T0197952

Get the eBook FREE!

(PDF, ePub, Kindle, and liveBook all included)

We believe that once you buy a book from us, you should be able to read it in any format we have available. To get electronic versions of this book at no additional cost to you, purchase and then register this book at the Manning website.

Go to https://www.manning.com/freebook and follow the instructions to complete your pBook registration.

That's it!
Thanks from Manning!

Math for Programmers

3D graphics, machine learning and simulations with Python

Math for Programmers

3D GRAPHICS, MACHINE LEARNING AND SIMULATIONS WITH PYTHON

PAUL ORLAND

MANNING
SHELTER ISLAND

For online information and ordering of this and other Manning books, please visit www.manning.com. The publisher offers discounts on this book when ordered in quantity. For more information, please contact

Special Sales Department
Manning Publications Co.
20 Baldwin Road
PO Box 761
Shelter Island, NY 11964
Email: orders@manning.com

Manning Publications Co.
20 Baldwin Road
PO Box 761
Shelter Island, NY 11964

Development editor: Jenny Stout
Technical development editor: Kris Athi
Review editor: Aleks Dragosavljević
Production editor: Lori Weidert
Copy editor: Frances Buran
Proofreader: Jason Everett
Technical proofreader: Mike Shepard
Typesetter and cover designer: Marija Tudor

ISBN 9781617295355
Printed in the United States of America

To my first math teacher and
my first programming teacher—Dad.

brief contents

contents

preface

I started working on this book in 2017, when I was CTO of Tachyus, a company I founded that builds predictive analytics software for oil and gas companies. By that time, we had finished building our core product: a fluid-flow simulator powered by physics and machine learning, along with an optimization engine. These tools let our customers look into the future of their oil reservoirs and helped them to discover hundreds of millions of dollars of optimization opportunities.

My task as CTO was to productize and scale-out this software as some of the biggest companies in the world began to use it. The challenge was that this was not only a complex software project, but the code was very mathematical. Around that time, we started hiring for a position called "scientific software engineer," with the idea that we needed skilled professional software engineers who also had solid backgrounds in math, physics, and machine learning. In the process of searching for and hiring scientific software engineers, I realized that this combination was both rare and in high demand. Our software engineers realized this as well and were eager to hone their math skills to contribute to our specialized back-end components of our stack. With eager math learners on our team already, as well as in our hiring pipeline, I started to think about the best way to train a strong software engineer to become a formidable math user.

I realized there were no books with the right math content, presented at the right level. While there are probably hundreds of books and thousands of free online articles on topics like linear algebra and calculus, I'm not aware of any I could hand to a typical professional software engineer, and expect them to come back in a few months having mastered the material. I don't say this to disparage software engineers, I just mean that reading and understanding math books is a difficult skill to learn on its own. To do so, you often need to figure out what specific topics you need to learn (which is hard if you don't know anything about the material yet!), read them, and then choose some high quality exercises to practice applying those topics. If you were less discerning, you could read every word of a textbook and solve *all* of its exercises, but it could take months of full-time study to do that!

With *Math for Programmers*, I hope to offer an alternative. I believe it's possible to read this book cover-to-cover in a reasonable amount of time, including completing all the exercises, and then to walk away having mastered some key math concepts.

How this book was designed

In the fall of 2017, I got in touch with Manning and learned that they were interested in publishing this book. That started a long process of converting my vision for this book into a concrete plan, which was much more difficult than I imagined, being a first-time author. Manning asked some hard questions of my original table of contents, like

- Will anyone be interested in this topic?
- Will this be too abstract?
- Can you really teach a semester of calculus in one chapter?

All of these questions forced me to think a lot more carefully about what was achievable. I'll share some of the ways we answered these questions because they'll help you understand exactly how this book works.

First, I decided to focus this book around one core skill—expressing mathematical ideas in code. I think this is a great way to learn math, even if you aren't a programmer by trade. When I was in high school, I learned to program on my TI-84 graphing calculator. I had the grand idea that I could write programs to do my math and science homework for me, giving me the right answer *and* outputting the steps along the way. As you might expect, this was more difficult than just doing my homework in the first place, but it gave me some useful perspective. For any kind of problem I wanted to program, I had to clearly understand the inputs and outputs, and what happened in each of the steps of the solution. By the end, I was sure I knew the material, and I had a working program to prove it.

That's the experience I'll try to share with you in this book. Each chapter is organized around a tangible example program, and to get it working, you need to put all the mathematical pieces together correctly. Once you're done, you'll have confidence that you've understood the concept and can apply it again in the future. I've included plenty of exercises to help you check your understanding on the math and code I've included, as well as mini-projects which invite you to experiment with new variations on the material.

Another question I discussed with Manning was what programming language I should use for the examples. Originally, I wanted to write the book in a functional programming language because math is a functional language itself. After all, the concept of a "function" originated in math, long before computers even existed. In various parts of math, you have functions that return other functions like integrals and derivatives in calculus. However, asking readers to learn an unfamiliar language like LISP, Haskell, or F# *while* learning new math concepts would make the book more difficult and less accessible. Instead, we settled on Python, a popular, easy-to-learn language with great mathematical libraries. Python also happens to be a favorite for "real world" users of math in academia and in industry.

The last major question that I had to answer with Manning was what specific math topics I would include and which ones wouldn't make the cut. This was a difficult decision, but at least we agreed on the title *Math for Programmers*, the broadness of which gave us some flexibility for what to include. My main criterion became the following: this was going to be "Math for Programmers," not "Math for Computer Scientists." With that in mind, I could leave out topics like discrete math, combinatorics, graphs, logic, Big O notation, and so on, that are covered in computer science classes and mostly used to *study* programs.

Even with that decision made, there was still plenty of math to choose from. Ultimately, I chose to focus on linear algebra and calculus. I have some strong pedagogical views on these subjects, and there are plenty of good example applications in both that can be visual and interactive. You can write a big textbook on either linear algebra *or* calculus alone, so I had to get even more specific. To do that, I decided the book would build up to some applications in the trendy field of machine learning. With those decisions made, the contents of the book became clearer.

Mathematical ideas we cover

This book covers a lot of mathematical topics, but there are a few major themes. Here are a few that you can keep an eye out for as you start reading:

- *Multi-dimensional spaces*—Intuitively, you probably have a sense what the words two-dimensional (2D) and three-dimensional (3D) mean. We live in a 3D world, while a 2D world is flat like a piece of paper or a computer screen. A location in 2D can be described by two numbers (often called x and y-coordinates), while you need three numbers to identify a location in 3D. We can't picture a 17-dimensional space, but we can describe its points by lists of 17 numbers. Lists of numbers like these are called *vectors*, and vector math helps illuminate the notion of "dimension."
- *Spaces of functions*—Sometimes a list of numbers can specify a function. With two numbers like $a = 5$ and $b = 13$, you can create a (linear) function of the form $f(x) = ax + b$, and in this case, the function would be $f(x) = 5x + 13$. For every point in 2D space, labeled by coordinates (a, b), there's a linear function that goes with it. So we can think of the set of all linear functions as a 2D space.
- *Derivatives and gradients*—These are calculus operations that measure the rates of change of functions. The *derivative* tells you how rapidly a function $f(x)$ is increasing or decreasing as you increase the input value x. A function in 3D might look like $f(x, y)$ and can increase or decrease as you change the values of either x or y. Thinking of (x, y) pairs as points in a 2D space, you could ask what

direction you could go in this 2D space to make f increase most rapidly. The gradient answers this question.

- *Optimizing a function*—For a function of the form $f(x)$ or $f(x, y)$, you could ask an even broader version of the previous question: what inputs to the function yield the biggest output? For $f(x)$, the answer would be some value x, and for $f(x, y)$, it would be a point in 2D. In the 2D case, the gradient can help us. If the gradient tells us $f(x, y)$ is increasing in some direction, we can find a maximum value of $f(x, y)$ if we explore in that direction. A similar strategy applies if you want to find a minimum value of a function.
- *Predicting data with functions*—Say you want to predict a number, like the price of a stock at a given time. You could create a function $p(t)$ that takes a time t and outputs a price p. The measure of predictive quality of your function is how close it comes to actual data. In that sense, finding a predictive function means minimizing the error between your function and real data. To do that, you need to explore a space of functions and find a minimum value. This is called *regression*.

I think this is a useful collection of mathematical concepts for anyone to have in their toolbelt. Even if you're not interested in machine learning, these concepts—and others in this book—have plenty of other applications.

The subjects I'm saddest to leave out of the book are probability and statistics. Probability and the concept of quantifying uncertainty in general is important in machine learning as well. This is a big book already, so there just wasn't time or room to squeeze a meaningful introduction for these topics. Stay tuned for a sequel to this book. There's a lot more fun and useful math out there, beyond what I've been able to cover in these pages, and I hope to be able to share it with you in the future.

acknowledgments

From start to finish, this book has taken about three years to create. I have gotten a lot of help in that time, and so I have quite a few people to thank and acknowledge.

First and foremost, I want to thank Manning for making this book happen. I'm grateful they bet on me to write a big, challenging book as a first-time author and had a lot of patience with me as the book fell behind schedule a few times. In particular, I want to thank Marjan Bace and Michael Stephens for pushing the project forward and for helping define what exactly it would be. My original development editor, Richard Wattenbarger, was also critical to keeping the book alive as we iterated on the content. I think he reviewed six total drafts of chapters 1 and 2 before we settled on how the book would be structured.

I wrote most of the book in 2019 under the expert guidance of my second editor, Jennifer Stout, who both got the project over the finish line and taught me a lot about technical writing. My technical editor, Kris Athi, and technical reviewer, Mike Shepard, also made it to the end with us, and thanks to them reading every word and line of code, we've caught and fixed countless errors. Outside of Manning, I got a lot of editing help from Michaela Leung, who also reviewed the whole book for grammatical and technical accuracy. I'd also like to thank the marketing team at Manning. With the MEAP program, we've been able to validate that this is a book people are interested in. It's been a great motivator to know a book will be at least a modest commercial success while working on the intensive final steps to get it published.

My current and former coworkers at Tachyus have taught me a lot about programming, and many of those lessons have made their way into this book. I credit Jack Fox for first getting me to think about the connections between functional programming and math, which comes up in chapters 4 and 5. Will Smith taught me about video game design, and we have had many good discussions about vector geometry for 3D rendering. Most notably, Stelios Kyriacou taught me most of what I know about optimization algorithms and helped me get some of the code in this book to work. He also introduced me to the philosophy that "everything is an optimization problem," a theme that you should pick up on in the latter half of the book.

To all the reviewers: Adhir Ramjiawan, Anto Aravinth, Christopher Haupt, Clive Harber, Dan Sheikh, David Ong, David Trimm, Emanuele Piccinelli, Federico Bertolucci, Frances Buontempo, German Gonzalez-Morris, James Nyika, Jens Christian B. Madsen, Johannes Van Nimwegen, Johnny Hopkins, Joshua Horwitz, Juan Rufes, Kenneth Fricklas, Laurence Giglio, Nathan Mische, Philip Best, Reka Horvath, Robert Walsh, Sébastien Portebois, Stefano Paluello, and Vincent Zhu, your suggestions helped make this a better book.

I'm by no means a machine learning expert, so I consulted a number of resources to make sure I introduced it correctly and effectively. I was most influenced by Andrew Ng's "Machine Learning" course on Coursera and the "Deep Learning" series by 3Blue1Brown on YouTube. These are great resources, and if you've seen them, you'll notice that part 3 of this book is influenced by the way they introduce the subject. I also need to thank Dan Rathbone, whose handy website CarGraph.com was the source of the data for many of my examples.

I also want to thank my wife Margaret, an astronomer, for introducing me to Jupyter notebooks. Switching the code for this book to Jupyter has made it much easier to follow. My parents have also been very supportive as I've written this book; on a few occasions, I've scrambled to get a chapter finished during a holiday visit with them. They also personally guaranteed that I would sell at least one copy (thanks, Mom!).

Finally, this book is dedicated to my Dad, who first showed me how to do math in code when he taught me how to program in APL when I was in fifth grade. If there's a second edition of this book, I might enlist his help to rewrite all of the Python in a single line of APL code!

about this book

Math for Programmers teaches you how to solve mathematical problems with code using the Python programming language. Math skills are more and more important for professional software developers, especially as companies are staffing up teams for data science and machine learning. Math also plays an integral role in other modern applications like game development, computer graphics and animation, image and signal processing, pricing engines, and stock market analysis.

The book starts by introducing 2D and 3D vector geometry, vector spaces, linear transformations, and matrices; these are the bread and butter of the subject of linear algebra. In part 2, it introduces calculus with a focus on a few particularly useful subjects for programmers: derivatives, gradients, Euler's method, and symbolic evaluation. Finally, in part 3, all the pieces come together to show you how some important machine learning algorithms work. By the last chapter of the book, you'll have learned enough math to code-up your own neural network from scratch.

This isn't a textbook! It's designed to be a friendly introduction to material that can often seem intimidating, esoteric, or boring. Each chapter features a complete, real-world application of a mathematical concept, complemented by exercises to help you check your understanding as well as mini-projects to help you continue your exploration.

Who should read this book?

This book is for anyone with a solid programming background who wants to refresh their math skills or to learn more about applications of math in software. It doesn't require any previous exposure to calculus or linear algebra, just high-school level algebra and geometry (even if that feels long ago!). This book is designed to be read at your keyboard. You'll get the most out of it if you follow along with the examples and try all the exercises.

How this book is organized

Chapter 1 invites you into the world of math. It covers some of the important applications of mathematics in computer programming, introduces some of the topics that appear in the book, and explains how programming can be a valuable tool to a math learner. After that, this book is divided into three parts:

- Part 1 focuses on vectors and linear algebra.
  - Chapter 2 covers vector math in 2D with an emphasis on using coordinates to define 2D graphics. It also contains a review of some basic trigonometry.
  - Chapter 3 extends the material of the previous chapter to 3D, where points are labeled by three coordinates instead of two. It introduces the dot product and cross product, which are helpful to measure angles and render 3D models.
  - Chapter 4 introduces linear transformations, functions that take vectors as inputs and return vectors as outputs and that have specific geometric effects like rotation or reflection.
  - Chapter 5 introduces matrices, which are arrays of numbers that can encode a linear vector transformation.
  - Chapter 6 extends the ideas from 2D and 3D so you can work with collections of vectors of *any* dimension. These are called vector spaces. As a main example, it covers how to process images using vector math.
  - Chapter 7 focuses on the most important computational problem in linear algebra: solving systems of linear equations. It applies this to a collision-detection system in a simple video game.
- Part 2 introduces calculus and applications to physics.
  - Chapter 8 introduces the concept of the rate of change of a function. It covers derivatives, which calculate a functions rate of change, and integrals, which recover a function from its rate of change.
  - Chapter 9 covers an important technique for approximate integration called Euler's method. It expands the game from chapter 7 to include moving and accelerating objects.
  - Chapter 10 shows how to manipulate algebraic expressions in code, including automatically finding the formula for the derivative of a function. It introduces symbolic programming, a different approach to doing math in code than used elsewhere in the book.
  - Chapter 11 extends the calculus topics to two-dimensions, defining the gradient operation and showing how it can be used to define a force field.
  - Chapter 12 shows how to use derivatives to find the maximum or minimum values of functions.

  - Chapter 13 shows how to think of sound waves as functions, and how to decompose them into sums of other simpler functions, called Fourier series. It covers how to write Python code to play musical notes and chords.
- Part 3 combines the ideas from the first two parts to introduce some important ideas in machine learning.
  - Chapter 14 covers how to fit a line to 2D data, a process referred to as linear regression. The example we explore is finding a function to best predict the price of a used car based on its mileage.
  - Chapter 15 addresses a different machine learning problem: figuring out what model a car is based on some data about it. Figuring out what kind of object is represented by a data point is called classification.
  - Chapter 16 shows how to design and implement a neural network, a special kind of mathematical function, and use it to classify images. This chapter combines ideas from almost every preceding chapter.

Each chapter should be accessible if you've read and understand the previous ones. The cost of keeping all of the concepts in order is that the applications may seem eclectic. Hopefully the variety of examples make it an entertaining read, and show you the broad range of applications of the math we cover.

About the code

This book presents ideas in (hopefully) logical order. The ideas you learn in chapter 2 apply to chapter 3, then ideas in chapters 2 and 3 appear in chapter 4, and so on. Computer code is not always written "in order" like this. That is, the simplest ideas in a finished computer program are not always in the first lines of the first file of the source code. This difference makes it challenging to present source code for a book in an intelligible way.

My solution to this is to include a "walkthrough" code file in the form of a Jupyter notebook for each chapter. A Jupyter notebook is something like a recorded Python interactive session, with visuals like graphs and images built in. In a Jupyter notebook, you enter some code, run it, and then perhaps overwrite it later in your session as you develop your ideas. The notebook for each chapter has code for each section and subsection, run in the same order as it appears in the book. Most importantly, this means you can run the code for the book as you read. You don't need to get to the end of a chapter before your code is complete enough to work. Appendix A shows you how to set up Python and Jupyter, and appendix B includes some handy Python features if you're new to the language.

This book contains many examples of source code both in numbered listings and in line with normal text. In both cases, source code is formatted in a `fixed-width font like this` to separate it from ordinary text.

Additionally, comments in the source code have often been removed from the listings when the code is described in the text. Code annotations accompany many of the listings, highlighting important concepts. If errata or bugs are fixed in the source code online, I'll include notes there to reconcile any differences from the code printed in the text.

In a few cases, the code for an example consists of a standalone Python script, rather than cells of the walkthrough Jupyter notebook for the chapter. You can either run it on its own as, for instance, `python script.py` or run it from within Jupyter notebook cell as `!python script.py`. I've included references to standalone scripts in some Jupyter notebooks, so you can follow along section-by-section and find the relevant source files.

One convention I've used throughout the book is to represent evaluation of individual Python commands with the `>>>` prompt symbol you'd see in a Python interactive session. I suggest you use Jupyter instead of Python interactive, but in any case, lines with `>>>` represent inputs and lines without represent outputs. Here's an example of a code block representing an interactive evaluation of a piece of Python code, "`2 + 2`":

```
>>> 2 + 2
4
```

By contrast, this next code block doesn't have any `>>>` symbols, so it's ordinary Python code rather than a sequence of inputs and outputs:

```
def square(x):
    return x * x
```

This book has hundreds of exercises, which are intended to be straightforward applications of material already covered, as well as mini-projects, which either are more involved, require more creativity, or introduce new concepts. Most exercises and mini-projects in this book invite you to solve some math problem with working Python code. I've included solutions to almost all of them, excluding some of the more open-ended mini-projects. You can find the solution code in the corresponding chapter's walkthrough Jupyter notebook.

The code for the examples in this book is available for download from the Manning website at https://www.manning.com/books/math-for-programmers and from GitHub at https://github.com/orlandpm/math-for-programmers.

liveBook discussion forum

Purchase of *Math for Programmers* includes free access to a private web forum run by Manning Publications where you can make comments about the book, ask technical questions, and receive help from the author and from other users. To access the forum, go to https://livebook.manning.com/#!/book/math-for-programmers/discussion. You can also learn more about Manning's forums and the rules of conduct at https://livebook.manning.com/#!/discussion.

Manning's commitment to our readers is to provide a venue where a meaningful dialogue between individual readers and between readers and the author can take place. It is not a commitment to any specific amount of participation on the part of the author, whose contribution to the forum remains voluntary (and unpaid). We suggest you try asking the author some challenging questions lest his interest stray! The forum and the archives of previous discussions will be accessible from the publisher's website as long as the book is in print.

about the author

PAUL ORLAND is an entrepreneur, programmer, and math enthusiast. After a stint as a software engineer at Microsoft, he co-founded Tachyus, a start-up company building predictive analytics to optimize energy production in the oil and gas industry. As founding CTO of Tachyus, Paul led the productization of machine learning and physics-based modeling software, and later as CEO, he expanded the company to serve customers on five continents. Paul has a B.S. in math from Yale and an M.S. in physics from the University of Washington. His spirit animal is the lobster.

about the cover illustration

The figure on the cover of *Math for Progammers* is captioned "Femme Laponne," or a woman from Lapp, now Sapmi, which includes parts of northern Norway, Sweden, Finland, and Russia. The illustration is taken from a collection of dress costumes from various countries by Jacques Grasset de Saint-Sauveur (1757–1810), titled *Costumes de Différents Pays,* published in France in 1797. Each illustration is finely drawn and colored by hand. The rich variety of Grasset de Saint-Sauveur's collection reminds us vividly of how culturally apart the world's towns and regions were just 200 years ago. Isolated from each other, people spoke different dialects and languages. In the streets or in the countryside, it was easy to identify where they lived and what their trade or station in life was just by their dress.

The way we dress has changed since then and the diversity by region, so rich at the time, has faded away. It is now hard to tell apart the inhabitants of different continents, let alone different towns, regions, or countries. Perhaps we have traded cultural diversity for a more varied personal life—certainly for a more varied and fast-paced technological life.

At a time when it is hard to tell one computer book from another, Manning celebrates the inventiveness and initiative of the computer business with book covers based on the rich diversity of regional life of two centuries ago, brought back to life by Grasset de Saint-Sauveur's pictures.

1 Learning math with code

This chapter covers

- Solving lucrative problems with math and software
- Avoiding common pitfalls in learning math
- Building on intuition from programming to understand math
- Using Python as a powerful and extensible calculator

Math is like baseball, or poetry, or fine wine. Some people are so fascinated by math that they devote their whole lives to it, while others feel like they just don't get it. You've probably already been forced into one camp or another by twelve years of compulsory math education in school.

What if we learned about fine wine in school like we learned math? I don't think I'd like wine at all if I got lectured on grape varietals and fermentation techniques for an hour a day, five days a week. Maybe in such a world, I'd need to consume three or four glasses for homework as assigned by the teacher. Sometimes this would be a delicious educational experience, but sometimes I might not feel like getting loaded on a school night. My experience in math class went something like that, and it turned me off of the subject for a while. Like wine, mathematics is an acquired taste, and a daily grind of lectures and assignments is no way to refine one's palate.

It's easy to think you're either cut out for math or you aren't. If you already believe in yourself, and you're excited to start learning, that's great! Otherwise, this chapter is designed for those less optimistic. Feeling intimidated by math is so common, it has a name: *math anxiety*. I hope to dispel any anxiety you might have and show you that math can be a stimulating experience rather than a frightening one. All you need are the right tools and the right mindset.

The main tool for learning in this book is the Python programming language. I'm guessing that when you learned math in high school, you saw it written on the blackboard and not in computer code. That's a shame, because a high-level programming language is far more powerful than a blackboard and far more versatile than whatever overpriced calculator you may have used. An advantage of meeting math in code is that the ideas have to be precise enough for a computer to understand, and there's never any hand-waving about what new symbols mean.

As with learning any new subject, the best way to set yourself up for success is to *want* to learn. There are plenty of good reasons for this. You could be intrigued by the beauty of mathematical concepts or enjoy the "brain-teaser" feel of math problems. Maybe there's an app or game that you dream of building, and you need to write some mathematical code to make it work. For now, I'll focus on a more pragmatic kind of motivation—solving mathematical problems with software can make you a lot of money.

1.1 Solving lucrative problems with math and software

A classic criticism you hear in high school math class is, "When am I ever going to use this stuff in real life?" Our teachers told us that math would help us succeed professionally and make money. I think they were right about this, even though their examples were off. For instance, I don't calculate my compounding bank interest by hand (and neither does my bank). Maybe if I became a construction site surveyor as my trigonometry teacher suggested, I'd be using sines and cosines every day to earn my paycheck.

It turns out the "real world" applications from high school textbooks aren't that useful. Still, there are real applications of math out there, and some of them are mind-bogglingly lucrative. Many are solved by translating the right mathematical idea into usable software. I'll share some of my favorite examples.

1.1.1 Predicting financial market movements

We've all heard legends of stock traders making millions of dollars by buying and selling the right stocks at the right time. Based on the movies I've seen, I always picture a trader as a middle-aged man in a suit yelling at his broker over a cell phone while driving around in a sports car. Maybe this stereotype was spot-on at one point, but the situation is different today.

Holed up in back offices of skyscrapers all over Manhattan are thousands of people called *quants*. Quants, otherwise known as quantitative analysts, design mathematical algorithms to automatically trade stocks and earn a profit. They don't wear suits and they don't spend time yelling on their cell phones, but I'm sure many of them own very nice sports cars.

So how does a quant write a program that automatically makes money? The best answers to that question are closely-guarded trade secrets, but you can be sure they involve a lot of math. We can look at a brief example to get a sense of how an automated trading strategy might work.

Stocks are types of financial assets that represent ownership stakes in companies. When the market perceives a company is doing well, its stock price goes up—buying the stock becomes more costly and selling it becomes more rewarding. Stock prices change erratically and in real time. Figure 1.1 shows how a graph of a stock price over a day of trading might look.

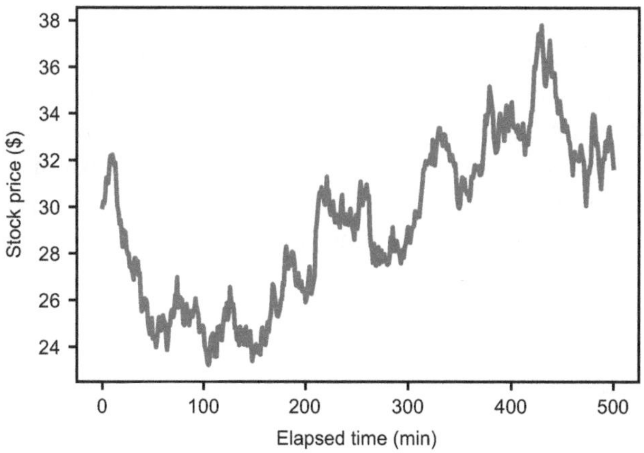

Figure 1.1 Typical graph of a stock price over time

If you bought a thousand shares of this stock for $24 around minute 100 and sold them for $38 at minute 400, you would make $14,000 for the day. Not bad! The challenge is that you'd have to know in advance that the stock was going up, and that minutes 100 and 400 were the best times to buy and sell, respectively. It may not be possible to predict the exact lowest or highest price points, but maybe you can find relatively good times to buy and sell throughout the day. Let's look at a way to do this mathematically.

We could measure whether the stock is going up or down by finding a line of "best fit" that approximately follows the direction the price is moving. This process is called *linear regression*, and we cover it in part 3 of this book. Based on the variability of data, we can calculate two more lines above and below the "best fit" line that show the region in which the price is wobbling up and down. Overlaid on the price graph, figure 1.2 shows that the lines follow the trend nicely.

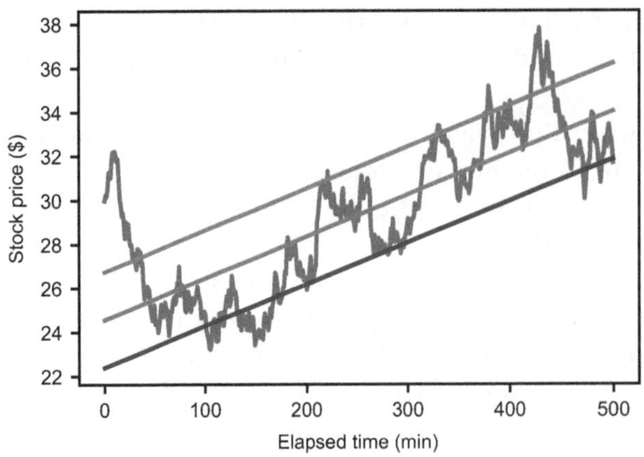

Figure 1.2 Using linear regression to identify a trend in changing stock prices

With a mathematical understanding of the price movement, we can then write code to automatically buy when the price is going through a low fluctuation relative to its trend and to sell when the price goes back up. Specifically, our program could connect to the stock exchange over the network and buy 100 shares when the price crosses the bottom line and sell 100 shares when the price crosses the top line. Figure 1.3 illustrates one such profitable trade: buying at around $27.80 and selling at around $32.60 makes you $480 in an hour.

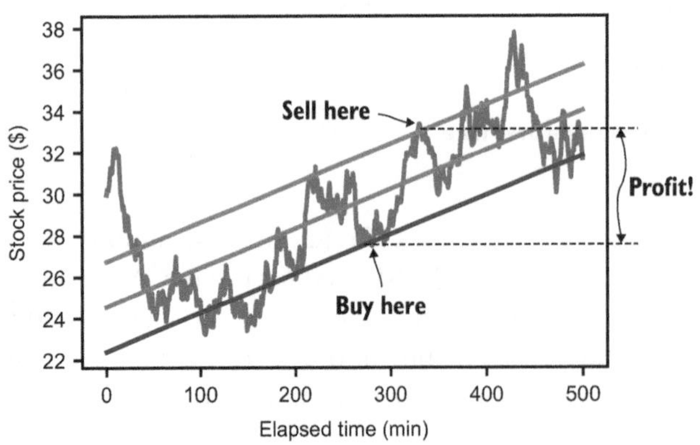

Figure 1.3 Buying and selling according to our rules-based software to make a profit

I don't claim I've shown you a complete or viable strategy here, but the point is that with the right mathematical model, you can make a profit automatically. At this moment, some unknown number of programs are building and updating models measuring the predicted trend of stocks and other financial instruments. If you write such a program, you can enjoy some leisure time while it makes money for you!

1.1.2 Finding a good deal

Maybe you don't have deep enough pockets to consider risky stock trading. Math can still help you make and save money in other transactions like buying a used car, for example. New cars are easy-to-understand commodities. If two dealers are selling the same car, you obviously want to buy from the dealer that has the lowest cost. But used cars have more numbers associated with them: an asking price, as well as mileage and model year. You can even use the duration that a particular used car has been on the market to assess its quality: the longer the duration, the more suspicious you might be.

In mathematics, objects you can describe with ordered lists of numbers are called *vectors*, and there is a whole field (called *linear algebra*) dedicated to studying them. For example, a used car might correspond to a *four-dimensional* vector, meaning a four-tuple of numbers:

(2015, 41429, 22.27, 16980)

These numbers represent the model year, mileage, days on the market, and asking price, respectively. A friend of mine runs a site called CarGraph.com that aggregates data on used cars for sale. At the time of writing, it shows 101 Toyota Priuses for sale, and it gives some or all of these four pieces of data for each one. The site also lives up to its name and visually presents the data in a graph (figure 1.4). It's hard to visualize four-dimensional objects, but if you choose two of the dimensions like price and mileage, you can graph them as points on a scatter plot.

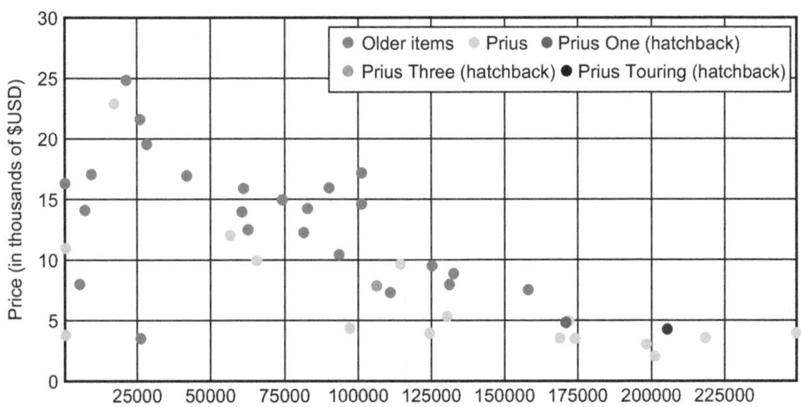

Figure 1.4 A graph of price vs. mileage for used Priuses from CarGraph.com

We might be interested in drawing a trend line here too. Every point on this graph represents someone's opinion of a fair price, so the trend line would aggregate these opinions together into a more reliable price at any mileage. In figure 1.5, I decided to

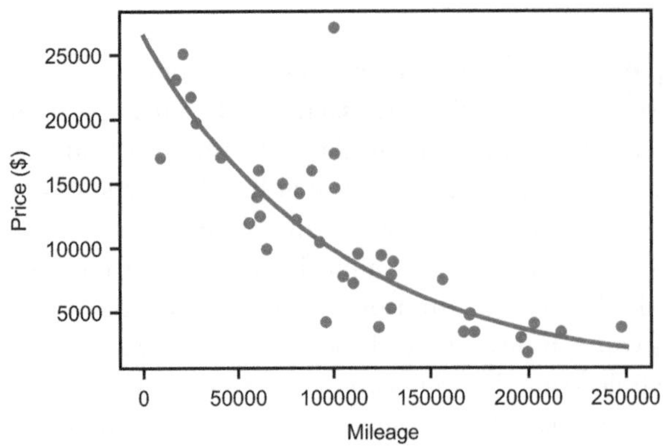

Figure 1.5 Fitting an exponential decline curve to price vs. mileage data for used Toyota Priuses

fit to an *exponential* decline curve rather than a line, and I omitted some of the nearly new cars selling for below retail price.

To make the numbers more manageable, I converted the mileage values to tens of thousands of miles, so a mileage of 5 represents 50,000 miles. Calling p the price and m the mileage, the equation for the curve of best fit is as follows:

$$p = \$26{,}500 \cdot (0.905)^m$$

Equation 1.1

Equation 1.1 shows that the best fit price is $26,500 times 0.905 raised to the power of the mileage. Plugging the values into the equation, I find that if my budget is $10,000, then I should buy a Prius with about 97,000 miles on it (figure 1.6). If I believe the curve indicates a *fair* price, then cars below the line should typically be good deals.

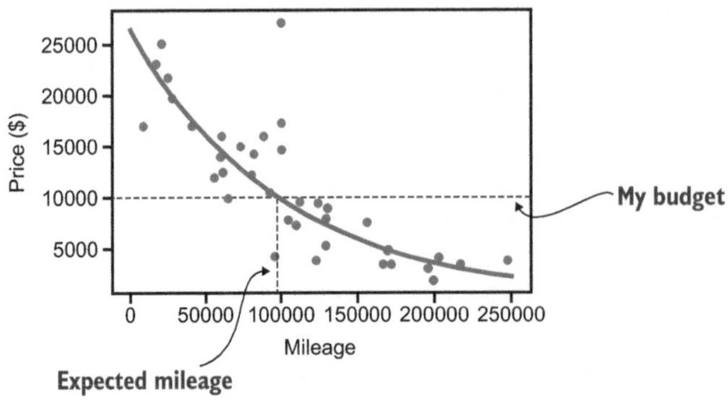

Figure 1.6 Finding the mileage I should expect on a used Prius for my $10,000 budget

But we can learn more from equation 1.1 than just how to find a good deal. It tells a story about how cars depreciate. The first number in the equation is $26,500, which is the exponential function's understanding of the price at zero mileage. This is an impressively close match to the retail price of a new Prius. If we use a line of best fit, it implies a Prius loses a fixed amount of value with each mile driven. This exponential function says, instead, that it loses a fixed *percentage* of its value with each mile driven. After driving 10,000 miles, a Prius is only worth 0.905 or 90.5% of its original price according to this equation. After 50,000 miles, we multiply its price by a factor of $(0.905)^5 = 0.607$. That tells us that it's worth about 61% of what it was originally.

To make the graph in figure 1.6, I implemented a `price(mileage)` function in Python, which takes a mileage as an input (measured in 10,000s of miles) and returns the best-fit price as an output. Calculating `price(0) - price(5)` and `price(5) - price(10)` tells me that the first and second 50,000 miles driven cost about $10,000 and $6,300, respectively.

If we use a line of best fit instead of an exponential curve, it implies that the car depreciated at a fixed rate of $0.10 per mile. This suggests that every 50,000 miles of driving leads to the same depreciation of $5,000. Conventional wisdom says that the first miles you drive a new car are the most expensive, so the exponential function (equation 1.1) agrees with this, while a linear model does not.

Remember, this is only a *two-dimensional* analysis. We only built a mathematical model to relate two of the four numerical dimensions describing each car. In part 1, you learn more about vectors of various dimensions and how to manipulate higher-dimensional data. In part 2, we cover different kinds of functions like linear functions and exponential functions, and we compare them by analyzing their rates of change. Finally, in part 3, we look at how to build mathematical models that incorporate *all* the dimensions of a data set to give us a more accurate picture.

1.1.3 Building 3D graphics and animations

Many of the most famous and financially successful software projects deal with multi-dimensional data, specifically *three-dimensional* or *3D* data. Here I'm thinking of 3D animated movies and 3D video games that gross in the billions of dollars. For example, Pixar's 3D animation software has helped them rake in over $13 billion at box offices. Activision's *Call of Duty* franchise of 3D action games has earned over $16 billion, and Rockstar's *Grand Theft Auto V* alone has brought in $6 billion.

Every one of these acclaimed projects is based on an understanding of how to do computations with 3D vectors, or triples of numbers of the form $v = (x, y, z)$. A triple of numbers is sufficient to locate a point in 3D space relative to a reference point called the *origin*. Figure 1.7 shows how each of the three numbers tells you how far to go in one of three perpendicular directions.

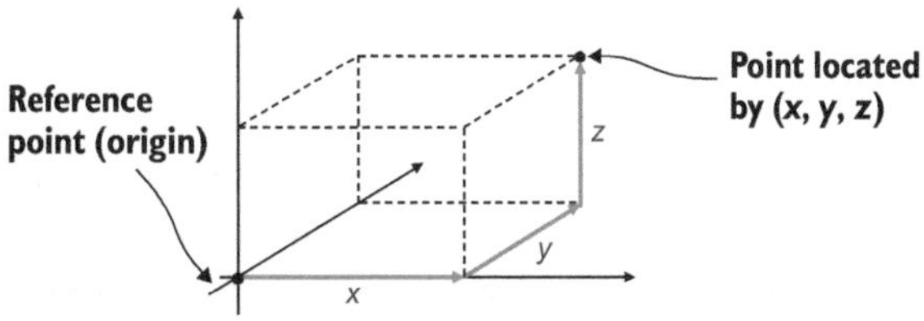

Figure 1.7 Labeling a point in 3D with a vector of three numbers, *x*, *y*, and *z*

Any 3D object from a clownfish in *Finding Nemo* to an aircraft carrier in *Call of Duty* can be defined for a computer as a collection of 3D vectors. In code, each of these objects looks like a list of triples of `float` values. With three triples of floats, we have three points in space that can define a triangle (figure 1.8). For instance,

```
triangle = [(2.3,1.1,0.9), (4.5,3.3,2.0), (1.0,3.5,3.9)]
```

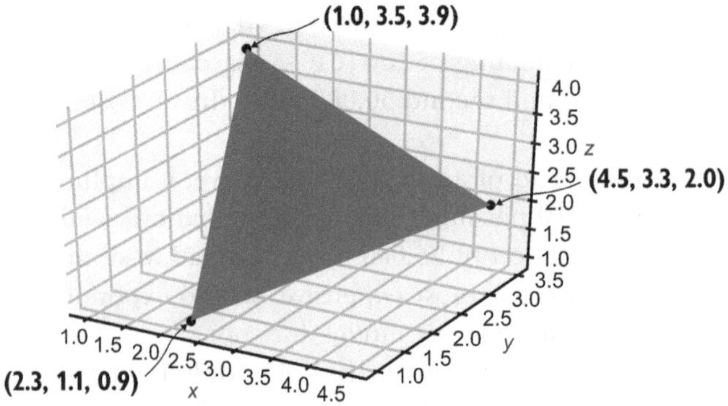

Figure 1.8 Building a 3D triangle using a triple of float values for each of its corners

Combining many triangles, you can define the surface of a 3D object. Using more, smaller triangles, you can even make the result look smooth. Figure 1.9 shows six renderings of a 3D sphere using an increasing number of smaller and smaller triangles.

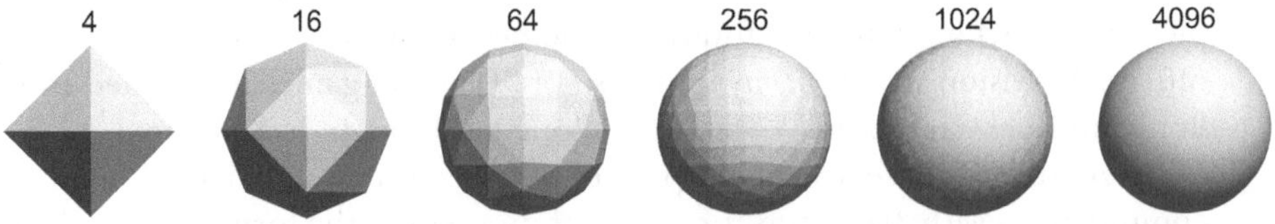

Figure 1.9 Three-dimensional (3D) spheres built out of the specified number of triangles.

In chapters 3 and 4, you learn how to use 3D vector math to turn 3D models into shaded 2D images like the ones in figure 1.9. You also need to make your 3D models smooth to make them realistic in a game or movie, and you need them to move and change in realistic ways. This means that your objects should obey the laws of physics, which are also expressed in terms of 3D vectors.

Suppose you're a programmer for *Grand Theft Auto V* and want to enable a basic use case like shooting a bazooka at a helicopter. A projectile coming out of a bazooka starts at the protagonist's location and then its position changes over time. You can use numeric subscripts to label the various positions it has over its flight, starting with $v_0 = (x_0, y_0, z_0)$. As time elapses, the projectile arrives at new positions labeled by vectors $\mathbf{v}_1 = (x_1, y_1, z_1)$, $\mathbf{v}_2 = (x_2, y_2, z_2)$, and so on. The rates of change for the *x*, *y*, and *z* values are decided by the direction and speed of the bazooka. Moreover, the rates can change over time—the projectile increases its *z* position at a decreasing rate because of the continuous downward pull of gravity (figure 1.10).

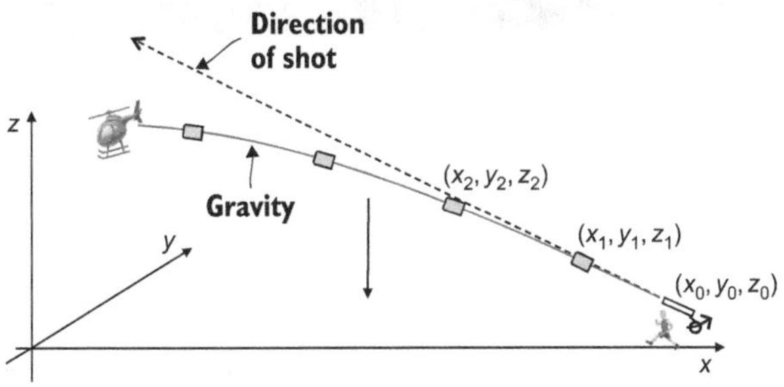

Figure 1.10 The position vector of the projectile changes over time due to its initial speed and the pull of gravity.

As any experienced action gamer will tell you, you need to aim slightly above the helicopter to hit it! To simulate physics, you have to know how forces affect objects and cause continuous change over time. The math of continuous change is called *calculus*, and the laws of physics are usually expressed in terms of objects from calculus called *differential equations*. You learn how to animate 3D objects in chapters 4 and 5, and then how to simulate physics using ideas from calculus in part 2.

1.1.4 Modeling the physical world

My claim that mathematical software produces real financial value isn't just speculation; I've seen the value in my own career. In 2013, I founded a company called Tachyus that builds software to optimize oil and gas production. Our software uses mathematical models to understand the flow of oil and gas underground to help producers extract it more efficiently and profitably. Using the insight it generates, our customers have achieved millions of dollars a year in cost savings and production increases.

To explain how our software works, you need to know a few pieces of oil terminology. Holes called *wells* are drilled into the ground until they reach the target layer of porous (sponge-like) rock containing oil. This layer of oil-rich rock underground is called a *reservoir*. Oil is pumped to the surface and is then sold to refiners who convert it into the products we use every day. A schematic of an oilfield (not to scale!) is shown in figure 1.11.

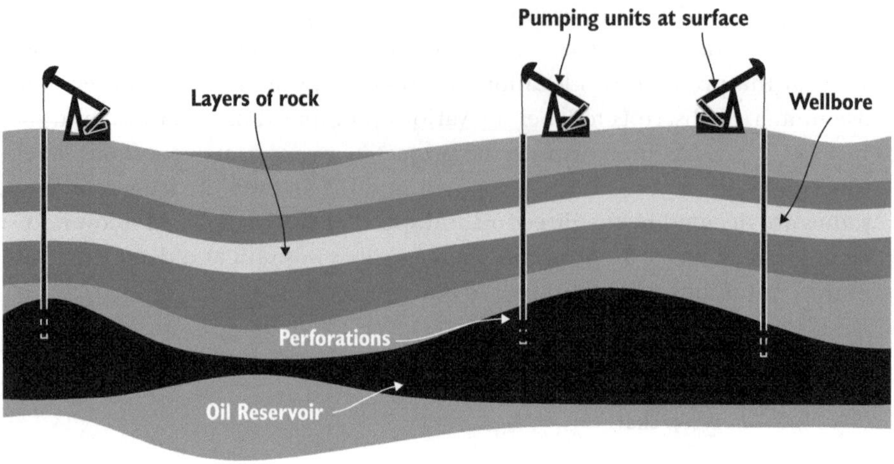

Figure 1.11 A schematic diagram of an oilfield

Over the past few years, the price of oil has varied significantly, but for our purposes, let's say it's worth $50 a barrel, where a barrel is a unit of volume equal to 42 gallons or about 159 liters. If by drilling wells and pumping effectively, a company is able to extract 1,000 barrels of oil per day (the volume of a few backyard swimming pools), it will have annual revenues in the tens of millions of dollars. Even a few percentage points of increased efficiency can mean a sizable amount of money.

The underlying question is what's going on underground: where is the oil now and how is it moving? This is a complicated question, but it can also be answered by solving differential equations. The changing quantities here are not positions of a projectile, but rather locations, pressures, and flow rates of fluids underground. Fluid flow rate is a special kind of function that returns a vector, called a *vector field*. This means that fluid can flow at any rate in any three-dimensional direction, and that direction and rate can vary across different locations within the reservoir.

With our best guess for some of these parameters, we can use a differential equation called *Darcy's law* to predict flow rate of liquid through a porous rock medium like sandstone. Figure 1.12 shows Darcy's law, but don't worry if some symbols are unfamiliar! The function named **q** representing flow rate is bold to indicate it returns a vector value.

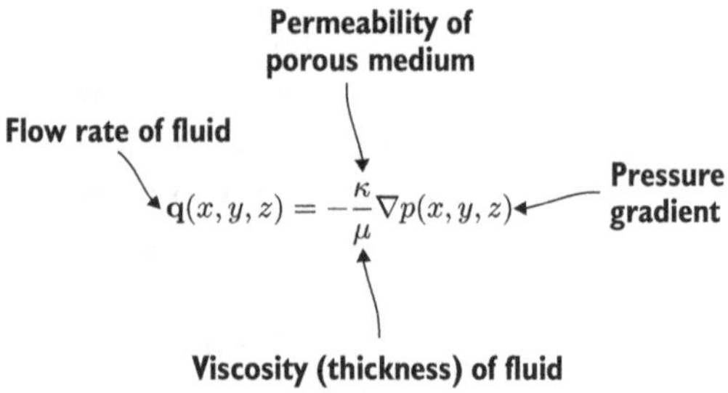

Figure 1.12 Darcy's law annotated for a physics equation, governing how fluid flows within a porous rock.

The most important part of this equation is the symbol that looks like an upside-down triangle, which represents the *gradient operator* in vector calculus. The gradient of the pressure function $p(x, y, z)$ at a given spatial point (x, y, z) is the 3D vector $\mathbf{q}(x, y, z)$, indicating the direction of increasing pressure and the rate of increase in pressure at that point. The negative sign tells us that the 3D vector of flow rate is in the *opposite* direction. This equation states, in mathematical terms, that fluid flows from areas of high pressure to areas of low pressure.

Negative gradients are common in the laws of physics. One way to think of this is that nature is always seeking to move toward lower potential energy states. The potential energy of a ball on a hill depends on the altitude h of the hill at any lateral point x. If the height of a hill is given by a function $h(x)$, the gradient points uphill while the ball rolls in the exact opposite direction (figure 1.13).

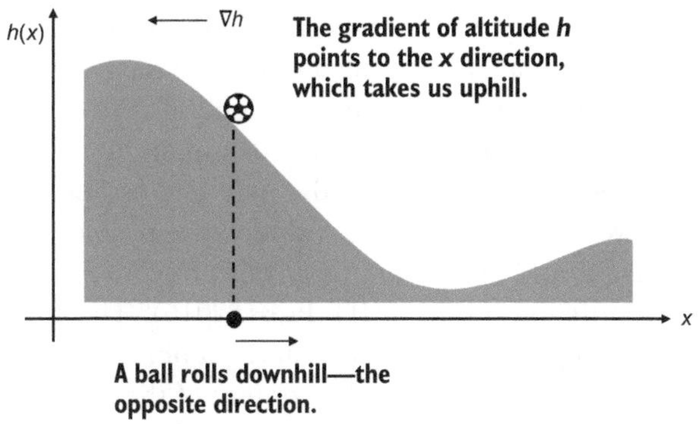

Figure 1.13 The positive gradient points uphill, while the negative gradient points downhill.

In chapter 11, you learn how to calculate gradients. There, I show you how to apply gradients to simulate physics and also to solve other mathematical problems. The gradient happens to be one of the most important mathematical concepts in machine learning as well.

I hope these examples have been more compelling and realistic than the real-world applications you heard in high school math class. Maybe, at this point, you're convinced these math concepts are worth learning, but you're worried that they might be too difficult. It's true that learning math can be hard, especially on your own. To make it as smooth as possible, let's talk about some of the pitfalls you can face as a math student and how I'll help you avoid them in this book.

1.2 How not to learn math

There are plenty of math books out there, but not all of them are equally useful. I have quite a few programmer friends who tried to learn mathematical concepts like the ones in the previous section, either motivated by intellectual curiosity or by career ambitions. When they use traditional math textbooks as their main resource, they often get stuck and give up. Here's what a typical *unsuccessful* math-learning story looks like.

1.2.1 Jane wants to learn some math

My (fictional) friend Jane is a full-stack web developer working at a medium-sized tech company in San Francisco. In college, Jane didn't study computer science or any mathematical subjects in depth, and she started her career as a product manager. Over the last ten years, she picked up coding in Python and JavaScript and was able to transition into software engineering. Now, at her new job, she is one of the most capable programmers on the team, able to build the databases, web services, and user interfaces required to deliver important new features to customers. Clearly, she's pretty smart!

Jane realizes that learning data science could help her design and implement better features at work, using data to improve the experience for her customers. Most days on the train to work, Jane reads blogs and articles about new technologies, and recently, she's been amazed by a few about a topic called "deep learning." One article talks about Google's AlphaGo, powered by deep learning, which beat the top-ranked human players in the world in a board game. Another article showed stunning impressionist paintings generated from ordinary images, again using a deep learning system.

After reading these articles, Jane overheard that her friend-of-a-friend Marcus got a deep learning research job at a big tech company. Marcus supposedly gets paid over $400,000 a year in salary and stock. Thinking about the next step in her career, what more could Jane want than to work on a fascinating and lucrative problem?

Jane did some research and found an authoritative (and free!) resource online: the book *Deep Learning* by Goodfellow, et al., (MIT Press, 2016). The introduction read much like the technical blog posts she was used to and got her even more excited about learning the topic. But as she kept reading, the content of the book got harder. The first chapter covered the required math concepts and introduced a lot of terminology and notation that Jane had never seen. She skimmed it and tried to get on to the meat of the book, but it continued to get more difficult.

Jane decided she needed to pause her study of AI and deep learning until she learned some math. Fortunately, the math chapter of *Deep Learning* listed a reference on linear algebra for students who had never seen the topic before. She tracked down this textbook, *Linear Algebra* by Georgi Shilov (Dover, 1977), and discovered that it was 400 pages long and equally as dense as *Deep Learning*.

After spending an afternoon reading abstruse theorems about concepts like number fields, determinants, and cofactors, she called it quits. She had no idea how these concepts were going to help her write a program to win a board game or to generate artwork, and she no longer cared to spend dozens of hours with this dry material to find out.

Jane and I met to catch up over a cup of coffee. She told me about her struggles reading real AI literature because she didn't know linear algebra. Recently, I'm hearing a lot of the same form of lamentation:

> *I'm trying to read about [new technology] but it seems like I need to learn [math topic] first.*

Her approach was admirable: she tracked down the best resource for the subject she wanted to learn and sought out resources for prerequisites she was missing. But in taking that approach to its logical conclusion, she found herself in a nauseating "depth-first" search of technical literature.

1.2.2 Slogging through math textbooks

College-level math books like the linear algebra book Jane picked up tend to be very formulaic. Every section follows the same format: it defines some new terminology, states some facts (called *theorems*) using that terminology, and then proves that those theorems are true.

This sounds like a good, logical order: you introduce the concept you're talking about, state some conclusions that can be drawn, and then justify them. Then why is it so hard to read advanced mathematical textbooks?

The problem is that this is not how math is actually created. When you're coming up with new mathematical ideas, there can be a long period of experimentation before you even find the right definitions. I think most professional mathematicians would describe their steps like this:

1. Invent a *game*. For example, start playing with some mathematical objects by trying to list all of them, find patterns among them, or find one with a particular property.
2. Form some *conjectures*. Speculate about some general facts you can state about your game and, at least, convince yourself these must be true.
3. Develop some *precise language* to describe your game and your conjectures. After all, your conjectures won't mean anything until you can communicate them.
4. Finally, with some determination and luck, find a *proof* for your conjecture, showing why it *needs* to be true.

The main lesson to learn from this process is that you should start by thinking about big ideas, and the formalism can wait. Once you have a rough idea of how the math works, the vocabulary and notation become an asset for you rather than a distraction. Math textbooks usually work in the opposite order, so I recommend using textbooks as references rather than as introductions to new subjects.

Instead of reading traditional textbooks, the best way to learn math is to explore ideas and draw your own conclusions. However, you don't have enough hours in the day to reinvent everything yourself. What is the right balance to strike? I'll give you my humble opinion, which guides how I've written this non-traditional book about math.

1.3 Using your well-trained left brain

This book is designed for people who are either experienced programmers or for those who are excited to learn programming as they work through it. It's great to write about math for an audience of programmers, because if you can write code, you've already trained your analytical left brain. I think the best way to learn math is with the help of a high-level programming language, and I predict that in the not-so-distant future, this will be the norm in math classrooms.

There are several specific ways programmers like you are well equipped to learn math. I list those here, not only to flatter you, but also to remind you what skills you already have that you can lean on in your mathematical studies.

1.3.1 Using a formal language

One of the first hard lessons you learn in programming is that you can't write your code like you write simple English. If your spelling or grammar is slightly off when writing a note to a friend, they can probably still understand what you're trying to say. But any syntactic error or misspelled identifier in code causes your program to fail. In some languages, even forgetting a semicolon at the end of an otherwise correct statement prevents the program from running. As another example, consider the two statements:

```
x = 5
5 = x
```

I could read either of these to mean that the symbol x has the value 5. But that's not *exactly* what either of these means in Python, and in fact, only the first one is correct. The Python statement `x = 5` is an instruction to set the variable x to have the value 5. On the other hand, you can't set the number 5 to have the value x. This may seem pedantic, but you need to know it to write a correct program.

Another example that trips up novice programmers (and experienced ones as well!) is reference equality. If you define a new Python class and create two identical instances of it, they are not equal!

```
>>> class A(): pass
...
>>> A() == A()
False
```

You might expect two identical expressions to be equal, but that's evidently not a rule in Python. Because these are different instances of the `A` class, they are not considered equal.

Be on the lookout for new mathematical objects that look like ones you know but don't behave the same way. For instance, if the letters A and B represent numbers, then $A \cdot B = B \cdot A$. But, as you'll learn in chapter 5, this is not necessarily the case if A and B are *not* numbers. If, instead, A and B are matrices, then the products $A \cdot B$ and $B \cdot A$ are different. In fact, it's possible that only one of the products is even doable or that neither product is correct.

When you're writing code, it's not enough to write statements with correct syntax. The ideas that your statements represent need to make sense to be valid. If you apply the same care when you're writing mathematical statements, you'll catch your mistakes faster. Even better, if you write your mathematical statements in code, you'll have the computer to help check your work.

1.3.2 Build your own calculator

Calculators are prevalent in math classes because it's useful to check your work. You need to know how to multiply 6 by 7 without using your calculator, but it's good to confirm that your answer of 42 is correct by consulting your calculator. The calculator also helps you save time once you've mastered mathematical concepts. If you're doing trigonometry, and you need to know the answer to 3.14159 / 6, the calculator is there to handle it so you can instead think about what the answer means. The more a calculator can do out-of-the-box, the more useful it should theoretically be.

But sometimes our calculators are too complicated for our own good. When I started high school, I was required to get a graphing calculator and I got a TI-84. It had about 40 buttons, each with 2 to 3 different modes. I only knew how to use maybe 20 of them, so it was a cumbersome tool to learn how to use. The story was the same when I got my first ever calculator in first grade. There were only 15 buttons or so, but I didn't know what some of them did. If I had to invent a first calculator for students, I would make it look something like the one in figure 1.14.

Figure 1.14 A calculator for students learning to count

This calculator only has two buttons. One of them resets the value to 1, and the other advances to the next number. Something like this would be the right "no-frills" tool for children learning to count. (My example may seem silly, but you can actually buy calculators like this! They are usually mechanical and sold as tally counters.)

Soon after you master counting, you want to practice writing numbers and adding them. The perfect calculator at that stage of learning might have a few more buttons (figure 1.15).

Figure 1.15 A calculator capable of writing whole numbers and adding them

There's no need for buttons like -, *, or ÷ to get in your way at this phase. As you solve subtraction problems like 5 - 2, you can still check your answer of 3 with this calculator by confirming the sum 3 + 2 = 5. Likewise, you can solve multiplication problems by adding numbers repeatedly. You could upgrade to a calculator that does all of the operations of arithmetic when you're done exploring with this one.

I think an ideal calculator would be extensible, meaning that you could add more functionality to it as needed. For instance, you could add a button to your calculator for every new mathematical operation you learn. Once you got to algebra, maybe you could enable it to understand symbols like x or y in addition to numbers. When you learned calculus, you could further enable it to understand and manipulate mathematical functions.

Extensible calculators that can hold many types of data seem far-fetched, but that's exactly what you get when you use a high-level programming language. Python comes with arithmetic operations, a `math` module, and numerous third-party mathematical libraries you can pull in to make your programming environment more powerful whenever you want. Because Python is *Turing complete*, you can (in principle) compute anything that can be computed. You only need a powerful enough computer, a clever enough implementation, or both.

In this book, we implement each new mathematical concept in reusable Python code. Working through the implementation yourself can be a great way of cementing your understanding of a new concept, and by the end, you've added a new tool to your toolbelt. After trying it yourself, you can always swap in a polished, mainstream library if you like. Either way, the new tools you build or import lay the groundwork to explore even bigger ideas.

1.3.3 Building abstractions with functions

In programming, the process I just described is called *abstraction*. For example, when you get tired of repeated counting, you create the abstraction of addition. When you get tired of doing repeated addition, you create the abstraction of multiplication, and so on.

Of all the ways that you can make abstractions in programming, the most important one to carry over to math is the *function*. A function in Python is a way of repeating some task that can take one or more inputs or that can produce an output. For example,

```
def greet(name):
    print("Hello %s!" % name)
```

allows me to issue multiple greetings with short, expressive code like this:

```
>>> for name in ["John","Paul","George","Ringo"]:
...     greet(name)
...
Hello John!
Hello Paul!
Hello George!
Hello Ringo!
```

This function can be useful, but it's not like a mathematical function. Mathematical functions always take input values, and they always return output values with no side effects.

In programming, we call the functions that behave like mathematical functions *pure functions*. For example, the square function $f(x) = x^2$ takes a number and returns the product of the number with itself. When you evaluate $f(3)$, the result is 9. That doesn't mean that the number 3 has now changed and becomes 9. Rather, it means 9 is the corresponding output for the input 3 for the function f. You can picture this squaring function as a machine that takes numbers in an input slot and produces results (numbers) in its output slot (figure 1.16).

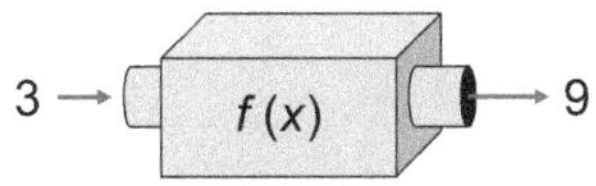

Figure 1.16 A function as a machine with an input slot and an output slot

This is a simple and useful mental model, and I'll return to it throughout the book. One of the things I like most about it is that you can picture a function as an object in and of itself. In math, as in Python, functions are data that you can manipulate independently and even pass to other functions.

Math can be intimidating because it is abstract. Remember, as in any well-written software, the abstraction is introduced for a reason: it helps you organize and communicate bigger and more powerful ideas. When you grasp these ideas and translate them into code, you'll open up some exciting possibilities.

If you didn't already, I hope you now believe there are many exciting applications of math in software development. As a programmer, you already have the right mindset and tools to learn some new mathematical ideas. The ideas in this book provided me with professional and personal enrichment, and I hope they will for you as well. Let's get started!

Summary

- There are interesting and lucrative applications of math in many software engineering domains.
- Math can help you quantify a trend for data that changes over time, for instance, to predict the movement of a stock price.
- Different types of functions convey different kinds of qualitative behavior. For instance, an exponential depreciation function means that a car loses a percentage of its resale value with each mile driven rather than a fixed amount.
- Tuples of numbers (called *vectors*) represent multidimensional data. Specifically, 3D vectors are triples of numbers and can represent points in space. You can build complex 3D graphics by assembling triangles specified by vectors.
- *Calculus* is the mathematical study of continuous change, and many of the laws of physics are written in terms of calculus equations that are called *differential equations*.
- It's hard to learn math from traditional textbooks! You learn math by exploration, not as a straightforward march through definitions and theorems.
- As a programmer, you've already trained yourself to think and communicate precisely; this skill will help you learn math as well.

Part 1

Vectors and graphics

In the first part of this book, we dig into the branch of math called *linear algebra*. At a very high level, linear algebra is the branch of math dealing with computations on multi-dimensional data. The concept of "dimension" is a geometric one; you probably intuitively know what I mean when I say "a square is 2-dimensional" while "a cube is 3-dimensional." Among other things, linear algebra lets us turn geometric ideas about dimension into things we can compute concretely.

The most basic concept in linear algebra is that of a *vector*, which you can think of as a data point in some multi-dimensional space. For instance, you've probably heard of the 2-dimensional (2D) coordinate plane in high school geometry and algebra. As we'll cover in chapter 2, vectors in 2D correspond to points in the plane, which can be labeled by ordered pairs of numbers of the form (x, y). In chapter 3, we'll consider 3-dimensional (3D) space, whose vectors (points) can be labeled by triples of numbers in the form (x, y, z). In both cases, we see we can use collections of vectors to define geometric shapes, which can, in turn, be converted into interesting graphics.

Another key concept in linear algebra is that of a *linear transformation*, which we introduce in chapter 4. A linear transformation is a kind of function that takes a vector as input and returns a vector as output, while preserving the geometry (in a special sense) of the vectors involved. For instance, if a collection of vectors (points) lie on a straight line in 2D, after applying a linear transformation, they will still lie on a straight line. In chapter 5, we introduce *matrices*, which are rectangular arrays of numbers that can represent linear transformations. Our culminating application of linear transformations is to apply them sequentially over time to graphics in a Python program, resulting in some animated graphics in 3D.

While we can only picture vectors and linear transformations in 2D and 3D, it's possible to define vectors with any number of dimensions. In n dimensions, a vector can be identified as an ordered n-tuple of numbers of the form $(x_1, x_2, ..., x_n)$. In chapter 6, we reverse-engineer the concepts of 2D and 3D space to define the general concept of a *vector space* and to define the concept of *dimension* more concretely. In particular, we'll see that digital images made of pixels can be thought of as vectors in a high-dimensional vector space and that we can do image manipulation with linear transformations.

Finally, in chapter 7, we look at the most ubiquitous computational tool in linear algebra: solving *systems of linear equations*. As you may remember from high school algebra, the solution to two linear equations in two variables like x and y tell us where two lines meet in the plane. In general, linear equations tell us where lines, planes, or higher-dimensional generalizations intersect in a vector space. Being able to automatically solve this problem in Python, we'll use it to build a first version of a video game engine.

2 Drawing with 2D vectors

This chapter covers

- Creating and manipulating 2D drawings as collections of vectors
- Thinking of 2D vectors as arrows, locations, and ordered pairs of coordinates
- Using vector arithmetic to transform shapes in the plane
- Using trigonometry to measure distances and angles in the plane

You probably already have some intuition for what it means to be two-dimensional or three-dimensional. A *two-dimensional* (2D) object is flat like an image on a piece of paper or a computer screen. It has only the dimensions of height and width. A *three-dimensional* (3D) object in our physical world, however, has not only height and width but also depth.

Models of 2D and 3D entities are important in programming. Anything that shows up on the screen of your phone, tablet, or PC is a 2D object, occupying some width and height of pixels. Any simulation, game, or animation that represents the physical world is stored as 3D data and eventually projected to the two dimensions of the screen. In virtual and augmented reality applications, the 3D models must be paired with real, measured 3D data about the user's position and perspective.

Even though our everyday experience takes place in three dimensions, it's useful to think of some data as higher dimensional. In physics, it's common to consider time as the fourth dimension. While an object exists at a location in 3D space, an event occurs at a 3D location and at a specified moment. In data science problems, it's common for data sets to have far more dimensions. For instance, a user tracked on a website can have hundreds of measurable attributes, which describe usage patterns. Grappling with these problems in graphics, physics, and data analysis requires a framework for dealing with data in higher dimensions. This framework is *vector mathematics.*

Vectors are objects that live in multi-dimensional spaces. These have their own notions of arithmetic (adding, multiplying, and so on). We start by studying 2D vectors, which are easy to visualize and compute with. We use a lot of 2D vectors in this book, and we also use them as a mental model when reasoning about higher-dimensional problems.

2.1 Picturing 2D vectors

The 2D world is flat like a piece of paper or a computer screen. In the language of math, a flat, 2D space is referred to as a *plane.* An object living in a 2D plane has the two dimensions of height and width but no third dimension of depth. Likewise, you can describe locations in 2D by two pieces of information: their vertical and horizontal positions. To describe the location of points in the plane, you need a reference point. We call that special reference point the *origin.* Figure 2.1 shows this relationship.

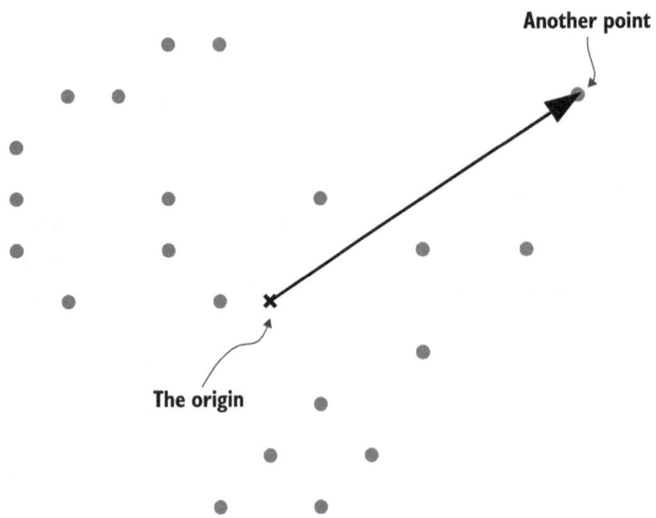

Figure 2.1 Locating one of several points in the plane, relative to the origin

There are many points to choose from, but we have to fix one of them as our origin. To distinguish it, we mark the origin with an *x* instead of with a dot as in figure 2.1. From the origin, we can draw an arrow (like the solid one in figure 2.1) to show the relative location of another point.

A *two-dimensional vector* is a point in the plane relative to the origin. Equivalently, you can think of a vector as a straight arrow in the plane; any arrow can be placed to start at the origin, and it indicates a particular point (figure 2.2).

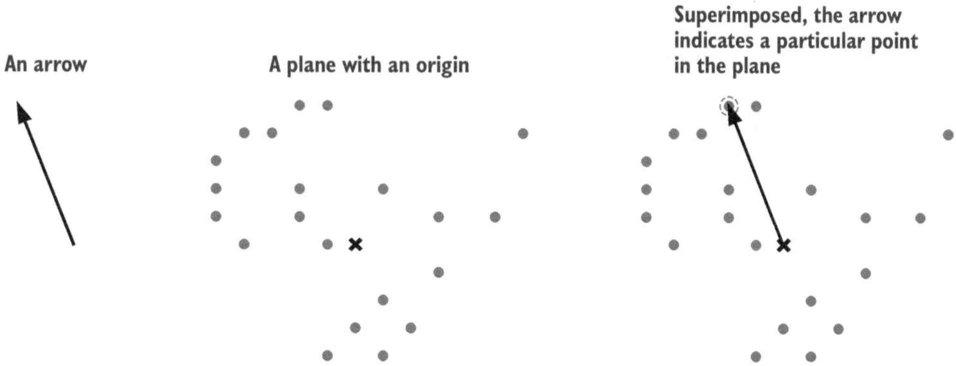

Figure 2.2 Superimposing an arrow on the plane indicates a point relative to the origin.

We'll use both arrows and points to represent vectors in this chapter and beyond. Points are useful to work with because we can build more interesting drawings out of them. If I connect the points in figure 2.2 as in figure 2.3, I get a drawing of a dinosaur:

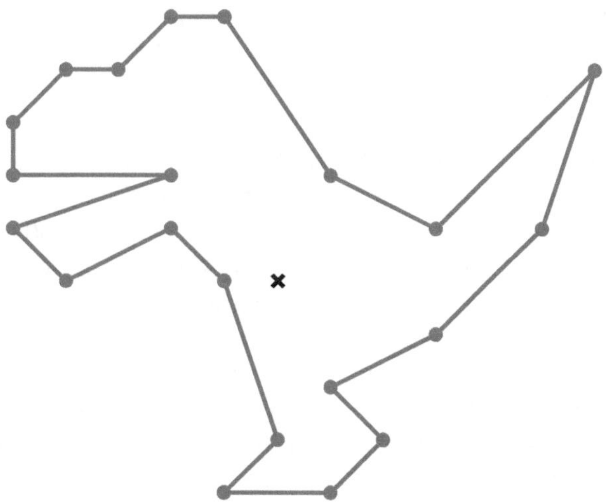

Figure 2.3 Connecting points in the plane to draw a shape

Any time a 2D or 3D drawing is displayed by a computer, from my modest dinosaur to a feature-length Pixar movie, it is defined by points—or vectors—connected to show the desired shape. To create the drawing you want, you need to pick vectors in the right places, requiring careful measurement. Let's take a look at how to measure vectors in the plane.

2.1.1 Representing 2D vectors

With a ruler, we can measure one dimension such as the length of an object. To measure in two dimensions, we need two rulers. These rulers are called *axes* (the singular is *axis*), and we lay them out in the plane perpendicular to one another, intersecting at the origin. Drawn with axes, figure 2.4 shows that our dinosaur has the notions of up and down as well as left and right. The horizontal axis is called the *x-axis* and the vertical one is called the *y-axis*.

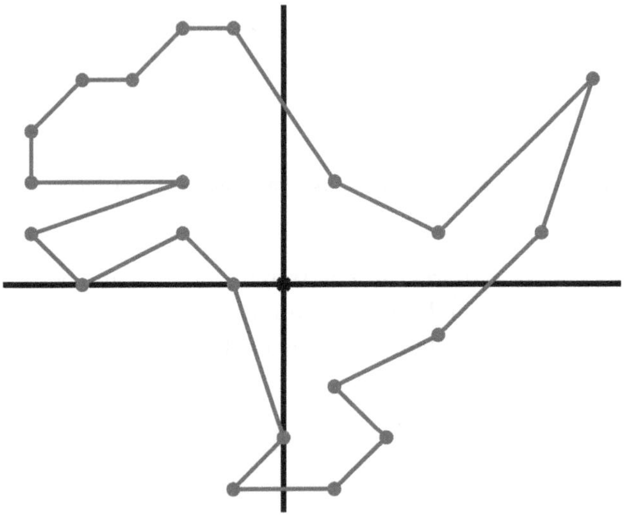

Figure 2.4 The dinosaur drawn with an *x*-axis and a *y*-axis.

With axes to orient us, we can say things like, "Four of the points are above and to the right of the origin." But we'll want to get more quantitative than that. A ruler has tick marks that show how many units along it we've measured. Likewise, in our 2D drawing, we can add grid lines perpendicular to the axes that show where points lie relative to them. By convention, we place the origin at tick 0 on both the *x*- and *y*-axes (figure 2.5).

In the context of this grid, we can measure vectors in the plane. For example, in figure 2.5, the tip of the dinosaur's tail lines up with positive 6 on the *x*-axis and positive 4 on the *y*-axis. We could think of these distances as centimeters, inches, pixels, or any other unit of length, but usually we leave the units unspecified unless we have a particular application in mind.

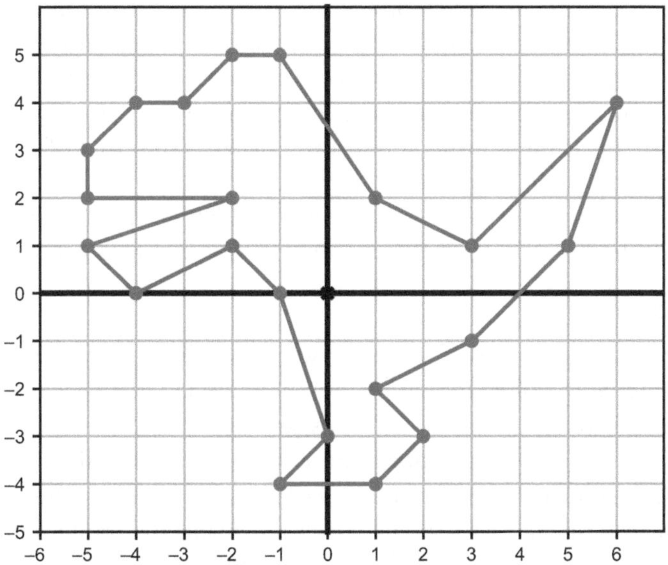

Figure 2.5 Grid lines let us measure the location of points relative to the axes.

The numbers 6 and 4 are called the *x- and y-coordinates* of the point, and this is enough to tell us exactly what point we're talking about. We typically write coordinates as an *ordered pair* (or *tuple*) with the *x*-coordinate first and the *y*-coordinate second, for example, (6, 4). Figure 2.6 shows how we can now describe the same vector in three ways.

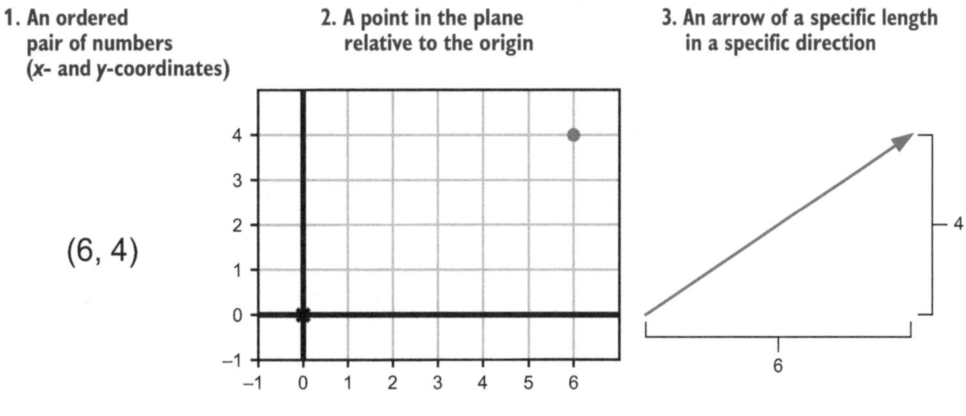

Figure 2.6 Three mental models describing the same vector.

From another pair of coordinates like (–3, 4.5), we can find the point in the plane or the arrow that represents them. To get to the point in the plane with these coordinates, start at the origin and then travel three grid lines to the left (because the *x*-coordinate is –3) and then four and a half grid lines up (where the *y*-coordinate is 4.5). The point won't lie at the intersection of two grid lines, but that's fine; any pair of real numbers gives us some point on the plane. The corresponding arrow will be the straight-line path from the origin to that location, which points up and to the left (northwest, if you prefer). Try drawing this example for yourself as practice!

2.1.2 2D drawing in Python

When you produce an image on a screen, you're working in a 2D space. The pixels on the screen are the available points in that plane. These are labeled by whole number coordinates rather than real number coordinates, and you can't illuminate the space between pixels. That said, most graphics libraries let you work with floating-point coordinates and automatically handle translating graphics to pixels on the screen.

We have plenty of language choices and libraries to specify graphics and to get them on the screen: OpenGL, CSS, SVG, and so on. Python has libraries like Pillow and Turtle that are well equipped for creating drawings with vector data. In this chapter, I use a small set of custom-built functions to create drawings, built on top of another Python library called Matplotlib. This lets us focus on using Python to build images with vector data. Once you understand this process, you'll be able to pick up any of the other libraries easily.

The most important function I've included, called `draw`, takes inputs representing geometric objects and keyword arguments specifying how you want your drawing to look. The Python classes listed in table 2.1 represent each kind of drawable geometric object.

Table 2.1 Some Python classes representing geometric figures, usable with the `draw` function.

Class	Constructor example	Description
`Polygon`	`Polygon(*vectors)`	Draws a polygon whose vertices (corners) are represented by a list of vectors
`Points`	`Points(*vectors)`	Represents a list of points (dots) to draw, one at each of the input vectors
`Arrow`	`Arrow(tip)` `Arrow(tip, tail)`	Draws an arrow from the origin to the `tip` vector or from the `tail` vector to the `head` vector if a tail is specified
`Segment`	`Segment(start,end)`	Draws a line segment from the start to the vector end

You can find these functions implemented in the file vector_drawing.py in the source code. At the end of the chapter, I'll say a bit more about how these are implemented.

> **NOTE** For this chapter (and each subsequent chapter), there is a Jupyter notebook in the source code folder showing how to run (in order) all of the code in the chapter, including importing the functions from the `vector_drawing` module. If you haven't already, you can consult appendix A to get set up with Python and Jupyter.

With these drawing functions in hand, we can draw the points outlining the dinosaur (figure 2.5):

```
from vector_drawing import *
     dino_vectors = [(6,4), (3,1), (1,2), (-1,5), (-2,5), (-3,4), (-4,4),
     # insert 16 remaining vectors here
]

draw(
    Points(*dino_vectors)
)
```

I didn't write out the complete list of `dino_vectors`, but with the suitable collection of vectors, the code gives you the points shown in figure 2.7 (matching figure 2.5 as well).

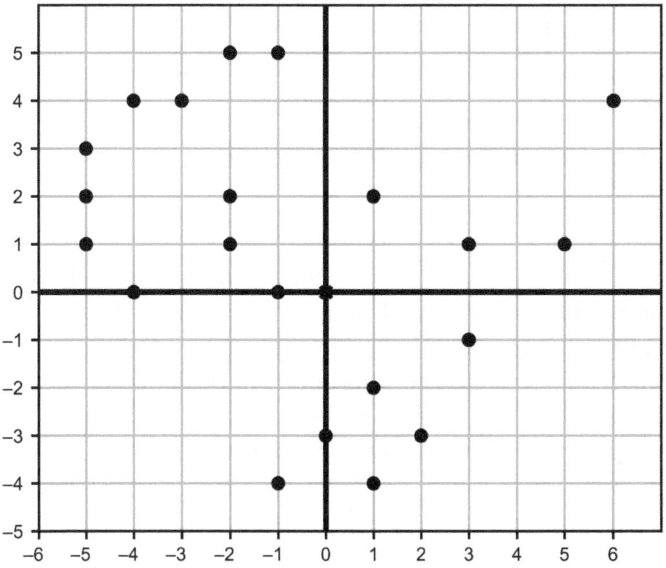

Figure 2.7 Plotting the dinosaur's points with the `draw` function in Python

As a next step in our drawing process, we can connect some dots. A first segment might connect the point (6, 4) with the point (3, 1) on the dinosaur's tail. We can draw the points along with this new segment using this function call, and figure 2.8 shows the results:

```
draw(
    Points(*dino_vectors),
    Segment((6,4),(3,1))
)
```

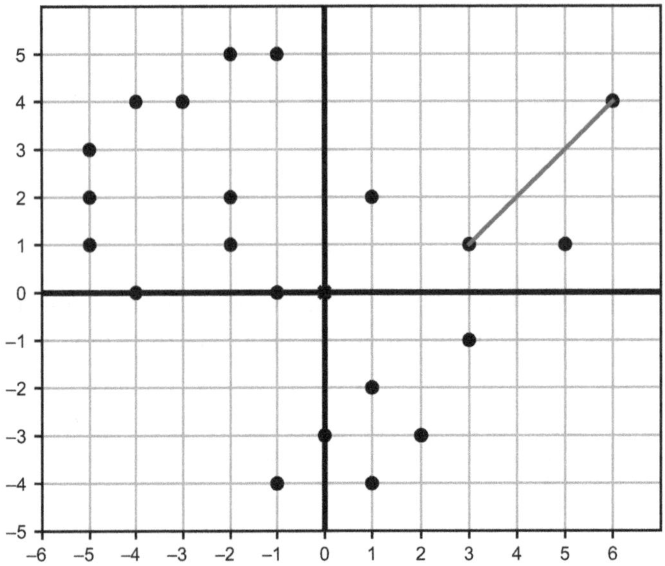

Figure 2.8 The dinosaur's points with a line segment connecting the first two points (6, 4) and (3, 1)

The line segment is actually the collection consisting of the points (6, 4) and (3, 1) as well as all of the points lying on the straight line between them. The `draw` function automatically colors all of the pixels at those points blue. The `Segment` class is a useful abstraction because we don't have to build every segment from the points that make up our geometric object (in this case, the dinosaur). Drawing 20 more segments, we get the complete outline of the dinosaur (figure 2.9).

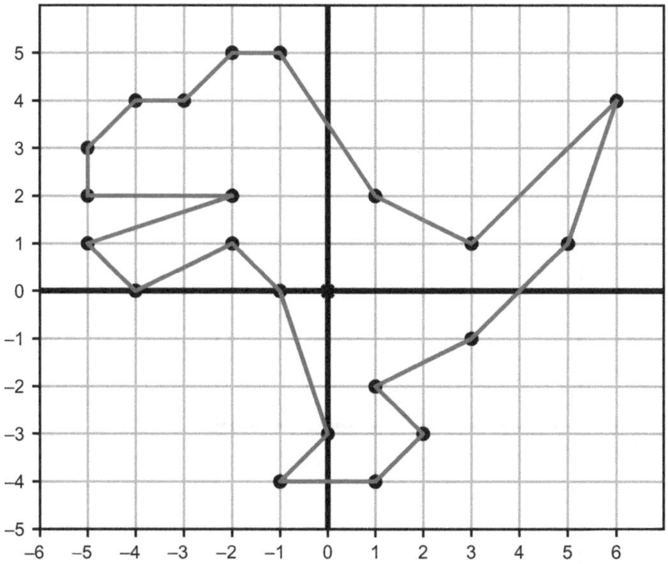

Figure 2.9 A total of 21 function calls give us 21 line segments, completing the outline of the dinosaur.

In principle, we can now outline any kind of 2D shape we want, provided we have all of the vectors to specify it. Coming up with all of the coordinates by hand can be tedious, so we'll start to look at ways to do computations with vectors to find their coordinates automatically.

2.1.3 Exercises

Exercise 2.1 What are the *x*- and *y*-coordinates of the point at the tip of the dinosaur's toe?

Solution (–1, –4)

Exercise 2.2 Draw the point in the plane and the arrow corresponding to the point (2, –2).

Solution Represented as a point on the plane and an arrow, (2, –2) looks like this:

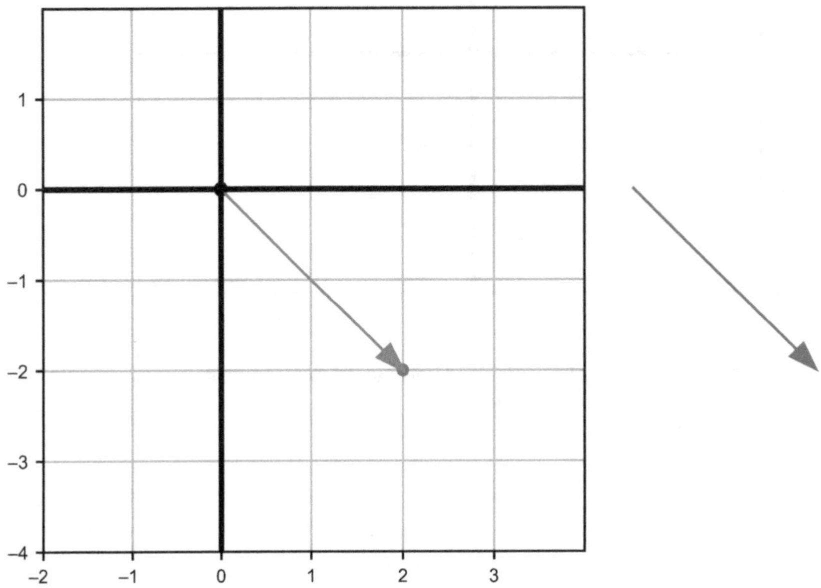

The point and arrow representing (2, –2)

Exercise 2.3 By looking at the locations of the dinosaur's points, infer the remaining vectors not included in the `dino_vectors` list. For instance, I already included (6, 4), which is the tip of the dinosaur's tail, but I didn't include the point (–5, 3), which is a point on the dinosaur's nose. When you're done, `dino_vectors` should be a list of 21 vectors represented as coordinate pairs.

Solution The complete set of vectors outlining the dinosaur is as follows:

```
dino_vectors = [(6,4), (3,1), (1,2), (-1,5), (-2,5), (-3,4), (-4,4),
    (-5,3), (-5,2), (-2,2), (-5,1), (-4,0), (-2,1), (-1,0), (0,-3),
    (-1,-4), (1,-4), (2,-3), (1,-2), (3,-1), (5,1)
]
```

Exercise 2.4 Draw the dinosaur with the dots connected by constructing a `Polygon` object with the `dino_vectors` as its vertices.

Solution

```
draw(
    Points(*dino_vectors),
    Polygon(*dino_vectors)
)
```

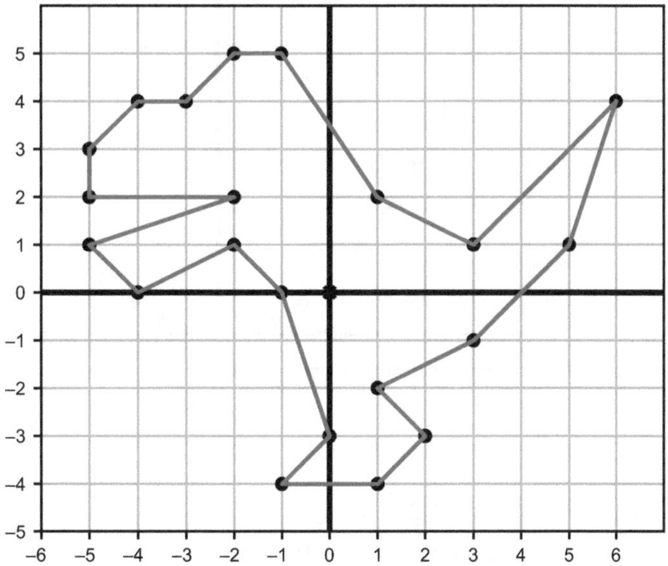

The dinosaur drawn as a polygon.

Exercise 2.5 Draw the vectors `(x,x**2)` for x in the range from $x = -10$ to $x = 11$) as points (dots) using the `draw` function. What is the result?

Solution The pairs draw the graph for the function $y = x^2$, plotted for the integers from 10 to 10:

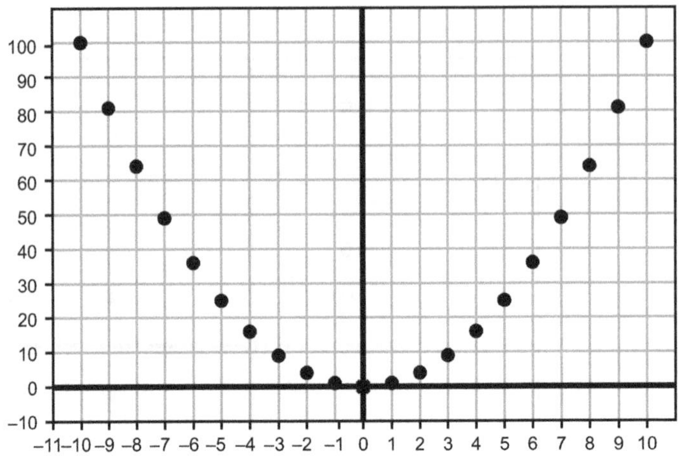

Points on the graph for $y = x^2$

To make this graph, I used two keyword arguments for the `draw` function. The `grid` keyword argument of (1, 10) draws vertical grid lines every one unit and horizontal grid lines every ten units. The `nice_aspect_ratio` keyword argument set to `False` tells the graph it doesn't have to keep the x-axis and the y-axis scales the same:

```
draw(
    Points(*[(x,x**2) for x in range(-10,11)]),
    grid=(1,10),
    nice_aspect_ratio=False
)
```

2.2 Plane vector arithmetic

Like numbers, vectors have their own kind of arithmetic; we can combine vectors with operations to make new vectors. The difference with vectors is that we can visualize the results. Operations from vector arithmetic all accomplish useful geometric transformations, not just algebraic ones. We'll start with the most basic operation: *vector addition.*

Vector addition is simple to calculate: given two input vectors, you add their x-coordinates to get the resulting x-coordinate and then you add their y-coordinates to get the resulting y-coordinate. Creating a new vector with these summed coordinates gives you the *vector sum* of the original vectors. For instance, (4, 3) + (–1, 1) = (3, 4) because 4 + (–1) = 3 and 3 + 1 = 4. Vector addition is a one-liner to implement in Python:

```
def add(v1,v2):
    return (v1[0] + v2[0], v1[1] + v2[1])
```

Because we can interpret vectors as arrows or as points in the plane, we can visualize the result of the addition in both ways (figure 2.10). As a point in the plane, you can reach (–1, 1) by starting at the origin, which is (0, 0), and move one unit to the left and one unit up. You reach the vector sum of (4, 3) + (–1, 1) by starting instead at (4, 3) and moving one unit to the left and one unit up. This is the same as saying you traverse one arrow and then traverse the second.

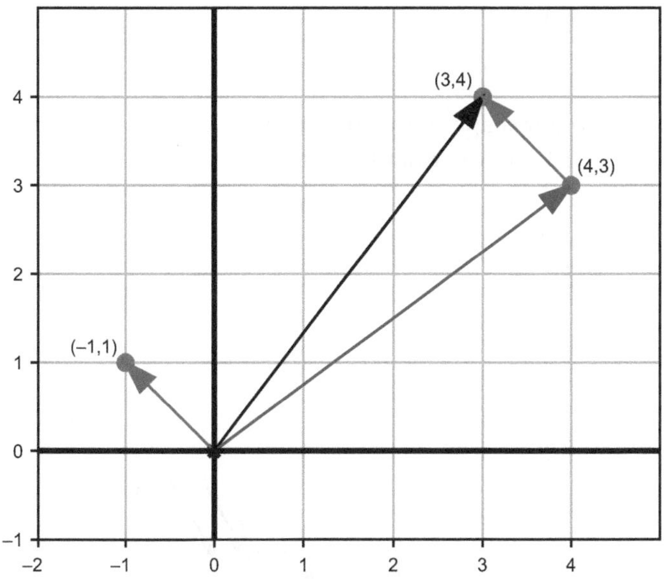

Figure 2.10 Picturing the vector sum of (4, 3) and (–1, 1)

The rule for vector addition of arrows is sometimes called *tip-to-tail* addition. That's because if you move the tail of the second arrow to the tip of the first (without changing its length or direction!), then the sum is the arrow from the start of the first to the end of the second (figure 2.11).

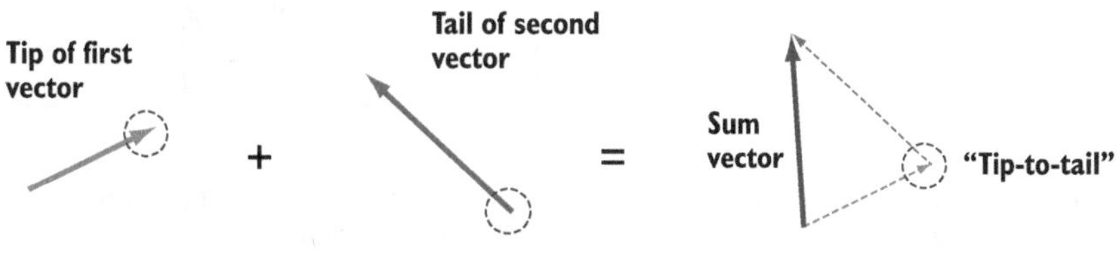

Figure 2.11 Tip-to-tail addition of vectors

When we talk about arrows, we really mean "a specific distance in a specific direction." If you walk one distance in one direction and another distance in another direction, the vector sum tells you the overall distance and direction you traveled (figure 2.12).

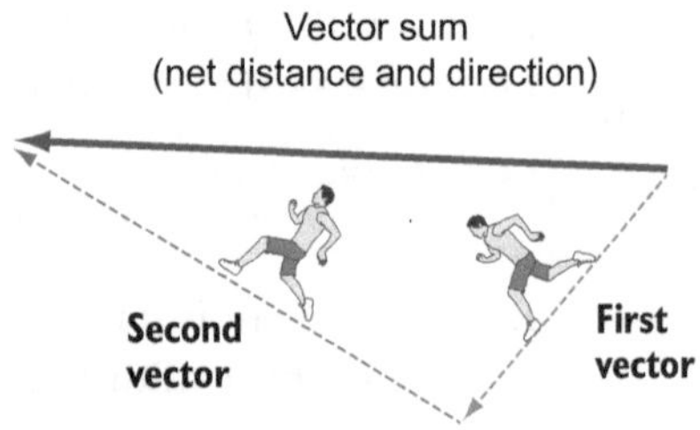

Figure 2.12 The vector sum as an overall distance and direction traveled in the plane.

Adding a vector has the effect of moving or *translating* an existing point or collection of points. If we add the vector (–1.5, –2.5) to every vector of dino_vectors, we get a new list of vectors, each of which is 1.5 units left and 2.5 units down from one of the original vectors. Here's the code for that:

```
dino_vectors2 = [add((-1.5,-2.5), v) for v in dino_vectors]
```

The result is the same dinosaur shape shifted down and to the left by the vector (–1.5, –2.5). To see this (figure 2.13), we can draw both dinosaurs as polygons:

```
draw(
    Points(*dino_vectors, color=blue),
    Polygon(*dino_vectors, color=blue),
    Points(*dino_vectors2, color=red),
    Polygon(*dino_vectors2, color=red)
)
```

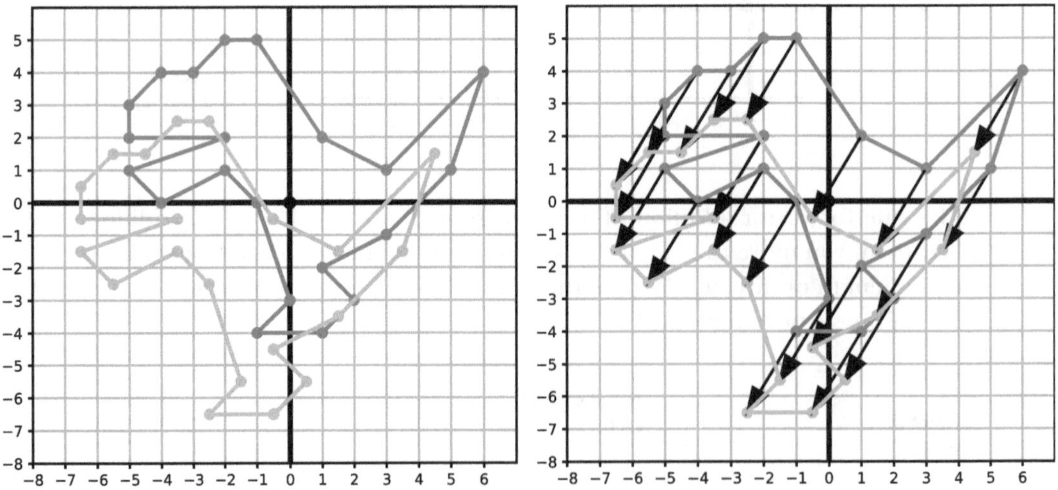

Figure 2.13 The original dinosaur (blue) and the translated copy (red). Each point on the translated dinosaur is moved by (–1.5, –2.5) down and to the left from its location on the original dinosaur.

The arrows in the copy on the right show that each point moved down and to the left by the same vector: (–1.5, –2.5). A translation like this is useful if, for instance, we want to make the dinosaur a moving character in a 2D computer game. Depending on the button pressed by the user, the dinosaur could translate in the corresponding direction on the screen. We'll implement a real game like this with moving vector graphics in chapters 7 and 9.

2.2.1 Vector components and lengths

Sometimes it's useful to take a vector we already have and decompose it as a sum of smaller vectors. For example, if I were asking for walking directions in New York City, it would be much more useful to hear "go four blocks east and three blocks north" rather than "go 800 meters northeast." Similarly, it can be useful to think of vectors as a sum of a vector pointing in the *x* direction and a vector pointing in the *y* direction.

As an example, figure 2.14 shows the vector (4, 3) rewritten as the sum (4, 0) + (0, 3). Thinking of the vector (4, 3) as a navigation path in the plane, the sum (4, 0) + (0, 3) gets us to the same point along a different path.

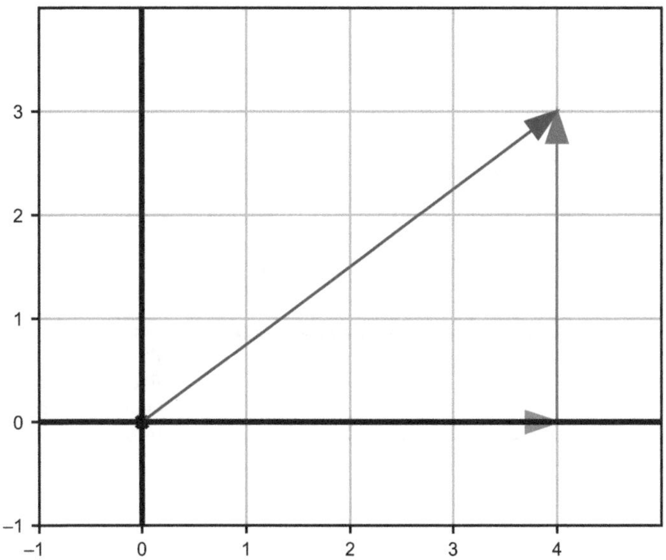

Figure 2.14 Breaking the vector (4, 3) into the sum (4, 0) + (0, 3)

The two vectors (4, 0) and (0, 3) are called the *x and y components*, respectively. If you couldn't walk diagonally in this plane (as if it were New York City), you would need to walk four units to the right and then three units up to get to the same destination, a total of seven units.

The *length* of a vector is the length of the arrow that represents it, or equivalently, the distance from the origin to the point that represents it. In New York City, this could be the distance between two intersections "as the crow flies." The length of a vector in the *x* or *y* direction can be measured immediately as a number of ticks passed on the corresponding axis: (4, 0) or (0, 4) are both vectors of the same length, 4, albeit in different directions. In general, though, vectors can lie diagonally, and we need to do a calculation to get their lengths.

You may recall the relevant formula: the *Pythagorean theorem.* For a right triangle (a triangle having two sides meeting at a 90° angle), the Pythagorean theorem says that the square of the length of the longest side is the sum of squares of the lengths of the other two sides. The longest side is called the *hypotenuse,* and its length is denoted by c in the memorable formula $a^2 + b^2 = c^2$, where a and b are the lengths of the other two sides. With $a = 4$ and $b = 3$, we can find c as the square root of $4^2 + 3^2$ (figure 2.15).

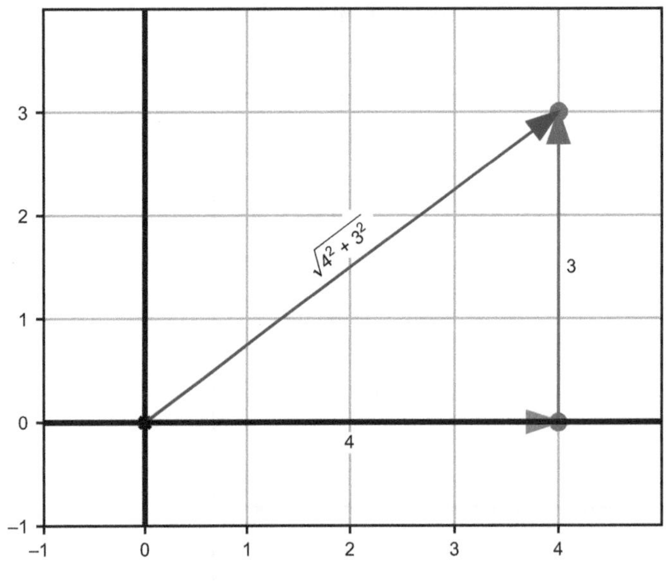

Figure 2.15 Using the Pythagorean theorem to find the length of a vector from the lengths of its *x*- and *y*- components

Breaking a vector into components is handy because it always gives us a right triangle. If we know the lengths of the components, we can compute the length of the hypotenuse, which is the length of the vector. Our vector (4, 3) is equal to (4, 0) + (0, 3), a sum of two perpendicular vectors whose sides are 4 and 3, respectively. The length of the vector (4, 3) is the square root of $4^2 + 3^2$, which is the square root of 25, or 5. In a city with perfectly square blocks, traveling 4 blocks east and 3 blocks north would take us the equivalent of 5 blocks northeast.

This is a special case where the distance turns out to be an integer, but typically, lengths that come out of the Pythagorean theorem are not whole numbers. The length of (–3, 7) is given in terms of the lengths of its components 3 and 7 by the following computation:

$$\sqrt{3^2 + 7^2} = \sqrt{9 + 49} = \sqrt{58} = 7.61577\ldots$$

We can translate this formula into a `length` function in Python, which takes a 2D vector and returns its floating-point length:

```
from math import sqrt
def length(v):
    return sqrt(v[0]**2 + v[1]**2)
```

2.2.2 Multiplying vectors by numbers

Repeated addition of vectors is unambiguous; you can keep stacking arrows tip-to-tail as long as you want. If a vector named **v** has coordinates (2, 1), then the fivefold sum **v** + **v** + **v** + **v** + **v** would look like that shown in figure 2.16.

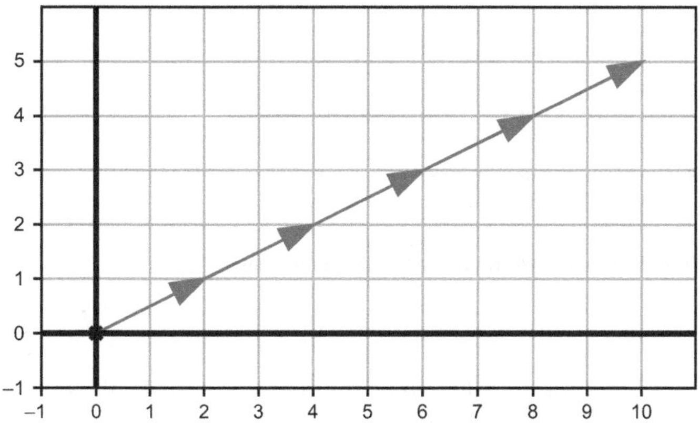

Figure 2.16 Repeated addition of the vector v = (2, 1) with itself

If **v** were a number, we wouldn't bother writing **v** + **v** + **v** + **v** + **v**. Instead, we'd write the simpler product 5 · **v**. There's no reason we can't do the same for vectors. The result of adding **v** to itself 5 times is a vector in the same direction but with 5 times the length. We can run with this definition, which lets us multiply a vector by any whole or fractional number.

The operation of multiplying a vector by a number is called *scalar multiplication*. When working with vectors, ordinary numbers are often called *scalars*. It's also an appropriate term because the effect of this operation is *scaling* the target vector by the given factor. It doesn't matter if the scalar is a whole number; we can easily draw a vector that is 2.5 times the length of another (figure 2.17).

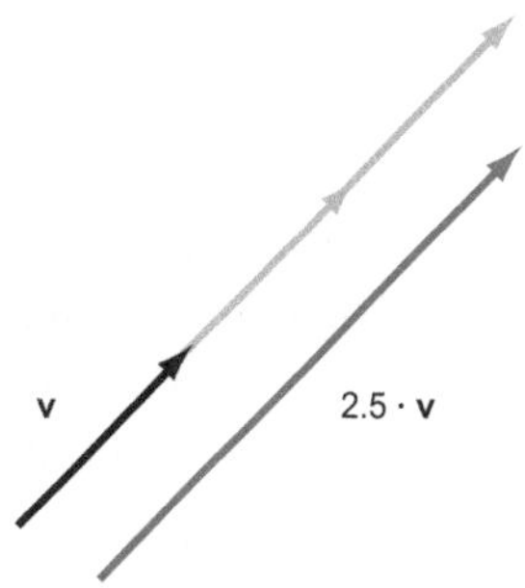

Figure 2.17 Scalar multiplication of a vector v by 2.5

The result on the vector components is that each component is scaled by the same factor. You can picture scalar multiplication as changing the size of the right triangle defined by a vector and its components, but not affecting its aspect ratio. Figure 2.18 superimposes a vector **v** and its scalar multiple 1.5 · **v**, where the scalar multiple is 1.5 times as long. Its components are also 1.5 times the length of the original components of **v**.

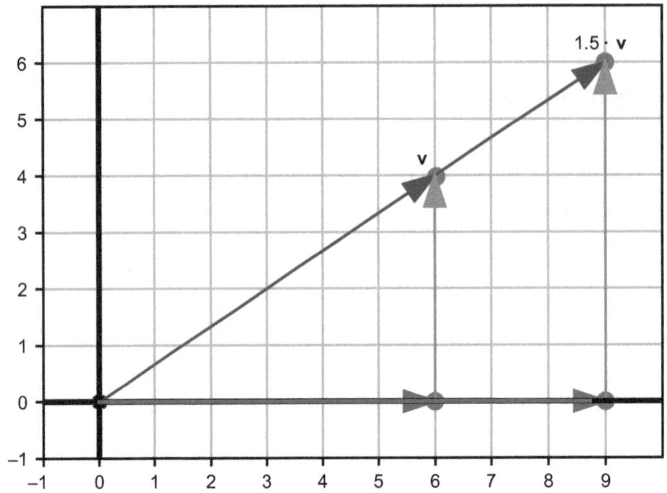

Figure 2.18 Scalar multiplication of a vector scales both components by the same factor.

In coordinates, the scalar multiple of 1.5 times the vector **v** = (6, 4) gives us a new vector (9, 6), where each component is 1.5 times its original value. Computationally, we execute any scalar multiplication on a vector by multiplying each coordinate of the vector by the scalar. As a second example, scaling a vector **w** = (1.2, –3.1) by a factor 6.5 can be accomplished like this:

$$6.5 \cdot \mathbf{w} = 6.5 \cdot (1.2, -3.1) = (6.5 \cdot 1.2, 6.5 \cdot -3.1) = (7.8, -20.15)$$

We tested this method for a fractional number as the scalar, but we should also test a negative number. If our original vector is (6, 4), what is –½ times that vector? Multiplying the coordinates, we expect the answer to be (–3, –2). Figure 2.19 shows that this vector is half the length of the original and points in the opposite direction.

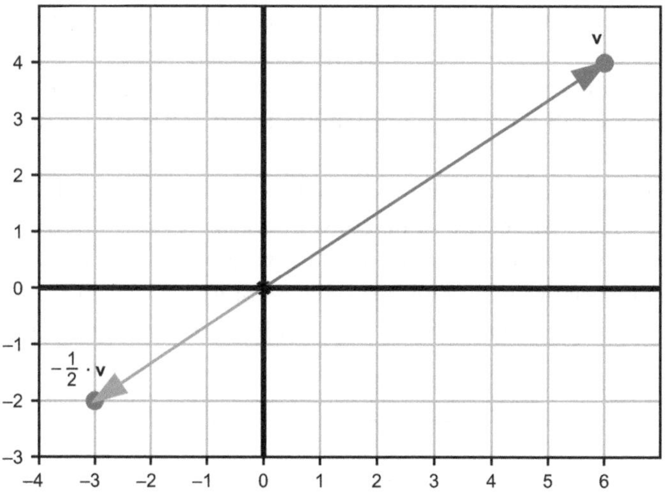

Figure 2.19 Scalar multiplication of a vector by a negative number, –½

2.2.3 *Subtraction, displacement, and distance*

Scalar multiplication agrees with our intuition for multiplying numbers. A whole number multiple of a number is the same as a repeated sum, and the same holds for vectors. We can make a similar argument for negative vectors and vector subtraction.

Given a vector **v**, the *opposite* vector, –**v**, is the same as the scalar multiple –1 · **v**. If **v** is (–4, 3), its opposite, –**v**, is (4, –3) as shown in figure 2.20. We get this by multiplying each coordinate by –1, or in other words, changing the sign of each.

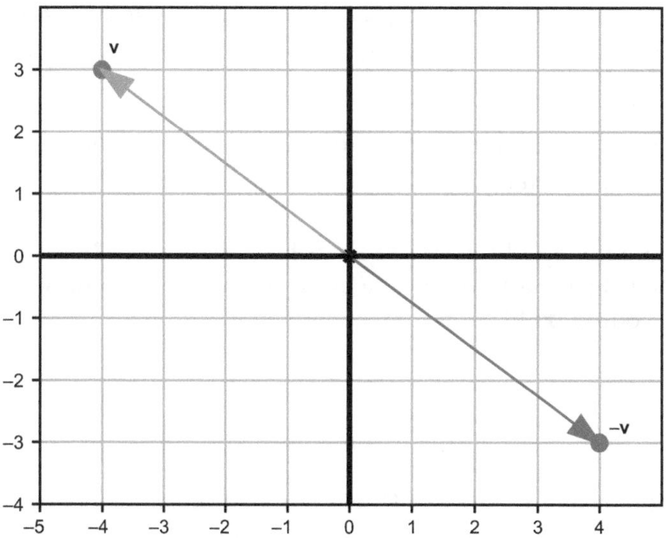

Figure 2.20 The vector v = (–4, 3) and its opposite –v = (4, –3).

On the number line, there are only two directions from zero: positive and negative. In the plane, there are many directions (infinitely many, in fact), so we can't say that one of **v** and −**v** is positive while the other is negative. What we can say is that for any vector **v**, the opposite vector −**v** will have the same length, but it will point in the opposite direction.

Having a notion of negating a vector, we can define *vector subtraction*. For numbers, $x - y$ is the same as $x + (-y)$. We set the same convention for vectors. To subtract a vector **w** from a vector **v**, you add the vector −**w** to **v**. Thinking of vectors **v** and **w** as points, **v** − **w** is the position of **v** relative to **w**. Thinking instead of **v** and **w** as arrows beginning at the origin, figure 2.21 shows that **v** − **w** is the arrow from the tip of **w** to the tip of **v**.

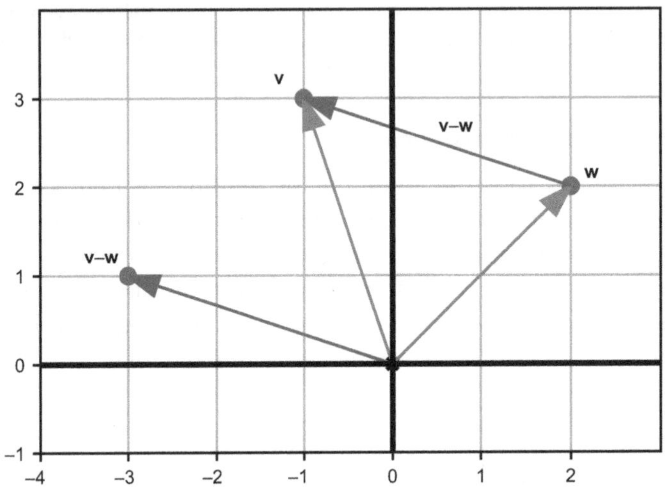

Figure 2.21 The result of subtracting v − w is an arrow from the tip of w to the tip of v.

The coordinates of **v** − **w** are the differences of the coordinates **v** and **w**. In figure 2.21, **v** = (−1, 3) and **w** = (2, 2). The difference for **v** − **w** has the coordinates (−1 − 2, 3 − 2) = (−3, 1).

Let's look at the difference of the vectors **v** = (−1, 3) and **w** = (2, 2) again. You can use the `draw` function I gave you to plot the points **v** and **w** and to draw a segment between them. The code looks like this:

```
draw(
    Points((2,2), (-1,3)),
    Segment((2,2), (-1,3), color=red)
)
```

The difference for the vectors **v** − **w** = (−3, 1) tells us that if we start at point *w*, we need to go three units left and one unit up to get to point *v*. This vector is sometimes called the *displacement* from **w** to **v**. The straight, red line segment from **w** to **v** in figure 2.22, drawn by this Python code, shows the *distance* between the two points.

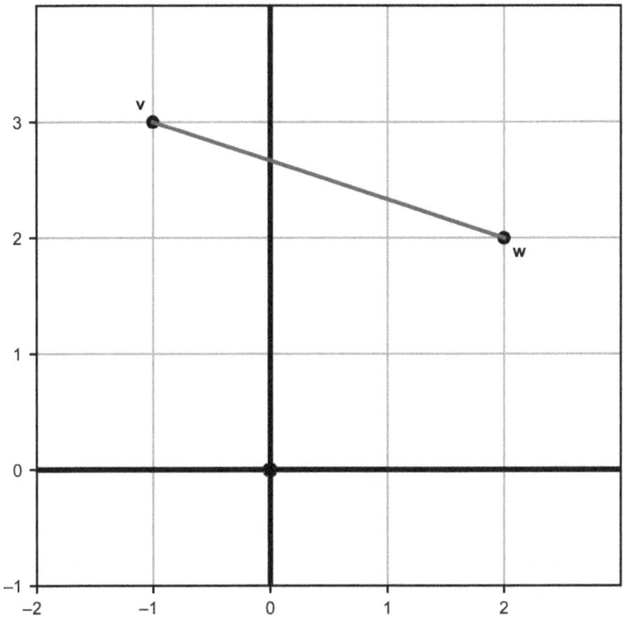

Figure 2.22 The distance between two points in the plane

The length of the line segment is computed with the Pythagorean theorem as follows:

$$\sqrt{(-3)^2 + 1^2} = \sqrt{9+1} = \sqrt{10} = 3.162\ldots$$

While the displacement is a vector, the distance is a scalar (a single number). The distance on its own is not enough to specify how to get from **w** to **v**; there are plenty of points that have the same distance from **w**. Figure 2.23 shows a few others with whole number coordinates.

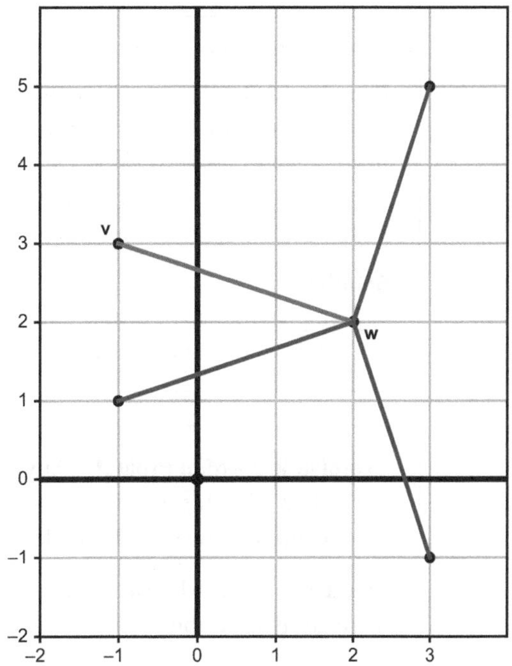

Figure 2.23 Several points equidistant from *w* = (2, 2)

2.2.4 Exercises

Exercise 2.6 If the vector **u** = (–2, 0), the vector **v** = (1.5, 1.5), and the vector **w** = (4, 1), what are the results of **u** + **v**, **v** + **w**, and **u** + **w**? What is the result of **u** + **v** + **w**?

Solution With the vector **u** = (–2, 0), the vector **v** = (1.5, 1.5), and the vector **w** = (4, 1), the results are as follows:

u + **v** = (–0.5, 1.5)
v + **w** = (5.5, 2.5)
u + **w** = (2, 1)
u + **v** + **w** = (3.5, 2.5)

Exercise 2.7—Mini Project You can add any number of vectors together by summing *all* of their *x*-coordinates and *all* of their *y*-coordinates. For instance, the fourfold sum (1, 2) + (2, 4) + (3, 6) + (4, 8) has *x* component 1 + 2 + 3 + 4 = 10 and *y* component 2 + 4 + 6 + 8 = 20, making the result (10, 20). Implement a revised add function that takes any number of vectors as arguments.

Solution

```
def add(*vectors):
    return (sum([v[0] for v in vectors]), sum([v[1] for v in vectors]))
```

Exercise 2.8 Write a function `translate(translation, vectors)` that takes a translation vector and a list of input vectors, and returns a list of the input vectors all translated by the translation vector. For instance, `translate ((1,1), [(0,0), (0,1,), (-3,-3)])` should return `[(1,1),(1,2),(-2, -2)]`.

Solution

```
def translate(translation, vectors):
    return [add(translation, v) for v in vectors]
```

Exercise 2.9—Mini Project Any sum of vectors **v** + **w** gives the same result as **w** + **v**. Explain why this is true using the definition of the vector sum on coordinates. Also, draw a picture to show why it is true geometrically.

Solution If you add two vectors **u** = (*a*, *b*) and **v** = (*c*, *d*), the coordinates *a*, *b*, *c*, and *d* are all real numbers. The result of the vector addition is **u** + **v** = (*a* + *c*, *b* + *d*). The result of **v** + **u** is $(c + a, d + b)$, which is the same pair of coordinates because order doesn't matter when adding real numbers. Tip-to-tail addition in either order yields the same sum vector. Visually, we can see this by adding an example pair of vectors tip-to-tail:

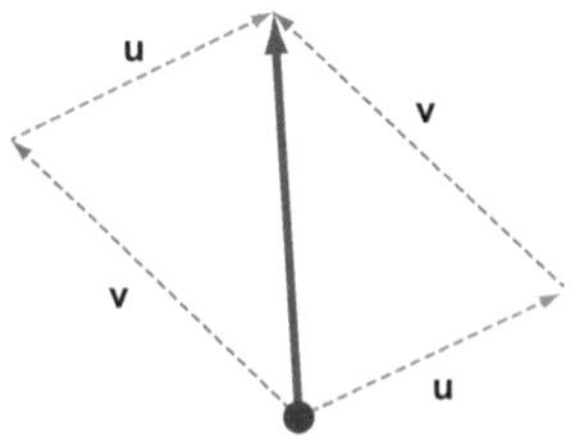

Tip-to-tail addition in either order yields the same sum vector.

It doesn't matter whether you add **u** + **v** or **v** + **u** (dashed lines), you get the same result vector (solid line). In geometric terms, **u** and **v** define a parallelogram, and the vector sum is the length of the diagonal.

Exercise 2.10 Among the following three arrow vectors (labeled **u**, **v**, and **w**), which pair has the sum that gives the *longest* arrow? Which pair sums to give the *shortest* arrow?

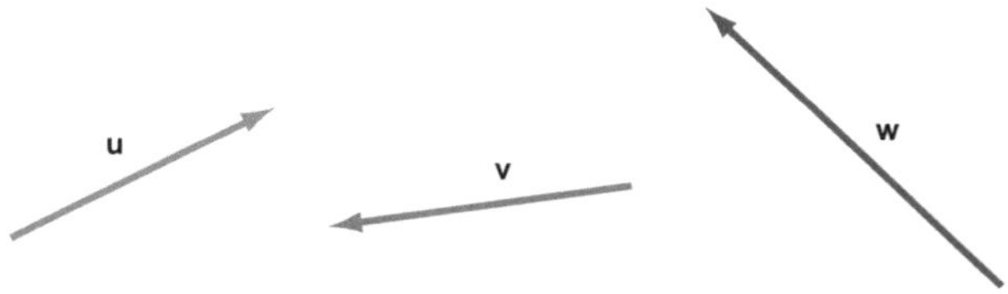

Which pair sums to the longest or shortest arrow?

Solution We can measure each of the vector sums by placing the vectors tip-to-tail:

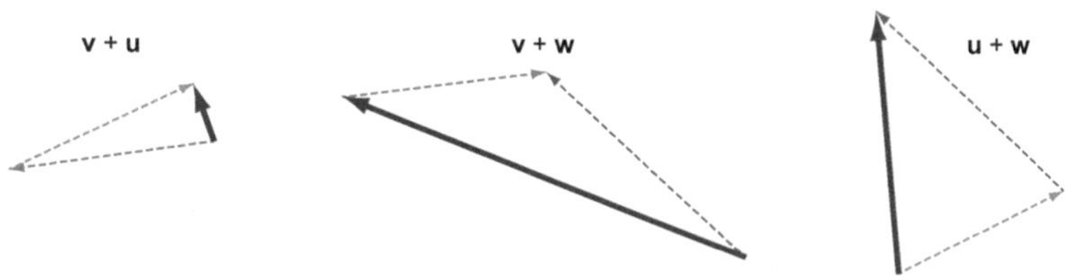

Tip-to-tail addition of the vectors in question

Inspecting the results, we can see that **v** + **u** is the shortest vector (**u** and **v** are in nearly opposite directions and come close to canceling each other out). The longest vector is **v** + **w**.

Exercise 2.11—Mini Project Write a Python function using vector addition to show 100 simultaneous and non-overlapping copies of the dinosaur. This shows the power of computer graphics; imagine how tedious it would be to specify all 2,100 coordinate pairs by hand!

Solution With some trial and error, you can translate the dinosaurs in the vertical and horizontal direction so that they don't overlap, and set the boundaries appropriately. I decided to leave out the grid lines, axes, origin, and points to make the drawing clearer. My code looks like this:

```
def hundred_dinos():
    translations = [(12*x,10*y)
                    for x in range(-5,5)
                    for y in range(-5,5)]
    dinos = [Polygon(*translate(t, dino_vectors),color=blue)
                for t in translations]
    draw(*dinos, grid=None, axes=None, origin=None)

hundred_dinos()
```

The result is as follows:

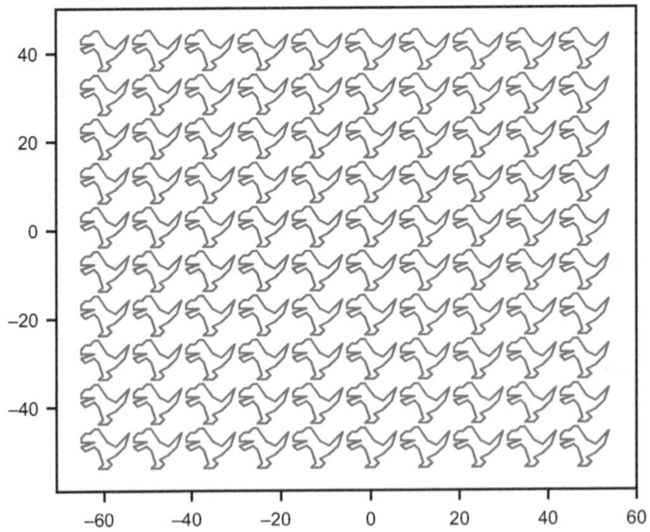

100 dinosaurs. Run for your life!

Exercise 2.12 Which is longer, the *x* or *y* component of (3, –2) + (1, 1) + (–2, –2)?

Solution The result of the vector sum (3, –2) + (1, 1) + (–2, –2) is (2, –3). The *x* component is (2, 0) and the *y* component is (0, –3). The *x* component has a length of 2 units (to the right), while the *y* component has a length of 3 units (downward because it is negative). This makes the *y* component longer.

Exercise 2.13 What are the components and lengths of the vectors (–6, –6) and (5, –12)?

Solution The components of (–6, –6) are (–6, 0) and (0, –6), both having length 6. The length of (–6, –6) is the square root of $6^2 + 6^2$, which is approximately 8.485.

The components of (5, –12) are (5, 0) and (0, –12), having lengths of 5 and 12, respectively. The length of (5, –12) is given by the square root of $5^2 + 12^2 = 25 + 144 = 169$. The result of the square root is exactly 13.

Exercise 2.14 Suppose I have a vector **v** that has a length of 6 and an *x* component (1, 0). What are the possible coordinates of **v**?

Solution The *x* component of (1, 0) has length 1 and the total length is 6, so the length *b* of the *y* component must satisfy the equation $1^2 + b^2 = 6^2$, or $1 + b^2 = 36$. Then $b^2 = 35$ and the length of the *y* component is approximately 5.916. This doesn't tell us the direction of the *y* component, however. The vector **v** could either be (1, 5.916) or (1, –5.916).

Exercise 2.15 What vector in the `dino_vectors` list has the longest length? Use the `length` function we wrote to compute the answer quickly.

Solution

```
>>> max(dino_vectors, key=length)
(6, 4)
```

Exercise 2.16 Suppose a vector **w** has the coordinates $(\sqrt{2}, \sqrt{3})$. What are the approximate coordinates of the scalar multiple $\pi \cdot \mathbf{w}$? Draw an approximation of the original vector and the new vector.

Solution The value of $(\sqrt{2}, \sqrt{3})$ is approximately

(1.4142135623730951, 1.7320508075688772)

Scaling each coordinate by a factor of π (pi), we get

(4.442882938158366, 5.441398092702653)

The scaled vector is longer than the original as shown here:

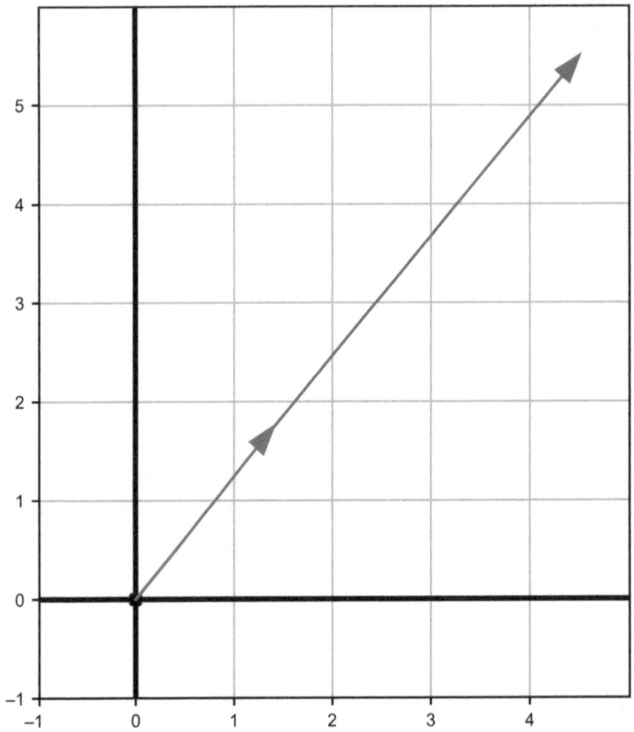

The original vector (shorter) and its scaled version (longer)

Exercise 2.17 Write a Python function `scale(s,v)` that multiplies the input vector **v** by the input scalar *s*.

Solution

```
def scale(scalar,v):
            return (scalar * v[0], scalar * v[1])
```

Exercise 2.18—Mini Project Convince yourself algebraically that scaling the coordinates by a factor also scales the length of the vector by the same factor. Suppose a vector of length c has the coordinates (a, b). Show that for any non-negative real number s, the length of $(s \cdot a, s \cdot b)$ is $s \cdot c$. (This can't work for a negative value of s because a vector can't have a negative length.)

Solution We use the notation $|(a, b)|$ to denote the length of a vector (a, b). So, the premise of the exercise tells us:

$$c = \sqrt{a^2 + b^2} = |(a, b)|$$

From that, we can compute the length of (sa, sb):

$$\begin{aligned} |(sa, sb)| &= \sqrt{(sa)^2 + (sb)^2} \\ &= \sqrt{s^2a^2 + s^2b^2} \\ &= \sqrt{s^2 \cdot (a^2 + b^2)} \\ &= |s| \cdot \sqrt{a^2 + b^2} \\ &= |s| \cdot c \end{aligned}$$

As long as s isn't negative, it's the same as its absolute value: $s = |s|$. Then the length of the scaled vector is sc as we hoped to show.

Exercise 2.19—Mini Project Suppose $\mathbf{u} = (-1, 1)$ and $\mathbf{v} = (1, 1)$, and suppose r and s are real numbers. Specifically, let's assume $-3 < r < 3$ and $-1 < s < 1$. Where are the possible points on the plane where the vector $r \cdot \mathbf{u} + s \cdot \mathbf{v}$ could end up?

Note that the order of operations is the same for vectors as it is for numbers. We assume scalar multiplication is carried out first and then vector addition (unless parentheses specify otherwise).

Solution If $\mathbf{r} = 0$, the possibilities lie on the line segment from $(-1, -1)$ to $(1, 1)$. If $\mathbf{r}$ is not zero, the possibilities can leave that line segment in the direction of $(-1, 1)$ or $-(-1, 1)$ by up to three units. The region of possible results is the parallelogram with vertices at $(2, 4)$, $(4, 2)$, $(2, -4)$, and $(4, -2)$. We can test many random, allowable values of $\mathbf{r}$ and $\mathbf{s}$ to validate this:

```
from random import uniform
u = (-1,1)
v = (1,1)
def random_r():
    return uniform(-3,3)
def random_s():
    return uniform(-1,1)
```

(continued)

```
possibilities = [add(scale(random_r(), u), scale(random_s(), v))
                 for i in range(0,500)]
draw(
    Points(*possibilities)
)
```

If you run this code, you get a picture like the following, showing the possible points where $r \cdot \mathbf{u} + s \cdot \mathbf{v}$ could end up given the constraints:

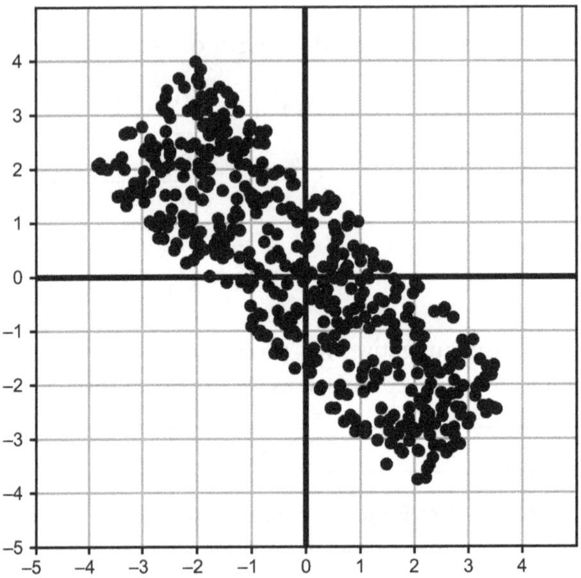

Location of possible points for r •u + s •v given the constraints.

Exercise 2.20 Show algebraically why a vector and its opposite have the same length.

Hint Plug the coordinates and their opposites into the Pythagorean theorem.

Solution The opposite vector of (a, b) has coordinates $(-a, -b)$, but this doesn't affect the length:

$$\sqrt{(-a)^2 + (-b)^2} = \sqrt{(-a) \cdot (-a) + (-b) \cdot (-b)} = \sqrt{a^2 + b^2}$$

The vector (–*a*, –*b*) has the same length as (*a*, *b*).

Exercise 2.21 Of the following seven vectors, represented as arrows, which two are a pair of opposite vectors?

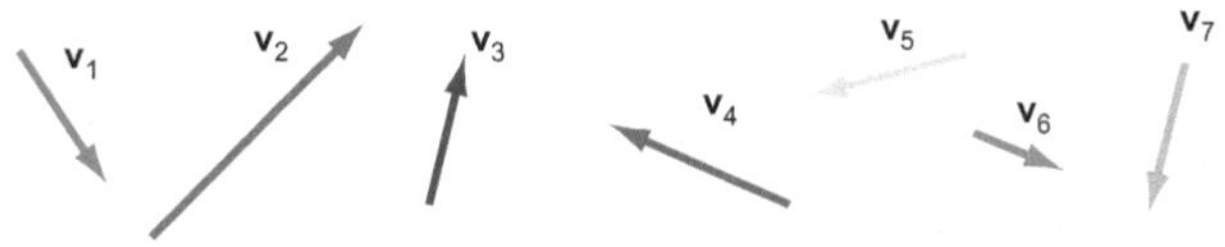

Solution Vectors v_3 and v_7 are the pair of opposite vectors.

Exercise 2.22 Suppose **u** is any 2D vector. What are the coordinates of **u** + –**u**?

Solution A 2D vector **u** has some coordinates (a, b). Its opposite has coordinates $(-a, -b)$, so:

$$u + (-u) = (a, b) + (-a, -b) = (a - a, b - b) = (0, 0)$$

The answer is (0, 0). Geometrically, this means that if you follow a vector and then its opposite, you end up back at the origin, (0, 0).

Exercise 2.23 For vectors $\mathbf{u} = (-2, 0)$, $\mathbf{v} = (1.5, 1.5)$, and $\mathbf{w} = (4, 1)$, what are the results of the vector subtractions $\mathbf{v} - \mathbf{w}$, $\mathbf{u} - \mathbf{v}$, and $\mathbf{w} - \mathbf{v}$?

Solution With $\mathbf{u} = (-2, 0)$, $\mathbf{v} = (1.5, 1.5)$, and $\mathbf{w} = (4, 1)$, we have

$\mathbf{v} - \mathbf{w} = (-2.5, 0.5)$
$\mathbf{u} - \mathbf{v} = (-3.5, -1.5)$
$\mathbf{w} - \mathbf{v} = (2.5, -0.5)$

Exercise 2.24 Write a Python function `subtract(v1,v2)` that returns the result of `v1` – `v2`, taking two 2D vectors as inputs and returning a 2D vector as an output.

Solution

```
def subtract(v1,v2):
    return (v1[0] - v2[0], v1[1] - v2[1])
```

Exercise 2.25 Write a Python function `distance(v1,v2)` that returns the *distance* between two input vectors. (Note that the `subtract` function from the previous exercise already gives the *displacement*.)

Write another Python function `perimeter(vectors)` that takes a list of vectors as an argument and returns the sum of distances from each vector to the next, including the distance from the last vector to the first. What is the perimeter of the dinosaur defined by `dino_vectors`?

Solution The distance is just the length of the difference of the two input vectors:

```
def distance(v1,v2):
    return length(subtract(v1,v2))
```

For the perimeter, we sum the distances of every pair of subsequent vectors in the list, as well as the pair of the first and the last vectors:

```
def perimeter(vectors):
    distances = [distance(vectors[i], vectors[(i+1)%len(vectors)])
                 for i in range(0,len(vectors))]
    return sum(distances)
```

We can use a square with side length of one as a sanity check:

```
>>> perimeter([(1,0),(1,1),(0,1),(0,0)])
4.0
```

Then we can calculate the perimeter of the dinosaur:

```
>>> perimeter(dino_vectors)
44.77115093694563
```

Exercise 2.26—Mini Project Let **u** be the vector (1, –1). Suppose there is another vector **v** with positive integer coordinates (n, m) such that $n > m$ and has a distance of 13 from **u**. What is the displacement from **u** to **v**?

Hint You can use Python to search for the vector **v**.

Solution We only need to search possible integer pairs (n, m) where n is within 13 units of 1 and m is within 13 units of –1:

```
for n in range(-12,15):
    for m in range(-14, 13):
        if distance((n,m), (1,-1)) == 13 and n > m > 0:
            print((n,m))
```

There is one result: (13, 4). It is 12 units to the right and 5 units up from (1, –1), so the displacement is (12, 5).

The length of a vector is not enough to describe it, nor is the distance between two vectors enough information to get from one to the other. In both cases, the missing ingredient is *direction*. If you know how long a vector is and you know what direction it is pointing, you can identify it and find its coordinates. To a large extent, this is what *trigonometry* is about, and we'll review that subject in the next section.

2.3 Angles and trigonometry in the plane

So far, we've used two "rulers" (called the x-axis and the y-axis) to measure vectors in the plane. An arrow from the origin covers some measurable displacement in the horizontal and vertical directions, and these values uniquely specify the vector. Instead of using two rulers, we could just as well use a ruler and a protractor. Starting with the vector (4, 3), we can measure or calculate its length to be 5 units, and then use our protractor to identify the direction as shown in figure 2.24.

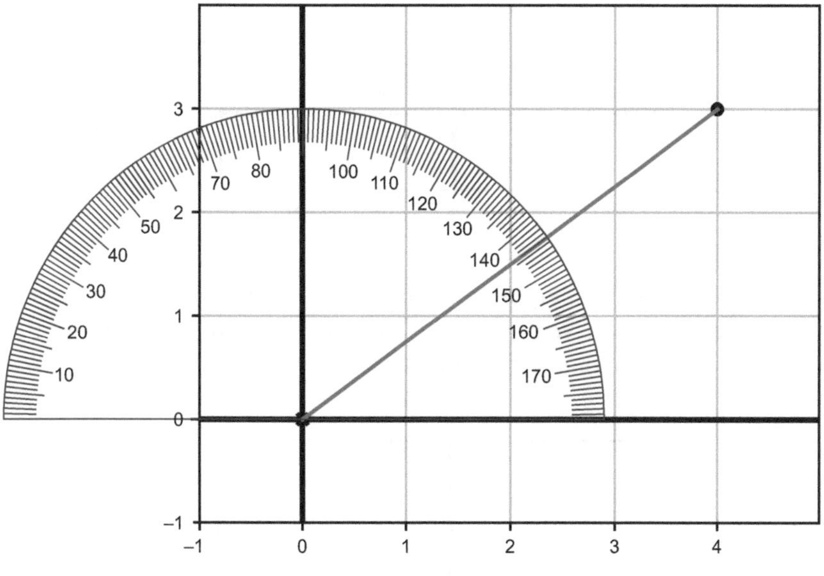

Figure 2.24 Using a protractor to measure the angle at which a vector points

This vector has a length of 5 units, and it points in a direction approximately 37° counterclockwise from the positive half of the x-axis. This gives us a new pair of numbers (5, 37°) that, like our original coordinates, uniquely specify the vector. These numbers are called *polar coordinates* and are just as good at describing points in the plane as the ones we've worked with so far, called *Cartesian coordinates.*

Sometimes, like when we're adding vectors, it's easier to use Cartesian coordinates. Other times, polar coordinates are more useful; for instance, when we want to look at vectors rotated by some angle. In code, we don't have literal rulers or protractors available, so we use trigonometric functions to convert back and forth instead.

2.3.1 From angles to components

Let's look at the reverse problem: imagine we already have an angle and a distance, say, 116.57° and 3. These define a pair of polar coordinates (3, 116.57°). How can we find the Cartesian coordinates for this vector geometrically?

First, we can position our protractor at the origin to find the right direction. We measure 116.57° counterclockwise from the positive *x*-axis and draw a line in that direction (figure 2.25). Our vector (3, 116.57°) lies somewhere on this line.

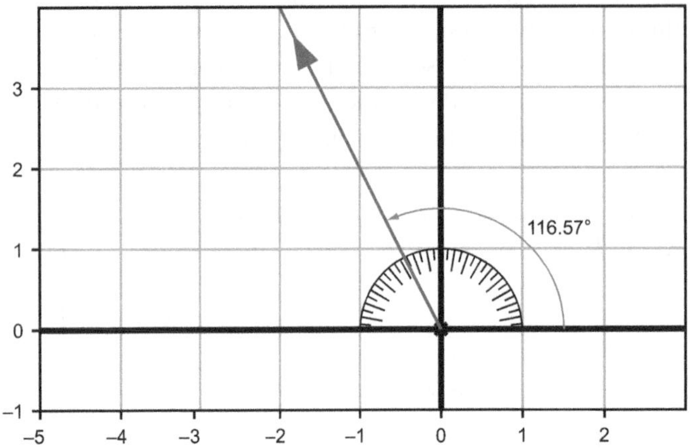

Figure 2.25 Measuring 116.57° from the positive *x*-axis using a protractor

The next step is to take a ruler and measure a point that is three units from the origin in this direction. Once we've found it, as in figure 2.26, we can measure the components and get our approximate coordinates (–1.34, 2.68).

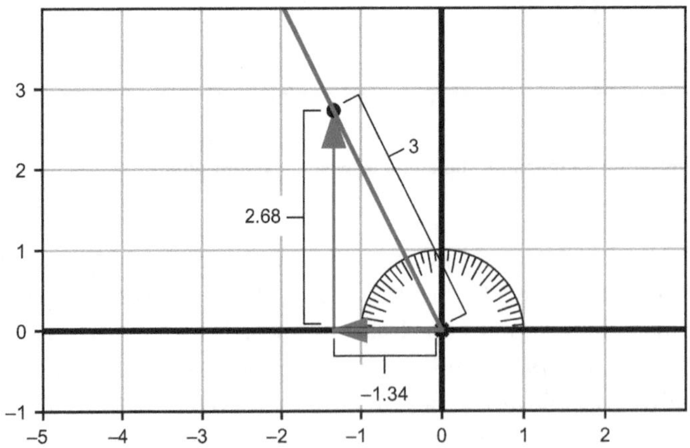

Figure 2.26 Using a ruler to measure the coordinates of the point that is three units from the origin

It may look like the angle 116.57° was a random choice, but it has a useful property. Starting from the origin and moving in that direction, you go up two units every time you go one unit to the left. Vectors that approximately lie along that line include (–1, 2), (–3, 6) and, of course, (–1.34, 2.68); the *y*-coordinates are –2 times their *x*-coordinates (figure 2.27).

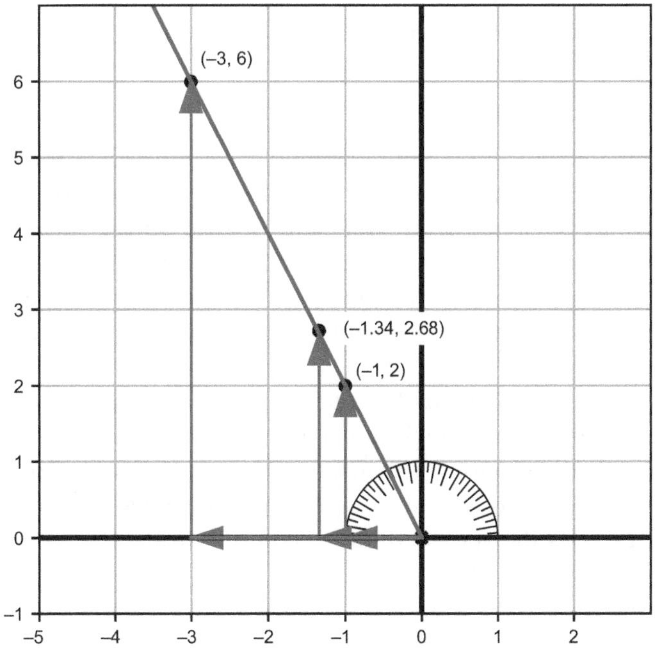

Figure 2.27 Traveling in the direction of 116.57°, you travel two units up for every unit you travel to the left.

The strange angle 116.57° happens to give us a nice round ratio of –2. We won't always be lucky enough to get a whole number ratio, but every angle does give us a *constant* ratio. The angle 45° gives us one vertical unit for every one horizontal unit or a ratio of 1. Figure 2.28 shows another angle, 200°. This gives us a constant ratio of –0.36 vertical units for every –1 horizontal unit covered or a ratio of 0.36.

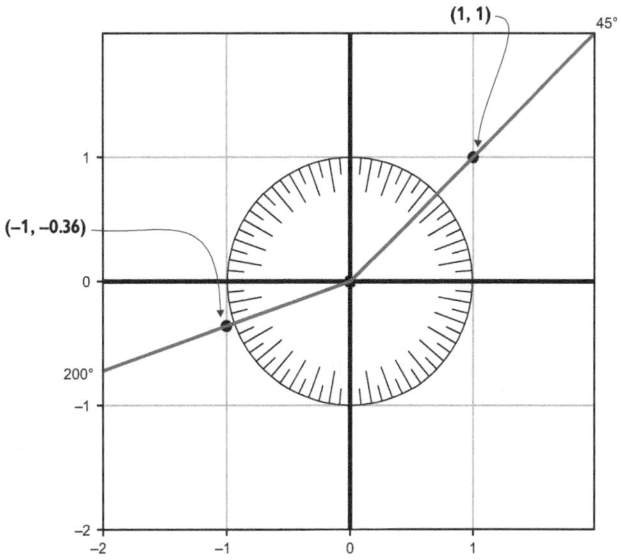

Figure 2.28 How much vertical distance is covered per unit of horizontal distance at different angles?

Given an angle, the coordinates of vectors along that angle will have a constant ratio. This ratio is called the *tangent* of the angle, and the tangent function is written as *tan*. You've seen a few of its approximate values so far:

$$\tan(37°) \approx 3/4$$
$$\tan(116.57°) \approx -2$$
$$\tan(45°) = 1$$
$$\tan(200°) \approx 0.36$$

Here, to denote *approximate* equality, I use the symbol ≈ as opposed to the symbol =. The tangent function is a *trigonometric* function because it helps us measure triangles. (The "trigon" in "trigonometry" means triangle and "metric" means measurement.) Note that I haven't told you *how* to calculate the tangent yet, only what a few of its values are. Python has a built-in tangent function that I'll show you how to use shortly. You almost never have to worry about calculating (or measuring) the tangent of an angle yourself.

The tangent function is clearly related to our original problem of finding Cartesian coordinates for a vector given an angle and a distance. But it doesn't actually provide the coordinates, only their ratio. For that, two other trigonometric functions are helpful: *sine* and *cosine*. If we measure some distance at some angle (figure 2.29), the tangent of the angle gives us the vertical distance covered divided by the horizontal distance.

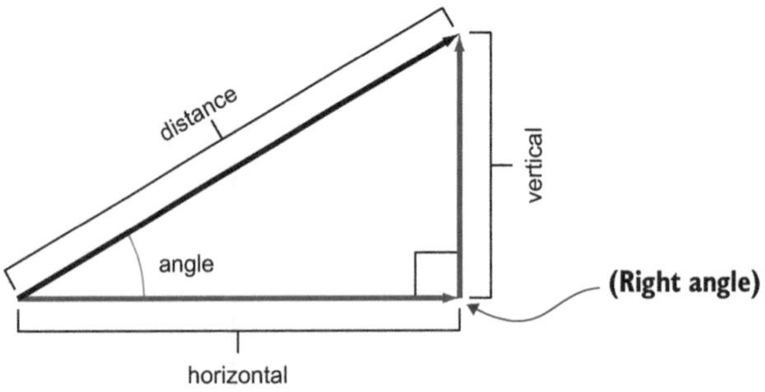

Figure 2.29 Schematic of distances and angles for a given vector

By comparison, the sine and cosine give us the vertical and horizontal distance covered relative to the overall distance. These are written *sin* and *cos* for short, and this equation shows the definitions for both:

$$\sin(\text{angle}) = \frac{\text{vertical}}{\text{distance}} \qquad \cos(\text{angle}) = \frac{\text{horizontal}}{\text{distance}}$$

Let's look at the angle 37° for a concrete example (figure 2.30). We saw that the point (4, 3) lies at a distance of 5 units from the origin at this angle.

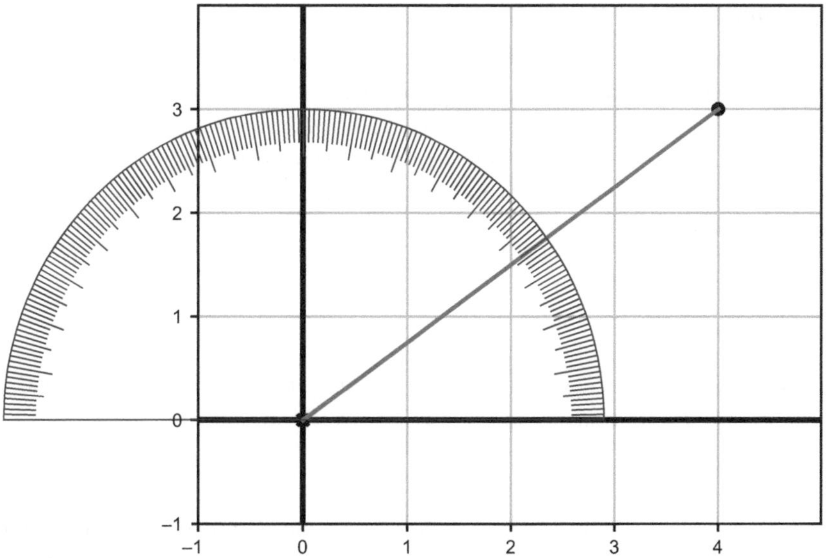

Figure 2.30 Measuring the angle to the point (4, 3) with a protractor

For every 5 units you travel at 37°, you cover approximately 3 vertical units. Therefore, we can write:

$$\sin(37°) \approx 3/5$$

Similarly, for every 5 units you travel at 37°, you cover approximately 4 horizontal units, so we can write:

$$\cos(37°) \approx 4/5$$

This is a general strategy for converting a vector in polar coordinates to corresponding Cartesian coordinates. If you know the sine and cosine of an angle θ (the Greek letter theta, commonly used for angles) and a distance r traveled in that direction, the Cartesian coordinates are given by $(r \cdot \cos(\theta), r \cdot \sin(\theta))$ and shown in figure 2.31.

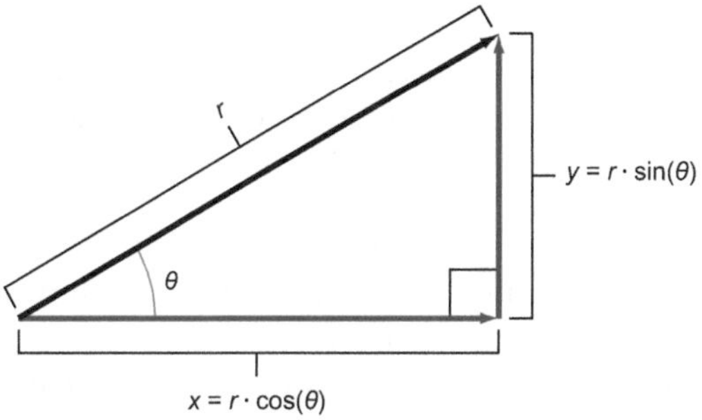

Figure 2.31 Picturing the conversion from polar coordinates to Cartesian coordinates for a right triangle

2.3.2 Radians and trigonometry in Python

Let's turn what we've reviewed about trigonometry into Python code. Specifically, let's build a function that takes a pair of polar coordinates (a length and an angle) and outputs a pair of Cartesian coordinates (lengths of x and y components).

The main hurdle is that Python's built-in trigonometric functions use different units than the ones we've used. We expect $\tan(45°) = 1$, for instance, but Python gives us a much different result:

```
>>> from math import tan
>>> tan(45)
1.6197751905438615
```

Python doesn't use degrees, and neither do most mathematicians. Instead, they use units called *radians* to measure angles. The conversion factor is

$$1 \text{ radian} \approx 57.296°$$

This may seem like an arbitrary conversion factor. Some more suggestive relationships between degrees and radians are given in terms of the special number π (pi), whose value is approximately 3.14159. Here are a few examples:

$$\pi \text{ radians} = 180°$$
$$2\pi \text{ radians} = 360°$$

In radians, half a trip around a circle is an angle of π and a whole revolution is 2π. These respectively agree with the half and whole circumference of a circle of radius 1 (figure 2.32).

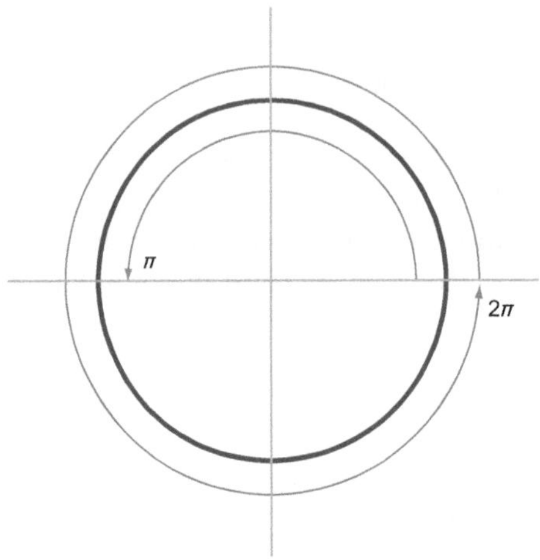

Figure 2.32 A half revolution is π radians, while a whole revolution is 2π radians.

You can think of radians as another kind of ratio: for a given angle, its measurement in radians tells you how many radiuses you've gone around the circle. Because of this special property, angle measurements without units are assumed to be radians. Noting that $45° = \pi/4$ (radians), we can get the correct result for the tangent of this angle:

```
>>> from math import tan, pi
>>> tan(pi/4)
0.9999999999999999
```

We can now make use of Python's trigonometric functions to write a `to_cartesian` function, taking a pair of polar coordinates and returning corresponding Cartesian coordinates:

```
from math import sin, cos
def to_cartesian(polar_vector):
    length, angle = polar_vector[0], polar_vector[1]
    return (length*cos(angle), length*sin(angle))
```

Using this, we can verify that 5 units at an angle of 37° gets us close to the point (4, 3):

```
>>> from math import pi
>>> angle = 37*pi/180
>>> to_cartesian((5,angle))
(3.993177550236464, 3.0090751157602416)
```

Now that we can convert from polar coordinates to Cartesian coordinates, let's see how to convert in the other direction.

2.3.3 From components back to angles

Given a pair of Cartesian coordinates like (–2, 3), we know how to find the length with the Pythagorean theorem. In this case, it is $\sqrt{13}$, which is the first of the two polar coordinates we are looking for. The second is the angle, which we can call θ (theta), indicating the direction of this vector (figure 2.33).

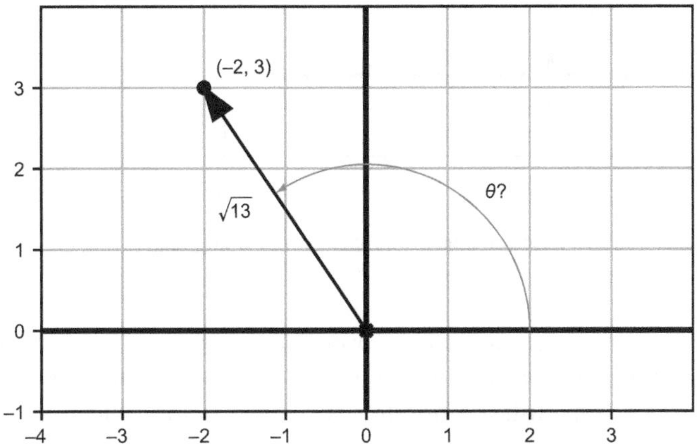

Figure 2.33 In what angle does the vector (–2, 3) point?

We can say some facts about the angle θ that we're looking for. Its tangent, $\tan(\theta)$, is $3/2$, while $\sin(\theta) = 3/\sqrt{13}$ and $\cos(\theta) = -2/\sqrt{13}$. All that's left is finding a value of θ that satisfies these. If you like, you can pause and try to approximate this angle yourself by guessing and checking.

Ideally, we'd like a more efficient method than this. It would be great if there were a function that took the value of $\sin(\theta)$, for instance, and gave you back θ. This turns out to be easier said than done, but Python's `math.asin` function makes a good attempt. This is an implementation of the *inverse trigonometric function* called the *arcsine*, and it returns a satisfactory value of θ:

```
>>> from math import asin
>>> sin(1)
0.8414709848078965
>>> asin(0.8414709848078965)
1.0
```

So far, so good. But what about the sine of our angle $3/\sqrt{13}$?

```
>>> from math import sqrt
>>> asin(3/sqrt(13))
0.9827937232473292
```

This angle is roughly 56.3°, and as figure 2.34 shows, that's the wrong direction!

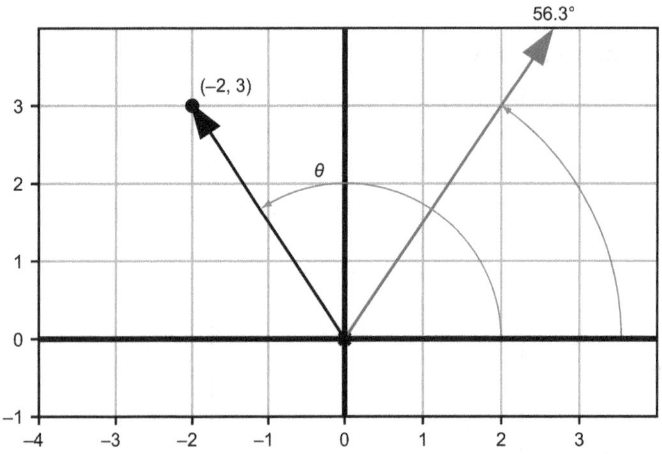

Figure 2.34 Python's `math.asin` function appears to give us the wrong angle.

It's not wrong that `math.asin` gives us this answer; another point (2, 3) *does* lie in this direction. It is at length $\sqrt{13}$ from the origin, so the sine of this angle is also $3\sqrt{13}$. This is why `math.asin` is not a full solution for us. There are multiple angles that can have the same sine.

The inverse trigonometric function, called *arccosine* and implemented in Python as `math.acos`, happens to give us the right value:

```
>>> from math import acos
>>> acos(-2/sqrt(13))
2.1587989303424644
```

This many radians is about the same as 123.7°, which we can confirm to be correct using a protractor. But this is only by happenstance; there are other angles that could have given us the same cosine. For instance, (−2, −3) also has distance $\sqrt{13}$ from the origin, so it lies at an angle with the same cosine as θ: $-2\sqrt{13}$. To find the value of θ that we actually want, we'll have to make sure the sine *and* cosine agree with our expectation. The angle returned by Python, which is approximately 2.159, satisfies this:

```
>> cos(2.1587989303424644)
-0.5547001962252293
>>> -2/sqrt(13)
-0.5547001962252291
>>> sin(2.1587989303424644)
0.8320502943378435
>>> 3/sqrt(13)
0.8320502943378437
```

None of the arcsine, arccosine, or arctangent functions are sufficient to find the angle to a point in the plane. It *is* possible to find the correct angle by a tricky geometric argument you probably learned in high school trigonometry class. I'll leave that as an exercise and cut to the chase—Python can do the work for you! The `math.atan2` function takes the Cartesian coordinates of a point in the plane (in reverse order!) and gives you back the angle at which it lies. For example,

```
>>> from math import atan2
>>> atan2(3,-2)
2.158798930342464
```

I apologize for burying the lede, but I did so because it's worth knowing the potential pitfalls of using inverse trigonometric functions. In summary, trigonometric functions are tricky to do in reverse; multiple different inputs can produce the same output, so an output can't be traced back to a unique input. This lets us complete the function we set out to write: a converter from Cartesian to polar coordinates:

```
def to_polar(vector):
    x, y = vector[0], vector[1]
    angle = atan2(y,x)
    return (length(vector), angle)
```

We can verify some simple examples: `to_polar((1,0))` should be one unit in the positive *x* direction or an angle of zero degrees. Indeed, the function gives us an angle of zero and a length of one:

```
>>> to_polar((1,0))
(1.0, 0.0)
```

(The fact that the input and the output are the same is coincidental; they have different geometric meanings.) Likewise, we get the expected answer for (−2, 3):

```
>>> to_polar((-2,3))
(3.605551275463989, 2.158798930342464)
```

2.3.4 Exercises

Exercise 2.27 Confirm that the vector given by Cartesian coordinates (–1.34, 2.68) has a length of approximately 3 as expected.

Solution

```
>>> length((-1.34,2.68))
2.9963310898497184
```

Close enough!

Exercise 2.28 The figure shows a line that makes a 22° angle in the counterclockwise direction from the positive *x*-axis. Based on the following picture, what is the approximate value of tan(22°)?

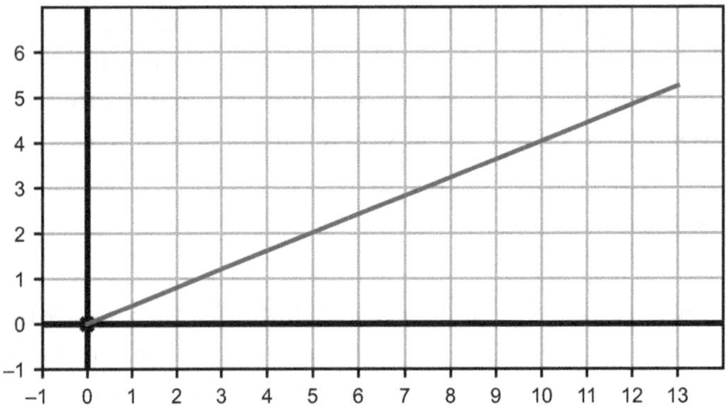

Solution The line passes close to the point (10, 4), so 4 / 10 = 0.4 is a reasonable approximation of tan(22°) as shown here:

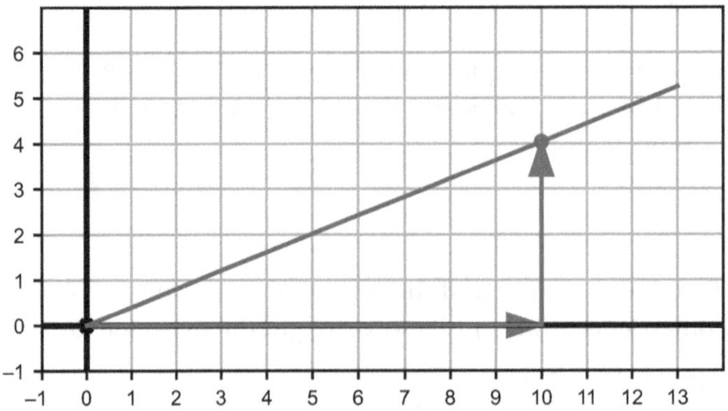

Exercise 2.29 Turning the question around, suppose we know the length and direction of a vector and want to find its components. What are the x and y components of a vector with length 15 pointing at a 37° angle?

Solution

The sine of 37° is roughly 3/5, which tells us that every 5 units of distance covered at this angle takes us 3 units upward. So, 15 units of distance give us a vertical component of $3/5 \cdot 15$, or 9.

The cosine of 37° is roughly 4/5, which tells us that each 5 units of distance in this direction take us 4 units to the right, so the horizontal component is $4/5 \cdot 15$ or 12. In summary, the polar coordinates (15, 37°) correspond approximately to the Cartesian coordinates (12, 9).

Exercise 2.30 Suppose I travel 8.5 units from the origin at an angle of 125°, measured counterclockwise from the positive x-axis. Given that $\sin(125°) = 0.819$ and $\cos(125°) = -0.574$, what are my final coordinates? Draw a picture to show the angle and path traveled.

Solution

$$x = r \cdot \cos(\theta) = 8.5 \cdot -0.574 = -4.879$$
$$y = r \cdot \sin(\theta) = 8.5 \cdot 0.819 = 6.962$$

The following figure shows the final position, (–4.879, 6.962):

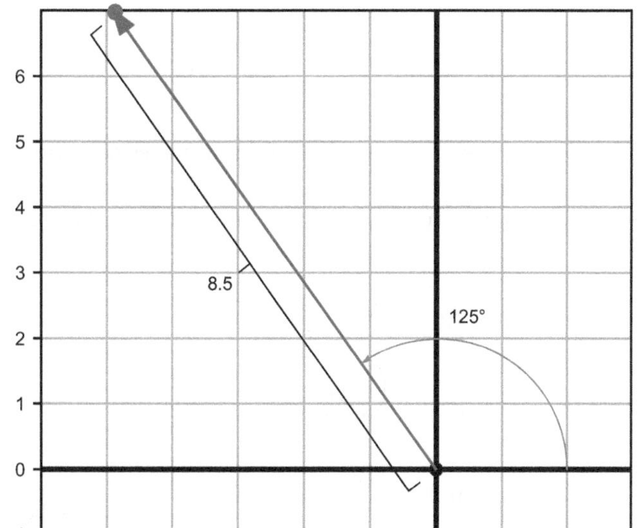

Exercise 2.31 What are the sine and cosine of 0°? Of 90°? Of 180°? In other words, how many vertical and horizontal units are covered per unit distance in any of these directions?

Solution At 0°, no vertical distance is covered, so sin(0°) = 0; rather, every unit of distance traveled is a unit of horizontal distance, so cos(0°) = 1.

For 90° (a quarter turn counterclockwise), every unit traveled is a positive vertical unit, so sin(90°) = 1, while cos(90°) = 0.

Finally, at 180°, every unit of distance traveled is a negative unit in the *x* direction, so cos(180°) = −1 and sin(180°) = 0.

Exercise 2.32 The following diagram gives some exact measurements for a right triangle:

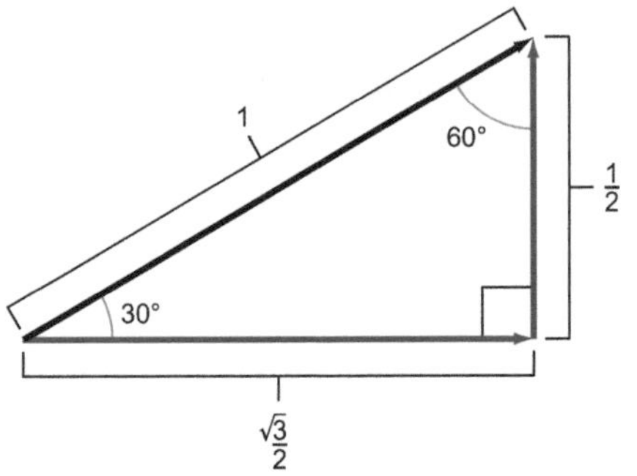

First, confirm that these lengths are valid for a right triangle because they satisfy the Pythagorean theorem. Then, calculate the values of sin(30°), cos(30°), and tan(30°) to three decimal places using the measurements in the diagram.

Solution These side lengths indeed satisfy the Pythagorean theorem:

$$\sqrt{\left(\frac{1}{2}\right)^2 + \left(\frac{\sqrt{3}}{2}\right)^2} = \sqrt{\frac{1}{4} + \frac{3}{4}} = \sqrt{\frac{4}{4}} = 1$$

Plugging the side lengths into the Pythagorean theorem

The trigonometric function values are given by the appropriate ratios of side lengths:

$$\sin(30°) = \frac{\left(\frac{1}{2}\right)}{1} = 0.500$$

$$\cos(30°) = \frac{\left(\frac{\sqrt{3}}{2}\right)}{1} \approx 0.866$$

$$\tan(30°) = \frac{\left(\frac{1}{2}\right)}{\left(\frac{\sqrt{3}}{2}\right)} \approx 0.577$$

Calculating the sine, cosine, and tangent by their definitions

Exercise 2.33 Looking at the triangle from the previous exercise from a different perspective, use it to calculate the values of $\sin(60°)$, $\cos(60°)$, and $\tan(60°)$ to three decimal places.

Solution Rotating and reflecting the triangle from the previous exercise has no effect on its side lengths or angles.

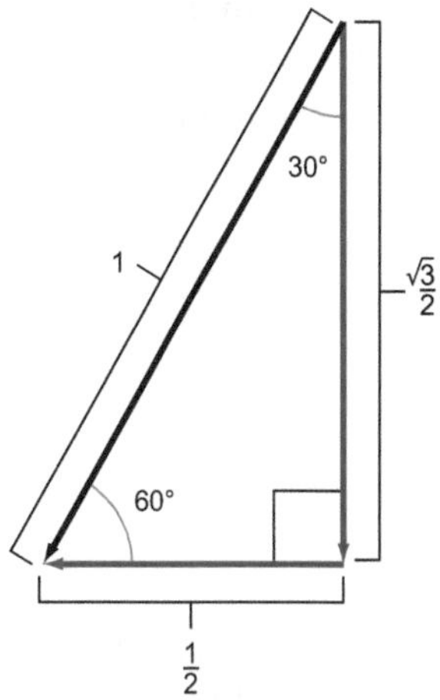

A rotated copy of the triangle from the previous exercise

The ratios of the side lengths give the trigonometric function values for 60°:

$$\sin(60°) = \frac{\left(\frac{\sqrt{3}}{2}\right)}{1} \approx 0.866$$

$$\cos(60°) = \frac{\left(\frac{1}{2}\right)}{1} = 0.500$$

$$\tan(60°) = \frac{\left(\frac{\sqrt{3}}{2}\right)}{\left(\frac{1}{2}\right)} \approx 1.732$$

Calculating the defining ratios when horizontal and vertical components have switched

Exercise 2.34 The cosine of 50° is 0.643. What is $\sin(50°)$ and what is $\tan(50°)$? Draw a picture to help you calculate the answer.

Solution Given that the cosine of 50° is 0.643, the following triangle is valid:

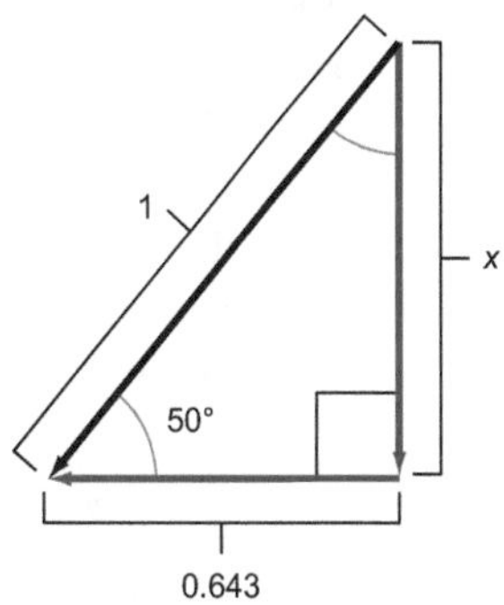

(continued)

That is, it has the right ratio of the two known side lengths: 0.643 / 1 = 0.643. To find the unknown side length, we can use the Pythagorean theorem:

$$\sqrt{0.643^2 + x^2} = 1$$
$$0.643^2 + x^2 = 1$$
$$0.413 + x^2 = 1$$
$$x^2 = 0.587$$
$$x = 0.766$$

With the known side lengths, $\sin(50°) = 0.766/1 = 0.766$. Also, $\tan(50°) = 0.766/0.643 = 1.192$.

Exercise 2.35 What is 116.57° in radians? Use Python to compute the tangent of this angle and confirm that it is close to −2 as we saw previously.

Solution $116.57° \cdot (1 \text{ radian}/57.296°) = 2.035$ radians:

```
>>> from math import tan
>>> tan(2.035)
-1.9972227673316139
```

Exercise 2.36 Locate the angle $10\pi/6$. Do you expect the values of $\cos(10\pi/6)$ and $\sin(10\pi/6)$ to be positive or negative? Use Python to calculate their values and confirm.

Solution A whole circle is 2π radians, so the angle $\pi/6$ is one twelfth of a circle. You can picture cutting a pizza in 12 slices, and counting counterclockwise from the positive *x*-axis; the angle $10\pi/6$ is two slices short of a full rotation. This means that it points down and to the right. The cosine should be positive, and the sine should be negative because the distance in this direction corresponds with a positive horizontal displacement and a negative vertical displacement:

```
>>> from math import pi, cos, sin
>>> sin(10*pi/6)
-0.8660254037844386
>>> cos(10*pi/6)
0.5000000000000001
```

Exercise 2.37 The following list comprehension creates 1,000 points in polar coordinates:

```
[(cos(5*x*pi/500.0), 2*pi*x/1000.0) for x in range(0,1000)]
```

In Python code, convert these to Cartesian coordinates and connect them in a closed loop with line segments to draw a picture.

Solution Including the setup and the original list of data, the code is as follows:

```
polar_coords = [(cos(x*pi/100.0), 2*pi*x/1000.0) for x in range(0,1000)]
vectors = [to_cartesian(p) for p in polar_coords]
draw(Polygon(*vectors, color=green))
```

And the result is a five-leafed flower:

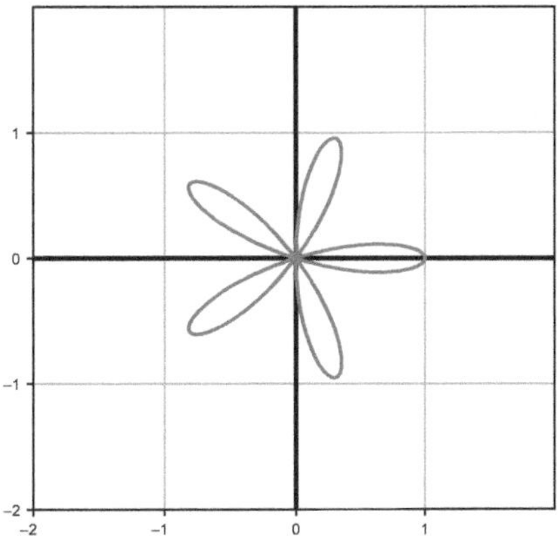

The plot of the 1,000 connected points is a flower shape.

Exercise 2.38 Find the angle to get to the point (−2, 3) by "guess-and-check."

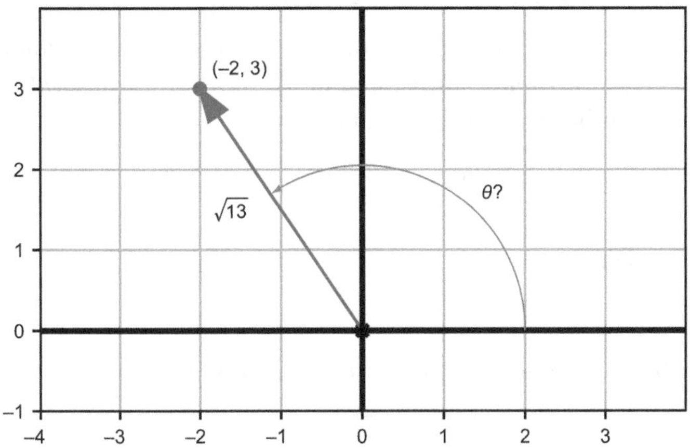

What is the angle to get to the point (−2, 3)?

(continued)

Hint We can tell visually that the answer is between $\pi/2$ and π. On that interval, the values of sine and cosine always decrease as the angle increases.

Solution Here's an example of guessing and checking between $\pi/2$ and π, looking for an angle with tangent close to $-3/2 = -1.5$:

```
>>> from math import tan, pi
>>> pi, pi/2
(3.141592653589793, 1.5707963267948966)
>>> tan(1.8)
-4.286261674628062
>>> tan(2.5)
-0.7470222972386603
>>> tan(2.2)
-1.3738230567687946
>>> tan(2.1)
-1.7098465429045073
>>> tan(2.15)
-1.5289797578045665
>>> tan(2.16)
-1.496103541616277
>>> tan(2.155)
-1.5124173422757465
>>> tan(2.156)
-1.5091348993879299
>>> tan(2.157)
-1.5058623488727219
>>> tan(2.158)
-1.5025996395625054
>>> tan(2.159)
-1.4993467206361923
```

The value must be between 2.158 and 2.159.

Exercise 2.39 Find another point in the plane with the same tangent as θ, namely $-3/2$. Use Python's implementation of the *arctangent* function, `math.atan`, to find the value of this angle.

Solution Another point with tangent $-3/2$ is $(3, -2)$. Python's `math.atan` finds the angle to this point:

```
>>> from math import atan
>>> atan(-3/2)
-0.982793723247329
```

This is slightly less than a quarter turn in the clockwise direction.

Exercise 2.40 Without using Python, what are the polar coordinates corresponding to the Cartesian coordinates (1, 1) and (1, –1)? Once you've found the answers, use `to_polar` to check your work.

Solution In polar coordinates, (1, 1) becomes $(\sqrt{2}, \pi/4)$ and (1, –1) becomes $(\sqrt{2}, -\pi/4)$.

With some care, you can find any angle on a shape made up of known vectors. The angle between two vectors is either a sum or difference of angles these make with the *x*-axis. You measure some trickier angles in the next mini-project.

Exercise 2.41—Mini Project What is the angle of the Dinosaur's mouth? What is the angle of the dinosaur's toe? Of the point of its tail?

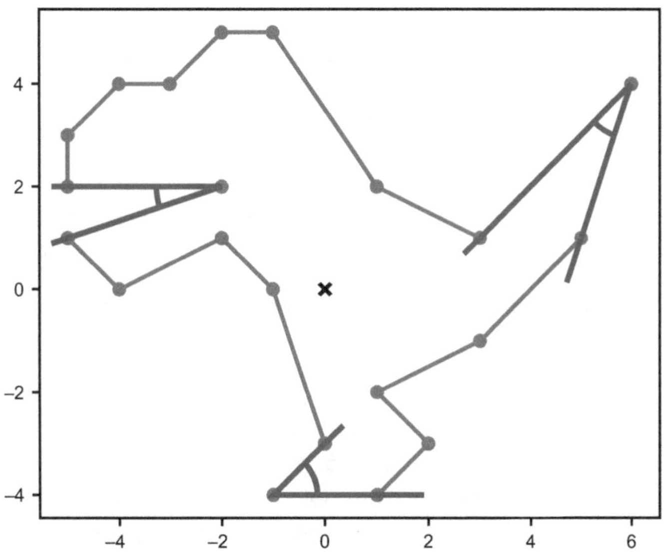

Some angles we can measure or calculate on our dinosaur.

2.4 Transforming collections of vectors

Collections of vectors store spatial data like drawings of dinosaurs regardless of what coordinate system we use: polar or Cartesian. It turns out that when we want to manipulate vectors, one coordinate system can be better than another. We already saw that moving (or translating) a collection of vectors is easy with Cartesian coordinates. It turns out to be much less natural in polar coordinates. Because polar coordinates have angles built in, these make it simple to carry out rotations.

In polar coordinates, adding to the angle rotates a vector further counterclockwise, while subtracting from it rotates the vector clockwise. The polar coordinate (1, 2) is at distance 1 and at an angle of 2 radians. (Remember that we are working in radians if there is no degree symbol!) Starting with the angle 2 and adding or subtracting 1 takes the vector either 1 radian counterclockwise or clockwise, respectively (figure 2.35).

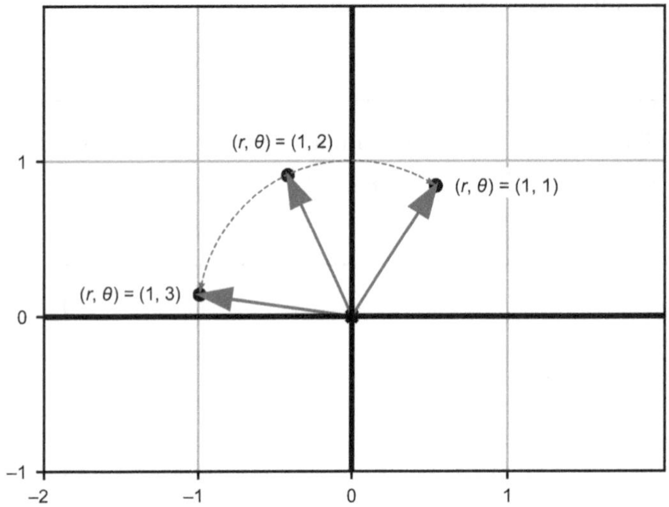

Figure 2.35 Adding or subtracting from the angle rotates the vector about the origin.

Rotating a number of vectors simultaneously has the effect of rotating the figure these represent about the origin. The `draw` function only understands Cartesian coordinates, so we need to convert from polar to Cartesian before using it. Likewise, we have only seen how to rotate vectors in polar coordinates, so we need to convert Cartesian coordinates to polar coordinates before executing a rotation. Using this approach, we can rotate the dinosaur like this:

```
rotation_angle = pi/4
dino_polar = [to_polar(v) for v in dino_vectors]
dino_rotated_polar = [(l,angle + rotation_angle) for l,angle in dino_polar]
dino_rotated = [to_cartesian(p) for p in dino_rotated_polar]
draw(
    Polygon(*dino_vectors, color=gray),
    Polygon(*dino_rotated, color=red)
)
```

The result of this code is a gray copy of the original dinosaur, plus a superimposed red copy that's rotated by $\pi/4$, or an eighth of a full revolution counterclockwise (figure 2.36).

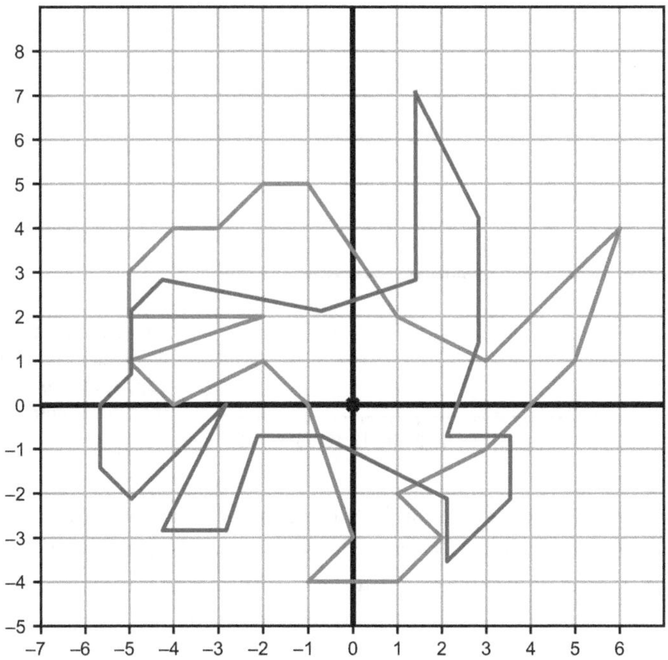

Figure 2.36 The original dinosaur in gray and a rotated copy in red

As an exercise at the end of this section, you can write a general-purpose `rotate` function that rotates a list of vectors by the same specified angle. I'm going to use such a function in the next few examples, and you can either use the implementation I provide in the source code or one you come up with yourself.

2.4.1 *Combining vector transformations*

So far, we've seen how to translate, rescale, and rotate vectors. Applying any of these transformations to a collection of vectors achieves the same effect on the shape that these define in the plane. The full power of these vector transformations comes when we apply them in sequence.

For instance, we could first rotate and *then* translate the dinosaur. Using the `translate` function from the exercise in section 2.2.4 and the `rotate` function, we can write such a transformation concisely (see the result in figure 2.37):

```
new_dino = translate((8,8), rotate(5 * pi/3, dino_vectors))
```

The rotation comes first, turning the dinosaur counterclockwise by $5\pi/3$, which is most of a full counterclockwise revolution. Then the dinosaur is translated up and to

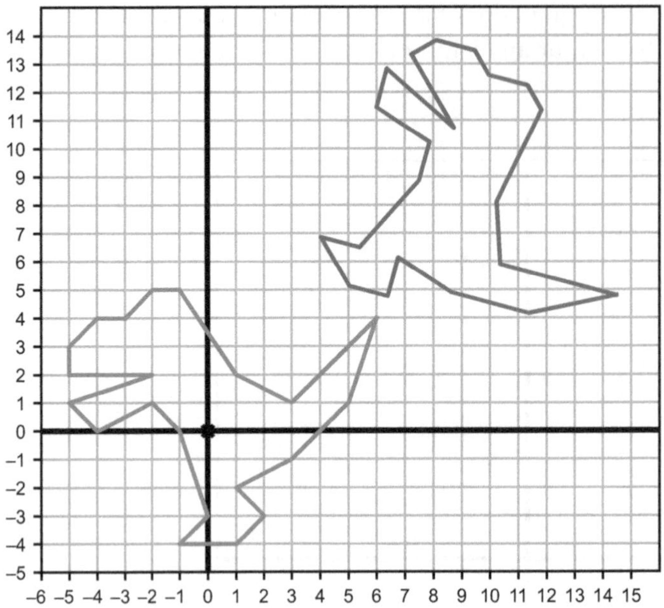

Figure 2.37 The original dinosaur in gray and a red copy that's rotated and then translated

the right by 8 units each. As you can imagine, combining rotations and translations appropriately can move the dinosaur (or any shape) to any desired location and orientation in the plane. Whether we're animating our dinosaur in a movie or in a game, the flexibility to move it around with vector transformations lets us give it life programmatically.

Our applications will soon take us past cartoon dinosaurs; there are plenty of other operations on vectors and many generalize to higher dimensions. Real-world data sets often live in dozens or hundreds of dimensions, so we'll apply the same kinds of transformations to these as well. It's often useful to both translate and rotate data sets to make their important features clearer. We won't be able to picture rotations in 100 dimensions, but we can always think of two dimensions as a trusty metaphor.

2.4.2 Exercises

Exercise 2.42 Create a `rotate(angle, vectors)` function that takes an array of input vectors in Cartesian coordinates and rotates those by the specified angle (counterclockwise or clockwise, according to whether the angle is positive or negative).

Solution

```
def rotate(angle, vectors):
    polars = [to_polar(v) for v in vectors]
    return [to_cartesian((l, a+angle)) for l,a in polars]
```

Exercise 2.43 Create a function `regular_polygon(n)` that returns Cartesian coordinates for the vertices of a regular n-sided polygon (that is, having all angles and side lengths equal). For instance, `polygon(7)` produces vectors defining the following heptagon:

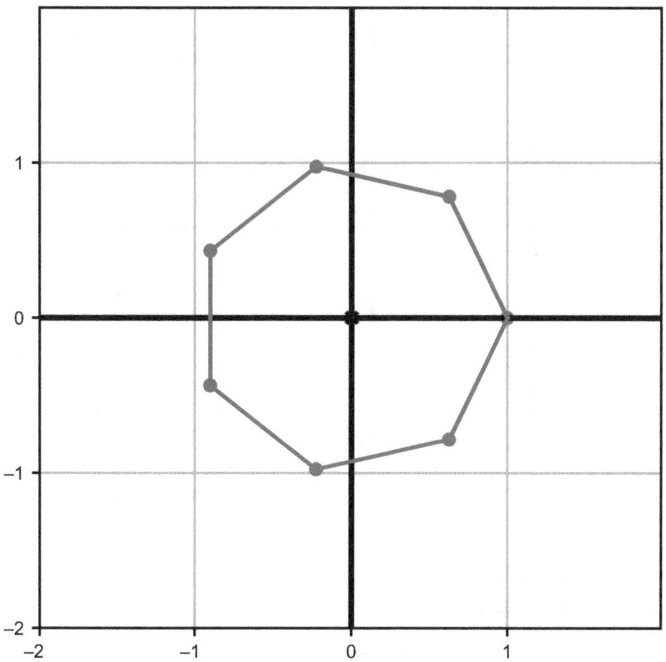

A regular heptagon with points at seven evenly-spaced angles around the origin

Hint In this picture, I used the vector (1, 0) and copies that are rotated by seven evenly-spaced angles about the origin.

Solution

```
def regular_polygon(n):
    return [to_cartesian((1, 2*pi*k/n)) for k in range(0,n)]
```

Exercise 2.44 What is the result of first translating the dinosaur by the vector (8, 8) and then rotating it by $5\pi/3$? Is the result the same as rotating and then translating?

(continued)

Solution

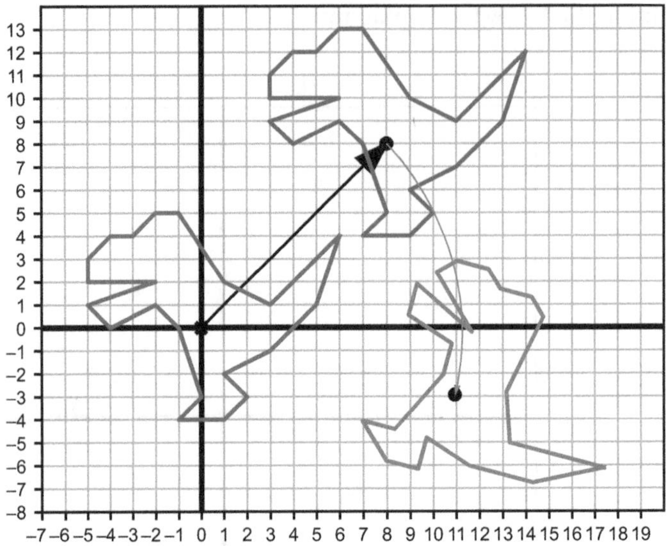

First translating and then rotating the dinosaur

The result is *not* the same. In general, applying rotations and translations in different orders yields different results.

2.5 Drawing with Matplotlib

As promised, I'll conclude by showing you how to build "from scratch" the drawing functions used in this chapter from the Matplotlib library. After installing Matplotlib with pip, you can import it (and some of its submodules); for example,

```
import matplotlib
from matplotlib.patches import Polygon
from matplotlib.collections import PatchCollection
```

The `Polygon`, `Points`, `Arrow`, and `Segment` classes are not that interesting; they simply hold the data passed to them in their constructors. For instance, the `Points` class contains only a constructor that receives and stores a list of vectors and a color keyword argument:

```
class Points():
    def __init__(self, *vectors, color=black):
        self.vectors = list(vectors)
        self.color = color
```

The `draw` function starts by figuring out how big the plot should be and then draws each of the objects it is passed one-by-one. For instance, to draw dots on the plane represented by a `Points` object, `draw` uses Matplotlib's scatter-plotting functionality:

```
def draw(*objects, ...
    # ...
    for object in objects:
    # ...
        elif type(object) == Points:
            xs = [v[0] for v in object.vectors]
            ys = [v[1] for v in object.vectors]
            plt.scatter(xs, ys, color=object.color)

        # ...
```

Some setup happens here, which is not shown.

Iterates over the objects passed in

If the current object is an instance of the Points class, draws dots for all of its vectors using Matplotlib's scatter function

Arrows, segments, and polygons are handled in much the same way using different pre-built Matplotlib functions to make the geometric objects appear on the plot. You can find all of these implemented in the source code file vector_drawing.py. We'll use Matplotlib throughout this book to plot data and mathematical functions, and I'll provide periodic refreshers on its functionality as we use it.

Now that you've mastered two dimensions, you're ready to add another one. With the third dimension, we can fully describe the world we live in. In the next chapter, you'll see how to model three-dimensional objects in code.

Summary

- Vectors are mathematical objects that live in multi-dimensional spaces. These can be geometric spaces like the two-dimensional (2D) plane of a computer screen or the three-dimensional (3D) world we inhabit.
- You can think of vectors equivalently as arrows having a specified length and direction, or as points in the plane relative to a reference point called the *origin*. Given a point, there is a corresponding arrow that shows how to get to that point from the origin.
- You can connect collections of points in the plane to form interesting shapes like a dinosaur.
- In 2D, coordinates are pairs of numbers that help us measure the location of points in the plane. Written as a tuple (x, y), the x and y values tell us how far horizontally and vertically to travel to get to the point.
- We can store points as coordinate tuples in Python and choose from a number of libraries to draw the points on the screen.
- Vector addition has the effect of translating (or moving) a first vector in the direction of a second added vector. Thinking of a collection of vectors as paths to travel, their vector sum gives the overall direction and distance traveled.
- Scalar multiplication of a vector by a numeric factor yields a vector that is longer by that factor and points in the same direction as the original.

- Subtracting one vector from a second gives the relative position of the second vector from the first.
- Vectors can be specified by their length and direction (as an angle). These two numbers define the polar coordinates of a given 2D vector.
- The trigonometric functions sine, cosine, and tangent are used to convert between ordinary (Cartesian) coordinates and polar coordinates.
- It's easy to rotate shapes defined by collections of vectors in polar coordinates. You only need to add or subtract the given rotation angle from the angle of each vector. Rotating and translating shapes in the plane lets us place them anywhere and in any orientation.

3 Ascending to the 3D world

This chapter covers

- Building a mental model for 3D vectors
- Doing 3D vector arithmetic
- Using the dot product and cross product to measure lengths and directions
- Rendering a 3D object in 2D

The 2D world is easy to visualize, but the real world has three dimensions. Whether we are using software to design a building, animate a movie, or run an action game, our programs need to be aware of the three spatial dimensions in which we live.

In a 2D space, like a page of this book, we have a vertical and a horizontal direction. Adding a third dimension, we could also talk about points outside of the page or arrows perpendicular to the page. But even when programs simulate three dimensions, most computer displays are two-dimensional. Our mission in this chapter is to build the tools we need to take 3D objects measured by 3D vectors and convert them to 2D so our objects can show up on the screen.

A sphere is one example of a 3D shape. A successfully drawn 3D sphere could look like the one shown in figure 3.1. Without the shading, it would just look like a circle.

Figure 3.1 Shading on a 2D circle makes it look like a 3D sphere.

The shading shows that light hits our sphere at a certain angle in 3D and gives it an illusion of depth. Our general strategy is not to draw a perfectly round sphere, but an approximation made up of polygons. Each polygon can be shaded according to the precise angle it makes with a light source. Believe it or not, figure 3.1 is not a picture of a round ball, but of 8,000 triangles in varying shades. Figure 3.2 shows another example with fewer triangles.

Figure 3.2 Drawing a shaded sphere using many small, solid-colored triangles

We have the mathematical machinery to define a triangle on a 2D screen: we only need the three 2D vectors defining the corners. But we can't decide how to shade them unless we also think of them as having a life in three dimensions. For this, we need to learn to work with 3D vectors.

Of course, this is already a solved problem, and we start by using a pre-built library to draw our 3D shapes. Once we have the feel for the world of 3D vectors, we can build our own renderer and show how to draw the sphere.

3.1 Picturing vectors in 3D space

In the 2D plane, we worked with three interchangeable mental models of a vector: coordinate pairs, arrows of fixed length and direction, and points positioned relative to the origin. Because the pages of this book have a finite size, we limited our view to a small portion of the plane—a rectangle of fixed height and width like the one shown in figure 3.3.

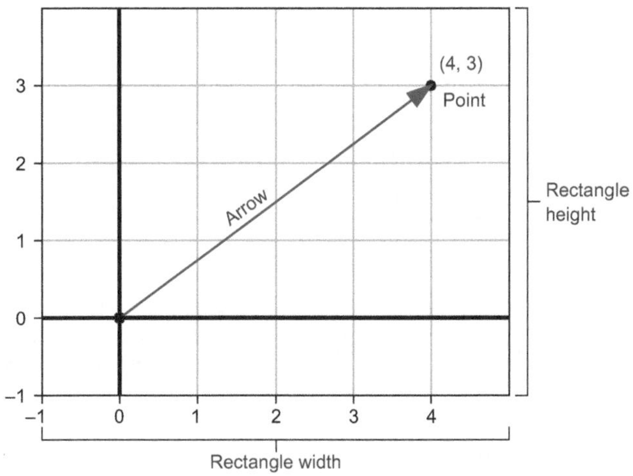

Figure 3.3 The height and width of a small segment of the 2D plane

We can interpret a 3D vector in similar ways. Instead of viewing a rectangular portion of the plane, we start with a finite box of 3D space. Such a 3D box, pictured in figure 3.4, has a finite height, width, and depth. In 3D, we keep the notions of *x* and *y* directions, and we add a *z* direction with which to measure the depth.

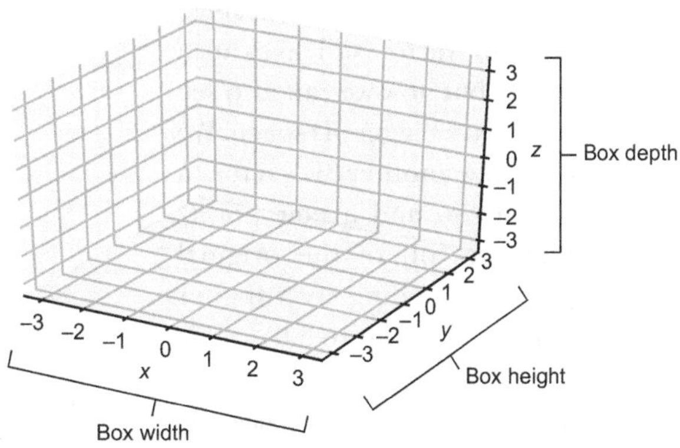

Figure 3.4 A small finite box of 3D space has a width (*x*), a height (*y*), and a depth (*z*).

We can think of any 2D vectors as living in 3D space,

having the same size and orientation but fixed to a plane where the depth *z* is zero. Figure 3.5 shows the 2D drawing of the vector (4, 3) embedded in 3D space with all the same features it had before. The second drawing (on the bottom) annotates all of the features that are still included.

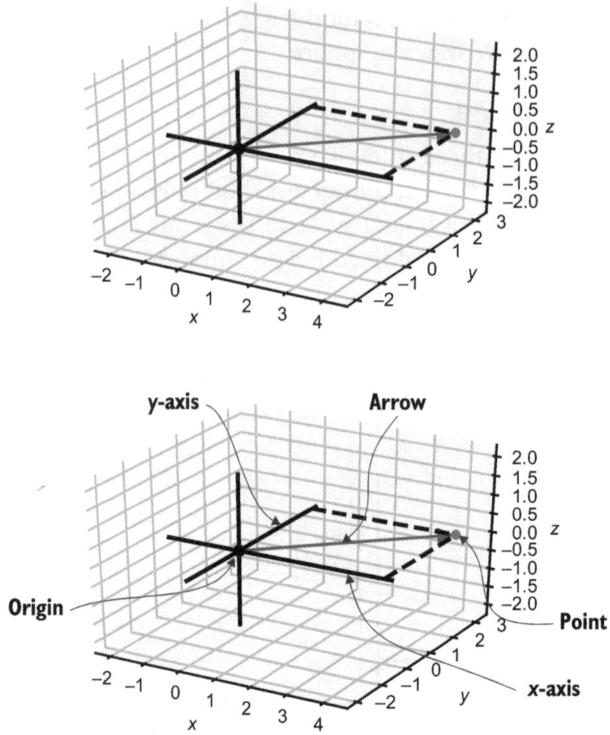

Figure 3.5 The 2D world and inhabitant vector (4, 3) contained in the 3D world

The dashed lines form a rectangle in the 2D plane where depth is zero. It's helpful to draw dashed lines meeting at right angles to help us locate points in 3D. Otherwise, our perspective might deceive us and a point may not be where we think it is.

Our vector still lives in a plane, but now we can also see it lives in a bigger 3D space. We can draw another 3D vector (a new arrow and a new point) that lives off of the original plane, extending to a higher depth value (figure 3.6).

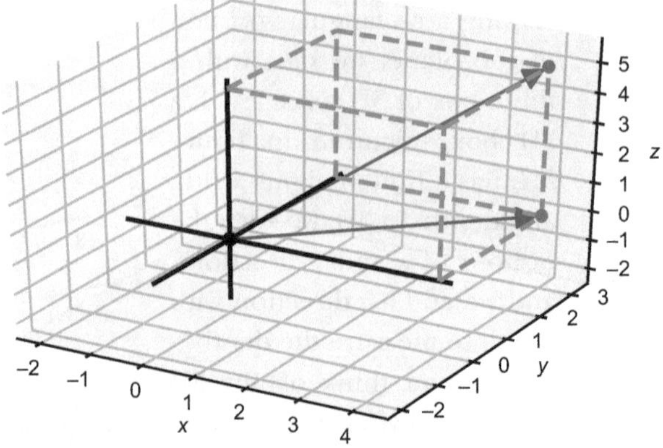

Figure 3.6 A vector extending into the third dimension as compared to the 2D world and its inhabitant vector (4, 3) of figure 3.5

To make the location of this second vector clear, I drew a dashed box instead of a dashed rectangle as in figure 3.5. In figure 3.6, this box shows the length, width, and depth the vector covers in 3D space. Arrows and points work as mental models for vectors in 3D just like in 2D, and we can measure them similarly with coordinates.

3.1.1 Representing 3D vectors with coordinates

The pair of numbers (4, 3) is enough to specify a single point or arrow in 2D, but in 3D, there are numerous points with an *x*-coordinate of 4 and a *y*-coordinate of 3. In fact, as figure 3.7 shows, there is a whole line of points in 3D with these coordinates, each having different positions in the *z* (or depth) direction.

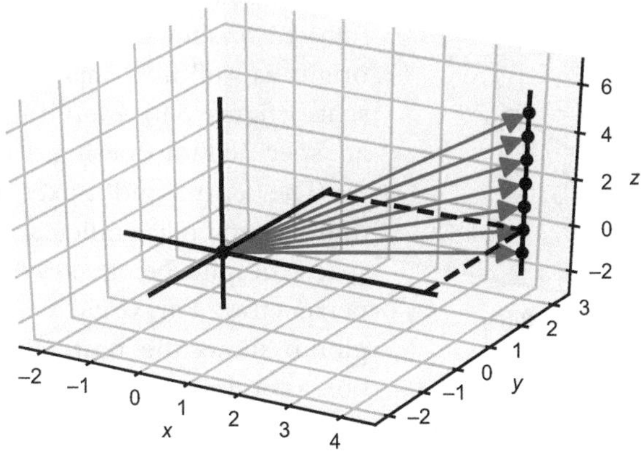

Figure 3.7 Several vectors with the same *x*- and *y*-coordinates but with different *z*-coordinates

To specify a unique point in 3D, we need three numbers in total. A triple of numbers like (4, 3, 5) are called the *x*-, *y*-, and *z*-coordinates for a vector in 3D. As before, we can read these as instructions to find the desired point. As shown in figure 3.8, to get to the point (4, 3, 5), we first go +4 units in the *x* direction, then go +3 units in the *y* direction, and then finally, go +5 units in the *z* direction.

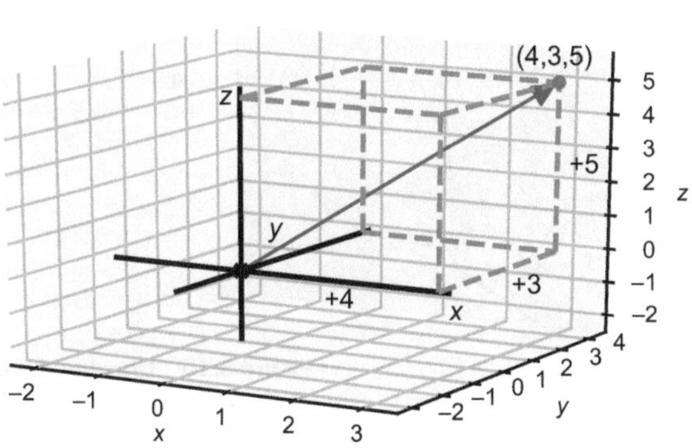

Figure 3.8 The three coordinates (4, 3, 5) give us directions to a point in 3D.

3.1.2 3D drawing with Python

As in the previous chapter, I use a wrapper around Python's Matplotlib library to make vector drawings in 3D. You can find the implementation in the source code for this book, but I'll stick with the wrapper to focus on the conceptual process of drawing rather than the details of Matplotlib.

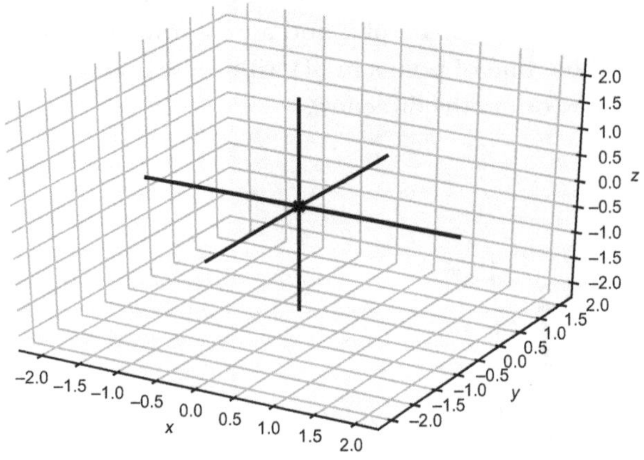

Figure 3.9 Drawing an empty 3D region with Matplotlib's `draw3d()`

My wrapper uses new classes like `Points3D` and `Arrow3D` to distinguish 3D objects from their 2D counterparts. A new function, `draw3d`, knows how to interpret and render these objects so as to make these look three-dimensional. By default, `draw3d()` shows the axes and the origin as well as a small box of 3D space (figure 3.9), even if no objects are specified for drawing.

The *x*-, *y*-, and *z*-axes that are drawn are perpendicular in the space, despite being skewed by our perspective. For visual clarity, Matplotlib shows the units outside the box, but the origin and the axes themselves are displayed within the box. The origin is the coordinate (0, 0, 0), and the axes emanate from it in the positive and negative *x*, *y*, and *z* directions.

The `Points3D` class stores a collection of vectors we want to think of as points and, therefore, draw as dots in 3D space. For instance, we could plot the vectors (2, 2, 2) and (1, –2, –2) with the following code that produces figure 3.10:

```
draw3d(
    Points3D((2,2,2),(1,-2,-2))
)
```

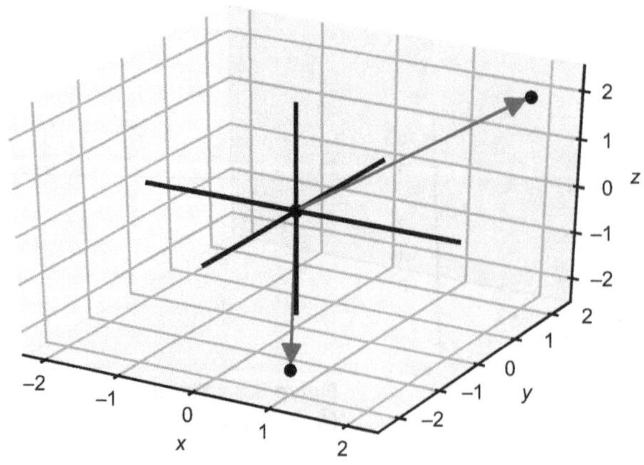

Figure 3.10 Drawing the points (2, 2, 2) and (1, –2, –2)

To visualize these vectors instead as arrows, we can represent the vectors as `Arrow3D` objects. We can also connect the tips of arrows with a `Segment3D` object as follows, producing figure 3.11:

```
draw3d(
    Points3D((2,2,2),(1,-2,-2)),
    Arrow3D((2,2,2)),
    Arrow3D((1,-2,-2)),
    Segment3D((2,2,2), (1,-2,-2))
)
```

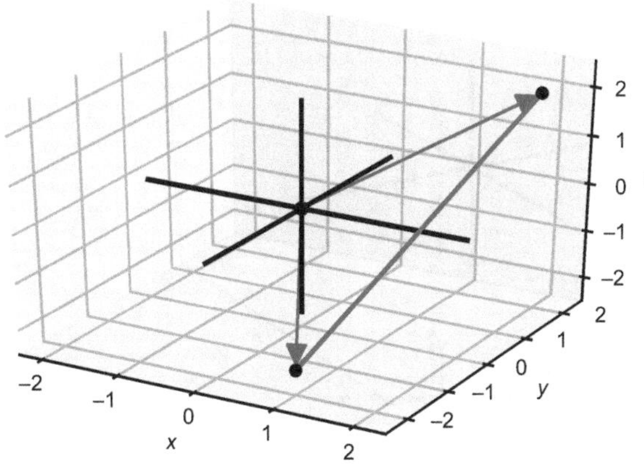

Figure 3.11 Drawing 3D arrows

It's a bit hard to see which direction the arrows are pointing in figure 3.11. To make it clearer, we can draw dashed boxes around the arrows to make them look more three-dimensional. Because we'll draw these boxes so frequently, I created a `Box3D` class to represent a box with one corner at the origin and the opposite one at a given point. Figure 3.12 illustrates the 3D box, but first, here's the code:

```
draw3d(
    Points3D((2,2,2),(1,-2,-2)),
    Arrow3D((2,2,2)),
    Arrow3D((1,-2,-2)),
    Segment3D((2,2,2), (1,-2,-2)),
    Box3D(2,2,2),
    Box3D(1,-2,-2)
)
```

In this chapter, I use a number of (hopefully self-explanatory) keyword arguments without introducing them explicitly. For instance, a `color` keyword argument can be passed to most of these constructors, controlling the color of the object that shows up in the drawing.

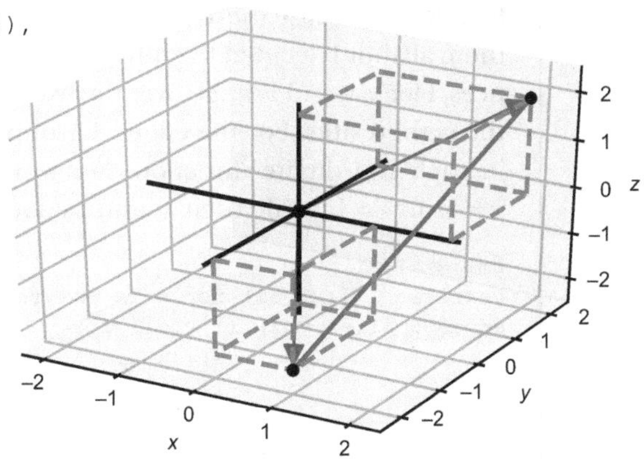

Figure 3.12 Drawing boxes to make our arrows look 3D

3.1.3 Exercises

Exercise 3.1 Draw the 3D arrow and point representing the coordinates (–1, –2, 2) as well as the dashed box that makes the arrow look 3D. Do this drawing by hand for practice, but from now on, we'll use Python to draw for us.

Solution

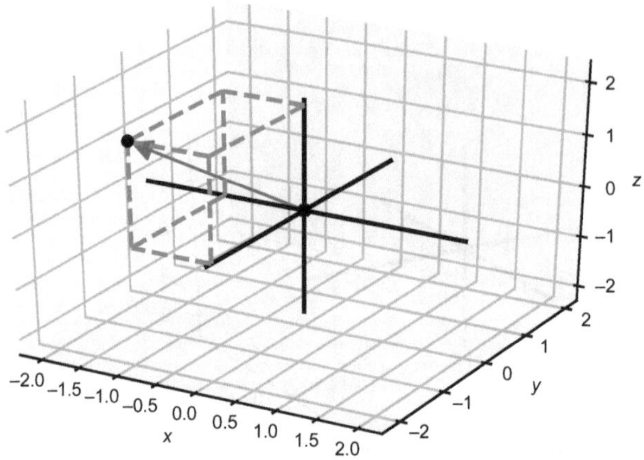

The vector (–1, –2, 2) and the box that makes it look 3D

Exercise 3.2—Mini Project There are exactly eight 3D vectors whose coordinates are all either +1 or –1. For instance, (1, –1, 1) is one of these. Plot all of these eight vectors as points. Then figure out how to connect them with line segments using `Segment3D` objects to form the outline of a cube.

Hint You'll need 12 segments in total.

Solution Because there are only 8 vertices and 12 edges, it's not too tedious to list them all, but I decided to enumerate them with a list comprehension. For the vertices, I let *x*, *y*, and *z* range over the list of two possible values `[1,-1]` and collected the eight results. For the edges, I grouped them into three sets of four that point in each coordinate direction. For instance, there are four edges that go from $x = -1$ to $x = 1$, while their *y*- and *z*-coordinates are the same at both endpoints:

```
pm1 = [1,-1]
vertices = [(x,y,z) for x in pm1 for y in pm1 for z in pm1]
edges = [((-1,y,z),(1,y,z)) for y in pm1 for z in pm1] +\
          [((x,-1,z),(x,1,z)) for x in pm1 for z in pm1] +\
          [((x,y,-1),(x,y,1)) for x in pm1 for y in pm1]
draw3d(
    Points3D(*vertices,color=blue),
    *[Segment3D(*edge) for edge in edges]
)
```

(continued)

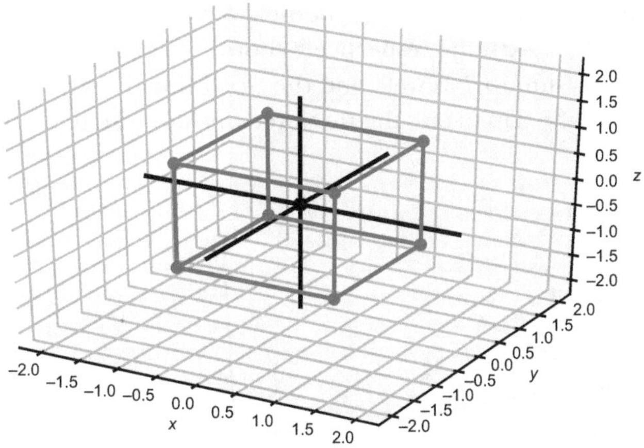

The cube with all vertex coordinates equal to +1 or –1

3.2 Vector arithmetic in 3D

With these Python functions in hand, it's easy to visualize the results of vector arithmetic in three dimensions. All of the arithmetic operations we saw in 2D have analogies in 3D, and the geometric effects of each are similar.

3.2.1 Adding 3D vectors

In 3D, vector addition can still be accomplished by adding coordinates. The vectors (2, 1, 1) and (1, 2, 2) sum to (2 + 1, 1 + 2, 1 + 2) = (3, 3, 3). We can start at the origin and place the two input vectors tip-to-tail in either order to get to the sum point (3, 3, 3) (figure 3.13).

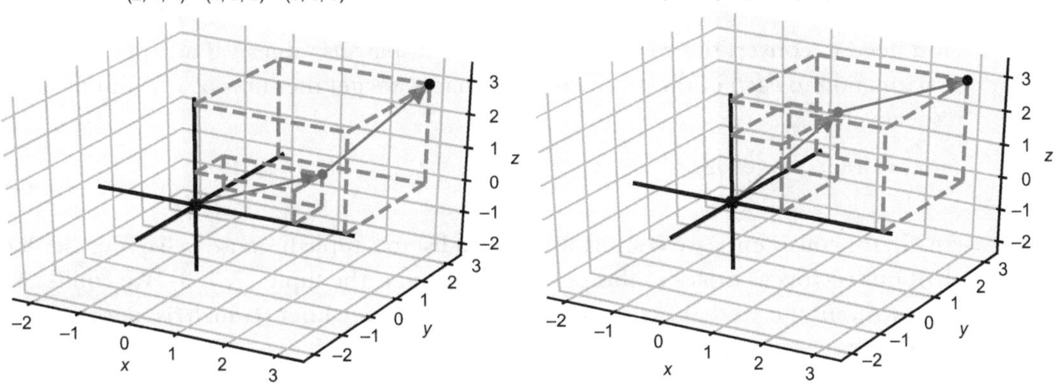

Figure 3.13 Two visual examples of vector addition in 3D

As in 2D, we can add any number of 3D vectors together by summing all of their *x*-coordinates, all of their *y*-coordinates, and all of their *z*-coordinates. These three sums give us the coordinates of the new vector. For instance, in the sum (1, 1, 3) + (2, 4, –4) + (4, 2, –2), the respective *x*-coordinates are 1, 2, and 4, which sum to 7. The *y*-coordinates sum to 7 as well, and the *z*-coordinates sum to –3; therefore, the vector sum is (7, 7, –3). Tip-to-tail, the three vectors look like those in figure 3.14.

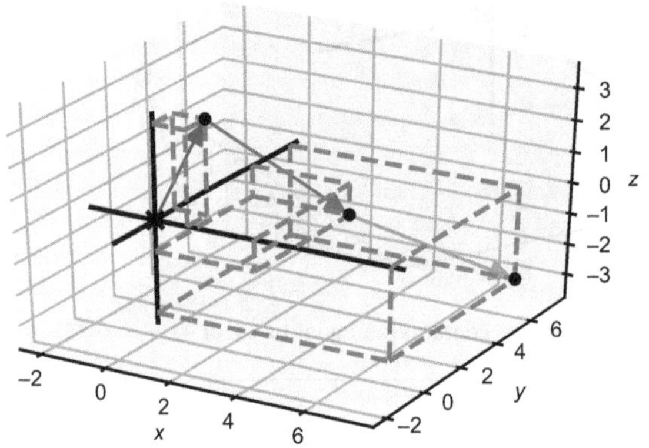

Figure 3.14 Adding three vectors tip-to-tail in 3D

In Python, we can write a concise function to sum any number of input vectors, and that works in two or three dimensions (or an even higher number of dimensions as we'll see later). Here it is:

```
def add(*vectors):
    by_coordinate = zip(*vectors)
    coordinate_sums = [sum(coords) for coords in by_coordinate]
    return tuple(coordinate_sums)
```

Let's break it down. Calling Python's `zip` function on the input vectors extracts their *x*-coordinates, *y*-coordinates, and *z*-coordinates. For instance,

```
>>> list(zip(*[(1,1,3),(2,4,-4),(4,2,-2)]))
[(1, 2, 4), (1, 4, 2), (3, -4, -2)]
```

(You need to convert the `zip` result to a list to display its values.) If we apply Python's `sum` function to each of the grouped coordinates, we get the sums of *x*, *y*, and *z* values, respectively:

```
[sum(coords) for coords in [(1, 2, 4), (1, 4, 2), (3, -4, -2)]]
[7, 7, -3]
```

Finally, for consistency, we convert this from a list to a tuple because we've represented all of our vectors as tuples to this point. The result is the tuple (7, 7, 3). We could also have written the `add` function as the following one-liner (which is perhaps less Pythonic):

```
def add(*vectors):
    return tuple(map(sum,zip(*vectors)))
```

3.2.2 *Scalar multiplication in 3D*

To multiply a 3D vector by a scalar, we multiply all of its components by the scalar factor. For example, the vector (1, 2, 3) multiplied by the scalar 2 gives us (2, 4, 6). This resulting vector is twice as long but points in the same direction as in the 2D case. Figure 3.15 shows **v** = (1, 2, 3) and its scalar multiple 2 · **v** = (2, 4, 6).

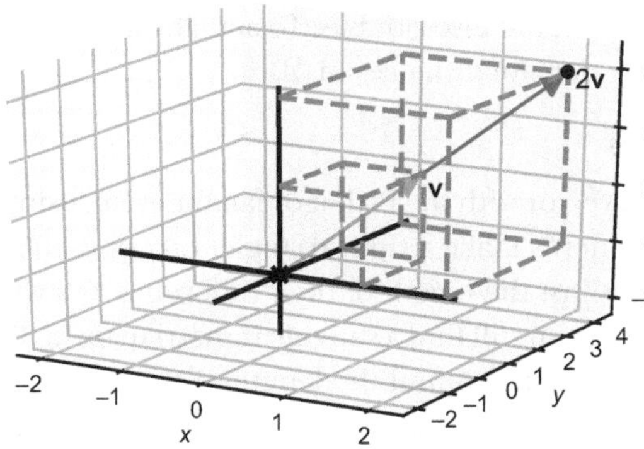

Figure 3.15 Scalar multiplication by 2 returns a vector pointing in the same direction, which is twice as long as the original vector.

3.2.3 *Subtracting 3D vectors*

In 2D, the difference of the two vectors **v** – **w** is the vector "from **w** to **v**," which is called the *displacement*. In 3D, the story is the same; in other words, **v** – **w** is the displacement from **w** to **v**, which is the vector you can add to **w** to get **v**. Thinking of **v** and **w** as arrows from the origin, the difference **v** – **w** is an arrow that can be positioned to have its tip at the tip of **v** and its tail at the tip of **w**. Figure 3.16 shows the difference of **v** = (–1, –3, 3) and **w** = (3, 2, 4), both as an arrow from **w** to **v** and as a point in its own right.

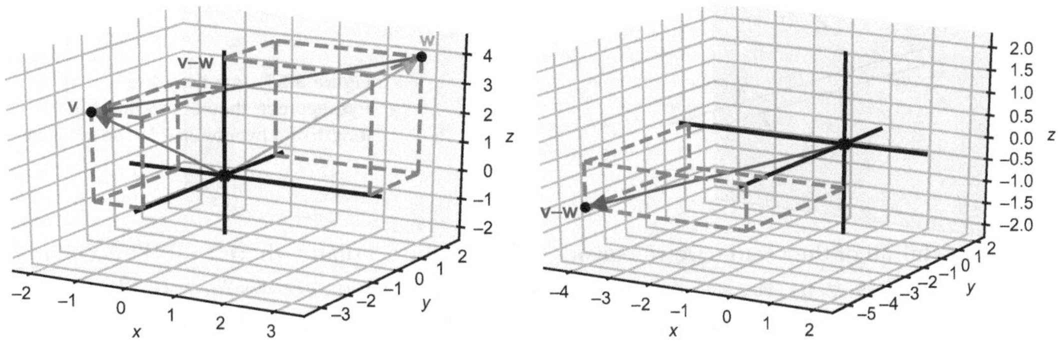

Figure 3.16 Subtracting the vector w from the vector v gives the displacement from w to v.

Subtracting a vector **w** from a vector **v** is accomplished in coordinates by taking the difference of the coordinates of **v** and **w**. For instance, **v** – **w** gives us (–1 –3, –3 – 2, 3 – 4) = (–4, –5, –1) as a result. These coordinates agree with the picture of **v** – **w** in figure 3.16, which shows it's a vector pointing in the negative *x*, negative *y*, and negative *z* directions.

When I claim scalar multiplication by two makes a vector "twice as long," I'm thinking in terms of geometric similarity. If each of the three components of **v** are doubled, corresponding to doubling the length, width, and depth of the box, the diagonal distance from one corner to the other should also double. To actually measure and confirm this, we need to know how to calculate distances in 3D.

3.2.4 Computing lengths and distances

In 2D, we calculated the length of a vector with the Pythagorean theorem, using the fact that an arrow vector and its components make a right triangle. Likewise, the distance between two points in the plane was just the length of their difference as a vector.

We have to look a bit closer, but we can still find a suitable right triangle in 3D to help us calculate the length of a vector. Let's try to find the length of the vector (4, 3, 12). The *x* and *y* components still give us a right triangle lying in the plane where $z = 0$. This triangle's hypotenuse, or diagonal side, has length $\sqrt{(4^2 + 3^2)} = \sqrt{25} = 5$. If this were a 2D vector, we'd be done, but the *z* component of 12 makes this vector quite a bit longer (figure 3.17).

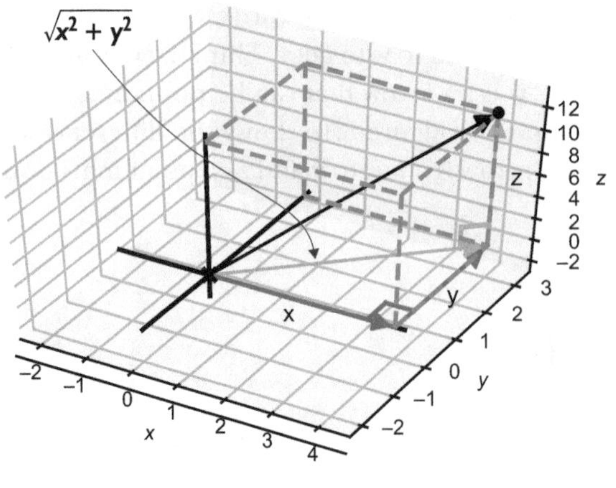

Figure 3.17 Applying the Pythagorean theorem to find the length of a hypotenuse in the *x*,*y* plane

So far all of the vectors we've considered lie in the *x*,*y* plane where $z = 0$. The *x* component is (4, 0, 0), the *y* component is (0, 3, 0), and their vector sum is (4, 3, 0). The *z* component of (0, 0, 12) is perpendicular to all three of these. That's useful because it gives us a second right triangle in the diagram: the one formed by (4, 3, 0) and (0, 0, 12) and placed tip-to-tail. The hypotenuse of this triangle is our original vector (4, 3, 12), whose length we want to find. Let's focus on this second right triangle and invoke the Pythagorean theorem again to find the hypotenuse length (shown in figure 3.18).

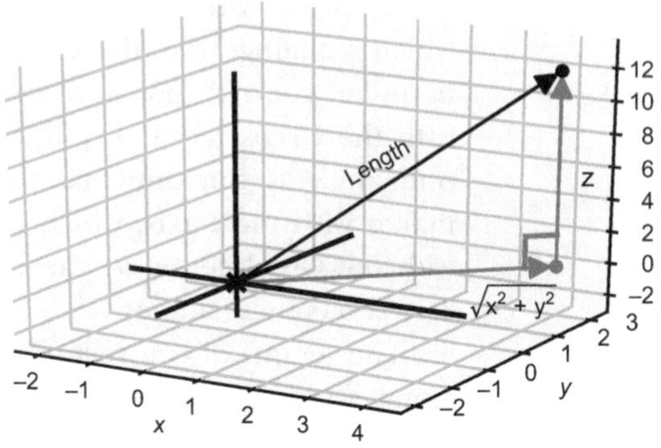

Figure 3.18 A second application of the Pythagorean theorem gives us the length of the 3D vector.

Squaring both known sides and taking the square root should give us the length. Here, the lengths are 5 and 12 so the result is $\sqrt{5^2 + 12^2} = 13$. In general, the following equation shows the formula for the length of a vector in 3D:

$$\text{length} = \sqrt{\left(\sqrt{x^2 + y^2}\right)^2 + z^2} = \sqrt{x^2 + y^2 + z^2}$$

This is conveniently similar to the 2D length formula. In either 2D or 3D, the length of a vector is the square root of the sum of squares of its components. Because we don't explicitly reference the length of the input tuple anywhere in the following `length` function, it will work on either 2D or 3D vectors:

```
from math import sqrt
    def length(v):
        return sqrt(sum([coord ** 2 for coord in v]))
```

So, for instance, `length((3,4,12))` returns 13.

3.2.5 *Computing angles and directions*

As in 2D, you can think of a 3D vector as an arrow or a displacement of a certain length in a certain direction. In 2D, this means that two numbers—a length and an angle making a pair of polar coordinates—are sufficient to specify any 2D vector. In 3D, one angle is not sufficient to specify a direction but two angles are.

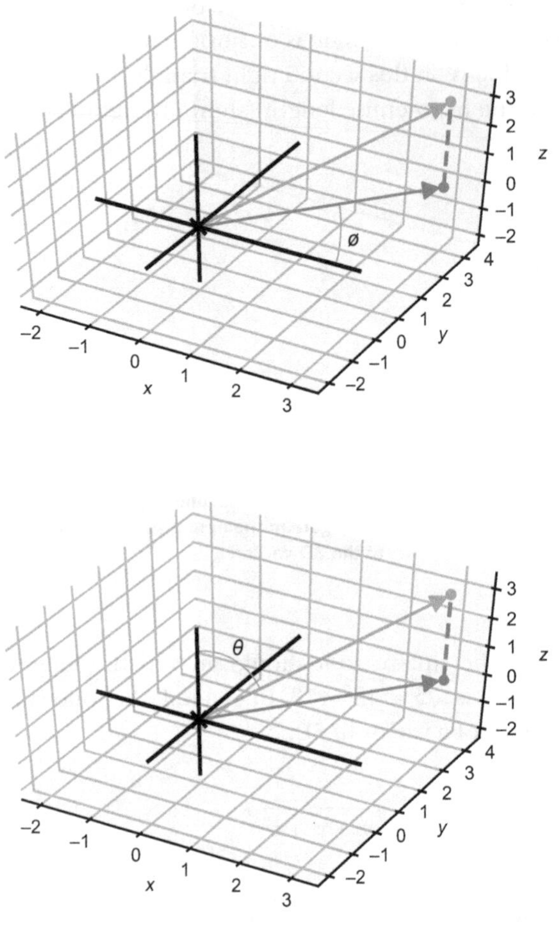

Figure 3.19 Two angles that together measure the direction of a 3D vector

For the first angle, we again think of the vector without its z-coordinate, as if it still lived in the x,y plane. Another way of thinking of this is as the shadow cast by the vector from a light at a very high z position. This shadow makes some angle with the positive x-axis, which is analogous to the angle we used in polar coordinates, and we label it with the Greek letter ϕ (phi). The second angle is the one that the vector makes with the z-axis, which is labeled with the Greek letter θ (theta). Figure 3.19 shows these angles.

The length of the vector, labeled **r**, along with the angles ϕ and θ can describe any vector in three dimensions. Together, the three numbers r, ϕ, and θ are called *spherical coordinates* as opposed to the Cartesian coordinates x, y, and z. Calculating spherical coordinates from Cartesian coordinates is a doable exercise with only the trigonometry we've covered, but we won't go into it here. In fact, we won't use spherical coordinates again in this book, but I want to briefly compare them with polar coordinates.

Polar coordinates were useful because they allowed us to perform any rotation of a collection of plane vectors by simply adding or subtracting from the angle. We were also able to read the angle between two vectors by taking the difference of their angles in polar coordinates. In three dimensions, neither of the angles ϕ and θ lets us immediately decide the angle between two vectors. And while we could rotate vectors easily around the z-axis by adding or subtracting from the angle ϕ, it's not convenient to rotate about any other axis in spherical coordinates.

We need some more general tools to handle angles and trigonometry in 3D. We'll cover two such tools, called *vector products*, in the next section.

3.2.6 Exercises

Exercise 3.3 Draw (4, 0, 3) and (–1, 0, 1) as `Arrow3D` objects, such that they are placed tip-to-tail in both orders in 3D. What is their vector sum?

Solution We can find the vector sum using the `add` function we built:

```
>>> add((4,0,3),(-1,0,1))
(3, 0, 4)
```

Then to draw these tip-to-tail, we draw arrows from the origin to each point and from each point to the vector sum (3, 0, 4). Like the 2D `Arrow` object, `Arrow3D` takes the tip vector of the arrow first and then, optionally, the tail vector if it is not the origin:

```
draw3d(
    Arrow3D((4,0,3),color=red),
    Arrow3D((-1,0,1),color=blue),
    Arrow3D((3,0,4),(4,0,3),color=blue),
    Arrow3D((-1,0,1),(3,0,4),color=red),
    Arrow3D((3,0,4),color=purple)
)
```

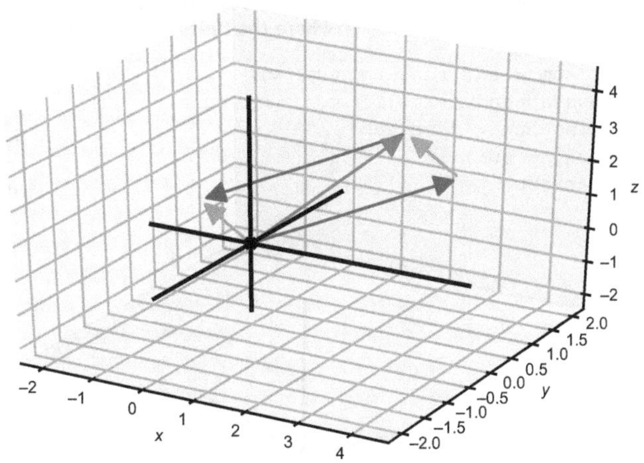

Tip-to-tail addition shows (4, 0, 3) + (–1, 0, 1) = (–1, 0, 1) + (4, 0, 3) = (3, 0, 4).

Exercise 3.4 Suppose we set `vectors1=[(1,2,3,4,5),(6,7,8,9,10)]` and `vectors2=[(1,2),(3,4),(5,6)]`. Without evaluating in Python, what are the lengths of `zip(*vectors1)` and `zip(*vectors2)`?

Solution The first `zip` has length 5. Because there are five coordinates in each of the two input vectors, `zip(*vectors1)` contains five tuples, having two

(continued)

elements each. Likewise, `zip(*vectors2)` has length 2; the two entries of `zip(*vectors2)` are tuples containing all of the *x* components and all of the *y* components, respectively.

Exercise 3.5—Mini Project The following comprehension creates a list of 24 Python vectors:

```
from math import sin, cos, pi
vs = [(sin(pi*t/6), cos(pi*t/6), 1.0/3) for t in range(0,24)]
```

What is the sum of the 24 vectors? Draw all 24 of them tip-to-tail as `Arrow3D` objects.

Solution Drawing these vectors tip-to-tail ends up producing a helix shape:

```
from math import sin, cos, pi
vs = [(sin(pi*t/6), cos(pi*t/6), 1.0/3) for t in range(0,24)]

running_sum = (0,0,0)
arrows = []
for v in vs:
    next_sum = add(running_sum, v)
    arrows.append(Arrow3D(next_sum, running_sum))
    running_sum = next_sum
print(running_sum)
draw3d(*arrows)
```

Begins a running sum at (0, 0, 0), where the tip-to-tail addition starts

To draw each subsequent vector tip-to-tail, we add it to the running sum. The latest arrow connects the previous running sum to the next.

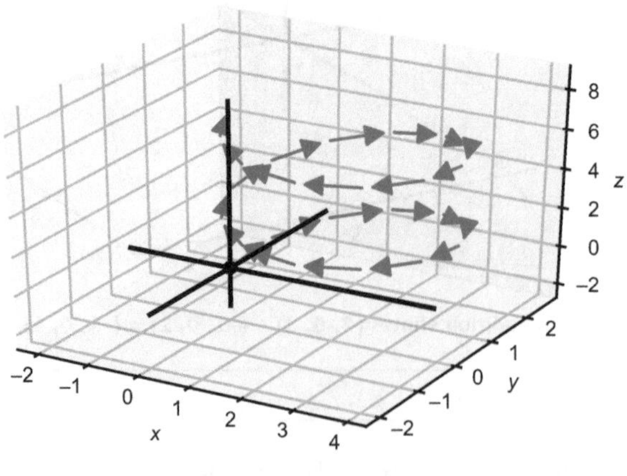

Finding the vector sum of 24 vectors in 3D

The sum is

```
(-4.440892098500626e-16, -7.771561172376096e-16, 7.9999999999999964)
```

which is approximately (0, 0, 8).

Exercise 3.6 Write a function `scale(scalar,vector)` that returns the input scalar times the input vector. Specifically, write it so it works on 2D or 3D vectors, or vectors of any number of coordinates.

Solution With a comprehension, we multiply each coordinate in the vector by the scalar. This is a generator comprehension that is converted to a tuple:

```
def scale(scalar,v):
    return tuple(scalar * coord for coord in v)
```

Exercise 3.7 Let $\mathbf{u} = (1, -1, -1)$ and $\mathbf{v} = (0, 0, 2)$. What is the result of $\mathbf{u} + \frac{1}{2} \cdot (\mathbf{v} - \mathbf{u})$?

Solution With $\mathbf{u} = (1, -1, -1)$ and $\mathbf{v} = (0, 0, 2)$, we can first compute $(\mathbf{v} - \mathbf{u}) = (0 - 1, 0 - (-1), 2 - (-1)) = (-1, 1, 3)$. Then $\frac{1}{2} \cdot (\mathbf{v} - \mathbf{u})$ is $(-\frac{1}{2}, \frac{1}{2}, 3/2)$. The final desired result of $\mathbf{u} + \frac{1}{2} \cdot (\mathbf{v} - \mathbf{u})$ is then $(\frac{1}{2}, -\frac{1}{2}, \frac{1}{2})$. Incidentally, this is the point exactly halfway between the point u and the point v.

Exercise 3.8 Try to find the answers for this exercise without using code and then check your work. What is the length of the 2D vector (1, 1)? What is the length of the 3D vector (1, 1, 1)? We haven't yet talked about 4D vectors, but these have four coordinates instead of two or three. If you had to guess, what is the length of the 4D vector with coordinates (1, 1, 1, 1)?

Solution The length of (1, 1) is $\sqrt{1^2 + 1^2} = \sqrt{2}$. The length of (1, 1, 1) is $\sqrt{1^2 + 1^2 + 1^2} = \sqrt{3}$. As you might guess, we use the same distance formula for higher dimensional vectors as well. The length of (1, 1, 1, 1) follows the same pattern: it is $\sqrt{1^2 + 1^2 + 1^2 + 1^2} = \sqrt{4}$, which is 2.

Exercise 3.9—Mini Project The coordinates 3, 4, 12 in any order create a vector of length 13, a whole number. This is unusual because most numbers are not perfect squares, so the square root in the length formula typically returns an irrational number. Find a different triple of whole numbers that define coordinates of a vector with a whole number length.

Solution The following code searches for triples of descending whole numbers less than 100 (an arbitrary choice):

```
def vectors_with_whole_number_length(max_coord=100):
    for x in range(1,max_coord):
        for y in range(1,x+1):
```

(continued)

```
        for z in range(1,y+1):
            if length((x,y,z)).is_integer():
                yield (x,y,z)
```

It finds 869 vectors with whole number coordinates and whole number lengths. The shortest of these is (2, 2, 1) with length exactly 3, and the longest is (99, 90, 70) with length exactly 150.

Exercise 3.10 Find a vector in the same direction as (–1, –1, 2) but which has length 1.

Hint Find the appropriate scalar to multiply the original vector to change its length appropriately.

Solution The length of (–1, –1, 2) is about 2.45, so we'll have to take a scalar multiple of this vector by (1/2.45) to make its length 1:

```
>>> length((-1,-1,2))
2.449489742783178
>>> s = 1/length((-1,-1,2))
>>> scale(s,(-1,-1,2))
(-0.4082482904638631, -0.4082482904638631, 0.8164965809277261)
>>> length(scale(s,(-1,-1,2)))
1.0
```

Rounding to the nearest hundredth in each coordinate, the vector is (–0.41, –0.41, 0.82).

3.3 The dot product: Measuring vector alignment

One kind of multiplication we've already seen for vectors is scalar multiplication, combining a scalar (a real number) and a vector to get a new vector. We haven't yet talked about any ways to multiply one vector with another. It turns out there are two important ways to do this, and these both give important geometric insights. One is called the *dot product* and we write it with a dot operator (for example, $\mathbf{u} \cdot \mathbf{v}$), while the other is called the *cross product* (for example, $\mathbf{u} \times \mathbf{v}$). For numbers, these symbols mean the same thing, so for instance $3 \cdot 4 = 3 \times 4$. For two vectors, the operations $\mathbf{u} \cdot \mathbf{v}$ and $\mathbf{u} \times \mathbf{v}$ aren't just different notations, these mean completely different things.

The dot product takes two vectors and returns a scalar (a number), while the cross product takes two vectors and returns another vector. Both, however, are operations that help us reason about lengths and directions of vectors in 3D. Let's start by focusing on the dot product.

3.3.1 Picturing the dot product

The dot product (also called the *inner product*) is an operation on two vectors that returns a scalar. In other words, given two vectors **u** and **v**, the result of **u** · **v** is a real number. The dot product works on vectors in 2D, 3D, or any number of dimensions. You can think of it as measuring "how aligned" the pair of input vectors are. Let's first look at some vectors in the *x*,*y* plane and show their dot products to give you some intuition for how this operation works.

The vectors **u** and **v** have lengths 4 and 5, respectively, and they point in nearly the same direction. Their dot product is positive, meaning they are aligned (figure 3.20).

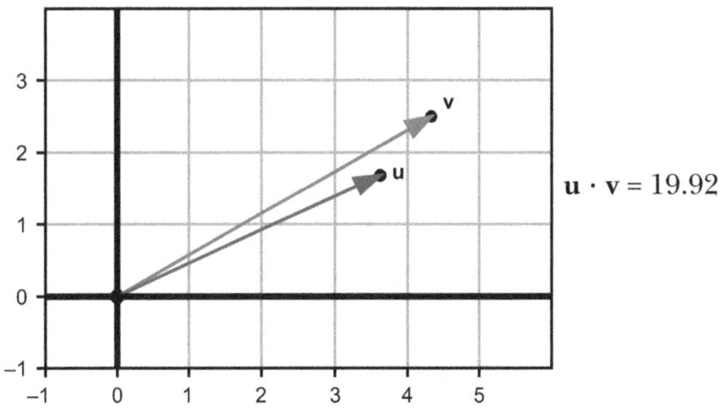

Figure 3.20 Two vectors that are relatively aligned give a large positive dot product.

Two vectors that are pointing in similar directions have a positive dot product, and the larger the vectors, the larger the product. Smaller vectors that are similarly aligned have a smaller but still positive dot product. The new vectors **u** and **v** both have a length of 2 (figure 3.21).

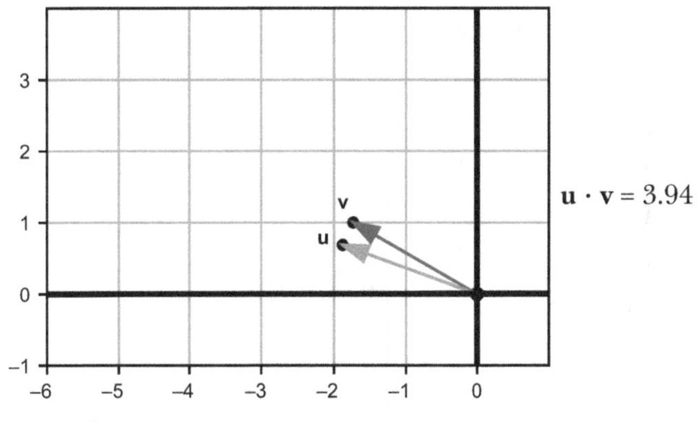

Figure 3.21 Two shorter vectors pointing in similar directions give a smaller but still positive dot product.

By contrast, if two vectors point in opposite or near opposite directions, their dot product is negative (figures 3.22 and 3.23). The bigger the magnitude of the vectors, the more negative their dot product.

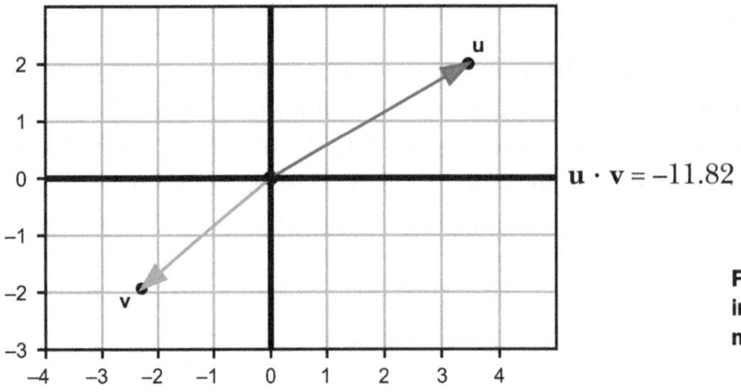

Figure 3.22 Vectors pointing in opposing directions have a negative dot product.

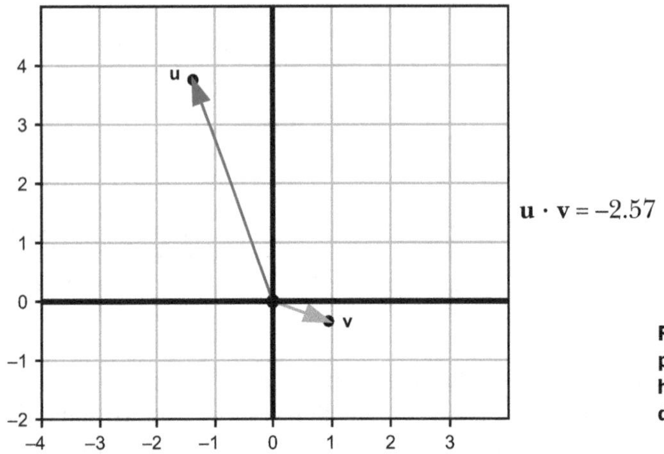

Figure 3.23 Shorter vectors pointing in opposing directions have a smaller but still negative dot product.

Not all pairs of vectors clearly point in similar or opposite directions, and the dot product detects this. As figure 3.24 shows, if two vectors point in exactly perpendicular directions, their dot product is zero regardless of their lengths.

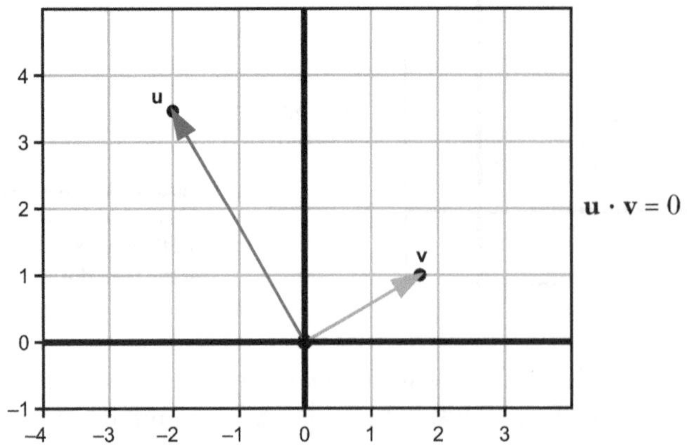

Figure 3.24 Perpendicular vectors always have a dot product of zero.

This turns out to be one of the most important applications of the dot product: it lets us compute whether two vectors are perpendicular without doing any trigonometry. This perpendicular case also serves to separate the other cases: if the angle between two vectors is less than 90°, the vectors then have a positive dot product. If the angle is greater than 90°, they have a negative dot product. While I haven't yet told you how to compute a dot product, you now know how to interpret the value. We move on to computing it next.

3.3.2 Computing the dot product

Given the coordinates for two vectors, there's a simple formula to compute the dot product: multiply the corresponding coordinates and then add the products. For instance in the dot product $(1, 2, -1) \cdot (3, 0, 3)$, the product of the *x*-coordinate is 3, the product of the *y*-coordinate is 0, and the product of the *z*-coordinate is –3. The sum is $3 + 0 + (-3) = 0$, so the dot product is zero. If my claim is correct, these two vectors should be perpendicular. Drawing them (figure 3.25) demonstrates this, if you look at them from the right perspective!

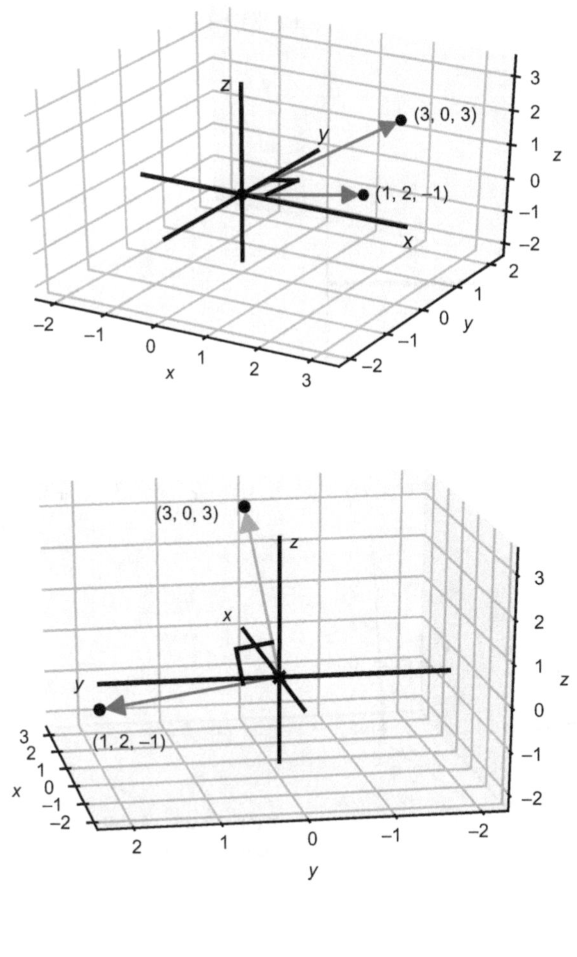

Figure 3.25 Two vectors with a dot product of zero are indeed perpendicular in 3D.

Our perspective can be misleading in 3D, making it all the more valuable to be able to *compute* relative directions rather than eyeballing them. As another example, figure 3.26 shows that the 2D vectors (2, 3) and (4, 5) lie in similar directions in the x,y plane. The product of the x-coordinates is $2 \cdot 4 = 8$, while the product of the y-coordinates is $3 \cdot 5 = 15$. The sum $8 + 15 = 23$ is the dot product. As a positive number, this result confirms that the vectors are separated by less than 90°. These vectors have the same relative geometry whether we consider them in 2D or in 3D as the vectors (2, 3, 0) and (4, 5, 0) that happen to lie in the plane where $z = 0$.

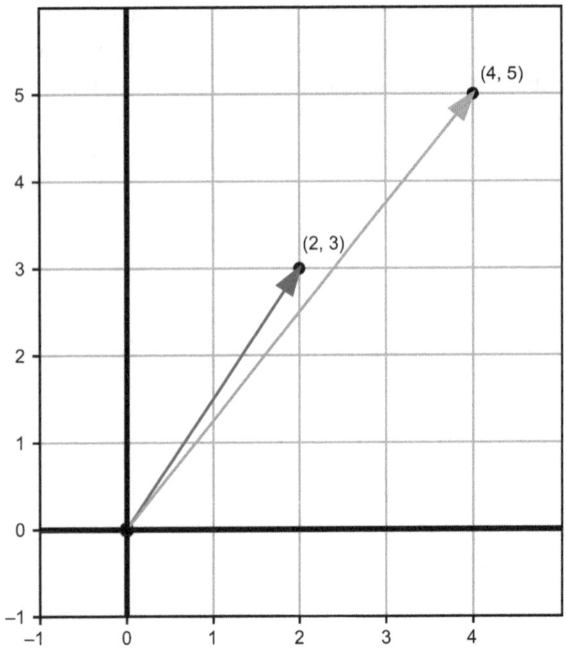

$$(2, 3) \cdot (4, 5) = 2 \cdot 4 + 3 \cdot 5 = 23$$

$$(2, 3, 0) \cdot (4, 5, 0) = 2 \cdot 4 + 3 \cdot 5 + 0 \cdot 0 = 23$$

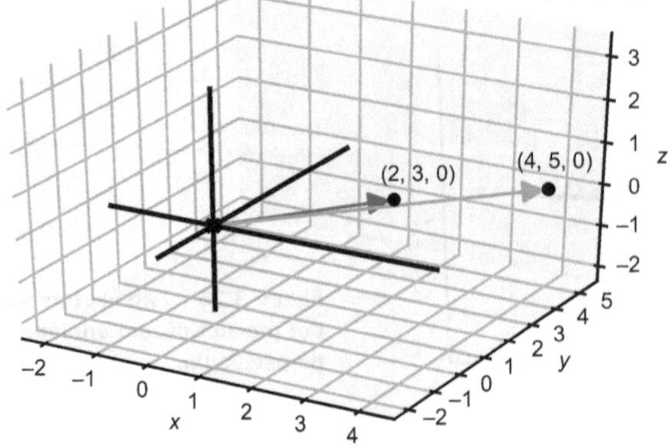

Figure 3.26 Another example of computing a dot product

In Python, we can write a dot product function that handles any pair of input vectors as long as they all have the same number of coordinates. For example,

```
def dot(u,v):
    return sum([coord1 * coord2 for coord1,coord2 in zip(u,v)])
```

This code uses Python's `zip` function to pair the appropriate coordinates, then multiplies each pair in a comprehension, and adds the resulting list. Let's use this to further explore how the dot product behaves.

3.3.3 Dot products by example

It's not surprising that two vectors lying on different axes have zero dot product. We know they are perpendicular:

```
>>> dot((1,0),(0,2))
0
>>> dot((0,3,0),(0,0,-5))
0
```

We can also confirm that longer vectors give longer dot products. For instance, scaling either input vector by a factor of 2 doubles the output of the dot product:

```
>>> dot((3,4),(2,3))
18
>>> dot(scale(2,(3,4)),(2,3))
36
>>> dot((3,4),scale(2,(2,3)))
36
```

It turns out the dot product is proportional to each of the lengths of its input vectors. If you take the dot product of two vectors in the same direction, the dot product is precisely equal to the product of the lengths. For instance, (4, 3) has a length of 5 and (8, 6) has a length of 10. The dot product is equal to $5 \cdot 10$:

```
>>> dot((4,3),(8,6))
50
```

Of course, the dot product is not always equal to the product of the lengths of its inputs. The vectors (5, 0), (–3, 4), (0, –5), and (–4, –3) all have the same length of 5 but different dot products with the original vector (4, 3) as shown in figure 3.27.

The dot product of two vectors of length 5 ranges from $5 \cdot 5 = 25$ when they are aligned to –25, when they point in opposite directions. In the next set of exercises, I invite you to convince yourself that the dot product of two vectors can range from the product of the lengths down to the opposite of that value.

3.3.4 Measuring angles with the dot product

We've seen that the dot product varies based on the angle between two vectors. Specifically, the dot product $\mathbf{u} \cdot \mathbf{v}$ ranges from 1 to –1 times the product of the lengths of $\mathbf{u}$ and $\mathbf{v}$ as the angle ranges from 0 to 180°. We already saw a function that behaves that

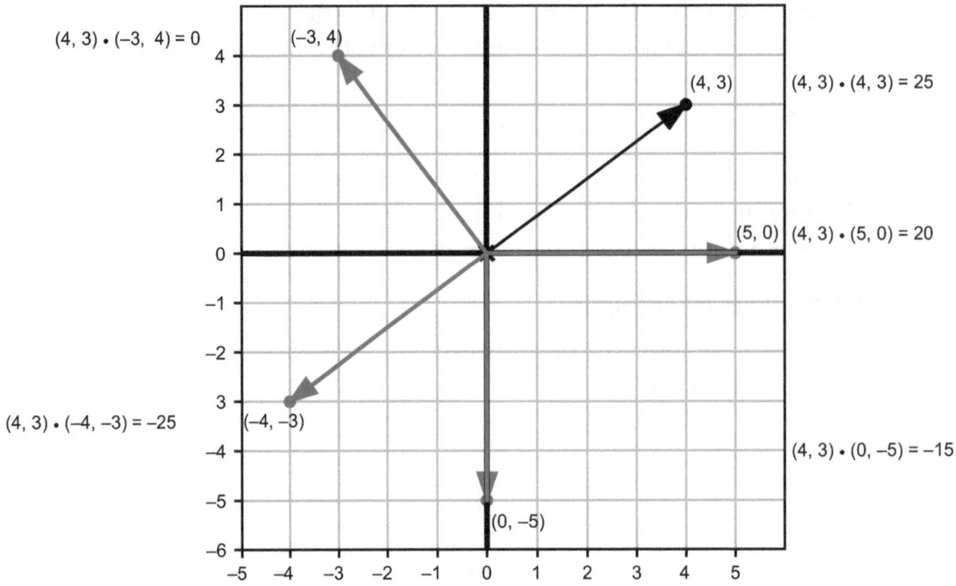

Figure 3.27 Vectors of the same length have different dot products with the vector (4, 3), depending on their direction.

way, namely the cosine function. It turns out that the dot product has an alternate formula. If $|\mathbf{u}|$ and $|\mathbf{v}|$ denote the lengths of vectors **u** and **v**, the dot product is given by

$$\mathbf{u} \cdot \mathbf{v} = |\mathbf{u}| \cdot |\mathbf{v}| \cdot \cos(\theta)$$

where θ is the angle between the vectors **u** and **v**. In principle this gives us a new way to compute a dot product. We could measure the lengths of two vectors and then measure the angle between them to get the result. Suppose, as in figure 3.28, we have two vectors of known lengths 3 and 2, respectively, and using our protractor, discovered that they are 75° apart.

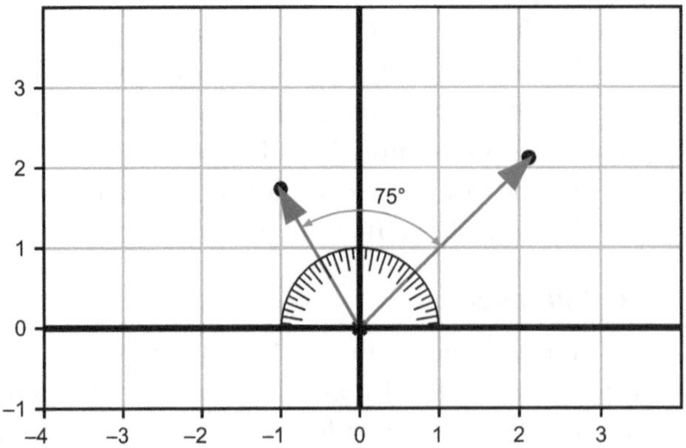

Figure 3.28 Two vectors of lengths 3 and 2, respectively, at 75° apart

The dot product for the two vectors in figure 3.28 is $3 \cdot 2 \cdot \cos(75°)$. With the appropriate conversion to radians, we can compute this in Python to be about 1.55:

```
>>> from math import cos,pi
>>> 3 * 2 * cos(75 * pi / 180)
1.5529142706151244
```

When doing computations with vectors, it's more common to start with coordinates and to compute angles from them. We can combine both of our formulas to recover an angle: first we compute the dot product and lengths using coordinates, then we solve for the angle.

Let's find the angle between the vectors (3, 4) and (4, 3). Their dot product is 24, and each of their lengths is 5. Our new dot product formula tells us that:

$$(3, 4) \cdot (4, 3) = 24 = 5 \cdot 5 \cdot \cos(\theta) = 25 \cdot \cos(\theta)$$

From $24 = 25 \cdot \cos(\theta)$, we can simplify it to $\cos(\theta) = 24/25$. Using Python's `math.acos`, we find that a θ value of 0.284 radians or 16.3° gives us a cosine of 24/25.

This exercise reminds us why we don't need the dot product in 2D. In chapter 2, we showed how to get the angle of a vector from the positive *x*-axis. Using that formula creatively, we could find any angle we want in the plane. The dot product really starts to shine in 3D, where a change of coordinates can't help us as much.

For instance, we can use the same formula to find the angle between (1, 2, 2) and (2, 2, 1). The dot product is $1 \cdot 2 + 2 \cdot 2 + 2 \cdot 1 = 8$ and the lengths are both 3. This means $8 = 3 \cdot 3 \cdot \cos(\theta)$, so $\cos(\theta) = 8/9$ and $\theta = 0.476$ radians or 27.3°.

This process is the same in 2D or 3D, and it's one we'll use over and over. We can save some effort by implementing a Python function to find the angle between two vectors. Because neither our dot function nor our length function has a hard-coded number of dimensions, this new function won't either. We can make use of the fact that $\mathbf{u} \cdot \mathbf{v} = |\mathbf{u}| \cdot |\mathbf{v}| \cdot \cos(\theta)$ and, therefore,

$$\cos(\theta) = \frac{\mathbf{u} \cdot \mathbf{v}}{|\mathbf{u}| \cdot |\mathbf{v}|}$$

and

$$\theta = \arccos\left(\frac{\mathbf{u} \cdot \mathbf{v}}{|\mathbf{u}| \cdot |\mathbf{v}|}\right)$$

This formula translates neatly to Python code as follows:

```
def angle_between(v1,v2):
    return acos(
                dot(v1,v2) /
                (length(v1) * length(v2))
            )
```

Nothing in this Python code depends on the number of dimensions of the vectors $\mathbf{v}_1$ and $\mathbf{v}_2$. These could both be tuples of 2 coordinates or tuples of 3 coordinates (or, in fact, tuples of 4 or more coordinates, which we discuss in the coming chapters). By contrast, the next vector product we meet (the cross product) only works in three dimensions.

3.3.5 Exercises

Exercise 3.11 Based on the following picture, rank $\mathbf{u} \cdot \mathbf{v}$, $\mathbf{u} \cdot \mathbf{w}$, and $\mathbf{v} \cdot \mathbf{w}$ from largest to smallest:

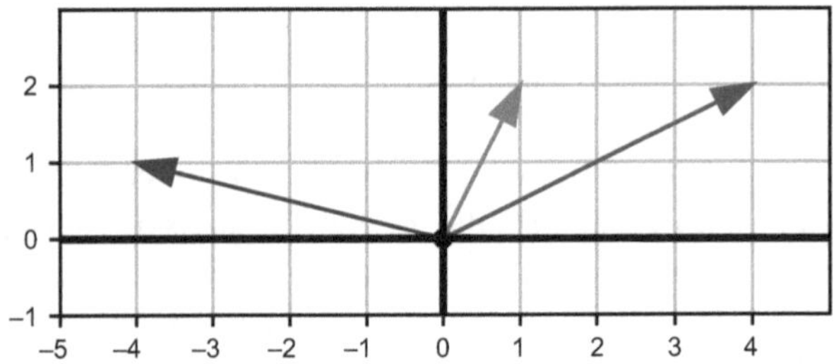

Solution The product $\mathbf{u} \cdot \mathbf{v}$ is the only positive dot product because **u** and **v** are the only pair with less than a right angle between them. Further, $\mathbf{u} \cdot \mathbf{w}$ is smaller (more negative) than $\mathbf{v} \cdot \mathbf{w}$ because **u** is both bigger and further from **w**, so $\mathbf{u} \cdot \mathbf{v} > \mathbf{v} \cdot \mathbf{w} > \mathbf{u} \cdot \mathbf{w}$.

Exercise 3.12 What is the dot product of (–1, –1, 1) and (1, 2, 1)? Are these two 3D vectors separated by more than 90°, less than 90°, or exactly 90°?

Solution (–1, –1, 1) and (1, 2, 1) have the dot product $-1 \cdot 1 + -1 \cdot 2 + 1 \cdot 1 = -2$. Because this is a negative number, the two vectors are more than 90° apart.

Exercise 3.13—Mini Project For two 3D vectors **u** and **v**, the values of $(2\mathbf{u}) \cdot \mathbf{v}$ and $\mathbf{u} \cdot (2\mathbf{v})$ are both equal to $2(\mathbf{u} \cdot \mathbf{v})$. In this case, $\mathbf{u} \cdot \mathbf{v} = 18$ and both $(2\mathbf{u}) \cdot \mathbf{v}$ and $\mathbf{u} \cdot (2\mathbf{v})$ are 36, twice the original result. Show that this works for any real number s, not just 2. In other words, show that for any s the values of $(s\mathbf{u}) \cdot \mathbf{v}$ and $\mathbf{u} \cdot (s\mathbf{v})$ are both equal to $s(\mathbf{u} \cdot \mathbf{v})$.

Solution Let's name the coordinates of **u** and **v**, say $\mathbf{u} = (a, b, c)$ and $\mathbf{v} = (d, e, f)$. Then $\mathbf{u} \cdot \mathbf{v} = ad + be + cf$. Because $s\mathbf{u} = (sa, sb, sc)$ and $s\mathbf{v} = (sd, se, sf)$, we can show both of the results by expanding the dot products:

Write out the coordinates

$$\begin{aligned}(s\mathbf{u}) \cdot \mathbf{v} &= (sa, sb, sc) \cdot (d, e, f) \\ &= sad + sbe + scf \\ &= s(ad + be + cf) \\ &= s(\mathbf{u} \cdot \mathbf{v})\end{aligned}$$

Do the dot product

Factor out s; recognize the original dot product

Proving that scalar multiplication scales the result of the dot product accordingly

And the other product works the same way:

$$\begin{aligned}\mathbf{u} \cdot (s\mathbf{v}) &= (a, b, c) \cdot (sd, se, sf) \\ &= asd + bse + csf \\ &= s(ad + be + cf) \\ &= s \cdot (\mathbf{u} \cdot \mathbf{v})\end{aligned}$$

Proving the same fact holds for the second vector input to the dot product.

Exercise 3.14—Mini Project Explain algebraically why the dot product of a vector with itself is the square of its length.

Solution If a vector has coordinates (a, b, c), then the dot product with itself is $a \cdot a + b \cdot b + c \cdot c$. Its length is $\sqrt{a \cdot a + b \cdot b + c \cdot c}$, so this is indeed the square.

Exercise 3.15—Mini Project Find a vector **u** of length 3 and a vector **v** of length 7 such that $\mathbf{u} \cdot \mathbf{v} = 21$. Find another pair of vectors **u** and **v** such that $\mathbf{u} \cdot \mathbf{v} = -21$. Finally, find three more pairs of vectors of respective lengths 3 and 7 and show that all of their lengths lie between –21 and 21.

Solution Two vectors in the same direction (for instance, along the positive *x*-axis) will have the highest possible dot product:

```
>>> dot((3,0),(7,0))
21
```

(continued)

Two vectors in the opposite direction (for instance, the positive and negative *y* directions) will have the lowest possible dot product:

```
>>> dot((0,3),(0,-7))
-21
```

Using polar coordinates, we can easily generate some more vectors of length 3 and 7 with random angles:

```
from vectors import to_cartesian
from random import random
from math import pi

def random_vector_of_length(l):
    return to_cartesian((l, 2*pi*random()))

pairs = [(random_vector_of_length(3), random_vector_of_length(7))
            for i in range(0,3)]
for u,v in pairs:
    print("u = %s, v = %s" % (u,v))
    print("length of u: %f, length of v: %f, dot product :%f" %
            (length(u), length(v), dot(u,v)))
```

Exercise 3.16 Let **u** and **v** be vectors, with $|\mathbf{u}| = 3.61$ and $|\mathbf{v}| = 1.44$. If the angle between **u** and **v** is 101.3°, what is $\mathbf{u} \cdot \mathbf{v}$?

1. 5.198
2. 5.098
3. –1.019
4. 1.019

Solution Again, we can plug these values into the new dot product formula and, with the appropriate conversion to radians, evaluate the result in Python:

```
>>> 3.61 * 1.44 * cos(101.3 * pi / 180)
-1.0186064362303022
```

Rounding to three decimal places, the answer agrees with *c*.

Exercise 3.17—Mini Project Find the angle between (3, 4) and (4, 3) by converting them to polar coordinates and taking the difference of the angles. The answer is

1. 1.569
2. 0.927

3. 0.643
4. 0.284

Hint The result should agree with the value from the dot product formula.

Solution The vector (3, 4) is further from the positive *x*-axis counterclockwise, so we subtract the angle of (4, 3) from the angle of (3, 4) to get our answer. It matches answer *d* exactly:

```
>>> from vectors import to_polar
>>> r1,t1 = to_polar((4,3))
>>> r2,t2 = to_polar((3,4))
>>> t1-t2
-0.2837941092083278
>>> t2-t1
0.2837941092083278
```

Exercise 3.18 What is the angle between (1, 1, 1) and (–1, –1, 1) in degrees?

1. 180°
2. 120°
3. 109.5°
4. 90°

Solution The lengths of both vectors are $\sqrt{3}$ or approximately 1.732. Their dot product is $1 \cdot (-1) + 1 \cdot (-1) + 1 \cdot 1 = -1$, so $-1 = \sqrt{3} \cdot \sqrt{3} \cdot \cos(\theta)$. Therefore, $\cos(\theta) = -1/3$. This makes the angle approximately 1.911 radians or 109.5° (answer *c*).

3.4 The cross product: Measuring oriented area

As previously introduced, the cross product takes two 3D vectors **u** and **v** as inputs, and its output **u** × **v** is another 3D vector. It is similar to the dot product in that the lengths and relative directions of the input vectors determine the output, but is different in that the output has not only a magnitude but also a direction. We need to think carefully about the concept of direction in 3D to understand the power of the cross product.

3.4.1 Orienting ourselves in 3D

When I introduced the *x*-, *y*-, and *z*-axes at the beginning of this chapter, I made two clear assertions. First, I promised that the familiar *x*,*y* plane exists within the 3D world. Second, I set the *z* direction to be perpendicular to the *x*,*y* plane with the *x*,*y* plane living where $z = 0$. What I didn't announce clearly was that the positive *z* direction was up instead of down.

In other words, if we look at the *x*,*y* plane from the usual perspective, we would see the positive *z*-axis emerging out of the plane toward us. The other choice we could make is sending the positive *z*-axis away from us (figure 3.29).

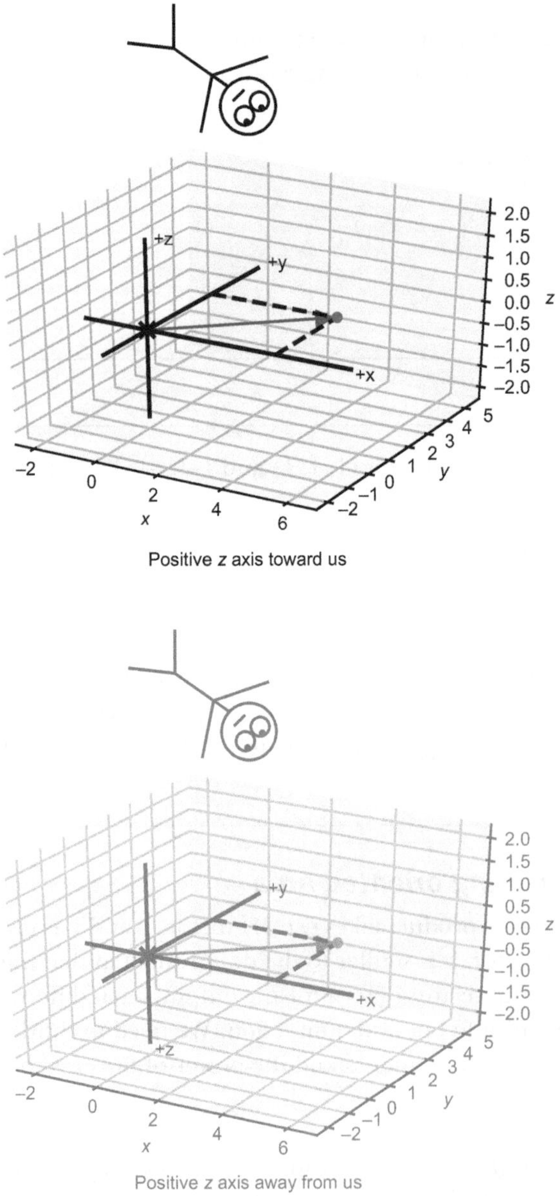

Figure 3.29 Positioning ourselves in 3D to see the *x*,*y* plane as we saw it in chapter 2. When looking at the *x*,*y* plane, we chose the positive *z*-axis to point toward us as opposed to away from us.

The difference here is not a matter of perspective; the two choices represent different orientations of 3D space, and they are distinguishable from any perspective. Suppose we are floating at some positive *z*-coordinate like the stick figure on the left in figure 3.29. We should see the positive *y*-axis positioned a quarter-turn counterclockwise from the positive *x*-axis; otherwise, the axes are arranged in the wrong orientation.

Plenty of things in the real world have orientations and don't look identical to their mirror images. For instance, left and right shoes have identical size and shape but different orientations. A plain coffee mug does not have an orientation; we cannot look at two pictures of an unmarked coffee mug and decide if they are different. But as figure 3.30 shows, two coffee mugs with graphics on opposite sides are distinguishable.

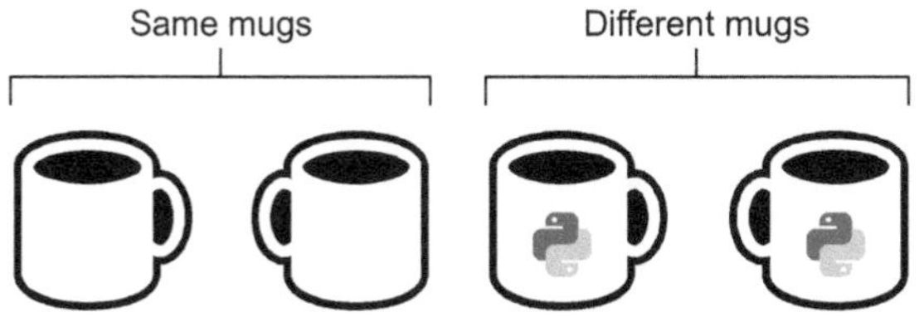

Figure 3.30 A mug with no image is the same object as its mirror image. A mug with an image on one side is *not* the same as its mirror image.

Figure 3.31 Is this a right or left hand?

The readily available object most mathematicians use to detect orientation is a hand. Our hands are oriented objects, so we can tell right hands from left hands even if they were unluckily detached from our bodies. Can you tell if the hand in figure 3.31 is a right or left hand?

Clearly, it's a right hand: we don't have fingernails on our left-hand fingertips! Mathematicians can use their hands to distinguish the two possible orientations of coordinate axes, and they call the two possibilities right-handed and left-handed orientations. Here's the rule as illustrated in figure 3.32: if you point your right index finger along the positive *x*-axis and curl your remaining fingers toward the positive *y*-axis, your thumb tells you the direction of the positive *z*-axis.

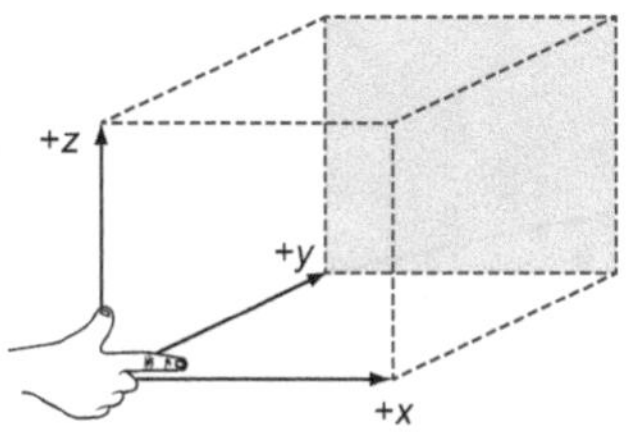

Figure 3.32 The right-hand rule helps us remember the orientation we've chosen.

This is called the *right-hand rule*, and if it agrees with your axes, then you are (correctly!) using the right-handed orientation. Orientation matters! If you are writing a program to steer a drone or control a laparoscopic surgery robot, you need to keep your ups, downs, lefts, rights, forwards, and backwards consistent. The cross product is an oriented machine, so it can help us keep track of orientation throughout all of our computations.

3.4.2 Finding the direction of the cross product

Again, before I tell you how to compute the cross product, I want to show you what it looks like. Given two input vectors, the cross product outputs a result that is perpendicular to both. For instance, if **u** = (1,0,0) and **v** = (0,1,0), then it happens that the cross product **u** × **v** is (0, 0, 1) as shown in figure 3.33.

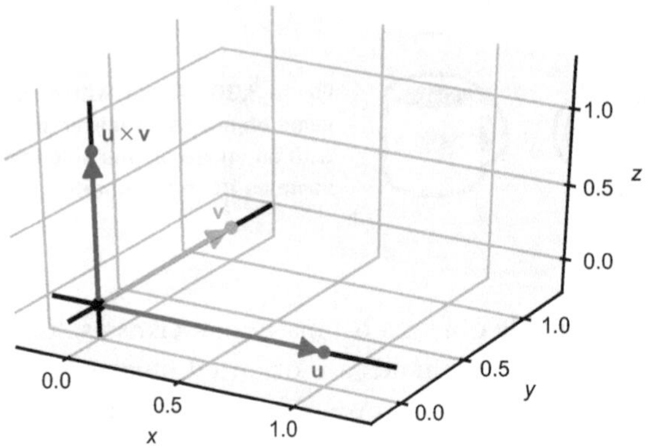

Figure 3.33 The cross product of u = (1, 0, 0) and v = (0, 1, 0)

In fact, as figure 3.34 shows, any two vectors in the *x*,*y* plane have a cross product that lies along the *z*-axis.

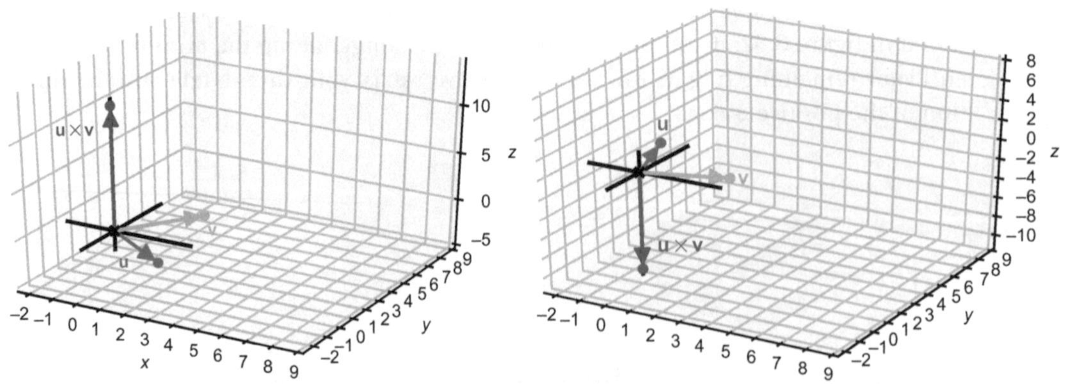

Figure 3.34 The cross product of any two vectors in the *x*,*y* plane lies on the *z*-axis.

This makes it clear why the cross product doesn't work in 2D: it returns a vector that lies outside of the plane containing the two input vectors. We can see the output of the cross product is perpendicular to both inputs even if they don't lie in the *x*,*y* plane (figure 3.35).

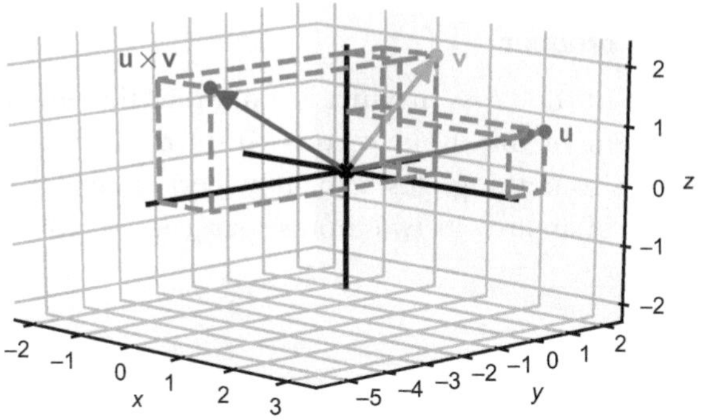

Figure 3.35 The cross product always returns a vector that is perpendicular to both inputs.

But there are two possible perpendicular directions, and the cross product selects only one. For instance, the result of (1, 0, 0) × (0, 1, 0) happens to be (0, 0, 1), pointing in the positive *z* direction. Any vector on the *z*-axis, positive or negative, would be perpendicular to both of these inputs. Why does the result point in the positive direction?

Here's where orientation comes in: the cross product obeys the right-hand rule as well. Once you've found the direction perpendicular to two input vectors **u** and **v**, the cross product **u** × **v** lies in a direction that puts the three vectors **u**, **v**, and **u** × **v** in a right-handed configuration. That is, we can point our right index finger in the direction of **u**, curl our other fingers toward **v**, and our thumb points in the direction of **u** × **v** (figure 3.36).

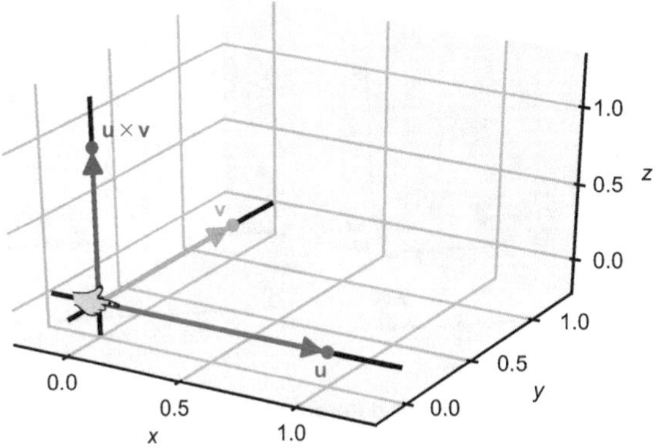

Figure 3.36 The right-hand rule tells us *which* perpendicular direction the cross product points toward.

When input vectors lie on two coordinate axes, it's not too hard to find the exact direction their cross product will point: it's one of the two directions along the remaining axis. In general, it's hard to describe a direction perpendicular to two vectors without computing their cross product. This is one of the features that make it so useful once we see how to compute it. But a vector doesn't just specify a direction; it also specifies a length. The length of the cross product encodes useful information as well.

3.4.3 Finding the length of the cross product

Like the dot product, the length of the cross product is a number that gives us information about the relative position of the input vectors. Instead of measuring how aligned two vectors are, it tells us something closer to "how perpendicular they are." More precisely, it tells us how big of an area its two inputs span (figure 3.37).

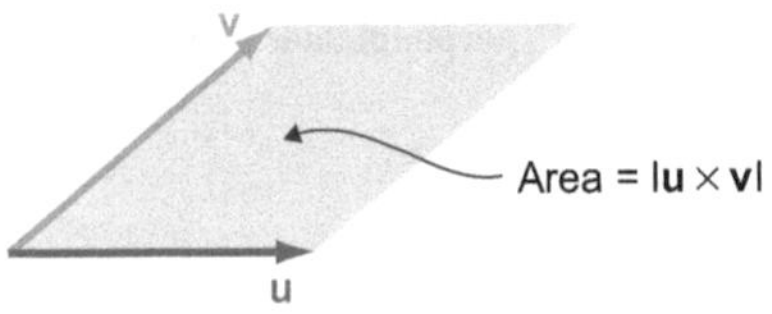

Figure 3.37 The length of the cross product is equal to the area of a parallelogram.

The parallelogram bounded by **u** and **v** as in figure 3.37 has an area that is the same as the length of the cross product **u** × **v**. For two vectors of given lengths, they span the most area if they are perpendicular. On the other hand, if **u** and **v** are in the same direction, they don't span any area; the cross product has zero length. This is convenient; we can't choose a unique perpendicular direction if the two input vectors are parallel.

Paired with the direction of the result, the length of the result gives us an exact vector. Two vectors in the plane are guaranteed to have a cross product pointing in the $+z$ or $-z$ direction. We can see in figure 3.38 that the bigger the parallelogram that the plane vectors span, the longer the cross product.

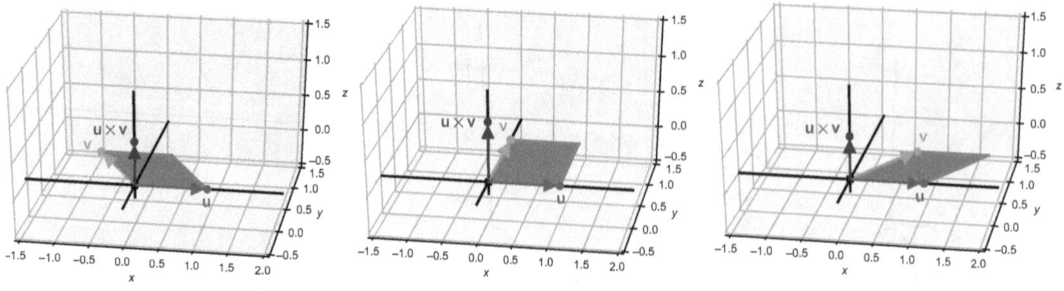

Figure 3.38 Pairs of vectors in the *x*,*y* plane have cross products of different sizes based on the area of the parallelogram these span.

There's a trigonometric formula for the area of this parallelogram: if **u** and **v** are separated by an angle θ, the area is $|\mathbf{u}| \cdot |\mathbf{v}| \cdot \sin(\theta)$. We can put the length and direction together to see some simple cross products. For instance, what is the cross product of (0, 2, 0) and (0, 0, –2)? These vectors lie on the *y*- and *z*-axes, respectively, so to be perpendicular to both, the cross product must lie on the *x*-axis. Let's find the direction of the result using the right-hand rule.

Pointing in the direction of the first vector with our index finger (the positive *y* direction) and bending our fingers in the direction of the second vector (the negative *z* direction), we find our thumb is in the negative *x* direction. The magnitude of the cross product is $2 \cdot 2 \cdot \sin(90°)$ because the *y*- and *z*-axes meet at a 90° angle. (The parallelogram happens to be a square in this case, having a side length of 2). This comes out to 4, so the result is (–4, 0, 0): a vector of length 4 in the $-x$ direction.

It's nice to convince ourselves that the cross product is a well-defined operation by computing it geometrically. But that's not practical, in general, when vectors don't always lie on an axis and it's not obvious what coordinates you need to find a perpendicular result. Fortunately, there's an explicit formula for the coordinates of the cross product in terms of the coordinates of its inputs.

3.4.4 Computing the cross product of 3D vectors

The formula for the cross product looks hairy at first glance, but we can quickly wrap it in a Python function and compute it with no sweat. Let's start with coordinates for two vectors **u** and **v**. We could name the coordinates $\mathbf{u} = (a, b, c)$ and $\mathbf{v} = (d, e, f)$, but it's clearer if we use better symbols: $\mathbf{u} = (u_x, u_y, u_z)$ and $\mathbf{v} = (v_x, v_y, v_z)$. It's easier to remember that the number called v_x is the *x*-coordinate of **v** than if we called it an arbitrary letter like *d*. In terms of these coordinates, the formula for the cross product is

$$\mathbf{u} \times \mathbf{v} = (u_y v_z - u_z v_y, u_z v_x - u_x v_z, u_x v_y - u_y v_x)$$

Or, in Python:

```
def cross(u, v):
    ux,uy,uz = u
    vx,vy,vz = v
    return (uy*vz - uz*vy, uz*vx - ux*vz, ux*vy - uy*vx)
```

You can test-drive this formula in the exercises. Note that in contrast to most of the formulas we used so far, this one doesn't appear to generalize well to other dimensions. It requires that the input vectors have exactly three components.

This algebraic procedure agrees with the geometric description we built in this chapter. Because it tells us area and direction, the cross product helps us decide whether an occupant of 3D space would see a polygon floating in space with them. For instance, as figure 3.39 shows, an observer standing on the *x*-axis would *not* see the parallelogram spanned by $\mathbf{u} = (1, 1, 0)$ and $\mathbf{v} = (-2, 1, 0)$.

In other words, the polygon in figure 3.39 is parallel to the observer's line of sight. Using the cross product, we could tell this without drawing the picture. Because the

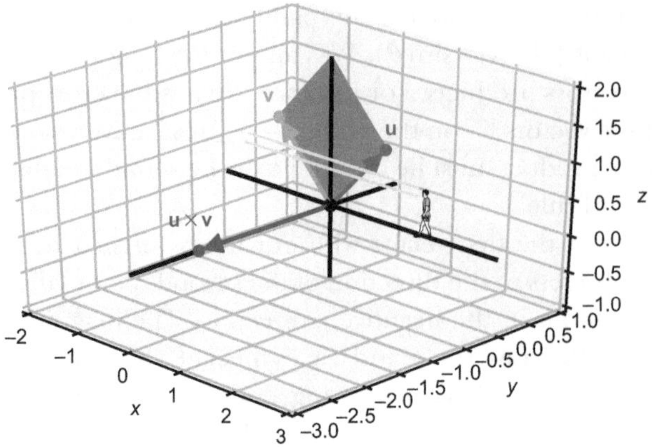

Figure 3.39 The cross product can indicate whether a polygon is visible to an observer.

cross product is perpendicular to the person's line of sight, none of the polygon is visible.

Now it's time for our culminating project: building a 3D object out of polygons and drawing it on a 2D canvas. You'll use all of the vector operations you've seen so far. In particular, the cross product will help you to decide which polygons are visible.

3.4.5 Exercises

Exercise 3.19 Each of the following diagrams show three mutually perpendicular arrows indicating positive *x*, *y*, and *z* directions. A 3D box is shown for perspective with the back of the box colored gray. Which of the four diagrams is compatible with the one we chose? That is, which shows the *x*-, *y*-, and *z*-axes as we've drawn them, even if from a different perspective?

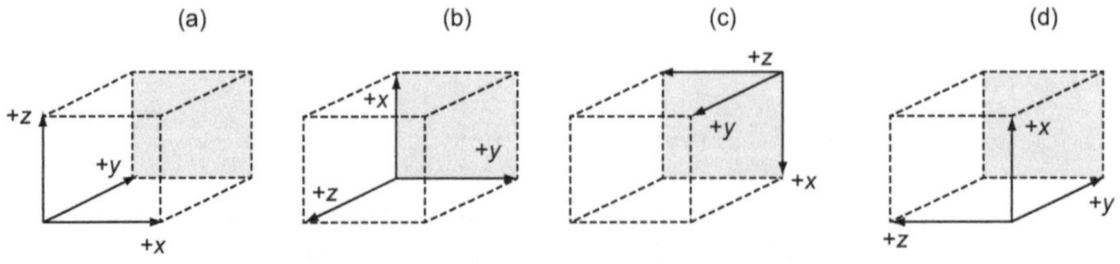

Which of these axes agrees with our orientation convention?

Solution Looking down on diagram *a* from above, we'd see the *x*- and *y*-axis as usual, with the *z*-axis pointing toward us. The diagram that agrees with our orientation is *a*.

In diagram *b*, the *z*-axis is coming toward us, while the +*y* direction is 90° clockwise from the +*x* direction. This does not agree with our orientation.

If we looked at diagram *c* from a point in the positive *z* direction (from the left side of the box), we would see the +*y* direction 90° counterclockwise from the +*x* direction. Diagram *c* also agrees with our orientation.

Looking at diagram *d* from the left of the box, the +*z* direction would be toward us and the +*y* direction would again be counterclockwise from the +*x* direction. This agrees with our orientation as well.

Exercise 3.20 If you held up three coordinate axes in front of a mirror, would the image in the mirror have the same orientation or a different one?

Solution The mirror image has reversed orientation. From this perspective, the *z*- and *y*-axes stay pointing in the same directions. The *x*-axis is clockwise from the *y*-axis in the original, but in the mirror image, it moves to counterclockwise:

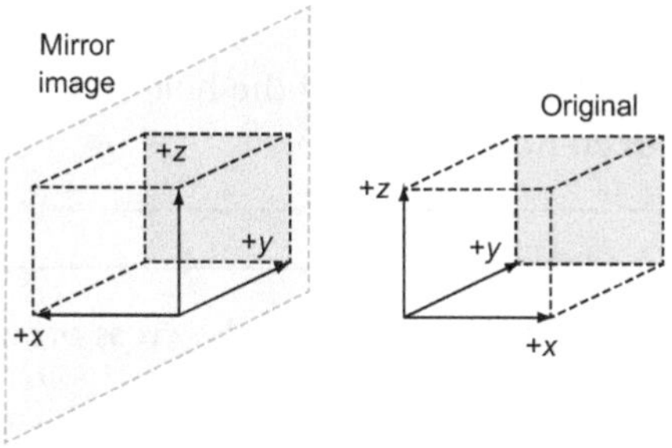

The *x*-, *y*-, and *z*-axes and their mirror image

Exercise 3.21 In what direction does the result of (0, 0, 3) × (0, –2, 0) point?

Solution If we point our right index finger in the direction of (0, 0, 3), the positive *z* direction, and curl our other fingers in the direction of (0, –2, 0), the negative *y* direction, our thumb points in the positive *x* direction. Therefore, (0, 0, 3) × (0, –2, 0) points in the positive *x* direction.

Exercise 3.22 What are the coordinates of the cross product of (1, –2, 1) and (–6, 12, –6)?

Solution As negative scalar multiples of one another, these vectors point in opposite directions and don't span any area. The length of the cross product is, therefore, zero. The only vector of length zero is (0, 0, 0), so that is the answer.

Exercise 3.23—Mini Project The area of a parallelogram is equal to the length of its base times its height as shown here:

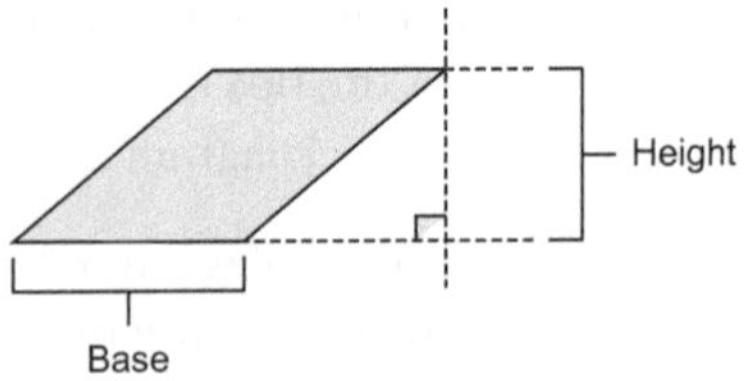

Given that, explain why the formula $|\mathbf{u}| \cdot |\mathbf{v}| \cdot \sin(\phi)$ makes sense.

Solution In the diagram, the vector **u** defines the base, so the base length is $|\mathbf{u}|$. From the tip of **v** to the base, we can draw a right triangle. The length of **v** is the hypotenuse, and the vertical leg of the triangle is the height we are looking for. By the definition of the sine function, the height is $|\mathbf{v}| \cdot \sin(\phi)$.

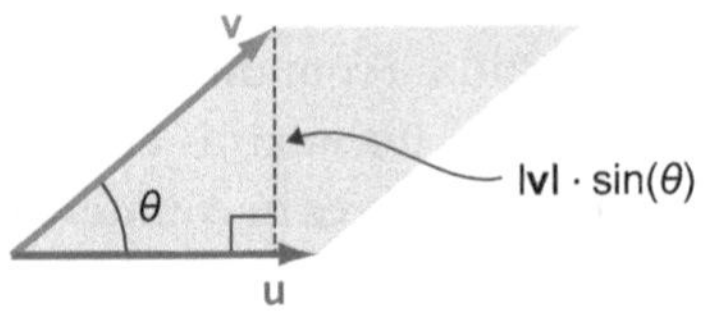

The formula for the area of a parallelogram in terms of the sine of one of its angles

Because the base length is $|\mathbf{u}|$ and the height is $|\mathbf{v}| \cdot \sin(\phi)$, the area of the parallelogram is indeed $|\mathbf{u}| \cdot |\mathbf{v}| \cdot \sin(\phi)$.

Exercise 3.24 What is the result of the cross product (1, 0, 1) × (–1, 0, 0)?

- **a** (0, 1, 0)
- **b** (0, –1, 0)
- **c** (0, –1, –1)
- **d** (0, 1, –1)

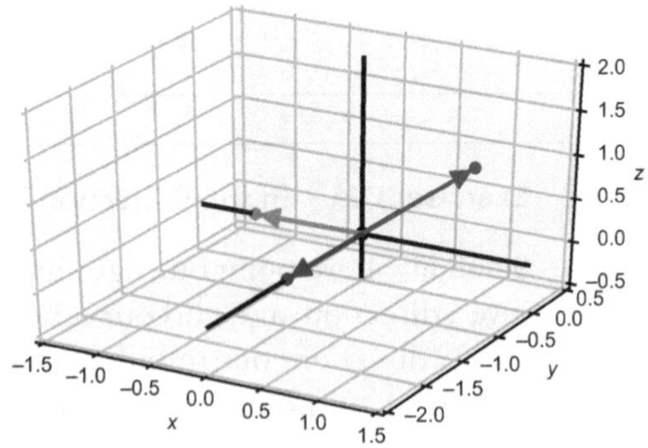

Computing the cross product of (1, 0, 1) and (–1, 0, 0) geometrically

Solution These vectors lie in the *x,z* plane, so their cross product lies on the *y*-axis. Pointing our right index finger in the direction of (1, 0, 1) and curling our fingers toward (–1, 0, 0) requires our thumb to point in the –*y* direction.

We could find the lengths of the vectors and the angle between them to get the size of the cross product, but we already have the base and height from the coordinates. These are both 1, so the length is 1. The cross product is, therefore, (0, –1, 0), a vector of length 1 in the –*y* direction; the answer is *b*.

Exercise 3.25 Use the Python `cross` function to compute $(0, 0, 1) \times \mathbf{v}$ for a few different values of a second vector $\mathbf{v}$. What is the z-coordinate of each result, and why?

Solution No matter what vector $\mathbf{v}$ is chosen, the z-coordinate is zero:

```
>>> cross((0,0,1),(1,2,3))
(-2, 1, 0)
>>> cross((0,0,1),(-1,-1,0))
(1, -1, 0)
>>> cross((0,0,1),(1,-1,5))
(1, 1, 0)
```

Because $\mathbf{u} = (0,0,1)$, both u_x and u_y are zero. This means the term $u_x v_y - u_y v_x$ in the cross product formula is zero, regardless of the values v_x and v_y. Geometrically this makes sense: the cross product should be perpendicular to both inputs, and to be perpendicular to $(0, 0, 1)$, the z component must be zero.

Exercise 3.26—Mini Project Show algebraically that $\mathbf{u} \times \mathbf{v}$ is perpendicular to both $\mathbf{u}$ and $\mathbf{v}$ regardless of the coordinates of $\mathbf{u}$ and $\mathbf{v}$.

Hint Show $(\mathbf{u} \times \mathbf{v}) \cdot \mathbf{u}$ and $(\mathbf{u} \times \mathbf{v}) \cdot \mathbf{v}$ by expanding these into coordinates.

Solution Let $\mathbf{u} = (u_x, u_y, u_z)$ and $\mathbf{v} = (v_x, v_y, v_z)$ in the following equations. We can write $(\mathbf{u} \times \mathbf{v}) \cdot \mathbf{u}$ in terms of coordinates as follows:

$$\mathbf{u} \times \mathbf{v} = (u_y v_z - u_z v_y, u_z v_x - u_x v_z, u_x v_y - u_y v_x) \cdot (u_x, u_y, u_z)$$

Expanding the dot product of a cross product

After we expand the dot product, we see that there are 6 terms. Each of these cancels out with one of the others.

$$= (u_y v_z - u_z v_y)u_x + (u_z v_x - u_x v_z)u_y + (u_x v_y - u_y v_x)u_z$$

$$= u_y v_z u_x - u_z v_y u_x + u_z v_x u_y - u_x v_z u_y + u_x v_y u_z - u_y v_x u_z$$

After fully expanding, all the terms cancel out.

Because all of the terms cancel out, the result is zero. To save "ink," I won't show the result of $(\mathbf{u} \times \mathbf{v}) \cdot \mathbf{v}$, but the same thing happens: six terms appear and cancel each other out, resulting in zero. This means that $(\mathbf{u} \times \mathbf{v})$ is perpendicular to both $\mathbf{u}$ and $\mathbf{v}$.

3.5 Rendering a 3D object in 2D

Let's try using what we've learned to render a simple 3D shape called an octahedron. Whereas a cube has six faces, all of which are squares, an octahedron has eight faces, all of which are triangles. You can think of an octahedron as two, four-sided pyramids stacked on top of each other. Figure 3.40 shows the skeleton of an octahedron.

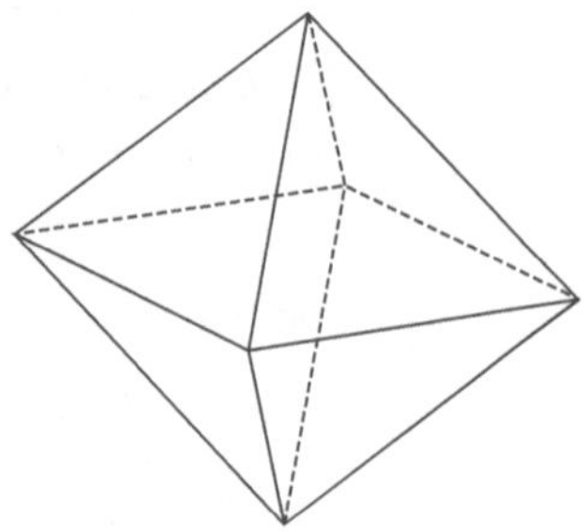

Figure 3.40 The skeleton of an octahedron, a shape with eight faces and six vertices. The dotted lines show the edges of the octahedron on the opposite side from us.

If this were a solid, we wouldn't be able to see the opposite sides. Instead, we'd see four of the eight triangular faces as shown in figure 3.41.

Rendering the octahedron comes down to identifying the four triangles we need to show and shading them appropriately. Let's see how to do that.

3.5.1 Defining a 3D object with vectors

An octahedron is an easy example because it has only six corners or vertices. We can give them simple coordinates: (1, 0, 0), (0, 1, 0), (0, 0, 1) and their three opposite vectors as shown in figure 3.42.

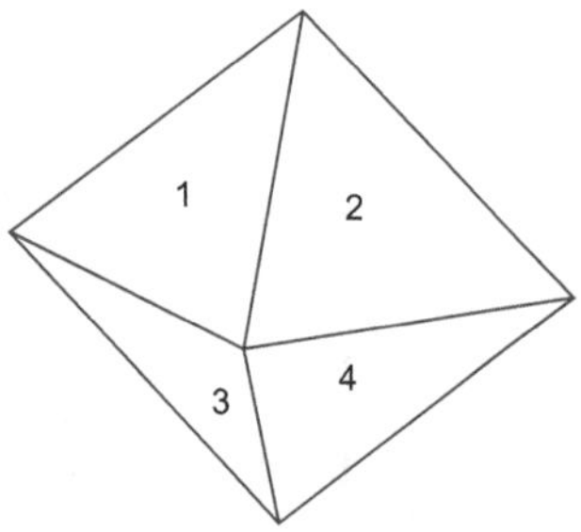

Figure 3.41 Four numbered faces of the octahedron that are visible to us in its current position

These six vectors define the boundaries of the shape but don't provide all the information we need to draw it. We also need to decide which of these vertices connect to form edges of the shape. For instance, the top point in figure 3.42 is (0, 0, 1) and it connects by an edge to all four points in the *x*,*y* plane (figure 3.43).

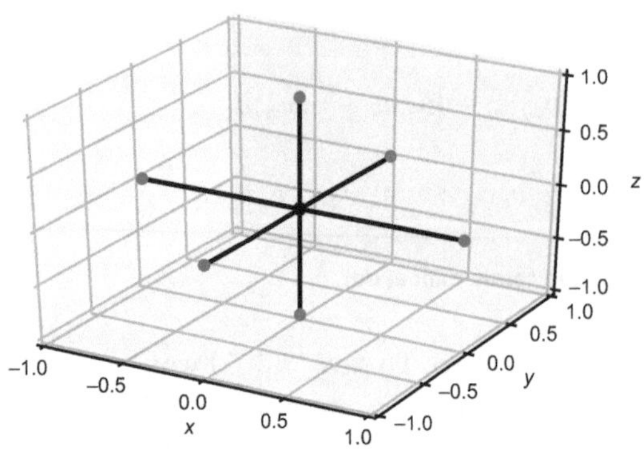

Figure 3.42 Vertices of an octahedron

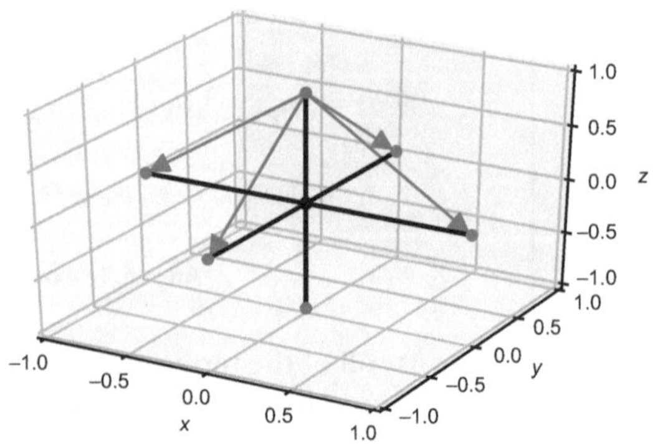

Figure 3.43 Four edges of the octahedron indicated by arrows

These edges outline the top pyramid of the octahedron. Note that there is no edge from (0, 0, 1) to (0, 0, –1) because that segment would lie within the octahedron, not on its outside. Each edge is defined by a pair of vectors: the start and end points of the edge as a line segment. For instance, (0, 0, 1) and (1, 0, 0) define one of the edges.

Edges still aren't enough data to complete the drawing. We also need to know which triples of vertices and edges define the triangular faces we want to fill with a solid, shaded color. Here's where orientation comes in: we want to know not only which segments define faces of the octahedron, but also whether they face toward us or away from us.

Here's the strategy: we'll model a triangular face as three vectors $\mathbf{v}_1$, $\mathbf{v}_2$, and $\mathbf{v}_3$, defining its edges. (Note that here I use subscripts 1, 2, and 3 to distinguish three different vectors, not components of the same vector.) Specifically, we'll order $\mathbf{v}_1$, $\mathbf{v}_2$, and $\mathbf{v}_3$ such that $(\mathbf{v}_2 - \mathbf{v}_1) \times (\mathbf{v}_3 - \mathbf{v}_1)$ points outside the octahedron (figure 3.44). If an outward-pointing vector is aimed toward us, it means the face is visible from our perspective. Otherwise, the face is obscured and we won't need to draw it.

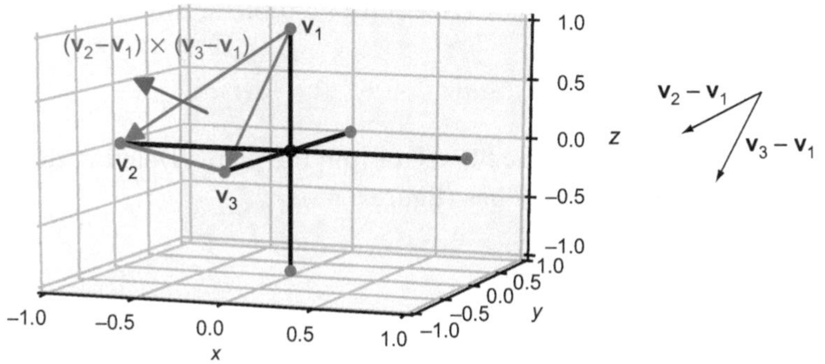

Figure 3.44 A face of the octahedron. The three points defining the face are ordered so that $(\mathbf{v}_2 \bullet \mathbf{v}_1) \times (\mathbf{v}_3 \bullet \mathbf{v}_1)$ points outside of the octahedron.

We can define the eight triangular faces as triples of three vectors $\mathbf{v}_1$, $\mathbf{v}_2$, and $\mathbf{v}_3$ as follows:

```
octahedron = [
    [(1,0,0), (0,1,0), (0,0,1)],
    [(1,0,0), (0,0,-1), (0,1,0)],
    [(1,0,0), (0,0,1), (0,-1,0)],
    [(1,0,0), (0,-1,0), (0,0,-1)],
    [(-1,0,0), (0,0,1), (0,1,0)],
    [(-1,0,0), (0,1,0), (0,0,-1)],
    [(-1,0,0), (0,-1,0), (0,0,1)],
    [(-1,0,0), (0,0,-1), (0,-1,0)],
]
```

The faces are actually the only data we need to render the shape; these contain the edges and vertices implicitly. For instance, we can get the vertices from the faces with the following function:

```
def vertices(faces):
    return list(set([vertex for face in faces for vertex in face]))
```

3.5.2 Projecting to 2D

To turn 3D points into 2D points, we must choose what 3D direction we are observing from. Once we have two 3D vectors defining "up" and "right" from our perspective, we can *project* any 3D vector onto them and get two components instead of three. The `component` function extracts the part of any 3D vector pointing in a given direction using the dot product:

```
def component(v,direction):
    return (dot(v,direction) / length(direction))
```

With two directions hard-coded (in this case, (1, 0, 0) and (0, 1, 0)), we can establish a way to project from three coordinates down to two. This function takes a 3D vector or a tuple of three numbers and returns a 2D vector or a tuple of two numbers:

```
def vector_to_2d(v):
    return (component(v,(1,0,0)), component(v,(0,1,0)))
```

We can picture this as "flattening" the 3D vector into the plane. Deleting the *z* component takes away any depth the vector has (figure 3.45).

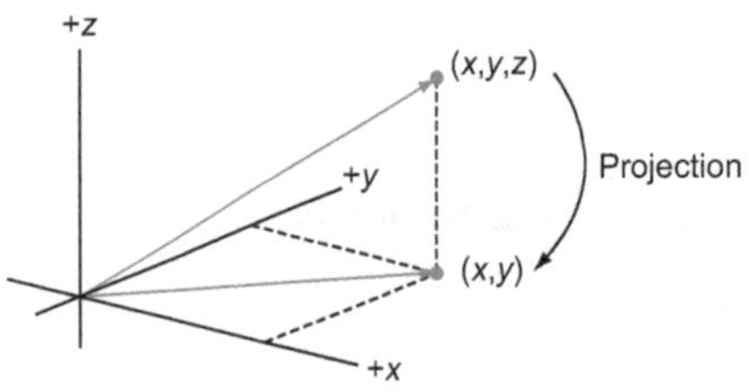

Figure 3.45 Deleting the *z* component of a 3D vector flattens it into the *x,y* plane.

Finally, to take a triangle from 3D to 2D, we need only apply this function to all of the vertices defining a face:

```
def face_to_2d(face):
    return [vector_to_2d(vertex) for vertex in face]
```

3.5.3 Orienting faces and shading

To shade our 2D drawing, we pick a fixed color for each triangle according to how much it faces a given light source. Let's say our light source lies at a vector of (1, 2, 3) from the origin. Then the brightness of a triangular face is decided by how close to perpendicular it is to the light. Another way to measure this is by how aligned a perpendicular vector to the face is with the light source. We don't have to worry about computing colors; Matplotlib has a built-in library to do that for us. For instance,

```
blues = matplotlib.cm.get_cmap('Blues')
```

gives us a function called `blues` that maps numbers from 0 to 1 onto a spectrum of darker to brighter blue values. Our task is to find a number from 0 to 1 that indicates how bright a face should be.

Given a vector perpendicular (or *normal*) to each face and a vector pointing to the light source, their dot product tells us how aligned they are. Moreover, because we're only considering directions, we can choose vectors with length 1. Then, if the face is pointing toward the light source at all, the dot product will lie between 0 and 1. If it is further than 90° from the light source, it will not be illuminated at all. This helper function takes a vector and returns another in the same direction but with length 1:

```
def unit(v):
    return scale(1./length(v), v)
```

This second helper function takes a face and gives us a vector perpendicular to it:

```
def normal(face):
    return(cross(subtract(face[1], face[0]), subtract(face[2], face[0])))
```

Putting it all together, we have a function that draws all the triangles we need to render a 3D shape using our `draw` function. (I've renamed `draw` to `draw2d` and renamed the classes accordingly to distinguish them from their 3D counterparts.)

```
def render(faces, light=(1,2,3), color_map=blues, lines=None):
    polygons = []
    for face in faces:
        unit_normal = unit(normal(face))
        if unit_normal[2] > 0:
            c = color_map(1 - dot(unit(normal(face)),
                                  unit(light)))
            p = Polygon2D(*face_to_2d(face),
                                  fill=c, color=lines)
            polygons.append(p)
    draw2d(*polygons,axes=False, origin=False, grid=None)
```

For each face, computes a vector of length 1 perpendicular to it

Only proceeds if the z component of this vector is positive, or in other words, if it points toward the viewer

The larger the dot product between the normal vector and the light source vector, the less shading

Specifies an optional lines argument for the edges of each triangle, revealing the skeleton of the shape we're drawing

With the following `render` function, it only takes a few lines of code to produce an octahedron. Figure 3.46 shows the result.

```
render(octahedron, color_map=matplotlib.cm.get_cmap('Blues'), lines=black)
```

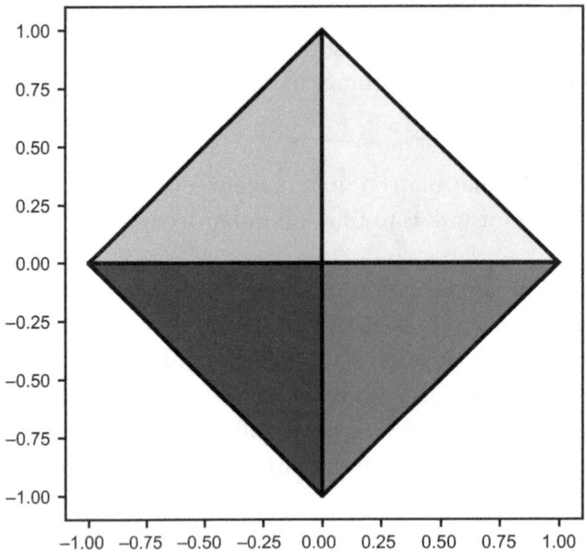

Figure 3.46 Four visible faces of the octahedron in shades of blue

The shaded octahedron doesn't look that special from the side, but adding more faces, we can tell that the shading is working (figure 3.47). You can find pre-built shapes with more faces in the source code for this book.

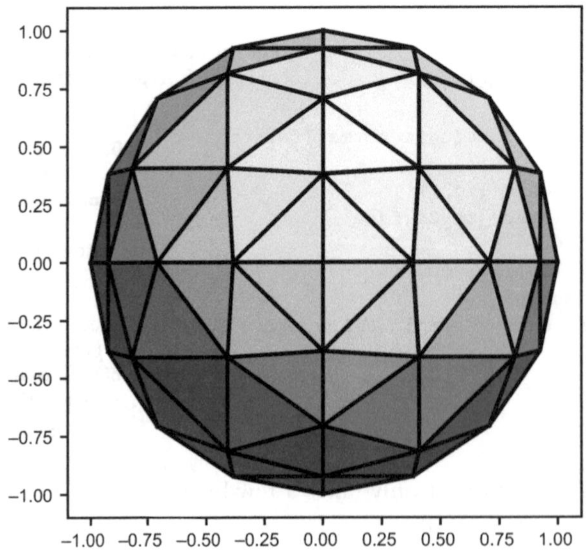

Figure 3.47 A 3D shape with many triangular sides. The effect of the shading is more apparent.

3.5.4 Exercises

Exercise 3.27—Mini Project Find pairs of vectors defining each of the 12 edges of the octahedron and draw all of the edges in Python.

Solution The top of the octahedron is (0, 0, 1). It connects to all four points in the *x,y* plane via four edges. Likewise, the bottom of the octahedron is (0, 0, –1) and it also connects to all four points in the *x,y* plane. Finally, the four points in the *x,y* plane connect to each other in a square:

```
top = (0,0,1)
bottom = (0,0,-1)
xy_plane = [(1,0,0),(0,1,0),(-1,0,0),(0,-1,0)]
edges = [Segment3D(top,p) for p in xy_plane] +\
            [Segment3D(bottom, p) for p in xy_plane] +\
            [Segment3D(xy_plane[i],xy_plane[(i+1)%4]) for i in
range(0,4)]
draw3d(*edges)
```

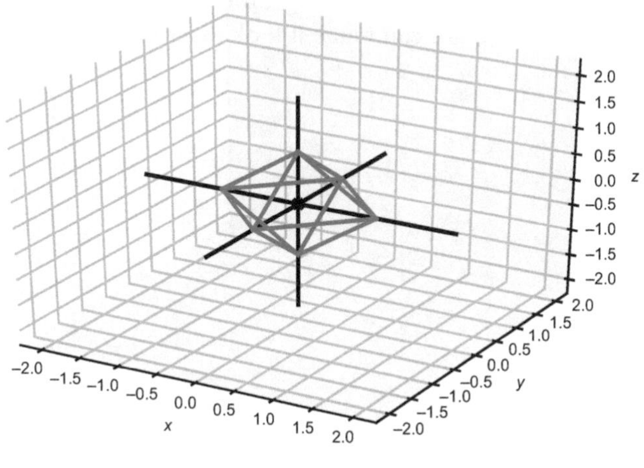

The resulting edges of the octahedron

Exercise 3.28 The first face of the octahedron is [(1, 0, 0), (0, 1, 0), (0, 0, 1)]. Is that the only valid order to write the vertices for this face?

Solution No, for instance [(0, 1, 0), (0, 0, 1), (1, 0, 0)] is the same set of three points, and the cross product still points in the same direction in this order.

Summary

- Whereas vectors in 2D have lengths and widths, vectors in 3D also have depths.
- 3D vectors are defined with triples of numbers called x-, y-, and z-coordinates. They tell us how far from the origin we need to travel in each direction to get to a 3D point.
- As with 2D vectors, 3D vectors can be added, subtracted, and multiplied by scalars. We can find their lengths using a 3D analogy of the Pythagorean theorem.
- The dot product is a way to multiply two vectors and get a scalar. It measures how aligned two vectors are, and we can use its value to find the angle between two vectors.
- The cross product is a way to multiply two vectors and get a third vector that is perpendicular to both input vectors. The magnitude of the output of the cross product is the area of the parallelogram spanned by the two input vectors.
- We can represent the surface of any 3D object as a collection of triangles, where each triangle is respectively defined by three vectors representing its vertices.
- Using the cross product, we can decide which direction a triangle is visible from in 3D. This can tell us whether a viewer can see it or how illuminated it is by a given light source. By drawing and shading all of the triangles defining an object's surface, we can make it look three-dimensional.

4 Transforming vectors and graphics

This chapter covers

- Transforming and drawing 3D objects by applying mathematical functions
- Creating computer animations using transformations to vector graphics
- Identifying linear transformations, which preserve lines and polygons
- Computing the effects of linear transformations on vectors and 3D models

With the techniques from the last two chapters and a little creativity, you can render any 2D or 3D figure you can think of. Whole objects, characters, and worlds can be built from line segments and polygons defined by vectors. But, there's still one thing standing in between you and your first feature-length, computer-animated film or life-like action video game—you need to be able to draw objects that *change* over time.

Animation works the same way for computer graphics as it does for film: you render static images and then display dozens of them every second. When we see that many snapshots of a moving object, it looks like the image is continuously changing. In chapters 2 and 3, we looked at a few mathematical operations that take in existing vectors and transform them geometrically to output new ones. By chaining together sequences of small transformations, we can create the illusion of continuous motion.

As a mental model for this, you can keep in mind our examples of rotating 2D vectors. You saw that you could write a Python function, `rotate`, that took in a 2D vector and rotated it by, say, 45° in the counterclockwise direction. As figure 4.1 shows, you can think of the `rotate` function as a machine that takes in a vector and outputs an appropriately transformed vector.

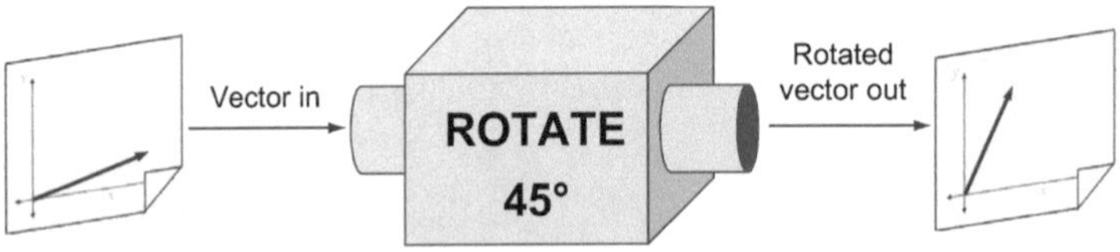

Figure 4.1 Picturing a vector function as a machine with an input slot and output slot

If we apply a 3D analogy of this function to every vector of every polygon defining a 3D shape, we can see the whole shape rotate. This 3D shape could be the octahedron from the previous chapter or a more interesting one like a teapot. In figure 4.2, this rotation machine takes a teapot as input and returns a rotated copy as its output.

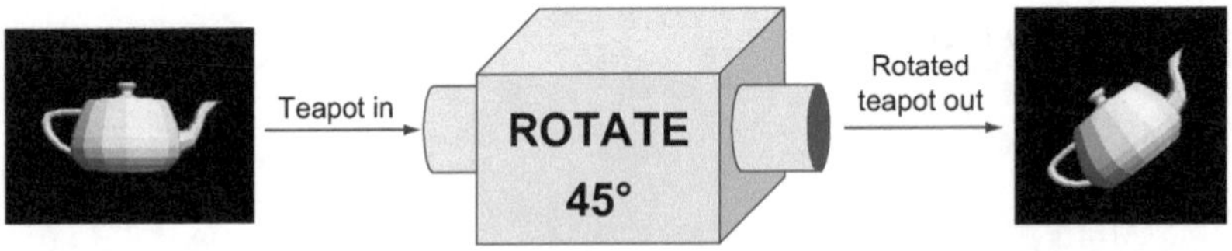

Figure 4.2 A transformation can be applied to every vector making up a 3D model, thereby transforming the whole model in the same geometric way.

If instead of rotating by 45° once, we rotated by one degree 45 times, we could generate frames of a movie showing a rotating teapot (figure 4.3).

Rotations turn out to be great examples to work with because when we rotate every point on a line segment by the same angle about the origin, we still have a line segment of the same length. As a result, when you rotate all the vectors outlining a 2D or 3D object, you can still recognize the object.

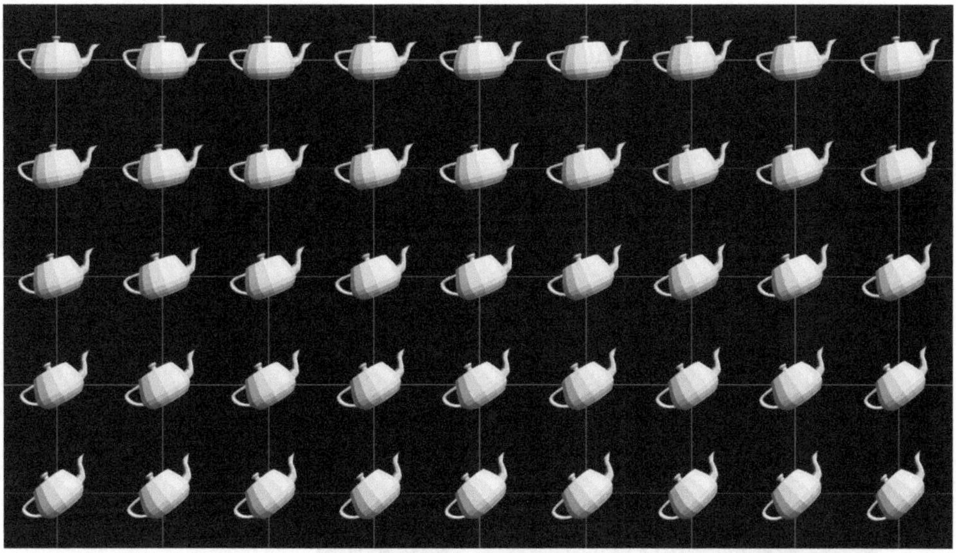

Figure 4.3 Rotating the teapot by 1° at a time, 45 times in a row, beginning with the upper left-hand corner

I'll introduce you to a broad class of vector transformations called *linear transformations* that, like rotations, send vectors lying on a straight line to new vectors that also lie on a straight line. Linear transformations have numerous applications in math, physics, and data analysis. It's helpful to know how to picture them geometrically when you meet them again in these contexts.

To visualize rotations, linear transformations, and other vector transformations in this chapter, we'll upgrade to more powerful drawing tools. We'll swap out Matplotlib for OpenGL, which is an industry standard library for high-performance graphics. Most OpenGL programming is done in C or C++, but we'll use a friendly Python wrapper called PyOpenGL. We'll also use a video game development library in Python called PyGame. Specifically, we'll use the features in PyGame that make it easy to render successive images into an animation. The set up for all of these new tools is covered in appendix C, so we can jump right in and focus on the math of transforming vectors. If you want to follow along with the code for this chapter (which I strongly recommend!), then you should skip to appendix C and return here once you get the code working.

4.1 Transforming 3D objects

Our main goal in this chapter is taking a 3D object (like the teapot) and changing it to create a new 3D object that is visually different. In chapter 2, we already saw that we could translate or scale each vector in a 2D dinosaur and the whole dinosaur shape would move or change in size accordingly. We take the same approach here. Every transformation we look at takes a vector as input and returns a vector as output, something like this pseudocode:

```
def transform(v):
    old_x, old_y, old_z = v
    # ... do some computation here ...
    return (new_x, new_y, new_z)
```

Let's start by adapting the familiar examples of translation and scaling from 2D to 3D.

4.1.1 *Drawing a transformed object*

If you've installed the dependencies described in appendix C, you should be able to run the file draw_teapot.py in the source code for chapter 4 (see appendix A for instructions to run a Python script from the command line). If it runs successfully, you should see a PyGame window that shows the image in figure 4.4.

Figure 4.4 The result of running draw_teapot.py

In the next few examples, we modify the vectors defining the teapot and then re-render it so that we can see the geometric effect. As a first example, we can scale all of the vectors by the same factor. The following function, `scale2`, multiplies an input vector by the scalar 2.0 and returns the result:

```
from vectors import scale
def scale2(v):
    return scale(2.0, v)
```

This `scale2(v)` function has the same form as the `transform(v)` function given at the top of this section; when passed a 3D vector as input, `scale2` returns a new 3D vector as output. To execute this transformation on the teapot, we need to transform each of its vertices. We can do this triangle by triangle. For each triangle that we use to build the teapot, we create a new triangle with the result of applying `scale2` to the original vertices:

```
original_triangles = load_triangles()
scaled_triangles = [
    [scale2(vertex) for vertex in triangle]
    for triangle in original_triangles
]
```

Loads the triangles using the code from appendix C

Applies scale2 to each vertex in a given triangle to get new vertices

Does this for each triangle in the list of original triangles

Now that we've got a new set of triangles, we can draw them by calling `draw_model(scaled_triangles)`. Figure 4.5 shows the teapot after this call, and you can reproduce this by running the file scale_teapot.py in the source code.

Figure 4.5 Applying `scale2` to each vertex of each triangle gives us a teapot that is twice as big.

This teapot looks bigger than the original, and in fact, it is twice as big because we multiplied each vector by 2. Let's apply another transformation to each vector: translation by the vector (–1, 0, 0).

Recall that "translating by a vector" is another way of saying "adding the vector," so what I'm really talking about is adding (–1, 0, 0) to every vertex of the teapot. This should move the whole teapot one unit in the negative *x* direction, which is to the left from our perspective. This function accomplishes the translation for a single vertex:

```
from vectors import add
def translate1left(v):
    return add((-1,0,0), v)
```

Starting with the original triangles, we now want to scale each of their vertices as before and then apply the translation. Figure 4.6 shows the result. You can reproduce it with the source file scale_translate_teapot.py:

```
scaled_translated_triangles = [
    [translate1left(scale2(vertex)) for vertex in triangle]
    for triangle in original_triangles
]
draw_model(scaled_translated_triangles)
```

Figure 4.6 The teapot is bigger and moved to the left as we hoped!

Different scalar multiples change the size of the teapot by different factors, and different translation vectors move the teapot to different positions in space. In the exercises that follow, you'll have a chance to try different scalar multiples and translations, but for now, let's focus on combining and applying more transformations.

4.1.2 Composing vector transformations

Applying any number of transformations sequentially defines a new transformation. In the previous section, for instance, we transformed the teapot by scaling it and then translating it. We can package this new transformation as its own Python function:

```
def scale2_then_translate1left(v):
    return translate1left(scale2(v))
```

This is an important principle! Because vector transformations take vectors as inputs and return vectors as outputs, we can combine as many of them as we want by *composition of functions.* If you haven't heard this term before, it means defining new functions by applying two or more existing ones in a specified order. If we picture the functions `scale2` and `translate1left` as machines that take in 3D models and output new ones (figure 4.7), we can combine them by passing the outputs of the first machine as inputs to the second.

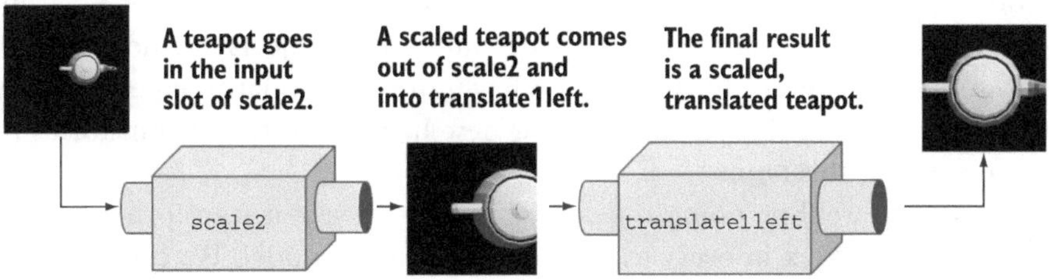

Figure 4.7 Calling `scale2` and then `translate1left` on a teapot to output a transformed version

We can imagine hiding the intermediate step by welding the output slot of the first machine to the input slot of the second machine (figure 4.8).

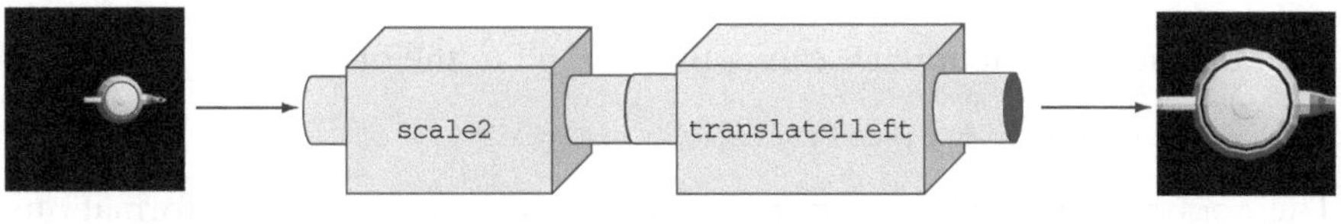

Figure 4.8 Welding the two function machines together to get a new one, which performs both transformations in one step

We can think of the result as a new machine that does the work of both the original functions in one step. This "welding" of functions can be done in code as well. We can write a general-purpose `compose` function that takes two Python functions (for vector transformations, for instance) and returns a new function, which is their composition:

```
def compose(f1,f2):
    def new_function(input):
        return f1(f2(input))
    return new_function
```

Instead of defining `scale2_then_translate1left` as its own function, we could write

```
scale2_then_translate1left = compose(translate1left, scale2)
```

You might have heard of the idea that Python treats functions as "first-class objects." What is usually meant by this slogan is that Python functions can be assigned to variables, passed as inputs to other functions, or created on-the-fly and returned as output values. These are *functional programming* techniques, meaning that they help us build complex programs by combining existing functions to make new ones.

There is some debate about whether functional programming is kosher in Python (or as a Python fan would say, whether or not functional programming is "Pythonic"). I won't opine about coding style, but I use functional programming because functions, namely vector transformations, are our central objects of study. With the `compose` function covered, I'll show you a few more functional "recipes" that justify this digression. Each of these is added in a new helper file called transforms.py in the source code for this book.

Something we'll be doing repeatedly is taking a vector transformation and applying it to every vertex in every triangle defining a 3D model. We can write a reusable function for this rather than writing a new list comprehension each time. The following `polygon_map` function takes a vector transformation and a list of polygons (usually triangles) and applies the transformation to each vertex of each polygon, yielding a new list of new polygons:

```
def polygon_map(transformation, polygons):
    return [
        [transformation(vertex) for vertex in triangle]
        for triangle in polygons
    ]
```

With this helper function, we can apply `scale2` to the original teapot in one line:

```
draw_model(polygon_map(scale2, load_triangles()))
```

The `compose` and `polygon_map` functions both take vector transformations as arguments, but it's also useful to have functions that return vector transformations. For instance, it might have bothered you that we named a function `scale2` and hardcoded the number two into its definition. A replacement for this could be a `scale_by` function that returns a scaling transformation for a specified scalar:

```
def scale_by(scalar):
    def new_function(v):
        return scale(scalar, v)
    return new_function
```

With this function, we can write `scale_by(2)` and the return value would be a new function that behaves identically to `scale2`. While we're picturing functions as machines with input and output slots, you can picture `scale_by` as a machine that takes numbers in its input slot and outputs new function machines from its output slot as shown in figure 4.9.

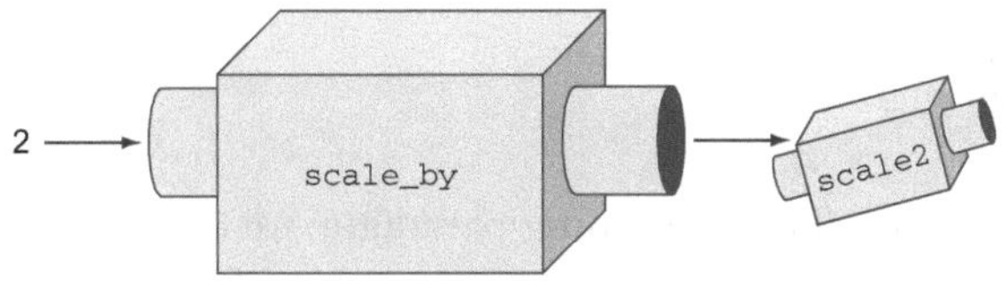

Figure 4.9 A function machine that takes numbers as inputs and produces new function machines as outputs

As an exercise, you can write a similar `translate_by` function that takes a translation vector as input and returns a translation function as output. In the terminology of functional programming, this process is called *currying*. Currying takes a function that accepts multiple inputs and refactors it to a function that returns another function.

The result is a programmatic machine that behaves identically but is invoked differently; for instance, `scale_by(s)(v)` gives the same result as `scale(s,v)` for any inputs *s* and *v*. The advantage is that `scale(...)` and `add(...)` accept different kinds of arguments, so the resulting functions, `scale_by(s)` and `translate_by(w)`, are interchangeable. Next, we'll think similarly about rotations: for any given angle, we can produce a vector transformation that rotates our model by that angle.

4.1.3 Rotating an object about an axis

You already saw how to do rotations in 2D in chapter 2: you convert the Cartesian coordinates to polar coordinates, increase or decrease the angle by the rotation factor, and then convert back. Even though this is a 2D trick, it is helpful in 3D because all 3D vector rotations are, in a sense, isolated in planes. Picture, for instance, a single point in 3D being rotated about the *z*-axis. Its *x*- and *y*-coordinates change, but its *z*-coordinate remains the same. If a given point is rotated around the *z*-axis, it stays in a circle with a constant *z*-coordinate, regardless of the rotation angle (figure 4.10).

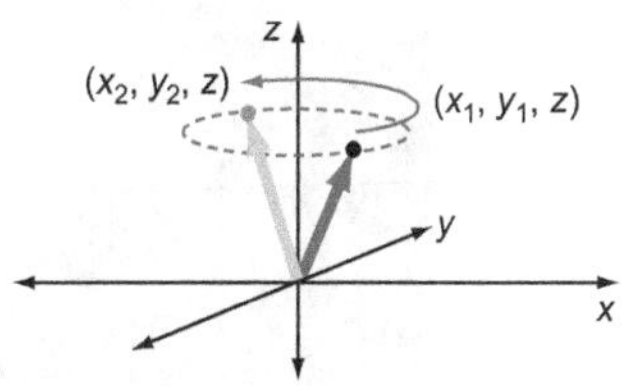

Figure 4.10 Rotating a point around the *z*-axis

What this means is that we can rotate a 3D point around the *z*-axis by holding the *z*-coordinate constant and applying our 2D rotation function only to the *x*- and *y*-coordinates. We'll work through the code here, and you can also find it in rotate_teapot.py in the source code. First, we write a 2D rotation function adapted from the strategy we used in chapter 2:

```
def rotate2d(angle, vector):
    l,a = to_polar(vector)
    return to_cartesian((l, a+angle))
```

This function takes an angle and a 2D vector and returns a rotated 2D vector. Now, let's create a function, `rotate_z`, that applies this function only to the *x* and *y* components of a 3D vector:

```
def rotate_z(angle, vector):
    x,y,z = vector
    new_x, new_y = rotate2d(angle, (x,y))
    return new_x, new_y, z
```

Continuing to think in the functional programming paradigm, we can curry this function. Given any angle, the curried version produces a vector transformation that does the corresponding rotation:

```
def rotate_z_by(angle):
    def new_function(v):
        return rotate_z(angle,v)
    return new_function
```

Let's see it in action. The following line yields the teapot in figure 4.11, which is rotated by $\pi/4$ or 45°:

```
draw_model(polygon_map(rotate_z_by(pi/4.), load_triangles()))
```

Figure 4.11 The teapot is rotated 45° counterclockwise about the *z*-axis.

We can write a similar function to rotate the teapot about the *x*-axis, meaning the rotation affects only the *y* and *z* components of the vector:

```
def rotate_x(angle, vector):
    x,y,z = vector
    new_y, new_z = rotate2d(angle, (y,z))
    return x, new_y, new_z
```

```
def rotate_x_by(angle):
    def new_function(v):
        return rotate_x(angle,v)
    return new_function
```

In the function `rotate_x_by`, a rotation about the *x*-axis is achieved by fixing the *x* coordinate and executing a 2D rotation in the *y*,*z* plane. The following code draws a 90° or $\pi/2$ radian rotation (counterclockwise) about the *x*-axis, resulting in the upright teapot shown in figure 4.12:

```
draw_model(polygon_map(rotate_x_by(pi/2.), load_triangles()))
```

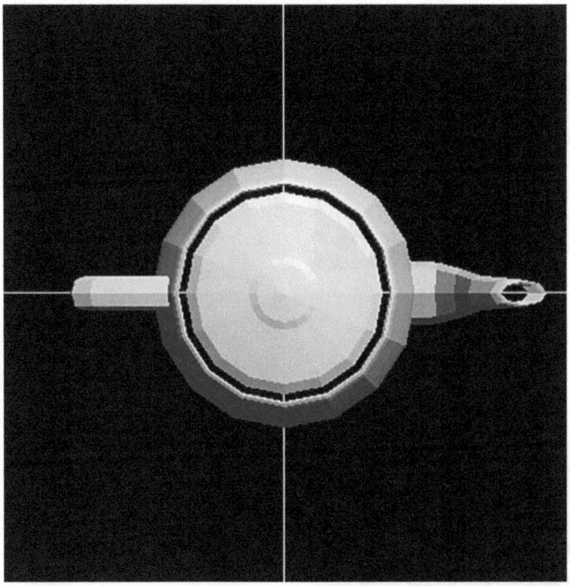

Figure 4.12 The teapot rotated by π/2 about the *x*-axis.

You can reproduce figure 4.12 with the source file rotate_teapot_x.py. The shading is consistent among these rotated teapots; their brightest polygons are toward the top-right of the figures, which is expected because the light source remains at (1, 2, 3). This is a good sign that we are successfully moving the teapot and not just changing our OpenGL perspective as before.

It turns out that it's possible to get *any* rotation we want by composing rotations in the *x* and *z* directions. In the exercises at the end of the section, you can try your hand at some more rotations, but for now, we'll move on to other kinds of vector transformations.

4.1.4 Inventing your own geometric transformations

So far, I've focused on the vector transformations we already saw in some way in the preceding chapters. Now, let's throw caution to the wind and see what other interesting transformations we can come up with. Remember, the only requirement for a 3D vector transformation is that it accepts a single 3D vector as input and returns a new 3D vector as its output. Let's look at a few transformations that don't quite fall in any of the categories we've seen so far.

For our teapot, let's modify one coordinate at a time. This function stretches vectors by a (hard-coded) factor of four, but only in the *x* direction:

```
def stretch_x(vector):
    x,y,z = vector
    return (4.*x, y, z)
```

The result is a long, skinny teapot along the *x*-axis or in the handle-to-spout direction (figure 4.13). This is fully implemented in stretch_teapot.py.

Figure 4.13 A teapot stretched along the *x*-axis.

A similar `stretch_y` function elongates the teapot from top-to-bottom. You can implement `stretch_y` and apply it to a teapot yourself, and you should get the image in figure 4.14. Otherwise, you can look at the implementation in stretch_teapot_y.py in the source code.

We can get even more creative, stretching the teapot by cubing the *y*-coordinate rather than just multiplying it by a number. This transformation gives the teapot a disproportionately elongated lid as implemented in cube_teapot.py and shown in figure 4.15:

```
def cube_stretch_z(vector):
    x,y,z = vector
    return (x, y*y*y, z)
```

Figure 4.14 Stretching the teapot in the *y* direction instead

Figure 4.15 Cubing the vertical dimension of the teapot

If we selectively add two of the three coordinates in the formula for the transformation, for instance the *x* and *y* coordinates, we can cause the teapot to slant. This is implemented in slant_teapot.py and shown in figure 4.16:

```
def slant_xy(vector):
    x,y,z = vector
    return (x+y, y, z)
```

Figure 4.16 Adding the *y*-coordinate to the existing *x*-coordinate causes the teapot to slant in the *x* direction.

The point is not that any one of these transformations is important or useful, but that any mathematical transformation of the vectors constituting a 3D model have *some* geometric consequence on the appearance of the model. It is possible to go too crazy with the transformation, at which point the model can become too distorted to recognize or even to draw successfully. Indeed, some transformations are better-behaved in general, and we'll classify them in the next section.

4.1.5 Exercises

Exercise 4.1 Implement a `translate_by` function (referred to in section 4.1.2), taking a translation vector as input and returning a translation function as output.

Solution

```
def translate_by(translation):
    def new_function(v):
        return add(translation,v)
    return new_function
```

Exercise 4.2 Render the teapot translated by 20 units in the negative *z* direction. What does the resulting image look like?

Solution We can accomplish this by applying `translate_by((0,0,-20))` to every vector of every polygon with `polgyon_map`:

```
draw_model(polygon_map(translate_by((0,0,-20)), load_triangles()))
```

Remember, we are looking at the teapot from five units up the *z*-axis. This transformation brings the teapot 20 units further from us, so it looks much smaller than the original. You can find the complete implementation in translate_teapot_down_z.py in the source code.

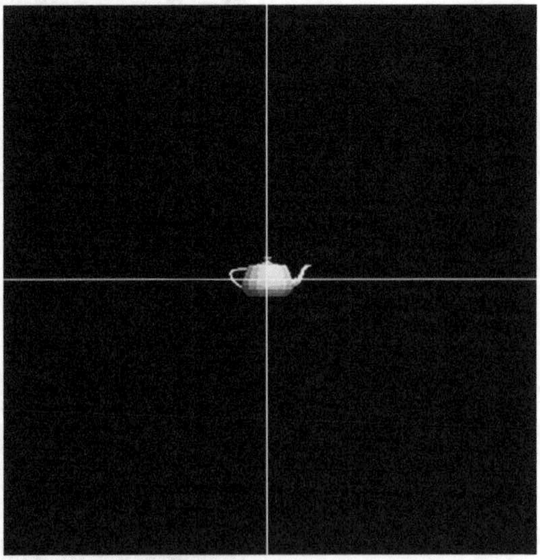

The teapot translated 20 units down the *z*-axis. It appears smaller because it is further from the viewpoint.

Exercise 4.3—Mini Project What happens to the teapot when you scale every vector by a scalar between 0 and 1? What happens when you scale it by a factor of –1?

Solution We can apply `scale_by(0.5)` and `scale_by(-1)` to see the results:

```
draw_model(polygon_map(scale_by(0.5), load_triangles()))
draw_model(polygon_map(scale_by(-1), load_triangles()))
```

Left-to-right, the original teapot, the teapot scaled by 0.5, and the teapot scaled by –1.

As you can see, `scale_by(0.5)` shrinks the teapot to half its original size. The action of `scale_by(-1)` seems to rotate the teapot by 180°, but the situation is a bit more complicated. It's actually turned inside-out as well! Each triangle has been reflected, so each normal vector now points into the teapot rather than outward from its surface.

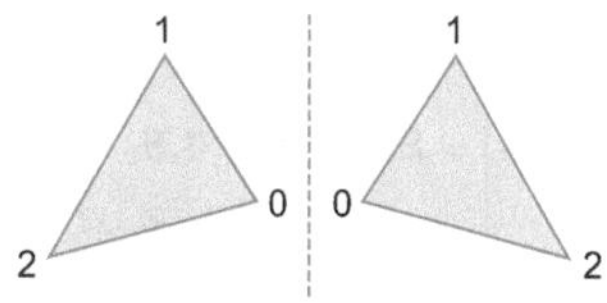

Reflection changes the orientation of a triangle. The indexed vertices are in counterclockwise order on the left and clockwise order in the reflection on the right. The normal vectors to these triangles point in opposite directions.

Rotating the teapot, you can see that it is not quite rendering correctly as a result. We should be careful with reflections of our graphics for this reason!

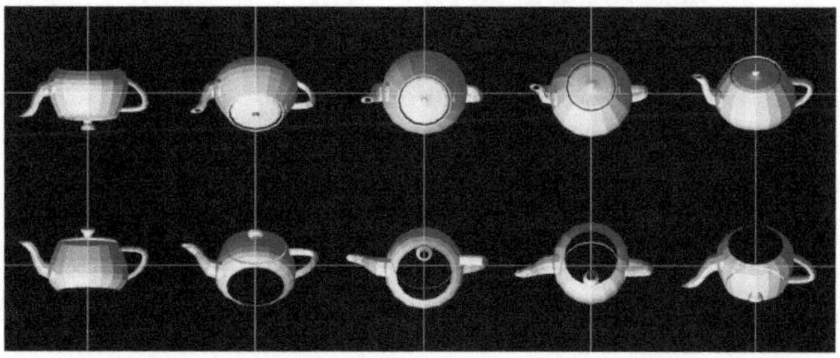

The rotated, reflected teapot does not look quite right. Some features appear but should be concealed. For instance, we can see both the lid and the hollow bottom in the bottom right frame.

Exercise 4.4 First apply `translate1left` to the teapot and then apply `scale2`. How is the result different from the opposite order of composition? Why?

Solution We can compose these two functions in the specified order and then apply them with `polygon_map`:

```
draw_model(polygon_map(compose(scale2, translate1left),
load_triangles()))
```

The result is that the teapot is still twice as large as the original, but this one is translated further to the left. This is because when a scaling factor of 2 is applied after a translation, the distance of the translation doubles as well. You can convince yourself by running the source files scale_translate_teapot.py and translate_scale_teapot.py and comparing the results.

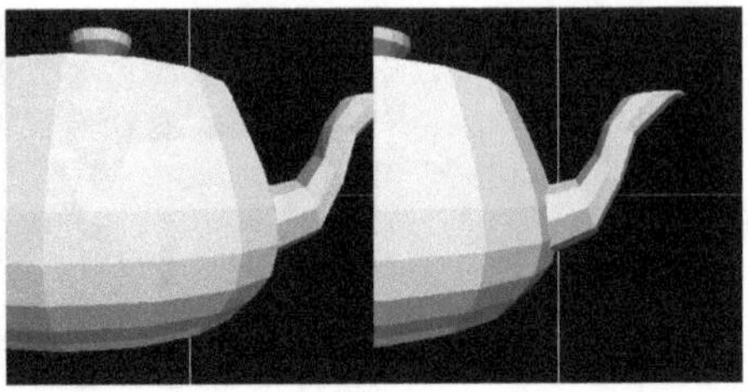

Scaling and then translating the teapot (left) vs. translating and then scaling (right)

Exercise 4.5 What is the effect of the transformation `compose(scale_by(0.4), scale_by(1.5))`?

Solution Applying this to a vector scales it by 1.5 and then by 0.4 for a net scaling factor of 0.6. The resulting figure will be 60% of the size of the original.

Exercise 4.6 Modify the `compose(f,g)` function to `compose(*args)`, which takes several functions as arguments and returns a new function that is their composition.

Solution

```
def compose(*args):
    def new_function(input):
        state = input
        for f in reversed(args):
            state = f(state)
        return state
    return new_function
```

- `def new_function(input):` — **Starts defining the function that compose returns**
- `state = input` — **Sets the current state equal to the input**
- `for f in reversed(args):` — **Iterates over the input functions in reverse order because the inner functions of a composition are applied first. For example, compose(f,g,h)(x) should equal f(g(h(x))), so the first function to apply is h.**
- `state = f(state)` — **At each step, updates the state by applying the next function. The final state has all the functions applied in the correct order.**

To check our work, we can build some functions and compose them:

```
def prepend(string):
    def new_function(input):
        return string + input
    return new_function

f = compose(prepend("P"), prepend("y"), prepend("t"))
```

Then running `f("hon")` returns the string `"Python"`. In general, the constructed function `f` appends the string `"Pyt"` to whatever string it is given.

Exercise 4.7 Write a `curry2(f)` function that takes a Python function `f(x,y)` with two arguments and returns a curried version. For instance, once you write `g = curry2(f)`, the two expressions `f(x,y)` and `g(x)(y)` should return the same result.

Solution The return value should be a new function that, in turn, produces a new function when called:

```
def curry2(f):
    def g(x):
        def new_function(y):
            return f(x,y)
        return new_function
    return g
```

As an example, we could have built the `scale_by` function like this:

```
>>> scale_by = curry2(scale)
>>> scale_by(2)((1,2,3))

(2, 4, 6)
```

Exercise 4.8 Without running it, what is the result of applying the transformation `compose(rotate_z_by(pi/2),rotate_x_by(pi/2))`? What if you switch the order of the composition?

Solution This composition is equivalent to a clockwise rotation by $\pi/2$ about the y-axis. Reversing the order gives a counterclockwise rotation by $\pi/2$ about the y-axis.

Exercise 4.9 Write a function `stretch_x(scalar,vector)` that scales the target vector by the given factor but only in the *x* direction. Also write a curried version `stretch_x_by` so that `stretch_x_by(scalar)(vector)` returns the same result.

Solution

```
def stretch_x(scalar,vector):
    x,y,z = vector
    return (scalar*x, y, z)

def stretch_x_by(scalar):
    def new_function(vector):
        return stretch_x(scalar,vector)
    return new_function
```

4.2 Linear transformations

The well-behaved vector transformations we're going to focus on are called *linear transformations.* Along with vectors, linear transformations are the other main objects of study in linear algebra. Linear transformations are special transformations where vector arithmetic looks the same before and after the transformation. Let's draw some diagrams to show exactly what that means.

4.2.1 Preserving vector arithmetic

The two most important arithmetic operations on vectors are addition and scalar multiplication. Let's return to our 2D pictures of these operations and see how they look before and after a transformation is applied.

We can picture the sum of two vectors as the new vector we arrive at when we place them tip-to-tail, or as the vector to the tip of the parallelogram they define. For instance, figure 4.17 represents the vector sum **u** + **v** = **w**.

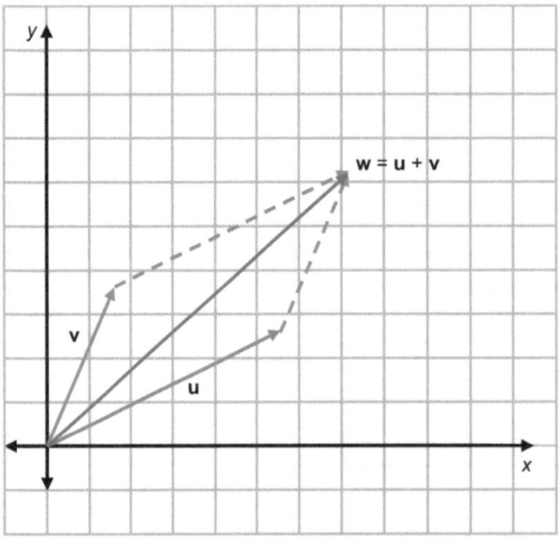

Figure 4.17 Geometric demonstration of the vector sum u + v = w

The question we want to ask is, if we apply the same vector transformation to all three of the vectors in this diagram, will it still look like a vector sum? Let's try a vector transformation, which is a counterclockwise rotation about the origin, and call this transformation R. Figure 4.18 shows **u**, **v**, and **w** rotated by the same angle by the transformation R.

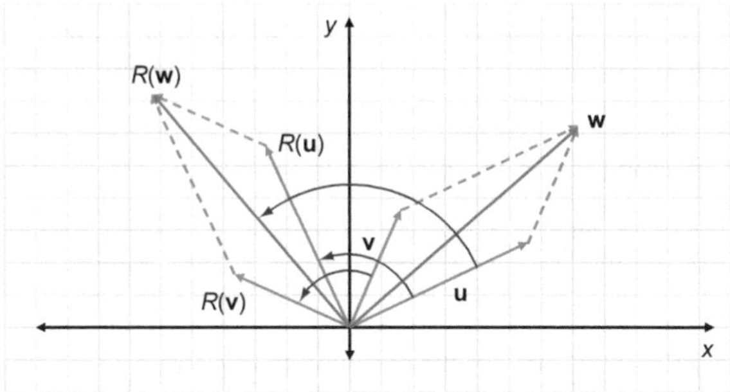

Figure 4.18 After rotating *u*, *v*, and *w* by the same rotation *R*, the sum still holds.

The rotated diagram is exactly the diagram representing the vector sum $R(\mathbf{u}) + R(\mathbf{v}) = R(\mathbf{w})$. You can draw the picture for any three vectors **u**, **v**, and **w**, and as long as $\mathbf{u} + \mathbf{v} = \mathbf{w}$ and if you apply the same rotation transformation R to each of the vectors, you find that $R(\mathbf{u}) + R(\mathbf{v}) = R(\mathbf{w})$ as well. To describe this property, we say that rotations *preserve* vector sums.

Similarly, rotations preserve scalar multiples. If **v** is a vector and $s\mathbf{v}$ is a multiple of **v** by a scalar s, then $s\mathbf{v}$ points in the same direction but is scaled by a factor of s. If we rotate **v** and $s\mathbf{v}$ by the same rotation R, we'll see that $R(s\mathbf{v})$ is a scalar multiple of $R(\mathbf{v})$ by the same factor s (figure 4.19).

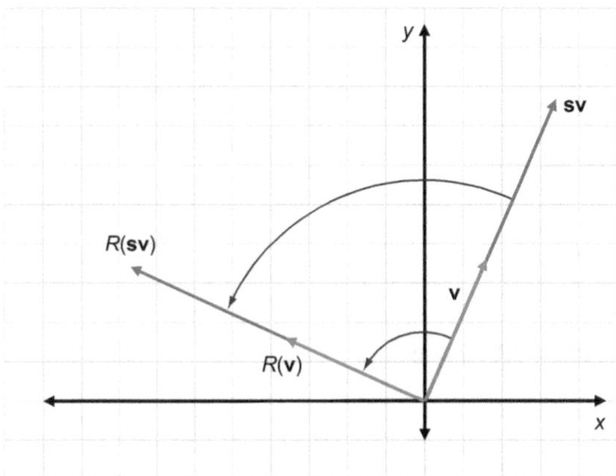

Figure 4.19 Scalar multiplication is preserved by rotation.

Again, this is only a visual example and not a proof, but you'll find that for any vector **v**, scalar s, and rotation R, the same picture holds. Rotations or any other vector transformations that preserve vector sums and scalar multiples are called *linear transformations.*

Linear transformation

A *linear transformation* is a vector transformation T that preserves vector addition and scalar multiplication. That is, for any input vectors **u** and **v**, we have

$$T(\mathbf{u}) + T(\mathbf{v}) = T(\mathbf{u} + \mathbf{v})$$

and for any pair of a scalar s and a vector **v**, we have

$$T(s\mathbf{v}) = sT(\mathbf{v})$$

Make sure you pause to digest this definition; linear transformations are so important that the whole subject of linear algebra is named after them. To help you recognize linear transformations when you see them, we'll look at a few more examples.

4.2.2 *Picturing linear transformations*

First, let's look at a counterexample: a vector transformation that's *not* linear. Such an example is a transformation $S(\mathbf{v})$ that takes a vector $\mathbf{v} = (x, y)$ and outputs a vector with both coordinates squared: $S(\mathbf{v}) = (x^2, y^2)$. As an example, let's look at the sum of $\mathbf{u} = (2, 3)$ and $\mathbf{v} = (1, -1)$. The sum is $(2, 3) + (1, -1) = (3, 2)$. This is shown with vector addition in figure 4.20.

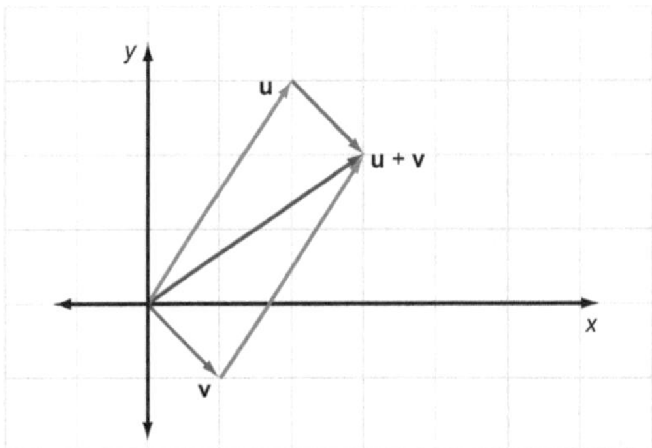

Figure 4.20 Picturing the vector sum of u = (2, 3) and v = (1, –1), u + v = (3, 2)

Now let's apply S to each of these: $S(\mathbf{u}) = (4, 9)$, $S(\mathbf{v}) = (1, 1)$, and $S(\mathbf{u} + \mathbf{v}) = (9, 4)$. Figure 4.21 clearly shows that $S(\mathbf{u}) + S(\mathbf{v})$ does *not* agree with $S(\mathbf{u} + \mathbf{v})$.

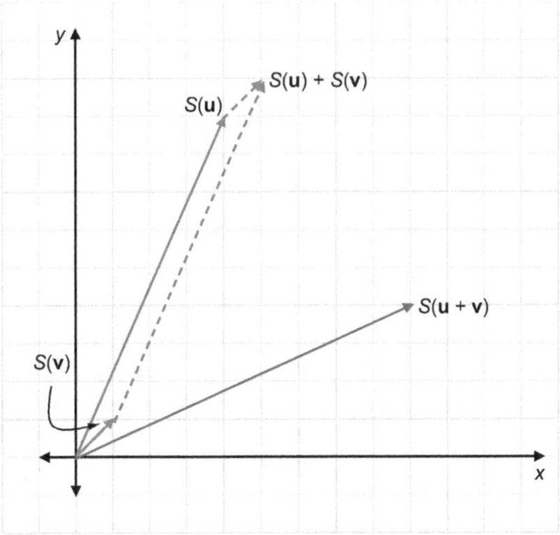

Figure 4.21 S does not respect sums! S(u) + S(v) is far from S(u + v).

As an exercise, you can try to find a counterexample demonstrating that S does not preserve scalar multiples either. For now, let's examine another transformation. Let $D(\mathbf{v})$ be the vector transformation that scales the input vector by a factor of 2. In other words, $D(\mathbf{v}) = 2\mathbf{v}$. This *does* preserve vector sums: if $\mathbf{u} + \mathbf{v} = \mathbf{w}$, then $2\mathbf{u} + 2\mathbf{v} = 2\mathbf{w}$ as well. Figure 4.22 provides a visual example.

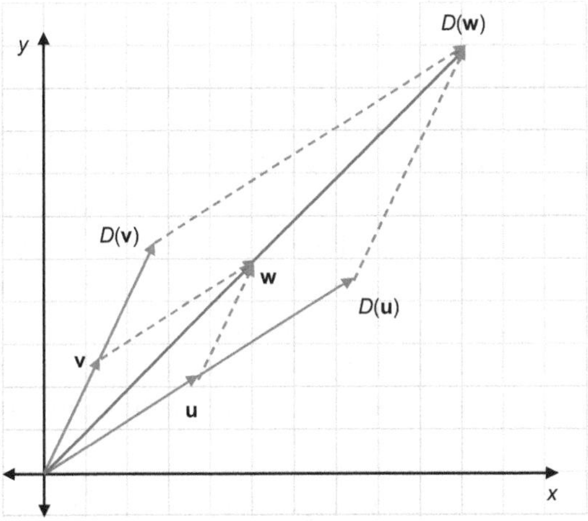

Figure 4.22 Doubling the lengths of vectors preserves their sums: if u + v = w, then *D*(u) + *D*(v) = *D*(w).

Likewise, $D(\mathbf{v})$ preserves scalar multiplication. This is a bit harder to draw, but you can see it algebraically. For any scalar s, $D(s\mathbf{v}) = 2(s\mathbf{v}) = s(2\mathbf{v}) = sD(\mathbf{v})$.

How about translation? Suppose $B(\mathbf{v})$ translates any input vector $\mathbf{v}$ by $(7, 0)$. Surprisingly, this is *not* a linear transformation. Figure 4.23 provides a visual counterexample where $\mathbf{u} + \mathbf{v} = \mathbf{w}$, but $B(\mathbf{v}) + B(\mathbf{w})$ is not the same as $B(\mathbf{v} + \mathbf{w})$.

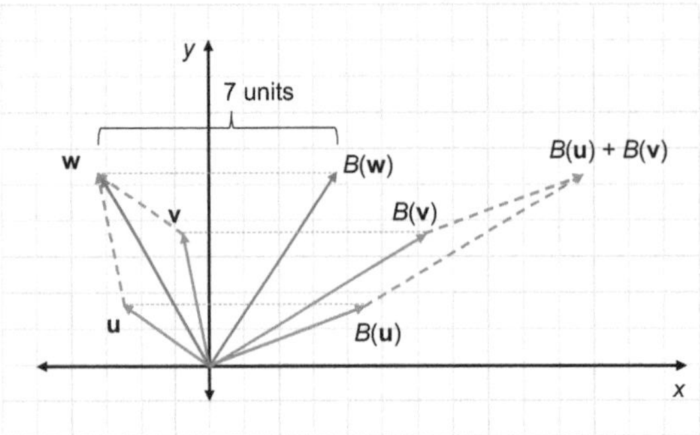

Figure 4.23 The translation transformation *B* does not preserve a vector sum because *B*(u) + *B*(v) is not equal to *B*(u + v).

It turns out that for a transformation to be linear, it must not move the origin (see why as an exercise later). Translation by any non-zero vector transforms the origin, which ends up at a different point, so it cannot be linear.

Other examples of linear transformations include reflection, projection, shearing, and any 3D analogy of the preceding linear transformations. These are defined in the exercises section and you should convince yourself with several examples that each of these transformations preserves vector addition and scalar multiplication. With practice, you can recognize which transformations are linear and which are not. Next, we'll look at why the special properties of linear transformations are useful.

4.2.3 *Why linear transformations?*

Because linear transformations preserve vector sums and scalar multiples, they also preserve a broader class of vector arithmetic operations. The most general operation is called a *linear combination.* A linear combination of a collection of vectors is a sum of scalar multiples of them. For instance, one linear combination of two vectors **u** and **v** would be $3\mathbf{u} - 2\mathbf{v}$. Given three vectors **u**, **v**, and **w**, the expression $0.5\mathbf{u} - \mathbf{v} + 6w$ *is a linear combination of* **u**, **v**, *and* **w**. Because linear transformations preserve vector sums and scalar multiples, these preserve linear combinations as well.

We can restate this fact algebraically. If you have a collection of n vectors, $\mathbf{v}_1, \mathbf{v}_2, \ldots, \mathbf{v}_n$, as well as any choice of n scalars, $s_1, s_2, s_3, \ldots, s_n$, a linear transformation T preserves the linear combination:

$$T(s_1\mathbf{v}_1 + s_2\mathbf{v}_2 + s_3\mathbf{v}_3 + \ldots + s_n\mathbf{v}_n) = s_1T(\mathbf{v}_1) + s_2T(\mathbf{v}_2) + s_3T(\mathbf{v}_3) + \ldots + s_nT(\mathbf{v}_n)$$

One easy-to-picture linear combination we've seen before is ½ **u** + ½ **v** *for vectors* **u** *and* **v**, which is equivalent to ½ (**u** + **v**). Figure 4.24 shows that this linear combination of two vectors gives us the midpoint of the line segment connecting them.

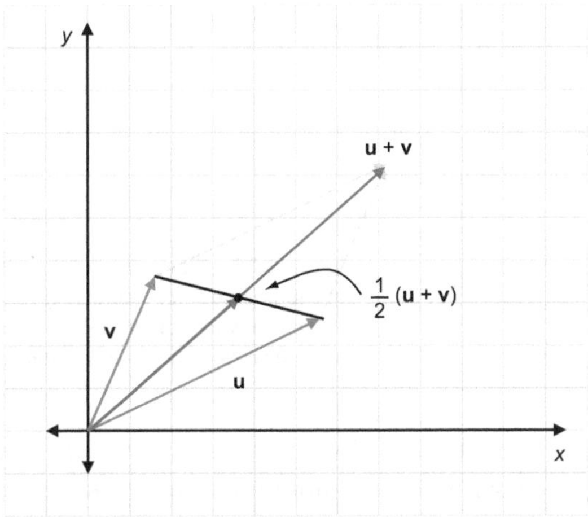

Figure 4.24 The midpoint between the tips of two vectors u and v can be found as the linear combination ½ u + ½ v = ½ (u + v).

This means linear transformations send midpoints to other midpoints: for example, *T*(½ **u** + ½ **v**) = ½ *T*(**u**) + ½ *T*(**v**), which is the midpoint of the segment connecting *T*(**u**) and *T*(**v**) as figure 4.25 shows.

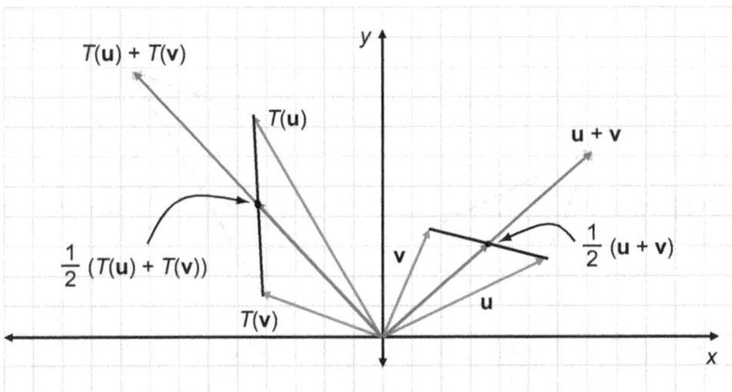

Figure 4.25 Because the midpoint between two vectors is a linear combination of the vectors, the linear transformation *T* sets the midpoint between u and v to the midpoint between *T*(u) and *T*(v).

It's less obvious, but a linear combination like 0.25**u** + 0.75**v** also lies on the line segment between **u** and **v** (figure 4.26). Specifically, this is the point 75% of the way from **u** to **v**. Likewise, 0.6**u** + 0.4**v** is 40% of the way from **u** to **v**, and so on.

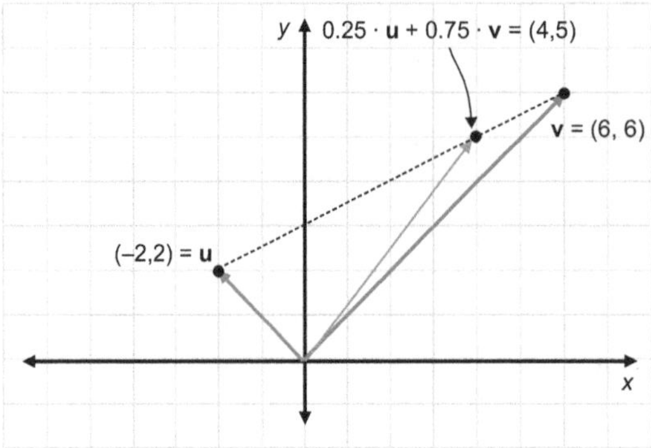

Figure 4.26 The point 0.25u + 0.75v lies on the line segment connecting u and v, 75% of the way from u to v. You can see this concretely with u = (–2, 2) and v = (6, 6).

In fact, *every* point on the line segment between two vectors is a "weighted average" like this, having the form $s\mathbf{u} + (1 - s)\mathbf{v}$ for some number s between 0 and 1. To convince you, figure 4.27 shows the vectors $s\mathbf{u} + (1 - s)\mathbf{v}$ for $\mathbf{u} = (-1, 1)$ and $\mathbf{v} = (3, 4)$ for 10 values of s between 0 and 1 and then for 100 values of s between 0 and 1.

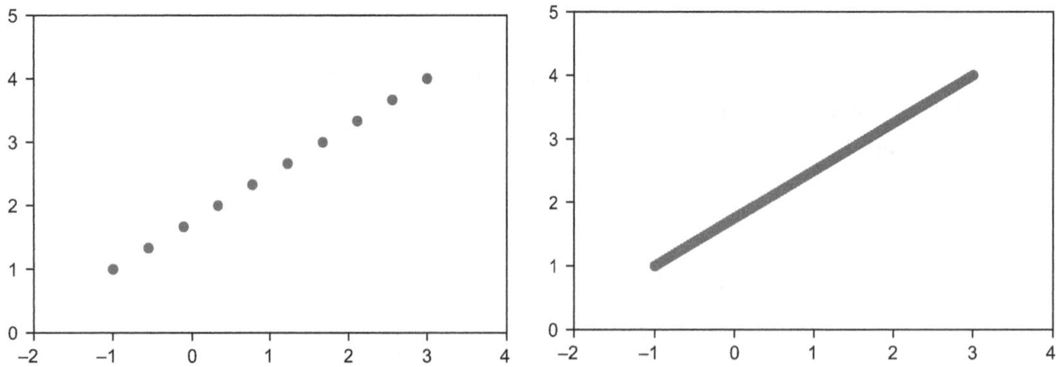

Figure 4.27 Plotting various weighted averages of (–1, 1) and (3, 4) with 10 values of s between 0 and 1 (left) and 100 values of s between 0 and 1 (right)

The key idea here is that every point on a line segment connecting two vectors **u** and **v** is a weighted average and, therefore, a linear combination of points **u** and **v**. With this in mind, we can think about what a linear transformation does to a whole line segment.

Any point on the line segment connecting **u** and **v** is a weighted average of **u** and **v**, so it has the form $s \cdot \mathbf{u} + (1 - s) \cdot \mathbf{v}$ for some value s. A linear transformation, T, transforms **u** and **v** to some new vectors **T**(**u**) and **T**(**v**). The point on the line segment is transformed to some new point $T(s \cdot \mathbf{u} + (1 - s) \cdot \mathbf{v})$ or $s \cdot T(\mathbf{u}) + (1 - s) \cdot T(\mathbf{v})$. This is, in turn, a weighted average of **T**(**u**) and **T**(**v**), so it is a point that lies on the segment connecting **T**(**u**) and **T**(**v**) as shown in figure 4.28.

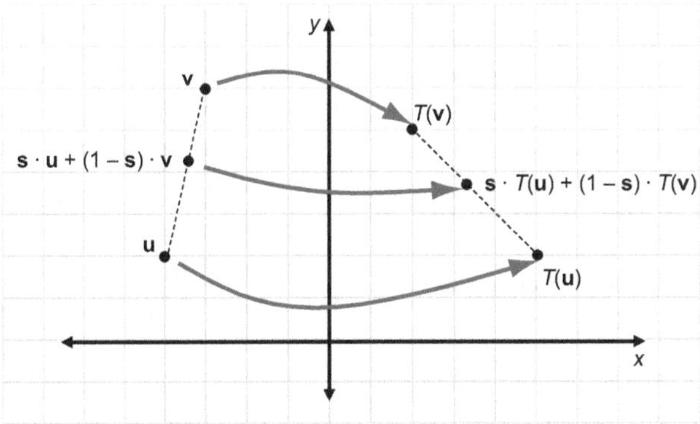

Figure 4.28 A linear transformation *T* transforms a weighted average of u and v to a weighted average of T(u) and T(v). The original weighted average lies on the segment connecting u and v, and the transformed one lies on the segment connecting T(u) and T(v).

Because of this, a linear transformation *T* takes every point on the line segment connecting **u** and **v** to a point on the line segment connecting **T**(**u**) and **T**(**v**). This is a key property of linear transformations: they send every existing line segment to a new line segment. Because our 3D models are made up of polygons and polygons are outlined by line segments, linear transformations can be expected to preserve the structure of our 3D models to some extent (figure 4.29).

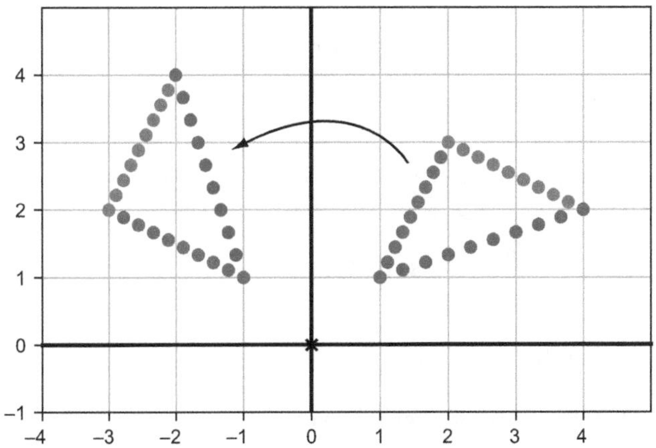

Figure 4.29 Applying a linear transformation (rotation by 60°) to points making up a triangle. The result is a rotated triangle (on the left).

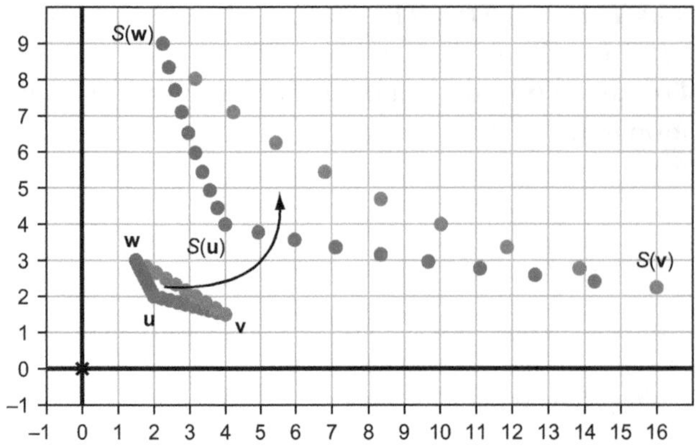

Figure 4.30 Applying the non-linear transformation S does *not* preserve the straightness of edges of the triangle.

By contrast, if we use the non-linear transformation $S(\mathbf{v})$ sending $\mathbf{v} = (x, y)$ to (x^2, y^2), we can see that line segments are distorted. This means that a triangle defined by vectors **u**, **v**, and **w** is not really sent to another triangle defined by $S(\mathbf{u})$, $S(\mathbf{v})$, and $S(\mathbf{w})$ as shown in figure 4.30.

In summary, linear transformations respect the algebraic properties of vectors, preserving sums, scalar multiples, and linear combinations. They also respect the geometric properties of collections of vectors, sending line segments and polygons defined by vectors to new ones defined by the transformed vectors. Next, we'll see that linear transformations are not only special from a geometric perspective; they're also easy to compute.

4.2.4 Computing linear transformations

In chapters 2 and 3, you saw how to break 2D and 3D vectors into components. For instance, the vector (4, 3, 5) can be decomposed as a sum (4, 0, 0) + (0, 3, 0) + (0, 0, 5). This makes it easy to picture how far the vector extends in each of the three dimensions of the space that we're in. We can decompose this even further into a linear combination (figure 4.31):

$$(4, 3, 5) = 4 \cdot (1, 0, 0) + 3 \cdot (0, 1, 0) + 5 \cdot (0, 0, 1)$$

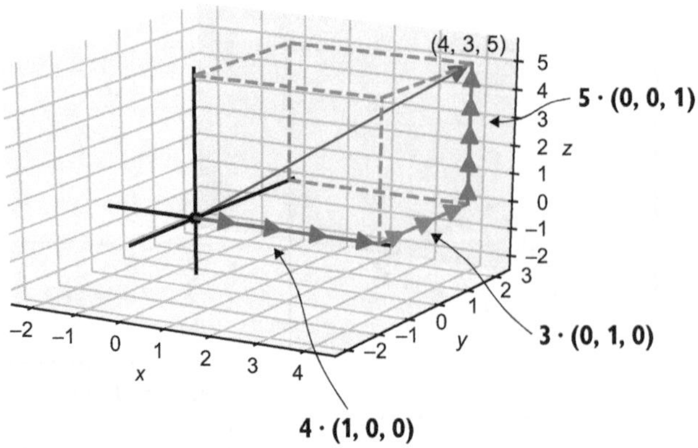

Figure 4.31 The 3D vector (4, 3, 5) as a linear combination of (1, 0, 0), (0, 1, 0), and (0, 0, 1)

This might seem like a boring fact, but it's one of the profound insights from linear algebra: any 3D vector can be decomposed into a linear combination of three vectors (1, 0, 0), (0, 1, 0), and (0, 0, 1). The scalars appearing in this decomposition for a vector **v** are exactly the coordinates of **v**.

The three vectors (1, 0, 0), (0, 1, 0), and (0, 0, 1) are called the *standard basis* for three-dimensional space. These are denoted $\mathbf{e}_1$, $\mathbf{e}_2$, and $\mathbf{e}_3$, so we could write the previous linear combination as $(3, 4, 5) = 3\ \mathbf{e}_1 + 4\ \mathbf{e}_2 + 5\ \mathbf{e}_3$. When we're working in 2D space, we call $\mathbf{e}_1 = (1, 0)$ and $\mathbf{e}_2 = (0, 1)$; so, for example, $(7, -4) = 7\ \mathbf{e}_1 - 4\ \mathbf{e}_2$ (figure 4.32). (When we say $\mathbf{e}_1$, we could mean (1, 0) or (1, 0, 0), but usually it's clear which one we mean once we've established whether we're working in two or three dimensions.)

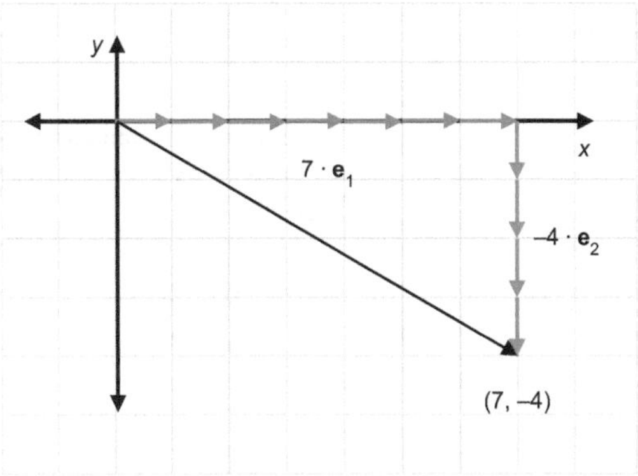

Figure 4.32 The 2D vector (7, –4) as a linear combination of the standard basis vectors $\mathbf{e}_1$ and $\mathbf{e}_2$

We've only written the same vectors in a slightly different way, but it turns out this change in perspective makes it easy to compute linear transformations. Because linear transformations respect linear combinations, all we need to know to compute a linear transformation is how it affects standard basis vectors.

Let's look at a visual example (figure 4.33). Say we know nothing about a 2D vector transformation T except that it is linear and we know what $T(\mathbf{e}_1)$ and $T(\mathbf{e}_2)$ are.

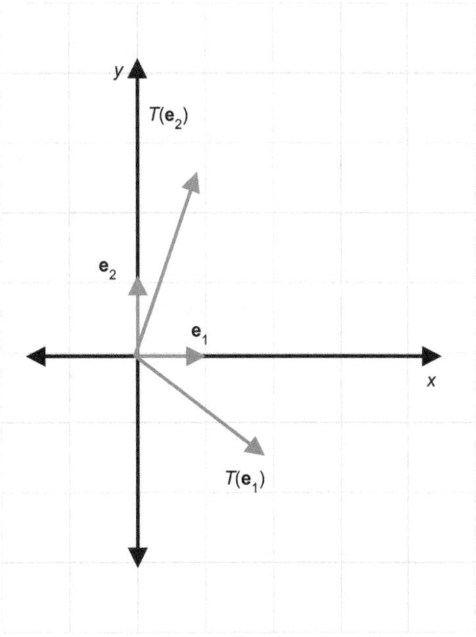

Figure 4.33 When a linear transformation acts on the two standard basis vectors in 2D, we get two new vectors as a result.

For any other vector **v**, we automatically know where $T(\mathbf{v})$ ends up. Say $\mathbf{v} = (3, 2)$, then we can assert:

$$T(\mathbf{v}) = T(3\mathbf{e}_1 + 2\mathbf{e}_2) = 3T(\mathbf{e}_1) + 2T(\mathbf{e}_2)$$

Because we already know where $T(\mathbf{e}_1)$ and $T(\mathbf{e}_2)$ are, we can locate $T(\mathbf{v})$ as shown in figure 4.34.

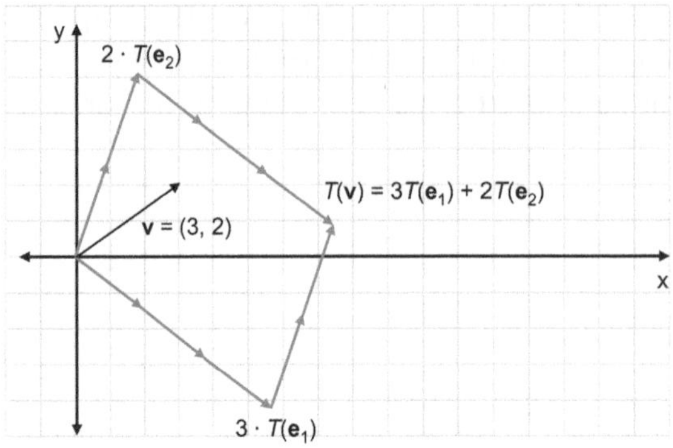

Figure 4.34 We can compute $T(\mathbf{v})$ for any vector v as a linear combination of $T(\mathbf{e}_1)$ and $T(\mathbf{e}_2)$.

To make this more concrete, let's do a complete example in 3D. Say A is a linear transformation, and all we know about A is that $A(\mathbf{e}_1) = (1, 1, 1)$, $A(\mathbf{e}_2) = (1, 0, -1)$, and $A(\mathbf{e}_3) = (0, 1, 1)$. If $\mathbf{v} = (-1, 2, 2)$, what is $A(\mathbf{v})$? Well, first we can expand **v** as a linear combination of the three standard basis vectors. Because $\mathbf{v} = (-1, 2, 2) = -\mathbf{e}_1 + 2\mathbf{e}_2 + 2\mathbf{e}_3$, we can make the substitution:

$$A(\mathbf{v}) = A(-\mathbf{e}_1 + 2\mathbf{e}_2 + 2\mathbf{e}_3)$$

Next, we can use the fact that A is linear and preserves linear combinations:

$$= -A(\mathbf{e}_1) + 2A(\mathbf{e}_2) + 2A(\mathbf{e}_3)$$

Finally, we can substitute in the known values of $A(\mathbf{e}_1)$, $A(\mathbf{e}_2)$, and $A(\mathbf{e}_3)$, and simplify:

$$= -(1, 1, 1) + 2 \cdot (1, 0, -1) + 2 \cdot (0, 1, 1)$$
$$= (1, 1, -1)$$

As proof we really know how A works, we can apply it to the teapot:

```
Ae1 = (1,1,1)
Ae2 = (1,0,-1)
Ae3 = (0,1,1)

def apply_A(v):
    return add(
        scale(v[0], Ae1),
        scale(v[1], Ae2),
        scale(v[2], Ae3)
    )

draw_model(polygon_map(apply_A, load_triangles()))
```

- **The known results of applying A to the standard basis vectors** (Ae1, Ae2, Ae3)
- **Builds a function apply_A(v) that returns the result of A on the input vector v**
- **The result should be a linear combination of these vectors, where the scalars are taken to be the coordinates of the target vector v.**
- **Uses polygon_map to apply A to every vector of every triangle in the teapot**

Figure 4.35 shows the result of this transformation.

Figure 4.35 In this rotated, skewed configuration, we see that the teapot does not have a bottom!

The takeaway here is that a 2D linear transformation T is defined completely by the values of $T(\mathbf{e}_1)$ and $T(\mathbf{e}_2)$; that's two vectors or four numbers in total. Likewise, a 3D linear transformation T is defined completely by the values of $T(\mathbf{e}_1)$, $T(\mathbf{e}_2)$, and $T(\mathbf{e}_3)$, which are three vectors or nine numbers in total. In any number of dimensions, the behavior of a linear transformation is specified by a list of vectors or an array-of-arrays of numbers. Such an array-of-arrays is called a *matrix*, and we'll see how to use matrices in the next chapter.

4.2.5 Exercises

Exercise 4.10 Considering S again, the vector transformation that squares all coordinates, show algebraically that $S(s\mathbf{v}) = sS(\mathbf{v})$ does not hold for all choices of scalars s and 2D vectors $\mathbf{v}$.

Solution Let $\mathbf{v} = (x, y)$. Then $s\mathbf{v} = (sx, sy)$ and $S(s\mathbf{v}) = (s^2x^2, s^2y^2) = s^2 \cdot (x^2, y^2) = s^2 \cdot S(\mathbf{v})$. For most values of s and most vectors $\mathbf{v}$, $S(s\mathbf{v}) = s^2 \cdot S(\mathbf{v})$ won't equal $s \cdot S(\mathbf{v})$. A specific counterexample is $s = 2$ and $\mathbf{v} = (1, 1, 1)$, where $S(s\mathbf{v}) = (4, 4, 4)$ while $s \cdot S(\mathbf{v}) = (2, 2, 2)$. This counterexample shows that S is not linear.

Exercise 4.11 Suppose T is a vector transformation and $T(\mathbf{0}) \neq 0$, where $\mathbf{0}$ represents the vector with all coordinates equal to zero. Why is T not linear according to the definition?

Solution For any vector $\mathbf{v}$, $\mathbf{v} + \mathbf{0} = \mathbf{v}$. For T to preserve vector addition, it should be that $T(\mathbf{v} + \mathbf{0}) = T(\mathbf{v}) + T(\mathbf{0})$. Because $T(\mathbf{v} + \mathbf{0}) = T(\mathbf{v})$, this requires that $T(\mathbf{v}) = T(\mathbf{v}) + T(\mathbf{0})$ or $\mathbf{0} = T(\mathbf{0})$. Given that this is not the case, T cannot be linear.

Exercise 4.12 The *identity transformation* is the vector transformation that returns the same vector it is passed. It is denoted with a capital I, so we could write its definition as $I(\mathbf{v}) = \mathbf{v}$ for all vectors $\mathbf{v}$. Explain why I is a linear transformation.

Solution For any vectors $\mathbf{v}$ and $\mathbf{w}$, $I(\mathbf{v} + \mathbf{w}) = \mathbf{v} + \mathbf{w} = I(\mathbf{v}) + I(\mathbf{w})$, and for any scalar s, $I(s\mathbf{v}) = s\mathbf{v} = s \cdot I(\mathbf{v})$. These equalities show that the identity transformation preserves vector sums and scalar multiples.

Exercise 4.13 What is the midpoint between (5, 3) and (–2, 1)? Plot all three of these points to see that you are correct.

Solution The midpoint is ½ (5, 3) + ½ (–2, 1) or (5/2, 3/2) + (–1, ½), which equals (3/2, 2). This is seen to be correct when drawn to scale in the diagram that follows:

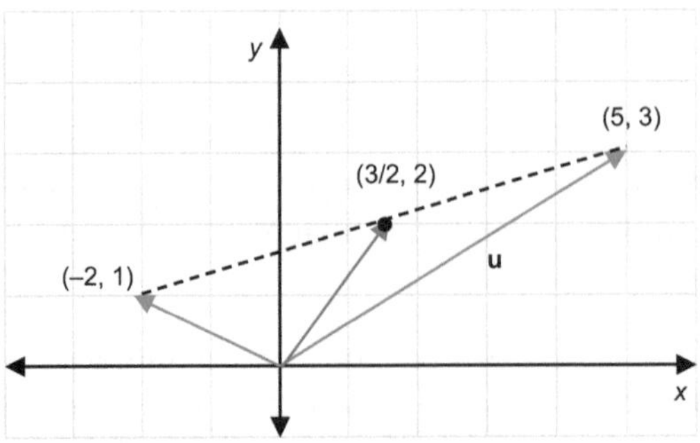

The midpoint of the segment connecting (5, 3) and (–2, 1) is (3/2, 2).

Exercise 4.14 Consider again the non-linear transformation $S(\mathbf{v})$ sending $\mathbf{v} = (x, y)$ to (x^2, y^2). Plot all 36 vectors $\mathbf{v}$ with integer coordinates 0 to 5 as points using the drawing code from chapter 2 and then plot $S(\mathbf{v})$ for each of them. What happens geometrically to vectors under the action of S?

Solution The space between points is uniform to begin with, but in the transformed picture, the spacing increases in the horizontal and vertical directions as the x- and y-coordinates increase, respectively.

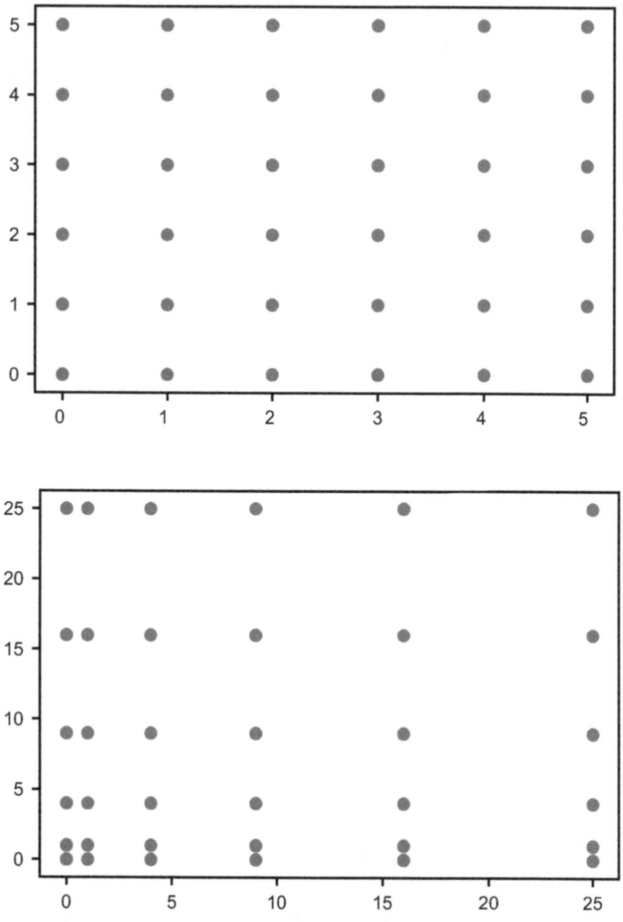

The grid of points is initially uniformly spaced, but after applying the transformation S, the spacing varies between points, even on the same lines.

Exercise 4.15—Mini Project *Property-based testing* is a type of unit testing that involves inventing arbitrary input data for a program and then checking that the outputs satisfy desired conditions. There are popular Python libraries like Hypothesis (available through pip) that make it easy to set this up. Using your library of choice, implement property-based tests that check if a vector transformation is linear.

Specifically, given a vector transformation T implemented as a Python function, generate a large number of pairs of random vectors and assert for all of those that their sum is preserved by T. Then, do the same thing for pairs of a scalar and a vector, and ensure that T preserves scalar multiples. You should find that linear transformations like `rotate_x_by(pi/2)` pass the test, but non-linear transformations like the coordinate-squaring transformation do not pass.

Exercise 4.16 One 2D vector transformation is *reflection* across the x-axis. This transformation takes a vector and returns another one, which is the mirror image with respect to the x-axis. Its x-coordinate should be unchanged, and its y-coordinate should change its sign. Denoting this transformation S_x, here is an image of a vector $\mathbf{v}$ = (3, 2) and the transformed vector $S_x(\mathbf{v})$.

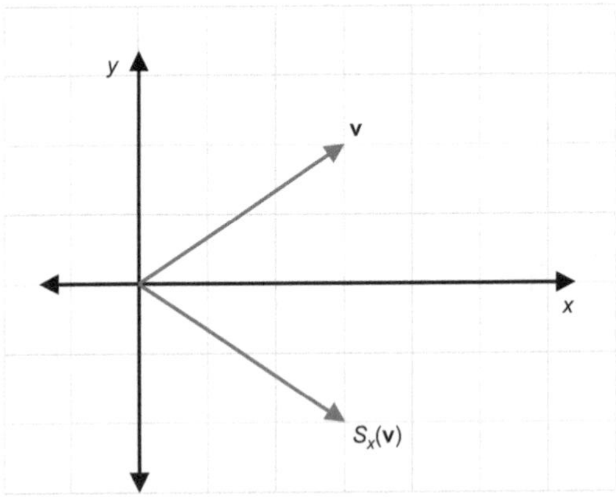

A vector v = (3, 2) and its reflection over the *x*-axis (3, –2)

Draw two vectors and their sum, as well as the reflection of these three vectors to demonstrate that this transformation preserves vector addition. Draw another diagram to show similarly that scalar multiplication is preserved, thereby demonstrating both criteria for linearity.

Solution Here's an example of reflection over the *x*-axis that preserves a vector sum:

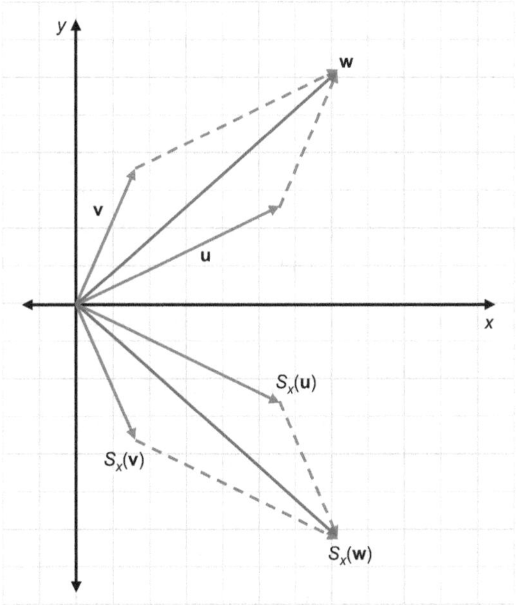

For u + v = w as shown, reflection over the *x*-axis preserves the sum; that is, $S_x(u) + S_x(v) = S_x(w)$.

Here's an example showing reflection preserving a scalar multiple: $S_x(s\mathbf{v})$ lies where $sS_x(\mathbf{v})$ is expected to be.

To *prove* that S_x is linear, you would need to show that you can draw analogous pictures for every vector sum and every scalar multiple. There are infinitely many of these, so it's better to use an algebraic proof. (Can you figure out how to show these two facts algebraically?)

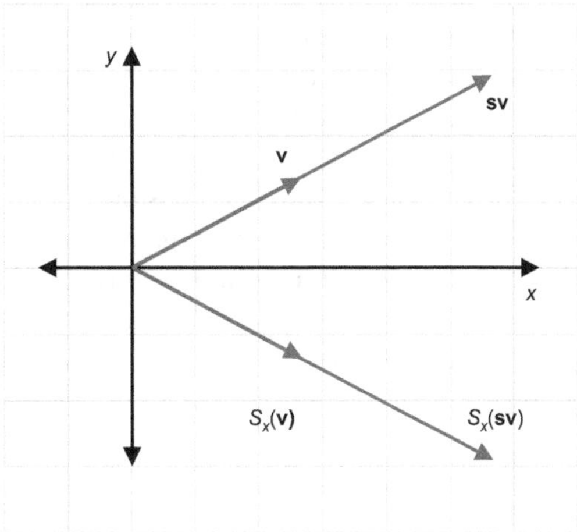

Reflection across the *x*-axis preserves this scalar multiple.

Exercise 4.17—Mini Project Suppose S and T are both linear transformations. Explain why the composition of S and T is also linear.

Solution The composition $S(T(\mathbf{v}))$ is linear if for any vector sum $\mathbf{u} + \mathbf{v} = \mathbf{w}$, we have $S(T(\mathbf{u})) + S(T(\mathbf{v})) = S(T(\mathbf{w}))$, and for any scalar multiple $s\mathbf{v}$, we have $S(T(s\mathbf{v})) = s \cdot S(T(\mathbf{v}))$. This is only a statement of the definition that must be satisfied.

Now let's see why it's true. Suppose first that $\mathbf{u} + \mathbf{v} = \mathbf{w}$ for any given input vectors $\mathbf{u}$ and $\mathbf{v}$. Then by the linearity of T, we also know that $T(\mathbf{u}) + T(\mathbf{v}) = T(\mathbf{w})$. Because this sum holds, the linearity of S tells us that the sum is preserved under S: $S(T(\mathbf{u})) + S(T(\mathbf{v})) = S(T(\mathbf{w}))$. That means that $S(T(\mathbf{v}))$ preserves vector sums.

Similarly, for any scalar multiple $s\mathbf{v}$, the linearity of T tells us that $s \cdot T(\mathbf{v}) = T(s\mathbf{v})$. By linearity of S, $s \cdot S(T(\mathbf{v})) = S(T(s\mathbf{v}))$ as well. This means $S(T(\mathbf{v}))$ preserves scalar multiplication and, therefore, that $S(T(\mathbf{v}))$ satisfies the full definition of linearity as previously stated. We can conclude that the composition of two linear transformations is linear.

Exercise 4.18 Let T be the linear transformation done by the Python function `rotate_x_by(pi/2)`, what are $T(\mathbf{e}_1)$, $T(\mathbf{e}_2)$, and $T(\mathbf{e}_3)$?

Solution Any rotation about an axis leaves points on the axis unaffected, so because $T(\mathbf{e}_1)$ is on the x-axis, $T(\mathbf{e}_1) = \mathbf{e}_1 = (1, 0, 0)$. A counterclockwise rotation of $\mathbf{e}_2 = (0, 1, 0)$ in the y,z plane takes this vector from the point one unit in the positive y direction to the point one unit in the positive z direction, so $T(\mathbf{e}_2) = \mathbf{e}_3 = (0, 0, 1)$. Likewise, $\mathbf{e}_3$ is rotated counterclockwise from the positive z direction to the negative y direction. $T(\mathbf{e}_3)$ still has length one in this direction, so it is $-\mathbf{e}_2$ or $(0, -1, 0)$.

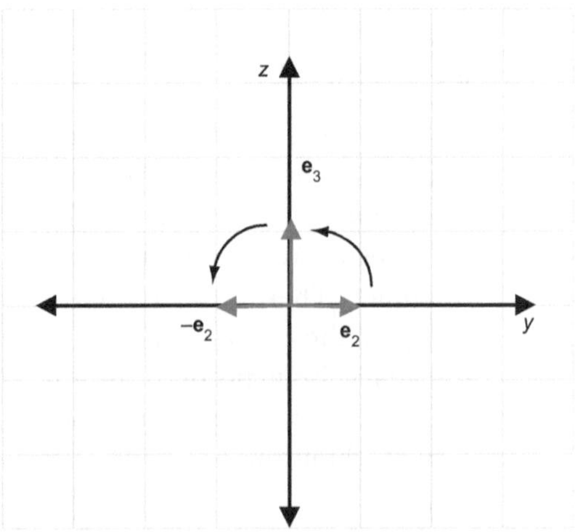

A quarter-turn counterclockwise in the y,z plane sends $\mathbf{e}_2$ to $\mathbf{e}_3$ and $\mathbf{e}_3$ to $-\mathbf{e}_2$.

Exercise 4.19 Write a `linear_combination(scalars, *vectors)` that takes a list of scalars and the same number of vectors, and returns a single vector. For example, `linear_combination([1,2,3], (1,0,0), (0,1,0), (0,0, 1))` should return $1 \cdot (1, 0, 0) + 2 \cdot (0, 1, 0) + 3 \cdot (0, 0, 1)$ or $(1, 2, 3)$.

Solution

```
from vectors import *
def linear_combination(scalars,*vectors):
    scaled = [scale(s,v) for s,v in zip(scalars,vectors)]
    return add(*scaled)
```

We can confirm this gives the expected result as previously described:

```
>>> linear_combination([1,2,3], (1,0,0), (0,1,0), (0,0,1))
(1, 2, 3)
```

Exercise 4.20 Write a function `transform_standard_basis(transform)` that takes a 3D vector transformation as an input and outputs the effect it has on the standard basis. It should output a tuple of 3 vectors that are the results of `transform` acting on $\mathbf{e}_1$, $\mathbf{e}_2$, and $\mathbf{e}_3$, respectively.

Solution As suggested, we just need to apply `transform` to each standard basis vector:

```
def transform_standard_basis(transform):
    return transform((1,0,0)), transform((0,1,0)), transform((0,0,1))
```

It confirms (within the floating-point error) our solution to a previous exercise, where we sought this output for `rotate_x_by(pi/2)`:

```
>>> from math import *
>>> transform_standard_basis(rotate_x_by(pi/2))
((1, 0.0, 0.0), (0, 6.123233995736766e-17, 1.0), (0, -1.0,
    1.2246467991473532e-16))
```

These vectors are approximately $(1, 0, 0)$, $(0, 0, 1)$, and $(0, -1, 0)$.

Exercise 4.21 Suppose B is a linear transformation, with $B(\mathbf{e}_1) = (0, 0, 1)$, $B(\mathbf{e}_2) = (2, 1, 0)$, $B(\mathbf{e}_3) = (-1, 0, -1)$, and $\mathbf{v} = (-1, 1, 2)$. What is $B(\mathbf{v})$?

Solution Because $\mathbf{v} = (-1, 1, 2) = -\mathbf{e}_1 + \mathbf{e}_2 + 2\mathbf{e}_3$, $B(\mathbf{v}) = B(-\mathbf{e}_1 + \mathbf{e}_2 + 2\mathbf{e}_3)$. Because B is linear, it preserves this linear combination: $B(\mathbf{v}) = -B(\mathbf{e}_1) + B(\mathbf{e}_2) + 2 \cdot B(\mathbf{e}_3)$. Now we have all the information we need: $B(\mathbf{v}) = -(0, 0, 1) + (2, 1, 0) + 2 \cdot (-1, 0, -1) = (0, 1, -3)$.

Exercise 4.22 Suppose A and B are both linear transformations with $A(\mathbf{e}_1) = (1, 1, 1)$, $A(\mathbf{e}_2) = (1, 0, -1)$, and $A(\mathbf{e}_3) = (0, 1, 1)$, and $B(\mathbf{e}_1) = (0, 0, 1)$, $B(\mathbf{e}_2) = (2, 1, 0)$, and $B(\mathbf{e}_3) = (-1, 0, -1)$. What is $A(B(\mathbf{e}_1))$, $A(B(\mathbf{e}_2))$, and $A(B(\mathbf{e}_3))$?

Solution $A(B(\mathbf{e}_1))$ is A applied to $B(\mathbf{e}_1) = (0, 0, 1) = \mathbf{e}_3$. We already know $A(\mathbf{e}_3) = (0, 1, 1)$, so $B(A(\mathbf{e}_1)) = (0, 1, 1)$.

$A(B(\mathbf{e}_2))$ is A applied to $B(\mathbf{e}_2) = (2, 1, 0)$. This is a linear combination of $A(\mathbf{e}_1)$, $A(\mathbf{e}_2)$, and $A(\mathbf{e}_3)$ with scalars $(2, 1, 0)$: $2 \cdot (1, 1, 1) + 1 \cdot (1, 0, -1) + 0 \cdot (0, 1, 1) = (3, 2, 1)$.

Finally, $A(B(\mathbf{e}_3))$ is A applied to $B(\mathbf{e}_3) = (-1, 0, -1)$. This is the linear combination $-1 \cdot (1, 1, 1) + 0 \cdot (1, 0, -1) + -1 \cdot (0, 1, 1) = (-1, -2, -2)$.

Note that now we know the result of the composition of A and B for all of the standard basis vectors, so we can calculate $A(B(\mathbf{v}))$ for any vector $\mathbf{v}$.

Linear transformations are both well-behaved and easy-to-compute because these can be specified with so little data. We explore this more in the next chapter when we compute linear transformations with *matrix* notation.

Summary

- Vector transformations are functions that take vectors as inputs and return vectors as outputs. Vector transformations can operate on 2D or 3D vectors.
- To effect a geometric transformation of the model, apply a vector transformation to every vertex of every polygon of a 3D model.
- You can combine existing vector transformations by composition of functions to create new transformations, which are equivalent to applying the existing vector transformations sequentially.
- Functional programming is a programming paradigm that emphasizes composing and, otherwise, manipulating functions.
- The functional operation of currying turns a function that takes multiple arguments into a function that takes one argument and returns a new function. Currying lets you turn existing Python functions (like `scale` and `add`) into vector transformations.
- Linear transformations are vector transformations that preserve vector sums and scalar multiples. In particular, points lying on a line segment still lie on a line segment after a linear transformation is applied.
- A linear combination is the most general combination of scalar multiplication and vector addition. Every 3D vector is a linear combination of the 3D standard basis vectors, which are denoted $\mathbf{e}_1 = (1, 0, 0)$, $\mathbf{e}_2 = (0, 1, 0)$, and $\mathbf{e}_3 = (0, 0, 1)$. Likewise, every 2D vector is a linear combination of the 2D standard basis vectors, which are $\mathbf{e}_1 = (1, 0)$ and $\mathbf{e}_2 = (0, 1)$.

- Once you know how a given linear transformation acts on the standard basis vectors, you can determine how it acts on any vector by writing the vector as a linear combination of the standard basis and using the fact that linear combinations are preserved.
  - In 3D, three vectors or nine total numbers specify a linear transformation.
  - In 2D, two vectors or four total numbers do the same.

  This last point is critical: linear transformations are both well-behaved and easy-to-compute with because they can be specified with so little data.

5 Computing transformations with matrices

This chapter covers

- Writing a linear transformation as a matrix
- Multiplying matrices to compose and apply linear transformations
- Operating on vectors of different dimensions with linear transformations
- Translating vectors in 2D or 3D with matrices

In the culmination of chapter 4, I stated a big idea: *any* linear transformation in 3D can be specified by just three vectors or nine numbers total. By correctly selecting these nine numbers, we can achieve rotation by any angle about any axis, reflection across any plane, projection onto any plane, scaling by any factor in any direction, or any other 3D linear transformation.

The transformation expressed as "a rotation counterclockwise by 90° about the *z*-axis" can equivalently be described by what it does to the standard basis vectors $\mathbf{e}_1 = (1, 0, 0)$, $\mathbf{e}_2 = (0, 1, 0)$, and $\mathbf{e}_3 = (0, 0, 1)$. Namely, the results are (0, 1, 0), (–1, 0, 0), and (0, 0, 1). Whether we think of this transformation geometrically or as described by these three vectors (or nine numbers), we're thinking of the same imaginary machine (figure 5.1) that operates on 3D vectors. The implementations might be different, but the machines still produce indistinguishable results.

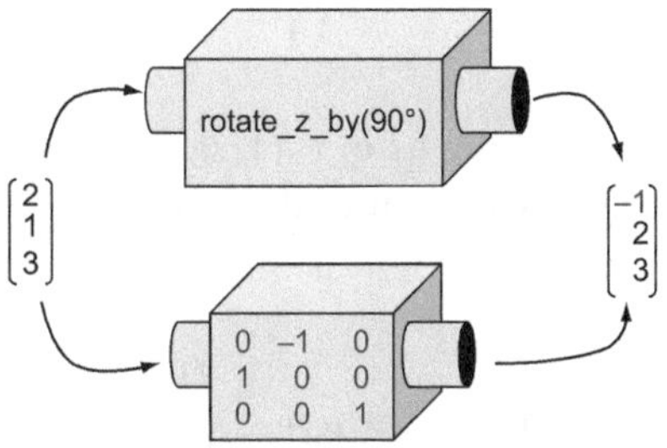

Figure 5.1 Two machines that do the same linear transformation. Geometric reasoning powers the machine on the top, while nine numbers power the one on the bottom.

When arranged appropriately in a grid, the numbers that tell us how to execute a linear transformation are called a *matrix*. This chapter focuses on using these grids of numbers as computational tools, so there's more number-crunching in this chapter than in the previous ones. Don't let this intimidate you! When it comes down to it, we're still just carrying out vector transformations.

A matrix lets us compute a given linear transformation using the data of what that transformation does to standard basis vectors. All of the notation in this chapter serves to organize that process, which we covered in section 4.2, not to introduce any unfamiliar ideas. I know it can feel like a pain to learn a new and complicated notation, but I promise, it will pay off. We are better off being able to think of vectors as geometric objects or as tuples of numbers. Likewise, we'll expand our mental model by thinking of linear transformations as matrices of numbers.

5.1 Representing linear transformations with matrices

Let's return to a concrete example of the nine numbers that specify a 3D linear transformation. Suppose A is a linear transformation, and we know $A(\mathbf{e}_1) = (1, 1, 1)$, $A(\mathbf{e}_2) = (1, 0, -1)$, and $A(\mathbf{e}_3) = (0, 1, 1)$. These three vectors having nine components in total contain all of the information required to specify the linear transformation A.

Because we reuse this concept over and over, it warrants a special notation. We'll adopt a new notation, called *matrix notation*, to work with these nine numbers as a representation of A.

5.1.1 Writing vectors and linear transformations as matrices

Matrices are rectangular grids of numbers, and their shapes tell us how to interpret them. For instance, we can interpret a matrix that is a single column of numbers as a vector with its entries being the coordinates, ordered top to bottom. In this form, the vectors are called *column vectors*. For example, the standard basis for three dimensions can be written as three column vectors like this:

$$\mathbf{e}_1 = \begin{pmatrix} 0 \\ 1 \\ 1 \end{pmatrix}, \quad \mathbf{e}_2 = \begin{pmatrix} 0 \\ 1 \\ 0 \end{pmatrix}, \quad \mathbf{e}_3 = \begin{pmatrix} 0 \\ 0 \\ 1 \end{pmatrix}$$

For our purposes, this notation means the same thing as $\mathbf{e}_1 = (1, 0, 0)$, $\mathbf{e}_2 = (0, 1, 0)$, and $\mathbf{e}_3 = (0, 0, 1)$. We can indicate how A transforms standard basis vectors with this notation as well:

$$A(\mathbf{e}_1) = \begin{pmatrix} 1 \\ 1 \\ 1 \end{pmatrix}, \quad A(\mathbf{e}_2) = \begin{pmatrix} 1 \\ 0 \\ -1 \end{pmatrix}, \quad A(\mathbf{e}_3) = \begin{pmatrix} 0 \\ 1 \\ 1 \end{pmatrix}$$

The matrix representing the linear transformation A is the 3-by-3 grid consisting of these vectors squashed together side by side:

$$A = \begin{pmatrix} 1 & 1 & 0 \\ 1 & 0 & 1 \\ 1 & -1 & 1 \end{pmatrix}$$

In 2D, a column vector consists of two entries, so 2 transformed vectors contain a total of 4 entries. We can look at the linear transformation D that scales input vectors by a multiple of 2. First, we write how it works on basis vectors:

$$D(\mathbf{e}_1) = \begin{pmatrix} 2 \\ 0 \end{pmatrix}, \quad D(\mathbf{e}_2) = \begin{pmatrix} 0 \\ 2 \end{pmatrix}$$

Then the matrix for D is obtained by putting these columns next to each other:

$$D = \begin{pmatrix} 2 & 0 \\ 0 & 2 \end{pmatrix}$$

Matrices can come in other shapes and sizes, but we'll focus on these two shapes for now: the single column matrices representing vectors and the square matrices representing linear transformations.

Remember, there are no new concepts here, only a new way of writing the core idea from section 4.2: a linear transformation is defined by its results acting on the standard basis vectors. The way to get a matrix from a linear transformation is to find the vectors it produces from all of the standard basis vectors and combine the results side by side. Now, we'll look at the opposite problem: how to evaluate a linear transformation given its matrix.

5.1.2 Multiplying a matrix with a vector

If a linear transformation B is represented as a matrix, and a vector $\mathbf{v}$ is also represented as a matrix (a column vector), we have all of the numbers required to evaluate $B(\mathbf{v})$. For instance, if B and $\mathbf{v}$ are given by

$$B = \begin{pmatrix} 0 & 2 & 1 \\ 0 & 1 & 0 \\ 1 & 0 & -1 \end{pmatrix}, \quad \mathbf{v} = \begin{pmatrix} 3 \\ -2 \\ 5 \end{pmatrix}$$

then the vectors $B(\mathbf{e}_1)$, $B(\mathbf{e}_2)$, and $B(\mathbf{e}_3)$ can be read off of B as the columns of its matrix. From that point, we use the same procedure as before. Because $\mathbf{v} = 3\mathbf{e}_1 - 2\mathbf{e}_2 + 5\mathbf{e}_3$, it follows that $B(\mathbf{v}) = 3B(\mathbf{e}_1) - 2\,B(\mathbf{e}_2) + 5B(\mathbf{e}_3)$. Expanding this, we get

$$B(\mathbf{v}) = 3 \cdot \begin{pmatrix} 0 \\ 0 \\ 1 \end{pmatrix} - 2 \cdot \begin{pmatrix} 2 \\ 1 \\ 0 \end{pmatrix} + 5 \cdot \begin{pmatrix} 1 \\ 0 \\ -1 \end{pmatrix} = \begin{pmatrix} 0 \\ 0 \\ 3 \end{pmatrix} + \begin{pmatrix} -4 \\ -2 \\ 0 \end{pmatrix} + \begin{pmatrix} 5 \\ 0 \\ -5 \end{pmatrix} = \begin{pmatrix} 1 \\ -2 \\ -2 \end{pmatrix}$$

and the result is the vector (1, –2, –2). Treating a square matrix as a function that operates on a column vector is a special case of an operation called *matrix multiplication*. Again, this has an impact on our notation and terminology, but we are simply doing the same thing: applying a linear transformation to a vector. Written as a matrix multiplication, it looks like this:

$$B\mathbf{v} = \begin{pmatrix} 0 & 2 & 1 \\ 0 & 1 & 0 \\ 1 & 0 & -1 \end{pmatrix} \begin{pmatrix} 3 \\ -2 \\ 5 \end{pmatrix} = \begin{pmatrix} 1 \\ -2 \\ -2 \end{pmatrix}$$

As opposed to multiplying numbers, the order matters when you multiply matrices by vectors. In this case, $B\mathbf{v}$ is a valid product but $\mathbf{v}B$ is not. Shortly, we'll see how to multiply matrices of various shapes and a general rule for the order in which matrices can be multiplied. For now, take my word for it and think of this multiplication as valid because it means applying a 3D linear operator to a 3D vector.

We can write Python code that multiplies a matrix by a vector. Let's say we encode the matrix B as a tuple-of-tuples and the vector $\mathbf{v}$ as a tuple as usual:

```
B = (
    (0,2,1),
    (0,1,0),
    (1,0,-1)
)

v = (3,-2,5)
```

This is a bit different from how we originally thought about the matrix B. We originally created it by combining three columns, but here B is created as a sequence of rows. The advantage of defining a matrix in Python as a tuple of rows is that the numbers are laid out in the same order as we would write them on paper. We can, however, get the columns any time we want by using Python's `zip` function (covered in appendix B):

```
>>> list(zip(*B))
[(0, 0, 1), (2, 1, 0), (1, 0, -1)]
```

The first entry of this list is $(0, 0, 1)$, which is the first column of B, and so on. What we want is the linear combination of these vectors, where the scalars are the coordinates of **v**. To get this, we can use the `linear_combination` function from the exercise in section 4.2.5. The first argument to `linear_combination` should be **v**, which serves as the list of scalars, and the subsequent arguments should be the columns of B. Here's the complete function:

```
def multiply_matrix_vector(matrix, vector):
    return linear_combination(vector, *zip(*matrix))
```

It confirms the calculation we did by hand with B and **v**:

```
>>> multiply_matrix_vector(B,v)
(1, -2, -2)
```

There are two other mnemonic recipes for multiplying a matrix by a vector, both of which give the same results. To see these, let's write a prototypical matrix multiplication:

$$\begin{pmatrix} a & b & c \\ d & e & f \\ g & h & i \end{pmatrix} \begin{pmatrix} x \\ y \\ z \end{pmatrix}$$

The result of this calculation is the linear combination of the columns of the matrix with the coordinates x, y, and z as the scalars:

$$= x \cdot \begin{pmatrix} a \\ d \\ g \end{pmatrix} + y \cdot \begin{pmatrix} b \\ e \\ h \end{pmatrix} + z \cdot \begin{pmatrix} c \\ f \\ i \end{pmatrix} = \begin{pmatrix} ax + by + cz \\ dx + ey + fz \\ gx + hy + iz \end{pmatrix}$$

This is an explicit formula for the product of a 3-by-3 matrix with a 3D vector. You can write a similar one for a 2D vector:

$$\begin{pmatrix} j & k \\ l & m \end{pmatrix} \begin{pmatrix} x \\ y \end{pmatrix} = x \cdot \begin{pmatrix} j \\ l \end{pmatrix} + y \cdot \begin{pmatrix} k \\ m \end{pmatrix} = \begin{pmatrix} jx + ky \\ lx + my \end{pmatrix}$$

The first mnemonic is that each coordinate of the output vector is a function of all the coordinates of the input vector. For instance, the first coordinate of the 3D output is a function $f(x, y, z) = ax + by + cz$. Moreover, this is a *linear function* (in the sense that you used the word in high school algebra); it is a sum of a number times each variable. We originally introduced the term "linear transformation" because linear transformations preserve lines. Another reason to use that term: a linear transformation is a collection of linear *functions* on the input coordinates that give the respective output coordinates.

The second mnemonic presents the same formula differently: the coordinates of the output vector are dot products of the rows of the matrix with the target vector. For instance, the first row of the 3-by-3 matrix is (a, b, c) and the multiplied vector is (x, y, z), so the first coordinate of the output is $(a, b, c) \cdot (x, y, z) = ax + by + cz$. We can combine our two notations to state this fact in a formula:

$$\begin{pmatrix} a & b & c \\ d & e & f \\ g & h & i \end{pmatrix} \begin{pmatrix} x \\ y \\ z \end{pmatrix} = \begin{pmatrix} (a,b,c) \cdot (x,y,z) \\ (d,e,f) \cdot (x,y,z) \\ (g,h,i) \cdot (x,y,z) \end{pmatrix} = \begin{pmatrix} ax + by + cz \\ dx + ey + fz \\ gx + hy + iz \end{pmatrix}$$

If your eyes are starting to glaze over from looking at so many letters and numbers in arrays, don't worry. The notation can be overwhelming at first, and it takes some time to connect it to your intuition. There are more examples of matrices in this chapter, and the next chapter provides more review and practice as well.

5.1.3 Composing linear transformations by matrix multiplication

Some of the examples of linear transformations we've seen so far are rotations, reflections, rescalings, and other geometric transformations. What's more, any number of linear transformations chained together give us a new linear transformation. In math terminology, the *composition* of any number of linear transformations is also a linear transformation.

Because any linear transformation can be represented by a matrix, any two composed linear transformations can be as well. In fact, if you want to compose linear transformations to build new ones, matrices are the best tools for the job.

> **NOTE** Let me take off my mathematician hat and put on my programmer hat for a moment. Suppose you want to compute the result of, say, 1,000 composed linear transformations operating on a vector. This can come up if you are animating an object by applying additional, small transformations within every frame of the animation. In Python, it would be computationally expensive to apply 1,000 sequential functions because there is computational overhead for every function call. However, if you were to find a matrix representing the composition of 1,000 linear transformations, you would boil the whole process down to a handful of numbers and a handful of computations.

Let's look at a composition of two linear transformations: $A(B(\mathbf{v}))$, where the matrix representations of A and B are known to be the following:

$$A = \begin{pmatrix} 1 & 1 & 0 \\ 1 & 0 & 1 \\ 1 & -1 & 1 \end{pmatrix}, \quad B = \begin{pmatrix} 0 & 2 & 1 \\ 0 & 1 & 0 \\ 1 & 0 & -1 \end{pmatrix}$$

Here's how the composition works step by step. First, the transformation B is applied to $\mathbf{v}$, yielding a new vector $B(\mathbf{v})$, or $B\mathbf{v}$ if we're writing it as a multiplication. Second, this vector becomes the input to the transformation A, yielding a final 3D vector as a result: $A(B\mathbf{v})$. Once again, we'll drop the parentheses and write $A(B\mathbf{v})$ as the product $AB\mathbf{v}$. Writing this product out for $\mathbf{v} = (x, y, z)$ gives us a formula that looks like this:

$$AB\mathbf{v} = \begin{pmatrix} 1 & 1 & 0 \\ 1 & 0 & 1 \\ 1 & -1 & 1 \end{pmatrix} \begin{pmatrix} 0 & 2 & 1 \\ 0 & 1 & 0 \\ 1 & 0 & -1 \end{pmatrix} \begin{pmatrix} x \\ y \\ z \end{pmatrix}$$

If we work right to left, we know how to evaluate this. Now I'm going to claim that we can work left to right as well and get the same result. Specifically, we can ascribe meaning to the product matrix AB on its own; it will be a new matrix (to be discovered) representing the composition of the linear transformations A and B:

$$AB = \begin{pmatrix} 1 & 1 & 0 \\ 1 & 0 & 1 \\ 1 & -1 & 1 \end{pmatrix} \begin{pmatrix} 0 & 2 & 1 \\ 0 & 1 & 0 \\ 1 & 0 & -1 \end{pmatrix} = \begin{pmatrix} ? & ? & ? \\ ? & ? & ? \\ ? & ? & ? \end{pmatrix}$$

Now, what should the entries of this new matrix be? Its purpose is to represent the composition of the transformations A and B, which give us a new linear transformation, AB. As we saw, the columns of a matrix are the results of applying its transformation to standard basis vectors. The columns of the matrix AB are the result of applying the transformation AB to each of $\mathbf{e}_1$, $\mathbf{e}_2$, and $\mathbf{e}_3$.

The columns of AB are, therefore, $AB(\mathbf{e}_1)$, $AB(\mathbf{e}_2)$ and $AB(\mathbf{e}_3)$. Let's look at the first column, for instance, which should be $AB(\mathbf{e}_1)$ or A applied to the vector $B(\mathbf{e}_1)$. In other words, to get the first column of AB, we multiply a matrix by a vector, an operation that we already practiced:

$$AB = \underbrace{\begin{pmatrix} 1 & 1 & 0 \\ 1 & 0 & 1 \\ 1 & -1 & 1 \end{pmatrix}}_{A} \underbrace{\begin{pmatrix} 0 & 2 & 1 \\ 0 & 1 & 0 \\ 1 & 0 & -1 \end{pmatrix}}_{B(\mathbf{e}_1)} = \underbrace{\begin{pmatrix} 0 & ? & ? \\ 1 & ? & ? \\ 1 & ? & ? \end{pmatrix}}_{AB(\mathbf{e}_1)}$$

Similarly, we find that $AB(\mathbf{e}_2) = (3, 2, 1)$ and $AB(\mathbf{e}_3) = (1, 0, 0)$, which are the second and third columns of AB:

$$AB = \begin{pmatrix} 0 & 3 & 1 \\ 1 & 2 & 0 \\ 1 & 1 & 0 \end{pmatrix}$$

That's how we do matrix multiplication. You can see there's nothing to it besides carefully composing linear operators. Similarly, you can use mnemonics instead of reasoning through this process each time. Because multiplying a 3-by-3 matrix by a column vector is the same as doing three dot products, multiplying two 3-by-3 matrices together is the same as doing nine dot products—all possible dot products of rows of the first matrix with columns of the second as shown in figure 5.2.

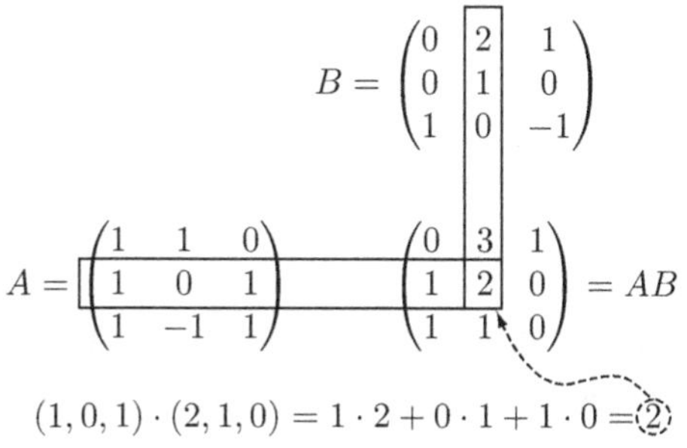

Figure 5.2 Each entry of a product matrix is a dot product of a row of the first matrix with a column of the second matrix.

Everything we've said about 3-by-3 matrix multiplication applies to 2-by-2 matrices as well. For instance, to find the product of these 2-by-2 matrices

$$\begin{pmatrix} 1 & 2 \\ 3 & 4 \end{pmatrix} \begin{pmatrix} 0 & -1 \\ 1 & 0 \end{pmatrix}$$

we can take the dot products of the rows of the first with the columns of the second. The dot product of the first row of the first matrix with the first column of the second matrix is $(1, 2) \cdot (0, 1) = 2$. This tells us that the entry in the first row and first column of the result matrix is 2:

$$\begin{pmatrix} 1 & 2 \\ 3 & 4 \end{pmatrix} \begin{pmatrix} 0 & -1 \\ 1 & 0 \end{pmatrix} = \begin{pmatrix} 2 & ? \\ ? & ? \end{pmatrix}$$

Repeating this procedure, we can find all the entries of the product matrix:

$$\begin{pmatrix} 1 & 2 \\ 3 & 4 \end{pmatrix} \begin{pmatrix} 0 & -1 \\ 1 & 0 \end{pmatrix} = \begin{pmatrix} 2 & -1 \\ 4 & 3 \end{pmatrix}$$

You can do some matrix multiplication as an exercise to get the hang of it, but you'll quickly prefer that your computer does the work for you. Let's implement matrix multiplication in Python to make this possible.

5.1.4 Implementing matrix multiplication

There are a few ways we could write our matrix multiplication function, but I prefer using the dot product trick. Because the result of matrix multiplication should be a tuple of tuples, we can write it as a nested comprehension. It takes in two nested tuples as well, called `a` and `b`, representing our input matrices A and B. The input matrix `a` is already a tuple of rows of the first matrix, and we can pair these up with `zip(*b)`, which is a tuple of columns of the second matrix. Finally, for each pair, we should take the dot product and yield it in the inner comprehension. Here's the implementation:

```
from vectors import *

def matrix_multiply(a,b):
    return tuple(
        tuple(dot(row,col) for col in zip(*b))
        for row in a
    )
```

The outer comprehension builds the rows of the result, and the inner one builds the entries of each row. Because the output rows are formed by the various dot products with rows of `a`, the outer comprehension iterates over `a`.

Our `matrix_multiply` function doesn't have any hard-coded dimensions. That means we can use it to do the matrix multiplications from the preceding 2D and 3D examples:

```
>>> a = ((1,1,0),(1,0,1),(1,-1,1))
>>> b = ((0,2,1),(0,1,0),(1,0,-1))
>>> matrix_multiply(a,b)
((0, 3, 1), (1, 2, 0), (1, 1, 0))
>>> c = ((1,2),(3,4))
>>> d = ((0,-1),(1,0))
>>> matrix_multiply(c,d)
((2, -1), (4, -3))
```

Equipped with the computational tool of matrix multiplication, we can now do some easy manipulations of our 3D graphics.

5.1.5 3D animation with matrix transformations

To animate a 3D model, we redraw a transformed version of the original model in each frame. To make the model appear to move or change over time, we need to use different transformations as time progresses. If these transformations are linear transformations specified by matrices, we need a new matrix for every new frame of the animation.

Because PyGame's built-in clock keeps track of time (in milliseconds), one thing we can do is to generate matrices whose entries depend on time. In other words, instead of thinking of every entry of a matrix as a number, we can think of it as a function that takes the current time, *t*, and returns a number (figure 5.3).

$$\begin{pmatrix} a & b & c \\ d & e & f \\ g & h & i \end{pmatrix} \rightarrow \begin{pmatrix} a(t) & b(t) & c(t) \\ d(t) & e(t) & f(t) \\ g(t) & h(t) & i(t) \end{pmatrix}$$

Figure 5.3 Thinking of matrix entries as functions of time allows the overall matrix to change as time passes.

For instance, we could use these nine expressions:

$$\begin{pmatrix} \cos(t) & 0 & -\sin(t) \\ 0 & 1 & 0 \\ \sin(t) & 0 & \cos(t) \end{pmatrix}$$

As we covered in chapter 2, cosine and sine are both functions that take a number and return another number as a result. The other five entries happen to not change over time, but if you crave consistency, you can think of these as constant functions (as in $f(t) = 1$ in the center entry). Given any value of *t*, this matrix represents the same linear transformation as `rotate_y_by(t)`. Time moves forward and the value of *t* increases, so if we apply this matrix transformation to each frame, we'll get a bigger rotation each time.

Let's give our `draw_model` function (covered in appendix C and used extensively in chapter 4) a `get_matrix` keyword argument, where the value passed to `get_matrix` is a function that takes time in milliseconds and returns the transformation matrix that should be applied at that time. In the source code file, animate_teapot.py, I call it like this to animate the rotating teapot from chapter 4:

```
from teapot import load_triangles
from draw_model import draw_model
from math import sin,cos

def get_rotation_matrix(t):
    seconds = t/1000
    return (
        (cos(seconds),0,-sin(seconds)),
        (0,1,0),
        (sin(seconds),0,cos(seconds))
    )
draw_model(load_triangles(),
           get_matrix=get_rotation_matrix)
```

Generates a new transformation matrix for any numeric input representing time (annotates `def get_rotation_matrix(t):`)

Converts the time to seconds so the transformation doesn't happen too quickly (annotates `seconds = t/1000`)

Passes the function as a keyword argument to draw_model (annotates `get_matrix=get_rotation_matrix)`)

Now, draw_model is passed the data required to transform the underlying teapot model over time, but we need to use it in the function's body. Before iterating over the teapot faces, we execute the appropriate matrix transformation:

```
def draw_model(faces, color_map=blues, light=(1,2,3),
               camera=Camera("default_camera",[]),
               glRotatefArgs=None,
               get_matrix=None):
    #...
    def do_matrix_transform(v):
        if get_matrix:
            m = get_matrix(pygame.time.get_ticks())
            return multiply_matrix_vector(m, v)
        else:
            return v
    transformed_faces = polygon_map(do_matrix_transform,
                                    faces)
    for face in transformed_faces:
    #...
```

Most of the function body is unchanged, so we don't print it here.

Creates a new function inside the main while loop that applies the matrix for this frame

Uses the elapsed milliseconds given by pygame.time.get_ticks() as well as the provided get_matrix function to compute a matrix for this frame

If no get_matrix is specified, doesn't carry out any transformation and returns the vector unchanged

Applies the function to every polygon using polygon_map

The rest of the draw_model is the same as described in appendix C.

With these changes, you can run the code and see the teapot rotate (figure 5.4).

Figure 5.4 The teapot is transformed by a new matrix in every frame, depending on the elapsed time when the frame is drawn.

Hopefully, I've convinced you with the preceding examples that matrices are entirely interchangeable with linear transformations. We've managed to transform and animate the teapot the same way, using only nine numbers to specify each transformation. You can practice your matrix skills some more in the following exercises and then I'll show you there's even more to learn from the matrix_multiply function we've already implemented.

5.1.6 Exercises

Exercise 5.1 Write a function `infer_matrix(n, transformation)` that takes a dimension (like 2 or 3) and a function that is a vector transformation assumed to be linear. It should return an n-by-n square matrix (an n-tuple of n-tuples of numbers, which is the matrix representing the linear transformation). Of course, the output is only meaningful if the input transformation is linear. Otherwise, it represents an entirely different function!

Solution

```
def infer_matrix(n, transformation):
    def standard_basis_vector(i):
        return tuple(1 if i==j else 0 for j in range(1,n+1))
    standard_basis = [standard_basis_vector(i) for i in range(1,n+1)]
    cols = [transformation(v) for v in standard_basis]
    return tuple(zip(*cols))
```

- Creates the ith standard basis vector as a tuple containing a one in the ith coordinate and zeroes in all other coordinates
- Creates the standard basis as a list of n vectors
- Defines the columns of a matrix to be the result of applying the corresponding linear transformation to the standard basis vectors
- Reshapes the matrix to be a tuple of rows instead of a list of columns, following our convention

We can test this on a linear transformation like `rotate_z_by(pi/2)`:

```
>>> from transforms import rotate_z_by
>>> from math import pi
>>> infer_matrix(3,rotate_z_by(pi/2))
((6.123233995736766e-17, -1.0, 0.0), (1.0, 1.2246467991473532e-16, 0.0),
(0, 0, 1))
```

Exercise 5.2 What is the result of the following product of a 2-by-2 matrix with a 2D vector?

$$\begin{pmatrix} 1.3 & 0.7 \\ 6.5 & 3.2 \end{pmatrix} \begin{pmatrix} -2.5 \\ 0.3 \end{pmatrix}$$

Solution The dot product of the vector with the first row of the matrix is $-2.5 \cdot 1.3 + 0.3 \cdot -0.7 = -3.46$. The dot product of the vector with the second row of the matrix is $-2.5 \cdot 6.5 + 0.3 \cdot 3.2 = -15.29$. These are the coordinates of the output vector, so the result is:

$$\begin{pmatrix} 1.3 & 0.7 \\ 6.5 & 3.2 \end{pmatrix} \begin{pmatrix} -2.5 \\ 0.3 \end{pmatrix} = \begin{pmatrix} -3.46 \\ -15.29 \end{pmatrix}$$

Exercise 5.3—Mini Project Write a `random_matrix` function that generates matrices of a specified size with random whole number entries. Use the function to generate five pairs of 3-by-3 matrices. Multiply each of the pairs together by hand (for practice) and then check your work with the `matrix_multiply` function.

Solution First, we give the `random_matrix` function arguments to specify the number of rows, the number of columns, and the minimum and maximum values for entries:

```
from random import randint
def random_matrix(rows,cols,min=-2,max=2):
    return tuple(
        tuple(
        randint(min,max) for j in range(0,cols))
        for i in range(0,rows)
    )
```

Next, we can generate a random 3-by-3 matrix with entries between 0 and 10 as follows:

```
>>> random_matrix(3,3,0,10)
((3, 4, 9), (7, 10, 2), (0, 7, 4))
```

Exercise 5.4 For each of your pairs of matrices from the previous exercise, multiply them in the opposite order. Do you get the same result?

Solution Unless you get very lucky, your results will all be different. Most pairs of matrices give different results when multiplied in different orders. In math jargon, we say an operation is *commutative* if it gives the same result regardless of the order of inputs. For instance, multiplying numbers is a commutative operation because $xy = yx$ for any choice of numbers x and y. However, matrix multiplication is *not* commutative because for two square matrices A and B, AB does not always equal BA.

Exercise 5.5 In either 2D or 3D, there is a boring but important vector transformation called the *identity transformation* that takes in a vector and returns the same vector as output. This transformation is linear because it takes any input vector sum, scalar multiple, or linear combination and returns the same thing as output. What are the matrices representing the identity transformation in 2D and 3D, respectively?

Solution In 2D or 3D, the identity transformation acts on the standard basis vectors and leaves them unchanged. Therefore, in either dimension, the matrix for this transformation has the standard basis vectors as its columns. In 2D and 3D, these *identity matrices* are denoted by I_2 and I_3, respectively, and look like this:

$$I_2 = \begin{pmatrix} 1 & 0 \\ 0 & 1 \end{pmatrix} \quad I_3 = \begin{pmatrix} 1 & 0 & 0 \\ 0 & 1 & 0 \\ 0 & 0 & 1 \end{pmatrix}$$

Exercise 5.6 Apply the matrix `((2,1,1),(1,2,1),(1,1,2))` to all the vectors defining the teapot. What happens to the teapot and why?

Solution The following function is included in the source file matrix_transform_teapot.py:

```
def transform(v):
    m = ((2,1,1),(1,2,1),(1,1,2))
    return multiply_matrix_vector(m,v)

draw_model(polygon_map(transform, load_triangles()))
```

Running the code, we see that the front of the teapot is stretched out into the region where x, y, and z are all positive.

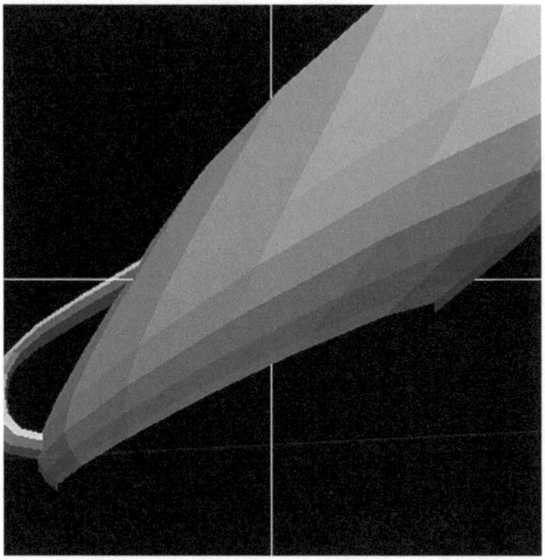

Applying the given matrix to all vertices of the teapot

(continued)

This is because all of the standard basis vectors are transformed to vectors with positive coordinates: (2, 1, 1), (1, 2, 1), and (1, 1, 2), respectively.

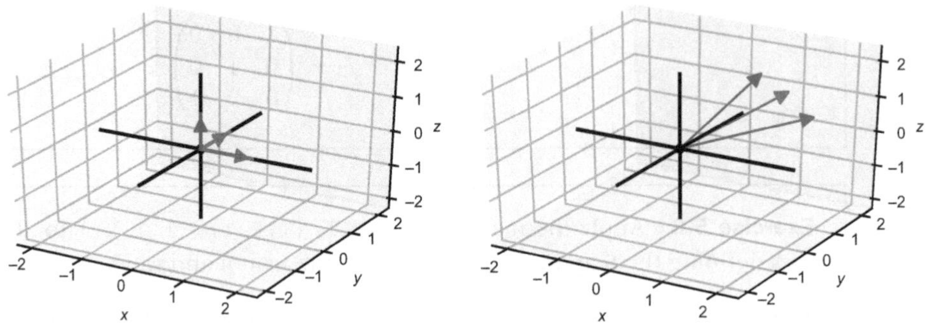

How the linear transformation defined by this matrix affects the standard basis vectors.

A linear combination of these new vectors with positive scalars is stretched further in the +*x*, +*y*, and +*z* directions than the same linear combination of the standard basis.

Exercise 5.7 Implement `multiply_matrix_vector` in a different way by using two nested comprehensions: one traversing the rows of the matrix and one traversing the entries of each row.

Solution

```
def multiply_matrix_vector(matrix,vector):
    return tuple(
        sum(vector_entry * matrix_entry
            for vector_entry, matrix_entry in zip(row,vector))
        for row in matrix
    )
```

Exercise 5.8 Implement `multiply_matrix_vector` yet another way using the fact that the output coordinates are the dot products of the input matrix rows with the input vector.

Solution This is a simplified version of the previous exercise solution:

```
def multiply_matrix_vector(matrix,vector):
    return tuple(
        dot(row,vector)
        for row in matrix
    )
```

Exercise 5.9—Mini Project I first told you what a linear transformation was and then showed you that any linear transformation can be represented by a matrix. Let's prove the converse fact now: all matrices represent linear transformations. Starting with the explicit formulas for multiplying a 2D vector by a 2-by-2 matrix or multiplying a 3D vector by a 3-by-3 matrix, prove that algebraically. That is, show that matrix multiplication preserves sums and scalar multiples.

Solution I'll show the proof for 2D; the 3D proof has the same structure but with a bit more writing. Suppose we have a 2-by-2 matrix called A with any four numbers a, b, c, and d as its entries. Let's see how A operates on two vectors **u** and **v**:

$$A = \begin{pmatrix} a & b \\ c & d \end{pmatrix} \quad \mathbf{u} = \begin{pmatrix} u_1 \\ u_2 \end{pmatrix} \quad \mathbf{v} = \begin{pmatrix} v_1 \\ v_2 \end{pmatrix}$$

You can do the matrix multiplications explicitly to find $A\mathbf{u}$ and $A\mathbf{v}$:

$$A\mathbf{u} = \begin{pmatrix} a & b \\ c & d \end{pmatrix} \begin{pmatrix} u_1 \\ u_2 \end{pmatrix} = \begin{pmatrix} au_1 + bu_2 \\ cu_1 + du_2 \end{pmatrix}$$

$$A\mathbf{v} = \begin{pmatrix} a & b \\ c & d \end{pmatrix} \begin{pmatrix} v_1 \\ v_2 \end{pmatrix} = \begin{pmatrix} av_1 + bv_2 \\ cv_1 + dv_2 \end{pmatrix}$$

And then we can compute $A\mathbf{u} + A\mathbf{v}$ and $A(\mathbf{u} + \mathbf{v})$ and see that the results match:

$$A\mathbf{u} + A\mathbf{v} = \begin{pmatrix} au_1 + bu_2 \\ cu_1 + du_2 \end{pmatrix} + \begin{pmatrix} av_1 + bv_2 \\ cv_1 + dv_2 \end{pmatrix} = \begin{pmatrix} au_1 + av_1 + bu_2 + bv_2 \\ cu_1 + cv_1 + du_2 + dv_2 \end{pmatrix}$$

$$A(\mathbf{u}+\mathbf{v}) = \begin{pmatrix} a & b \\ c & d \end{pmatrix} \begin{pmatrix} u_1 + v_1 \\ u_2 + v_2 \end{pmatrix} = \begin{pmatrix} a(u_1 + v_1) + b(u_2 + v_2) \\ c(u_1 + v_1) + d(u_2 + v_2) \end{pmatrix} = \begin{pmatrix} au_1 + av_1 + bu_2 + bv_2 \\ cu_1 + cv_1 + du_2 + dv_2 \end{pmatrix}$$

This tells us that the 2D vector transformation defined by multiplying *any* 2-by-2 matrix preserves vector sums. Likewise, for any number s, we have

$$s\mathbf{v} = \begin{pmatrix} sv_1 \\ sv_2 \end{pmatrix}$$

$$s(A\mathbf{v}) = \begin{pmatrix} s(av_1 + bv_2) \\ s(cv_1 + dv_2) \end{pmatrix} = \begin{pmatrix} sav_1 + sbv_2 \\ scv_1 + sdv_2 \end{pmatrix}$$

$$A(s\mathbf{v}) = \begin{pmatrix} a(sv_1) + b(sv_2) \\ c(sv_1) + d(sv_2) \end{pmatrix} = \begin{pmatrix} sav_1 + sbv_2 \\ scv_1 + sdv_2 \end{pmatrix}$$

(continued)

So $s \cdot (A\mathbf{v})$ and $A(s\mathbf{v})$ give the same results, and we see that multiplying by the matrix A preserves scalar multiples as well. These two facts mean that multiplying by any 2-by-2 matrix is a linear transformation of 2D vectors.

Exercise 5.10 Once again, let's use the two matrices from section 5.1.3:

$$A = \begin{pmatrix} 1 & 1 & 0 \\ 1 & 0 & 1 \\ 1 & -1 & 1 \end{pmatrix}, \quad B = \begin{pmatrix} 0 & 2 & 1 \\ 0 & 1 & 0 \\ 1 & 0 & -1 \end{pmatrix}$$

Write a function `compose_a_b` that executes the composition of the linear transformation for A and the linear transformation for B. Then use the `infer_matrix` function from a previous exercise in this section to show that `infer_matrix(3, compose_a_b)` is the same as the matrix product AB.

Solution First, we implement two functions `transform_a` and `transform_b` that do the linear transformations defined by the matrices A and B. Then, we combine these using our `compose` function:

```
from transforms import compose

a = ((1,1,0),(1,0,1),(1,-1,1))
b = ((0,2,1),(0,1,0),(1,0,-1))

def transform_a(v):
    return multiply_matrix_vector(a,v)

def transform_b(v):
    return multiply_matrix_vector(b,v)

compose_a_b = compose(transform_a, transform_b)
```

Now we can use our `infer_matrix` function to find the matrix corresponding to this composition of linear transformations and compare it to the matrix product AB:

```
>>> infer_matrix(3, compose_a_b)
((0, 3, 1), (1, 2, 0), (1, 1, 0))
>>> matrix_multiply(a,b)
((0, 3, 1), (1, 2, 0), (1, 1, 0))
```

Exercise 5.11—Mini Project Find two, 2-by-2 matrices, neither of which is the identity matrix I_2, but whose product *is* the identity matrix.

Solution One way to do this is to write two matrices and play with their entries until you get the identity matrix as a product. Another way is to think of the problem in terms of linear transformations. If two matrices multiplied together produce the identity matrix, then the composition of their corresponding linear transformations should produce the identity transformation.

With that in mind, what are two 2D linear transformations whose composition is the identity transformation? When applied in sequence to a given 2D vector, these linear transformations should return the original vector as output. One such pair of transformations is rotation by 90° clockwise, then rotation by 270° clockwise. Applying both of these executes a 360° rotation that brings any vector back to its original position. The matrices for a 270° rotation and a 90° rotation are as follows, and their product is the identity matrix:

$$\begin{pmatrix} 0 & 1 \\ -1 & 0 \end{pmatrix} \begin{pmatrix} 0 & -1 \\ 1 & 0 \end{pmatrix} = \begin{pmatrix} 1 & 0 \\ 0 & 1 \end{pmatrix}$$

Exercise 5.12 We can multiply a square matrix by itself any number of times. We can then think of successive matrix multiplications as "raising a matrix to a power." For a square matrix A, we can write AA as A^2; we can write AAA as A^3; and so on. Write a `matrix_power(power,matrix)` function that raises a matrix to the specified (whole number) power.

Solution Here is an implementation that works for whole number powers greater than or equal to 1:

```
def matrix_power(power,matrix):
    result = matrix
    for _ in range(1,power):
        result = matrix_multiply(result,matrix)
    return result
```

5.2 *Interpreting matrices of different shapes*

The `matrix_multiply` function doesn't hard-code the size of the input matrices, so we can use it to multiply either 2-by-2 or 3-by-3 matrices together. As it turns out, it can also handle matrices of other sizes as well. For instance, it can handle these two 5-by-5 matrices:

```
>>> a = ((-1, 0, -1, -2, -2), (0, 0, 2, -2, 1), (-2, -1, -2, 0, 1), (0, 2, -2,
-1, 0), (1, 1, -1, -1, 0))
```

```
>>> b = ((-1, 0, -1, -2, -2), (0, 0, 2, -2, 1), (-2, -1, -2, 0, 1), (0, 2, -2,
-1, 0), (1, 1, -1, -1, 0))
>>> matrix_multiply(a,b)
((-10, -1, 2, -7, 4), (-2, 5, 5, 4, -6), (-1, 1, -4, 2, -2), (-4, -5, -5, -9,
4), (-1, -2, -2, -6, 4))
```

There's no reason we shouldn't take this result seriously—our functions for vector addition, scalar multiplication, dot products, and, therefore, matrix multiplication don't depend on the dimension of the vectors we use. Even though we can't picture a 5D vector, we can do all the same algebra on five tuples of numbers that we did on pairs and triples of numbers in 2D and 3D, respectively. In this 5D product, the entries of the resulting matrix are still dot products of rows of the first matrix with columns of the second (figure 5.5):

Fourth column, second row

Fourth column

Second row

$$\begin{pmatrix} -1 & 0 & -1 & -1 & -1 \\ 0 & 0 & 2 & -2 & 1 \\ -2 & -1 & -2 & 0 & 1 \\ 0 & 2 & -2 & -1 & 0 \\ 1 & 1 & -1 & -1 & 0 \end{pmatrix} \begin{pmatrix} 2 & 0 & 0 & -1 & 2 \\ -1 & -2 & -1 & -2 & 0 \\ 0 & 1 & 2 & 2 & -2 \\ 2 & -1 & -1 & 1 & 0 \\ 2 & 1 & -1 & 2 & -2 \end{pmatrix} = \begin{pmatrix} -10 & -1 & 2 & -7 & 4 \\ -2 & 5 & 5 & 4 & -6 \\ -1 & 1 & -4 & 2 & -2 \\ -4 & -5 & -5 & -9 & 4 \\ -1 & -2 & -2 & -6 & 4 \end{pmatrix}$$

$$(0, 0, 2, -2, 1) \cdot (-1, -2, 2, 1, 2) = 4$$

Figure 5.5 The dot product of a row of the first matrix with a column of the second matrix produces one entry of the matrix product.

You can't visualize it in the same way, but you can show algebraically that a 5-by-5 matrix specifies a linear transformation of 5D vectors. We spend time talking about what kind of objects live in four, five, or more dimensions in the next chapter.

5.2.1 *Column vectors as matrices*

Let's return to the example of multiplying a matrix by a column vector. I already showed you how to do a multiplication like this, but we treated it as its own case with the `multiply_matrix_vector` function. It turns out `matrix_multiply` is capable of doing these products as well, but we have to write the column vector as a matrix. As an example, let's pass the following square matrix and single-column matrix to our `matrix_multiply` function:

$$C = \begin{pmatrix} -1 & -1 & 0 \\ -2 & 1 & 2 \\ 1 & 0 & -1 \end{pmatrix}, \quad D = \begin{pmatrix} 1 \\ 1 \\ 1 \end{pmatrix}$$

I claimed before that you can think of a vector and a single-column matrix interchangeably, so we might encode d as a vector `(1,1,1)`. But this time, let's force ourselves to think of it as a matrix, having three rows with one entry each. Note that we have to write (1,) instead of (1) to make Python think of it as a 1-tuple instead of as a number.

```
>>> c = ((-1, -1, 0), (-2, 1, 2), (1, 0, -1))
>>> d = ((1,),(1,),(1,))
>>> matrix_multiply(c,d)
((-2,), (1,), (0,))
```

The result has three rows with one entry each, so it is a single-column matrix as well. Here's what this product looks like in matrix notation:

$$\begin{pmatrix} -1 & -1 & 0 \\ -2 & 1 & 2 \\ 1 & 0 & -1 \end{pmatrix} \begin{pmatrix} 1 \\ 1 \\ 1 \end{pmatrix} = \begin{pmatrix} -2 \\ 1 \\ 0 \end{pmatrix}$$

Our `multiply_matrix_vector` function can evaluate the same product but in a different format:

```
>>> multiply_matrix_vector(c,(1,1,1))
(-2, 1, 0)
```

This demonstrates that multiplying a matrix and a column vector is a special case of matrix multiplication. We don't need a separate function `multiply_matrix_vector` after all. We can further see that the entries of the output are dot products of the rows of the first matrix with the single column of the second (figure 5.6).

$$\begin{pmatrix} -1 & -1 & 0 \\ -2 & 1 & 2 \\ 1 & 0 & -1 \end{pmatrix} \begin{pmatrix} 1 \\ 1 \\ 1 \end{pmatrix} = \begin{pmatrix} -2 \\ 1 \\ 0 \end{pmatrix}$$

Figure 5.6 An entry of the resulting vector computed as a dot product

On paper, you'll see vectors represented interchangeably as tuples (with commas) or as column vectors. But for the Python functions we've written, the distinction is critical. The tuple (–2, 1, 0) can't be used interchangeably with the tuple-of-tuples ((–2,), (1,), (0,)). Yet another way of writing the same vector would be as a *row vector*, or a matrix with one row. Here are the three notations for comparison:

Table 5.1 Comparison of mathematical notations for vectors with corresponding Python representations

Representation	In math notation	In Python
Ordered triple (ordered tuple)	$\mathbf{v} = (-2, 1, 0)$	`v = (-2,1,0)`
Column vector	$\mathbf{v} = \begin{pmatrix} -2 \\ 1 \\ 0 \end{pmatrix}$	`v = ((-2,),(1,),(0,))`
Row vector	$\mathbf{v} = (-2, 1, 0)$	`v = ((-2,1,0),)`

If you've seen this comparison in math class, you may have thought it was a pedantic notational distinction. Once we represent these in Python, however, we see that they are really three distinct objects that need to be treated differently. While these all represent the same geometric data, which is a 3D arrow or point in space, only one of these, the column vector, can be multiplied by a 3-by-3 matrix. The row vector doesn't work because, as shown in figure 5.7, we can't take the dot product of a row of the first matrix with a column of the second.

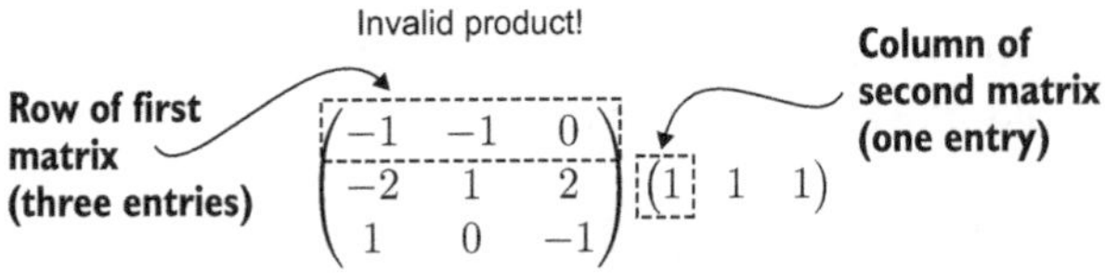

Figure 5.7 Two matrices that cannot be multiplied together

For our definition of matrix multiplication to be consistent, we can only multiply a matrix on the left of a *column* vector. This prompts the general question posed by the next section.

5.2.2 *What pairs of matrices can be multiplied?*

We can make grids of numbers of any dimension. When can our matrix multiplication formula work, and what does it mean when it does?

The answer is that the number of columns of the first matrix has to match the number of rows of the second. This is clear when we do the matrix multiplication in terms of dot products. For instance, we can multiply any matrix with three columns by a second matrix with three rows. This means that rows of the first matrix and columns of the second each have three entries, so we can take their dot products. Figure 5.8 shows the dot product of the first row of the first matrix with the first column of the second matrix gives us an entry of the product matrix.

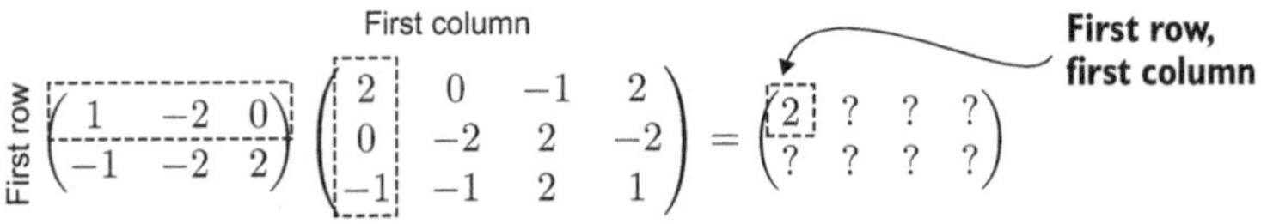

Figure 5.8 Finding the first entry of the product matrix

We can complete this matrix product by taking the remaining seven dot products. Figure 5.9 shows another entry, computed from a dot product.

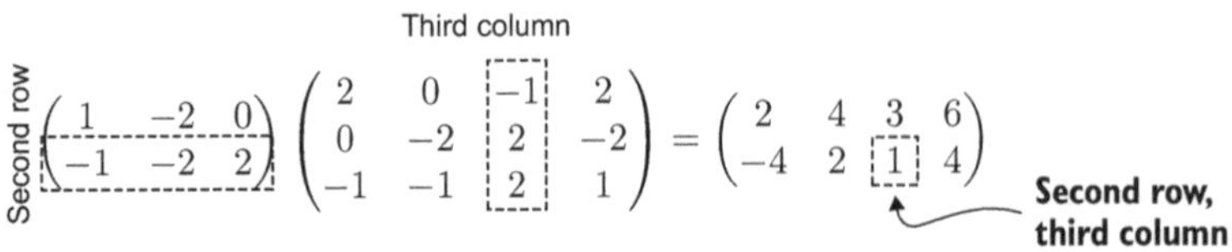

Figure 5.9 Finding another entry of the product matrix

This constraint also makes sense in terms of our original definition of matrix multiplication: the columns of the output are each linear combinations of the columns of the first matrix with scalars given by a row of the second matrix (figure 5.10).

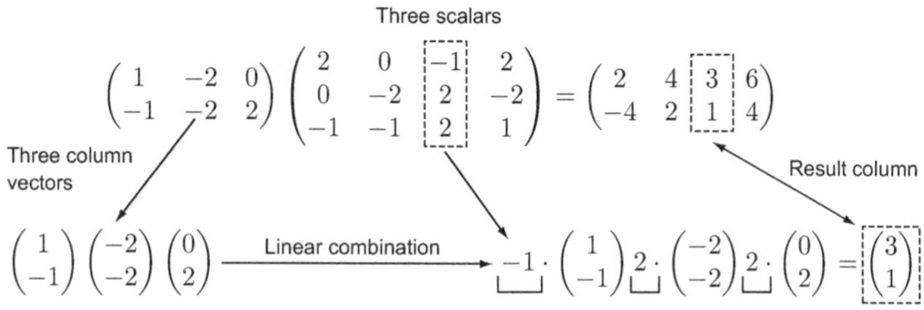

Figure 5.10 Each column of the result is a linear combination of the columns of the first matrix.

I was calling the previous square matrices 2-by-2 and 3-by-3 matrices. The last example (figure 5.10) was the product of a 2-by-3 and a 3-by-4 matrix. When we describe the *dimensions* of a matrix like this, we say the number of rows first and then the number of columns. For instance, a 3D column vector would be a 3-by-1 matrix.

> **NOTE** Sometimes you'll see matrix dimensions written with a multiplication sign as in a 3×3 matrix or a 3×1 matrix.

In this language, we can make a general statement about the shapes of matrices that can be multiplied: you can only multiply an n-by-m matrix by a p-by-q matrix if $m = p$. When that is true, the resulting matrix will be a n-by-q matrix. For instance, a 17×9 matrix cannot be multiplied by a 6×11 matrix. However, a 5×8 matrix can be multiplied by an 8×10 matrix. Figure 5.11 shows the result of the latter, which is a 5×10 matrix.

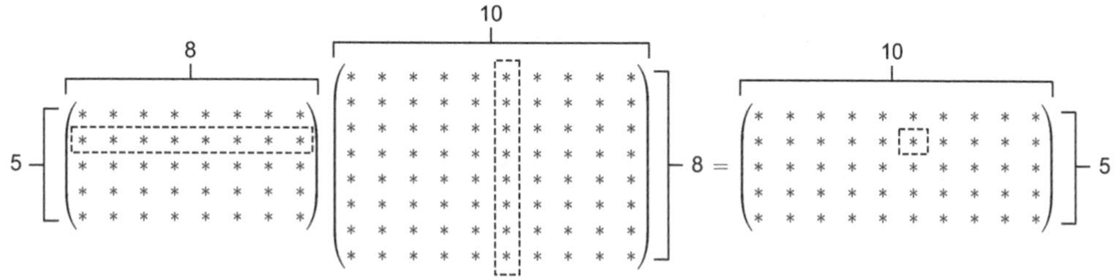

Figure 5.11 Each of the five rows of the first matrix can be paired with one of the ten columns of the second matrix to produce one of the 5 × 10 = 50 entries of the product matrix. I used stars instead of numbers to show you that *any* matrices of these sizes are compatible.

By contrast, you couldn't multiply these matrices in the opposite order: a 10×8 matrix can't be multiplied by a 5×8 matrix. Now it's clear how to multiply bigger matrices, but what do the results mean? It turns out we can learn something from the result: *all* matrices represent vector functions, and all valid matrix products can be interpreted as composition of these functions. Let's see how this works.

5.2.3 Viewing square and non-square matrices as vector functions

We can think of a 2×2 matrix as the data required to do a given linear transformation of a 2D vector. Pictured as a machine in figure 5.12, this transformation takes a 2D vector into its input slot and produces a 2D vector out of its output slot as a result.

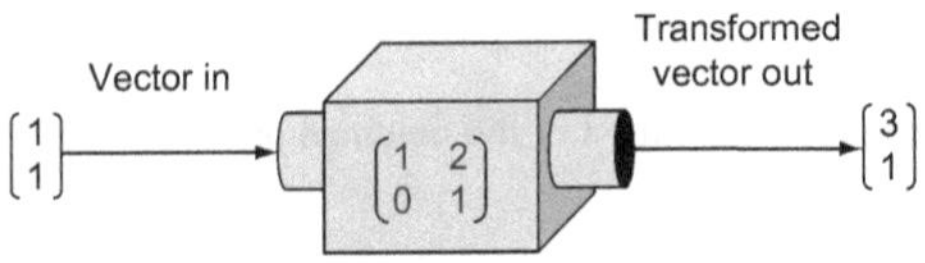

Figure 5.12 Visualizing a matrix as a machine that takes vectors as inputs and produces vectors as outputs

Under the hood, our machine does this matrix multiplication:

$$\begin{pmatrix}1 & 2\\0 & 1\end{pmatrix}\begin{pmatrix}1\\1\end{pmatrix} = \begin{pmatrix}3\\1\end{pmatrix}$$

It's fair to think of matrices as machines that take vectors as inputs and produce vectors as outputs. Figure 5.13, however, shows a matrix can't take just any vector as input; it is a 2×2 matrix so it does a linear transformation of 2D vectors. Correspondingly, this matrix can only be multiplied by a column vector with two entries. Let's split up the machine's input and output slots to suggest that these take and produce 2D vectors or pairs of numbers.

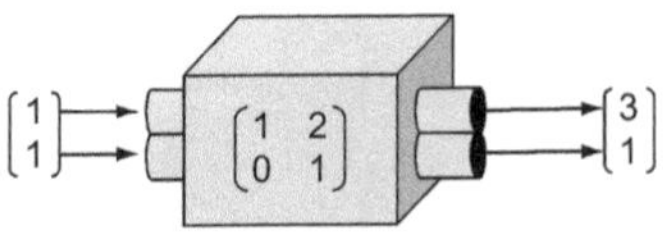

Figure 5.13 Refining our mental model by redrawing the machine's input and output slots to indicate that its inputs and outputs are pairs of numbers

Likewise, a linear transformation machine (figure 5.14) powered by a 3×3 matrix can only take in 3D vectors and produce 3D vectors as a result.

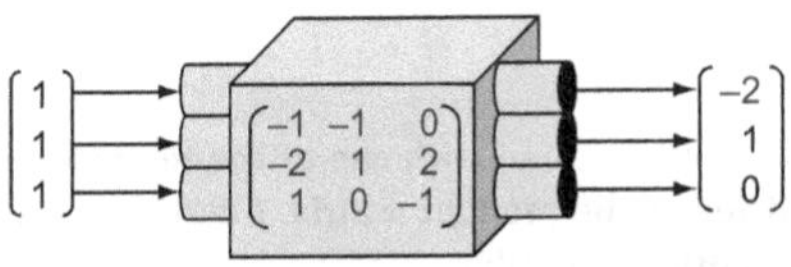

Figure 5.14 A linear transformation machine powered by a 3×3 matrix takes in 3D vectors and outputs 3D vectors.

Now we can ask ourselves, what would a machine look like if it were powered by a non-square matrix? Perhaps the matrix would look something like this:

$$\begin{pmatrix} -2 & -1 & -1 \\ 2 & -2 & 1 \end{pmatrix}$$

As a specific example, what kinds of vectors could this 2×3 matrix act on? If we're going to multiply this matrix with a column vector, the column vector must have three entries to match the size of the rows of this matrix. Multiplying our 2×3 matrix by a 3×1 column vector gives us a 2×1 matrix as a result, or a 2D column vector. For example,

$$\begin{pmatrix} -2 & -1 & -1 \\ 2 & -2 & 1 \end{pmatrix} \begin{pmatrix} 0 \\ -1 \\ 1 \end{pmatrix} = \begin{pmatrix} 0 \\ 3 \end{pmatrix}$$

This tells us that this 2×3 matrix represents a function taking 3D vectors to 2D vectors. If we were to draw it as a machine, like in figure 5.15, it would accept 3D vectors in its input slot and produce 2D vectors from its output slot.

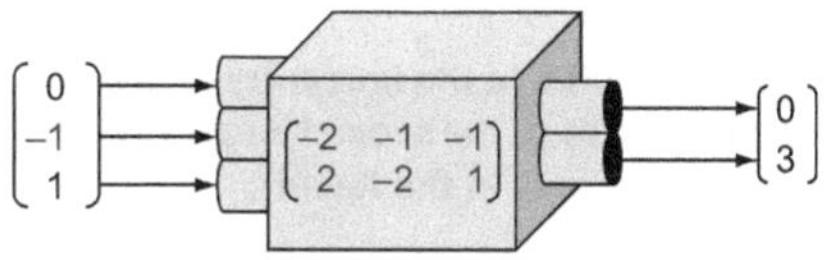

Figure 5.15 A machine that takes in 3D vectors and outputs 2D vectors, powered by a 2×3 matrix

In general, an m-by-n matrix defines a function taking n-dimensional vectors as inputs and returning m-dimensional vectors as outputs. Any such function is linear in the sense that it preserves vector sums and scalar multiples. It's not a transformation because it doesn't just modify input, it returns an entirely different kind of output: a vector living in a different number of dimensions. For this reason, we'll use a more general terminology; we'll call it a *linear function* or a *linear map*. Let's consider an in-depth example of a familiar linear map from 3D to 2D.

5.2.4 Projection as a linear map from 3D to 2D

We already saw a vector function that accepts 3D vectors and produces 2D vectors: a projection of a 3D vector onto the x,y plane (section 3.5.2). This transformation (we can call it P) takes vectors of the form (x, y, z) and returns these with their z component deleted: (x, y). I'll spend some time carefully showing why this is a linear map and how it preserves vector addition and scalar multiplication.

First of all, let's write P as a matrix. To accept 3D vectors and return 2D vectors, it should be a 2×3 matrix. Let's follow our trusty formula for finding a matrix by testing the action of P on standard basis vectors. Remember, in 3D the standard basis vectors are defined as $\mathbf{e}_1 = (1, 0, 0)$, $\mathbf{e}_2 = (0, 1, 0)$, and $\mathbf{e}_3 = (0, 0, 1)$, and when we apply the projection to these three vectors, we get (1, 0), (0, 1), and (0, 0), respectively. We can write these as column vectors

$$P(\mathbf{e}_1) = \begin{pmatrix} 1 \\ 0 \end{pmatrix} \quad P(\mathbf{e}_2) = \begin{pmatrix} 0 \\ 1 \end{pmatrix}, \quad P(\mathbf{e}_3) = \begin{pmatrix} 0 \\ 0 \end{pmatrix}$$

and then stick them together side by side to get a matrix:

$$\begin{pmatrix} 1 & 0 & 0 \\ 0 & 1 & 0 \end{pmatrix}$$

To check this, let's multiply it by a test vector (a, b, c). The dot product of (a, b, c) with (1, 0, 0) is a, so that's the first entry of the result. The second entry is the dot product of (a, b, c) with (0, 1, 0), or b. You can picture this matrix as grabbing a and b from (a, b, c) and ignoring c (figure 5.16).

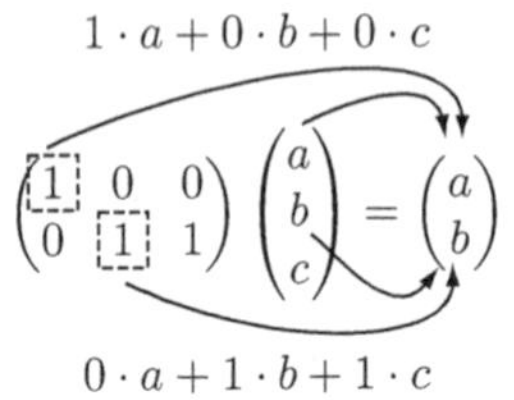

Figure 5.16 Only 1 · *a* contributes to the first entry of the product, and only 1 · *b* contributes to the second entry. The other entries are zeroed out (in gray in the figure).

This matrix does what we want; it deletes the third coordinate of a 3D vector, leaving us with only the first two coordinates. It's good news that we can write this projection as a matrix, but let's also give an algebraic *proof* that this is a linear map. To do this, we have to show that the two key conditions of linearity are satisfied.

PROVING THAT PROJECTION PRESERVES VECTOR SUMS

If P is linear, any vector sum $\mathbf{u} + \mathbf{v} = \mathbf{w}$ should be respected by P. That is, $P(\mathbf{u}) + P(\mathbf{v})$ should equal $P(\mathbf{w})$ as well. Let's confirm this using these equations: $\mathbf{u} = (u_1, u_2, u_3)$ and $\mathbf{v} = (v_1, v_2, v_3)$. Then $\mathbf{w} = \mathbf{u} + \mathbf{v}$ so that

$$\mathbf{w} = (u_1 + v_1, u_2 + v_2, u_3 + v_3)$$

Executing P on all of these vectors is simple because we only need to remove the third coordinate:

$$P(\mathbf{u}) = (u_1, u_2)$$
$$P(\mathbf{v}) = (v_1, v_2)$$

so

$$P(\mathbf{w}) = (u_1 + v_1, u_2 + v_2)$$

Adding $P(\mathbf{u})$ and $P(\mathbf{v})$, we get $(u_1 + v_1, u_2 + v_2)$, which is the same as $P(\mathbf{w})$. For any three 3D vectors $\mathbf{u} + \mathbf{v} = \mathbf{w}$, we, therefore, also have $P(\mathbf{u}) + P(\mathbf{v}) = P(\mathbf{w})$. This validates our first condition.

Proving projection preserves scalar multiples

The second thing we need to show is that P preserves scalar multiples. Letting s stand for *any* real number and letting $\mathbf{u} = (u_1, u_2, u_3)$, we want to demonstrate that $P(s\mathbf{u})$ is the same as $sP(\mathbf{u})$.

Deleting the third coordinate and doing the scalar multiplication give the same result regardless of which order these operations are carried out. The result of $s\mathbf{u}$ is (su_1, su_2, su_3), so $P(s\mathbf{u}) = (su_1, su_2)$. The result of $P(\mathbf{u})$ is $(\mathbf{u}_1, \mathbf{u}_2)$, so $sP(u) = (su_1, su_2)$. This validates the second condition and confirms that P satisfies the definition of linearity.

These kinds of proofs are usually easier to do than to follow, so I've given you another one as an exercise. In the exercise, you can check that a function from 2D to 3D, specified by a given matrix, is linear using the same approach.

More illustrative than an algebraic proof is an example. What does it look like when we project a 3D vector sum down to 2D? We can see it in three steps. First, we can draw a vector sum of two vectors **u** and **v** in 3D as shown in figure 5.17.

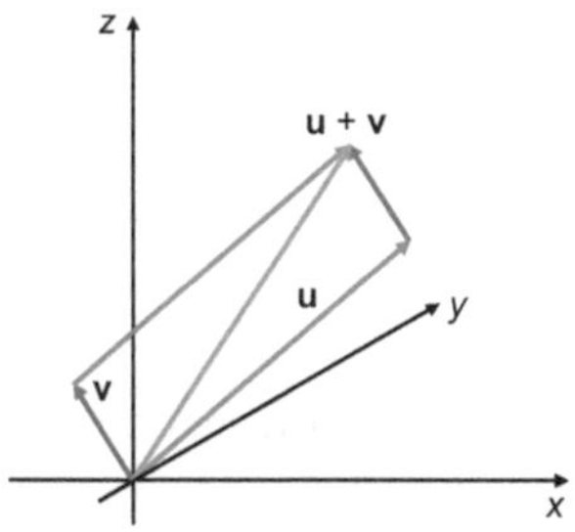

Figure 5.17 A vector sum of two arbitrary vectors u and v in 3D

Then, we can trace a line from each vector to the *x,y* plane to show where these vectors end up after projection (figure 5.18).

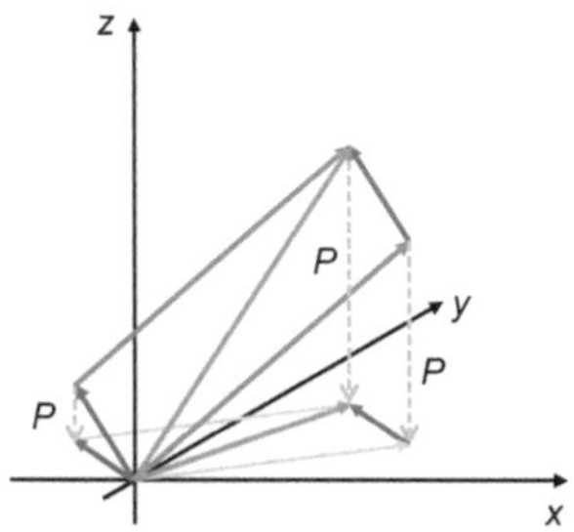

Figure 5.18 Visualizing where u, v, and u + v end up after projection to the *x,y* plane

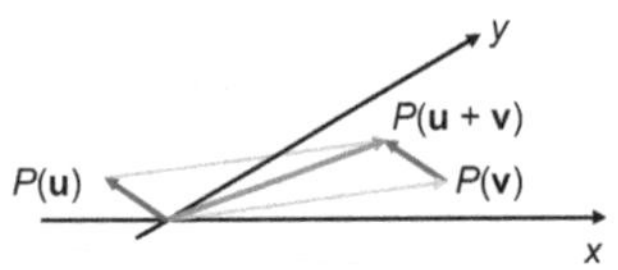

Figure 5.19 The projected vectors form a sum: P(v) + P(v) = P(u + v).

Finally, we can look at these new vectors and see that they *still* constitute a vector sum (figure 5.19).

In other words, if three vectors **u**, **v**, and **w** form a vector sum **u** + **v** = **w**, then their "shadows" in the *x*,*y* plane also form a vector sum. Now that you've got some insight into linear transformation from 3D to 2D and a matrix that represents it, let's return to our discussion of linear maps in general.

5.2.5 Composing linear maps

The beauty of matrices is that they store all of the data required to evaluate a linear function on a given vector. What's more, the dimensions of a matrix tell us the dimensions of input vectors and output vectors for the underlying function. We captured that visually in figure 5.20 by drawing machines for matrices of varying dimensions, whose input and output slots have different shapes. Here are four examples we've seen, labeled with letters so we can refer back to them.

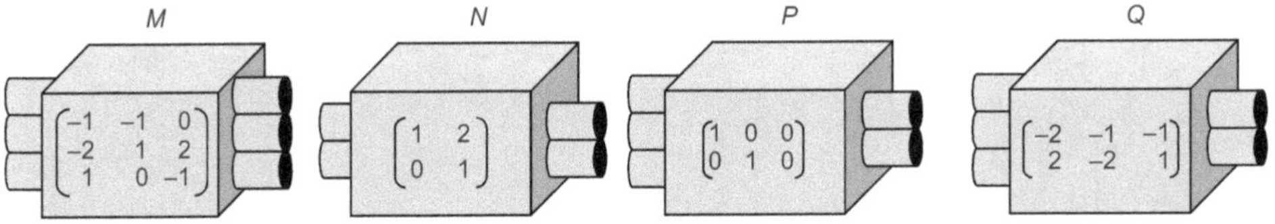

Figure 5.20 Four linear functions represented as machines with input and output slots. The shape of a slot tells us what dimension of vector it accepts or produces.

Drawn like this, it's easy to pick out which pairs of linear function machines could be welded together to build a new one. For instance, the output slot of *M* has the same shape as the input slot of *P*, so we could make the composition *P*(*M*(**v**)) for a 3D vector **v**. The output of *M* is a 3D vector that can be passed right along into the input slot of *P* (figure 5.21).

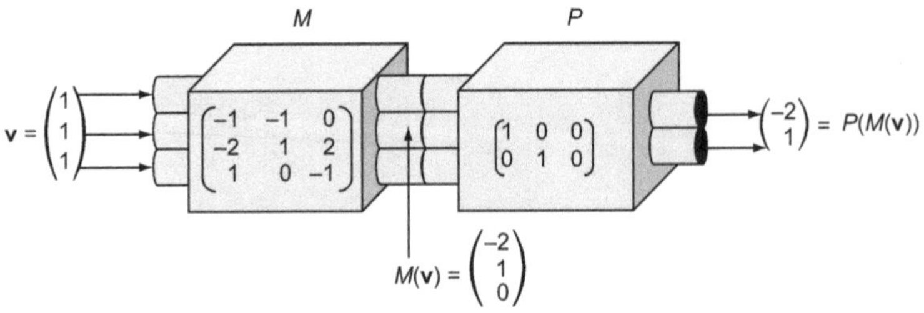

Figure 5.21 The composition of *P* and *M*. A vector is passed into the input slot of *M*, the output *M*(v) passes invisibly through the plumbing and into *P*, and the output *P*(*M*(v)) emerges from the other end.

By contrast, figure 5.22 shows that we can't compose *N* and *M* because *N* doesn't have enough output slots to fill every input of *M*.

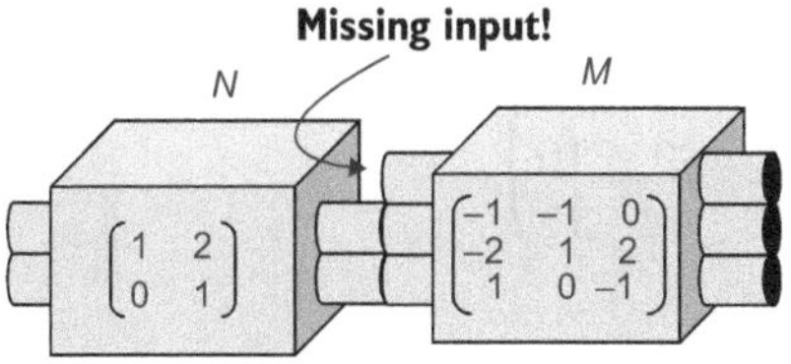

Figure 5.22 The composition of *N* and *M* is not possible because outputs of *N* are 2D vectors, while inputs to *M* are 3D vectors.

I'm making this idea visual now by talking about slots, but hidden underneath is the same reasoning we use to decide if two matrices can be multiplied together. The count of the columns of the first matrix has to match the count of rows of the second. When the dimensions match in this way, so do the slots, and we can compose the linear functions and multiply their matrices.

Thinking of *P* and *M* as matrices, the composition of *P* and *M* is written *PM* as a matrix product. (Remember, if *PM* acts on a vector **v** as *PM***v**, *M* is applied first and then *P*.) When **v** = (1, 1, 1), the product *PM***v** is a product of two matrices and a column vector, and it can be simplified into a single matrix times a column vector if we evaluate *PM* (figure 5.23).

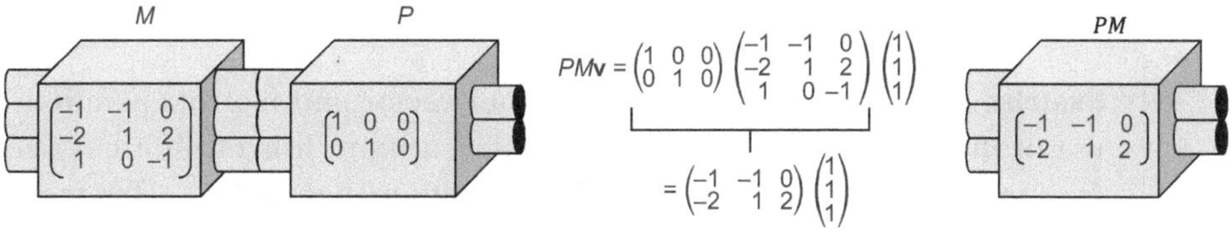

Figure 5.23 Applying *M* and then *P* is equivalent to applying the composition *PM*. We consolidate the composition into a single matrix by doing the matrix multiplication.

As a programmer, you're used to thinking of functions in terms of the types of data they consume and produce. I've given you a lot of notation and terminology to digest thus far in this chapter, but as long as you grasp this core concept, you'll get the hang of it eventually.

I strongly encourage you to work through the following exercises to make sure you understand the language of matrices. For the rest of this chapter and the next, there won't be many big new concepts, only applications of what we've seen so far. These applications will give you even more practice with matrix and vector computations.

5.2.6 Exercises

Exercise 5.13 What are the dimensions of this matrix?

$$\begin{pmatrix} 1 & 2 & 3 & 4 & 5 \\ 6 & 7 & 8 & 9 & 10 \\ 11 & 12 & 13 & 14 & 15 \end{pmatrix}$$

1. 5×3
2. 3×5

Solution This is a 3×5 matrix because it has three rows and five columns.

Exercise 5.14 What are the dimensions of a 2D column vector considered as a matrix? What about a 2D row vector? A 3D column vector? A 3D row vector?

Solution A 2D column vector has two rows and one column, so it is a 2×1 matrix. A 2D row vector has one row with two columns, so it is a 1×2 matrix. Likewise, a 3D column and row vector have the dimensions 3×1 and 1×3 as matrices, respectively.

Exercise 5.15—Mini Project Many of our vector and matrix operations make use of the Python `zip` function. When given input lists of different sizes, this function truncates the longer of the two rather than failing. This means that when we pass invalid inputs, we get meaningless results back. For instance, there is no such thing as a dot product between a 2D vector and a 3D vector, but our `dot` function returns something anyway:

```
>>> from vectors import dot
>>> dot((1,1),(1,1,1))
2
```

Add safeguards to all of the vector arithmetic functions so that they throw exceptions rather than returning values for vectors of invalid sizes. Once you've done that, show that `matrix_multiply` no longer accepts a product of a 3×2 and a 4×5 matrix.

Exercise 5.16 Which of the following are valid matrix products? For those that are valid, what dimension is the product matrix?

A.

$$\begin{pmatrix} 10 & 0 \\ 3 & 4 \end{pmatrix} \begin{pmatrix} 8 & 2 & 3 & 6 \\ 7 & 8 & 9 & 4 \\ 5 & 7 & 0 & 9 \\ 3 & 3 & 0 & 2 \end{pmatrix}$$

B.

$$\begin{pmatrix} 0 & 2 & 1 & -2 \\ -2 & 1 & -2 & -1 \end{pmatrix} \begin{pmatrix} -3 & -5 \\ 1 & -4 \\ -4 & -4 \\ -2 & -4 \end{pmatrix}$$

C.

$$\begin{pmatrix} 1 \\ 3 \\ 0 \end{pmatrix} \begin{pmatrix} 3 & 3 & 5 & 1 & 3 & 0 & 5 & 1 \end{pmatrix}$$

D.

$$\begin{pmatrix} 9 & 2 & 3 \\ 0 & 6 & 8 \\ 7 & 7 & 9 \end{pmatrix} \begin{pmatrix} 7 & 8 & 9 \\ 10 & 7 & 8 \end{pmatrix}$$

Solution

A. This product of a 2×2 matrix and a 4×4 matrix is not valid; the first matrix has two columns but the second matrix has four rows.

B. This product of a 2×4 matrix and a 4×2 matrix *is* valid; the four columns of the first matrix match the four rows of the second matrix. The result is a 2×2 matrix.

C. This product of a 3×1 matrix and a 1×8 matrix *is* valid; the single column of the first matrix matches the single row of the second. The result is a 3×8 matrix.

D. This product of a 3×3 matrix and a 2×3 matrix is not valid; the three columns of the first matrix do not match the two rows of the second.

Exercise 5.17 A matrix with 15 total entries is multiplied by a matrix with 6 total entries. What are the dimensions of the two matrices, and what is the dimension of the product matrix?

(continued)

Solution Let's call the dimensions of the matrices m-by-n and n-by-k because the number of columns of the first matrix has to match the number of rows of the second. Then $mn = 15$ and $nk = 6$. There are actually two possibilities:

- The first possibility is that $m = 5$, $n = 3$, and $k = 2$. Then this would be a 5×3 matrix multiplied by a 3×2 matrix resulting in a 5×2 matrix.
- The second possibility is that $m = 15$, $n = 1$, and $k = 6$. Then this would be a 15×1 matrix times a 1×6 matrix, resulting in a 15×6 matrix.

Exercise 5.18 Write a function that turns a column vector into a row vector, or vice versa. Flipping a matrix on its side like this is called *transposition* and the resulting matrix is called the *transpose* of the original.

Solution

```
def transpose(matrix):
    return tuple(zip(*matrix))
```

The call to `zip(*matrix)` returns a list of columns of the matrix and then we tuple them. This has the effect of swapping rows and columns of any input matrix, specifically turning column vectors into row vectors and vice versa:

```
>>> transpose(((1,),(2,),(3,)))
((1, 2, 3),)
>>> transpose(((1, 2, 3),))
((1,), (2,), (3,))
```

Exercise 5.19 Draw a picture that shows that a 10×8 and a 5×8 matrix *can't* be multiplied in that order.

Solution

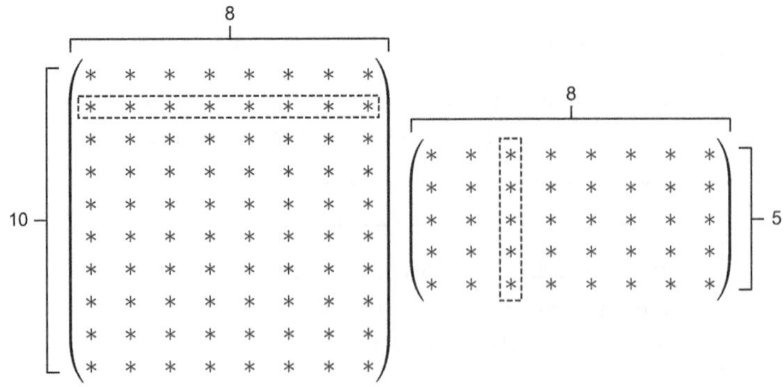

The rows of the first matrix have ten entries but the columns of the second have five, meaning we can't evaluate this matrix product.

Exercise 5.20 We want to multiply three matrices together: A is 5×7, B is 2×3, and C is 3×5. What order can they be multiplied in and what is the size of the result?

Solution One valid product is BC, a 2x3 times a 3×5 matrix yielding a 2×5 matrix. Another is CA, a 3×5 matrix times a 5×7 matrix yielding a 3×7 matrix. The product of three matrices, BCA, is valid regardless of the order you use. $(BC)A$ is a 2×5 matrix times a 5×7 matrix, while $B(CA)$ is a 2×3 matrix times a 3×7 matrix. Each yields the same 2×7 matrix as a result.

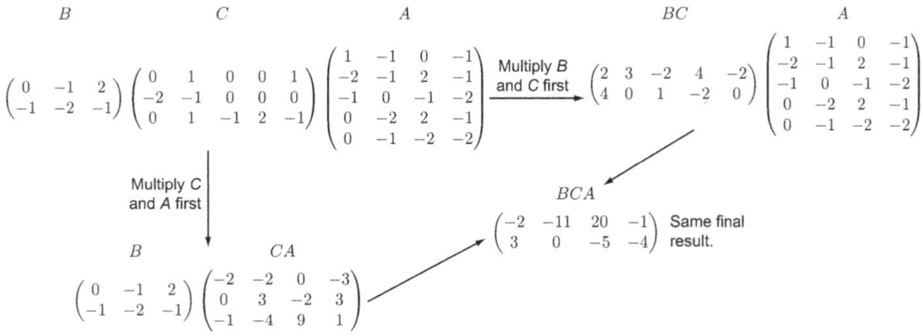

Multiplying three matrices in different orders

Exercise 5.21 Projection onto the *y,z* plane and onto the *x,z* plane are also linear maps from 3D to 2D. What are their matrices?

Solution Projection onto the *y,z* plane deletes the *x*-coordinate. The matrix for this operation is

$$\begin{pmatrix} 0 & 1 & 0 \\ 0 & 0 & 1 \end{pmatrix}$$

Likewise, projection onto the *x,z* plane deletes the *y*-coordinate:

$$\begin{pmatrix} 1 & 0 & 0 \\ 0 & 0 & 1 \end{pmatrix}$$

For example,

$$\begin{pmatrix} 1 & 0 & 0 \\ 0 & 0 & 1 \end{pmatrix} \begin{pmatrix} x \\ y \\ z \end{pmatrix} = \begin{pmatrix} x \\ z \end{pmatrix} \quad \text{and} \quad \begin{pmatrix} 0 & 1 & 0 \\ 0 & 0 & 1 \end{pmatrix} \begin{pmatrix} x \\ y \\ z \end{pmatrix} = \begin{pmatrix} y \\ z \end{pmatrix}$$

Exercise 5.22 Show by example that the `infer_matrix` function from a previous exercise can create matrices for linear functions whose inputs and outputs have different dimensions.

Solution One function we could test would be projection onto the x,y plane, which takes in 3D vectors and returns 2D vectors. We can implement this linear transformation as a Python function and then infer its 2×3 matrix:

```
>>> def project_xy(v):
...     x,y,z = v
...     return (x,y)
...
>>> infer_matrix(3,project_xy)
((1, 0, 0), (0, 1, 0))
```

Note that we had to supply the dimension of *input* vectors as an argument, so that we can build the correct standard basis vectors to test under the action of `project_xy`. Once `project_xy` is passed the 3D standard basis vectors, it automatically outputs 2D vectors to supply the columns of the matrix.

Exercise 5.23 Write a 4×5 matrix that acts on a 5D vector by deleting the third of its five entries, thereby producing a 4D vector. For instance, multiplying it with the column vector form of (1, 2, 3, 4, 5) should return (1, 2, 4, 5).

Solution The matrix is

$$\begin{pmatrix} 1 & 0 & 0 & 0 & 0 \\ 0 & 1 & 0 & 0 & 0 \\ 0 & 0 & 0 & 1 & 0 \\ 0 & 0 & 0 & 0 & 1 \end{pmatrix}$$

You can see that the first, second, fourth, and fifth coordinates of an input vector form the four coordinates of the output vector:

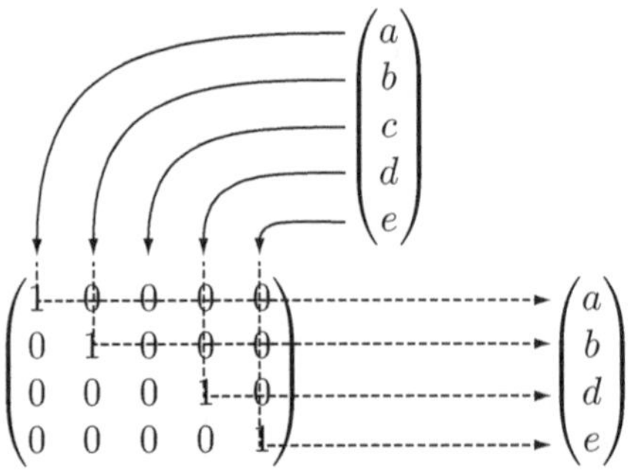

The 1s in the matrix indicate where coordinates of the input vector end up in the output vector.

Exercise 5.24—Mini Project Consider the vector of six variables (l, e, m, o, n, s). Find the matrix for the linear transformation that acts on this vector to produce the vector (s, o, l, e, m, n) as a result.

Hint The third coordinate of the output equals the first coordinate of the input, so the transformation must send the standard basis vector (1, 0, 0, 0, 0, 0) to (0, 0, 1, 0, 0, 0).

Solution

$$\begin{pmatrix} 0 & 0 & 0 & 0 & 0 & 1 \\ 0 & 0 & 0 & 1 & 0 & 0 \\ 1 & 0 & 0 & 0 & 0 & 0 \\ 0 & 1 & 0 & 0 & 0 & 0 \\ 0 & 0 & 1 & 0 & 0 & 0 \\ 0 & 0 & 0 & 0 & 1 & 0 \end{pmatrix} \begin{pmatrix} l \\ e \\ m \\ o \\ n \\ s \end{pmatrix} = \begin{pmatrix} 0+0+0+0+0+s \\ 0+0+0+o+0+0 \\ l+0+0+0+0+0 \\ 0+e+0+0+0+0 \\ 0+0+m+0+n+0 \\ 0+0+0+0+1+0 \end{pmatrix} = \begin{pmatrix} s \\ o \\ l \\ e \\ m \\ n \end{pmatrix}$$

This matrix reorders the entries of a 6D vector in the specified way.

Exercise 5.25 What valid products can be made from the matrices M, N, P, and Q from section 5.2.5? Include in your consideration the products of matrices with themselves. For those products that are valid, what are the dimensions of the matrix products?

Solution M is 3×3, N is 2×2, and P and Q are both 2×3. The product of M with itself, $MM = M^2$ is valid and a 3×3 matrix, so is $NN = N^2$ which is a 2×2 matrix. Apart from that, PM, QM, NP, and NQ are all 3×2 matrices.

5.3 Translating vectors with matrices

One advantage of matrices is that computations look the same in any number of dimensions. We don't need to worry about picturing the configurations of vectors in 2D or 3D; we can simply plug them into the formulas for matrix multiplication or use them as inputs to our Python `matrix_multiply`. This is especially useful when we want to do computations in more than three dimensions.

The human brain isn't wired to picture vectors in four or five dimensions, let alone 100, but we already saw we can do computations with vectors in higher dimensions. In this section, we'll cover a computation that *requires* doing computation in higher dimensions: translating vectors using a matrix.

5.3.1 Making plane translations linear

In the last chapter, we showed that translations are not linear transformations. When we move every point in the plane by a given vector, the origin moves and vector sums are not preserved. How can we hope to execute a 2D transformation with a matrix if it is not a linear transformation?

The trick is that we can think of our 2D points to translate as living in 3D. Let's return to our dinosaur from chapter 2. The dinosaur was composed of 21 points, and we could connect these in order to create the outline of the figure:

```
from vector_drawing import *

dino_vectors = [(6,4), (3,1), (1,2), (-1,5), (-2,5), (-3,4), (-4,4),
    (-5,3), (-5,2), (-2,2), (-5,1), (-4,0), (-2,1), (-1,0), (0,-3),
    (-1,-4), (1,-4), (2,-3), (1,-2), (3,-1), (5,1)
]

draw(
    Points(*dino_vectors),
    Polygon(*dino_vectors)
)
```

The result is the familiar 2D dinosaur (figure 5.24).

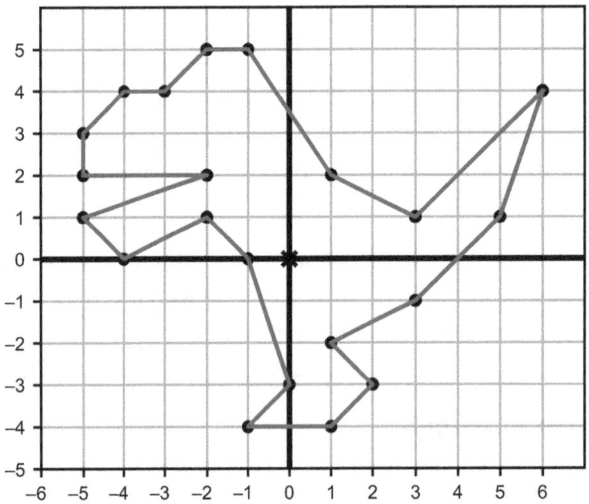

Figure 5.24 The familiar 2D dinosaur from chapter 2

If we want to translate the dinosaur to the right by 3 units and up by 1 unit, we could simply add the vector (3, 1) to each of the dinosaur's vertices. But this isn't a linear map, so we can't produce a 2×2 matrix that does this translation. If we think of the dinosaur as an inhabitant of 3D space instead of the 2D plane, it turns out we *can* formulate the translation as a matrix.

Bear with me for a moment while I show you the trick; I'll explain how it works shortly. Let's give every point of the dinosaur a *z*-coordinate of 1. Then we can draw it in 3D by connecting each of the points with segments and see that the resulting polygon lies on the plane where $z = 1$ (figure 5.25). I've created a helper function called `polygon_segments_3d` to get the segments of the dinosaur polygon in 3D.

```
from draw3d import *
def polygon_segments_3d(points,color='blue'):
    count = len(points)
    return [Segment3D(points[i], points[(i+1) % count],color=color) for i in
range(0,count)]

dino_3d = [(x,y,1) for x,y in dino_vectors]

draw3d(
    Points3D(*dino_3d, color='blue'),
    *polygon_segments_3d(dino_3d)
)
```

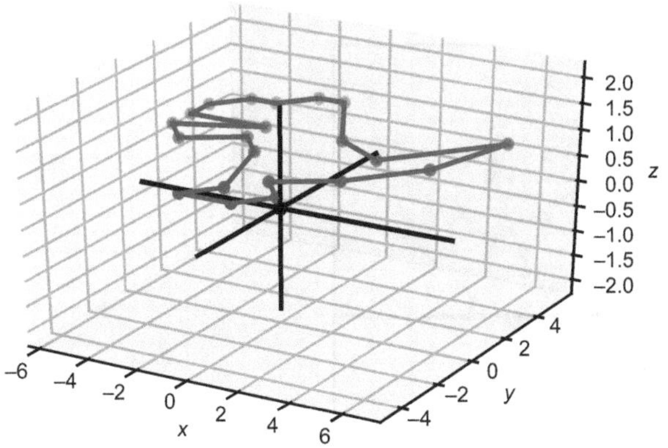

Figure 5.25 The same dinosaur with each of its vertices given a *z*-coordinate of 1

Figure 5.26 shows a matrix that "skews" 3D space, so that the origin stays put, but the plane where $z = 1$ is translated as desired. Trust me for now! I've highlighted the numbers relating to the translation that you should pay attention to.

$$\begin{pmatrix} 1 & 0 & 3 \\ 0 & 1 & 1 \\ 0 & 0 & 1 \end{pmatrix}$$

Figure 5.26 A magic matrix that moves the plane $z = 1$ by +3 in the *x* direction and by +1 in the *y* direction

We can apply this matrix to each vertex of the dinosaur and then voila! The dinosaur is translated by (3, 1) in its plane (figure 5.27).

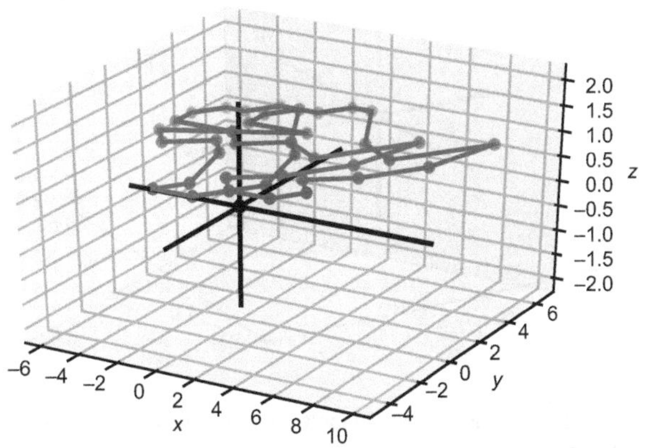

Figure 5.27 Applying the matrix to every point keeps the dinosaur in the same plane, but translates it within the plane by (3, 1).

Here's the code:

```
magic_matrix = (
    (1,0,3),
    (0,1,1),
    (0,0,1))

translated = [multiply_matrix_vector(magic_matrix, v) for v in
dino_vectors_3d]
```

For clarity, we could then delete the *z*-coordinates again and show the translated dinosaur in the plane with the original one (figure 5.28).

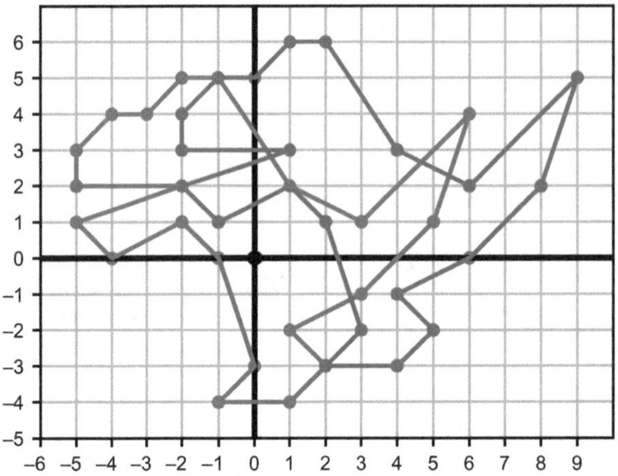

Figure 5.28 Dropping the translated dinosaur back into 2D

You can reproduce the code and check the coordinates to see that the dinosaur was indeed translated by (3, 1) in the final picture. Now let me show you how the trick works.

5.3.2 Finding a 3D matrix for a 2D translation

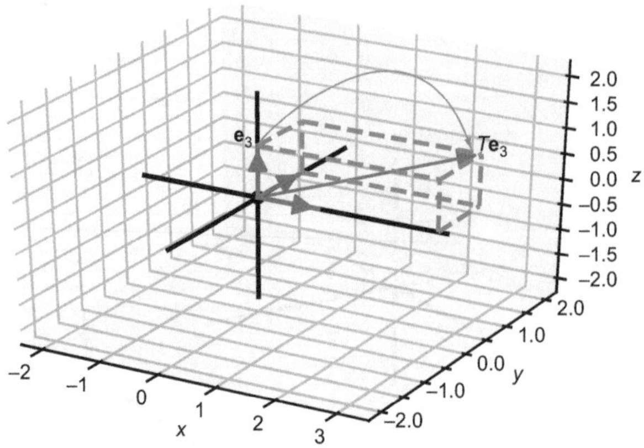

Figure 5.29 This matrix doesn't move $\mathbf{e}_1$ or $\mathbf{e}_2$, but it does move $\mathbf{e}_3$.

The columns of our "magic" matrix, like the columns of any matrix, tell us where the standard basis vectors end up after being transformed. Calling this matrix *T*, the vectors $\mathbf{e}_1$, $\mathbf{e}_2$, and $\mathbf{e}_3$ would be transformed into the vectors $T\mathbf{e}_1 = (1, 0, 0)$, $T\mathbf{e}_2 = (0, 1, 0)$, and $T\mathbf{e}_3 = (3, 1, 1)$. This means $\mathbf{e}_1$ and $\mathbf{e}_2$ are unaffected, and $\mathbf{e}_3$ changes only its *x*- and *y*-components (figure 5.29).

Any point in 3D and, therefore, any point on our dinosaur is built as a linear combination of $\mathbf{e}_1$, $\mathbf{e}_2$, and $\mathbf{e}_3$. For instance, the tip of the dinosaur's tail is at (6, 4, 1), which is $6\mathbf{e}_1 + 4\mathbf{e}_2 + \mathbf{e}_3$.

Because T doesn't move $\mathbf{e}_1$ or $\mathbf{e}_2$, only the effect on $\mathbf{e}_3$ moves the point, $T(\mathbf{e}_3) = \mathbf{e}_3 + (3, 1, 0)$, so that the point is translated by +3 in the x direction and +1 in the y direction. You can also see this algebraically. Any vector $(x, y, 1)$ is translated by $(3, 1, 0)$ by this matrix:

$$\begin{pmatrix} 1 & 0 & 3 \\ 0 & 1 & 1 \\ 0 & 0 & 1 \end{pmatrix} \begin{pmatrix} x \\ y \\ 1 \end{pmatrix} = \begin{pmatrix} 1 \cdot x + 0 \cdot y + 3 \cdot 1 \\ 0 \cdot x + 1 \cdot y + 1 \cdot 1 \\ 0 \cdot x + 0 \cdot y + 1 \cdot 1 \end{pmatrix} = \begin{pmatrix} x + 3 \\ y + 1 \\ 1 \end{pmatrix}$$

If you want to translate a collection of 2D vectors by some vector (a, b), the general steps are as follows:

1. Move the 2D vectors into the plane in 3D space, where $z = 1$ and each has a z-coordinate of 1.
2. Multiply the vectors by the matrix with your given choices of a and b plugged in:

$$\begin{pmatrix} 1 & 0 & a \\ 0 & 1 & b \\ 0 & 0 & 1 \end{pmatrix}$$

3. Delete the z-coordinate of all of the vectors so you are left with 2D vectors as a result.

Now that we can do translations with matrices, we can creatively combine them with other linear transformations.

5.3.3 *Combining translation with other linear transformations*

In the previous matrix, the first two columns are exactly $\mathbf{e}_1$ and $\mathbf{e}_2$, meaning that only the change in $\mathbf{e}_3$ moves a figure. We don't want $T(\mathbf{e}_1)$ or $T(\mathbf{e}_2)$ to have any z-component because that would move the figure out of the plane $z = 1$. But we can modify or interchange the other components (figure 5.30).

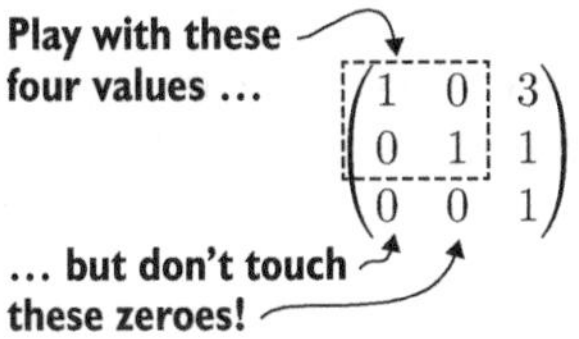

Figure 5.30 Let's see what happens when we move $T(\mathbf{e}_1)$ and $T(\mathbf{e}_2)$ in the x,y plane.

It turns out you can put any 2×2 matrix in the top left (as shown by figure 5.30) by doing the corresponding linear transformation *in addition* to the translation specified in the third column. For instance, this matrix

$$\begin{pmatrix} 0 & -1 \\ 1 & 0 \end{pmatrix}$$

produces a 90° counterclockwise rotation. Inserting it in the translation matrix, we get a new matrix that rotates the x,y plane by 90° and then translates it by (3, 1) as shown in figure 5.31.

$$\begin{pmatrix} 0 & -1 & 3 \\ 1 & 0 & 1 \\ 0 & 0 & 1 \end{pmatrix}$$

Figure 5.31 A matrix that rotates e_1 and e_3 by 90° *and* translates e_3 by (3, 1). Any figure in the plane where $z = 1$ experiences both transformations.

To show this works, we can carry out this transformation on all of the 3D dinosaur vertices in Python. Figure 5.32 shows the output of the following code:

```
rotate_and_translate = ((0,-1,3),(1,0,1),(0,0,1))
rotated_translated_dino = [
    multiply_matrix_vector(rotate_and_translate, v)

    for v in dino_vectors_3d]
```

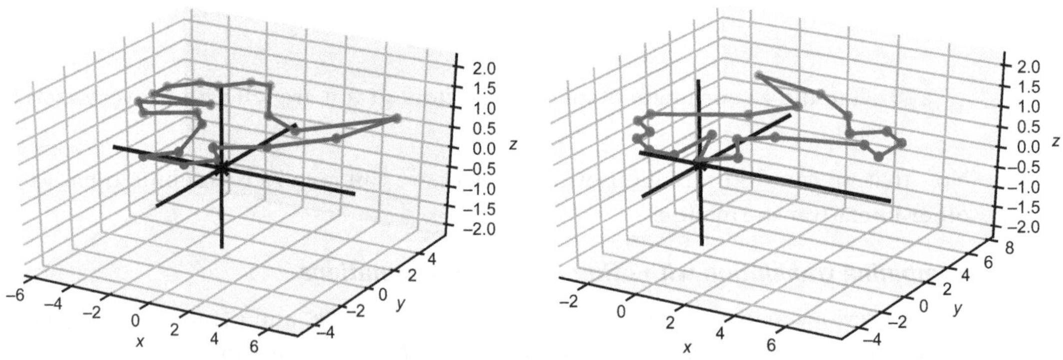

Figure 5.32 The original dinosaur (left) and a second dinosaur (right) that is both rotated and translated by a single matrix

Once you get the hang of doing 2D translations with a matrix, you can apply the same approach to doing a 3D translation. To do that, you'll have to use a 4×4 matrix and enter the mysterious world of 4D.

5.3.4 *Translating 3D objects in a 4D world*

What is the fourth dimension? A 4D vector would be an arrow with some length, width, depth, and one other dimension. When we built 3D space from 2D space, we added a z-coordinate. That means that 3D vectors can live in the x,y plane, where $z = 0$, or they can live in any other parallel plane, where z takes a different value. Figure 5.33 shows some of these parallel planes.

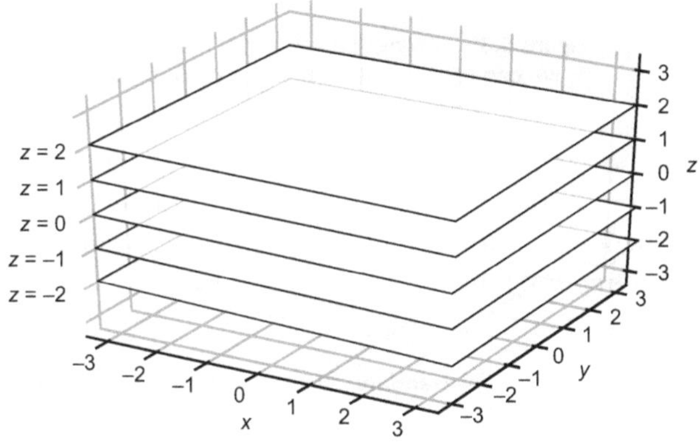

Figure 5.33 Building 3D space out of a stack of parallel planes, each looking like the *x,y* plane but at different *z*-coordinates

We can think of four dimensions in analogy to this model: a collection of 3D spaces that are indexed by some fourth coordinate. One way to interpret the fourth coordinate is "time." Each snapshot at a given time is a 3D space, but the collection of all of the snapshots is a fourth dimension called a *spacetime*. The origin of the spacetime is the origin of the space at the moment when time, *t*, is equal to 0 (figure 5.34).

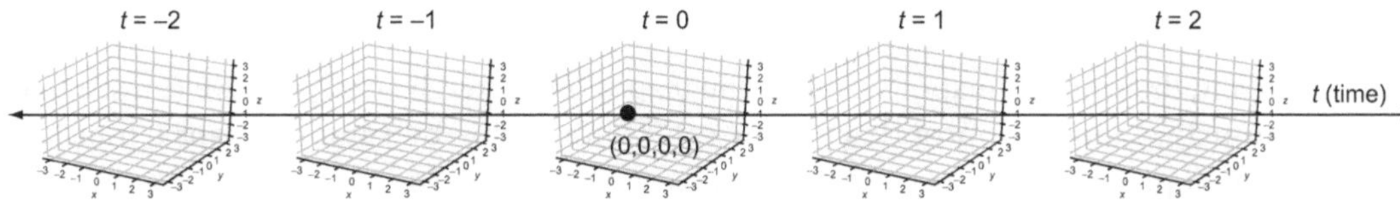

Figure 5.34 An illustration of 4D spacetime, similar to how a slice of 3D space at a given *z* value is a 2D plane and a slice of 4D spacetime at a given *t* value is a 3D space

This is the starting point for Einstein's theory of relativity. (In fact, you are now qualified to go read about this theory because it is based on 4D spacetime and linear transformations given by 4×4 matrices.)

Vector math is indispensable in higher dimensions because we quickly run out of good analogies. For five, six, seven, or more dimensions, I have a hard time picturing them, but the coordinate math is no harder than in two or three dimensions. For our current purposes, it's sufficient to think of a 4D vector as a four-tuple of numbers.

Let's replicate the trick that worked for translating 2D vectors in 3D. If we start with a 3D vector like (x, y, z) and we want to translate it by a vector (a, b, c), we can attach a fourth coordinate of 1 to the target vector and use an analogous 4D matrix to do the translation. Doing the matrix multiplication confirms that we get the desired result (figure 5.35).

$$\begin{pmatrix} 1 & 0 & 0 & a \\ 0 & 1 & 0 & b \\ 0 & 0 & 1 & c \\ 0 & 0 & 0 & 1 \end{pmatrix} \begin{pmatrix} x \\ y \\ z \\ 1 \end{pmatrix} = \begin{pmatrix} x+a \\ y+b \\ z+c \\ 1 \end{pmatrix}$$

Figure 5.35 Giving the vector (*x*, *y*, *z*) a fourth coordinate of 1, we can translate the vector by (*a*, *b*, *c*) using this matrix.

This matrix increases the *x*-coordinate by *a*, the *y*-coordinate by *b*, and the *z*-coordinate by *c*, so it does the transformation required to translate by the vector (*a*, *b*, *c*). We can package in a Python function the work of adding a fourth coordinate, applying this 4×4 matrix, and then deleting the fourth coordinate:

```
def translate_3d(translation):
    def new_function(target):
        a,b,c = translation
        x,y,z = target
        matrix = ((1,0,0,a),

            0,1,0,b),

            (0,0,1,c),

            (0,0,0,1))
        vector = (x,y,z,1)
        x_out, y_out, z_out, _ =\

          multiply_matrix_vector(matrix,vector)
        return (x_out,y_out,z_out)
    return new_function
```

The translate_3d function takes a translation vector and returns a new function that applies that translation to a 3D vector.

Builds the 4×4 matrix for the translation, and on the next line, turns (x, y, z) into a 4D vector with a fourth coordinate 1

Does the 4D matrix transformation

Finally, drawing the teapot as well as the teapot translated by (2, 2, –3), we can see that the teapot moves appropriately. You can confirm this by running matrix_translate_teapot.py. You should see the same image as in figure 5.36.

Figure 5.36 The untranslated teapot (left) and a translated teapot (right). As expected, the translated teapot moves up and to the right, and away from our viewpoint.

With translation packaged as a matrix operation, we can now combine that operation with other 3D linear transformations and do them in one step. It turns out you *can* interpret the artificial fourth-coordinate in this setup as time, t.

The two images in figure 5.36 could be snapshots of a teapot at $t = 0$ and $t = 1$, which is moving in the direction (2, 2, –3) at a constant speed. If you're looking for a fun challenge, you can replace the vector $(x, y, z, 1)$ in this implementation with vectors of the form (x, y, z, t), where the coordinate t changes over time. With $t = 0$ and $t = 1$, the teapot should match the frames in figure 5.36, and at the time between the two, it should move smoothly between the two positions. If you can figure out how this works, you'll catch up with Einstein!

So far, we've focused exclusively on vectors as points in space that we can render to a computer screen. This is clearly an important use case, but it only scratches the surface of what we can do with vectors and matrices. The study of how vectors and linear transformations work together in general is called *linear algebra*, and I'll give you a broader picture of this subject in the next chapter, along with some fresh examples that are relevant to programmers.

5.3.5 Exercises

Exercise 5.26 Show that the 3D "magic" matrix transformation does not work if you move a 2D figure such as the dinosaur we have been using to the plane $z = 2$. What happens instead?

Solution Using `[(x,y,2) for x,y in dino_vectors]` and applying the same 3×3 matrix, the dinosaur is translated twice as far by the vector (6, 2) instead of (3, 1). This is because the vector (0, 0, 1) is translated by (3, 1), and the transformation is linear.

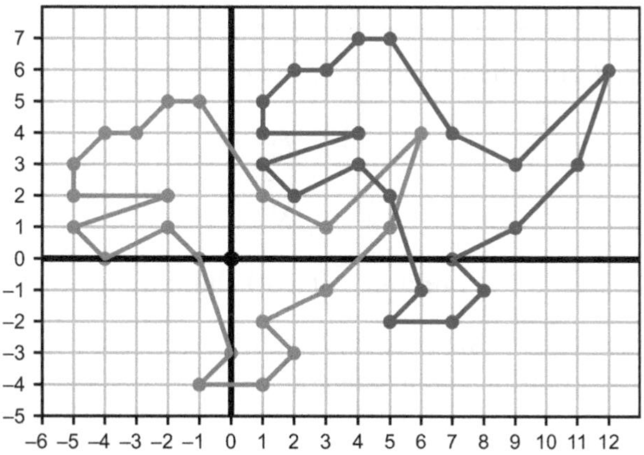

A dinosaur in the plane where z = 2 is translated twice as far by the same matrix.

Exercise 5.27 Come up with a matrix to translate the dinosaur by –2 units in the x direction and –2 units in the y direction. Execute the transformation and show the result.

Solution Replacing the values 3 and 1 in the original matrix with –2 and –2, we get

$$\begin{pmatrix} 1 & 0 & 2 \\ 0 & 1 & 2 \\ 0 & 0 & 1 \end{pmatrix}$$

The dinosaur, indeed, translates down and to the left by the vector (–2, –2).

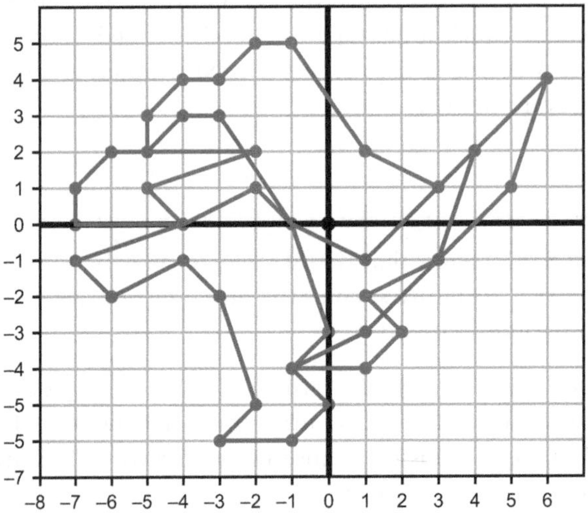

Exercise 5.28 Show that any matrix of the form

$$\begin{pmatrix} a & b & c \\ d & e & f \\ 0 & 0 & 1 \end{pmatrix}$$

doesn't affect the z-coordinate of a 3D column vector it is multiplied by.

Solution If the initial z-coordinate of a 3D vector is a number z, this matrix leaves that coordinate unchanged:

$$\begin{pmatrix} a & b & c \\ d & e & f \\ 0 & 0 & 1 \end{pmatrix} \begin{pmatrix} x \\ y \\ z \end{pmatrix} = \begin{pmatrix} ax + by + cz \\ dx + ey + fz \\ 0x + 0y + z \end{pmatrix}$$

Exercise 5.29—Mini Project Find a 3×3 matrix that rotates a 2D figure in the plane $z = 1$ by 45°, decreases its size by a factor of 2, and translates it by the vector (2, 2). Demonstrate that it works by applying it to the vertices of the dinosaur.

Solution First, let's find a 2×2 matrix for rotating a 2D vector by 45°:

```
>>> from vectors import rotate2d
>>> from transforms import *
>>> from math import pi
>>> rotate_45_degrees = curry2(rotate2d)(pi/4)
>>> rotation_matrix = infer_matrix(2,rotate_45_degrees)
>>> rotation_matrix
((0.7071067811865476, -0.7071067811865475), (0.7071067811865475,
0.7071067811865476))
```

Builds a function that executes rotate2d with an angle of 45° (or with 4 radians) for an input 2D vector

This matrix is approximately:

$$\begin{pmatrix} 0.707 & -0.707 \\ 0.707 & 0.707 \end{pmatrix}$$

Similarly, we can find a matrix to scale by a factor of ½:

$$\begin{pmatrix} 0.5 & 0 \\ 0 & 0.5 \end{pmatrix}$$

Multiplying these matrices together, we accomplish both transformations at once with this code:

```
>>> from matrices import *
>>> scale_matrix = ((0.5,0),(0,0.5))
>>> rotate_and_scale = matrix_multiply(scale_matrix,rotation_matrix)
>>> rotate_and_scale
((0.3535533905932738, -0.35355339059327373), (0.35355339059327373,
0.3535533905932738))
```

And this is a 3×3 matrix that translates the dinosaur by (2, 2) in the plane where $z = 1$:

$$\begin{pmatrix} 1 & 0 & 2 \\ 0 & 1 & 2 \\ 0 & 0 & 1 \end{pmatrix}$$

We can plug our 2×2 rotation and scaling matrix into the top left of this matrix, giving us the final matrix that we want:

```
>>> ((a,b),(c,d)) = rotate_and_scale
>>> final_matrix = ((a,b,2),(c,d,2),(0,0,1))
>>> final_matrix
((0.3535533905932738, -0.35355339059327373, 2), (0.35355339059327373,
0.3535533905932738, 2), (0, 0, 1))
```

(continued)

Moving the dinosaur to the plane $z = 1$, applying this matrix in 3D, and then projecting back to 2D gives us the rotated, scaled, and translated dinosaur, using only one matrix multiplication as shown here:

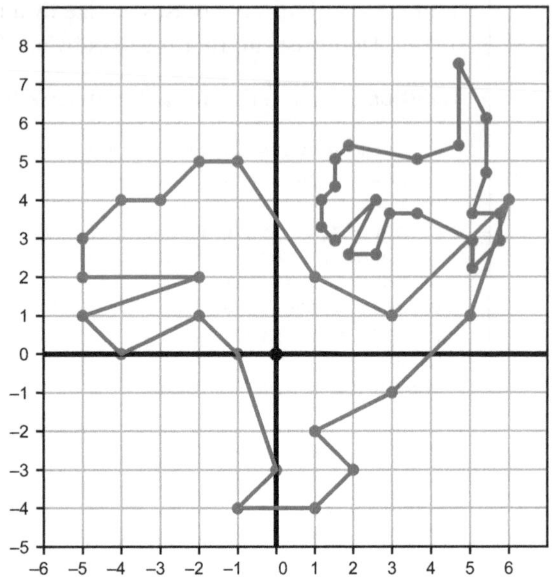

Exercise 5.30 The matrix in the preceding exercise rotates the dinosaur by 45° and then translates it by (3, 1). Using matrix multiplication, build a matrix that does this in the opposite order.

Solution If the dinosaur is in the plane where $z = 1$, then the following matrix does a rotation by 90° with *no* translation:

$$\begin{pmatrix} 0 & -1 & 0 \\ 1 & 0 & 0 \\ 0 & 0 & 1 \end{pmatrix}$$

We want to translate first and then rotate, so we multiply this rotation matrix by the translation matrix:

$$\begin{pmatrix} 0 & -1 & 0 \\ 1 & 0 & 0 \\ 0 & 0 & 1 \end{pmatrix} \begin{pmatrix} 1 & 0 & 3 \\ 0 & 1 & 1 \\ 0 & 0 & 1 \end{pmatrix} = \begin{pmatrix} 0 & -1 & -1 \\ 1 & 0 & 3 \\ 0 & 0 & 1 \end{pmatrix}$$

This is different from the other matrix, which rotates before the translation. In this case, we see that the translation vector (3, 1) is affected by the 90° rotation. The new effective translation is (–1, 3).

Exercise 5.31 Write a function analogous to `translate_3d` called `translate_4d` that uses a 5×5 matrix to translate a 4D vector by another 4D vector. Run an example to show that the coordinates are translated.

Solution The setup is the same, except that we lift the 4D vector to 5D by giving it a fifth coordinate of 1:

```
def translate_4d(translation):
    def new_function(target):
        a,b,c,d = translation
        x,y,z,w = target
        matrix = (
            (1,0,0,0,a),
            (0,1,0,0,b),
            (0,0,1,0,c),
            (0,0,0,1,d),
            (0,0,0,0,1))
        vector = (x,y,z,w,1)
        x_out,y_out,z_out,w_out,_ = multiply_matrix_vector(matrix,vector)
        return (x_out,y_out,z_out,w_out)
    return new_function
```

We can see that the translation works (the effect is the same as adding the two vectors):

```
>>> translate_4d((1,2,3,4))((10,20,30,40))
(11, 22, 33, 44)
```

In the previous chapters, we used visual examples in 2D and 3D to motivate vector and matrix arithmetic. As we've gone along, we've put more emphasis on computation. At the end of this chapter, we calculated vector transformations in higher dimensions where we didn't have any physical insight. This is one of the benefits of linear algebra: it gives you the tools to solve geometric problems that are too complicated to picture. We'll survey the broad range of this application in the next chapter.

Summary

- A linear transformation is defined by what it does to standard basis vectors. When you apply a linear transformation to the standard basis, the resulting vectors contain all the data required to do the transformation. This means that only nine numbers are required to specify a 3D linear transformation of any kind (the three coordinates of each of these three resulting vectors). For a 2D linear transformation, four numbers are required.
- In matrix notation, we represent a linear transformation by putting these numbers in a rectangular grid. By convention, you build a matrix by applying a transformation to the standard basis vectors and putting the resulting coordinate vectors side by side as columns.

- Using a matrix to evaluate the result of the linear transformation it represents on a given vector is called *multiplying the matrix by the vector.* When you do this multiplication, the vector is typically written as a column of its coordinates from top to bottom rather than as a tuple.
- Two square matrices can also be multiplied together. The resulting matrix represents the composition of the linear transformations of the original two matrices.
- To calculate the product of two matrices, you take the dot products of the rows of the first with the columns of the second. For instance, the dot product of row i of the first matrix and column j of the second matrix gives you the value in row i and column j of the product.
- As square matrices represent linear transformations, non-square matrices represent linear functions from vectors of one dimension to vectors of another dimension. That is, these functions send vector sums to vector sums and scalar multiples to scalar multiples.
- The dimension of a matrix tells you what kind of vectors its corresponding linear function accepts and returns. A matrix with m rows and n columns is called an m-by-n matrix (sometimes written $m \times n$). It defines a linear function from n-dimensional space to m-dimensional space.
- Translation is *not* a linear function, but it can be made linear if you perform it in a higher dimension. This observation allows us to do translations (simultaneously with other linear transformations) by matrix multiplication.

6 Generalizing to higher dimensions

This chapter covers

- Implementing a Python abstract base class for general vectors
- Defining vector spaces and listing their useful properties
- Interpreting functions, matrices, images, and sound waves as vectors
- Finding useful subspaces of vector spaces containing data of interest

Even if you're not interested in animating teapots, the machinery of vectors, linear transformations, and matrices can still be useful. In fact, these concepts are so useful there's an entire branch of math devoted to them: *linear algebra.* Linear algebra generalizes everything we know about 2D and 3D geometry to study data in any number of dimensions.

As a programmer, you're probably skilled at generalizing ideas. When writing complex software, it's common to find yourself writing similar code over and over. At some point, you catch yourself doing this, and you consolidate the code into one class or function capable of handling all of the cases you see. This saves you typing and often improves code organization and maintainability. Mathematicians follow the same process: after encountering similar patterns over and over, they can better state exactly what they see and refine their definitions.

In this chapter, we use this kind of logic to define *vector spaces*. Vector spaces are collections of objects we can treat like vectors. These can be arrows in the plane, tuples of numbers, or objects completely different from the ones we've seen so far. For instance, you can treat images as vectors and take a linear combination of them (figure 6.1).

Figure 6.1 A linear combination of two pictures produces a new picture.

The key operations in a vector space are vector addition and scalar multiplication. With these, you can make linear combinations (including negation, subtraction, weighted averages, and so on), and you can reason about which transformations are linear. It turns out these operations help us make sense of the word *dimension*. For instance, we'll see that the images used in figure 6.1 are 270,000-dimensional objects! We'll cover higher-dimensional and even infinite-dimensional spaces soon enough, but let's start by reviewing the 2D and 3D spaces we already know.

6.1 Generalizing our definition of vectors

Python supports object-oriented programming (OOP), which is a great framework for generalization. Specifically, Python classes support *inheritance*: you can create new classes of objects that inherit properties and behaviors of an existing parent class. In our case, we want to realize the 2D and 3D vectors we've already seen as instances of a more general class of objects simply called vectors. Then any other objects that inherit behaviors from the parent class can rightly be called vectors as well (figure 6.2).

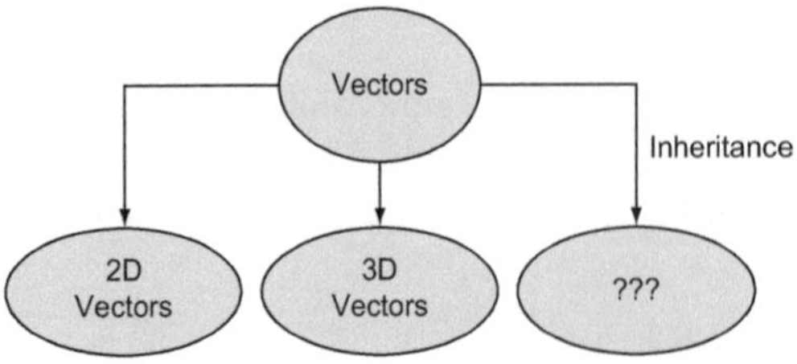

Figure 6.2 Treating 2D vectors, 3D vectors, and other objects as special cases of vectors using inheritance

If you haven't done object-oriented programming or you haven't seen it done in Python, don't worry. I stick to simple use cases in this chapter and will help you pick it up as we go. In case you want to learn more about classes and inheritance in Python before getting started, I've covered them in appendix B.

6.1.1 Creating a class for 2D coordinate vectors

In code, our 2D and 3D vectors have been *coordinate* vectors, meaning that they were defined as tuples of numbers, which are their coordinates. (We also saw that vector arithmetic can be defined geometrically in terms of arrows, but we can't translate that approach directly into Python code.) For 2D coordinate vectors, the data is the ordered pair of the *x*- and *y*-coordinates. A tuple is a great way to store this data, but we can equivalently use a class. We'll call the class representing 2D coordinate vectors `Vec2`:

```
class Vec2():
    def __init__(self,x,y):
        self.x = x
        self.y = y
```

We can initialize a vector like `v = Vec2(1.6,3.8)` and retrieve its coordinates as `v.x` and `v.y`. Next, we can give this class the methods required to do 2D vector arithmetic, specifically addition and scalar multiplication. The addition function, `add`, takes a second vector as an argument and returns a new `Vec2` object whose coordinates are the sum of the *x*- and *y*-coordinates, respectively:

```
class Vec2():
    ...
    def add(self, v2):
        return Vec2(self.x + v2.x, self.y + v2.y)
```

When adding to an existing class, I sometimes use ... as a placeholder for existing code.

Doing vector addition with `Vec2` could look like this:

```
v = Vec2(3,4)
w = v.add(Vec2(-2,6))
print(w.x)
```

Creates a new Vec2 called v with an x-coordinate 3 and y-coordinate 4

Adds a second Vec2 to v to produce a new Vec2 instance called w. This operation returns (3,4) + (–2,6) = (1,10).

Prints the x-coordinate of w. The result is 1.

Like our original implementation of vector addition, we do not perform the addition "in-place." That is, the two input vectors are not modified; a new `Vec2` object is created to store the sum. We can implement scalar multiplication in a similar way, taking a scalar as input and returning a new, scaled vector as output:

```
class Vec2():
    ...
    def scale(self, scalar):
        return Vec2(scalar * self.x, scalar * self.y)
```

`Vec(1,1).scale(50)` returns a new vector with the *x*-and *y*-coordinates both equal to 50. There's one more critical detail we need to take care of: currently the output of a comparison like `Vec2(3,4) == Vec2(3,4)` is `False`. This is problematic because these instances represent the same vector. By default, Python compares instances by their references (asking whether they are located in the same place in memory) rather than by their values. We can fix this by overriding the equality method, which causes Python to treat the `==` operator differently for objects of the `Vec2` class. (If you haven't seen this before, appendix B explains it in more depth.)

```
class Vec2():
    ...
    def __eq__(self,other):
        return self.x == other.x and self.y == other.y
```

We want two 2D coordinate vectors to be equal if their *x*- and *y*-coordinates agree, and this new definition of equality captures that. With this implemented, you'll find that `Vec2(3,4) == Vec2(3,4)`.

Our `Vec2` class now has the fundamental vector operations of addition and scalar multiplication, as well as an equality test that makes sense. We can now turn our attention to some syntactic sugar.

6.1.2 Improving the Vec2 class

As we changed the behavior of the `==` operator, we can also customize the Python operators `+` and `*` to mean vector addition and scalar multiplication, respectively. This is called *operator overloading*, and it is covered in appendix B:

```
class Vec2():
        ...
        def __add__(self, v2):
            return self.add(v2)
        def __mul__(self, scalar):
            return self.scale(scalar)
        def __rmul__(self,scalar):
            return self.scale(scalar)
```

The `__mul__` and `__rmul__` methods define both orders of multiplication, so we can multiply vectors by scalars on the left or the right. Mathematically, we consider both orders to mean the same thing.

We can now write a linear combination concisely. For instance, `3.0 * Vec2(1,0) + 4.0 * Vec2(0,1)` gives us a new `Vec2` object with *x*-coordinate 3.0 and *y*-coordinate 4.0. It's hard to read this in an interactive session though, because Python doesn't print `Vec2` nicely:

```
>>> 3.0 * Vec2(1,0) + 4.0 * Vec2(0,1)
<__main__.Vec2 at 0x1cef56d6390>
```

Python gives us the memory address of the resulting `Vec2` instance, but we already observed that's not what's important to us. Fortunately, we can change the string representation of `Vec2` objects by overriding the `__repr__` method:

```
class Vec2():
    ...
```

```
    def __repr__(self):
        return "Vec2({},{})".format(self.x,self.y)
```

This string representation shows the coordinates that are the most important data for a `Vec2` object. The results of `Vec2` arithmetic are much clearer now:

```
>>> 3.0 * Vec2(1,0) + 4.0 * Vec2(0,1)
Vec2(3.0,4.0)
```

We're doing the same math here as we did with our original tuple vectors but, in my opinion, this is a lot nicer. Building a class required some boilerplate, like the custom equality we wanted, but it also enabled operator overloading for vector arithmetic. The custom string representation also makes it clear that we're not just working with *any* tuples, but rather 2D vectors that we intend to use in a certain way. Now, we can implement 3D vectors represented by their own special class.

6.1.3 Repeating the process with 3D vectors

I'll call the 3D vector class `Vec3`, and it looks a lot like the 2D `Vec2` class except that its defining data will be three coordinates instead of two. In each method that explicitly references the coordinates, we need to make sure to properly use the *x*, *y*, and *z* values for `Vec3`.

```
class Vec3():
    def __init__(self,x,y,z): #1
        self.x = x
        self.y = y
        self.z = z
    def add(self, other):
        return Vec3(self.x + other.x, self.y + other.y, self.z + other.z)
    def scale(self, scalar):
        return Vec3(scalar * self.x, scalar * self.y, scalar * self.z)
    def __eq__(self,other):
        return (self.x == other.x
                and self.y == other.y
                and self.z == other.z)
    def __add__(self, other):
        return self.add(other)
    def __mul__(self, scalar):
        return self.scale(scalar)
    def __rmul__(self,scalar):
        return self.scale(scalar)
    def __repr__(self):
        return "Vec3({},{},{})".format(self.x,self.y, self.z)
```

We can now write 3D vector math in Python using the built-in arithmetic operators:

```
>>> 2.0 * (Vec3(1,0,0) + Vec3(0,1,0))
Vec3(2.0,2.0,0.0)
```

This `Vec3` class, much like the `Vec2` class, puts us in a good place to think about generalization. There are a few different directions we can go, and like many software design choices, the decision is subjective. We could, for example, focus on simplifying the arithmetic. Instead of implementing add differently for Vec2 and Vec3, they can both use the add function we built in chapter 3, which already handles coordinate vectors of any size. We could also store coordinates internally as a tuple or list, letting the constructor accept any number of coordinates and create a 2D, 3D, or other coordinate vector. I'll leave these possibilities as exercises for you, however, and take us in a different direction.

The generalization I want to focus on is based on how we *use* the vectors, not on how they work. This gets us to a mental model that both organizes the code well and aligns with the mathematical definition of a vector. For instance, we can write a generic average function that can be used on any kind of vector:

```
def average(v1,v2):
    return 0.5 * v1 + 0.5 * v2
```

We can insert either 3D vectors or 2D vectors; for instance, average(Vec2(9.0, 1.0), Vec2(8.0,6.0)) and average(Vec3(1,2,3), Vec3(4,5,6)) both give us correct and meaningful results. As a spoiler, we will soon be able to average pictures together as well. Once we've implemented a suitable class for images, we'll be able to write average(img1, img2) and get a new image back.

This is where we see the beauty and the economy that comes with generalization. We can write a single, generic function like average and use it for a wide variety of types of inputs. The only constraint on the input is that it needs to support multiplication by scalars and addition with one another. The implementation of arithmetic varies between Vec2 objects, Vec3 objects, images, or other kinds of data, but there's always an important overlap in *what* arithmetic we can do with them. When we separate the *what* from the *how*, we open the door for code reuse and far-reaching mathematical statements.

How can we best describe *what* we can do with vectors separately from the details of *how* we carry them out? We can capture this in Python using an abstract base class.

6.1.4 Building a vector base class

The basic things we can do with Vec2 or Vec3 include constructing a new instance, adding with other vectors, multiplying by a scalar, testing equality with another vector, and representing an instance as a string. Of these, only addition and scalar multiplication are distinctive vector operations. Any new Python class automatically includes the rest. This prompts a definition of a Vector base class:

```
from abc import ABCMeta, abstractmethod

class Vector(metaclass=ABCMeta):
    @abstractmethod
    def scale(self,scalar):
        pass
    @abstractmethod
    def add(self,other):
        pass
```

The abc module contains helper classes, functions, and method decorators that help define an *abstract base class*, a class that is not intended to be instantiated. Instead, it's designed to be used as a template for classes that inherit from it. The @abstractmethod decorator means that a method is not implemented in the base class and needs to be implemented for any child class. For instance, if you try to instantiate a vector with code like v = Vector(), you get the following TypeError:

```
TypeError: Can't instantiate abstract class Vector with abstract methods add,
scale
```

This makes sense; there is no such thing as a vector that is "just a vector." It needs to have some concrete manifestation such as a list of coordinates, an arrow in the plane, or something else. But this is still a useful base class because it forces any child class to include requisite methods. It is also useful to have this base class because we can equip it with all the methods that depend only on addition and scalar multiplication, like our operator overloads:

```
class Vector(metaclass=ABCMeta):
    ...
    def __mul__(self, scalar):
        return self.scale(scalar)
    def __rmul__(self, scalar):
        return self.scale(scalar)
    def __add__(self,other):
        return self.add(other)
```

In contrast to the abstract methods scale and add, these implementations are automatically available to any child class. We can simplify Vec2 and Vec3 to inherit from Vector. Here's a new implementation for Vec2:

```
class Vec2(Vector):
    def __init__(self,x,y):
        self.x = x
        self.y = y
    def add(self,other):
        return Vec2(self.x + other.x, self.y + other.y)
    def scale(self,scalar):
        return Vec2(scalar * self.x, scalar * self.y)
    def __eq__(self,other):
        return self.x == other.x and self.y == other.y
    def __repr__(self):
        return "Vec2({},{})".format(self.x, self.y)
```

This has indeed saved us from repeating ourselves! The methods that were identical between Vec2 and Vec3 now live in the Vector class. All remaining methods on Vec2 are specific to 2D vectors; they need to be modified to work for Vec3 (as you will see in the exercises) or for vectors with any other number of coordinates.

The Vector base class is a good representation of what we can do with vectors. If we can add any useful methods to it, chances are they will be useful for *any* kind of vector. For instance, we can add two methods to Vector:

```
class Vector(metaclass=ABCMeta):
    ...
    def subtract(self,other):
        return self.add(-1 * other)
    def __sub__(self,other):
        return self.subtract(other)
```

And without any modification of `Vec2`, we can automatically subtract them:

```
>>> Vec2(1,3) - Vec2(5,1)
Vec2(-4,2)
```

This abstract class makes it easier to implement general vector operations, and it also agrees with the mathematical definition of a vector. Let's switch languages from Python to English and see how the abstraction carries over from code to become a real mathematical definition.

6.1.5 Defining vector spaces

In math, a vector is defined by what it does rather than what it is, much like how we defined the abstract `Vector` class. Here's a first (incomplete) definition of a vector.

> **DEFINITION** A vector is an object equipped with a *suitable* way to add it to other vectors and multiply it by scalars.

Our `Vec2` or `Vec3` objects, or any other objects inheriting from the `Vector` class can be added to each other and multiplied by scalars. This definition is incomplete because I haven't said what "suitable" means, and that ends up being the most important part of the definition!

There are a few important rules outlawing weird behaviors, many of which you might have already assumed. It's not necessary to memorize all these rules. If you ever find yourself testing whether a new kind of object can be thought of as a vector, you can refer back to these rules. The first set of rules says that addition should be well-behaved. Specifically:

1. Adding vectors in any order shouldn't matter: $\mathbf{v} + \mathbf{w} = \mathbf{w} + \mathbf{v}$ for any vectors $\mathbf{v}$ and $\mathbf{w}$.
2. Adding vectors in any grouping shouldn't matter: $\mathbf{u} + (\mathbf{v} + \mathbf{w})$ should be the same as $(\mathbf{u} + \mathbf{v}) + \mathbf{w}$, meaning that a statement like $\mathbf{u} + \mathbf{v} + \mathbf{w}$ should be unambiguous.

A good counterexample is adding strings by concatenation. In Python, you can do the sum `"hot" + "dog"`, but this doesn't support the case that strings can be vectors because the sums `"hot" + "dog"` and `"dog" + "hot"` are not equal, violating rule 1.

Scalar multiplication also needs to be well-behaved and compatible with addition. For instance, a whole number scalar multiple should be equal to a repeated addition (like $3\mathbf{v} = \mathbf{v} + \mathbf{v} + \mathbf{v}$). Here are the specific rules:

3. Multiplying vectors by several scalars should be the same as multiplying by all the scalars at once. If a and b are scalars and $\mathbf{v}$ is a vector, then $a \cdot (b \cdot \mathbf{v})$ should be the same as $(a \cdot b) \cdot \mathbf{v}$.

4. Multiplying a vector by 1 should leave it unchanged: $1 \cdot \mathbf{v} = \mathbf{v}$.
5. Addition of scalars should be compatible with scalar multiplication: $a \cdot \mathbf{v} + b \cdot \mathbf{v}$ should be the same as $(a + b) \cdot \mathbf{v}$.
6. Addition of vectors should also be compatible with scalar multiplication: $a \cdot (\mathbf{v} + \mathbf{w})$ should be the same as $a \cdot \mathbf{v} + a \cdot \mathbf{w}$.

None of these rules should be too surprising. For instance, $3 \cdot \mathbf{v} + 5 \cdot \mathbf{v}$ could be translated to English as "3 of **v** added together plus 5 of **v** added together." Of course, this is the same as 8 of **v** added together, or $8 \cdot \mathbf{v}$, agreeing with rule 5.

The takeaway from these rules is that not all addition and multiplication operations are created equal. We need to verify each of the rules to ensure that addition and multiplication behave as expected. If so, the objects in question can rightly be called vectors.

A *vector space* is a collection of compatible vectors. Here's the definition:

DEFINITION A vector space is a collection of objects called vectors, equipped with suitable vector addition and scalar multiplication operations (obeying the rules above), such that every linear combination of vectors in the collection produces a vector that is also in the collection.

A collection like `[Vec2(1,0), Vec2(5,-3), Vec2(1.1,0.8)]` is a group of vectors that can be suitably added and multiplied, but it is not a vector space. For instance, `1 * Vec2(1,0) + 1 * Vec2(5,-3)` is a linear combination whose result is `Vec2(6,-3)`, which is not in the collection. One example of a vector space is the infinite collection of all possible 2D vectors. In fact, most vector spaces you meet are infinite sets; there are infinitely many linear combinations using infinitely many scalars after all!

There are two implications of the fact that vector spaces need to contain all their scalar multiples, and these implications are important enough to mention on their own. First, no matter what vector **v** you pick in a vector space, $0 \cdot \mathbf{v}$ gives you the same result, which is called the *zero vector* and denoted as **0** (bold, to distinguish it from the number 0). Adding the zero vector to any vector leaves that vector unchanged: $\mathbf{0} + \mathbf{v} = \mathbf{v} + \mathbf{0} = \mathbf{v}$. The second implication is that every vector **v** has an opposite vector, $-1 \cdot \mathbf{v}$, written as $-\mathbf{v}$. Due to rule #5, $\mathbf{v} + -\mathbf{v} = (1 + -1) \cdot \mathbf{v} = 0 \cdot \mathbf{v} = \mathbf{0}$. For every vector, there is another vector in the vector space that "cancels it out" by addition. As an exercise, you can improve the `Vector` class by adding a zero vector and a negation function as required members.

A class like `Vec2` or `Vec3` is not a collection per se, but it does describe a collection of values. In this way, we can think of the classes `Vec2` and `Vec3` as representing two different vector spaces, and their instances represent vectors. We'll see a lot more examples of vector spaces with classes that represent them in the next section, but first, let's look at how to validate that they satisfy the specific rules we've covered.

6.1.6 Unit testing vector space classes

It was helpful to use an abstract `Vector` base class to think about what a vector should be able to do, rather than how it's done. But even giving the base class an abstract `add` method doesn't guarantee every inheriting class will implement a suitable addition operation.

In math, the usual way we guarantee suitability is by *writing a proof*. In code, and especially in a dynamic language like Python, the best we can do is to write unit tests. For instance, we can check rule #6 from the previous section by creating two vectors and a scalar and making sure the equality holds:

```
>>> s = -3
>>> u, v = Vec2(42,-10), Vec2(1.5, 8)
>>> s * (u + v) == s * v + s * u
True
```

This is often how unit tests are written, but it's a pretty weak test because we're only trying one example. We can make it stronger by plugging in random numbers and ensuring that it works. Here I use the `random.uniform` function to generate evenly distributed floating-point numbers between –10 and 10:

```
from random import uniform

def random_scalar():
    return uniform(-10,10)

def random_vec2():
    return Vec2(random_scalar(),random_scalar())

a = random_scalar()
u, v = random_vec2(), random_vec2()
assert a * (u + v) == a * v + a * u
```

Unless you're lucky, this test will fail with an `AssertionError`. Here are the offending values of a, u, and v that caused the test to fail for me:

```
>>> a, u, v
(0.17952747449930084,
 Vec2(0.8353326458605844,0.2632539730989293),
 Vec2(0.555146137477196,0.34288853317521084))
```

And the expressions from the left and right of the equals sign in the `assert` call from the previous code have these values:

```
>>> a * (u + v), a * u + a * v
(Vec2(0.24962914431749222,0.10881923333807299),
 Vec2(0.24962914431749225,0.108819233338073))
```

These are two different vectors, but only because their components differ by a few quadrillionths (very, very small numbers). This doesn't mean that the math is wrong, just that floating-point arithmetic is approximate rather than exact.

To ignore such small discrepancies, we can use another notion of equality suitable for testing. Python's `math.isclose` function checks that two float values don't differ by a significant amount (by default, by more than one-billionth of the larger value). Using that function instead, the test passes 100 times in a row:

```
from math import isclose

def approx_equal_vec2(v,w):
    return isclose(v.x,w.x) and isclose(v.y,w.y)

for _ in range(0,100):
    a = random_scalar()
    u, v = random_vec2(), random_vec2()
    assert approx_equal_vec2(a * (u + v),
                             a * v + a * u)
```

Tests whether the x and y components are close (even if not equal)

Runs the test for 100 different randomly generated scalars and pairs of vectors

Replaces a strict equality check with the new function

With the floating-point error removed from the equation, we can test all six of the vector space properties in this way:

```
def test(eq, a, b, u, v, w):
    assert eq(u + v, v + u)
    assert eq(u + (v + w), (u + v) + w)
    assert eq(a * (b * v), (a * b) * v)
    assert eq(1 * v, v)
    assert eq((a + b) * v, a * v + b * v)
    assert eq(a * v + a * w, a * (v + w))

for i in range(0,100):
    a,b = random_scalar(), random_scalar()
        u,v,w = random_vec2(), random_vec2(), random_vec2()
        test(approx_equal_vec2,a,b,u,v,w)
```

Passes in the equality test function as eq. This keeps the test function agnostic as to the particular concrete vector implementation being passed in.

This test shows that all six rules (properties) hold for 100 different random selections of scalars and vectors. That 600 randomized unit tests pass is a good indication that our `Vec2` class satisfies the list of properties from the previous section. Once you implement the `zero()` property and the negation operator in the exercises, you can test a few more properties.

This setup isn't completely generic; we had to write special functions to generate random `Vec2` instances and to compare them. The important part is that the `test` function itself and the expressions within it are completely generic. As long as the class we're testing inherits from `Vector`, it can run expressions like `a * v + a * w` and `a * (v + w)` that we can then test for equality. Now, we can go wild exploring all the different objects that can be treated as vectors, and we know how to test them as we go.

6.1.7 Exercises

Exercise 6.1 Implement a `Vec3` class inheriting from `Vector`.

Solution

```
class Vec3(Vector):
    def __init__(self,x,y,z):
        self.x = x
        self.y = y
        self.z = z
    def add(self,other):
        return Vec3(self.x + other.x,
                    self.y + other.y,
                    self.z + other.z)
    def scale(self,scalar):
        return Vec3(scalar * self.x,
                    scalar * self.y,
                    scalar * self.z)
    def __eq__(self,other):
        return (self.x == other.x
                and self.y == other.y
                and self.z == other.z)
    def __repr__(self):
        return "Vec3({},{},{})".format(self.x, self.y, self.z)
```

Exercise 6.2—Mini Project Implement a `CoordinateVector` class inheriting from `Vector` with an abstract property representing the dimension. This should save repetitious work when implementing specific coordinate vector classes. Inheriting from `CoordinateVector` and setting the dimension to `6` should be all you need to do to implement a `Vec6` class.

Solution We can use the dimension-independent operations `add` and `scale` from chapters 2 and 3. The only thing not implemented in the following class is the dimension, and not knowing how many dimensions we're working with prevents us from instantiating a `CoordinateVector`:

```
from abc import abstractproperty
from vectors import add, scale

class CoordinateVector(Vector):
    @abstractproperty
    def dimension(self):
        pass
    def __init__(self,*coordinates):
        self.coordinates = tuple(x for x in coordinates)
    def add(self,other):
        return self.__class__(*add(self.coordinates, other.coordinates))
    def scale(self,scalar):
        return self.__class__(*scale(scalar, self.coordinates))
```

```
    def __repr__(self):
        return "{}{}".format(self.__class__.__qualname__,
self.coordinates)
```

Once we pick a dimension (say 6), we have a concrete class that we can instantiate:

```
class Vec6(CoordinateVector):
    def dimension(self):
        return 6
```

The definitions of addition, scalar multiplication, and so on are picked up from the `CoordinateVector` base class:

```
>>> Vec6(1,2,3,4,5,6) + Vec6(1, 2, 3, 4, 5, 6)
Vec6(2, 4, 6, 8, 10, 12)
```

Exercise 6.3 Add a `zero` abstract method to the `Vector` class to return the zero vector in a given vector space, as well as an implementation for the negation operator. These are useful because we're required to have a zero vector and negations of any vector in a vector space.

Solution

```
from abc import ABCMeta, abstractmethod, abstractproperty

class Vector(metaclass=ABCMeta):
    ...
    @classmethod
    @abstractproperty
    def zero():
        pass

    def __neg__(self):
        return self.scale(-1)
```

zero is a class method because there's only one zero value for any vector space.

It's also an abstract property because we haven't said what zero is yet.

Special method name for overloading negation

We don't need to implement `__neg__` for any child class because its definition is included in the parent class, based only on scalar multiplication. We do, however, need to implement `zero` for each class:

```
class Vec2(Vector):
    ...
    def zero():
        return Vec2(0,0)
```

Exercise 6.4 Write unit tests to show that the addition and scalar multiplication operations for `Vec3` satisfy the vector space properties.

Solution Because the test function is general, we only need to supply a new equality function for `Vec3` objects and 100 random sets of inputs:

```
def random_vec3():
    return Vec3(random_scalar(),random_scalar(),random_scalar())

def approx_equal_vec3(v,w):
    return isclose(v.x,w.x) and isclose(v.y,w.y) and isclose(v.z, w.z)

for i in range(0,100):
    a,b = random_scalar(), random_scalar()
    u,v,w = random_vec3(), random_vec3(), random_vec3()
    test(approx_equal_vec3,a,b,u,v,w)
```

Exercise 6.5 Add unit tests to check that $\mathbf{0} + \mathbf{v} = \mathbf{v}$, $0 \cdot \mathbf{v} = \mathbf{0}$, and $-\mathbf{v} + \mathbf{v} = \mathbf{0}$ for any vector **v**, where again 0 is the number zero and **0** is the zero vector.

Solution Because the zero vector is different, depending on which class we're testing, we need to pass it in as an argument to the function:

```
def test(zero,eq,a,b,u,v,w):
    ...
    assert eq(zero + v, v)
    assert eq(0 * v, zero)
    assert eq(-v + v, zero)
```

We can test any vector class with a `zero` method implemented (see exercise 6.3):

```
for i in range(0,100):
    a,b = random_scalar(), random_scalar()
    u,v,w = random_vec2(), random_vec2(), random_vec2()
    test(Vec2.zero(), approx_equal_vec2, a,b,u,v,w)
```

Exercise 6.6 As equality is implemented for `Vec2` and `Vec3`, it turns out that `Vec2(1,2) == Vec3(1,2,3)` returns `True`. Python's duck typing is too forgiving for its own good! Fix this by adding a check that classes must match before testing vector equality.

Solution It turns out, we need to do the check for addition as well!

```
class Vec2(Vector):
    ...
    def add(self,other):
        assert self.__class__ == other.__class__
        return Vec2(self.x + other.x, self.y + other.y)
```

```
    ...
    def __eq__(self,other):
        return (self.__class__ == other.__class__
            and self.x == other.x and self.y == other.y)
```

To be safe, you can add checks like this to other child classes of `Vector` as well.

Exercise 6.7 Implement a `__truediv__` function on `Vector` that allows you to divide vectors by scalars. You can divide vectors by a non-zero scalar by multiplying them by the reciprocal of the scalar (1.0/scalar).

Solution

```
class Vector(metaclass=ABCMeta):
    ...
    def __truediv__(self, scalar):
        return self.scale(1.0/scalar)
```

With this implemented, you can do division like `Vec2(1,2)/2`, getting back `Vec2(0.5,1.0)`.

6.2 Exploring different vector spaces

Now that you know what a vector space is, let's look at some examples. In each case, we take a new kind of object and implement it as a class that inherits from `Vector`. At that point, no matter what kind of object it is, we can do addition, scalar multiplication, or any other vector operation with it.

6.2.1 Enumerating all coordinate vector spaces

We've spent a lot of time on the coordinate vectors `Vec2` and `Vec3` so far, so coordinate vectors in 2D and 3D don't need much more explanation. It is worth reviewing, however, that a vector space of coordinate vectors can have *any* number of coordinates. `Vec2` vectors have two coordinates, `Vec3` vectors have three, and we could just as well have a `Vec15` class with 15 coordinates. We can't picture it geometrically, but `Vec15` objects represent points in a 15D space.

One special case worth mentioning is the class we might call `Vec1`, vectors with a single coordinate. The implementation looks like this:

```
class Vec1(Vector):
    def __init__(self,x):
        self.x = x
    def add(self,other):
        return Vec1(self.x + other.x)
    def scale(self,scalar):
        return Vec1(scalar * self.x)
    @classmethod
```

```
    def zero(cls):
        return Vec1(0)
    def __eq__(self,other):
        return self.x == other.x
    def __repr__(self):
        return "Vec1({})".format(self.x)
```

This is a lot of boilerplate to wrap a single number, and it doesn't give us any arithmetic we don't already have. Adding and multiplying `Vec1` scalar objects is just addition and multiplication of the underlying numbers:

```
>>> Vec1(2) + Vec1(2)
Vec1(4)
>>> 3 * Vec1(1)
Vec1(3)
```

For this reason, we probably will never need a `Vec1` class. But it is important to know that numbers on their own are vectors. The set of all real numbers (including integers, fractions, and irrational numbers like π) is denoted as $\mathbb{R}$, and it is a vector space in its own right. This is a special case where the scalars and the vectors are the same kind of objects.

Coordinate vector spaces are denoted $\mathbb{R}^n$, where n is the dimension or number of coordinates. For instance, the 2D plane is denoted as $\mathbb{R}^2$ and 3D space is denoted as $\mathbb{R}^3$. As long as you use real numbers as your scalars, any vector space you stumble across is some $\mathbb{R}^n$ in disguise.[1] This is why we need to mention the vector space $\mathbb{R}$, even if it is boring. The other vector space we need to mention is the *zero-dimensional* one, $\mathbb{R}^0$. This is the set of vectors with zero coordinates that we can describe as empty tuples or as a `Vec0` class inheriting from `Vector`:

```
class Vec0(Vector):
    def __init__(self):
        pass
    def add(self,other):
        return Vec0()
    def scale(self,scalar):
        return Vec0()
    @classmethod
    def zero(cls):
        return Vec0()
    def __eq__(self,other):
        return self.__class__ == other.__class__ == Vec0
    def __repr__(self):
        return "Vec0()"
```

No coordinates don't mean that there are no possible vectors; it means there is exactly one zero-dimensional vector. This makes zero-dimensional vector math stupidly easy; any result vector is always the same:

[1] That is, as long as you can guarantee your vector space has only finitely many dimensions! There is a vector space called $\mathbb{R}^\infty$, but it is not the only infinitely dimensional vector space.

```
>>> - 3.14 * Vec0()
Vec0()
>>> Vec0() + Vec0() + Vec0() + Vec0()
Vec0()
```

This is something like a singleton class from an OOP perspective. From a mathematical perspective, we know that every vector space has to have a zero vector, so we can think of `Vec0()` as being this zero vector.

That covers it for coordinate vectors of dimensions zero, one, two, three, or more. Now, when you see a vector in the wild, you'll be able to match it up with one of these vector spaces.

6.2.2 *Identifying vector spaces in the wild*

Let's return to an example from chapter 1 and look at a data set of used Toyota Priuses. In the source code, you'll see how to load the data set generously provided by my friend Dan Rathbone at CarGraph.com. To make the cars easy to work with, I've loaded them into a class:

```
class CarForSale():
    def __init__(self, model_year, mileage, price, posted_datetime,
                 model, source, location, description):
        self.model_year = model_year
        self.mileage = mileage
        self.price = price
        self.posted_datetime = posted_datetime
        self.model = model
        self.source = source
        self.location = location
        self.description = description
```

It would be useful to think of `CarForSale` objects as vectors. Then, for example, I could average them together as a linear combination to see what the typical Prius for sale looks like. To do that, I need to retrofit this class to inherit from `Vector`.

How can we add two cars? The numeric fields `model_year`, `mileage`, and `price` can be added like components of a vector, but the string properties can't be added in a meaningful way. (Remember, you saw that we can't think of strings as vectors.) When we do arithmetic on cars, the result is not a real car for sale but a *virtual* car defined by its properties. To represent this, I'll change all the string properties to the string `"(virtual)"` to remind us of this. Finally, we can't add datetimes, but we can add time spans. In figure 6.3, I use the day I retrieved the data as a reference point and add the time spans since the cars were posted for sale. The code for the entire process is shown in listing 6.1.

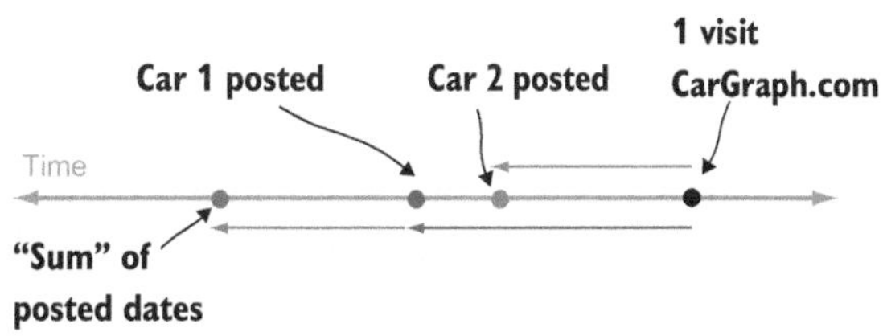

Figure 6.3 Timeline of cars posted for sale

All this applies to scalar multiplication as well. We can multiply the numeric properties and the time span since posting by a scalar. The string properties are no longer meaningful, however.

Listing 6.1 Making `CarForSale` behave like a `Vector` by implementing required methods

```
from datetime import datetime

class CarForSale(Vector):
    retrieved_date = datetime(2018,11,30,12)
    def __init__(self, model_year, mileage, price, posted_datetime,
                 model="(virtual)",
                         source="(virtual)",
                 location="(virtual)", description="(virtual)"):
        self.model_year = model_year
        self.mileage = mileage
        self.price = price
        self.posted_datetime = posted_datetime
        self.model = model
        self.source = source
        self.location = location
        self.description = description
    def add(self, other):
        def add_dates(d1, d2):
            age1 = CarForSale.retrieved_date - d1
            age2 = CarForSale.retrieved_date - d2
            sum_age = age1 + age2
            return CarForSale.retrieved_date - sum_age
        return CarForSale(
            self.model_year + other.model_year,
            self.mileage + other.mileage,
            self.price + other.price,
            add_dates(self.posted_datetime, other.posted_datetime)
        )
    def scale(self,scalar):
        def scale_date(d):
            age = CarForSale.retrieved_date - d
            return CarForSale.retrieved_date - (scalar * age)
        return CarForSale(
            scalar * self.model_year,
            scalar * self.mileage,
            scalar * self.price,
            scale_date(self.posted_datetime)
        )
    @classmethod
    def zero(cls):
        return CarForSale(0, 0, 0, CarForSale.retrieved_date)
```

I retrieved the data set from CarGraph.com on 11/30/2018 at noon.

To simplify construction of virtual cars, all of the string parameters are optional with a default value "(virtual)".

Helper function that adds dates by adding the time spans from the reference date

Adds CarForSale objects by adding underlying properties and constructing a new object

Helper function that scales a datetime by scaling the time span from the reference date

In the source code, you'll find the complete implementation of the class as well as the code to load a list of sample car data. With the list of cars loaded, we can try some vector arithmetic:

```
>>> (cars[0] + cars[1]).__dict__
{'model_year': 4012,
```

```
 'mileage': 306000.0,
 'price': 6100.0,
 'posted_datetime': datetime.datetime(2018, 11, 30, 3, 59),
 'model': '(virtual)',
 'source': '(virtual)',
 'location': '(virtual)',
 'description': '(virtual)'}
```

The sum of the first two cars is evidently a Prius from model year 4012 (maybe it can fly?) with 306,000 miles on it and going for an asking price of $6,100. It was posted for sale at 3:59 AM on the same day I looked at CarGraph.com. This unusual car doesn't look too helpful, but bear with me, averages (as shown in the following) look a lot more meaningful:

```
>>> average_prius = sum(cars, CarForSale.zero()) * (1.0/len(cars))
>>> average_prius.__dict__

{'model_year': 2012.5365853658536,
 'mileage': 87731.63414634147,
 'price': 12574.731707317074,
 'posted_datetime': datetime.datetime(2018, 11, 30, 9, 0, 49, 756098),
 'model': '(virtual)',
 'source': '(virtual)',
 'location': '(virtual)',
 'description': '(virtual)'}
```

We can learn real things from this result. The average Prius for sale is about 6 years old, has about 88,000 miles on it, is selling for about $12,500, and was posted at 9:49 AM the morning I accessed the website. (In Part 3, we spend a lot of time learning from data sets by treating them as vectors.)

Ignoring the text data, `CarForSale` behaves like a vector. In fact, it behaves like a 4D vector having dimensions of price, model year, mileage, and datetime of posting. It's not quite a coordinate vector because the posting date is not a number. Even though the data is not numeric, the class satisfies the vector space properties (you verify this with unit tests in the exercises), so its objects are vectors and can be manipulated as such. Specifically, they are 4D vectors, so it is possible to write a 1-to-1 mapping between `CarForSale` objects and `Vec4` objects (also an exercise for you). For our next example, we'll see some objects that look even less like coordinate vectors but still satisfy the defining properties.

6.2.3 *Treating functions as vectors*

It turns out that mathematical functions can be thought of as vectors. Specifically, I'm talking about functions that take in a single real number and return a single real number, though there are plenty of other types of mathematical functions. The mathematical shorthand to say that a function f takes any real number and returns a real number is $f: \mathbb{R} \to \mathbb{R}$. With Python, we'll think of functions that take `float` values in and return `float` values.

As with 2D or 3D vectors, we can do addition and scalar multiplication of functions visually or algebraically. To start, we can write functions algebraically; for instance, $f(x) = 0.5 \cdot x + 3$ or $g(x) = \sin(x)$. Alternatively, we can visualize these with a graph.

In the source code, I've written a simple `plot` function that draws the graph of one or more functions on a specified range of inputs (figure 6.4). For instance, the following code plots both of our functions $f(x)$ and $g(x)$ on x values between –10 and 10:

```
def f(x):
    return 0.5 * x + 3
def g(x):
    return sin(x)
plot([f,g],-10,10)
```

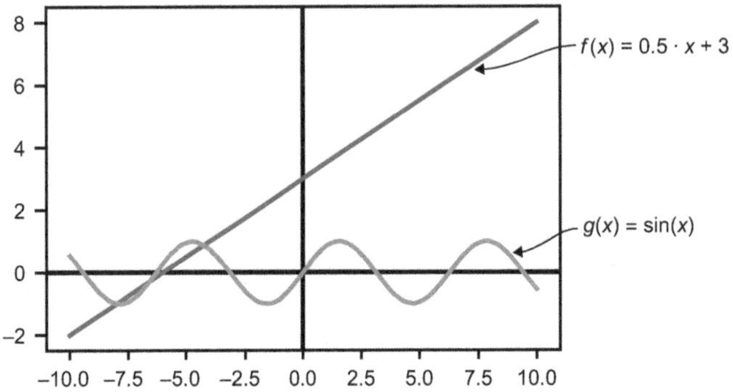

Figure 6.4 Graph of the functions $f(x) = 0.5 \cdot x + 3$ and $g(x) = \sin(x)$

Algebraically, we can add functions by adding the expressions that define them. This means $f + g$ is a function defined by $(f + g)(x) = f(x) + g(x) = 0.5 \cdot x + 3 + \sin(x)$. Graphically, the y values of each point are added, so it's something like stacking the two functions together as shown in figure 6.5.

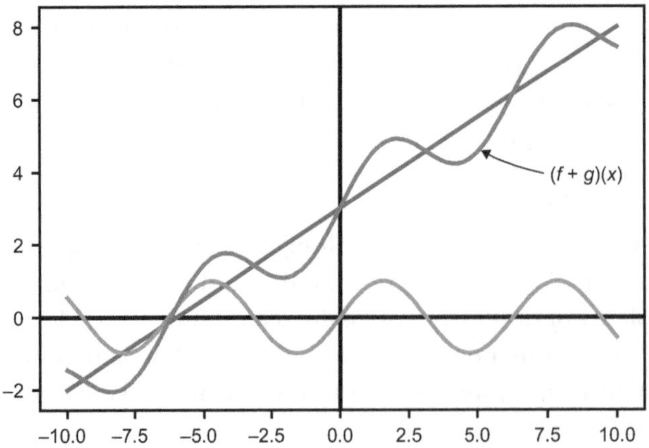

Figure 6.5 Visualizing the sum of two functions on a graph

To implement this sum, you can write some functional Python code. This code takes two functions as inputs and returns a new one, which is their sum:

```
def add_functions(f,g):
    def new_function(x):
        return f(x) + g(x)
    return new_function
```

Likewise, we can multiply a function by a scalar by multiplying its expression by the scalar. For instance, $3g$ is defined by $(3g)(x) = 3 \cdot g(x) = 3 \cdot \sin(x)$. This has the effect of stretching the graph of the function g in the y direction by a factor of 3 (figure 6.6).

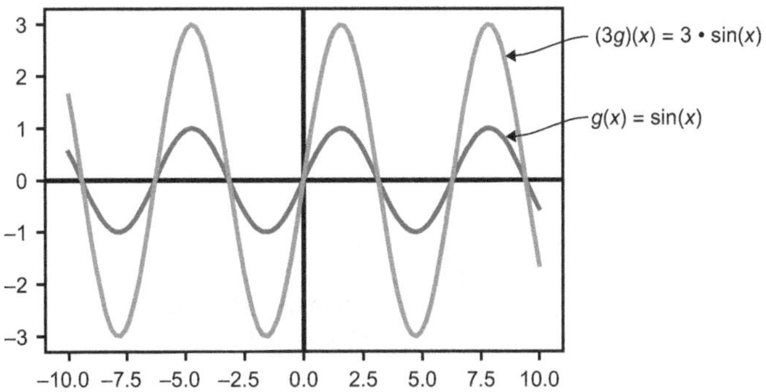

Figure 6.6 The function (3*g*) looks like the function *g* stretched by a factor of 3 in the *y* direction.

It's possible to nicely wrap Python functions in a class that inherits from vector, and I leave it as an exercise for you. After doing so, you can write satisfying function arithmetic expressions like $3 \cdot f$ or $2 \cdot f - 6 \cdot g$. You can even make the class *callable* or able to accept arguments as if it were a function to allow expressions like $(f + g)(6)$. Unfortunately, unit testing to determine if functions satisfy the vector space properties is much harder because it's difficult to generate random functions or to test whether two functions are equal. To really know if two functions are equal, you have to know that they return the same output for every single possible input. That means a test for every real number or at least every `float` value!

This brings us to another question: what is the *dimension* of the vector space of functions? Or, to be concrete, how many real number coordinates are needed to uniquely identify a function?

Instead of naming the coordinates of a `Vec3` object x, y, and z, you could index them from $i = 1$ to 3. Likewise, you could index the coordinates of a `Vec15` from $i = 1$ to 15. A function, however, has infinitely many numbers that define it; for instance, the values $f(x)$ for any value of x. In other words, you can think of the coordinates of f as being its values at every point, indexed by all real numbers instead of the first few integers. This means that the vector space of functions is *infinite dimensional.* This has important implications, but it mostly makes the vector space of all functions hard to work with. We'll return to this space later, specifically looking at some subsets that are simpler. For now, let's return to the comfort of finitely many dimensions and look at two more examples.

6.2.4 Treating matrices as vectors

Because an n-by-m matrix is a list of $n \cdot m$ numbers, albeit arranged in a rectangle, we can treat it as a $n \cdot m$-dimensional vector. The only difference between the vector space of, say, 5×3 matrices from the vector space of 15D coordinate vectors is that the coordinates are presented in a matrix. We still add and scalar multiply coordinate by coordinate. Figure 6.7 shows how this addition looks.

$$2 + (-3) = -1$$

$$\begin{pmatrix} 2 & 0 & 6 \\ 10 & 6 & -3 \\ 9 & -3 & -5 \\ -5 & 5 & 5 \\ -2 & 8 & -2 \end{pmatrix} + \begin{pmatrix} -3 & -4 & 3 \\ 1 & -1 & 8 \\ -1 & -3 & 8 \\ 4 & 7 & -4 \\ -3 & 10 & 6 \end{pmatrix} = \begin{pmatrix} -1 & -4 & 9 \\ 11 & 5 & 5 \\ 8 & -6 & 3 \\ -1 & 12 & 1 \\ -5 & 18 & 4 \end{pmatrix}$$

Figure 6.7 Adding two 5×3 matrices by adding their corresponding entries

Implementing a class for 5×3 matrices inheriting from `Vector` is more typing than simply implementing a `Vec15` class because you need two loops to iterate over a matrix. The arithmetic, however, is no more complicated than as that shown in this listing.

Listing 6.2 A class representing 5×3 matrices thought of as vectors

```
class Matrix5_by_3(Vector):
    rows = 5
    columns = 3
    def __init__(self, matrix):
        self.matrix = matrix
    def add(self, other):
        return Matrix5_by_3(tuple(
            tuple(a + b for a,b in zip(row1, row2))
            for (row1, row2) in zip(self.matrix, other.matrix)
        ))
    def scale(self,scalar):
        return Matrix5_by_3(tuple(
            tuple(scalar * x for x in row)
            for row in self.matrix
        ))
    @classmethod
```

You need to know the number of rows and columns to be able to construct the zero matrix.

```
    def zero(cls):
        return Matrix5_by_3(tuple(
            tuple(0 for j in range(0, cls.columns))
            for i in range(0, cls.rows)
        ))
```

The zero vector for 5×3 matrices is a 5×3 matrix consisting of all zeroes. Adding this to any other 5×3 matrix M returns M.

You could just as well create a `Matrix2_by_2` class or a `Matrix99_by_17` class to represent different vector spaces. In these cases, much of the implementation would be the same, but the dimensions would no longer be 15, they would be $2 \cdot 2 = 4$ or $99 \cdot 17 = 1{,}683$. As an exercise, you can create a `Matrix` class inheriting from `Vector` that includes all the data except for specified numbers of rows and columns. Then any `MatrixM_by_N` class could inherit from `Matrix`.

The interesting thing about matrices isn't that they are numbers arranged in grids, but rather that we can think of them as representing linear functions. We already saw that lists of numbers and functions are two cases of vector spaces, but it turns out that matrices are vectors in both senses. If a matrix A has n rows and m columns, it represents a linear function from m-dimensional space to n-dimensional space. (You can write $A : \mathbb{R}^m \to \mathbb{R}^n$ to say this same sentence in mathematical shorthand.)

Just as we added and scalar-multiplied functions from $\mathbb{R} \to \mathbb{R}$, so can we add and scalar multiply functions from $\mathbb{R}^m \to \mathbb{R}^n$. In a mini-project at the end of this section, you can try running the vector space unit tests on matrices to check they are vectors in both senses. That doesn't mean grids of numbers aren't useful in their own right; sometimes we don't care to interpret them as functions. For instance, we can use arrays of numbers to represent images.

6.2.5 Manipulating images with vector operations

On a computer, images are displayed as arrays of colored squares called *pixels*. A typical image can be a few hundred pixels tall by a few hundred pixels wide. In a color image, three numbers are needed to specify the red, green, and blue (RGB) content of the color of any given pixel (figure 6.8). In total, a 300×300 pixel image is specified by $300 \cdot 300 \cdot 3 = 270{,}000$ numbers. When thinking of images of this size as vectors, the pixels live in a 270,000-dimensional space!

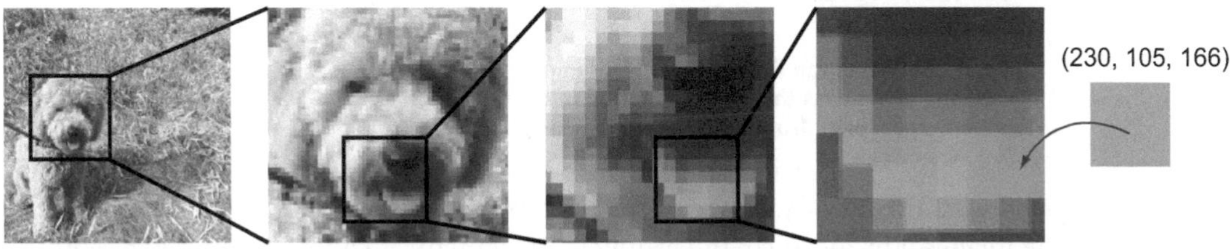

Figure 6.8 Zooming in on a picture of my dog, Melba, until we can pick out one pixel with red, green, and blue content (230, 105, 166, respectively)

Depending on what format you're reading this, you may or may not see the pink color of Melba's tongue. But because we'll represent color numerically rather than visually in this discussion, everything should still make sense. You can also see the pictures in full color in the source code for this book.

Python has a de-facto standard image manipulation library, PIL, which is distributed in pip under the package name `pillow`. You won't need to learn much about the library because we immediately encapsulate our use of it inside a new class (listing 6.3). This class, `ImageVector`, inherits from `Vector`, stores the pixel data of a 300×300 image, and supports addition and scalar multiplication.

Listing 6.3 A class representing an image as a vector

```
from PIL import Image
class ImageVector(Vector):
    size = (300,300)
    def __init__(self,input):
        try:
            img = Image.open(input).\
                  resize(ImageVector.size)
            self.pixels = img.getdata()
        except:
            self.pixels = input
    def image(self):
        img = Image.new('RGB', ImageVector.size)
        img.putdata([(int(r), int(g), int(b))
                     for (r,g,b) in self.pixels])
        return img
    def add(self,img2):
        return ImageVector([(r1+r2,g1+g2,b1+b2)
                            for ((r1,g1,b1),(r2,g2,b2))
                            in zip(self.pixels,img2.pixels)])
    def scale(self,scalar):
        return ImageVector([(scalar*r,scalar*g,scalar*b)
                      for (r,g,b) in self.pixels])
    @classmethod
    def zero(cls):
        total_pixels = cls.size[0] * cls.size[1]
        return ImageVector([(0,0,0) for _ in range(0,total_pixels)])
    def _repr_png_(self):
        return self.image()._repr_png_()
```

Handles images of a fixed size: 300×300 pixels, for example

The constructor accepts the name of an image file. We create an Image object with PIL, resize it to 300×300, and then extract its list of pixels with the getdata() method. Each pixel is a triple consisting of red, green, and blue values.

The constructor also accepts a list of pixels directly.

This method returns the underlying PIL image, reconstructed from the pixels stored as an attribute on the class. The values must be converted to integers to create a displayable image.

Performs vector addition for images by adding the respective red, green, and blue values for each pixel

Performs scalar multiplication by multiplying every red, green, and blue value for every pixel by the given scalar

The zero image has zero red, green, or blue content at any pixel.

Jupyter notebooks can display PIL images inline, as long as we pass the implementation of the function _repr_png_ from the underlying image.

Equipped with this library, we can load images by filename and do vector arithmetic with the images. For instance, the average of two pictures can be computed as a linear combination as follows with a result shown in figure 6.9:

```
0.5 * ImageVector("inside.JPG") + 0.5 * ImageVector("outside.JPG")
```

Figure 6.9 The average of two images of Melba as a linear combination

While any `ImageVector` is valid, the minimum and maximum color values that render as visually different are 0 and 255, respectively. Because of this, the negative of any image you import will be black, having gone below the minimum brightness at every pixel. Likewise, positive scalar multiples quickly become washed out with most pixels exceeding the maximum displayable brightness. Figure 6.10 shows these characteristics.

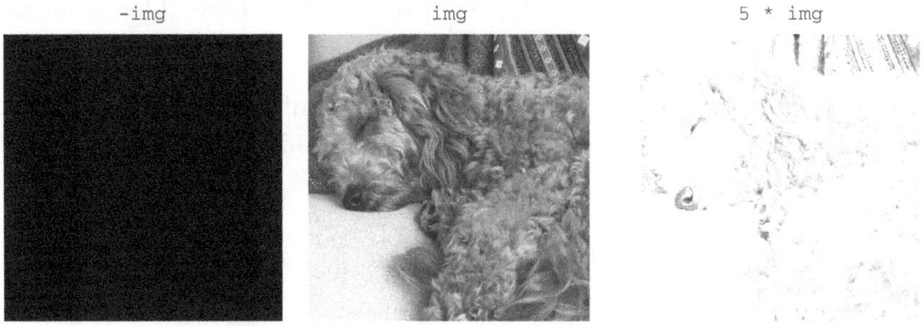

Figure 6.10 Negation and scalar multiplication of an image

To make visually interesting changes, you need to do operations that land you in the right brightness range for all colors. The zero vector (black) and the vector with all values equal to 255 (white) are good reference points. For instance, subtracting an image from an all white image has the effect of reversing the colors. As figure 6.11 shows, for the following white vector

```
white = ImageVector([(255,255,255) for _ in range(0,300*300)])
```

subtracting an image yields an eerily recolored picture. (The difference should be striking even if you're looking at the picture in black and white.)

Figure 6.11 Reversing the color of an image by subtracting it from a plain, white image

Vector arithmetic is clearly a general concept: the defining concepts of addition and scalar multiplication apply to numbers, coordinate vectors, functions, matrices, images, and many other kinds of objects. It's striking to see such visual results when we apply the same math across unrelated domains. We'll keep all of these examples of vector spaces in mind and continue to explore the generalizations we can make across them.

6.2.6 Exercises

Exercise 6.8 Run the vector space unit tests with float values for `u`, `v`, and `w`, rather than with objects inheriting from the `Vector` class. This demonstrates that real numbers are indeed vectors.

Solution With vectors as random scalars, the number zero as the zero vector, and `math.isclose` as the equality test, the 100 random tests pass:

```
for i in range(0,100):
    a,b = random_scalar(), random_scalar()
    u,v,w = random_scalar(), random_scalar(), random_scalar()
    test(0, isclose, a,b,u,v,w)
```

Exercise 6.9—Mini Project Run the vector space unit tests for `CarForSale` to show its objects form a vector space (ignoring their textual attributes).

Solution Most of the work is generating random data and building an approximate equality test that handles datetimes as shown here:

```
from math import isclose
from random import uniform, random, randint
```

```
from datetime import datetime, timedelta

def random_time():
    return CarForSale.retrieved_date - timedelta(days=uniform(0,10))

def approx_equal_time(t1, t2):
    test = datetime.now()
    return isclose((test-t1).total_seconds(), (test-t2).total_seconds())

def random_car():
    return CarForSale(randint(1990,2019), randint(0,250000),
              27000. * random(), random_time())

def approx_equal_car(c1,c2):
    return (isclose(c1.model_year,c2.model_year)
            and isclose(c1.mileage,c2.mileage)
            and isclose(c1.price, c2.price)
            and approx_equal_time(c1.posted_datetime,
c2.posted_datetime))

for i in range(0,100):
    a,b = random_scalar(), random_scalar()
    u,v,w = random_car(), random_car(), random_car()
    test(CarForSale.zero(), approx_equal_car, a,b,u,v,w)
```

Exercise 6.10 Implement the class `Function(Vector)` that takes a function of one variable as an argument to its constructor and implement a `__call__` method so you can treat it as a function. You should be able to run `plot([f,g,f+g,3*g],-10,10)`.

Solution

```
class Function(Vector):
    def __init__(self, f):
        self.function = f
    def add(self, other):
        return Function(lambda x: self.function(x) + other.function(x))
    def scale(self, scalar):
        return Function(lambda x: scalar * self.function(x))
    @classmethod
    def zero(cls):
        return Function(lambda x: 0)
    def __call__(self, arg):
        return self.function(arg)

f = Function(lambda x: 0.5 * x + 3)
g = Function(sin)

plot([f, g, f+g, 3*g], -10, 10)
```

(continued)

The result of the last line is shown in this plot:

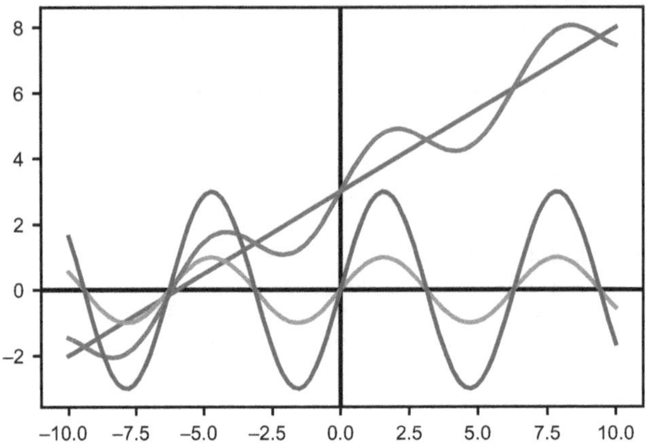

Our objects `f` and `g` behave like vectors, so we can add and scalar multiply them. Because they also behave like functions, we can plot them.

Exercise 6.11—Mini Project Testing equality of functions is difficult. Do your best to write a function to test whether two functions are equal.

Solution Because we're usually interested in well-behaved, continuous functions, it might be enough to check that their values are close for a few random input values as shown here:

```
def approx_equal_function(f,g):
    results = []
    for _ in range(0,10):
        x = uniform(-10,10)
        results.append(isclose(f(x),g(x)))
    return all(results)
```

Unfortunately, this can give us misleading results. The following returns `True`, even though the functions cannot be equal to zero:

```
approx_equal_function(lambda x: (x*x)/x, lambda x: x)
```

It turns out that computing equality of functions is an *undecidable* problem. That is, it has been proved there is no algorithm that can guarantee whether any two functions are equal.

Exercise 6.12—Mini Project Unit test your `Function` class to demonstrate that functions satisfy the vector space properties.

Solution It's difficult to test function equality, and it's also difficult to generate random functions. Here, I used a `Polynomial` class (that you'll meet in the next section) to generate some random polynomial functions. Using `approx_equal_function` from the previous mini-project, we can get the test to pass:

```
def random_function():
    degree = randint(0,5)
    p = Polynomial(*[uniform(-10,10) for _ in range(0,degree)])
    return Function(lambda x: p(x))

for i in range(0,100):
    a,b = random_scalar(), random_scalar()
    u,v,w = random_function(), random_function(), random_function()
    test(Function.zero(), approx_equal_function, a,b,u,v,w)
```

Exercise 6.13—Mini Project Implement a class `Function2(Vector)` that stores a function of *two* variables like $f(x, y) = x + y$.

Solution The definition is not much different than the `Function` class, but all functions are given two arguments:

```
class Function(Vector):
    def __init__(self, f):
        self.function = f
    def add(self, other):
        return Function(lambda x,y: self.function(x,y) +
other.function(x,y))
    def scale(self, scalar):
        return Function(lambda x,y: scalar * self.function(x,y))
    @classmethod
    def zero(cls):
        return Function(lambda x,y: 0)
    def __call__(self, *args):
        return self.function(*args)
```

For instance, the sum of $f(x, y) = x + y$ and $g(x, y) = x - y + 1$ should be $2x + 1$. We can confirm this:

```
>>> f = Function(lambda x,y:x+y)
>>> g = Function(lambda x,y: x-y+1)
>>> (f+g)(3,10)
7
```

Exercise 6.14 What is the dimension of the vector space of 9×9 matrices?

1. 9
2. 18
3. 27
4. 81

Solution A 9×9 matrix has 81 entries, so there are 81 independent numbers (or coordinates) that determine it. It, therefore, is an 81-dimensional vector space and answer *d* is correct.

Exercise 6.15—Mini Project Implement a `Matrix` class inheriting from `Vector` with abstract properties representing the number of rows and number of columns. You should not be able to instantiate a `Matrix` class, but you could make a `Matrix5_by_3` class by inheriting from `Matrix` and explicitly specifying the number of rows and columns.

Solution

```
class Matrix(Vector):
    @abstractproperty
    def rows(self):
        pass
    @abstractproperty
    def columns(self):
        pass
    def __init__(self,entries):
        self.entries = entries
    def add(self,other):
        return self.__class__(
            tuple(
                tuple(self.entries[i][j] + other.entries[i][j]
                        for j in range(0,self.columns()))
                for i in range(0,self.rows())))
    def scale(self,scalar):
        return self.__class__(
            tuple(
                tuple(scalar * e for e in row)
                for row in self.entries))
    def __repr__(self):
        return "%s%r" % (self.__class__.__qualname__, self.entries)
    def zero(self):
        return self.__class__(
            tuple(
                tuple(0 for i in range(0,self.columns()))
                for j in range(0,self.rows())))
```

We can now quickly implement any class representing a vector space of matrices of fixed size, for instance, 2×2:

```
class Matrix2_by_2(Matrix):
    def rows(self):
        return 2
    def columns(self):
        return 2
```

Then we can compute with 2×2 matrices as vectors:

```
>>> 2 * Matrix2_by_2(((1,2),(3,4))) + Matrix2_by_2(((1,2),(3,4)))
Matrix2_by_2((3, 6), (9, 12))
```

Exercise 6.16 Unit test the `Matrix5_by_3` class to demonstrate that it obeys the defining properties of a vector space.

Solution

```
def random_matrix(rows, columns):
    return tuple(
        tuple(uniform(-10,10) for j in range(0,columns))
        for i in range(0,rows)
    )

def random_5_by_3():
    return Matrix5_by_3(random_matrix(5,3))

def approx_equal_matrix_5_by_3(m1,m2):
    return all([
        isclose(m1.matrix[i][j],m2.matrix[i][j])
        for j in range(0,3)
        for i in range(0,5)
    ])

for i in range(0,100):
    a,b = random_scalar(), random_scalar()
    u,v,w = random_5_by_3(), random_5_by_3(), random_5_by_3()
    test(Matrix5_by_3.zero(), approx_equal_matrix_5_by_3, a,b,u,v,w)
```

Exercise 6.17—Mini Project Write a `LinearMap3d_to_5d` class inheriting from `Vector` that uses a 5×3 matrix as its data but implements `__call__` to act as a linear map from $\mathbb{R}^3$ to $\mathbb{R}^5$. Show that it agrees with `Matrix5_by_3` in its underlying computations and that it independently passes the defining properties of a vector space.

Exercise 6.18—Mini Project Write a Python function enabling you to multiply `Matrix5_by_3` objects by `Vec3` objects in the sense of matrix multiplication. Update your overloading of the `*` operator for the vector and matrix classes so you can multiply vectors on their left by either scalars or matrices.

Exercise 6.19 Convince yourself that the zero vector for the `ImageVector` class doesn't visibly alter any image when it is added.

Solution For any image of your choice, look at the result of `ImageVector ("my_ image.jpg") + ImageVector.zero()`.

Exercise 6.20 Pick two images and display 10 different weighted averages of them. These will be points on a line segment connecting the images in 270,000-dimensional space!

Solution I ran the following code with $s = 0.1, 0.2, 0.3, \ldots, 0.9, 1.0$:

```
s * ImageVector("inside.JPG") + (1-s) * ImageVector("outside.JPG")
```

When you put your images side-by-side, you'll get something like this:

Several different weighted averages of two images

Exercise 6.21 Adapt the vector space unit tests to images and run them. What do your randomized unit tests look like as images?

Solution One way to generate random images is to put random red, green, and blue values at every pixel, for example,

```
def random_image():
    return ImageVector([(randint(0,255), randint(0,255), randint(0,255))
                        for i in range(0,300 * 300)])
```

The result is a fuzzy mess, but that doesn't matter to us. The unit tests compare each pixel. With an approximate equality test such as the following, we can run the tests:

```
def approx_equal_image(i1,i2):
    return all([isclose(c1,c2)
        for p1,p2 in zip(i1.pixels,i2.pixels)
        for c1,c2 in zip(p1,p2)])

for i in range(0,100):
    a,b = random_scalar(), random_scalar()
    u,v,w = random_image(), random_image(), random_image()
    test(ImageVector.zero(), approx_equal_image, a,b,u,v,w)
```

6.3 Looking for smaller vector spaces

The vector space of 300×300 color images has a whopping 270,000 dimensions, meaning we need to list as many numbers to specify any image of that size. This isn't a problematic amount of data on its own, but when we have larger images, a large quantity of images, or thousands of images chained together to make a movie, the data can add up.

In this section, we look at how to start with a vector space and find smaller ones (having fewer dimensions) that retain most of the interesting data from the original space. With images, we can reduce the number of distinct pixels used in an image or convert it to black and white. The result may not be beautiful, but it can still be recognizable.

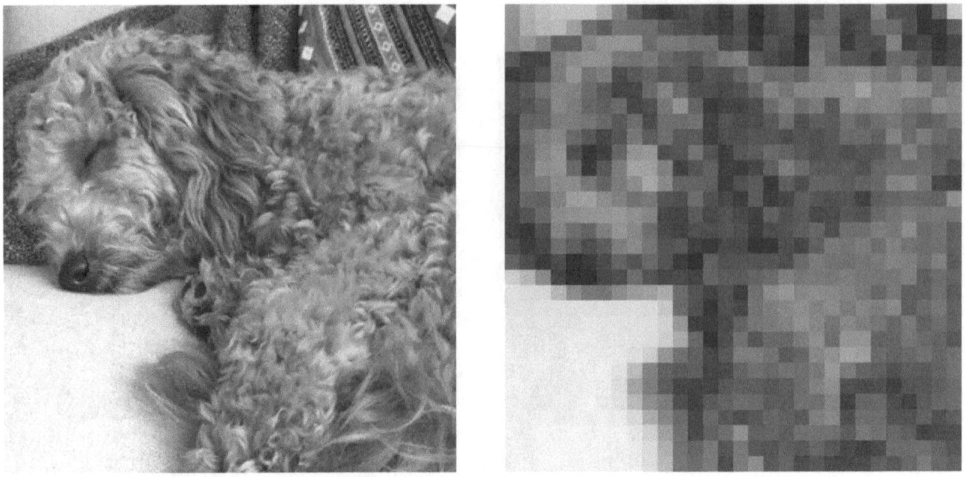

Figure 6.12 Converting from an image specified by 270,000 numbers (left) to another one specified by 900 numbers (right)

For instance, the image on the right in figure 6.12 takes 900 numbers to specify, compared to the 270,000 numbers to specify the image on the left.

Pictures that look like the one on the right live in a 900-dimensional *subspace* of a 270,000-dimensional space. That means that they are still 270,000-dimensional image vectors, but they can be represented or stored with only 900 coordinates. This is a starting point for a study of *compression*. We won't go too deep into the best practices of compression, but we will take a close look at subspaces of vector spaces.

6.3.1 Identifying subspaces

A vector *subspace*, or subspace for short, is just what it sounds like: a vector space that exists inside another vector space. One example we've looked at a few times already is the 2D x,y plane within 3D space as the plane where $z = 0$. To be specific, the subspace consists of vectors of the form $(x, y, 0)$. These vectors have three components, so they are veritable 3D vectors, but they form a subset that happens to be constrained to lie on a plane. For that reason, we say this is a 2D subspace of $\mathbb{R}^3$.

> **NOTE** At the risk of being pedantic, the 2D vector space $\mathbb{R}^2$, which consists of the ordered pairs (x, y), is not technically a subspace of 3D space $\mathbb{R}^3$. That's because vectors of the form (x, y) are not 3D vectors. However, it has a one-to-one correspondence with the set of vectors $(x, y, 0)$, and vector arithmetic looks the same whether or not the extra zero z-coordinate is present. For these reasons, I consider it okay to call $\mathbb{R}^2$ a subspace of $\mathbb{R}^3$.

Not every subset of 3D vectors is a subspace. The plane where $z = 0$ is special because the vectors (x, y, 0) form a self-contained vector space. There's no way to build a linear combination of vectors in this plane that somehow "escapes" it; the third coordinate always remains zero. In math lingo, the precise way to say that a subspace is self-contained is to say it is *closed* under linear combinations.

To get the feel for what a vector subspace looks like in general, let's search for subsets of vector spaces that are also subspaces (figure 6.13). What subsets of vectors in the plane can make a standalone vector space? Can we just draw any region in the plane and only take vectors that live within it?

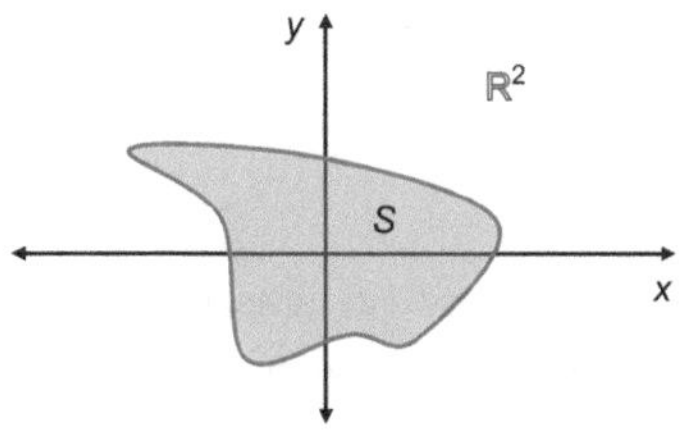

Figure 6.13 S is a subset of points (vectors) in the plane $\mathbb{R}^2$. Is S a subspace of $\mathbb{R}^2$?

The answer is no: the subset in figure 6.13 contains some vectors that lie on the x-axis and some that live on the y-axis. These can respectively be scaled to give us the standard basis vectors $\mathbf{e}_1 = (1, 0)$ and $\mathbf{e}_2 = (0, 1)$. From these vectors, we can make linear combinations to get to any point in the plane, not only the ones in S (figure 6.14).

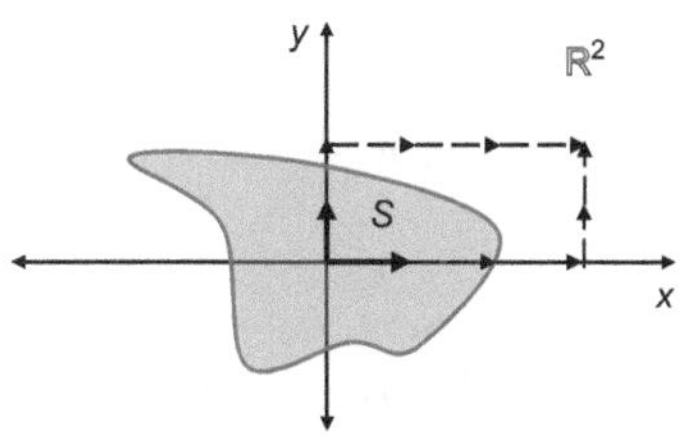

Figure 6.14 Linear combinations of two vectors in S give us an "escape route" from S. It cannot be a subspace of the plane.

Instead of drawing a random subspace, let's mimic the example of the plane in 3D. There is no z-coordinate, so let's instead choose the points where $y = 0$. This leaves us with the points on the x-axis, having the form $(x, 0)$. No matter how hard we try, we can't find a linear combination of vectors of this form that have a non-zero y-coordinate (figure 6.15).

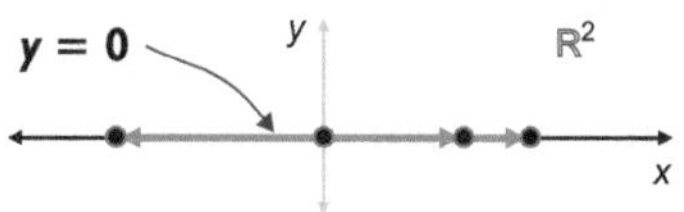

Figure 6.15 Focusing on the line where y = 0. This is a vector space, containing all linear combinations of its points.

This line, $y = 0$, is a vector subspace of $\mathbb{R}^2$. As we originally found a 2D subspace of 3D, we also have found a 1D subspace of 2D. Instead of a 3D *space* or a 2D *plane*, a 1D vector space like this is called a *line*. In fact, we can identify this subspace as the real number line $\mathbb{R}$.

The next step could be to set $x = 0$ as well. Once we've set both $x = 0$ and $y = 0$ to zero, there's only one point remaining: the zero vector. This is a vector subspace as well! No matter how you take linear combinations of the zero vector, the result is the zero vector. This is a *zero-dimensional subspace* of the 1D line, the 2D plane, and the 3D space. Geometrically, a zero-dimensional subspace is a point, and that point has to be zero. If it were some other point, v for instance, it would also contain $0 \cdot v = 0$ and an infinity of other different scalar multiples like $3 \cdot v$ and $-42 \cdot v$. Let's run with this idea.

6.3.2 Starting with a single vector

A vector subspace containing a non-zero vector **v** contains (at least) all of the scalar multiples of **v**. Geometrically, the set of all scalar multiples of a non-zero vector **v** lie on a line through the origin as shown in figure 6.16.

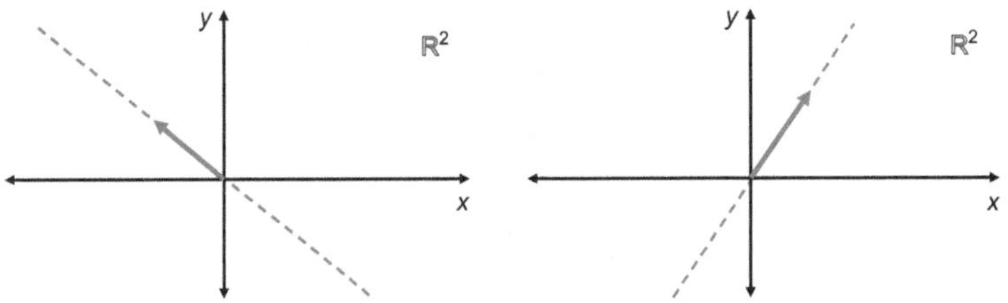

Figure 6.16 Two different vectors with dotted lines, showing where all of their scalar multiples will lie.

Each of these lines through the origin is a vector space. There's no way to escape any line like this by adding or scaling vectors that lie in it. This is true of lines through the origin in 3D as well: they are all of the linear combinations of a single 3D vector, and they form a vector space. This is the first example of a general way of building subspaces: picking a vector and seeing all of the linear combinations that must come with it.

6.3.3 Spanning a bigger space

Given a set of one or more vectors, their *span* is defined as the set of all linear combinations. The important part of the span is that it's automatically a vector subspace. To rephrase what we just discovered, the span of a single vector **v** is a line through the origin. We denote a set of objects by including them in curly braces, so the set containing only **v** is {**v**}, and the span of this set could be written span({**v**}).

As soon as we include another vector **w**, which is not parallel to **v**, the space gets bigger because we are no longer constrained to a single linear direction. The span of the set of two vectors {**v**, **w**} includes two lines, span({**v**}) and span({**w**}), as well as linear combinations including both **v** and **w**, which lie on neither line (figure 6.17).

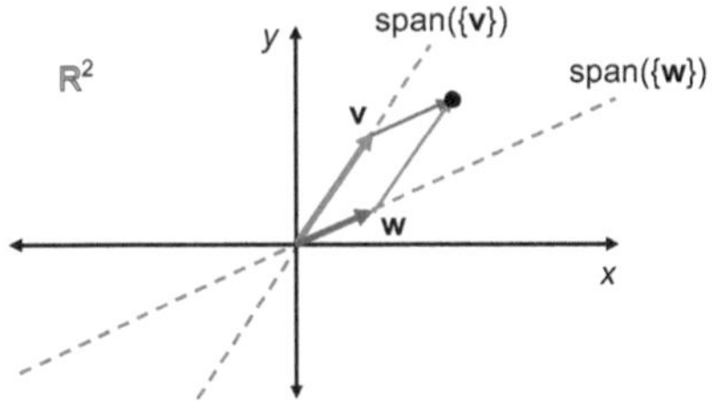

Figure 6.17 The span of two non-parallel vectors. Each individual vector spans a line, but together they span more points, for instance, v + w lies on neither line.

It might not be obvious, but the span of these two vectors is the entire plane. This is true of any pair of non-parallel vectors in the plane, but most strikingly for the standard basis vectors. Any point (x, y) can be reached as the linear combination $x \cdot (1, 0) + y \cdot (0, 1)$. The same is true for other pairs of non-parallel vectors like **v** = (1, 0) and **w** = (1, 1), but there's a bit more arithmetic to see it.

You can get any point like (4, 3) by taking the right linear combination of (1, 0) and (1, 1). The only way to get the *y*-coordinate of 3 is to have three of the vector (1, 1). That's (3, 3) instead of (4, 3), so you can correct the *x*-coordinate by adding one unit of (1, 0). That gets us a linear combination $3 \cdot (1, 1) + 1 \cdot (1, 0)$, which takes us to the point (4, 3) as shown in figure 6.18.

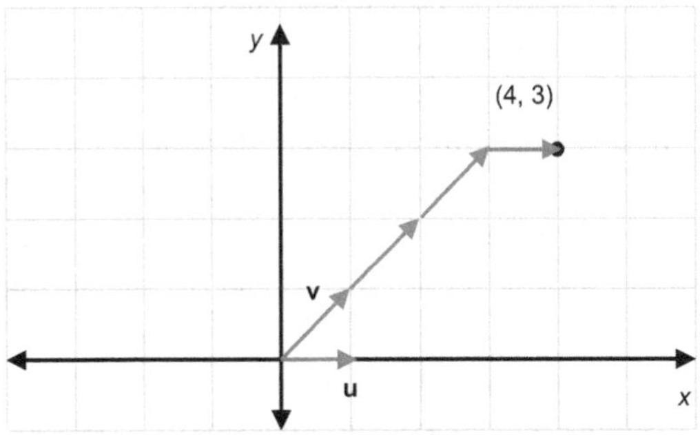

Figure 6.18 Getting to an arbitrary point (4, 3) by a linear combination of (1, 0) and (1, 1)

A single non-zero vector spans a line in 2D or 3D, and it turns out, two non-parallel vectors can span either the whole 2D plane or a plane passing through the origin in 3D space. A plane spanned by two 3D vectors could look like that shown in figure 6.19.

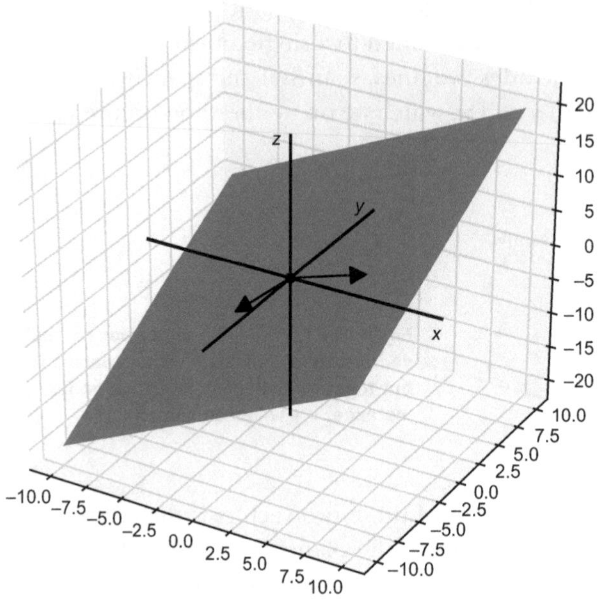

Figure 6.19 A plane spanned by two 3D vectors

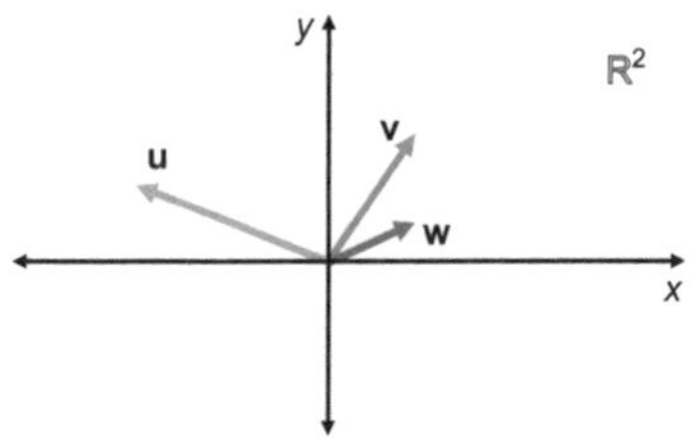

Figure 6.20 Three non-parallel vectors that only span a 2D space

It's slanted, so it doesn't look like the plane where $z = 0$, and it doesn't contain any of the three standard basis vectors. But it's still a plane and a vector subspace of 3D space. One vector spans a 1D space, and two non-parallel vectors span a 2D space. If we add a third non-parallel vector to the mix, do the three vectors span a 3D space? Figure 6.20 shows that clearly the answer is no.

No pair of the vectors **u**, **v**, and **w** is parallel, but these vectors don't span a 3D space. They all live in the 2D plane, so no linear combination of them can magically obtain a *z*-coordinate. We need a better generalization of the concept of "non-parallel" vectors.

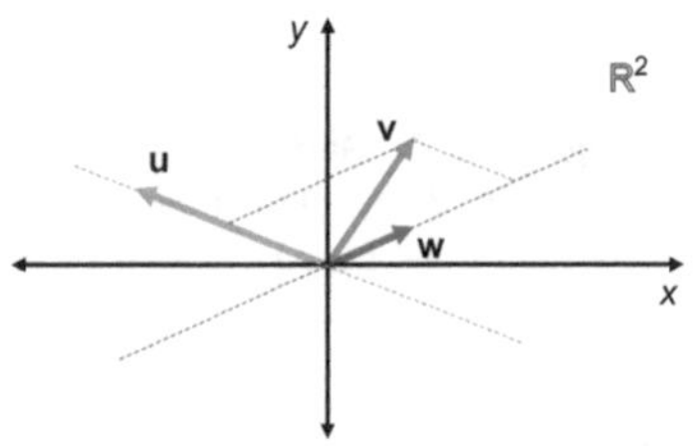

Figure 6.21 A linear combination of u and w returns v, so the span of u, v, and w should be no bigger than the span of u and w.

If we want to add a vector to a set and span a higher dimensional space, the new vector needs to point in a new direction that isn't included in the span of the existing ones. In the plane, three vectors always have some redundancy. For instance, as shown in figure 6.21, a linear combination of **u** and **w** gives us **v**.

The right generalization of "non-parallel" is *linearly independent*. A collection of vectors is *linearly dependent* if any of its members can be obtained as a linear combination of the others. Two parallel vectors are linearly dependent because they are scalar multiples of each other. Likewise, the set of three vectors {**u**, **v**, **w**} is linearly dependent because we can make **v** out of a linear combination of **u** and **w** (or **w** out of a linear combination of **u** and **v**, and so on). You should make sure to get a feel for this concept yourself. As one of the exercises at the end of this section, you can check that any of the three vectors (1, 0), (1, 1) and (–1, 1) can be written as a linear combination of the other two.

By contrast, the set {**u**, **v**} is *linearly independent* because the components are non-parallel and cannot be scalar multiples of one another. This means that **u** and **v** span a bigger space than either on its own. Similarly, the standard basis {$\mathbf{e}_1$, $\mathbf{e}_2$, $\mathbf{e}_3$} for $\mathbb{R}^3$ is a linearly independent set. None of these vectors can be built as a linear combination of the other two, and all three are required to span 3D space. We're starting to get at the properties of a vector space or subspace that indicate its dimension.

6.3.4 Defining the word dimension

Here's a motivational question: is the following set of 3D vectors linearly independent?

$$\{(1, 1, 1), (2, 0, -3), (0, 0, 1), (-1, -2, 0)\}$$

To answer this, you could draw these vectors in 3D or attempt to find a linear combination of three of them to get the fourth. But there's an easier answer: only three vectors are needed to span all of 3D space, so any list of four 3D vectors has to have some redundancy.

We know that a set with one or two 3D vectors will span a line or plane, respectively, rather than all of $\mathbb{R}^3$. Three is the magic number of vectors that can both span a 3D space and still be linearly independent. That's really *why* we call it three-dimensional: there are three independent directions after all.

A linearly independent set of vectors that spans a whole vector space like {$\mathbf{e}_1$, $\mathbf{e}_2$, $\mathbf{e}_3$} for $\mathbb{R}^3$ is called a *basis.* Any basis for a space has the same number of vectors, and that number is its *dimension.* For instance, we saw (1, 0) and (1, 1) are linearly independent and span the whole plane, so they are a basis for the vector space $\mathbb{R}^2$. Likewise (1, 0, 0) and (0, 1, 0) are linearly independent and span the plane where $z = 0$ in $\mathbb{R}^3$. That makes them a basis for this 2D subspace, albeit not a basis for all of $\mathbb{R}^3$.

I have already used the word *basis* in the context of the "standard basis" for $\mathbb{R}^2$ and for $\mathbb{R}^3$. These are called "standard" because they are such natural choices. It takes no computation to decompose a coordinate vector in the standard basis; the coordinates *are* the scalars in this decomposition. For instance, (3, 2) means the linear combination $3 \cdot (1, 0) + 2 \cdot (0, 1)$ or $3\mathbf{e}_1 + 2\mathbf{e}_2$.

In general, deciding whether vectors are linearly independent requires some work. Even if you know that a vector is a linear combination of some other vectors, finding that linear combination requires doing some algebra. In the next chapter, we cover how to do that; it ends up being a ubiquitous computational problem in linear algebra. But before that let's get in some more practice identifying subspaces and measuring their dimensions.

6.3.5 Finding subspaces of the vector space of functions

Mathematical functions from ℝ to ℝ contain an infinite amount of data, namely the output value when they are given any of infinitely many real numbers as inputs. That doesn't mean that it takes infinite data to describe a function though. For instance, a linear function requires only two real numbers. They are the values of a and b in this general formula that you've probably seen:

$$f(x) = ax + b$$

where a and b can be any real number. This is much more tractable than the infinite-dimensional space of all functions. Any linear function can be specified by two real numbers, so it looks like the subspace of linear functions will be 2D.

> **CAUTION** I've used the word *linear* in a lot of new contexts in the last few chapters. Here, I'm returning to a meaning you used in high school algebra: a *linear function* is a function whose graph is a straight line. Unfortunately, functions of this form are not linear in the sense we spent all of chapter 4 discussing, and you can prove it yourself in an exercise. Because of this, I'll try to be clear as to which sense of the word *linear* I'm using at any point.

We can quickly implement a `LinearFunction` class inheriting from `Vector`. Instead of holding a function as its underlying data, it can hold two numbers for the coefficients a and b. We can add these functions by adding coefficients because

$$(ax + b) + (cx + d) = (ax + cx) + (b + d) = (a + c)x + (b + d)$$

And we can scale the function by multiplying both coefficients by the scalar: $r(ax + b) = rax + rb$. Finally, it turns out the zero function $f(x) = 0$ is linear. It's the case where $a = b = 0$. Here's the implementation:

```
class LinearFunction(Vector):
    def __init__(self,a,b):
        self.a = a
        self.b = b
    def add(self,v):
        return LinearFunction(self.a + v.a, self.b + v.b)
    def scale(self,scalar):
        return LinearFunction(scalar * self.a, scalar * self.b)
    def __call__(self,x):
        return self.a * x + self.b
    @classmethod
    def zero(cls):
        return LinearFunction(0,0,0)
```

As figure 6.22 shows, the result is a linear function `plot([LinearFunction(-2,2)],-5,5)` gives us the straight line graph of $f(x) = -2x + 2$.

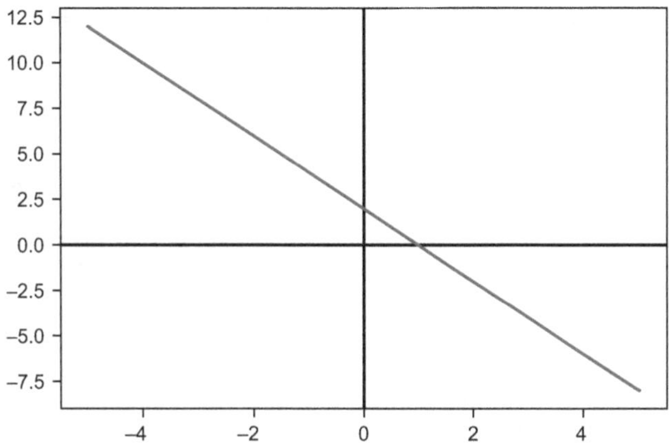

Figure 6.22 The graph of `LinearFunction(-2,2)` representing $f(x) = -2x + 2$

We can prove to ourselves that linear functions form a vector subspace of dimension 2 by writing a basis. The basis vectors should both be functions, they should span the whole space of linear functions, and they should be linearly independent (not multiples of one another). Such a set is $\{x, 1\}$ or, more specifically, $\{f(x) = x, g(x) = 1\}$. Named this way, functions of the form $ax + b$ can be written as a linear combination $a \cdot f + b \cdot g$.

This is as close as we can get to a standard basis for linear functions; $f(x) = x$ and $f(x) = 1$ are clearly different functions, not scalar multiples of one another. By contrast, $f(x) = x$ and $h(x) = 4x$ *are* scalar multiples of one another and would not be a linearly independent pair. But $\{x, 1\}$ is not the only basis we could have chosen; $\{4x + 1, x - 3\}$ is also a basis.

The same concept applies to *quadratic functions* having the form $f(x) = ax^2 + bx + c$. These form a 3D subspace of the vector space of functions with one choice of basis being $\{x^2, x, 1\}$. Linear functions form a vector subspace of the space of quadratic functions where the x^2 component is zero. Linear functions and quadratic functions are examples of *polynomial functions*, which are linear combinations of powers of x; for example,

$$f(x) = a_0 + a_1 x + a_2 x^2 + \ldots + a_n x^n$$

Linear and quadratic functions have *degree* 1 and 2, respectively, because those are the highest powers of x that appear in each. The polynomial written in the previous equation has degree n and $n + 1$ coefficients in total. In the exercises, you'll see that the space of polynomials of *any* degree forms another vector subspace of the space of functions.

6.3.6 *Subspaces of images*

Because our `ImageVector` objects are represented by 270,000 numbers, we could follow the standard basis formula and construct a basis of 270,000 images, each with one of the 270,000 numbers equal to 1 and all others equal to 0. The listing shows what the first basis vector would look like.

Listing 6.4 Pseudocode that builds a first standard basis vector

```
ImageVector([
    (1,0,0), (0,0,0), (0,0,0), ..., (0,0,0),
    (0,0,0), (0,0,0), (0,0,0), ..., (0,0,0),
    ...
])
```

Only the first pixel in the first row is non-zero: it has a red value of 1. All the other pixels have a value of (0,0,0).

The second row consists of 300 black pixels, each with a value (0,0,0).

I skipped the next 298 rows, but they are all identical to row 2; no pixels have any color values.

This single vector spans a 1D subspace consisting of the images that are black except for a single, red pixel in the top left corner. Scalar multiples of this image could have brighter or dimmer red pixels at this location, but no other pixels can be illuminated. In order to show more pixels, we need more basis vectors.

There's not too much to be learned from writing out these 270,000 basis vectors. Let's instead look for a small set of vectors that span an interesting subspace. Here's a single `ImageVector` consisting of dark gray pixels at every position:

```
gray = ImageVector([
    (1,1,1), (1,1,1), (1,1,1), ..., (1,1,1),
    (1,1,1), (1,1,1), (1,1,1), ..., (1,1,1),
    ...
])
```

More concisely, we could write this instead:

```
gray = ImageVector([(1,1,1) for _ in range(0,300*300)])
```

One way to picture the subspace spanned by the single vector gray is to look at some vectors that belong to it. Figure 6.23 shows scalar multiples of gray.

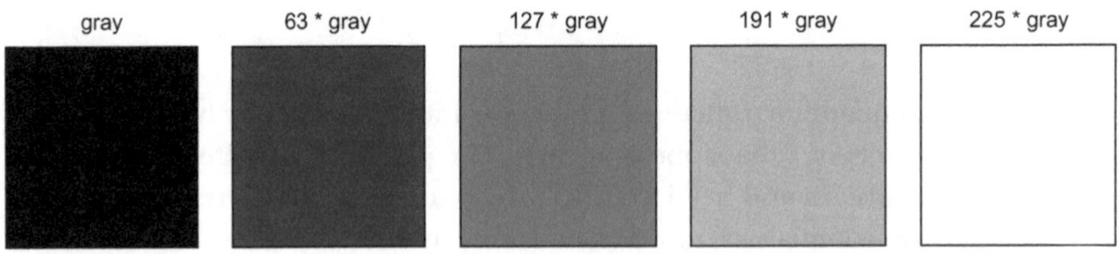

Figure 6.23 Some of the vectors in the 1D subspace of images spanned by the gray instance of `ImageVector`.

This collection of images is "one-dimensional" in the colloquial sense. There's only one thing changing about them, their brightness.

Another way we can look at this subspace is by thinking about the pixel values. In this subspace, any image has the same value at each pixel. For any given pixel, there is a 3D space of color possibilities measured by red, green, and blue coordinates. Gray pixels form a 1D subspace of this, containing points with all coordinates $s \cdot (1, 1, 1)$ for some scalar s (figure 6.24).

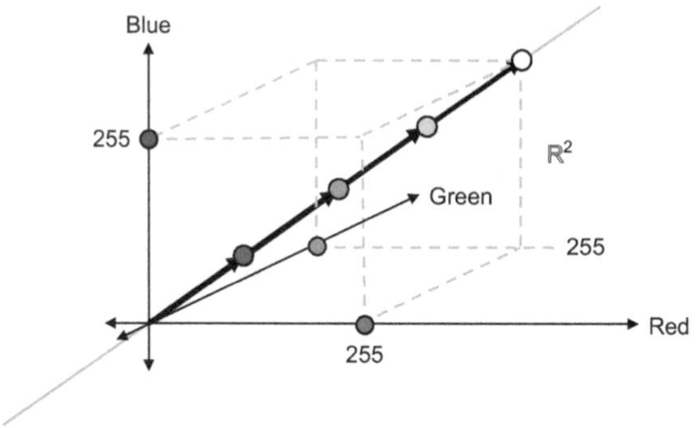

Figure 6.24 Gray pixels of varying brightness on a line. The gray pixels form a 1D subspace of the 3D vector space of pixel values.

Each of the images in the basis would be black, except for one pixel that would be a very dim red, green, or blue. Changing one pixel at a time doesn't yield striking results, so let's look for smaller and more interesting subspaces.

There are many subspaces of images you can explore. You could look at solid color images of any color. These would be images of the form:

```
ImageVector([
    (r,g,b), (r,g,b), (r,g,b), ..., (r,g,b),
    (r,g,b), (r,g,b), (r,g,b), ..., (r,g,b),
    ...
])
```

There are no constraints on the pixels themselves; the only constraint on a solid color image is that every pixel is the same. As a final example, you could consider a subspace consisting of low resolution, grayscale images like that shown in figure 6.25.

Each 10×10 pixel block has a constant gray value across its pixels, making it look like a 30×30 grid. There are only $30 \cdot 30 = 900$ numbers defining this image, so images like this one define a 900-dimensional subspace of the 270,000 dimensional space of images. It's a lot less data, but it's still possible to create recognizable images.

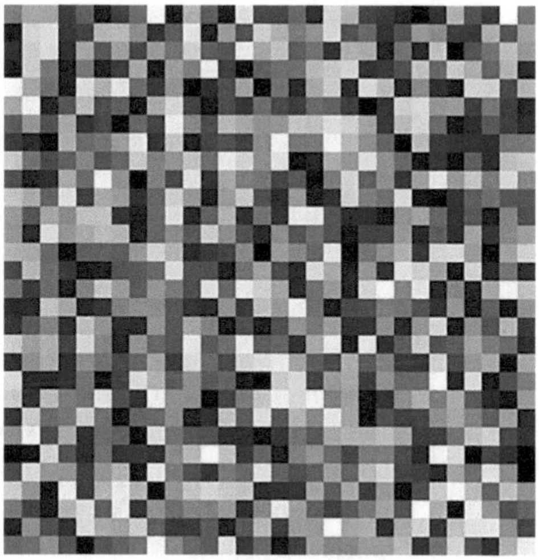

Figure 6.25 A low resolution grayscale image. Each 10×10 block of pixels has the same value.

One way to make an image in this subspace is to start with any image and average all red, green, and blue values in each 10×10 pixel block. This average gives you the brightness *b*, and you can set all pixels in the block to (*b*, *b*, *b*) to build your new image. This turns out to be a linear map (figure 6.26), and you can implement it later as a mini-project.

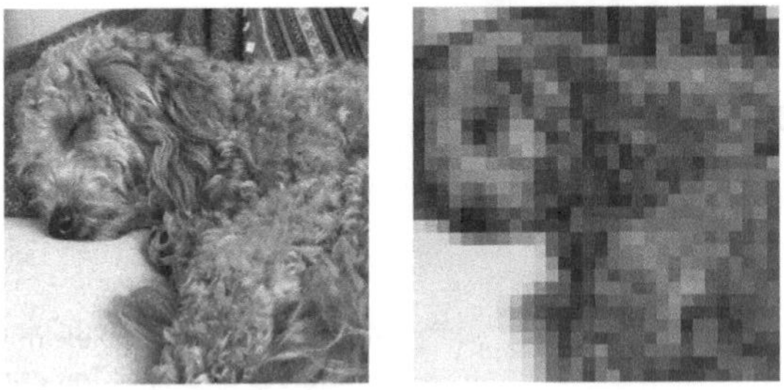

Figure 6.26 A linear map takes any image (left) and returns a new one (right) that lies in a 900-dimensional subspace.

My dog, Melba, isn't as photogenic in the second picture, but the picture is still recognizable. This is the example I mentioned at the beginning of the section, and the remarkable thing is that you can tell it's the same picture with only 0.3% of the data. There's clearly room for improvement, but the approach of mapping to a subspace is a starting point for more fruitful exploration. In chapter 13, we'll see how to compress audio data in this way.

6.3.7 Exercises

Exercise 6.22 Give a geometric argument for why the following region *S* of the plane can't be a vector subspace of the plane.

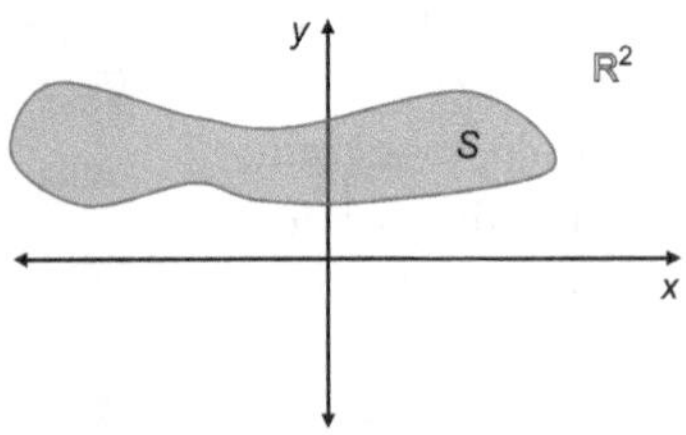

Solution There are many linear combinations of points in this region that don't end up in the region. More obviously, this region cannot be a vector space because it doesn't include the zero vector. The zero vector is a scalar multiple of any vector (by the scalar zero), so it must be included in any vector space or subspace.

Exercise 6.23 Show that the region of the plane where $x = 0$ forms a 1D vector space.

Solution These are the vectors that lie on the y-axis and have the form $(0, y)$ for a real number y. Addition and scalar multiplication of vectors of the form $(0, y)$ is the same as for real numbers; there just happens to be an extra 0 along for the ride. We can conclude that this is $\mathbb{R}$ in disguise and, therefore, a 1D vector space. If you want to be more rigorous, you can check all of the vector space properties explicitly.

Exercise 6.24 Show that three vectors (1, 0), (1, 1), and (–1, 1) are linearly dependent by writing each one as a linear combination of the other two.

Solution

$$(1, 0) = ½ \cdot (1, 1) - ½ \cdot (-1, 1)$$
$$(1, 1) = 2 \cdot (1, 0) + (-1, 1)$$
$$(-1, 1) = (1, 1) - 2 \cdot (1, 0)$$

Exercise 6.25 Show that you can get any vector (x, y) as a linear combination of (1, 0) and (1, 1).

Solution We know that (1, 0) can't contribute to the y-coordinate, so we need y times (1, 1) as part of the linear combination. To make the algebra work, we need $(x - y)$ units of (1, 0):

$$(x, y) = (x - y) \cdot (1, 0) + y(1, 1)$$

Exercise 6.26 Given a single vector **v**, explain why the set of all linear combinations of **v** is the same as the set of all scalar multiples of **v**.

Solution Linear combinations of a vector and itself reduce to scalar multiples according to one of the vector space laws. For instance, the linear combination $a \cdot \mathbf{v} + b \cdot \mathbf{v}$ is equal to $(a + b) \cdot \mathbf{v}$.

Exercise 6.27 From a geometric perspective, explain why a line that *doesn't* pass through the origin is not a vector subspace (of the plane or of the 3D space).

Solution One simple reason this cannot be a subspace is that it doesn't contain the origin (the zero vector). Another reason is that such a line will have two non-parallel vectors. Their span would be the whole plane, which is much bigger than the line.

Exercise 6.28 Any two of $\{\mathbf{e}_1, \mathbf{e}_2, \mathbf{e}_3\}$ will fail to span all of $\mathbb{R}^3$ and will instead span 2D subspaces of a 3D space. What are these subspaces?

Solution The span of the set $\{\mathbf{e}_1, \mathbf{e}_2\}$ consists of all linear combinations $a \cdot \mathbf{e}_1 + b \cdot \mathbf{e}_2$, or $a \cdot (1, 0, 0) + b \cdot (0, 1, 0) = (a, b, 0)$. Depending on the choice of a and b, this can be any point in the plane where $z = 0$, often called the x,y plane. By the same argument, the vectors $\{\mathbf{e}_2, \mathbf{e}_3\}$ span the plane where $x = 0$, called the y,z plane, and the vectors $\{\mathbf{e}_1, \mathbf{e}_3\}$ span the plane where $y = 0$, called the x,z plane.

Exercise 6.29 Write the vector (–5, 4) as a linear combination of (0, 3) and (–2, 1).

Solution Only (–2, 1) can contribute to the x-coordinate, so we need to have $2.5 \cdot (-2, 1)$ in the sum. That gets us to (–5, 2.5), so we need an additional 1.5 units on the x-coordinate or $0.5 \cdot (0, 3)$. The linear combination is

$$(-5, 4) = 0.5 \cdot (0, 3) + 2.5 \cdot (-2, 1)$$

Exercise 6.30—Mini Project Are (1, 2, 0), (5, 0, 5), and (2, –6, 5) linearly independent or linearly dependent vectors?

Solution It's not easy to find, but there is a linear combination of the first two vectors that yields the third:

$$-3 \cdot (1, 2, 0) + (5, 0, 5) = (2, -6, 5)$$

This means that the third vector is redundant, and the vectors *are* linearly dependent. They only span a 2D subspace of 3D rather than all of 3D space.

Exercise 6.31 Explain why the linear function $f(x) = ax + b$ is not a linear map from the vector space $\mathbb{R}$ to itself unless $b = 0$.

Solution We can turn directly to the definition: a linear map must preserve linear combinations. We see that f doesn't preserve linear combinations of real numbers. For instance, $f(1+1) = 2a + b$ while $f(1) + f(1) = (a + b) + (a + b) = 2a + 2b$. This won't hold unless $b = 0$.

As an alternative explanation, we know that linear functions $\mathbb{R} \to \mathbb{R}$ should be representable as 1-by-1 matrices. Matrix multiplication of a 1D column vector $[x]$ by a 1-by-1 matrix $[a]$ gives you $[ax]$. This is an unusual case of matrix multiplication, but your implementation from chapter 5 confirms this result. If a function $\mathbb{R} \to \mathbb{R}$ is going to be linear, it must agree with 1-by-1 matrix multiplication and, therefore, be multiplication by a scalar.

Exercise 6.32 Rebuild the `LinearFunction` class by inheriting from `Vec2` and implementing the `__call__` method.

Solution The data of a `Vec2` are called x and y instead of a and b; otherwise, the functionality is the same. All you need to do is implement `__call__`:

```
class LinearFunction(Vec2):
    def __call__(self,input):
        return self.x * input + self.y
```

Exercise 6.33 Prove (algebraically!) that the linear functions of the form $f(x) = ax + b$ make up a vector subspace of the vector space of all functions.

Solution To prove this, you need to be sure a linear combination of two linear functions is another linear function. If $f(x) = ax + b$ and $g(x) = cx + d$, then $r \cdot f + s \cdot g$ returns

$$r \cdot f + s \cdot g = r \cdot (ax + b) + s \cdot (cx + d) = rax + b + scx + d = (ra + sc) \cdot x + (b + d)$$

Because $(ra + sc)$ and $(b + d)$ are scalars, this has the form we want. We can conclude that linear functions are closed under linear combinations and, therefore, that they form a subspace.

Exercise 6.34 Find a basis for the set of 3-by-3 matrices. What is the dimension of this vector space?

Solution Here's a basis consisting of nine, 3-by-3 matrices:

$$\begin{pmatrix} 1 & 0 & 0 \\ 0 & 0 & 0 \\ 0 & 0 & 0 \end{pmatrix} \begin{pmatrix} 0 & 1 & 0 \\ 0 & 0 & 0 \\ 0 & 0 & 0 \end{pmatrix} \begin{pmatrix} 0 & 0 & 1 \\ 0 & 0 & 0 \\ 0 & 0 & 0 \end{pmatrix}$$

$$\begin{pmatrix} 0 & 0 & 0 \\ 1 & 0 & 0 \\ 0 & 0 & 0 \end{pmatrix} \begin{pmatrix} 0 & 0 & 0 \\ 0 & 1 & 0 \\ 0 & 0 & 0 \end{pmatrix} \begin{pmatrix} 0 & 0 & 0 \\ 0 & 0 & 1 \\ 0 & 0 & 0 \end{pmatrix}$$

$$\begin{pmatrix} 0 & 0 & 0 \\ 0 & 0 & 0 \\ 1 & 0 & 0 \end{pmatrix} \begin{pmatrix} 0 & 0 & 0 \\ 0 & 0 & 0 \\ 0 & 1 & 0 \end{pmatrix} \begin{pmatrix} 0 & 0 & 0 \\ 0 & 0 & 0 \\ 0 & 0 & 1 \end{pmatrix}$$

They are linearly independent; each contributes a unique entry to any linear combination. They also span the space because any matrix can be constructed as a linear combination of these; the coefficient on any particular matrix decides one entry of the result. Because these nine vectors provide a basis for the space of 3-by-3 matrices, the space has nine dimensions.

Exercise 6.35—Mini Project Implement a class `QuadraticFunction(Vector)` that represents the vector subspace of functions of the form $ax^2 + bx + c$. What is a basis for this subspace?

Solution The implementation looks a lot like `LinearFunction`, except there are three coefficients instead of two, and the `__call__` function has a square term:

```
class QuadraticFunction(Vector):
    def __init__(self,a,b,c):
        self.a = a
        self.b = b
        self.c = c
    def add(self,v):
        return QuadraticFunction(self.a + v.a,
                                 self.b + v.b,
                                 self.c + v.c)
    def scale(self,scalar):
        return QuadraticFunction(scalar * self.a,
                                 scalar * self.b,
                                 scalar * self.c)
    def __call__(self,x):
        return self.a * x * x + self.b * x + self.c
    @classmethod
    def zero(cls):
        return QuadraticFunction(0,0,0)
```

We can take note that $ax^2 + bx + c$ looks like a linear combination of the set $\{x^2, x, 1\}$. Indeed, these three functions span the space, and none of these three can be written as a linear combination of the others. There's no way to get a x^2 term by adding together linear functions, for example. Therefore, this is a basis. Because there are three vectors, we can conclude that this is a 3D subspace of the space of functions.

Exercise 6.36—Mini Project I claimed that $\{4x + 1, x - 2\}$ are a basis for the set of linear functions. Show that you can write $-2x + 5$ as a linear combination of these two functions.

Solution $(1/9) \cdot (4x + 1) - (22/9) \cdot (x - 2) = -2x + 5$. If your algebra skills aren't too rusty, you can figure this out by hand. Otherwise, don't worry; we cover how to solve tricky problems like this in the next chapter.

Exercise 6.37—Mini Project The vector space of all polynomials is an infinite-dimensional subspace. Implement that vector space as a class and describe a basis (which must be an infinite set!).

Solution

```
class Polynomial(Vector):
    def __init__(self, *coefficients):
        self.coefficients = coefficients
    def __call__(self,x):
        return sum(coefficient * x ** power
                   for (power,coefficient)
                   in enumerate(self.coefficients))
    def add(self,p):
        return Polynomial([a + b
                          for a,b
                          in zip(self.coefficients,
                                 p.coefficients)])
    def scale(self,scalar):
        return Polynomial([scalar * a
                          for a in self.coefficients])
        return "$ %s $" % (" + ".join(monomials))
    @classmethod
    def zero(cls):
        return Polynomial(0)
```

A basis for the set of all polynomials is the infinite set $\{1, x, x^2, x^3, x^4, \ldots\}$. Given all of the possible powers of x at your disposal, you can build any polynomial as a linear combination.

Exercise 6.38 I showed you pseudocode for a basis vector for the 270,000 dimensional space of images. What would the second basis vector look like?

Solution The second basis vector could be given by putting a one in the next possible place. It would yield a dim green pixel in the very top left of the image:

```
ImageVector([
    (0,1,0), (0,0,0), (0,0,0), ..., (0,0,0),
    (0,0,0), (0,0,0), (0,0,0), ..., (0,0,0),
    ...
])
```

For the second basis vector, the 1 has moved to the second possible slot.

All other rows remain empty

Exercise 6.39 Write a function `solid_color(r,g,b)` that returns a solid color `ImageVector` with the given red, green, and blue content at every pixel.

Solution

```
def solid_color(r,g,b):
    return ImageVector([(r,g,b) for _ in range(0,300*300)])
```

Exercise 6.40—Mini Project Write a linear map that generates an `ImageVector` from a 30×30 grayscale image, implemented as a 30×30 matrix of brightness values. Then, implement the linear map that takes a 300×300 image to a 30×30 grayscale image by averaging the brightness (average of red, green, and blue) at each pixel.

Solution

```
image_size = (300,300)
total_pixels = image_size[0] * image_size[1]
square_count = 30
square_width = 10

def ij(n):
    return (n // image_size[0], n % image_size[1])

def to_lowres_grayscale(img):

    matrix = [
        [0 for i in range(0,square_count)]
        for j in range(0,square_count)
    ]
    for (n,p) in enumerate(img.pixels):
        i,j = ij(n)
        weight = 1.0 / (3 * square_width * square_width)
        matrix[i // square_width][ j // square_width] += (sum(p) * weight)
    return matrix
```

Indicates that we're breaking the picture into a 30×30 grid

The function takes an ImageVector and returns an array of 30 arrays of 30 values each, giving grayscale values square by square.

> **The second function takes a 30×30 matrix and returns an image built from 10×10 pixel blocks, having a brightness given by the matrix values.**

```
def from_lowres_grayscale(matrix):
    def lowres(pixels, ij):
        i,j = ij
        return pixels[i // square_width][ j // square_width]
    def make_highres(limg):
        pixels = list(matrix)
        triple = lambda x: (x,x,x)
        return ImageVector([triple(lowres(matrix, ij(n))) for n in
range(0,total_pixels)])
    return make_highres(matrix)
```

Calling `from_lowres_grayscale(to_lowres_grayscale(img))` transforms the image `img` in the way I showed in the chapter.

Summary

- A vector space is a generalization of the 2D plane and 3D space: a collection of objects that can be added and multiplied by scalars. These addition and scalar multiplication operations must behave in certain ways (listed in section 6.1.5) to mimic the more familiar operations in 2D and 3D.
- You can generalize in Python by pulling common features of different data types into an abstract base class and inheriting from it.
- You can overload arithmetic operators in Python so that vector math looks the same in code, regardless of what kind of vectors you're using.
- Addition and scalar multiplication need to behave in certain ways to match your intuition, and you can verify these behaviors by writing unit tests involving random vectors.
- Real-world objects like used cars can be described by several numbers (coordinates) and, therefore, treated as vectors. This lets us think about abstract concepts like a "weighted average of two cars."
- Functions can be thought of as vectors. You add or multiply them by adding or multiplying the expressions that define them.
- Matrices can be thought of as vectors. The entries of an $m \times n$ matrix can be thought of as coordinates of an $(m \cdot n)$-dimensional vector. Adding or scalar multiplying matrices has the same effect as adding or scalar multiplying the linear functions they define.
- Images of a fixed height and width make up a vector space. They are defined by a red, green, and blue (RGB) value at each pixel, so the number of coordinates and, therefore, the dimension of the space is defined by three times the number of pixels.

- A subspace of a vector space is a subset of the vectors in a vector space, which is a vector space on its own. That is, linear combinations of vectors in the subspace stay in the subspace.
- For any line through the origin in 2D or 3D, the set vectors that lie on it form a 1D subspace. For any plane through the origin in 3D, the vectors that lie on it form a 2D subspace.
- The span of a set of vectors is the collection of all linear combinations of the vectors. It is guaranteed to be a subspace of whatever space the vectors live in.
- A set of vectors is linearly *independent* if you can't make any one of them as a linear combination of the others. Otherwise, the set is linearly *dependent.* A set of linearly independent vectors that span a vector space (or subspace) is called a *basis* for that space. For a given space, any basis will have the same number of vectors. That number defines the dimension of the space.
- When you can think of your data as living in a vector space, subspaces often consist of data with similar properties. For instance, the subset of image vectors that are solid colors forms a subspace.

7 Solving systems of linear equations

This chapter covers

- Detecting collisions of objects in a 2D video game
- Writing equations to represent lines and finding where lines intersect in the plane
- Picturing and solving systems of linear equations in 3D or beyond
- Rewriting vectors as linear combinations of other vectors

When you think of algebra, you probably think of problems that require "solving for *x*." For instance, you probably spent quite a bit of time in algebra class learning to solve equations like $3x^2 + 2x + 4 = 0$; that is, figuring out what value or values of x make the equation true.

Linear algebra, being a branch of algebra, has the same kinds of computational questions. The difference is that what you want to solve for may be a vector or matrix rather than a number. If you take a traditional linear algebra course, you might cover a lot of algorithms to solve these kinds of problems. But because you have Python at your disposal, you only need to know how to recognize the problem you're facing and choose the right library to find the answer for you.

I'm going to cover the most important class of linear algebra problems you'll see in the wild: *systems of linear equations*. These problems boil down to finding points where lines, planes, or their higher dimensional analogies intersect. One example is the infamous high school math problem involving two trains leaving Boston and New York at different times and speeds. But because I don't assume railroad operation interests you, I'll use a more entertaining example.

In this chapter, we build a simple remake of the classic Asteroids arcade game (figure 7.1). In this game, the player controls a triangle representing a spaceship and fires a laser at polygons floating around it, which represent asteroids. The player must destroy the asteroids to prevent them from hitting and destroying the spaceship.

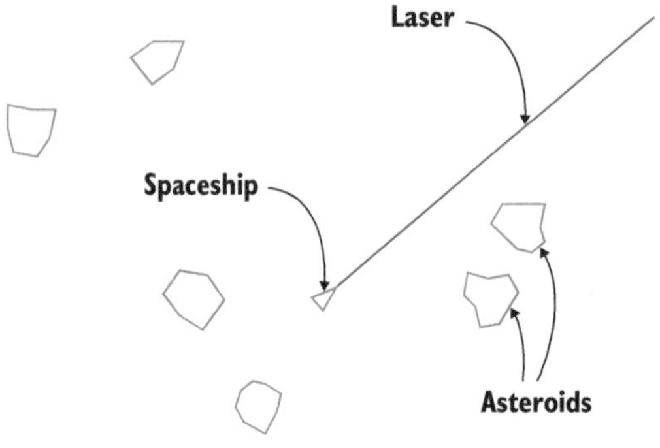

Figure 7.1 Setup of the classic Asteroids arcade game

One of the key mechanics in this game is deciding whether the laser hits an asteroid. This requires us to figure out whether the line defining the laser beam intersects with the line segments outlining the asteroids. If these lines intersect, the asteroid is destroyed. We'll set up the game first, and then we'll see how to solve the underlying linear algebra problem.

After we implement our game, I'll show you how this 2D example generalizes to 3D or any number of dimensions. The latter half of this chapter covers a bit more theory, but it will round out your linear algebra education. We'll have covered many of the major concepts you'd find in a college-level linear algebra class, albeit in less depth. After completing this chapter, you should be well prepared to crack open a denser textbook on linear algebra and fill in the details. But for now, let's focus on building our game.

7.1 Designing an arcade game

In this chapter, I focus on a simplified version of the asteroid game where the ship and asteroids are static. In the source code, you'll see that I already made the asteroids move, and we'll cover how to make them move according to the laws of physics in part

2 of this book. To get started, we model the entities of the game—the spaceship, the laser, and the asteroids—and show how to render them onscreen.

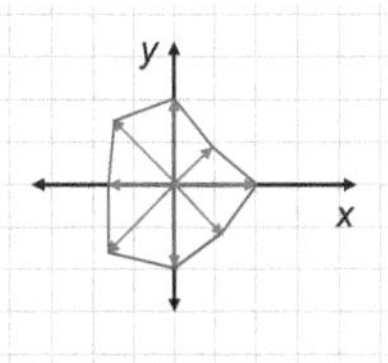

Figure 7.2 An eight-sided polygon representing an asteroid

7.1.1 Modeling the game

In this section, we display the spaceship and the asteroids as polygons in the game. As before, we model these as collections of vectors. For instance, we can represent an eight-sided asteroid by eight vectors (indicated by arrows in figure 7.2), and we can connect them to draw its outline.

The asteroid or spaceship translates or rotates as it travels through space, but its shape remains the same. Therefore, we store the vectors representing this shape separately from the *x*- and *y*-coordinates of its center, which can change over time. We also store an angle, indicating the rotation of the object at the current moment. The `PolygonModel` class represents a game entity (the ship or an asteroid) that keeps its shape but can translate or rotate. It's initialized with a set of vector points that define the outline of the asteroid, and by default, its center *x*- and *y*-coordinates and its angle of rotation are set to zero:

```
class PolygonModel():
    def __init__(self,points):
        self.points = points
        self.rotation_angle = 0
        self.x = 0
        self.y = 0
```

When the spaceship or asteroid moves, we need to apply the translation by `self.x,self.y` and the rotation by `self.rotation_angle` to find out its actual location. As an exercise, you can give `PolygonModel` a method to compute the actual, transformed vectors outlining it.

The spaceship and asteroids are specific cases of `PolygonModel` that initialize automatically with their respective shapes. For instance, the ship has a fixed triangular shape, given by three points:

```
class Ship(PolygonModel):
    def __init__(self):
        super().__init__([(0.5,0), (-0.25,0.25), (-0.25,-0.25)])
```

For the asteroid, we initialize it with somewhere between 5 and 9 vectors at equally spaced angles and random lengths between 0.5 and 1.0. This randomness gives the asteroids some character:

```
class Asteroid(PolygonModel):
    def __init__(self):
        sides = randint(5,9)
        vs = [vectors.to_cartesian((uniform(0.5,1.0), 2*pi*i/sides))
                for i in range(0,sides)]
        super().__init__(vs)
```

An asteroid has a random number of sides between 5 and 9.

Lengths are randomly selected between 0.5 and 1.0, and the angles are multiples of $2\pi/n$, where n is the number of sides.

With these objects defined, we can turn our attention to instantiating them and rendering them onscreen.

7.1.2 Rendering the game

For the initial state of the game, we need a ship and several asteroids. The ship can begin at the center of the screen, but the asteroids should be randomly spread out over the screen. We can show an area of the plane ranging from –10 to 10 in the *x* and *y* directions like this:

```
ship = Ship()

asteroid_count = 10
asteroids = [Asteroid() for _ in range(0,asteroid_count)]

for ast in asteroids:
    ast.x = randint(-9,9)
    ast.y = randint(-9,9)
```

Creates a list of a specified number of Asteroid objects, in this case, 10

Sets the position of each object to a random point with coordinates between –10 and 10 so it shows up onscreen

I use a 400×400 pixel screen, which requires transforming the *x*- and *y*-coordinates before rendering them. Using PyGame's built-in 2D graphics instead of OpenGL, the top left pixel on the screen has the coordinate (0, 0) and the bottom right has the coordinate (400, 400). These coordinates are not only bigger, they're also translated and upside down, so we need to write a `to_pixels` function (illustrated in figure 7.3) that does the transformation from our coordinate system to PyGame's pixels.

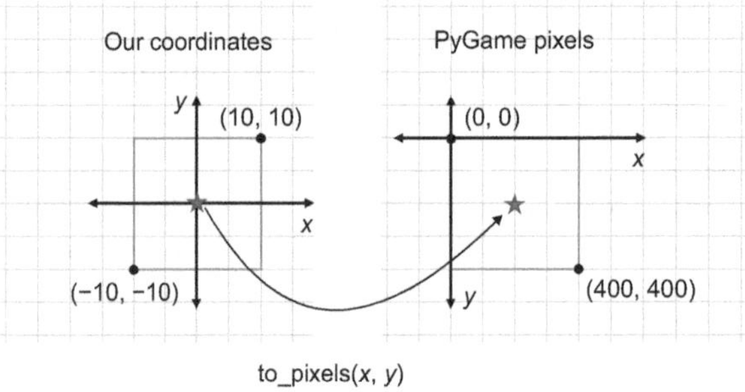

Figure 7.3 The `to_pixels` function maps an object from the center of our coordinate system to the center of the PyGame screen.

With the `to_pixels` function implemented, we can write a function to draw a polygon defined by points to the PyGame screen. First, we take the transformed points (translated and rotated) that define the polygon and convert them to pixels. Then we draw them with a PyGame function:

```
GREEN = (0, 255, 0)
def draw_poly(screen, polygon_model, color=GREEN):
    pixel_points = [to_pixels(x,y) for x,y in polygon_model.transformed()]
    pygame.draw.aalines(screen, color, True, pixel_points, 10)
```

Draws lines connecting given points to a specified PyGame object. The True parameter connects the first and last points to create a closed polygon.

You can see the whole game loop in the source code, but it basically calls `draw_poly` for the ship and each asteroid every time a frame is rendered. The result is our simple triangular spaceship surrounded by an asteroid field in a PyGame window (figure 7.4).

Figure 7.4 The game rendered in a PyGame window

7.1.3 Shooting the laser

Now it's time for the most important part: giving our ship a way to defend itself! The player should be able to aim the ship using the left and right arrow keys and then shoot a laser by pressing the spacebar. The laser beam should come out of the tip of the spaceship and extend to the edge of the screen.

In the 2D world we've invented, the laser beam should be a line segment starting at the *transformed* tip of the spaceship and extending in whatever direction the ship is pointed. We can make sure it reaches the end of the screen by making it sufficiently long. Because the laser's line segment is associated with the state of the `Ship` object, we can make a method on the `Ship` class to compute it:

```
class Ship(PolygonModel):
        ...
        def laser_segment(self):
        dist = 20. * sqrt(2)
        x,y = self.transformed()[0]
        return ((x,y),
            (x + dist * cos(self.rotation_angle),
             y + dist*sin(self.rotation_angle)))
```

Uses the Pythagorean theorem to find the longest segment that fits onscreen

Gets the value of the first of the definition points (the tip of the ship)

Uses trigonometry to find an endpoint for the laser if it extends dist units from the tip (x,y) at a self.rotation_angle (figure 7.5)

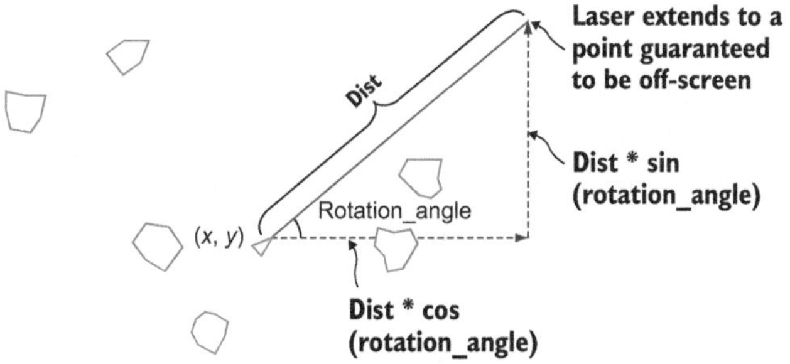

Figure 7.5 Using trigonometry to find the off-screen point where the laser beam ends

In the source code, you can see how to make PyGame respond to keystrokes and draw the laser as a line segment only if the spacebar is pressed. Finally, if the player fires the laser and hits an asteroid, we want to know something happened. In every iteration of the game loop, we want to check each asteroid to see if it is currently hit by the laser. We do this with a `does_intersect(segment)` method on the `PolygonModel` class, which computes whether the input segment intersects any segment of the given `PolygonModel`. The final code includes some lines like the following:

```
laser = ship.laser_segment()
keys = pygame.key.get_pressed()
    if keys[pygame.K_SPACE]:
    draw_segment(*laser)

    for asteroid in asteroids:
        if asteroid.does_intersect(laser):
            asteroids.remove(asteroid)
```

Calculates the line segment representing the laser beam based on the ship's current position and orientation

Detects which keys are pressed. If the spacebar is pressed, renders the laser beam to the screen with a helper function draw_segment (similar to draw_poly).

For every asteroid, checks whether the laser line segment intersects it. If so, destroys the given asteroid by removing it from the list of asteroids.

The work that remains is implementing the `does_intersect(segment)` method. In the next section, we cover the math to do so.

7.1.4 Exercises

Exercise 7.1 Implement a `transformed()` method on the `PolygonModel` that returns the points of the model translated by the object's x and y attributes and rotated by its `rotation_angle` attribute.

Solution Make sure to apply the rotation first; otherwise, the translation vector is rotated by the angle as well; for example,

```
class PolygonModel():
    ...
    def transformed(self):
        rotated = [vectors.rotate2d(self.rotation_angle, v) for v in
self.points]
        return [vectors.add((self.x,self.y),v) for v in rotated]
```

Exercise 7.2 Write a function `to_pixels(x,y)` that takes a pair of x- and y-coordinates in the square where $-10 < x < 10$ and $-10 < y < 10$ and maps them to the corresponding PyGame x and y pixel coordinates, each ranging from 0 to 400.

Solution

```
width, height = 400, 400
def to_pixels(x,y):
    return (width/2 + width * x / 20, height/2 - height * y / 20)
```

7.2 Finding intersection points of lines

The problem at hand is to decide whether the laser beam hits the asteroid. To do this, we'll look at each line segment defining the asteroid and decide whether it intersects with the segment defining the laser beam. There are a few algorithms we could use, but we'll solve this as a *system of linear equations in two variables.* Geometrically, this means looking at the lines defined by an edge of the asteroid and the laser beam and seeing where they intersect (figure 7.6).

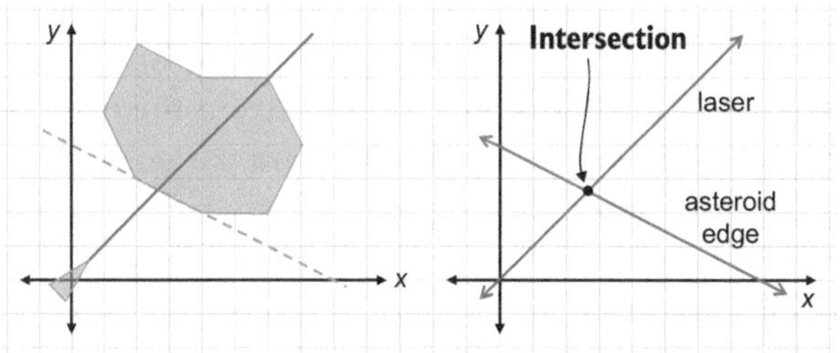

Figure 7.6 The laser hitting an edge of an asteroid (left) and the corresponding system of linear equations (right)

Once we know the location of the intersection, we can see whether it lies within the bounds of both segments. If so, the segments collide and the asteroid is hit. We first review equations for lines in the plane, then cover how to find where pairs of lines intersect. Finally, we write the code for the `does_intersect` method for our game.

7.2.1 Choosing the right formula for a line

In the previous chapter, we saw that 1D subspaces of the 2D plane are lines. These subspaces consist of all of the scalar multiples $t \cdot \mathbf{v}$ for a single chosen vector **v**. Because one such scalar multiple is $0 \cdot \mathbf{v}$, these lines always pass through the origin, so $t \cdot \mathbf{v}$ is not quite a general formula for any line we encounter.

If we start with a line through the origin and translate it by another vector **u**, we can get any possible line. The points on this line have the form $\mathbf{u} + t \cdot \mathbf{v}$ for some scalar t. For instance, take $\mathbf{v} = (2, -1)$. Points of the form $t \cdot (2, -1)$ lie on a line through the origin. But if we translate by a second vector, $\mathbf{u} = (2, 3)$, the points are now $(2, 3) + t \cdot (2, -1)$, which constitute a line that *doesn't* pass through the origin (figure 7.7).

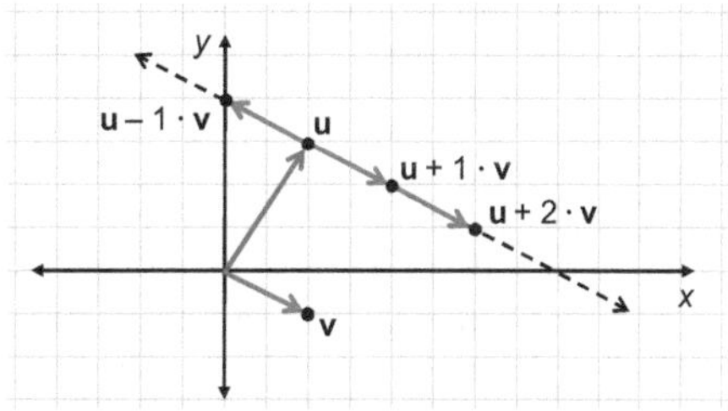

Figure 7.7 Vectors u = (2, 3) and V = (2, –1). Points of the form u + t · v lie on a straight line.

Any line can be described as the points $\mathbf{u} + t \cdot \mathbf{v}$ for some selection of vectors **u** and **v** and *all* possible scalar multiples t. This is probably not the general formula for a line you're used to. Instead of writing y as a function of x, we've given both the x- and y-coordinates of points on the line as functions of another parameter t. Sometimes,

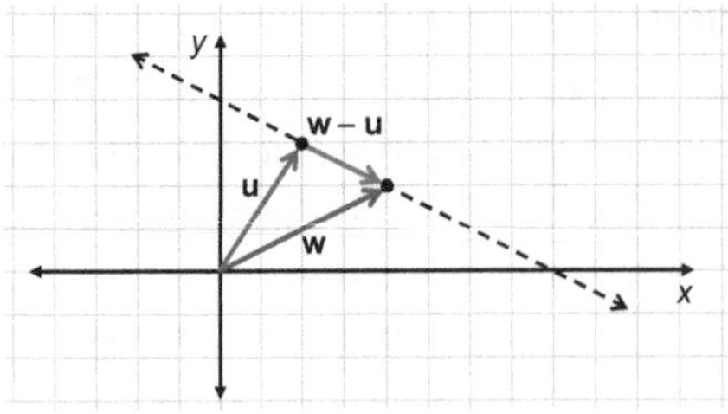

Figure 7.8 Given u and w, the line that connects them is r(*t*) = u + *t* · (w – u).

you'll see the line written $\mathbf{r}(t) = \mathbf{u} + t \cdot \mathbf{v}$ to indicate that this line is a vector valued function **r** of the scalar parameter *t*. The input *t* decides how many units of **v** you go from the starting point **u** to get the output $\mathbf{r}(t)$.

The advantage of this kind of formula for a line is that it's dead simple to find if you have two points on the line. If your points are **u** and **w**, then you can use **u** as the translation vector, and **w** – **u** as the vector that is scaled (figure 7.8).

The formula $\mathbf{r}(t) = \mathbf{u} + t \cdot \mathbf{v}$ also has its downside. As you'll see in the exercises, there are multiple ways to write the same line in this form. The extra parameter *t* also makes it harder to solve equations because there is one extra unknown variable. Let's look at some alternative formulas with other advantages.

If you recall any formula for a line from high school, it is probably $y = m \cdot x + b$. This formula is useful because it gives you a *y*-coordinate explicitly as a function of the *x*-coordinate. In this form, it's easy to graph a line; you go through a bunch of *x* values, compute the corresponding *y* values, and put dots at the resulting (x, y) points. But this formula also has some limitations. Most importantly, you can't represent a vertical line like $\mathbf{r}(t) = (3, 0) + t \cdot (0, 1)$. This is the line consisting of vectors where $x = 3$.

We'll continue to use the *parametric* formula $\mathbf{r}(t) = \mathbf{u} + t \cdot \mathbf{v}$ because it avoids this problem, but it would be great to have a formula with no extra parameter *t* that can represent any line. The one we use is $ax + by = c$. As an example, the line we're looking at in the last few images can be written as $x + 2y = 8$ (figure 7.9). It is the set of (x, y) points in the plane satisfying that equation.

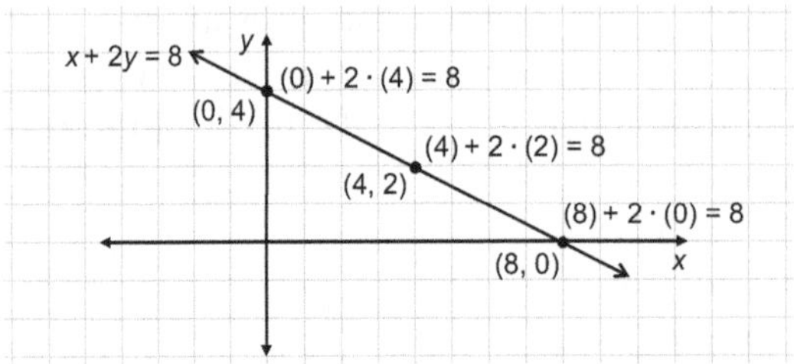

Figure 7.9 All (*x*, *y*) points on the line satisfy *x* + 2*y* = 8.

The form $ax + by = c$ has no extra parameters and can represent any line. Even a vertical line can be written in this form; for instance, $x = 3$ is the same as $1 \cdot x + 0 \cdot y = 3$. Any equation representing a line is called a *linear equation* and this, in particular, is called the *standard form* for a linear equation. We prefer to use it in this chapter because it makes it easy to organize our computations.

7.2.2 Finding the standard form equation for a line

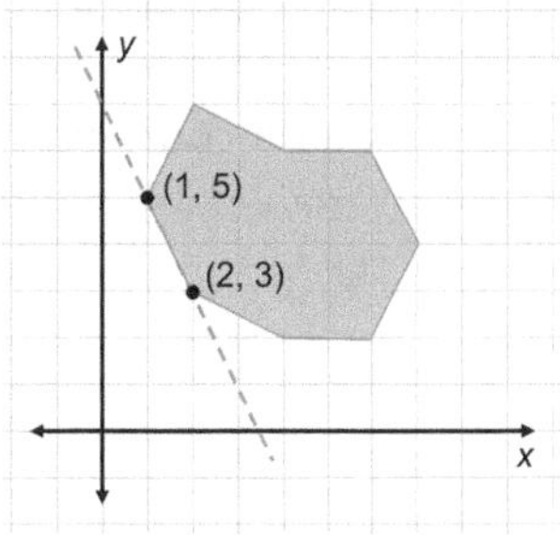

Figure 7.10 The points (1, 5) and (2, 3) define a second segment of the asteroid.

The formula $x + 2y = 8$ is the equation for a line containing one of the segments on the example asteroid. Next, we'll look at another one (figure 7.10) and then try to systematize finding the standard form for linear equations. Brace yourself for a bit of algebra! I'll explain each of the steps carefully, but it may be a bit dry to read. You'll have a better time if you follow along on your own with a pencil and paper.

The vector (1, 5) – (2, 3) is (–1, 2), which is parallel to the line. Because (2, 3) lies on the line, a parametric equation for the line is $r(t) = (2, 3) + t \cdot (-1, 2)$. Knowing that all points on the line have the form $(2, 3) + t \cdot (-1, 2)$ for some t, how can we rewrite this condition to be a standard form equation? We need to do some algebra and, particularly, get rid of t. Because $(x, y) = (2, 3) + t \cdot (-1, 2)$, we really have two equations to start with:

$$x = 2 - t$$
$$y = 3 + 2t$$

We can manipulate both of them to get two new equations that have the same value ($2t$):

$$4 - 2x = 2t$$
$$y - 3 = 2t$$

Because both of the expressions on the left-hand sides equal $2t$, they equal each other:

$$4 - 2x = y - 3$$

We've now gotten rid of t! Finally, pulling the x and y terms to one side, we get the standard form equation:

$$2x + y = 7$$

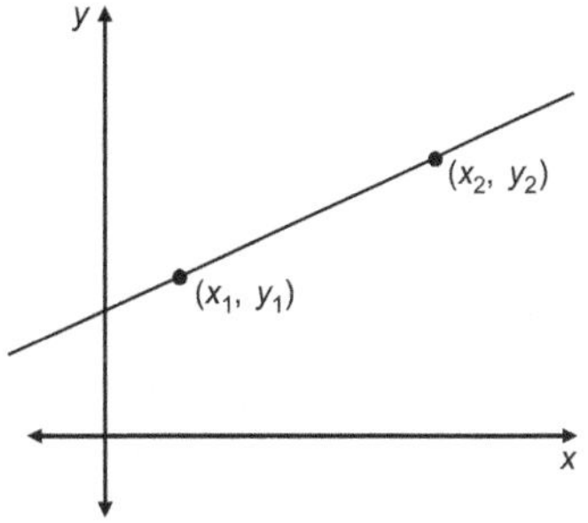

Figure 7.11 The general problem of finding the equation of the line that passes through two known points

This process isn't too hard, but we need to be more precise about how to do it if we want to convert it to code. Let's try to solve the general problem: given two points (x_1, y_1) and (x_2, y_2), what is the equation of the line that passes through them (see figure 7.11)?

Using the parametric formula, the points on the line have the following form:

$$(x, y) = (x_1, y_1) + t \cdot (x_2 - x_1, y_2 - y_1)$$

There are a lot of x and y variables here, but remember that x_1, x_2, y_1, and y_2 are all constants for the purpose of this discussion. We assume we have two points with known coordinates, and we could have called them (a, b) and (c, d) just as easily. The variables are x and y (with no subscripts), which stand for coordinates of *any* point on the line. As before, we can break this equation into two pieces:

$$x = x_1 + t \cdot (x_2 - x_1)$$
$$y = y_1 + t \cdot (y_2 - y_1)$$

We can move x_1 and y_1 to the left-hand side of their respective equations:

$$x - x_1 = t \cdot (x_2 - x_1)$$
$$y - y_1 = t \cdot (y_2 - y_1)$$

Our next goal is to make the right-hand side of both equations look the same, so we can set the left-hand sides equal to each other. Multiplying both sides of the first equation by $(y_2 - y_1)$ and both sides of the second equation by $(x_2 - x_1)$ gives us

$$(y_2 - y_1) \cdot (x - x_1) = t \cdot (x_2 - x_1) \cdot (y_2 - y_1)$$
$$(x_2 - x_1) \cdot (y - y_1) = t \cdot (x_2 - x_1) \cdot (y_2 - y_1)$$

Because the right-hand sides are identical, we know that the first and second equations' left-hand sides equal each other too. That lets us create a new equation with no t in it:

$$(y_2 - y_1) \cdot (x - x_1) = (x_2 - x_1) \cdot (y - y_1)$$

Remember, we want an equation of the form $ax + by = c$, so we need to get x and y on the same side and the constants on the other side. The first thing we can do is expand both sides:

$$(y_2 - y_1) \cdot x - (y_2 - y_1) \cdot x = (x_2 - x_1) \cdot y - (x_2 - x_1) \cdot y_1$$

Then we can move the constants to the left and the variables to the right:

$$(y_2 - y_1) \cdot x - (x_2 - x_1) \cdot y = (y_2 - y_1) \cdot x_1 - (x_2 - x_1) \cdot y_1$$

Expanding the right side, we see some of the terms cancel out:

$$(y_2 - y_1) \cdot x - (x_2 - x_1) \cdot y = y_2x_1 - y_1x_1 - x_2y_1 + x_1y_1 = x_1y_2 - x_2y_1$$

We've done it! This is the linear equation in standard form $ax + by = c$, where $a = (y_2 - y_1)$, $b = -(x_2 - x_1)$, or in other words, $(x_1 - x_2)$, and $c = (x_1y_2 - x_2y_1)$. Let's check this with the previous example we did, using the two points $(x_1, y_1) = (2, 3)$ and $(x_2, y_2) = (1, 5)$. In this case,

$$a = y_2 - y_1 = 5 - 3 = 2$$
$$b = -(x_2 - x_1) = -(1 - 2) = 1$$

and

$$c = x_1y_2 - x_2y_1 = 2 \cdot 5 - 3 \cdot 1 = 7$$

As expected, this means the standard form equation is $2x + y = 7$. This formula seems trustworthy! As one final application, let's find the standard form equation for the line defined by the laser. It looks like it passes through (2, 2) and (4, 4) as I drew it before (figure 7.12).

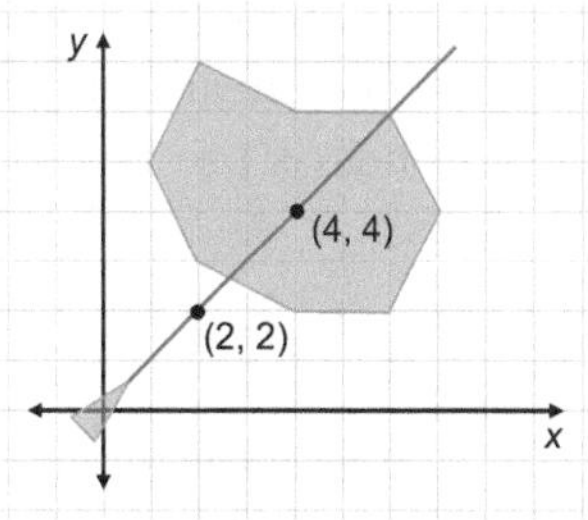

Figure 7.12 The laser passes through the points (2, 2) and (4, 4).

In our asteroid game, we have exact start and end points for the laser line segment, but these numbers are nice for an example. Plugging into the formula, we find

$$a = y_2 - y_1 = 4 - 2 = 2$$
$$b = -(x_2 - x_1) = -(4 - 2) = -2$$

and

$$c = x_1y_2 - x_2y_1 = 2 \cdot 4 - 2 \cdot 4 = 0$$

This means the line is $2y - 2x = 0$, which is equivalent to saying $x - y = 0$ (or simply $x = y$). To decide whether the laser hits the asteroid, we'll have to find where the line $x - y = 0$ intersects the line $x + 2y = 8$, the line $2x + y = 7$, or any of the other lines bounding the asteroid.

7.2.3 *Linear equations in matrix notation*

Let's focus on an intersection we can see: the laser clearly hits the closest edge of the asteroid, whose line has equation $x + 2y = 8$ (figure 7.13).

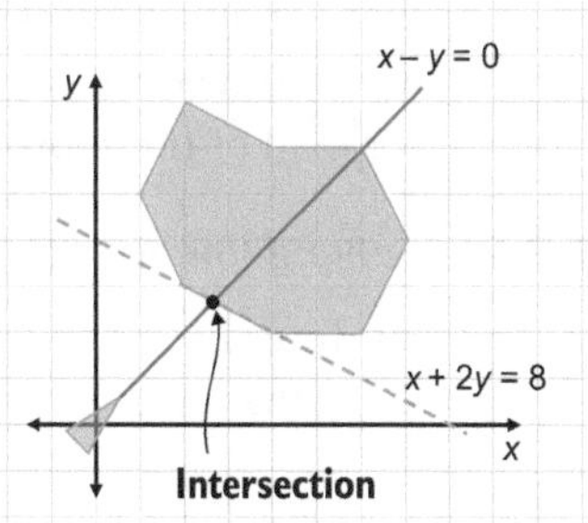

Figure 7.13 The laser hits the asteroid where the lines $x - y = 0$ and $x + 2y = 8$ intersect.

After quite a bit of build-up, we've met our first real system of linear equations. It's customary to write systems of linear equations in a grid like the following, so that the variables x and y line up:

$$\begin{aligned} x - y &= 0 \\ x + 2y &= 8 \end{aligned}$$

Thinking back to chapter 5, we can organize these two equations into a single matrix equation. One way to do this is to write a linear combination of column vectors, where x and y are coefficients:

$$x \begin{pmatrix} 1 \\ 1 \end{pmatrix} + y \begin{pmatrix} -1 \\ 2 \end{pmatrix} = \begin{pmatrix} 0 \\ 8 \end{pmatrix}$$

Another way is to consolidate this even further and write it as a matrix multiplication. The linear combination of (1,–1) and (–1,–2) with coefficients x and y is the same as a matrix product:

$$\begin{pmatrix} 1 & -1 \\ 1 & 2 \end{pmatrix} \begin{pmatrix} x \\ y \end{pmatrix} = \begin{pmatrix} 0 \\ 8 \end{pmatrix}$$

When we write it this way, the task of solving the system of linear equations looks like solving for a vector in a matrix multiplication problem. If we call the 2-by-2 matrix A, the problem becomes what vector (x, y) is multiplied by the matrix A to yield (0, 8)? In other words, we know that an output of the linear transformation A is (0, 8) and we want to know what input yields it (figure 7.14).

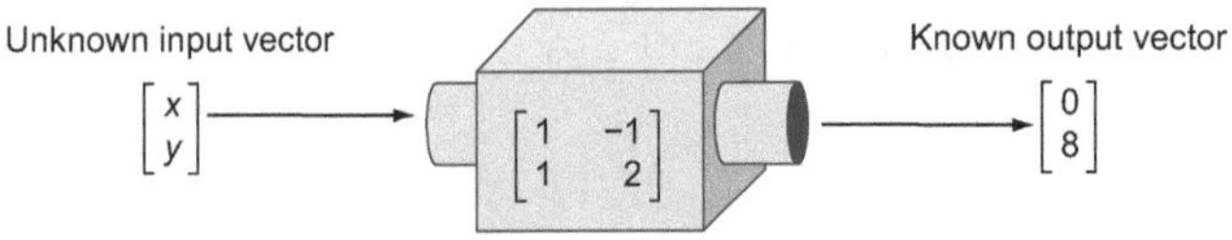

Figure 7.14 Framing the problem as finding an input vector that yields the desired output vector

These different notations show new ways to look at the same problem. Solving a system of linear equations is equivalent to finding a linear combination of some vectors that produces another given vector. It's also equivalent to finding an input vector to a linear transformation that produces a given output. Thus, we're about to see how to solve all of these problems at once.

7.2.4 *Solving linear equations with NumPy*

Finding the intersection of $x - y = 0$ and $x + 2y = 8$ is the same as finding the vector (x, y) that satisfies the matrix multiplication equation:

$$\begin{pmatrix} 1 & -1 \\ 1 & 2 \end{pmatrix} \begin{pmatrix} x \\ y \end{pmatrix} = \begin{pmatrix} 0 \\ 8 \end{pmatrix}$$

This is only a notational difference, but framing the problem in this form allows us to use pre-built tools to solve it. Specifically, Python's NumPy library has a linear algebra module and a function that solves this kind of equation. Here's an example:

```
>>> import numpy as np
>>> matrix = np.array(((1,-1),(1,2)))
>>> output = np.array((0,8))
```

Packages the matrix as a NumPy array object

Packages the output vector as a NumPy array (although it needn't be reshaped to a column vector)

```
>>> np.linalg.solve(matrix,output)
array([2.66666667, 2.66666667])
```

The numpy.linalg.solve function takes a matrix and an output vector and finds the input vector that produces it.

The result is (x, y) = (2.66..., 2.66...).

NumPy has told us that the x- and y-coordinates of the intersection are approximately 2⅔ or 8/3 each, which looks about right geometrically. Eyeballing the diagram, it looks like both coordinates of the intersection point should be between 2 and 3. We can check to see that this point lies on both lines by plugging it in to both equations:

$$1x - 1y = 1 \cdot (2.66666667) - 1 \cdot (2.66666667) = 0$$

$$1x + 2y = 1 \cdot (2.66666667) + 2 \cdot (2.66666667) = 8.00000001$$

These results are close enough to (0, 8) and, indeed, make an exact solution. This solution vector, roughly (8/3, 8/3) is also the vector that satisfies the matrix equation 7.1.

$$\begin{pmatrix} 1 & -1 \\ 1 & 2 \end{pmatrix} \begin{pmatrix} 8/3 \\ 8/3 \end{pmatrix} = \begin{pmatrix} 0 \\ 8 \end{pmatrix}$$

As figure 7.15 shows, we can picture (8/3, 8/3) as the vector we pass into the linear transformation machine defined by the matrix that gives us the desired output vector.

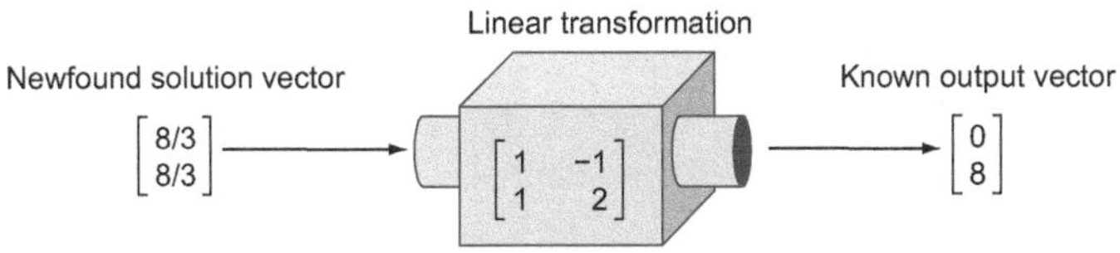

Figure 7.15 The vector (8/3, 8/3) when passed to the linear transformation produces the desired output (0, 8).

We can think of the Python function `numpy.linalg.solve` as a differently shaped machine that takes in matrices and output vectors, and returns the "solution" vectors for the linear equation they represent (figure 7.16).

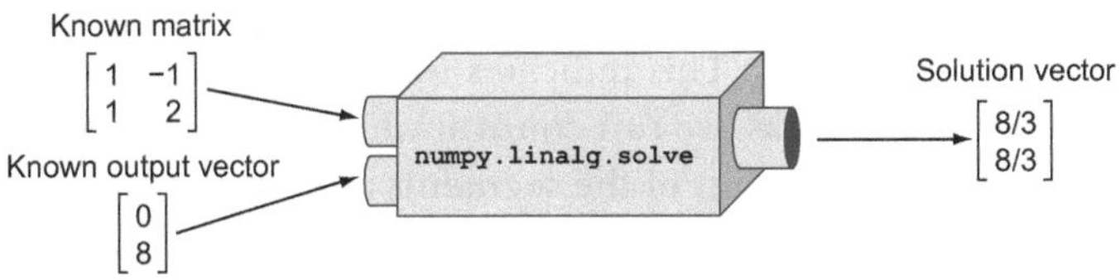

Figure 7.16 The `numpy.linalg.solve` function takes a matrix and a vector and outputs the solution vector to the linear system they represent.

This is perhaps the most important computational task in linear algebra; starting with a matrix A, and a vector $\mathbf{w}$, and finding the vector $\mathbf{v}$ such that $A\mathbf{v} = \mathbf{w}$. Such a vector gives the solution to a system of linear equations represented by A and $\mathbf{w}$. We're lucky to have a Python function that can do this for us so we don't have to worry about the tedious algebra required to do it by hand. We can now use this function to find out when our laser hits asteroids.

7.2.5 Deciding whether the laser hits an asteroid

The missing piece of our game was an implementation for the `does_intersect` method on the `PolygonModel` class. For any instance of this class, which represents a polygon-shaped object living in our 2D game world, this method should return `True` if an input line segment intersects any line segment of the polygon.

For this, we need a few helper functions. First, we need to convert the given line segments from pairs of endpoint vectors to linear equations in standard form. At the end of the section, I give you an exercise to implement the function `standard_form`, which takes two input vectors and returns a tuple (a, b, c) where $ax + by = c$ is the line on which the segment lies.

Next, given two segments, each represented by its pair of endpoint vectors, we want to find out where their lines intersect. If u_1 and u_2 are endpoints of the first segment, and v_1 and v_2 are endpoints of the second, we need to first find the standard form equations and then pass them to NumPy to solve. For example,

```
def intersection(u1,u2,v1,v2):
    a1, b1, c1 = standard_form(u1,u2)
    a2, b2, c2 = standard_form(v1,v2)
    m = np.array(((a1,b1),(a2,b2)))
    c = np.array((c1,c2))
    return np.linalg.solve(m,c)
```

The output is the point where the two lines on which the segments lie intersect. But this point might not lie on either of the segments as shown in figure 7.17.

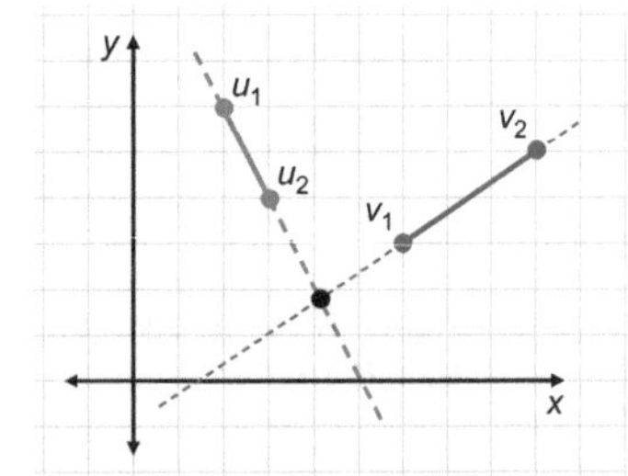

Figure 7.17 One segment connects u_1 and u_2 and the other connects points v_1 and v_2. The lines extending the segments intersect, but the segments themselves don't.

To detect whether the two segments intersect, we need to check that the intersection point of their lines lies between the two pairs of endpoints. We can check that using distances. In figure 7.17, the intersection point is further from point v_2 than point v_1. Likewise, it's further from u_1 than u_2. This indicates that the point is on neither segment. With four total distance checks, we can confirm whether the intersection point of the lines (x, y) is an intersection point of the segments as well:

```
def do_segments_intersect(s1,s2):
    u1,u2 = s1
    v1,v2 = s2
    d1, d2 = distance(*s1), distance(*s2)
```

Stores the lengths of the first and second segments as d1 and d2, respectively

```
    x,y = intersection(u1,u2,v1,v2)
    return (distance(u1, (x,y)) <= d1 and
            distance(u2, (x,y)) <= d1 and
            distance(v1, (x,y)) <= d2 and
            distance(v2, (x,y)) <= d2)
```

Finds the intersection point (x, y) of the lines on which the segments lie

Does four checks to ensure the intersection point lies between the four endpoints of the line segments, confirming the segments intersect

Finally, we can write the `does_intersect` method by checking whether `do_segments_intersect` returns `True` for the input segment and any of the edges of the (transformed) polygon:

```
class PolygonModel():
    ...
    def does_intersect(self, other_segment):
        for segment in self.segments():
            if do_segments_intersect(other_segment,segment):
                return True
        return False
```

If any of the segments of the polygon intersect other_segment, the method returns True.

In the exercises that follow, you can confirm that this actually works by building an asteroid with known coordinate points and a laser beam with a known start and end point. With `does_intersect` implemented as in the source code, you should be able to rotate the spaceship to aim at asteroids and destroy them.

7.2.6 *Identifying unsolvable systems*

Let me leave you with one final admonition: not every system of linear equations in 2D can be solved! It's rare in an application like the asteroid game, but some pairs of linear equations in 2D don't have unique solutions or even solutions at all. If we pass NumPy a system of linear equations with no solution, we get an exception, so we need to handle this case.

When a pair of lines in 2D are not parallel, they intersect somewhere. Even the two lines in figure 7.18 that are nearly parallel (but not quite) intersect somewhere off in the distance.

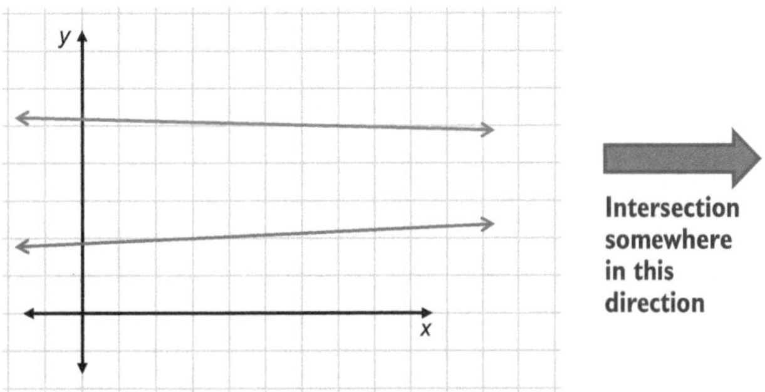

Figure 7.18 Two lines that are not quite parallel intersect somewhere in the distance.

Where we run into trouble is when the lines are parallel, meaning the lines never intersect (or they're the same line!) as shown in figure 7.19.

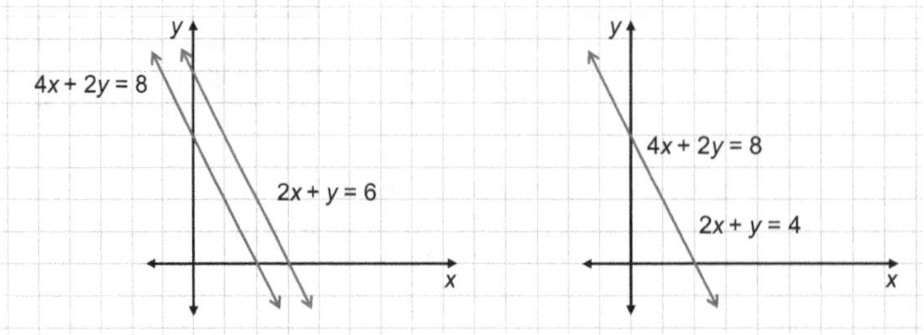

Figure 7.19 A pair of parallel lines that never intersect and a pair of parallel lines that are, in fact, the same line despite having different equations

In the first case, there are zero intersection points, while in the second, there are *infinitely* many intersection points—every point on the line is an intersection point. Both of these cases are problematic computationally because our code demands a single, unique result. If we try to solve either of these systems with NumPy, for instance, the system consisting of $2x + y = 6$ and $4x + 2y = 8$, we get an exception:

```
>>> import numpy as np
>>> m = np.array(((2,1),(4,2)))
>>> v = np.array((6,4))
>>> np.linalg.solve(m,v)
Traceback (most recent call last):
  File "<stdin>", line 1, in <module>
...
numpy.linalg.linalg.LinAlgError: Singular matrix
```

NumPy points to the matrix as the source of the error. The matrix

$$\begin{pmatrix} 2 & 1 \\ 4 & 2 \end{pmatrix}$$

is called a *singular* matrix, meaning there is no unique solution to the linear system. A system of linear equations is defined by a matrix and a vector, but the matrix on its own is enough to tell us whether the lines are parallel and whether the system has a unique solution. For any non-zero *w* we pick, there won't be a unique *v* that solves the system.

$$\begin{pmatrix} 2 & 1 \\ 4 & 2 \end{pmatrix} \mathbf{v} = \mathbf{w}$$

We'll philosophize more about singular matrices later, but for now you can see that the rows (2, 1) and (4, 2) and the columns (2, 4) and (1, 2) are both parallel and, therefore, linearly dependent. This is the key clue that tells us the lines are parallel and that the system does not have a unique solution. Solvability of linear systems is one of the central concepts in linear algebra; it closely relates to the notions of linear independence and dimension. We discuss that in the last two sections of this chapter.

For the purpose of our asteroid game, we can make the simplifying assumption that any parallel line segments don't intersect. Given that we're building the game with random floats, it's highly unlikely that any two segments are exactly parallel. Even if the laser lined up exactly with the edge of an asteroid, this would be a glancing hit and the player doesn't deserve to have the asteroid destroyed. We can modify `do_segments_intersect` to catch the exception and return the default result of `False`:

```
def do_segments_intersect(s1,s2):
    u1,u2 = s1
    v1,v2 = s2
    l1, l2 = distance(*s1), distance(*s2)
    try:
        x,y = intersection(u1,u2,v1,v2)
        return (distance(u1, (x,y)) <= l1 and
                distance(u2, (x,y)) <= l1 and
                distance(v1, (x,y)) <= l2 and
                distance(v2, (x,y)) <= l2)
    except np.linalg.linalg.LinAlgError:
        return False
```

7.2.7 Exercises

Exercise 7.3 It's possible that $\mathbf{u} + t \cdot \mathbf{v}$ can be a line through the origin. In this case, what can you say about the vectors **u** and **v**?

Solution One possibility is that $\mathbf{u} = \mathbf{0} = (0, 0)$; in which case, the line automatically passes through the origin. The point $\mathbf{u} + 0 \cdot \mathbf{v}$ is the origin in this case, regardless of what **v** is. Otherwise, if **u** and **v** are scalar multiples, say $\mathbf{u} = s \cdot \mathbf{v}$, then the line passes through the origin as well because $\mathbf{u} - s \cdot \mathbf{v} = \mathbf{0}$ is on the line.

Exercise 7.4 If $\mathbf{v} = \mathbf{0} = (0, 0)$, do points of the form $\mathbf{u} + t \cdot \mathbf{v}$ represent a line?

Solution No, regardless of the value of t, we have $\mathbf{u} + t \cdot \mathbf{v} = \mathbf{u} + t \cdot (0, 0) = \mathbf{u}$. Every point of this form is equal to **u**.

Exercise 7.5 It turns out that the formula $\mathbf{u} + t \cdot \mathbf{v}$ is not unique; that is, you can pick different values of **u** and **v** to represent the same line. What is another line representing $(2, 2) + t \cdot (-1, 3)$?

Solution One possibility is to replace $\mathbf{v} = (-1, 3)$ with a scalar multiple of itself such as $(2, -6)$. The points of the form $(2, 2) + t \cdot (-1, 3)$ agree with the points $(2, 2) + s \cdot (2, -6)$ when $t = -2 \cdot s$. You can also replace **u** with any point on the line. Because $(2, 2) + 1 \cdot (-1, 3) = (1, 5)$ is on the line, $(1, 5) + t \cdot (2, -6)$ is a valid equation for the same line as well.

Exercise 7.6 Does $a \cdot x + b \cdot y = c$ represent a line for *any* values of a, b, and c?

Solution No, if both a and b are zero, the equation does not describe a line. In that case, the formula would be $0 \cdot x + 0 \cdot y = c$. If $c = 0$, this would always be true, and if $c \neq 0$, it would never be true. Either way, it establishes no relationship between x and y and, therefore, it would not describe a line.

Exercise 7.7 Find another equation for the line $2x + y = 3$, showing that the choices of a, b, and c are not unique.

Solution One example of another equation is $6x + 3y = 9$. In fact, multiplying both sides of the equation by the same non-zero number gives you a different equation for the same line.

Exercise 7.8 The equation $ax + by = c$ is equivalent to an equation involving a dot product of two 2D vectors: $(a, b) \cdot (x, y) = c$. You could, therefore, say that a line is a set of vectors whose dot product with a given vector is constant. What is the geometric interpretation of this statement?

Solution See the discussion in section 7.3.1.

Exercise 7.9 Confirm that the vectors $(0, 7)$ and $(3.5, 0)$ both satisfy the equation $2x + y = 7$.

Solution $2 \cdot 0 + 7 = 7$ and $2 \cdot (3.5) + 0 = 7$.

Exercise 7.10 Draw a graph for $(3, 0) + t \cdot (0, 1)$ and convert it to the standard form using the formula.

Solution $(3, 0) + t \cdot (0, 1)$ yields a vertical line, where $x = 3$:

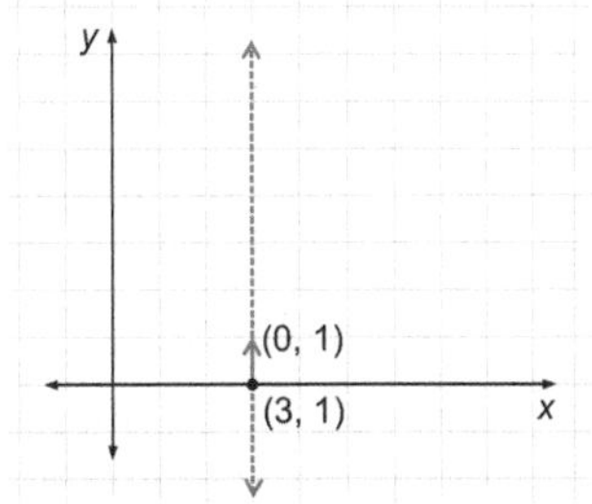

The formula $x = 3$ is already the equation of a line in standard form, but we can confirm this with the formulas. The first point on our line is already given: $(x_1, y_1) = (3, 0)$. A second point on the line is $(3, 0) + (0, 1) = (3, 1) = (x_2, y_2)$. We have $a = y_2 - y_1 = 1$, $b = x_1 - x_2 = 0$, and $c = x_1y_2 - x_2y_1 = 3 \cdot 1 - 1 \cdot 0 = 3$. This gives us $1 \cdot x + 0 \cdot y = 3$ or simply $x = 3$.

Exercise 7.11 Write a Python function `standard_form` that takes two vectors $\mathbf{v}_1$ and $\mathbf{v}_2$ and finds the line $ax + by = c$ passing through both of them. Specifically, it should output the tuple of constants (a, b, c).

Solution All you need to do is translate the formulas you wrote in Python:

```
def standard_form(v1, v2):
    x1, y1 = v1
    x2, y2 = v2
    a = y2 - y1
    b = x1 - x2
    c = x1 * y2 - y1 * x2
    return a,b,c
```

Exercise 7.12—Mini Project For each of the four distance checks in `do_segments_intersect`, find a pair of line segments that fail one of the checks but pass the other three checks.

Solution To make it easier to run experiments, we can create a modified version of `do_segments_intersect` that returns a list of the True/False values returned by each of the four checks:

```
def segment_checks(s1,s2):
    u1,u2 = s1
    v1,v2 = s2
    l1, l2 = distance(*s1), distance(*s2)
    x,y = intersection(u1,u2,v1,v2)
```

(continued)

```
    return [
        distance(u1, (x,y)) <= l1,
        distance(u2, (x,y)) <= l1,
        distance(v1, (x,y)) <= l2,
        distance(v2, (x,y)) <= l2
    ]
```

In general, these checks fail when one endpoint of a segment is closer to the other endpoint than to the intersection point.

Here are some other solutions I found using segments on the lines $y = 0$ and $x = 0$, which intersect at the origin. Each of these fails exactly one of the four checks. If in doubt, draw them yourself to see what's going on.

```
>>> segment_checks(((-3,0),(-1,0)),((0,-1),(0,1)))
[False, True, True, True]
>>> segment_checks(((1,0),(3,0)),((0,-1),(0,1)))
[True, False, True, True]
>>> segment_checks(((-1,0),(1,0)),((0,-3),(0,-1)))
[True, True, False, True]
>>> segment_checks(((-1,0),(1,0)),((0,1),(0,3)))
[True, True, True, False]
```

Exercise 7.13 For the example laser line and asteroid, confirm the `does_intersect` function returns `True`. (Hint: use grid lines to find the vertices of the asteroid and build a `PolygonModel` object representing it.)

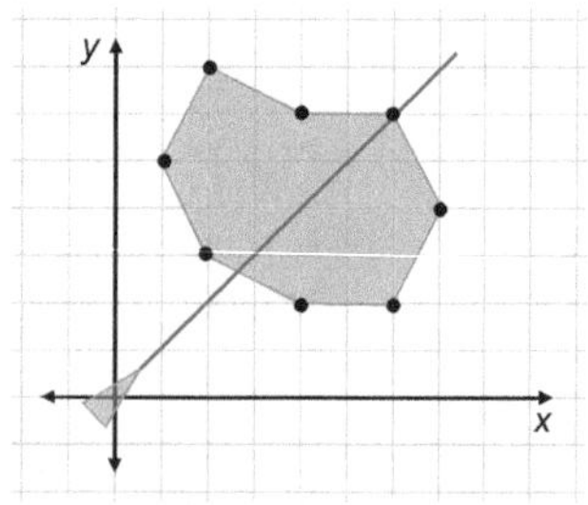

The laser hits the asteroid.

Solution In counterclockwise order, starting with the topmost point, the vertices are (2, 7), (1, 5), (2, 3), (4, 2), (6, 2), (7, 4), (6, 6), and (4, 6). We can assume the endpoints of the laser beam are (1, 1) and (7, 7):

```
>>> from asteroids import PolygonModel
>>> asteroid = PolygonModel([(2,7), (1,5), (2,3), (4,2), (6,2), (7,4),
(6,6), (4,6)])
>>> asteroid.does_intersect([(0,0),(7,7)])
True
```

This confirms the laser hits the asteroid! By contrast, a shot directly up the y-axis from (0, 0) to (0, 7) does not hit:

```
>>> asteroid.does_intersect([(0,0),(0,7)])
False
```

Exercise 7.14 Write a `does_collide(other_polygon)` method to decide whether the current `PolygonModel` object collides with another `other_polygon` by checking whether any of the segments that define the two are intersecting. This could help us decide whether an asteroid has hit the ship or another asteroid.

Solution First, it's convenient to add a `segments()` method to `PolygonModel` to avoid duplication of the work of returning the (transformed) line segments that constitute the polygon. Then, we can check every segment of the other polygon to see if it returns true for `does_intersect` with the current one:

```
class PolygonModel():
    ...
    def segments(self):
        point_count = len(self.points)
        points = self.transformed()
        return [(points[i], points[(i+1)%point_count])
                for i in range(0,point_count)]

    def does_collide(self, other_poly):
        for other_segment in other_poly.segments():
            if self.does_intersect(other_segment):
                return True
        return False
```

We can test this by building some squares that should and shouldn't overlap, and seeing whether the `does_collide` method correctly detects which is which. Indeed, it does:

```
>>> square1 = PolygonModel([(0,0), (3,0), (3,3), (0,3)])
>>> square2 = PolygonModel([(1,1), (4,1), (4,4), (1,4)])
>>> square1.does_collide(square2)
True
>>> square3 = PolygonModel([(-3,-3),(-2,-3),(-2,-2),(-3,-2)])
>>> square1.does_collide(square3)
False
```

Exercise 7.15—Mini Project We can't pick a vector **w** so that the following system has a unique solution **v**.

$$\begin{pmatrix} 2 & 1 \\ 4 & 2 \end{pmatrix} \mathbf{v} = \mathbf{w}$$

Find a vector **w** such that there are *infinitely* many solutions to the system; that is, infinitely many values of v that satisfy the equation.

(continued)

Solution If $\mathbf{w} = (0, 0)$, for example, the two lines represented by the system are identical. (Graph them if you are skeptical!) The solutions have the form $\mathbf{v} = (a, -2a)$ for any real number a. Here are some of the infinite possibilities for $\mathbf{v}$ when $\mathbf{w} = (0, 0)$:

$$\begin{pmatrix} 2 & 1 \\ 4 & 2 \end{pmatrix} \begin{pmatrix} 1 \\ -2 \end{pmatrix} = \begin{pmatrix} 2 & 1 \\ 4 & 2 \end{pmatrix} \begin{pmatrix} -4 \\ 8 \end{pmatrix} = \begin{pmatrix} 2 & 1 \\ 4 & 2 \end{pmatrix} \begin{pmatrix} 10 \\ -20 \end{pmatrix} = \begin{pmatrix} 0 \\ 0 \end{pmatrix}$$

7.3 Generalizing linear equations to higher dimensions

Now that we've built a functional (albeit minimal) game, let's broaden our perspective. We can represent a wide variety of problems as systems of linear equations, not just arcade games. Linear equations in the wild often have more than two "unknown" variables, x and y. Such equations describe collections of points in more than two dimensions. In more than three dimensions, it's hard to picture much of anything, but the 3D case can be a useful mental model. Planes in 3D end up being the analogy of lines in 2D, and they are also represented by linear equations.

7.3.1 Representing planes in 3D

To see why lines and planes are analogous, it's useful to think of lines in terms of dot products. As you saw in a previous exercise, or may have noticed yourself, the equation $ax + by = c$ is the set of points (x, y) in the 2D plane where the dot product with a fixed vector (a, b) is equal to a fixed number c. That is, the equation $ax + by = c$ is equivalent to the equation $(a, b) \cdot (x, y) = c$. In case you didn't figure out how to interpret this geometrically in the exercise, let's go through it here.

If we have a point and a (non-zero) vector in 2D, there's a unique line that is perpendicular to the vector and also passes through the point as shown in figure 7.20.

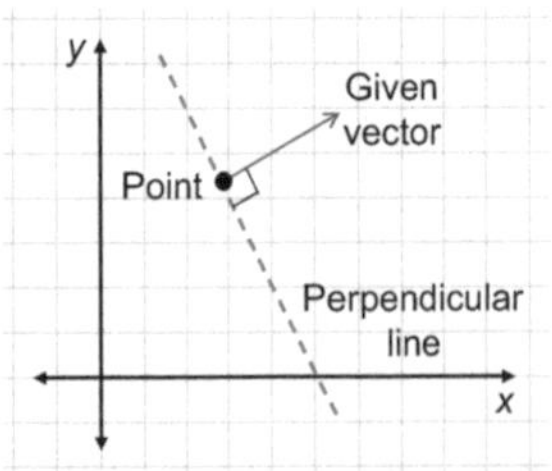

Figure 7.20 A unique line passing through a given point and perpendicular to a given vector

If we call the given point (x_0, y_0) and the given vector (a, b), we can write a criterion for a point (x, y) to lie on the line. Specifically, if (x, y) lies on the line, then $(x-x_0, y-y_0)$ is parallel to the line and perpendicular to (a, b) as shown in figure 7.21.

Because two perpendicular vectors have a zero dot product, that's equivalent to the algebraic statement:

$$(a, b) \cdot (x-x_0, y-y_0) = 0$$

That dot product expands to

$$a(x-x_0) + b(y-y_0) = 0$$

or

$$ax + by = ax_0 + by_0$$

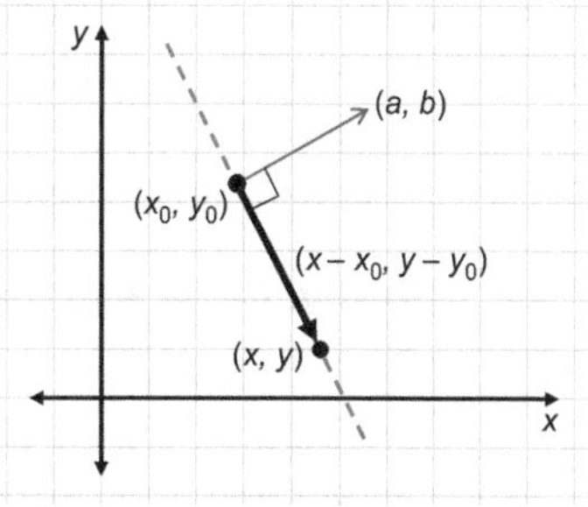

Figure 7.21 The vector $(x-x_0, y-y_0)$ is parallel to the line and, therefore, perpendicular to (a, b).

The quantity on the right-hand side of this equation is a constant, so we can rename it c, giving us the general form equation for a line: $ax + by = c$. This is a handy geometric interpretation of the formula $ax + by = c$, and one that we can generalize to 3D.

Given a point and a vector in 3D, there is a unique *plane* perpendicular to the vector and passing through that point. If the vector is (a, b, c) and the point is (x_0, y_0, z_0), we can conclude that if a vector (x, y, z) lies in the plane, then $(x-x_0, y-y_0, z-z_0)$ is perpendicular to (a, b, c). Figure 7.22 shows this logic.

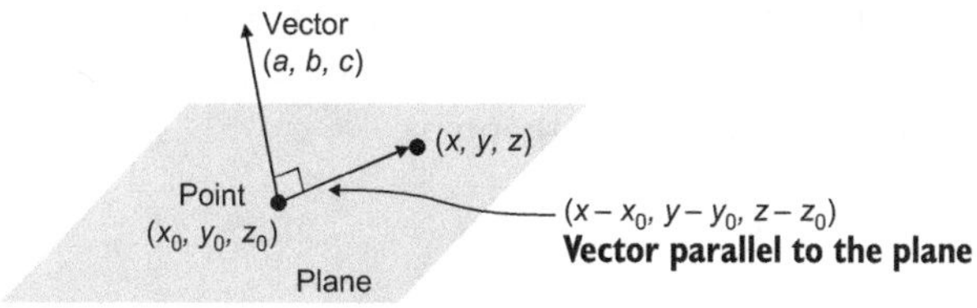

Figure 7.22 A plane parallel to the vector (a, b, c) passes through the point (x_0, y_0, z_0).

Every point on the plane gives us such a perpendicular vector to (a, b, c), and every vector perpendicular to (a, b, c) leads us to a point in the plane. We can express this perpendicularity as a dot product of the two vectors, so the equation satisfied by every point (x, y, z) in the plane is

$$(a, b, c) \cdot (x-x_0, y-y_0, z-z_0) = 0$$

This expands to

$$ax + by + cz = ax_0 + by_0 + cz_0$$

And because the right-hand side of the equation is a constant, we can conclude that every plane in 3D has an equation of the form $ax + by + cz = d$. In 3D, the computational problem is to decide where the planes intersect or which values of (x, y, z) simultaneously satisfy multiple linear equations like this.

7.3.2 Solving linear equations in 3D

A pair of non-parallel lines in the plane intersects at exactly one point. Is that single intersection point true for planes as well? If we draw a pair of intersecting planes, we can see that it's possible for non-parallel planes to intersect at many points. In fact, figure 7.23 shows there is a whole *line*, consisting of an infinite number of points where two non-parallel planes intersect.

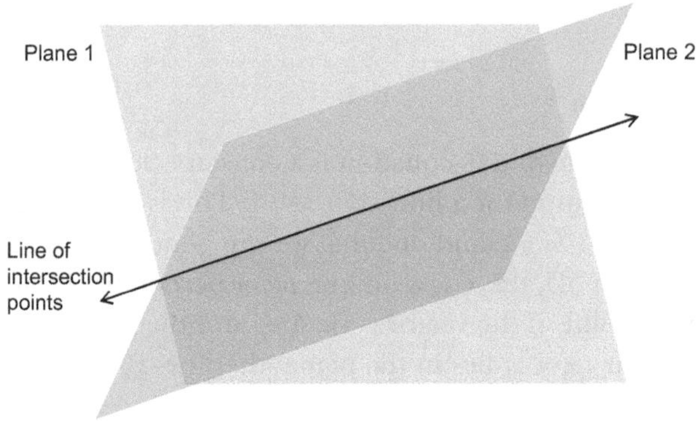

Figure 7.23 Two non-parallel planes intersect along a line.

If you add a third plane that is not parallel to this intersection line, you can find a unique intersection point. Figure 7.24 shows that each pair among the three planes intersects along a line and the lines share a single point.

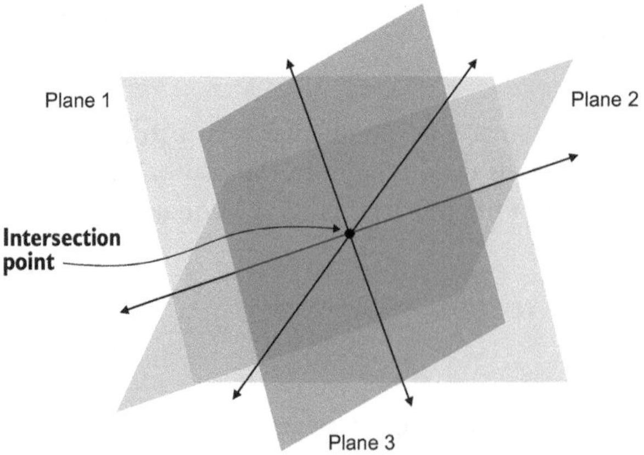

Figure 7.24 Two non-parallel planes intersect along a line.

Finding this point algebraically requires finding a common solution to three linear equations in three variables, each representing one of the planes and having the form $ax + by + cz = d$. Such a system of three linear equations would have the form:

$$a_1x + b_1y + c_1z = d_1$$
$$a_2x + b_2y + c_2z = d_2$$
$$a_3x + b_3y + c_3z = d_3$$

Each plane is determined by four numbers: a_i, b_i, c_i, and d_i, where $i = 1, 2$, or 3 and is the index of the plane we're looking at. Subscripts like this are useful for systems of linear equations where there can be a lot of variables that need to be named. These twelve numbers in total are enough to find the point (x, y, z) where the planes intersect, if there is one. To solve the system, we can convert the system into a matrix equation:

$$\begin{pmatrix} a_1 & b_1 & c_1 \\ a_2 & b_2 & c_2 \\ a_3 & b_3 & c_3 \end{pmatrix} \begin{pmatrix} x \\ y \\ z \end{pmatrix} = \begin{pmatrix} d_1 \\ d_2 \\ d_3 \end{pmatrix}$$

Let's try an example. Say our three planes are given by the following equations:

$$x + y - z = -1$$
$$2y - z = 3$$
$$x + z = 2$$

You can see how to plot these planes in Matplotlib in the source code for this book. Figure 7.25 shows the result.

It's not easy to see, but somewhere in there, the three planes intersect. To find that intersection point, we need the values of x, y, and z that simultaneously satisfy all three linear equations. Once again, we can convert the system to matrix form and use NumPy to solve it. The matrix equation equivalent to this linear system is

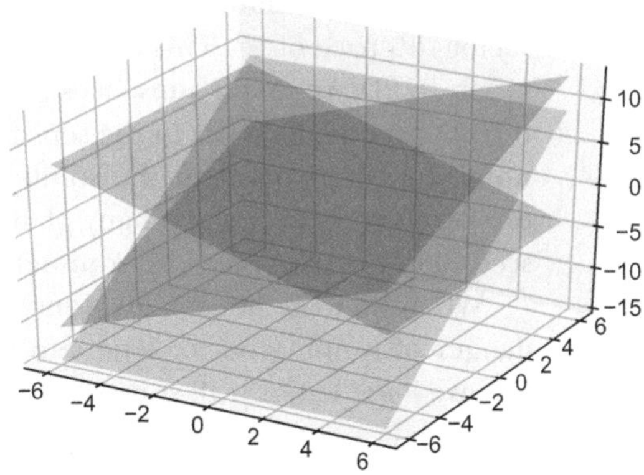

Figure 7.25 Three planes plotted in Matplotlib

$$\begin{pmatrix} 1 & 1 & -1 \\ 0 & 2 & -1 \\ 1 & 0 & 1 \end{pmatrix} \begin{pmatrix} x \\ y \\ z \end{pmatrix} = \begin{pmatrix} -1 \\ 3 \\ 2 \end{pmatrix}$$

Converting the matrix and vector to NumPy arrays in Python, we can quickly find the solution vector:

```
>>> matrix = np.array(((1,1,-1),(0,2,-1),(1,0,1)))
>>> vector = np.array((-1,3,2))
>>> np.linalg.solve(matrix,vector)
array([-1., 3., 3.])
```

This tells us that (–1, 3, 3) is the (x, y, z) point where all three planes intersect and the point that simultaneously satisfies all three linear equations.

While this result was easy to compute with NumPy, you can see it's already a bit harder to visualize systems of linear equations in 3D. Beyond 3D, it's difficult (if not impossible) to visualize linear systems, but solving them is mechanically the same. The analogy to a line or a plane in any number of dimensions is called a *hyperplane*, and the problem boils down to finding the points where multiple hyperplanes intersect.

7.3.3 Studying hyperplanes algebraically

To be precise, a hyperplane in n dimensions is a solution to a linear equation in n unknown variables. A line is a 1D hyperplane living in 2D, and a plane is a 2D hyperplane living in 3D. As you might guess, a linear equation in standard form in 4D has the following form:

$$aw + bx + cy + dz = e$$

The set of solutions (w, x, y, z) form a region that is a 3D hyperplane living in 4D space. We need to be careful when we use the adjective 3D because it isn't necessarily a 3D vector subspace of $\mathbb{R}^4$. This is analogous to the 2D case: the lines passing through the origin in 2D are vector subspaces of $\mathbb{R}^2$, but other lines are not. Vector space or not, the 3D hyperplane is 3D in the sense that there are three linearly independent directions you could travel in the solution set, like there are two linearly independent directions you can travel on any plane. I've included a mini-project at the end of this section to help you check your understanding of this.

When we write linear equations in even higher numbers of dimensions, we're in danger of running out of letters to represent coordinates and coefficients. To solve this, we'll use letters with subscript indices. For instance, in 4D, we could write a linear equation in standard form as:

$$a_1x_1 + a_2x_2 + a_3x_3 + a_4x_4 = b$$

Here, the coefficients are a_1, a_2, a_3, and a_4, and the 4D vector has the coordinates (x_1, x_2, x_3, x_4). We could just as easily write a linear equation in 10 dimensions:

$$a_1x_1 + a_2x_2 + a_3x_3 + a_4x_4 + a_5x_5 + a_6x_6 + a_7x_7 + a_8x_8 + a_9x_9 + a_{10}x_{10} = b$$

When the pattern of terms we're summing is clear, we sometimes use an ellipsis (...) to save space. You may see equations like the previous one written $a_1x_1 + a_2x_2 + ... + a_{10}x_{10} = b$. Another compact notation you'll see involves the summation symbol Σ, which is the Greek letter *sigma.* If I want to write the sum of terms of the form $a_i x_i$ with the index i ranging from $i = 1$ to $i = 10$, and I want to state that the sum is equal to some other number b, I can use the mathematical shorthand:

$$\sum_{i=1}^{10} a_i x_i = b$$

This equation means the same thing as the earlier one; it is merely a more concise way of writing it. Whatever number of dimensions n we're working in, the standard form of a linear equation has the same shape:

$$a_1 x_1 + a_2 x_2 + \ldots + a_n x_n = b$$

To represent a system of m linear equations in n dimensions, we need even more indices. Our array of constants on the left-hand side of the equals sign can be denoted a_{ij}, where the subscript i indicates which equation we're talking about and the subscript j indicates which coordinate (x_j) the constant is multiplied by. For example,

$$\begin{aligned} &a_{11} x_1 + a_{12} x_2 + \ldots + a_{1n} x_n = b_1 \\ &a_{21} x_1 + a_{22} x_2 + \ldots + a_{2n} x_n = b_2 \\ &\ldots \\ &a_{m1} x_1 + a_{m2} x_2 + \ldots + a_{mn} x_n = b_m \end{aligned}$$

You can see that I also used the ellipsis to skip equations three through m–1 in the middle. There are m equations and n constants in each equation, so there are mn constants of the form a_{ij} in total. On the right-hand side, there are m constants in total, one per equation: $b_1, b_2, \ldots, b_m$.

Regardless of the number of dimensions (the same as the number of unknown variables) and the number of equations, we can represent such a system as a linear equation. The previous system with n unknowns and m equations can be rewritten as shown in figure 7.26.

$$\begin{pmatrix} a_{11} & a_{12} & \cdots & a_{1n} \\ a_{21} & a_{22} & \cdots & a_{2n} \\ \vdots & \vdots & \ddots & \vdots \\ a_{m1} & a_{m2} & \cdots & a_{mn} \end{pmatrix} \begin{pmatrix} x_1 \\ x_2 \\ \vdots \\ x_n \end{pmatrix} = \begin{pmatrix} b_1 \\ b_2 \\ \vdots \\ b_m \end{pmatrix}$$

Figure 7.26 A system of *m* linear equations with *n* unknowns written in matrix form

7.3.4 *Counting dimensions, equations, and solutions*

We saw in both 2D and 3D that it's possible to write linear equations that don't have a solution, or at least not a unique one. How will we know if a system of m equations in n unknowns is solvable? In other words, how will we know if m hyperplanes in n-dimensions have a unique intersection point? We'll discuss this in detail in the last section of this chapter, but there's one important conclusion we can draw now.

In 2D, a pair of lines can intersect at a single point. They won't always (for instance, if the lines are parallel), but they can. The algebraic equivalent to this statement is that a system of two linear equations in two variables can have a unique solution.

In 3D, three planes can intersect at a single point. Likewise, this is not always the case, but three is the minimum number of planes (or linear equations) required to specify a point in 3D. With only two planes, you have at least a 1D space of possible solutions, which is the line of intersection. Algebraically, this means you need two linear equations to get a unique solution in 2D and three linear equations to get a unique solution in 3D. In general, you need n linear equations to be able to get a unique solution in n-dimensions.

Here's an example when working in 4D with the coordinates (x_1, x_2, x_3, x_4), which can seem overly simple but is useful because of how concrete it is. Let's take our first linear equation to be $x_4 = 0$. The solutions to this linear equation form a 3D hyperplane, consisting of vectors of the form $(x_1, x_2, x_3, 0)$. This is clearly a 3D space of solutions, and it turns out to be a vector subspace of $\mathbb{R}^4$ with basis (1, 0, 0, 0), (0, 1, 0, 0), (0, 0, 1, 0).

A second linear equation could be $x_2 = 0$. The solutions of this equation on its own are also a 3D hyperplane. The intersection of these two 3D hyperplanes is a 2D space, consisting of vectors of the form $(x_1, 0, x_3, 0)$, which satisfy both equations. If we could picture such a thing, we would see this as a 2D plane living in 4D space. Specifically, it is the plane spanned by (1, 0, 0, 0) and (0, 0, 1, 0).

Adding one more linear equation, $x_1 = 0$, which defines its own hyperplane, the solutions to all three equations are now a 1D space. The vectors in this 1D space lie on a line in 4D, and have the form $(0, 0, x_3, 0)$. This line is exactly the x_3-axis, which is a 1D subspace of $\mathbb{R}^4$.

Finally, if we impose a fourth linear equation, $x_3 = 0$, the only possible solution is (0, 0, 0, 0), a zero-dimensional vector space. The statements $x_4 = 0$, $x_2 = 0$, $x_1 = 0$, and $x_3 = 0$ are, in fact, linear equations, but these are so simple they describe the solution exactly: $(x_1, x_2, x_3, x_4) = (0, 0, 0, 0)$. Each time we add an equation, we reduced the dimension of the solution space by one, until we got a zero-dimensional space consisting of the single point (0, 0, 0, 0).

Had we chosen different equations, each step would not have been as clear; we would have to test whether each successive hyperplane truly reduces the dimension of the solution space by one. For instance, if we started with

$$x_1 = 0$$

and

$$x_2 = 0$$

we would have reduced the solution set to a 2D space, but then adding another equation to the mix

$$x_1 + x_2 = 0$$

there is no effect on the solution space. Because x_1 and x_2 are already constrained to be zero, the equation $x_1 + x_2 = 0$ is automatically satisfied. This third equation, therefore, adds no more specificity to the solution set.

In the first case, four dimensions with three linear equations to satisfy left us with a $4 - 3 = 1$ dimensional solution space. But in the second case, three equations described a less specific 2D solution space. If you have n dimensions (n unknown variables) and n linear equations, it's possible there's a unique solution—a zero-dimensional solution space—but this is not always the case. More generally, if you're working in n dimensions, the lowest dimensional solution space you can get with m linear equations is $n - m$. In that case, we call the system of linear equations *independent*.

Every basis vector in a space gives us a new independent direction we can move in the space. Independent directions in a space are sometimes called *degrees of freedom*; the z direction, for instance, "freed" us from the plane into larger 3D space. By contrast, every independent linear equation we introduce is a constraint; it removes a degree of freedom and restricts the space of solutions to have a smaller number of dimensions. When the number of independent degrees of freedom (dimensions) equals the number of independent constraints (linear equations), there are no longer any degrees of freedom, and we are left with a unique point.

This is a major philosophical point in linear algebra, and one you can explore more in some mini-projects that follow. In the final section of this chapter, we'll connect the concepts of independent equations and (linearly) independent vectors.

7.3.5 Exercises

Exercise 7.16 What's the equation for a line that passes through (5, 4) and that is perpendicular to (–3, 3)?

Solution Here's the set up:

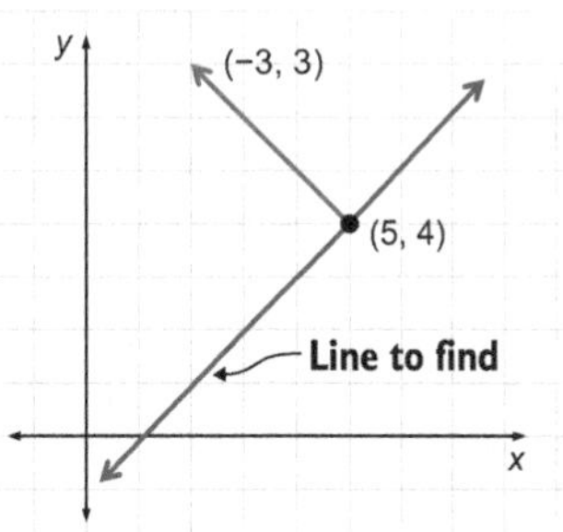

For every point (x, y) on the line, the vector $(x - 5, y - 4)$ is parallel to the line and, therefore, perpendicular to $(-3, 3)$. That means that the dot product $(x - 5, y - 4) \cdot (-3, 3)$ is zero for any (x, y) on the line. This equation expands to $-3x + 15 + 3y - 12 = 0$, which rearranges to give $-3x + 3y = -3$. We can divide both sides by -3 to get a simpler, equivalent equation: $x - y = 1$.

Exercise 7.17—Mini Project Consider a system of two linear equations in 4D:

$$x_1 + 2x_2 + 2x_3 + x_4 = 0$$
$$x_1 - x_4 = 0$$

Explain algebraically (rather than geometrically) why the solutions form a vector subspace of 4D.

Solution We can show that if (a_1, a_2, a_3, a_4) and (b_1, b_2, b_3, b_4) are two solutions, then a linear combination of those is a solution as well. That would imply that the solution set contains all linear combinations of its vectors, making it a vector subspace.

Let's start with the assumption that (a_1, a_2, a_3, a_4) and (b_1, b_2, b_3, b_4) are solutions to both linear equations, which explicitly means:

$$a_1 + 2a_2 + 2a_3 + a_4 = 0$$
$$b_1 + 2b_2 + 2b_3 + b_4 = 0$$
$$a_1 - a_4 = 0$$
$$b_1 - b_4 = 0$$

Picking scalars c and d, the linear combination $c(a_1, a_2, a_3, a_4) + d(b_1, b_2, b_3, b_4)$ is equal to $(ca_1 + db_1, ca_2 + db_2, ca_3 + db_3, ca_4 + db_4)$. Is this a solution to the two equations? We can find out by plugging the four coordinates in for x_1, x_2, x_3, and x_4. In the first equation,

$$x_1 + 2x_2 + 2x_3 + x_4$$

becomes

$$(ca_1 + db_1) + 2(ca_2 + db_2) + 2(ca_3 + db_3) + (ca_4 + db_4)$$

That expands to give us

$$ca_1 + db_1 + 2ca_2 + 2db_2 + 2ca_3 + 2db_3 + ca_4 + db_4$$

which rearranges to

$$c(a_1 + 2a_2+2a_3 + a_4) + d(b_1 + 2b_2 + 2b_3 + b_4)$$

Because $a_1 + 2a_2 + 2a_3 + a_4$ and $b_1 + 2b_2 + 2b_3 + b_4$ are both zero, this expression is zero:

$$c(a_1 + 2a_2 + 2a_3 + a_4) + d(b_1 + 2b_2 + 2b_3 + b_4) = c \cdot 0 + d \cdot 0 = 0$$

That means the linear combination is a solution to the first equation. Similarly, plugging the linear combination into the second equation, we see it's a solution to that equation as well:

$$(ca_1 + db_1) - (ca_4 + db_4) = c(a_1 - a_4) + d(b_1 - b_4) = c \cdot 0 + d \cdot 0 = 0$$

Any linear combination of any two solutions is also a solution, so the solution set contains all of its linear combinations. That means the solution set is a vector subspace of 4D.

Exercise 7.18 What is the standard form equation for a plane that passes through the point (1, 1, 1) and is perpendicular to the vector (1, 1, 1)?

Solution For any point (x, y, z) in the plane, the vector $(x - 1, y - 1, z - 1)$ is perpendicular to (1, 1, 1). That means that the dot product $(x - 1, y - 1, z - 1) \cdot (1, 1, 1)$ is zero for any x, y, and z values giving a point in the plane. This expands to give us $(x - 1) + (y - 1) + (z - 1) = 0$ or $x + y + z = 3$, the standard form equation for the plane.

Exercise 7.19—Mini Project Write a Python function that takes three 3D points as inputs and returns the standard form equation of the plane that they lie in. For instance, if the standard form equation is $ax + by + cz = d$, the function could return the tuple (a, b, c, d).

Hint Differences of any pairs of the three vectors are parallel to the plane, so cross products of the differences are perpendicular.

Solution If the points given are p_1, p_2, and p_3, then the vector differences like $p_3 - p_1$ and $p_2 - p_1$ are parallel to the plane. The cross product $(p_2 - p_1) \times (p_3 - p_1)$ is then perpendicular to the plane. (All is well as long as the points p_1, p_2, and p_3 form a triangle, so the differences are not parallel.) With a point in the plane (for instance, p_1) and a perpendicular vector, we can repeat the process of finding the standard form of the solution as in the previous two exercises:

```
from vectors import *

def plane_equation(p1,p2,p3):
    parallel1 = subtract(p2,p1)
    parallel2 = subtract(p3,p1)
    a,b,c = cross(parallel1, parallel2)
    d = dot((a,b,c), p1)
    return a,b,c,d
```

(continued)

For example, these are three points from the plane $x + y + z = 3$ from the preceding exercise:

```
>>> plane_equation((1,1,1), (3,0,0), (0,3,0))
(3, 3, 3, 9)
```

The result is (3, 3, 3, 9), meaning $3x + 3y + 3z = 9$, which is equivalent to $x + y + z = 3$. That means we got it right!

Exercise 7.20 How many total constants a_{ij} are in the following matrix equation? How many equations are there? How many unknowns? Write the full matrix equation (no dots) and the full system of linear equations (no dots).

$$\begin{pmatrix} a_{11} & a_{12} & \cdots & a_{17} \\ a_{21} & a_{22} & \cdots & a_{27} \\ \vdots & \vdots & \ddots & \vdots \\ a_{51} & a_{52} & \cdots & a_{57} \end{pmatrix} \begin{pmatrix} x_1 \\ x_2 \\ \vdots \\ x_7 \end{pmatrix} = \begin{pmatrix} b_1 \\ b_2 \\ \vdots \\ b_5 \end{pmatrix}$$

An abbreviated system of linear equations in matrix form

Solution To be clear, we can write out the full matrix equation first:

$$\begin{pmatrix} a_{11} & a_{12} & a_{13} & a_{14} & a_{15} & a_{16} & a_{17} \\ a_{21} & a_{22} & a_{23} & a_{24} & a_{25} & a_{26} & a_{27} \\ a_{31} & a_{32} & a_{33} & a_{34} & a_{35} & a_{36} & a_{37} \\ a_{41} & a_{42} & a_{43} & a_{44} & a_{45} & a_{46} & a_{47} \\ a_{51} & a_{52} & a_{53} & a_{54} & a_{55} & a_{56} & a_{57} \end{pmatrix} \begin{pmatrix} x_1 \\ x_2 \\ x_3 \\ x_4 \\ x_5 \\ x_6 \\ x_7 \end{pmatrix} = \begin{pmatrix} b_1 \\ b_2 \\ b_3 \\ b_4 \\ b_5 \end{pmatrix}$$

The unabbreviated version of the matrix equation

In total, there are $5 \cdot 7 = 35$ entries in this matrix and 35 a_{ij} constants on the left-hand side of the equations in the linear system. There are 7 unknown variables: $x_1, x_2, \ldots, x_7$ and 5 equations (one per row of the matrix). You can get the full linear system by carrying out the matrix multiplication:

$$\begin{aligned}
a_{11}x_1 + a_{12}x_2 + a_{13}x_3 + a_{14}x_4 + a_{15}x_5 + a_{16}x_6 + a_{17}x_7 &= b_1 \\
a_{21}x_1 + a_{22}x_2 + a_{23}x_3 + a_{24}x_4 + a_{25}x_5 + a_{26}x_6 + a_{27}x_7 &= b_2 \\
a_{31}x_1 + a_{32}x_2 + a_{33}x_3 + a_{34}x_4 + a_{35}x_5 + a_{36}x_6 + a_{37}x_7 &= b_3 \\
a_{41}x_1 + a_{42}x_2 + a_{43}x_3 + a_{44}x_4 + a_{45}x_5 + a_{46}x_6 + a_{47}x_7 &= b_4 \\
a_{51}x_1 + a_{52}x_2 + a_{53}x_3 + a_{54}x_4 + a_{55}x_5 + a_{56}x_6 + a_{57}x_7 &= b_5
\end{aligned}$$

The full system of linear equations represented by this matrix equation

You can see why we avoid this tedious writing with abbreviations!

Exercise 7.21 Write the following linear equation without summation shorthand. Geometrically, what does the set of solutions look like?

$$\sum_{i=1}^{3} x_i = 1$$

Solution The left-hand side of this equation is a sum of terms of the form x_i for i, ranging from 1 to 3. That gives us $x_1 + x_2 + x_3 = 1$. This is the standard form of a linear equation in three variables, so its solutions form a plane in 3D space.

Exercise 7.22 Sketch three planes, none of which are parallel and do not have a single point of intersection. (Better yet, find their equations and graph them!)

Solution Here are three planes: $z + y = 0$, $z - y = 0$, and $z = 3$ and the graph:

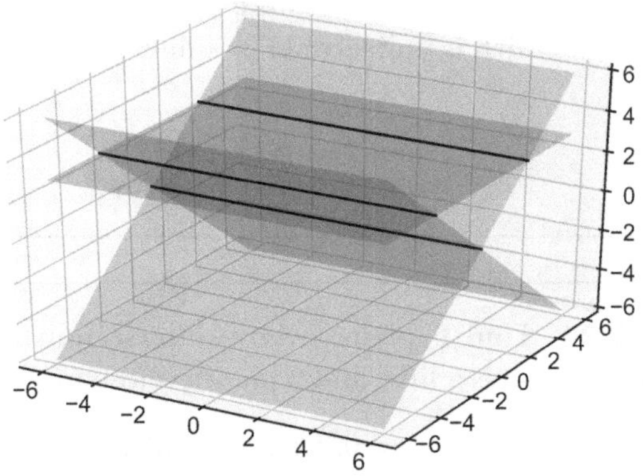

Three non-parallel planes that don't share an intersection point

I've drawn the intersections of the three pairs of planes, which are parallel lines. Because these lines never meet, there is no single point of intersection for all three planes. This is like the example you saw in chapter 6: three vectors can be linearly dependent even when no pair among them is parallel.

Exercise 7.23 Suppose we have m linear equations and n unknown variables. What do the following values of m and n say about whether there is a unique solution?

1. $m = 2, n = 2$
2. $m = 2, n = 7$
3. $m = 5, n = 5$
4. $m = 3, n = 2$

(continued)

Solution

1 With two linear equations and two unknowns, there *can* be a unique solution. The two equations represent lines in the plane, and they will intersect at a unique point unless they are parallel.

2 With two linear equations and seven unknowns, there *cannot* be a unique solution. Assuming the 6D hyperplanes defined by these equations are not parallel, there will be a 5D space of solutions.

3 With five linear equations and five unknowns, there *can* be a unique solution, as long as the equations are independent.

4 With three linear equations and two unknowns, there *can* be a unique solution, but it requires some luck. This would mean that the third line happens to pass through the intersection point of the first two lines, which is unlikely but possible.

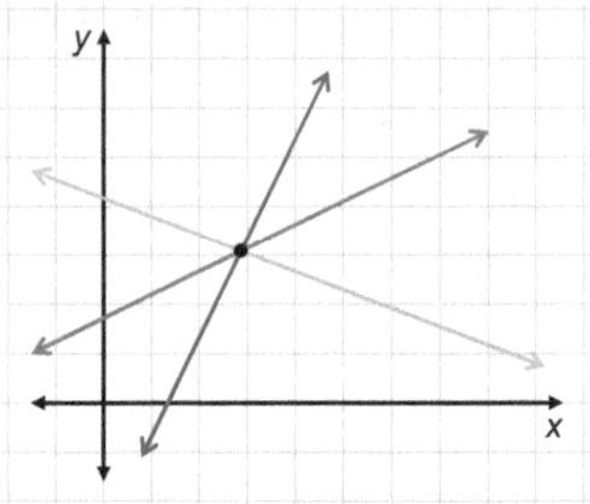

Three lines in the plane that happen to intersect at a point

Exercise 7.24 Find 3 planes whose intersection is a single point, 3 planes whose intersection is a line, and 3 planes whose intersection is a plane.

Solution The planes $z - y = 0$, $z + y = 0$, and $z + x = 0$ intersect at the single point (0, 0, 0). Most randomly selected planes will intersect at a unique point like this.

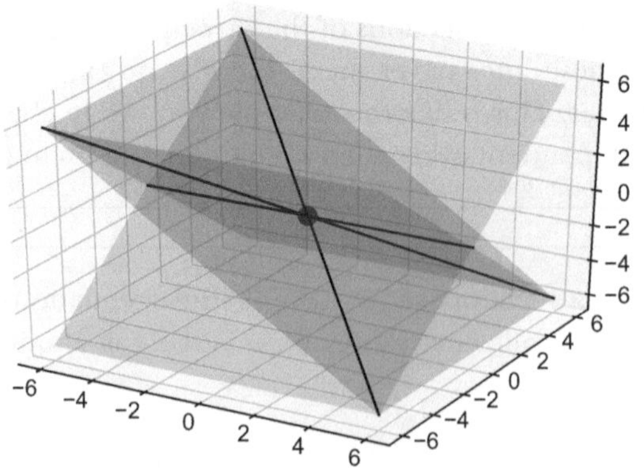

Three planes intersecting at a single point

The planes $z - y = 0$, $z + y = 0$, and $z = 0$ intersect on a line, specifically the x-axis. If you play with these equations, you'll find both y and z are constrained to be zero, but x doesn't even appear, so it has no constraints. Any vector $(x, 0, 0)$ on the x-axis is, therefore, a solution.

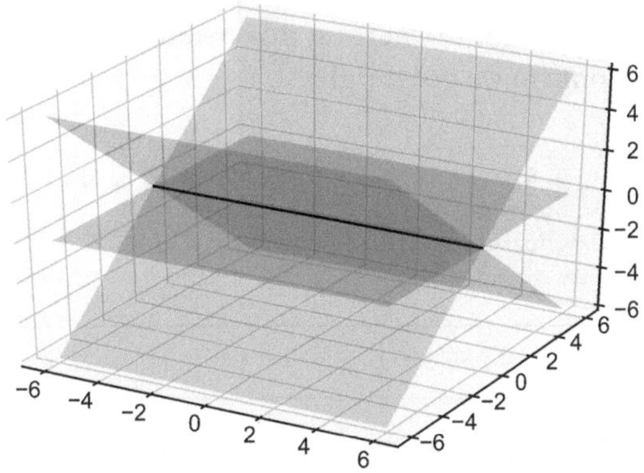

Three planes whose intersection points form a line

Finally, if all three equations represent the same plane, then that whole plane is a set of solutions. For instance, $z - y = 0$, $2z - 2y = 0$, and $3z - 3y = 0$ all represent the same plane.

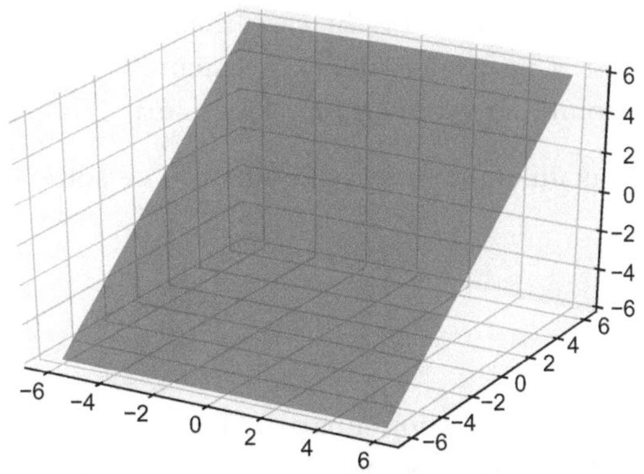

Three identical planes overlaid; their set of solutions is the whole plane.

Exercise 7.25 Without using Python, what is the solution of the system of linear equations in 5D? $x_5 = 3$, $x_2 = 1$, $x_4 = -1$, $x_1 = 0$, and $x_1 + x_2 + x_3 = -2$? Confirm the answer with NumPy.

(continued)

Solution Because four of these linear equations specify the value of a coordinate, we know the solution has the form $(0,1, x_3, -1,3)$. We need to do some algebra using the final equation to find out the value of x_3. Because $x_1 + x_2 + x_3 = -2$, we know $0 + 1 + x_3 = -2$, and x_3 must be -3. The unique solution point is, therefore, $(0, 1, -3, -1, 3)$. Converting this system to matrix form, we can solve it with NumPy to confirm we got it right:

```
>>> matrix =
np.array(((0,0,0,0,1),(0,1,0,0,0),(0,0,0,1,0),(1,0,0,0,0),(1,1,1,0,0)))
>>> vector = np.array((3,1,-1,0,-2))
>>> np.linalg.solve(matrix,vector)
array([ 0.,  1., -3., -1.,  3.])
```

Exercise 7.26—Mini Project In any number of dimensions, there is an identity matrix that acts as the identity map. That is, when you multiply the n-dimensional identity matrix I by any vector $\mathbf{v}$, you get the same vector $\mathbf{v}$ as a result; therefore, $I\mathbf{v} = \mathbf{v}$.

This means that $I\mathbf{v} = \mathbf{w}$ is an easy system of linear equations to solve: one possible answer for $\mathbf{v}$ is $\mathbf{v} = \mathbf{w}$. The idea of this mini-project is that you can start with a system of linear equations, $A\mathbf{v} = \mathbf{w}$, and multiply both sides by another matrix B such that $(BA) = I$. If that is the case, then you have $(BA)\mathbf{v} = B\mathbf{w}$ and $I\mathbf{v} = B\mathbf{w}$ or $\mathbf{v} = B\mathbf{w}$. In other words, if you have a system $A\mathbf{v} = \mathbf{w}$, and a suitable matrix B, then $B\mathbf{w}$ is the solution to your system. This matrix B is called the *inverse* matrix of A.

Let's look again at the system of equations we solved in section 7.3.2:

$$\begin{pmatrix} 1 & 1 & -1 \\ 0 & 2 & -1 \\ 1 & 0 & 1 \end{pmatrix} \begin{pmatrix} x \\ y \\ z \end{pmatrix} = \begin{pmatrix} -1 \\ 3 \\ 2 \end{pmatrix}$$

Use the NumPy function `numpy.linalg.inv(matrix)`, which returns the inverse of the matrix it is given to find the inverse of the matrix on the left-hand side of the equation. Then, multiply both sides by this matrix to find the solution to the linear system. Compare your results with the results we got from NumPy's solver.

Hint You might also want to use NumPy's built-in matrix multiplication routine, `numpy.matmul`, to make computations simpler.

Solution First, we can compute the inverse of the matrix using NumPy:

```
>>> matrix = np.array(((1,1,-1),(0,2,-1),(1,0,1)))
>>> vector = np.array((-1,3,2))
>>> inverse = np.linalg.inv(matrix)
>>> inverse
array([[ 0.66666667, -0.33333333,  0.33333333],
       [-0.33333333,  0.66666667,  0.33333333],
       [-0.66666667,  0.33333333,  0.66666667]])
```

The product of the inverse matrix with the original matrix gives us the identity matrix, having 1's on the diagonal and 0's elsewhere, albeit with some numerical error:

```
>>> np.matmul(inverse,matrix)
array([[ 1.00000000e+00,  1.11022302e-16, -1.11022302e-16],
       [ 0.00000000e+00,  1.00000000e+00,  0.00000000e+00],
       [ 0.00000000e+00,  0.00000000e+00,  1.00000000e+00]])
```

The trick is to multiply both sides of the matrix equation by this inverse matrix. Here I've rounded the values in the inverse matrix for the sake of readability. We already know that the first product on the left is a matrix and its inverse, so we can simplify accordingly:

$$\begin{pmatrix} 0.667 & -0.333 & 0.333 \\ -0.333 & 0.667 & 0.333 \\ -0.667 & 0.333 & 0.667 \end{pmatrix} \begin{pmatrix} 1 & -1 & 0 \\ 0 & -1 & -1 \\ 1 & 0 & 2 \end{pmatrix} \begin{pmatrix} x \\ y \\ z \end{pmatrix} = \begin{pmatrix} 0.667 & -0.333 & 0.333 \\ -0.333 & 0.667 & 0.333 \\ -0.667 & 0.333 & 0.667 \end{pmatrix} \begin{pmatrix} 1 \\ 3 \\ 2 \end{pmatrix}$$

$$\begin{pmatrix} 1 & 0 & 0 \\ 0 & 1 & 0 \\ 0 & 0 & 1 \end{pmatrix} \begin{pmatrix} x \\ y \\ z \end{pmatrix} = \begin{pmatrix} 0.667 & -0.333 & 0.333 \\ -0.333 & 0.667 & 0.333 \\ -0.667 & 0.333 & 0.667 \end{pmatrix} \begin{pmatrix} 1 \\ 3 \\ 2 \end{pmatrix}$$

$$\begin{pmatrix} x \\ y \\ z \end{pmatrix} = \begin{pmatrix} 0.667 & -0.333 & 0.333 \\ -0.333 & 0.667 & 0.333 \\ -0.667 & 0.333 & 0.667 \end{pmatrix} \begin{pmatrix} 1 \\ 3 \\ 2 \end{pmatrix}$$

Multiplying both sides of the system by the inverse matrix and simplifying

This gives us an explicit formula for the solution (x, y, z); all we need to do is to carry out the matrix multiplication. It turns out `numpy.matmul` also works for matrix vector multiplication:

```
>>> np.matmul(inverse, vector)
array([-1., 3., 3.])
```

This is the same as the solution we got earlier from the solver.

7.4 Changing basis by solving linear equations

The notion of linear independence of vectors is clearly related to the notion of independence of linear equations. The connection comes from the fact that solving a system of linear equations is the equivalent of rewriting vectors in a different basis. Let's explore what this means in 2D. When we write coordinates for a vector like (4, 3), we are implicitly writing the vector as a linear combination of the standard basis vectors:

$$(4, 3) = 4\mathbf{e}_1 + 3\mathbf{e}_2$$

In the last chapter, you learned that the standard basis consisting of $\mathbf{e}_1 = (1, 0)$ and $\mathbf{e}_2 = (0, 1)$ is not the only basis available. For instance, a pair of vectors like $\mathbf{u}_1 = (1, 1)$ and $\mathbf{u}_2 = (-1, 1)$ form a basis for $\mathbb{R}^2$. As any 2D vector can be written as a linear combination of $\mathbf{e}_1$ and $\mathbf{e}_2$, so can any 2D vector be written as a linear combination of this $\mathbf{u}_1$ and $\mathbf{u}_2$. For some c and d, we can make the following equation true, but it's not immediately obvious what the values of c and d are:

$$c \cdot (1, 1) + d \cdot (-1, 1) = (4, 2)$$

Figure 7.27 shows a visual representation of this.

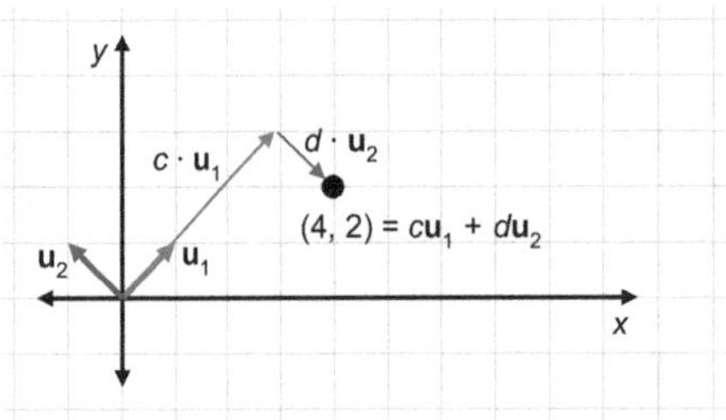

Figure 7.27 Writing (4, 2) as a linear combination of $u_1 = (1, 1)$ and $u_2 = (-1, 1)$

As a linear combination, this equation is equivalent to a matrix equation, namely:

$$\begin{pmatrix} 1 & -1 \\ 1 & 1 \end{pmatrix} \begin{pmatrix} c \\ d \end{pmatrix} = \begin{pmatrix} 4 \\ 2 \end{pmatrix}$$

This, too, is a system of linear equations! In this case, the unknown vector is written (c, d) rather than (x, y), and the linear equations hidden in the matrix equation are $c - d = 4$ and $c + d = 2$. There is a 2D space of vectors (c, d) that define different linear combinations of $\mathbf{u}_1$ and $\mathbf{u}_2$, but only one simultaneously satisfies these two equations.

Any choice of the pair (c, d) defines a different linear combination. As an example, let's look at an arbitrary value of (c, d), say $(c, d) = (3, 1)$. The vector (3, 1) doesn't live in the same vector space as $\mathbf{u}_1$ and $\mathbf{u}_2$; it lives in a vector space of (c, d) pairs, each of which describe a different linear combination of $\mathbf{u}_1$ and $\mathbf{u}_2$. The point $(c, d) = (3, 1)$ describes a specific linear combination in our original 2D space: $3\mathbf{u}_1 + 1\mathbf{u}_2$ gets us to the point $(x, y) = (2, 4)$ (figure 7.28).

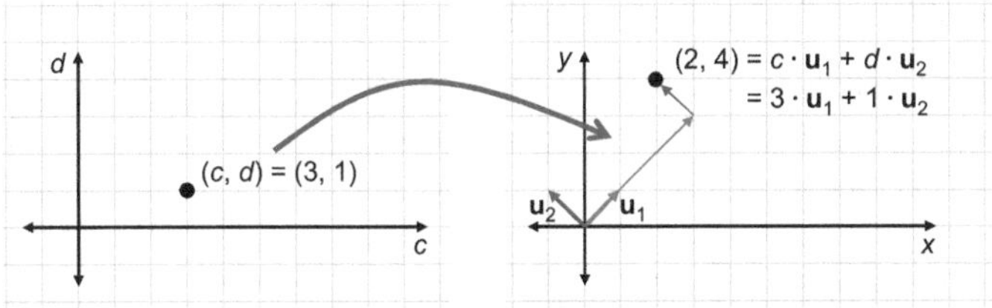

Figure 7.28 There is a 2D space of values (*c*, *d*), where (*c*, *d*) = (3, 1) and yields the linear combination $3u_1 + 1u_2 = (2, 4)$.

Recall that we're trying to make (4, 2) as a linear combination of $\mathbf{u}_1$ and $\mathbf{u}_2$, so this isn't the linear combination we were looking for. For $c\mathbf{u}_1 + d\mathbf{u}_2$ to equal (4, 2), we need to satisfy $c - d = 4$ and $c + d = 2$, as we saw previously.

Let's draw the system of linear equations in the *c*, *d* plane. Visually, we can tell that (3, –1) is a point that satisfies both $c + d = 2$ and $c - d = 4$. This gives us the pair of scalars to use in a linear combination to make (4, 2) out of $\mathbf{u}_1$ and $\mathbf{u}_2$ as shown in figure 7.29.

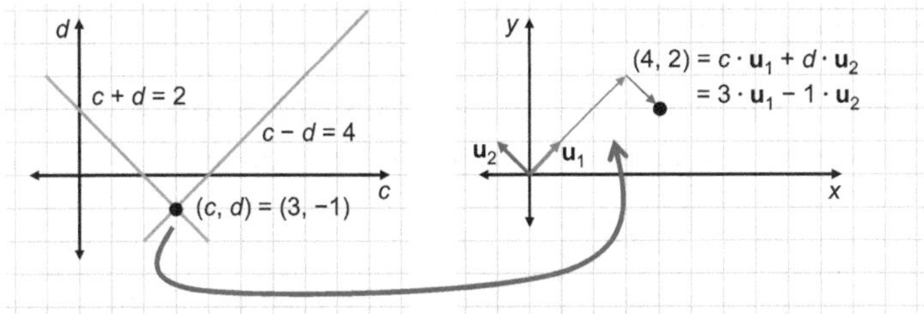

Figure 7.29 The point (*c*, *d*) = (3, –1) satisfies both *c* + *d* = 2 and *c* – *d* = 4. Therefore, it describes the linear combination we were looking for.

Now we can write (4, 2) as a linear combination of two different pairs of basis vectors: $(4, 2) = 4\mathbf{e}_1 + 2\mathbf{e}_2$ and $(4, 2) = 3\mathbf{u}_1 - 1\mathbf{u}_2$. Remember, the coordinates (4, 2) are exactly the scalars in the linear combination $4\mathbf{e}_1 + 2\mathbf{e}_2$. If we had drawn our axes differently, $\mathbf{u}_1$ and $\mathbf{u}_2$ could just as well have been our standard basis; our vector would be $3\mathbf{u}_1 - \mathbf{u}_2$ and we would say its coordinates were (3, 1). To emphasize that coordinates are determined by our choice of basis, we can say that the vector has coordinates (4, 2) with respect to the standard basis, but it has coordinates (3, –1) with respect to the basis consisting of $\mathbf{u}_1$ and $\mathbf{u}_2$.

Finding the coordinates of a vector with respect to a different basis is an example of a computational problem that is really a system of linear equations in disguise. It's an important example because every system of linear equations can be thought of in this way. Let's try another example, this time in 3D, to see what I mean.

7.4.1 Solving a 3D example

Let's start by writing an example of a system of linear equations in 3D and then we'll work on interpreting it. Instead of a 2-by-2 matrix and a 2D vector, we can start with a 3-by-3 matrix and a 3D vector:

$$\begin{pmatrix} 1 & -1 & 0 \\ 0 & -1 & -1 \\ 1 & 0 & 2 \end{pmatrix} \begin{pmatrix} x \\ y \\ z \end{pmatrix} = \begin{pmatrix} 1 \\ 3 \\ -7 \end{pmatrix}$$

The unknown here is a 3D vector; we need to find three numbers to identify it. Doing the matrix multiplication, we can break this up into three equations:

$$1 \cdot x - 1 \cdot y + 0 \cdot z = 1$$
$$0 \cdot x - 1 \cdot y - 1 \cdot z = 3$$
$$1 \cdot x + 0 \cdot y + 2 \cdot z = -7$$

This is a system of three linear equations with three unknowns, and $ax + by + cz = d$ is the standard form for a linear equation in 3D. In the next section, we look at the geometric interpretation for 3D linear equations. (It turns out they represent planes in 3D as opposed to lines in 2D.)

For now, let's look at this system as a linear combination with coefficients to be determined. The previous matrix equation is equivalent to the following:

$$x \begin{pmatrix} 1 \\ 0 \\ 1 \end{pmatrix} + y \begin{pmatrix} -1 \\ -1 \\ 0 \end{pmatrix} + z \begin{pmatrix} 0 \\ -1 \\ 2 \end{pmatrix} = \begin{pmatrix} 1 \\ 3 \\ -7 \end{pmatrix}$$

Solving this equation is equivalent to asking the question: What linear combination of (1, 0, 1), (–1, –1, 0), and (0, –1, 2) yields the vector (1, 3, –7)? This is harder to picture than the 2D example, and it is harder to compute the answer by hand as well. Fortunately, we know NumPy can handle systems of linear equations in three unknowns, so we simply pass a 3-by-3 matrix and 3D vector as inputs to the solver like this:

```
>>> import numpy as np
>>> w = np.array((1,3,-7))
>>> a = np.array(((1,-1,0),(0,-1,-1),(1,0,2)))
>>> np.linalg.solve(a,w)
array([ 3.,  2., -5.])
```

The values that solve our linear system are, therefore, $x = 3$, $y = 2$, and $z = -5$. In other words, these are the coefficients that build our desired linear combination. We can say that the vector (1, 3, –7) has coordinates (3, 2, –5) with respect to the basis (1, 0, 1), (–1, –1, 0), (0, –1, 2).

The story is the same in higher dimensions; as long as it is possible to do so, we can write a vector as a linear combination of other vectors by solving a corresponding system of linear equations. But, it's not always possible to write a linear combination, and not every system of linear equations has a unique solution or even a solution at all. The question of whether a collection of vectors forms a basis is computationally equivalent to the question of whether a system of linear equations has a unique solution.

This profound connection is a good place to bookend part 1 with its focus on linear algebra. There will be plenty of more linear algebra nuggets throughout the book, but they are even more useful when we pair them with the core topic of part 2: calculus.

7.4.2 Exercises

Exercise 7.27 How can you write the vector (5, 5) as a linear combination of (10, 1) (3, 2)?

Solution This is equivalent to asking what numbers *a* and *b* satisfy the equation

$$a\begin{pmatrix}10\\1\end{pmatrix} + b\begin{pmatrix}3\\2\end{pmatrix} = \begin{pmatrix}5\\5\end{pmatrix}$$

or what vector (*a*, *b*) satisfies the matrix equation:

$$\begin{pmatrix}10 & 3\\1 & 2\end{pmatrix}\begin{pmatrix}a\\b\end{pmatrix} = \begin{pmatrix}5\\5\end{pmatrix}$$

We can find a solution with NumPy:

```
>>> matrix = np.array(((10,3),(1,2)))
>>> vector = np.array((5,5))
>>> np.linalg.solve(matrix,vector)
array([-0.29411765, 2.64705882])
```

This means the linear combination (which you can check!) is as follows:

$$-0.29411765 \cdot \begin{pmatrix}10\\1\end{pmatrix} + 2.64705882 \cdot \begin{pmatrix}3\\2\end{pmatrix} = \begin{pmatrix}5\\5\end{pmatrix}$$

Exercise 7.28 Write the vector (3, 0, 6, 9) as a linear combination of the vectors (0, 0, 1, 1), (0, –2, –1, –1), (1, –2, 0, 2), and (0, 0, –2, 1).

(continued)

Solution The linear system to solve is

$$\begin{pmatrix} 0 & 0 & 1 & 0 \\ 0 & -2 & -2 & 0 \\ 1 & -1 & 0 & -2 \\ 1 & -1 & 2 & 1 \end{pmatrix} \begin{pmatrix} a \\ b \\ c \\ d \end{pmatrix} = \begin{pmatrix} 3 \\ 0 \\ 6 \\ 9 \end{pmatrix}$$

where the columns of the 4-by-4 matrix are the vectors we want to build the linear combination from. NumPy gives us the solution to this system:

```
>>> matrix = np.array(((0, 0, 1, 0), (0, -2, -2, 0), (1, -1, 0, -2), (1,
-1, 2, 1)))
>>> vector = np.array((3,0,6,9))
>>> np.linalg.solve(matrix,vector)
array([ 1., -3., 3., -1.])
```

This means that the linear combination is

$$1 \cdot \begin{pmatrix} 0 \\ 0 \\ 1 \\ 1 \end{pmatrix} - 3 \cdot \begin{pmatrix} 0 \\ -2 \\ -1 \\ -1 \end{pmatrix} + 3 \cdot \begin{pmatrix} 1 \\ -2 \\ 0 \\ 2 \end{pmatrix} - \begin{pmatrix} 0 \\ 0 \\ -2 \\ 1 \end{pmatrix} = \begin{pmatrix} 3 \\ 0 \\ 6 \\ 9 \end{pmatrix}$$

Summary

- Model objects in a 2D video game can be done as polygonal shapes built out of line segments.
- Given two vectors **u** and **v**, the points of the form $\mathbf{u} + t\mathbf{v}$ for any real number t lie on a straight line. In fact, any line can be described with this formula.
- Given real numbers a, b, and c, where at least one of a or b is non-zero, the points (x, y) in the plane satisfying $ax + by = c$ lie on a straight line. This is called the *standard form* for the equation of a line, and any line can be written in this form for some choice of a, b, and c. Equations for lines are called *linear equations.*
- Finding the point where two lines intersect in the plane is equivalent to finding the values (x, y) that simultaneously satisfy two linear equations. A collection of linear equations that we seek to solve simultaneously is called a *system of linear equations.*
- Solving a system of two linear equations is equivalent to finding what vector can be multiplied by a known 2-by-2 matrix to yield a known vector.
- NumPy has a built-in function, `numpy.linalg.solve`, that takes a matrix and a vector and solves the corresponding system of linear equations automatically, if possible.

- Some systems of linear equations cannot be solved. For instance, if two lines are parallel, they can have either no intersection points or infinitely many (which would mean they are the same line). This means there is no (x, y) value that simultaneously satisfies both lines' equations. A matrix representing such a system is called *singular.*
- Planes in 3D are the analogs of lines in 2D. They are the sets of points (x, y, z) satisfying equations of the form $ax + by + cz = d$.
- Two non-parallel planes in 3D intersect at infinitely many points, and specifically, the set of points they share form a 1D line in 3D. Three planes can have a unique point of intersection that can be found by solving the system of three linear equations representing the planes.
- Lines in 2D and planes in 3D are both cases of *hyperplanes,* sets of points in n-dimensions that are solutions to a single linear equation.
- In n-dimensions, you need a system of at least n linear equations to find a unique solution. If you have exactly n linear equations and they have a unique solution, those are called *independent equations.*
- Figuring out how to write a vector as a linear combination of a given set of vectors is computationally equivalent to solving a system of linear equations. If the set of vectors is a basis for the space, this is always possible.

Part 2

Calculus and physical simulation

In part 2 of this book, we embark on an overview of calculus. Broadly speaking, calculus is the study of continuous change, so we talk a lot about how to measure rates of change of different quantities and what these rates of change can tell us.

In my opinion, calculus gets a bad rap as a difficult subject because of how much algebra is required, not because the concepts are unfamiliar. If you've ever owned or driven a car, you have an intuitive understanding of rates and cumulative values: a speedometer measures your *rate* of movement over time, while an odometer measures the *cumulative* miles driven. To some extent their measurements must agree. If your speedometer reads a higher value over a period of time, your odometer should increase by a larger amount, and vice versa.

In calculus, we learn that if we have a function giving a cumulative value at any time, we can calculate its rate of change, also as a function of time. This operation of taking a "cumulative" function and returning a "rate" function is called a *derivative.* Similarly, if we start with a rate function, we can reconstruct a cumulative function that agrees with it, which is an operation called an *integral.* We spend all of chapter 8 making sure these conversions make conceptual sense, applying it to measured fluid volume (a cumulative function) and fluid flow rate (a corresponding rate function). In chapter 9, we extend these ideas to multiple dimensions. To simulate a moving object in a video game engine, we need to consider the relationship between speed and position in each coordinate independently.

Once you get a conceptual understanding of calculus in chapters 8 and 9, we'll cover the mechanics in chapter 10. We'll have more fun with this than in an ordinary calculus class, because Python will do most of the formula crunching for us. We model mathematical expressions like little computer programs, which we can parse and transform to find their derivatives and integrals. Chapter 10, therefore, shows quite a different approach to doing math in code, and this approach is called *symbolic programming*.

In chapter 11, we return to calculus in multiple dimensions. While speed on a speedometer or fluid flow rate through a pipe are functions that vary over time, we can also have functions that vary over space. These functions take vectors as inputs and return numbers or vectors as outputs. For instance, representing the strength of gravity as a function over a 2D space allows us to add some interesting physics to our video game from chapter 7. A key calculus operation for functions that vary over space is the *gradient*, an operation that tells us the spatial direction that a function increases at the highest rate. Because it measures a rate, a gradient is like a vector version of an ordinary derivative. In chapter 12, we use the gradient to *optimize* a function or to find the input for which it returns the largest output. By following the direction of the gradient vector, we can find increasingly large outputs, and eventually, we can converge on a maximum value for the whole function.

In chapter 13, we cover a completely different application of calculus. It turns out that the integral of a function tells us a lot about the geometry of the graph of a function. In particular, integrating the product of two functions tells us about how similar their graphs are. We'll apply this kind of analysis to sound waves. A *sound wave* is a graph of a function describing a sound, and the graph tells us whether the sound is loud or soft, high or low pitched, and so on. Comparing a sound wave with different musical notes, we can find out the musical notes it contains. Thinking of a sound wave as a function corresponds to an important mathematical concept called a *Fourier series*.

As compared to part 1, part 2 is more of a smorgasbord of topics, but there are two main themes you should keep your eye on. The first is the concept of the rate of change of a function; whether a function is increasing or decreasing at a point tells us how to find bigger or smaller values. The second is the idea of an operation that takes functions as inputs and returns functions as outputs. In calculus, the answer to many questions comes in the form of a function. These two ideas will be key to our machine learning applications in part 3.

8 Understanding rates of change

This chapter covers

- Calculating the average rate of change in a mathematical function
- Approximating the instantaneous rate of change at a point
- Picturing how the rate of change is itself changing
- Reconstructing a function from its rate of change

In this chapter, I introduce you to two of the most important concepts from calculus: the derivative and the integral. Both of these are operations that work with functions. The *derivative* takes a function and gives you another function measuring its rate of change. The *integral* does the opposite; it takes a function representing a rate of change and gives you back a function measuring the original, cumulative value.

I focus on a simple example from my own work in data analysis for oil production. The set up we'll picture is a pump lifting crude oil out of a well, which then flows through a pipe into a tank. The pipe is equipped with a meter that continuously measures the rate of fluid flow, and the tank is equipped with a sensor that

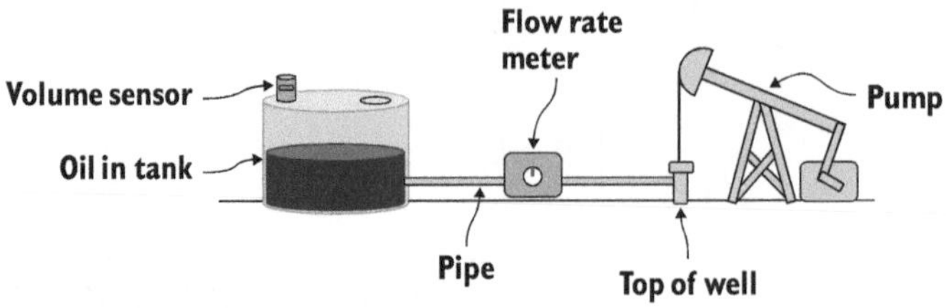

Figure 8.1 Schematic diagram of a pump lifting oil from a well and pumping it into a tank

detects the height of fluid in the tank and reports the volume of oil stored within (figure 8.1).

The volume sensor measurements tell us the volume of oil in the tank as a function of time, while the flow meter measurements tell us the volume flowing into the tank per hour, also as a function of time. In this example, the volume is the cumulative value and the flow rate is its rate of change.

In this chapter, we solve two main problems. First, in our example, we start with known, cumulative volumes over time and calculate the flow rate as a function of time using the derivative. Second, we do the opposite task, starting with the flow rate as a function of time and calculating the cumulative volume of oil in the tank over time using the integral. Figure 8.2 shows this process.

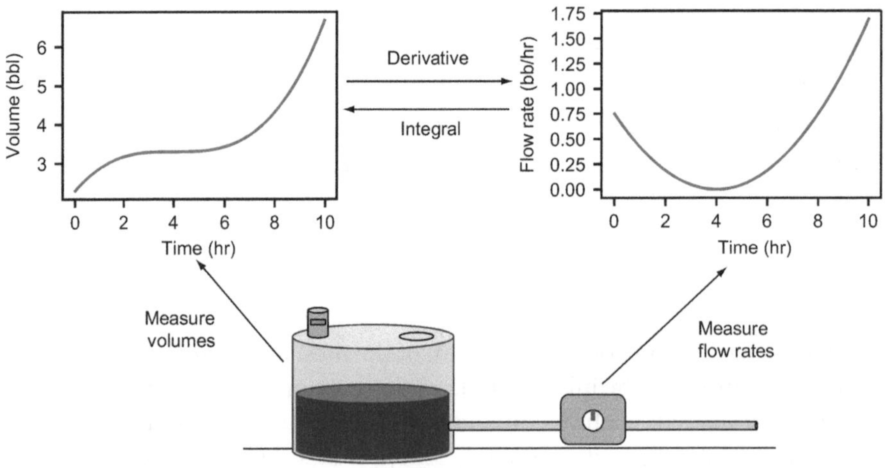

Figure 8.2 Finding the flow rate over time from the volume using the derivative and then finding the volume over time from the flow rate using the integral

We'll write a function called `get_flow_rate(volume_function)` that takes the volume function as an input and returns a new Python function that gives the flow rate at any time. Then we'll write a second function, `get_volume(flow_rate_function)`, that takes the flow rate function and returns a Python function giving volume over time. I intersperse a few smaller examples along the way as a warm up to help you think about rates of change.

Even though its big ideas aren't that complicated or foreign, calculus gets a bad reputation because it requires so much tedious algebra. For that reason, I focus on introducing new ideas in this chapter but not a lot of new techniques. Most of the examples require only the linear function math that we covered in chapter 7. Let's get started!

8.1 Calculating average flow rate from volume

Let's start by assuming we know the volume in the tank over time, which is encoded as a Python function called `volume`. This function takes as an argument, the time in hours after a predefined starting point, and returns the volume of oil in the tank at that time, measured in a unit called barrels (abbreviated "bbl"). To keep the focus on the ideas rather than the algebra, I won't even tell you the formula for the `volume` function (though you can see it in the source code if you're curious). All you need to do for now is to call it and to plot it. When you plot it, you'll see something like figure 8.3.

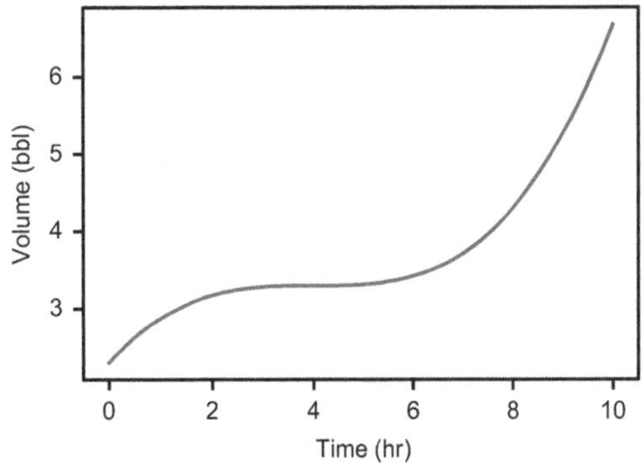

Figure 8.3 A plot of the `volume` function shows the volume of oil in the tank over time.

We want to move in the direction of finding the flow rate into the tank at any point in time, so for our first baby step, let's calculate this in an intuitive way. In this example, let's write a function `average_flow_rate(v, t1, t2)` that takes a volume function `v`, a start time `t1`, and an end time `t2`, and returns a number that is the *average flow rate* into the tank on the time interval. That is, it tells us the overall number of barrels per hour entering the tank.

8.1.1 Implementing an average_flow_rate function

The word *per* in "barrels per hour" suggests that we're going to do some division to get our answer. The way to calculate the average flow rate is to take the total change in volume divided by the elapsed time:

$$\text{average flow rate} = \frac{\text{change in volume}}{\text{elapsed time}}$$

The elapsed time between the starting time t_1 and the ending time t_2 measured in hours is $t_2 - t_1$. If we have a function $V(t)$ that tells us volume as a function of time, the overall change in volume is the volume at t_2 minus the volume at t_1, or $V(t_2) - V(t_1)$. That gives us a more specific equation to work with:

$$\text{average flow rate from } t_1 \text{ to } t_2 = \frac{V(t_2) - V(t_1)}{t_2 - t_1}$$

This is how we calculate rates of change in different contexts. For instance, your speed when driving a car is the rate at which you cover distance with respect to time. To calculate your average speed for a drive, you divide your total distance traveled in miles by the elapsed time in hours to get a result in miles per hour (mph). To know the distance traveled and time elapsed, you need to check your clock and odometer at the beginning and end of the trip.

Our formula for average flow rate depends on the volume function V and the starting and ending times t_1 and t_2, which are the parameters we'll pass to the corresponding Python function. The body of the function is a direct translation of this mathematical formula to Python:

```
def average_flow_rate(v,t1,t2):
    return (v(t2) - v(t1))/(t2 - t1)
```

This function is simple, but important enough to walk through as an example calculation. Let's use the `volume` function (plotted in figure 8.3 and included in the source code) and say we want to know the average flow rate into the tank between the 4-hr mark and the 9-hr mark. In this case, `t1 = 4` and `t2 = 9`. To find the starting and ending volumes, we can evaluate the `volume` function at these times:

```
>>> volume(4)
3.3
>>> volume(9)
5.253125
```

Rounding for simplicity, the difference between the two volumes is 5.25 bbl – 3.3 bbl = 1.95 bbl, and the total elapsed time is 9 hr – 4 hr = 5 hr. Therefore, the average flow rate into the tank is roughly 1.95 bbl divided by 5 hr or 0.39 bbl/hr. Our function confirms we got this right:

```
>>> average_flow_rate(volume,4,9)
0.390625
```

This completes our first basic example of finding the rate of change of a function. That wasn't too bad! Before we move on to some more interesting examples, let's spend a bit more time interpreting what the volume function does.

8.1.2 Picturing the average flow rate with a secant line

Another useful way to think about the average rate of change in volume over time is to look at the volume graph. Let's focus on the two points on the volume graph between which we calculated the average flow rate. In figure 8.4, the points are shown as dots

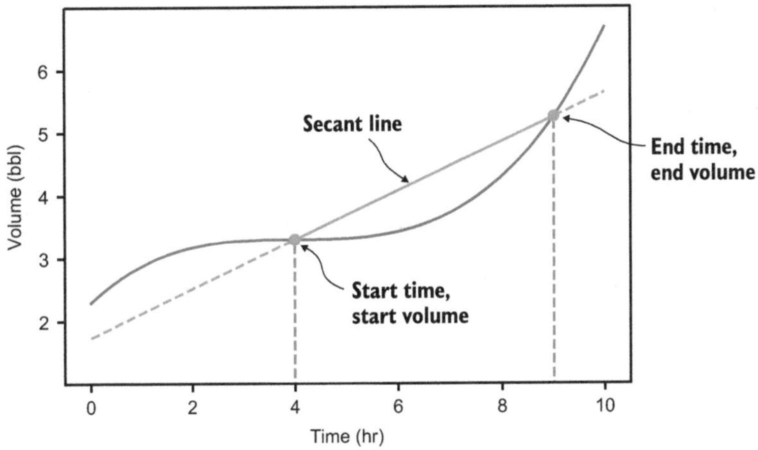

Figure 8.4 A secant line connects the starting and ending points on the volume graph.

on the graph, and I've drawn a line passing through them. A line passing through two points on a graph like this is called a *secant line.*

As you can see, the graph is higher at 9 hrs than at 4 hrs because the volume of oil in the tank increased during this period. This causes the secant line connecting the starting and ending points to slope upward. It turns out the slope of the secant tells us *exactly* what the average flow rate is on the time interval.

Here's why. Given two points on a line, the slope is the change in the vertical coordinate divided by the change in the horizontal coordinate. In this case, the vertical coordinate goes from $V(t_1)$ to $V(t_2)$ for a change of $V(t_2) - V(t_1)$, and the horizontal coordinate goes from t_1 to t_2 for a change of $t_2 - t_1$. The slope is then $(V(t_2) - V(t_1))$ divided by (t_2-t_1), exactly the same calculation as the average flow rate (figure 8.5)!

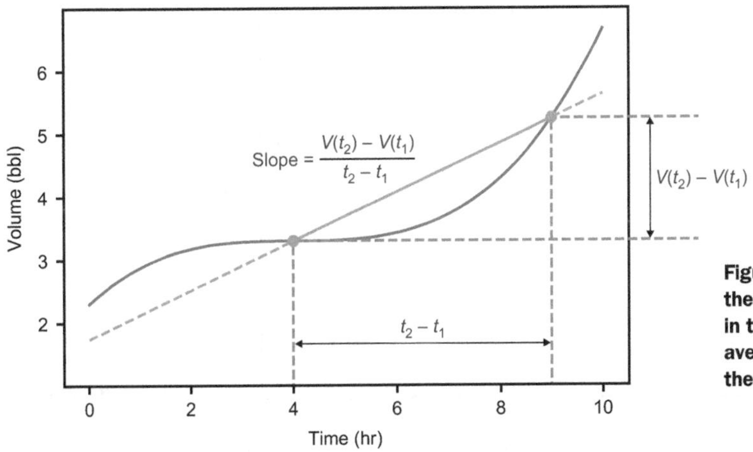

Figure 8.5 We calculate the slope of a secant line in the same way as the average rate of change of the `volume` function.

As we continue, you can picture secant lines on graphs to reason about the average rate of change in a function.

8.1.3 Negative rates of change

One case worth a brief mention is that the secant line can have a *negative* slope. Figure 8.6 shows the graph of a different `volume` function, which you can find implemented as `decreasing_volume` in the source code for this book. Figure 8.6 plots the volume in the tank decreasing over time.

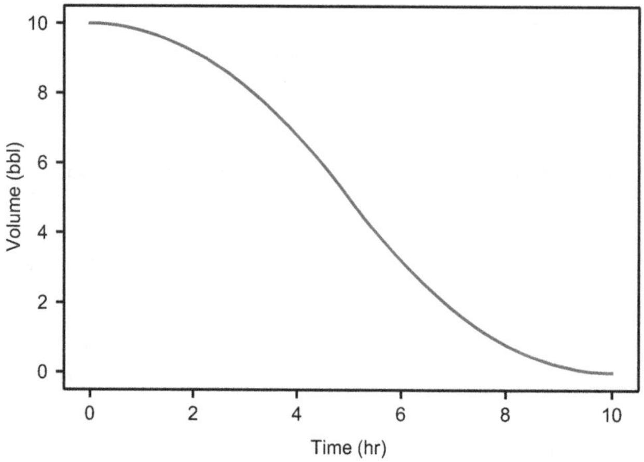

Figure 8.6 A different `volume` function shows that the volume in the tank decreases over time.

This example isn't compatible with our previous example because we don't expect oil to be flowing out of the tank back into the ground. But it does illustrate that a secant line can go downward, for instance, from $t = 0$ to $t = 4$. On this time interval, the change in volume is –3.2 bbl (figure 8.7).

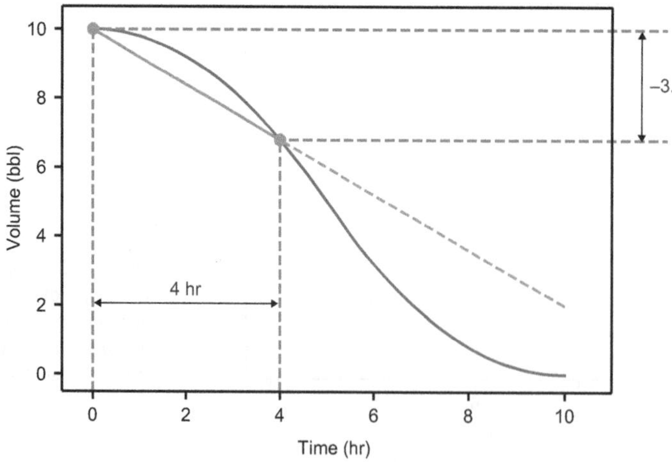

Figure 8.7 Two points on a graph that define a secant line with a negative slope

In this case, the slope is –3.2 bbl divided by 4 hr or –0.8 bbl/hr. That means that the rate at which oil is entering the tank is –0.8 bbl/hr. A more sensible way to say this is that oil is *leaving* the tank at a rate of 0.8 bbl/hr. Regardless of whether the `volume` function is increasing or decreasing, our `average_flow_rate` function is reliable. In this case,

```
>>> average_flow_rate(decreasing_volume,0,4)
-0.8
```

Equipped with this function to measure the average flow rate, we can go a step further in the next section—figuring out how the flow rate changes over time.

8.1.4 Exercises

Exercise 8.1 Suppose you start a road trip at noon when your odometer reads 77,641 miles, and you end your road trip at 4:30 in the afternoon with your odometer reading 77,905 miles. What was your average speed during the trip?

Solution The total distance traveled is 77,905 – 77,641 = 264 miles covered over 4.5 hrs. The average speed is 264 mi / 4.5 hr or about 58.7 mph.

Exercise 8.2 Write a Python function `secant_line(f,x1,x2)` that takes a function `f(x)` and two values, `x1` and `x2`, and that returns a new function representing a secant line over time. For instance, if you ran `line = secant_line (f,x1,x2)`, then `line(3)` would give you the y value of the secant line at $x = 3$.

Solution

```
def secant_line(f,x1,x2):
    def line(x):
        return f(x1) + (x-x1) * (f(x2)-f(x1))/(x2-x1)
    return line
```

Exercise 8.3 Write a function that uses the code from the previous exercise to plot a secant line of a function `f` between two given points.

Solution

```
def plot_secant(f,x1,x2,color='k'):
    line = secant_line(f,x1,x2)
    plot_function(line,x1,x2,c=color)
    plt.scatter([x1,x2],[f(x1),f(x2)],c=color)
```

8.2 Plotting the average flow rate over time

One of our big objectives for this chapter is to start with the volume function and recover the flow rate function. To find the flow rate as a function of time, we need to ask how rapidly the volume of the tank is changing at different points in time. For starters, we can see in figure 8.8 that the flow rate is changing over time—different secant lines on the volume graph have different slopes.

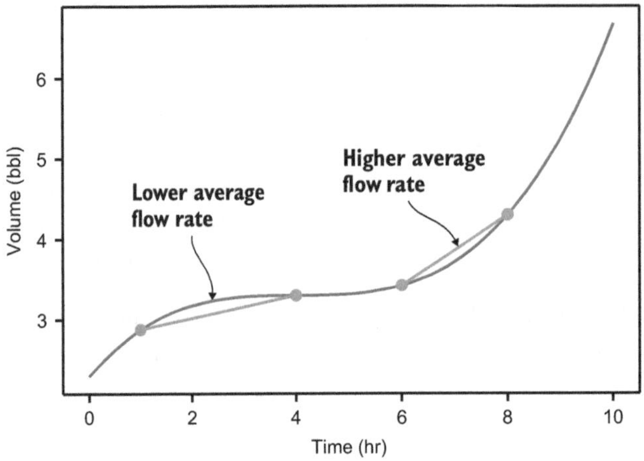

Figure 8.8 Different secant lines on the volume graph have different slopes, indicating that the flow rate is changing.

In this section, we get closer to finding the flow rate as a function of time by calculating the average flow rate on different intervals. We break up the 10-hr period into a number of smaller intervals of a fixed duration (for example, ten, 1-hr intervals) and calculate the average flow rate for each one.

We package this work in a function called `interval_flow_rates(v,t1,t2,dt)`, where `v` is the volume function, `t1` and `t2` are the starting and ending times, and `dt` is the fixed duration of the time intervals. This function returns a list of pairs of time and flow rate. For instance, if we break the 10 hrs into 1-hr segments, the result should look like this:

```
[(0,...), (1,...), (2,...), (3,...), (4,...), (5,...), (6,...), (7,...),
     (8,...), (9,...)]
```

Where each `...` would be replaced by the flow rate in the corresponding hour. Once we get these pairs, we can draw them as a scatter plot alongside the flow rate function from the beginning of the chapter and compare the results.

8.2.1 Finding the average flow rate in different time intervals

As a first step to implementing `interval_flow_rates()`, we need to find the starting points for each time interval. This means finding a list of time values from the starting time `t1` to the ending time `t2` in increments of the interval length `dt`.

There's a handy function in Python's NumPy library called `arange` that does this for us. For instance, starting from time zero and going to time 10 in 0.5-hr increments gives us the following interval start times:

```
>>> import numpy as np
>>> np.arange(0,10,0.5)
array([0. , 0.5, 1. , 1.5, 2. , 2.5, 3. , 3.5, 4. , 4.5, 5. , 5.5, 6. ,
       6.5, 7. , 7.5, 8. , 8.5, 9. , 9.5])
```

Note that the end time of 10 hrs isn't included in the list. This is because we list the *start* time for each half hour, and the half hour from t=10 to t=10.5 isn't part of the overall time interval we're considering.

For each of these interval start times, adding `dt` returns the corresponding interval end times. For instance, the interval starting at 3.5 hrs in the preceding list ends at 3.5 + 0.5 = 4.0 hrs. To implement the `interval_flow_rates` function, we just need to use our `average_flow_rate` function on each of the intervals. Here's how the complete function looks:

```
def interval_flow_rates(v,t1,t2,dt):
    return [(t,average_flow_rate(v,t,t+dt))
                for t in np.arange(t1,t2,dt)]
```

For every interval start time t, finds the average flow rate from t to t+dt. (We want the list of pairs of t with the corresponding rate.)

If we pass in our `volume` function with 0 hrs and 10 hrs as the start and end times, and 1 hr as the interval length, we get a list telling us the flow rate in each hour:

```
>>> interval_flow_rates(volume,0,10,1)
[(0, 0.578125),
 (1, 0.296875),
 (2, 0.109375),
 (3, 0.015625),
 (4, 0.015625),
 (5, 0.109375),
 (6, 0.296875),
 (7, 0.578125),
 (8, 0.953125),
 (9, 1.421875)]
```

We can tell a few things by looking at this list. The average flow rate is always positive, meaning that there is a net addition of oil into the tank in each hour. The flow rate decreases to its lowest value around hours 3 and 4 and then increases to its highest value in the final hour. This is even clearer if we plot it on a graph.

8.2.2 Plotting the interval flow rates

We can use Matplotlib's `scatter` function to quickly make a plot of these flow rates over time. This function plots a set of points on a graph, given a list of horizontal coordinates followed by a list of vertical coordinates. We need to pull out the times and flow rates as two separate lists of 10 numbers and then pass them to the function. To avoid repeating this process, we can build it all into one function:

```
def plot_interval_flow_rates(volume,t1,t2,dt):
    series = interval_flow_rates(volume,t1,t2,dt)
    times = [t for (t,_) in series]
    rates = [q for (_,q) in series]
    plt.scatter(times,rates)
```

Calling `plot_interval_flow_rates(volume,0,10,1)` generates a scatter plot of the data produced by `interval_flow_rates`. Figure 8.9 shows the result of plotting the `volume` function from zero to 10 hrs in increments of 1 hr.

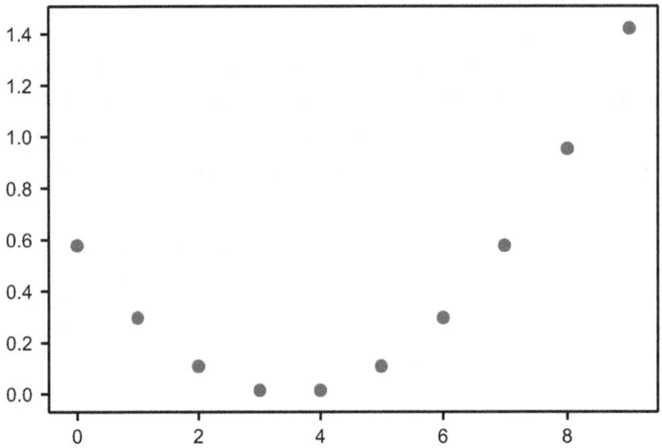

Figure 8.9 A plot of the average flow rate in each hour

This confirms what we saw in the data: the average flow rate decreases to its lowest value around hours 3 and 4 and then increases again after that to a highest rate of nearly 1.5 bbl/hr. Let's compare these average flow rates with the actual flow rate function. Again, I don't want you to worry about the formula for flow rate as a function of time. I include a `flow_rate` function in the source code for this book that we can plot (figure 8.10), along with the scatter plot.

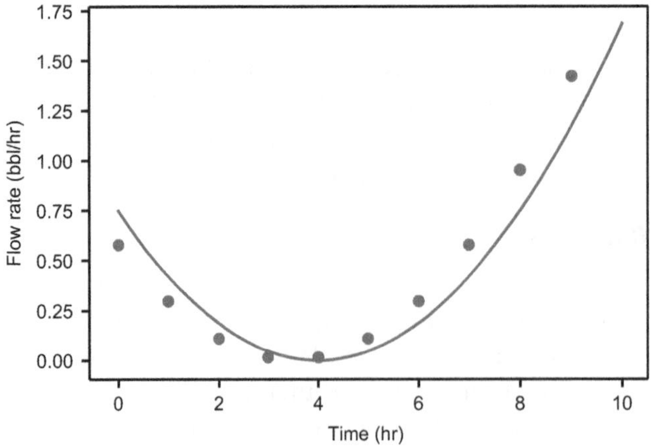

Figure 8.10 A plot of the average flow rate in each hour (dots) and the actual flow rate (smooth curve) per hour

These two plots tell the same story, but they don't quite line up. The difference is that the dots measure average flow rates, whereas the `flow_rate` function shows the *instantaneous* value of the flow rate at any point in time.

To understand this, it's helpful to think of the road trip example again. If you cover 60 miles in 1 hr, your average speed is 60 mph. However, it's unlikely your speedometer read exactly 60 mph at every instant of the hour. At some point on the open road, your *instantaneous speed* might have been 70 mph, while at another time in traffic, you might have slowed down to 50 mph.

Similarly, the flow rate meter on the pipeline needn't agree with the average flow rate on the subsequent hour. It turns out that if you make the time intervals smaller, the graphs are in closer agreement. Figure 8.11 shows the plot of the average flow rates at 20-min intervals (1/3 hrs) next to the flow rate function.

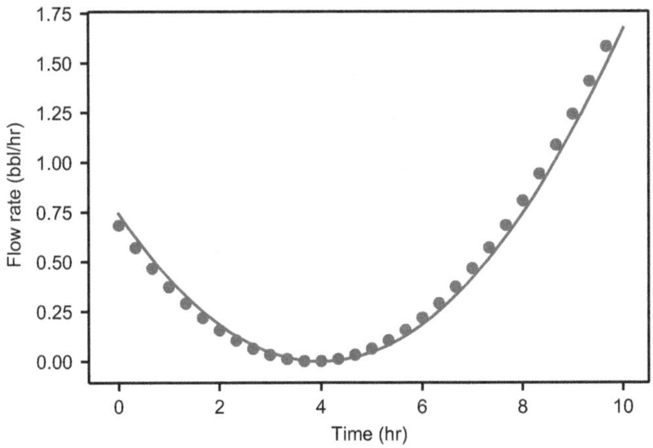

Figure 8.11 The graph of the flow rate over time compared with the average flow rates at 20-min intervals

The average flow rates are still not a perfect match to the instantaneous flow rates, but they're a lot closer. In the next section, we'll run with this idea and calculate the flow rates on extremely small intervals, where the difference between average and instantaneous rates is imperceptible.

8.2.3 Exercises

Exercise 8.4 Plot the `decreasing_volume` flow rates over time at 0.5-hr intervals. When is its flow rate the lowest? That is, when is oil leaving the tank at the fastest rate?

Solution Running `plot_interval_flow_rates(decreasing_volume,0,10,0.5)`, we can see that the rate is the lowest (most negative) just before the 5-hr mark.

(continued)

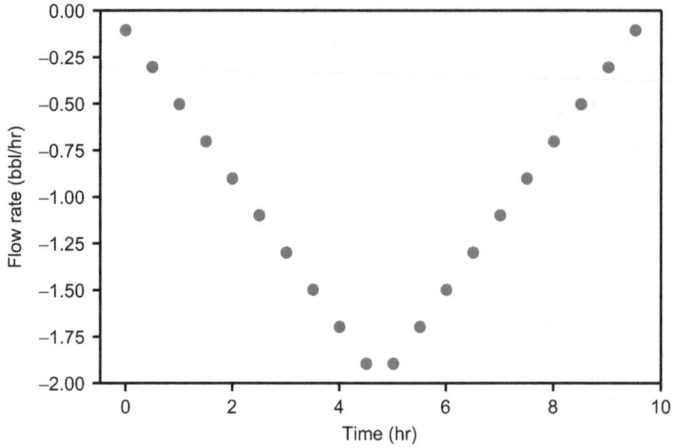

Exercise 8.5 Write a `linear_volume_function` and plot the flow rate over time to show that it is constant.

Solution A `linear_volume_function(t)` has the form $V(t) = at + b$ for the constants a and b. For instance,

```
def linear_volume_function(t):
    return 5*t + 3

plot_interval_flow_rates(linear_volume_function,0,10,0.25)
```

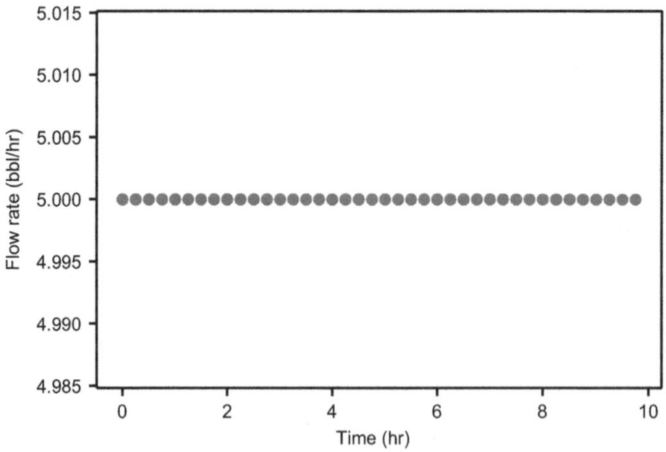

This graph shows that for a linear volume function, the flow rate is constant over time.

8.3 Approximating instantaneous flow rates

As we calculate the average rate of change in our `volume` function over smaller and smaller time intervals, we get closer and closer to measuring what's going on in a single instant. But if we try to measure the average rate of change in volume at a single instant, meaning an interval whose start and end times are the same, we run into trouble. At a time *t*, the formula for average flow rate would read:

$$\text{average flow rate at } t = \frac{V(t) - V(t)}{t - t} = \frac{0}{0}$$

Dividing 0/0 is undefined, so this method doesn't work. This is where algebra no longer helps us, and we need to turn to reasoning from calculus. In calculus, there's an operation called the *derivative* that sidesteps this undefined division problem to tell you the instantaneous rate of change in a function.

In this section, I explain why the instantaneous flow rate function, which in calculus is called *the derivative* of the volume function, is well-defined and how to approximate it. We'll write a function `instantaneous_flow_rate(v,t)` that takes a volume function *v* and a single point in time *t*, and returns an approximation of the instantaneous rate at which oil is flowing into the tank. This result is the number of barrels per hour, which should match the value of the `instantaneous_flow_rate` function exactly.

Once we do that, we'll write a second function `get_flow_rate_function(v)`, which is the curried version of `instantaneous_flow_rate()`. Its argument is a volume function, and it returns a function that takes a time and returns an instantaneous flow rate. This function completes our first of two major objectives for this chapter: starting with a volume function and producing a corresponding flow rate function.

8.3.1 Finding the slope of small secant lines

Before we do any coding, I want to convince you that it makes sense to talk about an "instantaneous flow rate" in the first place. To do that, let's zoom in on the volume graph around a single instant and see what's going on (figure 8.12). Let's pick the point where $t = 1$ hour and look at a smaller window around it.

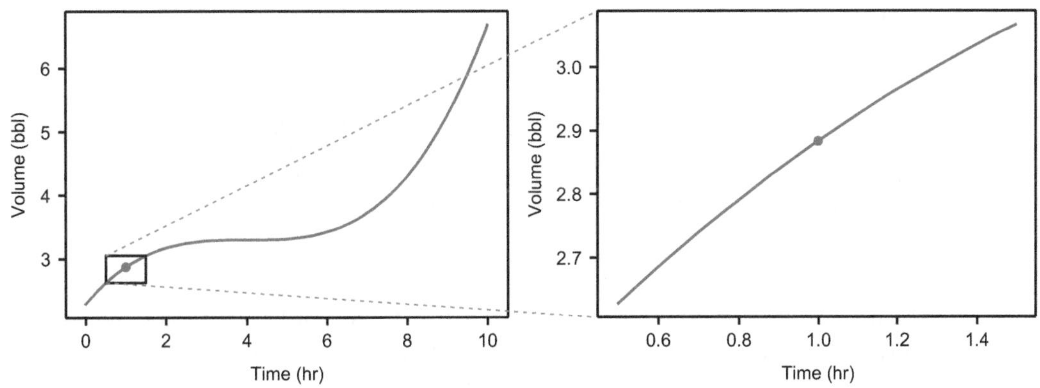

Figure 8.12 Zooming in on the 1-hr window around *t* = 1 hr

On this smaller time interval, we no longer see much of the curviness of the volume graph. That is, the steepness of the graph has less variability than on the whole 10-hr window. We can measure this by drawing some secant lines and seeing that their slopes are fairly close (figure 8.13).

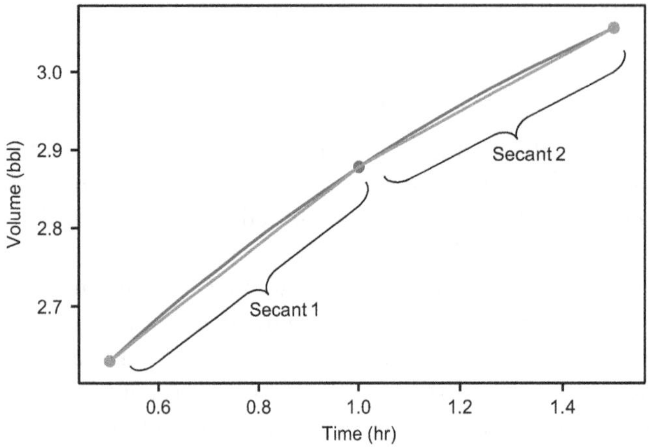

Figure 8.13 Two secant lines around *t* = 1 hr have similar slopes, meaning that the flow rate doesn't change much during this time interval.

If we zoom in even further, the steepness of the graph looks more and more constant. Zooming in to the interval between 0.9 hrs and 1.1 hrs, the volume graph is almost a straight line. If you draw a secant line over this interval, you can barely see the graph rise above the secant line (figure 8.14).

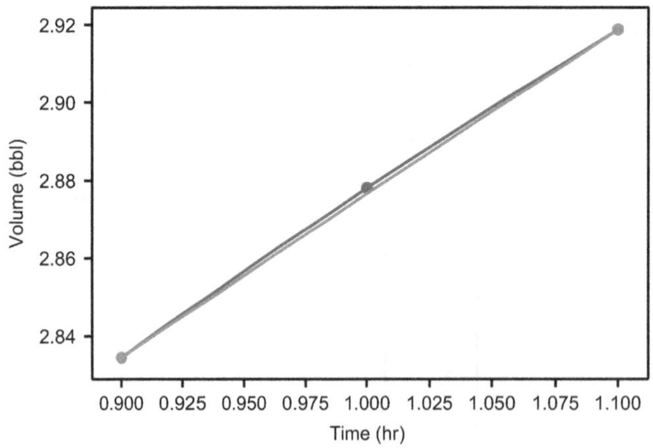

Figure 8.14 The volume graph looks nearly straight at a smaller interval around *t* = 1 hr.

Finally, if we zoom in to the window between $t = 0.99$ hrs and $t = 1.01$ hrs, the volume graph is indistinguishable from a straight line (figure 8.15). At this level, a secant line appears to overlap exactly with the graph of the function appearing like one line.

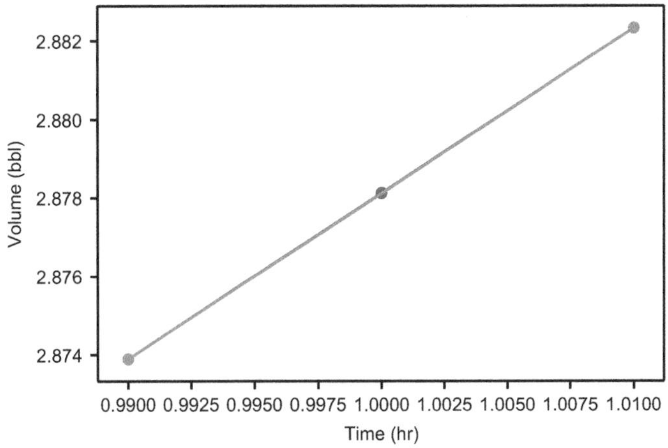

Figure 8.15 Zooming in even closer, the volume graph is visually indistinguishable from a straight line.

If you keep zooming in, the graph continues to look like a line. It's not that the graph *is* a line at this point, it's that it gets closer and closer to looking like a line the closer you zoom in. The leap in reasoning that we can make in calculus is that there's a single, best line approximating a smooth graph like the volume graph at any point. Here are a few calculations showing that the slopes of smaller and smaller secant lines *converge* to a single value, suggesting we really are approaching a single "best" approximation of the slope:

```
>>> average_flow_rate(volume,0.5,1.5)
0.42578125
>>> average_flow_rate(volume,0.9,1.1)
0.4220312499999988
>>> average_flow_rate(volume,0.99,1.01)
0.42187656249998945
>>> average_flow_rate(volume,0.999,1.001)
0.42187501562509583
>>> average_flow_rate(volume,0.9999,1.0001)
0.42187500015393936
>>> average_flow_rate(volume,0.99999,1.00001)
0.4218750000002602
```

Unless those zeroes are a big coincidence, the number we're approaching is 0.421875 bbl/hr. We can conclude that the line of best approximation for the volume function at the point $t = 1$ hr has a slope of 0.421875. If we zoom out again (figure 8.16), we can see what this line of best approximation looks like.

This line is called the *tangent line* to the volume graph at the point $t = 1$, and it's distinguished by the fact that it lies flat against the volume graph at that point. Because the tangent line is the line that best approximates the volume graph, its slope is the

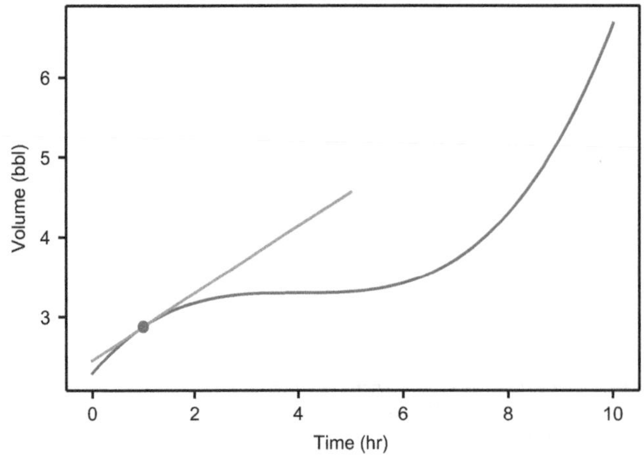

Figure 8.16 A line with slope 0.421875 is the best approximation of the volume function at time t = 1 hr.

best measure of the instantaneous slope of that graph and, therefore, the instantaneous flow rate at $t = 1$. Lo and behold, the `flow_rate` function I've provided in the source code gives us exactly the same number that the smaller and smaller secant line slopes approach:

```
>>> flow_rate(1)
0.421875
```

To have a tangent line, a function needs to be "smooth." As a mini-project at the end of this section, you can try repeating this exercise with a function that's not smooth, and you'll see that there's no line of best approximation. When we can find a tangent line to the graph of a function at a point, its slope is called the *derivative of the function at the point.* For instance, the derivative of the volume function at the point $t = 1$ is equal to 0.421875 (barrels per hour).

8.3.2 Building the instantaneous flow rate function

Now that we've seen how to calculate instantaneous rates of change of the volume function, we have what we need to implement the `instantaneous_flow_rate` function. There's one major obstacle to automating the procedure we previously used, which is that Python can't "eyeball" the slopes of several small secant lines and decide what number they're converging to. To get around this, we can calculate slopes of smaller and smaller secant lines until they stabilize to some fixed number of decimal digits.

For instance, we could have decided that we were going to find the slopes of a series of secant lines, each a tenth as wide as the previous, until the values stabilized to four decimal places. The following table shows the slopes once again.

Secant line interval	Secant line slope
0.5 to 1.5	0.42578125
0.9 to 1.1	0.4220312499999988
0.99 to 1.01	0.42187656249998945
0.999 to 1.001	0.42187501562509583

In the last two rows, the slopes agree to four decimal places (they differ by less than 10^{-4}), so we could round the final result to 0.4219 and call that our result. This isn't the exact result of 0.421875, but it's a solid approximation to the specified number of digits.

Fixing the number of digits of the approximation, we now have a way to know if we are done. If after some large number of steps, we still haven't converged to the specified number of digits, we can say that there is no line of best approximation and, therefore, no derivative at the point. Here's how this procedure translates to Python:

```
def instantaneous_flow_rate(v,t,digits=6):
    tolerance = 10 ** (-digits)
    h = 1
    approx = average_flow_rate(v,t-h,t+h)
    for i in range(0,2*digits):
        h = h / 10
        next_approx = average_flow_rate(v,t-h,t+h)
        if abs(next_approx - approx) < tolerance:
            return round(next_approx,digits)
        else:
            approx = next_approx
    raise Exception("Derivative did not converge")
```

If two numbers differ by less than a tolerance of 10^{-d}, then they agree to d decimal places.

Calculates a first secant line slope on an interval spanning h=1 units on either side of the target point t

As a crude approximation, we only try 2·digits iterations before giving up on the convergence.

At each step, calculates the slope of a new secant line around the point t on a 10 times smaller interval

If the last two approximations differ by less than the tolerance, rounds the result and returns it

Otherwise, runs the loop again with a smaller interval

If we exceed the maximum number of iterations, the procedure has not converged to a result.

I arbitrarily chose six digits as the default precision, so this function matches our result for the instantaneous flow rate at the 1-hr mark:

```
>>> instantaneous_flow_rate(volume,1)
0.421875
```

We can now compute the instantaneous flow rate at any point in time, which means we have the complete data of the flow rate function. Next, we can plot it and confirm it matches the `flow_rate` function I provide in the source code.

8.3.3 Currying and plotting the instantaneous flow rate function

For a function that behaves like the `flow_rate` function in the source code, taking a time variable and returning a flow rate, we need to curry the `instantaneous_flow_rate` function. The curried function takes a volume function (`v`) and returns a flow rate function:

```
def get_flow_rate_function(v):
    def flow_rate_function(t):
        instantaneous_flow_rate(v,t)
    return flow_rate_function
```

The output of `get_flow_rate_function(v)` is a function that should be identical to the function `flow_rate` in the source code. We can plot these both over the 10-hr period to confirm and, indeed, figure 8.17 shows that their graphs are indistinguishable:

```
plot_function(flow_rate,0,10)
plot_function(get_flow_rate_function(volume),0,10)
```

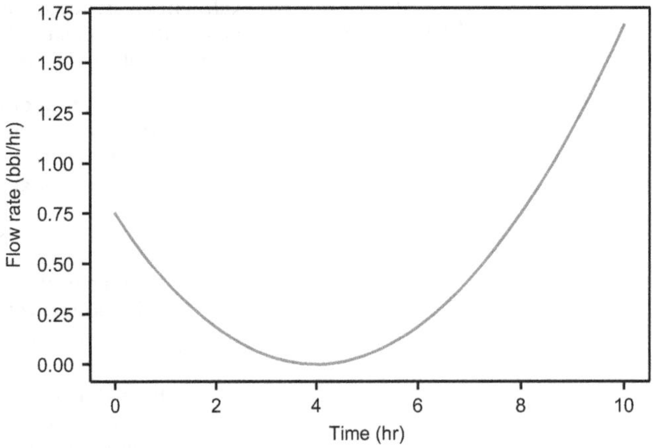

Figure 8.17 Plotting the `flow_rate` function alongside the `get_flow_rate` function shows that the graphs are indistinguishable.

We've completed our first major goal of this chapter, producing the flow rate function from the volume function. As I mentioned at the beginning of the chapter, this procedure is called *taking a derivative.*

Given a function like the `volume` function, another function giving its instantaneous rate of change at any given point is called its *derivative.* You can think of the derivative as an operation that takes one (sufficiently smooth) function and returns

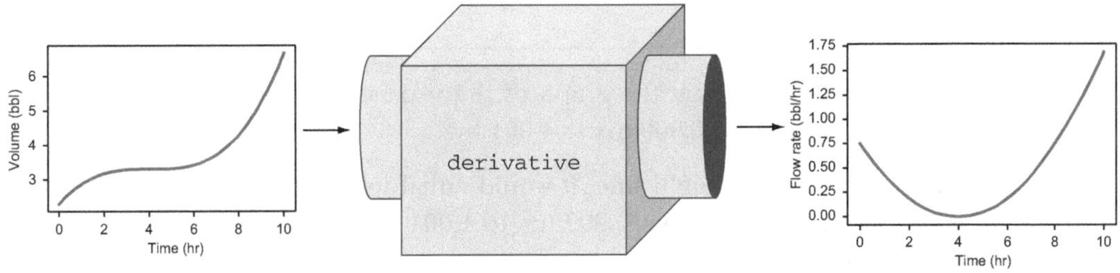

Figure 8.18 You can think of the derivative as a machine that takes a function and returns another function, measuring the rate of change of the input function.

another function measuring the rate of change of the original (figure 8.18). In this case, it would be correct to say that the flow rate function is the derivative of the volume function.

The derivative is a general procedure that works on *any* function $f(x)$, which is smooth enough to have tangent lines at every point. The derivative of a function f is written f' (and reads "f prime"), so $f'(x)$ means the instantaneous rate of change in f with respect to x. Specifically, $f'(5)$ is the derivative of $f(x)$ at $x = 5$, measuring the slope of a tangent line to f at $x = 5$. There are some other common notations for the derivative of a function including:

$$f'(x) = \frac{df}{dx} = \frac{d}{dx}f(x)$$

The df and dx are meant to signify infinitesimal (infinitely small) changes in f and x, respectively, and their quotient gives the slope of an infinitesimal secant line. The last notation of the three is nice because it makes d/dx look like an operation applied to $f(x)$. In many contexts, you'll see standalone derivative operators like d/dx. This, in particular, means "the operation of taking the derivative with respect to x." Figure 8.19 shows a schematic of how these notations fit together.

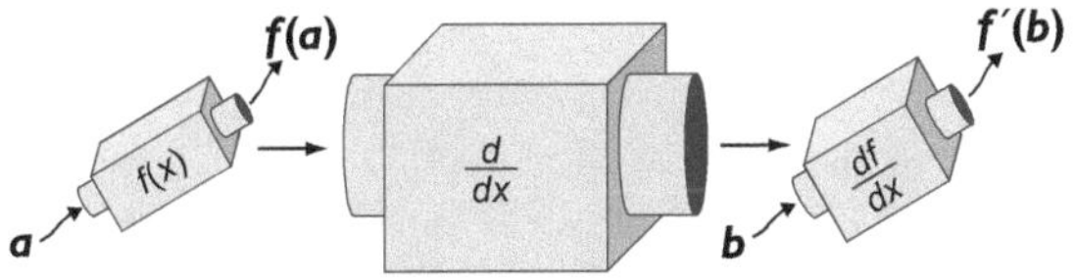

Figure 8.19 The "derivative with respect to x" as an operation that takes a function and returns a new function

We make more use of derivatives throughout the rest of this book, but for now, let's turn to the counterpart operation—the integral.

8.3.4 Exercises

Exercise 8.6 Confirm that the graph of the `volume` function is *not* a straight line on the interval from 0.999 hrs to 1.001 hrs.

Solution If it were a straight line, it would equal its secant line at every point. However, the secant line from 0.999 hrs to 1.001 hrs has a different value than the `volume` function at $t = 1$ hr:

```
>>> volume(1)
2.878125
>>> secant_line(volume,0.999,1.001)(1)
2.8781248593749997
```

Exercise 8.7 Approximate the slope of a tangent line to the volume graph at $t = 8$ by computing the slopes of smaller and smaller secant lines around $t = 8$.

Solution

```
>>> average_flow_rate(volume,7.9,8.1)
0.7501562500000007
>>> average_flow_rate(volume,7.99,8.01)
0.750001562499996
>>> average_flow_rate(volume,7.999,8.001)
0.7500000156249458
>>> average_flow_rate(volume,7.9999,8.0001)
0.7500000001554312
```

It appears that the instantaneous rate of change at $t = 8$ is 0.75 bbl/hr.

Exercise 8.8 For the `sign` function defined in Python, convince yourself that it doesn't have a derivative at $x = 0$:

```
def sign(x):
    return x / abs(x)
```

Solution On smaller and smaller intervals, the slope of a secant line gets bigger and bigger rather than converging on a single number:

```
>>> average_flow_rate(sign,-0.1,0.1)
10.0
>>> average_flow_rate(sign,-0.01,0.01)
100.0
>>> average_flow_rate(sign,-0.001,0.001)
1000.0
>>> average_flow_rate(sign,-0.000001,0.000001)
1000000.0
```

This is because the `sign` function jumps from –1 to 1 immediately at $x = 0$, and it doesn't look like a straight line when you zoom in on it.

8.4 Approximating the change in volume

For the rest of the chapter, I'm going to focus on our second major objective: starting with a known flow rate function and recovering the volume function. This is the reverse of the process of finding a derivative because we assume we know the rate of change of a function, and we want to recover the original function. In calculus, this is called *integration*.

I'll break the task of recovering the volume function into a few smaller examples, which will help you get a sense of how integration works. For the first example, we write two Python functions to help us find the change in volume in the tank over a specified period of time.

We call the first function `brief_volume_change(q,t,dt)`, taking a flow rate function `q`, a time `t`, and a short time duration `dt`, which returns the approximate change in volume from time t to time $t + dt$. This function calculates its result by assuming the time interval is so short that the flow rate does not change by much.

We call the second function `volume_change(q,t1,t2,dt)` and, as the difference in naming suggests, we use it to calculate the volume change on any time interval, not just a brief one. Its arguments are the flow rate function `q`, a start time `t1`, an end time `t2`, and a small time interval `dt`. The function breaks the time interval down into increments of duration `dt`, which are short enough to use the `brief_volume_change` function. The total volume change returned is the sum of all of the volume changes on the short time intervals.

8.4.1 Finding the change in volume for a short time interval

To understand the rationale behind the `brief_volume_change` function, let's return to the familiar example of a speedometer on a car. If you glance at your speedometer and it reads exactly 60 mph, you might predict that in the next 2 hrs, you'll travel 120 miles, which is 2 hrs times 60 mph. That estimate could be correct if you're lucky, but it's also possible that the speed limit increases or that you exit the freeway and park the car. The point is, one glance at a speedometer won't help you estimate the distance traveled over a long period.

On the other hand, if you used the value of 60 mph to calculate how far you traveled in a single *second* after looking at the speedometer, you'd probably get a very accurate answer; your speed is not going to change that much over a single second. A second is 1/3,600 of an hour, so 60 mph times 1/3,600 per hour gives you 1/60 of a mile, or 88 feet. Unless you're actively slamming on the brakes or flooring the gas pedal, this is probably a good estimate.

Returning to flow rates and volumes, let's assume that we're working with a short enough duration that the flow rate is roughly constant. In other words, the flow rate on the time interval is close to the average flow rate on the time interval, so we can apply our original equation:

$$\text{flow rate} \approx \text{average flow rate} = \frac{\text{change in volume}}{\text{elapsed time}}$$

Rearranging this equation, we can get an approximation for the change in volume:

$$\text{change in volume} \approx \text{flow rate} \times \text{elapsed time}$$

Our `small_volume_change` function is just a translation of this assumed formula into Python code. Given a flow rate function `q`, we can find the flow rate at the input time `t` as `q(t)`, and we just need to multiply it by the duration `dt` to get the change in volume:

```
def small_volume_change(q,t,dt):
    return q(t) * dt
```

Because we have an actual pair of volume and flow rate functions, we can now test how good our approximation is. As expected, the prediction is not great for a whole hour interval:

```
>>> small_volume_change(flow_rate,2,1)
0.1875
>>> volume(3) - volume(2)
0.109375
```

That approximation is off by about 70%. By comparison, we get a great approximation on a time interval of 0.01 hrs. The result is within 1% of the actual volume change:

```
>>> small_volume_change(flow_rate,2,0.01)
0.001875
>>> volume(2.01) - volume(2)
0.0018656406250001645
```

Because we can get good approximations for the volume change on small time intervals, we can piece them together to get the volume change on a longer interval.

8.4.2 Breaking up time into smaller intervals

To implement the function `volume_change(q,t1,t2,dt)`, we split the time from `t1` to `t2` into intervals of duration `dt`. For simplicity, we'll only handle values of `dt` that evenly divide `t2 - t1` so that we break the time period into a whole number of intervals.

Once again, we can use the NumPy `arange` function to get the starting time for each of the intervals. The function call `np.arange(t1,t2,dt)` gives us an array of times from `t1` to `t2` in increments of `dt`. For each time value `t` in this array, we can find the volume change in the ensuing time interval using `small_volume_change`. Finally, we need to sum the results to get the total volume change over all of the intervals. This can be done in roughly one line:

```
def volume_change(q,t1,t2,dt):
    return sum(small_volume_change(q,t,dt)
               for t in np.arange(t1,t2,dt))
```

With this function, we can break up the time from 0 to 10 hrs into 100 time intervals of duration 0.1 hrs and sum the volume changes during each. The result matches the actual volume change to one decimal place:

```
>>> volume_change(flow_rate,0,10,0.1)
4.32890625
>>> volume(10) - volume(0)
4.375
```

If we break the time into smaller and smaller intervals, the results get better and better. For instance:

```
>>> volume_change(flow_rate,0,10,0.0001)
4.3749531257812455
```

As with the process of taking a derivative, we can make the intervals smaller and smaller, and our results will converge to the expected answer. Calculating the overall change in a function on some interval from its rate of change is called a *definite integral*. We'll return to the definition of the definite integral in the last section, but for now, let's focus on how to picture it.

8.4.3 Picturing the volume change on the flow rate graph

Suppose we're breaking the 10-hr period into 1-hr intervals, even though we know this won't give us very accurate results. The only 10 points on the flow rate graph we care about are the beginning times of each interval: 0 hrs, 1 hrs, 2 hrs, 3 hrs, and so on, up to 9 hrs. Figure 8.20 shows these points marked on a graph.

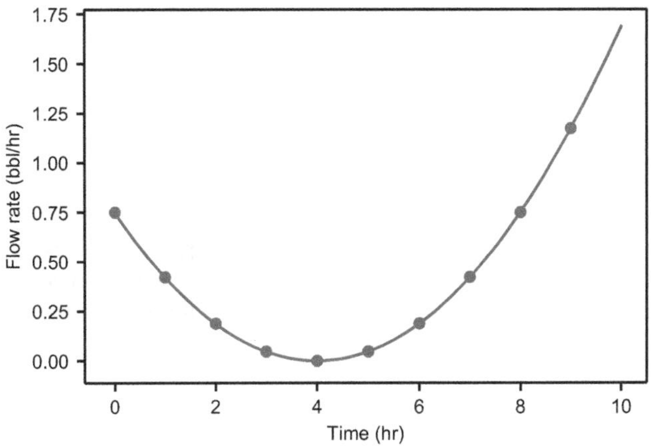

Figure 8.20 Plotting the points used to calculate `volume_change(flow_rate,0,10,1)`

Our calculation assumed the flow rates in each of the intervals remained constant, which is clearly not the case. Within each of these intervals, the flow rate visibly changes. In our assumption, it's as if we're working with a different flow rate function,

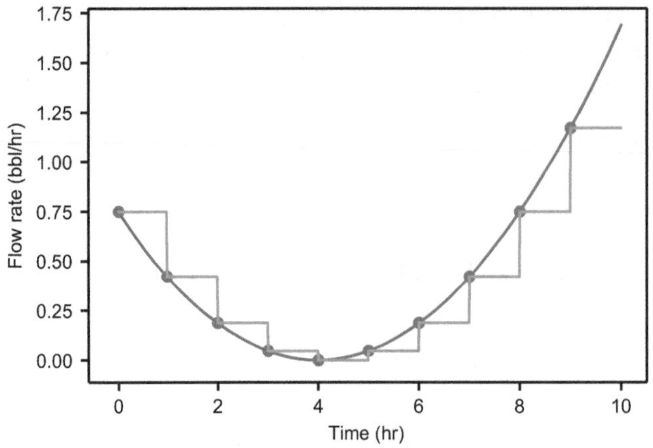

Figure 8.21 If we assumed flow rate were constant on each interval, its graph would look like a staircase going down and back up.

whose graph is constant during every hour. Figure 8.21 shows what these intervals look like next to the original.

In each interval, we calculate the flow rate (which is the height of each flat graph segment) times the elapsed time of 1 hr (which is the width of each graph segment). Each small volume we calculate is a height multiplied by a width on the graph, or the area of an imaginary rectangle. Figure 8.22 shows the rectangles filled in.

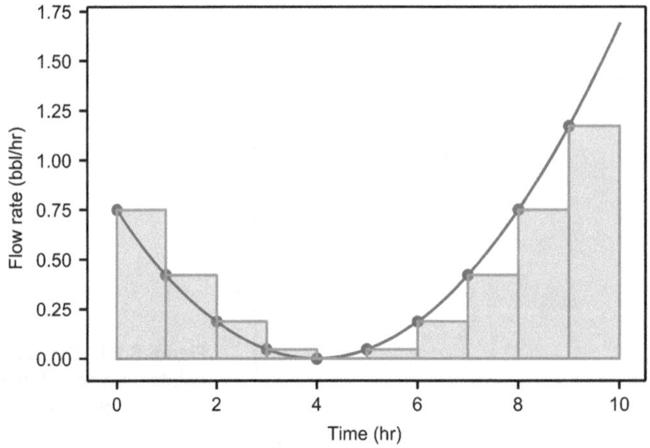

Figure 8.22 The overall change in volume as a sum of the areas of 10 rectangles

As we shorten the time intervals, we see our results improve. Visually, that corresponds with more rectangles that can hug the graph more closely. Figure 8.23 shows what the rectangles look like using 30 intervals of ⅓ hrs (20 mins) each, or 100 intervals of 0.1 hrs each.

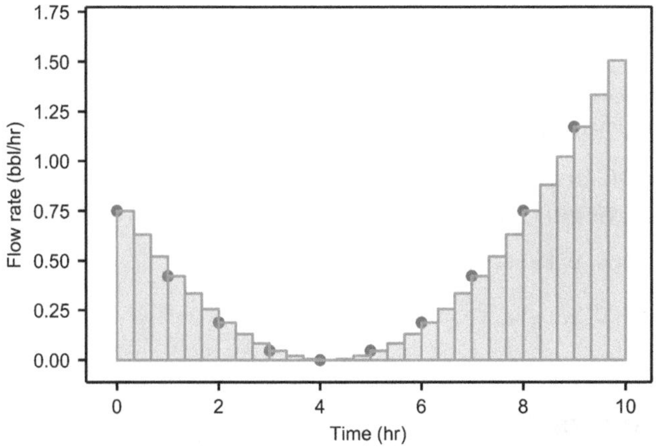

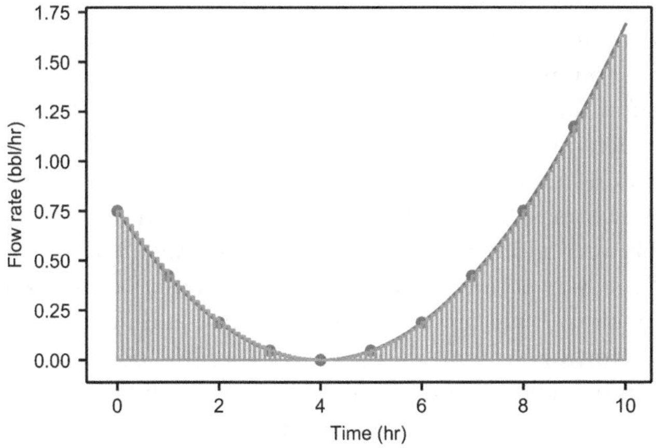

Figure 8.23 The volume as a sum of the area of 30 rectangles (top) or 100 rectangles (bottom) under the flow rate graph (figure 8.20)

From these pictures, you can see that as our intervals get smaller and our computed result approaches the actual change in volume, the rectangles come closer and closer to filling the space under the flow rate graph. The insight here is that the area under the flow rate graph on a given time interval *is (approximately) equal to* the volume added to the tank on the same interval.

A sum of the areas of rectangles approximating the area under a graph is called a *Riemann sum.* Riemann sums made of skinnier and skinnier rectangles converge to the area under a graph, much the same way as slopes of smaller and smaller secant lines converge to the slope of a tangent line. We'll return to the convergence of Riemann sums and definite integrals, but first let's make some more progress toward finding the volume over time.

8.4.4 Exercises

Exercise 8.9 Approximately how much oil is added to the tank in the first 6 hrs? In the last 4 hrs? During which time interval is more added?

Solution In the first 6 hrs, about 1.13 bbls of oil are pumped into the tank, which is less than the roughly 3.24 bbls pumped into the tank in the last 4 hrs:

```
>>> volume_change(flow_rate,0,6,0.01)
1.1278171874999996
>>> volume_change(flow_rate,6,10,0.01)
3.2425031249999257
```

8.5 Plotting the volume over time

In the previous section, we were able to start with the flow rate and come up with approximations for the *change* in volume over a given time interval. Our main goal is to come up with the *total* volume in the tank at any given point in time.

Here's a trick question for you: if oil flows into the tank at a constant rate of 1.2 bbl/hr for 3 hrs, how much oil is in the tank after 3 hrs? The answer is: we don't know because I didn't tell you how much was in the tank to begin with! Fortunately, if I tell you, then the answer is easy to figure out. For instance, if 0.5 bbls were in the tank to begin with, then 3.6 bbls were added during this period, and 0.5 + 3.6 = 4.1 bbls are in the tank at the end of the 3-hr period. Adding the initial volume at time zero to the change in volume to any time T, we can find the total volume at time T.

In our last examples for this section, we turn this idea into code to reconstruct a volume function. We implement a function called `approximate_volume(q,v0, dt,T)`, which takes a flow rate `q`, an initial volume of oil in the tank `v0`, a small time interval `dt`, and a time `T` in question. The output of the function is an approximation of the total volume in the tank at time `T`, by adding the starting volume `v0` to the change in volume from time zero to time `T`.

Once we do that, we can curry it to get a function called `approximate_volume_function(q,v0,dt)`, which produces a function giving the approximate volume as a function of time. The function returned by `approximate_volume_function` is a volume function we can plot alongside our original volume function for comparison.

8.5.1 Finding the volume over time

The basic formula we'll use is as follows:

volume at time T = (volume at time 0) + (change in volume from time 0 to time T)

We need to provide the first term of the sum, the volume in the tank at time zero, because there's no way to infer it from the flow rate function. Then we can use our `volume_change` function to find the volume from time zero to time T. Here's what the implementation looks like:

```
def approximate_volume(q,v0,dt,T):
    return v0 + volume_change(q,0,T,dt)
```

To curry this function, we can define a new function taking the first three arguments as parameters and returning a new function that takes the last parameter, `T`:

```
def approximate_volume_function(q,v0,dt):
    def volume_function(T):
        return approximate_volume(q,v0,dt,T)
    return volume_function
```

This function directly produces a plottable volume function from our `flow_rate` function. Because the `volume` function I provide in the source code has `volume(0)` equal to 2.3, let's use that value for `v0`. Finally, let's try a `dt` value of 0.5, meaning we're calculating our changes in volume in half-hour (30 mins) intervals. Let's see how this looks plotted against the original volume function (figure 8.24):

```
plot_function(approximate_volume_function(flow_rate,2.3,0.5),0,10)
plot_function(volume,0,10)
```

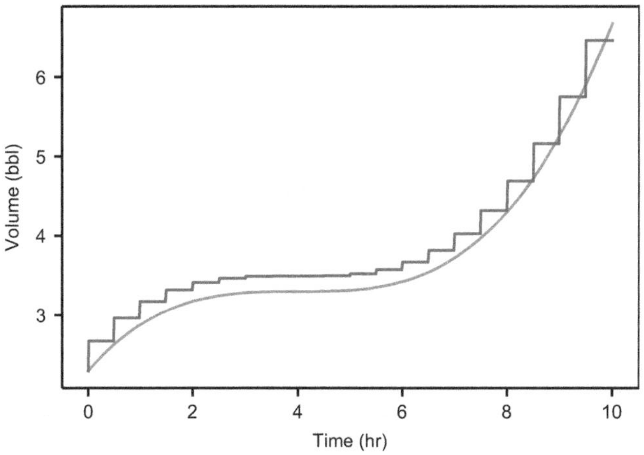

Figure 8.24 A plot of the output of `approximate_volume_function` (jagged line) alongside the original `volume` function (smooth line)

The good news is that the output is pretty close to our original volume function! But the result produced by `approximate_volume_function` is jagged, having steps every 0.5 hrs. You might guess that this has to do with our `dt` value of 0.5 and that we'll get a better approximation if we reduce this value. This is correct, but let's dig in to how the volume change is computed to see exactly why the graph looks like this, and why a smaller time interval will improve it.

8.5.2 *Picturing Riemann sums for the volume function*

At any point in time, the volume in the tank computed by our approximate `volume` function is the sum of the initial volume in the tank plus the change in volume to that point. For $t = 4$ hrs, the equation looks like this:

volume at 4 hrs = (volume at 0 hrs) + (change in volume from 0 hr to 4 hrs)

The result of this sum gives us one point on the graph at the 4-hr mark. The value at any other time is computed the same way. In this case, the sum consists of the 2.3 bbls at time zero plus a Riemann sum, giving us the change from 0 hrs to 4 hrs. This is the sum of eight rectangles, each having a width of 0.5 hrs, which fit evenly into the 4-hr window. The result is approximately 3.5 bbls (figure 8.25).

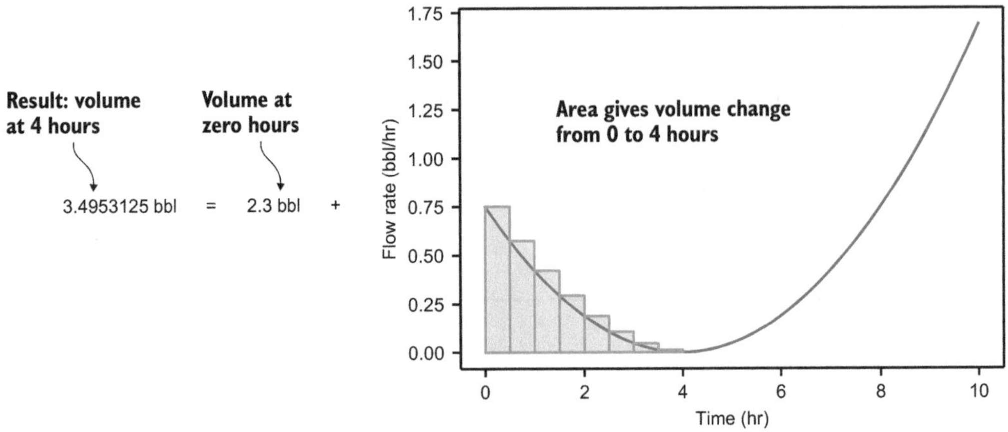

Figure 8.25 The volume in the tank at 4 hrs using a Riemann sum

We could do the same thing for any other point in time. For example, figure 8.26 shows the result for 8 hrs in.

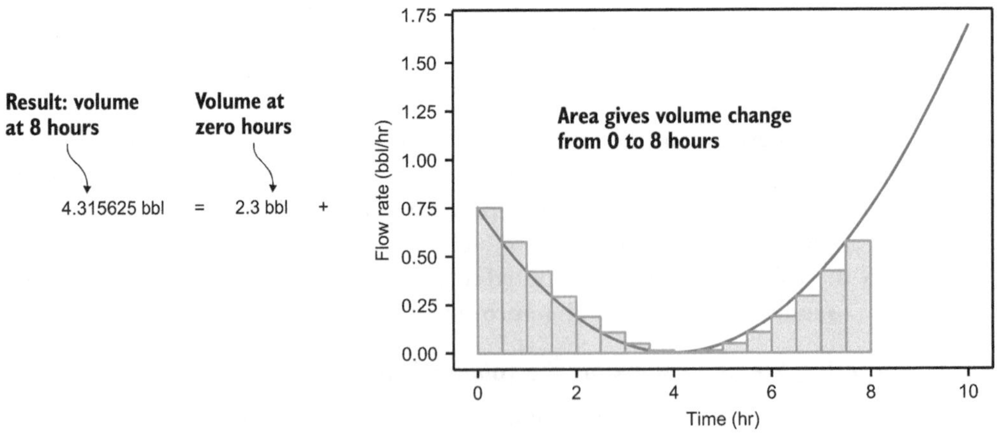

Figure 8.26 The volume in the tank at 8 hrs using a Riemann sum

In this case, the answer is approximately 4.32 bbls in the tank at the 8-hr mark. This required summing 8/0.5 = 16 rectangle areas. These two values show up as points on the graph we produced (figure 8.27):

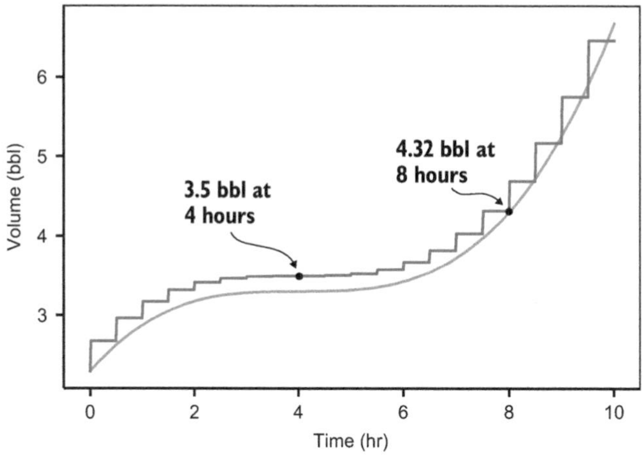

Figure 8.27 The two previous results shown on the approximate volume graph

In both of these cases, we could get from zero to the point in time in question using a whole number of timesteps. To produce this graph, our Python code computes a lot of Riemann sums, corresponding to whole number hours and half hours, as well as all the points plotted in between.

How does our code get the approximate volume at 3.9 hrs, which isn't divisible by the *dt* value of 0.5 hrs? Looking back at the implementation of `volume_change (q,t1,t2,dt)`, we made one small change in the volume calculation, corresponding to the area of one rectangle for every start time in `np.arange(t1,t2,dt)`. When we find the volume change from 0 to 3.9 hrs with a *dt* of 0.5, our rectangles are given by:

```
>>> np.arange(0,3.9,0.5)
array([0. , 0.5, 1. , 1.5, 2. , 2.5, 3. , 3.5])
```

Even though the eight rectangles of width 0.5 hr go past the 3.9-hr mark, we calculate the area of all eight! To be completely clean, we should have probably shortened our last time interval to 0.4 hrs, lasting from the end of the 7th time interval at 3.5 hrs to the end time of 3.9 hrs, and no further. As a mini-project at the end of this section, you can try updating the `volume_change` function to use a smaller duration for the last time interval, if necessary. For now, I'll ignore this oversight.

In the last section, we saw that we got better results by shrinking the *dt* value and, therefore, the widths of the rectangles. In addition to fitting the graph better, smaller rectangles are likely to have less error even if they slightly overshoot the end of a time interval. For instance, while 0.5-hr intervals can only add up to 3.5 hrs or 4.0 hrs but not 3.9 hrs, 0.1-hr intervals can add up evenly to 3.9 hrs.

8.5.3 Improving the approximation

Let's try using smaller values of *dt*, corresponding to smaller rectangle sizes, and see the improvements we get. Here's the approximation with `dt = 0.1` hrs (figure 8.28 plots the results). The steps on the graph are barely visible, but they are smaller, and the graph stays closer to the actual volume graph than it did with 0.5-hr intervals:

```
plot_function(approximate_volume_function(flow_rate,2.3,0.1),0,10)
plot_function(volume,0,10)
```

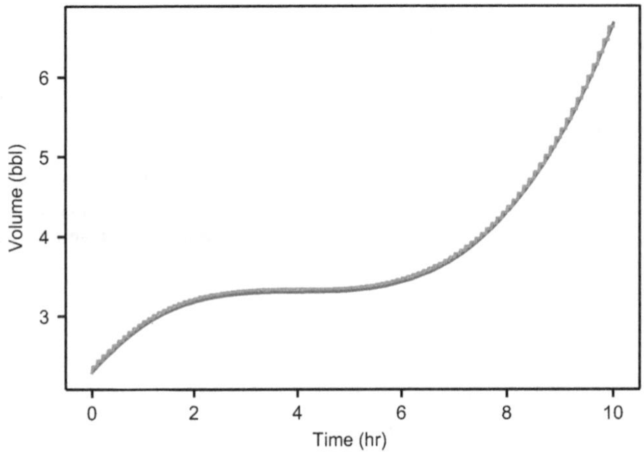

Figure 8.28 With *dt* = 0.1 hr, the graphs nearly match.

With even smaller timesteps, like `dt=0.01` hrs, the graphs are nearly indistinguishable (figure 8.29):

```
plot_function(approximate_volume_function(flow_rate,2.3,0.01),0,10)
plot_function(volume,0,10)
```

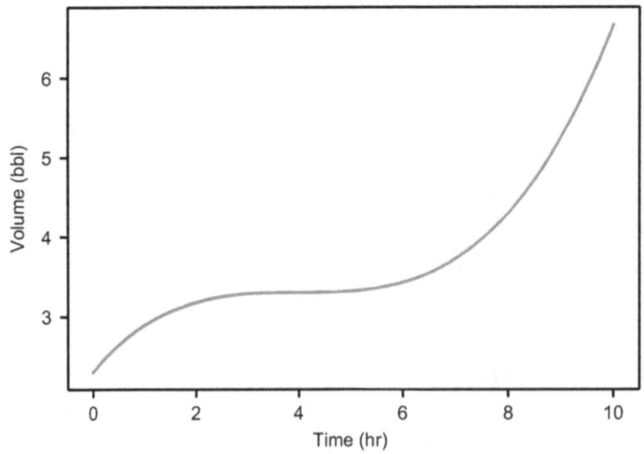

Figure 8.29 With 0.01-hr timesteps, the graph of the approximate `volume` function is indistinguishable from the actual `volume` function.

Even though the graphs appear to match exactly, we can ask the question of how accurate is this approximation. The graphs of the approximate `volume` functions with smaller and smaller values of `dt` get closer and closer to the actual volume graph at every point, so we could say the values are *converging* to the actual volume values. But at each step, the approximation still might disagree with the actual volume measurement.

Here's a way we could find the volume at any point to an arbitrary precision (within any tolerance we want). For any point `t` in time, we can recalculate `volume_change(q,0,t,dt)` with smaller and smaller values of `dt` until the outputs stop changing by more than the tolerance value. This looks a lot like our function to make repeated approximations of the derivative until they stabilize:

```
def get_volume_function(q,v0,digits=6):
    def volume_function(T):
        tolerance = 10 ** (-digits)
        dt = 1
        approx = v0 + volume_change(q,0,T,dt)
        for i in range(0,digits*2):
            dt = dt / 10
            next_approx = v0 + volume_change(q,0,T,dt)
            if abs(next_approx - approx) < tolerance:
                return round(next_approx,digits)
            else:
                approx = next_approx
        raise Exception("Did not converge!")
    return volume_function
```

For instance, the volume $v(1)$ is exactly 2.878125 bbls, and we can ask for any precision estimation of this result that we want. For example, for three digits, we get

```
>>> v = get_volume_function(flow_rate,2.3,digits=3)
>>> v(1)
2.878
```

and for six digits, we get the exact answer:

```
>>> v = get_volume_function(flow_rate,2.3,digits=6)
>>> v(1)
2.878125
```

If you run this code yourself, you'll see the second computation takes quite a while. This is because it has to run a Riemann sum consisting of millions of small volume changes to get the answer to this precision. There's probably no realistic use for this function, which computes volume values to an arbitrary precision, but it illustrates the point that with smaller and smaller *dt* values, our volume approximation *converges* to the exact value of the `volume` function. The result it is converging to is called the *integral* of the flow rate.

8.5.4 Definite and indefinite integrals

In the last two sections, we *integrated* the flow rate function to obtain the volume function. Like taking a derivative, finding an integral is a general procedure that you can do with functions. We can integrate any function specifying a rate of change to get a function giving a compatible, cumulative value. If we know the speed of a car as a function of time, for example, we can integrate it to get the distance traveled as a function of time. In this section, we look at two types of integrals: definite integrals and indefinite integrals.

A *definite integral* tells you the total change in a function on some interval from its derivative function. The function and a pair of start and end values for the argument, which in our case is time, specify the definite integral. The output is a single number, which gives the cumulative change. For instance, if $f(x)$ is our function of interest and $f'(x)$ is the derivative of $f(x)$, then the change in f from $x = a$ to $x = b$ is $f(b) - f(a)$, and it can be found by taking a definite integral (figure 8.30).

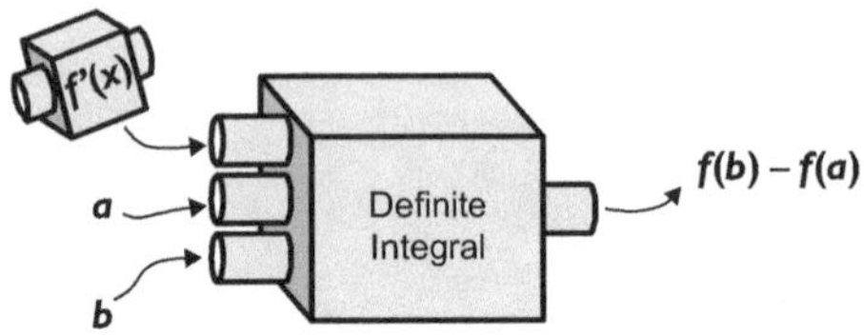

Figure 8.30 The definite integral takes the rate of change (derivative) of a function and a specified interval and recovers the cumulative change in the function on that interval.

In calculus, the definite integral of $f(x)$ from $x = a$ to $x = b$ is written like this:

$$\int_a^b f'(x)dx$$

and its value is $f(b) - f(a)$. The big $\int$ symbol is the integral symbol, a and b are called the *bounds of integration*, $f'(x)$ is the function being integrated, and dx indicates that the integral is being taken with respect to x.

Our `volume_change` function approximates definite integrals, and as we saw in section 8.4.3, it also approximates the area under the flow rate graph. It turns out that the definite integral of a function on an interval is equal to the area under the rate graph on that interval. For most rate functions you meet in the wild, the graphs will be nice enough that you can approximate the area underneath them with skinnier and skinnier rectangles, and your approximations will converge to a single value.

After taking a definite integral, let's look at an indefinite integral. The *indefinite integral* takes the derivative of a function and recovers the original function. For instance, if you know that $f'(x)$ is the derivative of $f(x)$, then to reconstruct $f(x)$, you have to find the indefinite integral of $f(x)$.

The catch is that the derivative $f(x)$ on its own is not enough to reconstruct the original function $f(x)$. As we saw with `get_volume_function`, which computed a definite integral, you need to know an initial value of $f(x)$, like $f(0)$ for instance. The value of $f(x)$ can then be found by adding a definite integral to $f(0)$. Because

$$f(b) - f(a) = \int_a^b f'(x)dx$$

we can get any value of $f(x)$ as:

$$f(x) - f(0) = \int_0^x f'(t)dt$$

Note we have to use a different name t for the argument of f because x becomes a bound of integration here. The indefinite integral of a function $f(x)$ is written as

$$f(x) = \int f'(x)dx$$

which looks like a definite integral but without specified bounds. If, for example, $g(x) = \int f(x)\ dx$, then $g(x)$ is said to be an *antiderivative* of $f(x)$. Antiderivatives are not unique, and in fact, there is a different function $g(x)$ whose derivative is $f(x)$ for any initial value $g(0)$ that you choose.

This is a lot of terminology to absorb in a short time, but fortunately, we spend the rest of the second part of this book reviewing it. We'll continue to work with functions and their rates of change using derivatives and integrals to switch between them interchangeably.

Summary

- The average rate of change in a function, say $f(x)$, is the change in the value of f on some x interval divided by the length of the interval. For instance, the average rate of change in $f(x)$ from $x = a$ to $x = b$ is

$$\frac{f(b) - f(a)}{b - a}$$

- The average rate of change in a function can be pictured as the steepness of a *secant line*, a line passing through the graph of the function at two points.
- Zooming in on the graph of a smooth function, it appears indistinguishable from a straight line. The line it looks like is the best linear approximation for the function in that area, and its slope is called the *derivative* of the function.
- You can approximate the derivative by taking the slopes of secant lines on successively smaller intervals containing the point. This approximates the instantaneous rate of change in the function at the point of interest.

- The *derivative* of a function is another function that tells you the instantaneous rate of change at every point. You can plot the derivative of a function to see its rate of change over time.
- Starting with a derivative of a function, you can figure out how it changes over time by breaking it into brief intervals and assuming the rate is constant on each. If each interval is short enough, the rate will be approximately constant and summed to find the total. This approximates the definite integral of a function.
- Knowing the initial value of a function and taking the definite integral of its rate on various intervals, you can reconstruct the function. This is called the *indefinite integral* of the function.

9 Simulating moving objects

This chapter covers

- Implementing Newton's laws of motion in code to simulate realistic motion
- Calculating velocity and acceleration vectors
- Using Euler's method to approximate the position of a moving object
- Finding the exact trajectory of a moving object with calculus

Our asteroid game from chapter 7 was functional but not that challenging. In order to make it more interesting, we need the asteroids to actually move! And, to give the player a chance to avoid the moving asteroids, we need to make it possible to move and steer the spaceship as well.

To implement motion in the asteroid game, we'll use many of the same calculus concepts from chapter 8. The numerical quantities we'll consider are the x and the y positions of the asteroids and of the spaceship. If we want the asteroids to move, these values are different at different points in time, so we can consider them to be functions of time: $x(t)$ and $y(t)$. The derivative of a position function with respect to time is called *velocity*, and the derivative of velocity with respect to time is called *acceleration*. Because we have two position functions, we have two velocity functions and two acceleration functions. This allows us to think of velocities and accelerations as vectors, as well.

Our first goal is to get the asteroids moving. For that, we'll provide random, constant velocity functions for the asteroids. Then we'll integrate these velocity functions in "real time" to get the position of each asteroid in each frame using an algorithm called *Euler's method*. Euler's method is mathematically similar to the integration we did in chapter 8, but it has the advantage that we can carry it out as the game runs.

After that, we can allow the user to control the spaceship. When the user presses the up arrow on their keyboard, the spaceship should accelerate in the direction it's pointing. That means the derivative of the derivative of each of $x(t)$ and $y(t)$ becomes non-zero; the velocity begins to change, and the position starts to change as well. Again, we'll use Euler's method to integrate the acceleration function and the velocity function in real time.

Euler's method is merely an approximation of the integral, and in this application, it's analogous to the Riemann sums from chapter 8. It is possible to calculate the exact positions of the asteroids and of the spaceship over time, and I conclude the chapter with a brief comparison of the Euler's method results and the exact solutions.

9.1 Simulating a constant velocity motion

In everyday usage, the word *velocity* is a synonym for the word *speed*. In math and physics, velocity has a special meaning; it includes the concepts of both speed and direction of motion. Therefore, velocity will be the concept that we focus on, and we'll think of it as a vector.

What we want to do is to give each of the asteroid objects a random velocity vector, meaning a pair of numbers (v_x, v_y), and interpret these to be the constant values of the derivatives of position with respect to time. That is, we assume $x'(t) = vx$ and $y'(t) = v_y$. With that information encoded, we can update the game engine so that the asteroids actually move with those velocities as the game progresses.

Because our game is two-dimensional, we work with pairs of positions and pairs of velocities. I switch back and forth from talking about $x(t)$ and $y(t)$ as a pair of position functions and $x'(t)$ and $y'(t)$ as a pair of velocity functions, and writing them as *vector-valued* functions: $\mathbf{s}(t) = (x(t), y(t))$ and $\mathbf{v}(t) = (x'(t), y'(t))$. This notation just means that $\mathbf{s}(t)$ and $\mathbf{v}(t)$ are both functions that take a time value and return a vector, representing position and velocity, respectively, at that time.

The asteroids already have position vectors, indicated by their x and y properties, but we need to give them velocity vectors as well, indicating how fast they are moving in the x and y directions. That's our first step to get them moving frame-by-frame.

9.1.1 Adding velocities to the asteroids

To give each asteroid a velocity vector, we can add the two components of the vectors `vx` and `vy` as properties on the `PolygonModel` object (in the chapter 9 version of `asteroids.py` in the source code):

```
class PolygonModel():
    def __init__(self,points):
        self.points = points
        self.angle = 0
        self.x = 0
        self.y = 0
        self.vx = 0
        self.vy = 0
```

The first four properties are kept from the original implementation of this class in chapter 7.

These vx and vy properties store the current values of $v_x = x'(t)$ and $v_y = y'(t)$. By default, they are set to 0, meaning the object is not moving.

Next, to make our asteroids move erratically, we can give them random values for the two components of their velocities. This means adding two lines at the bottom of the `Asteroid` constructor:

Up to this line, the code is unchanged from chapter 7; it initializes the asteroid's shape as a polygon with randomly positioned vertices.

```
class Asteroid(PolygonModel):
    def __init__(self):
        sides = randint(5,9)
        vs = [vectors.to_cartesian((uniform(0.5,1.0), 2 * pi * i / sides))
                for i in range(0,sides)]
        super().__init__(vs)
        self.vx = uniform(-1,1)
        self.vy = uniform(-1,1)
```

In the last two lines, the x and y velocities are set to random values between –1 and 1.

Remember, a negative derivative means that a function is decreasing, while a positive value means that a function is increasing. The fact that the *x* and *y* velocities could be positive or negative means that the *x* and *y* positions could each either be increasing or decreasing. That means our asteroids could be moving to the right or left and upward or downward.

9.1.2 Updating the game engine to move the asteroids

The next thing we need to do is use the velocity to update the position. Regardless of whether we're talking about the spaceship, the asteroids, or some other `PolygonModel` objects, the velocity components v_x and v_y tell us how to update the position components *x* and *y*.

If some time Δt elapses between frames, we update *x* by $v_x \cdot \Delta t$ and *y* by $v_y \cdot \Delta t$. (The symbol Δ is the capital Greek letter delta, often used to indicate a change in a variable). This is the same approximation we use to find a small change in volume from a small change in flow rate in chapter 8. In this case, it is better than an approximation because the velocities are constant, the velocity times the elapsed time gives the change in position.

We can add a `move` method to the `PolygonModel` class that updates an object's position based on this formula. The only thing that the object won't be intrinsically aware of is the elapsed time, so we pass that in (in milliseconds):

```
class PolygonModel():
    ...
    def move(self, milliseconds):
        dx, dy = (self.vx * milliseconds / 1000.0,
                  self.vy * milliseconds / 1000.0
        self.x, self.y = vectors.add((self.x,self.y),
                                     (dx,dy))
```

The change in x position is called dx, and the change in y position is called dy. Both are calculated by multiplying the asteroid's velocity by the elapsed time in seconds.

Completes the movement for the frame, updating the positions by adding the respective changes dx and dy

This is a first, simple application of the Euler's method algorithm. The algorithm consists of keeping track of the value of one or more functions (in our case, the positions $x(t)$ and $y(t)$ as well as their derivatives $x'(t) = v_x$ and $y'(t) = v_y$) and updating the functions according to their derivatives in each step. This works perfectly if the derivatives are constant, but it is still a fairly good approximation if the derivatives are themselves changing. When we turn our attention to the spaceship, we'll deal with changing velocity values and update our implementation of Euler's method.

9.1.3 Keeping the asteroids on the screen

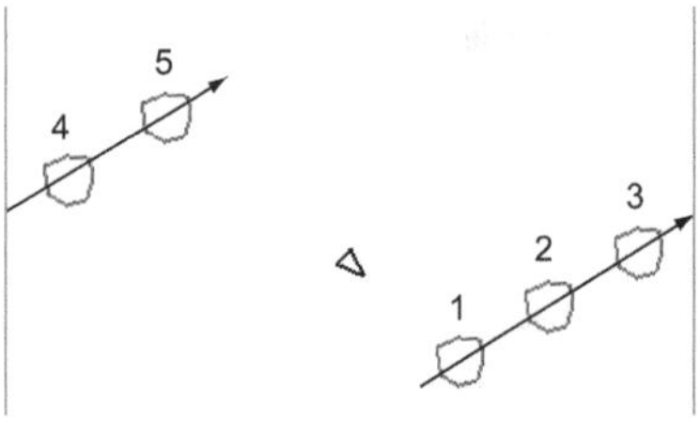

Figure 9.1 Keeping all objects' coordinates between –10 and 10 by "teleporting" the objects across the screen when they are about to leave it

We can add one more small feature to improve the gameplay experience. An asteroid with a random velocity is bound to drift off the screen at some point. To keep the asteroids within the screen area, we can add some logic to keep both coordinates between the minimum and maximum values of –10 and 10. When, for instance, the *x* property increases from 10.0 to 10.1, we subtract 20 so it becomes an acceptable value of –9.9. This has the effect of "teleporting" the asteroid from the right side of the screen to the left. This game mechanic has nothing to do with physics, but makes the game more interesting by keeping the asteroids in play (figure 9.1).

Here's the teleportation code:

```
class PolygonModel():
    ...
    def move(self, milliseconds):
        dx, dy = (self.vx * milliseconds / 1000.0,
                  self.vy * milliseconds / 1000.0)
        self.x, self.y = vectors.add((self.x,self.y),
                                     (dx,dy))
```

```
if self.x < -10:
    self.x += 20
if self.y < -10:
    self.y += 20
if self.x > 10:
    self.x -= 20
if self.y > 10:
    self.y -=20
```

If x < –10, the asteroid drifts off the left side of the screen, so we add 20 units to the x position to teleport it to the right side of the screen.

If y < –10, the asteroid drifts off the bottom of the screen, so we add 20 units to the y position to teleport it to the top of the screen.

Finally, we need to call the `move` method for every asteroid in play. To do that, we need the following lines within our game loop before the drawing begins:

```
milliseconds = clock.get_time()
for ast in asteroids:
    ast.move(milliseconds)
```

Figures out how many milliseconds have elapsed since the last frame

Signals all of the asteroids to update their position based on their velocity

It's unremarkable when printed on this page, but when you run the code yourself, you'll see the asteroids move randomly about the screen, each in a random direction. But if you focus on an asteroid, you'll see that its motion isn't random; it changes position by the same distance in the same direction in each passing second (figure 9.2).

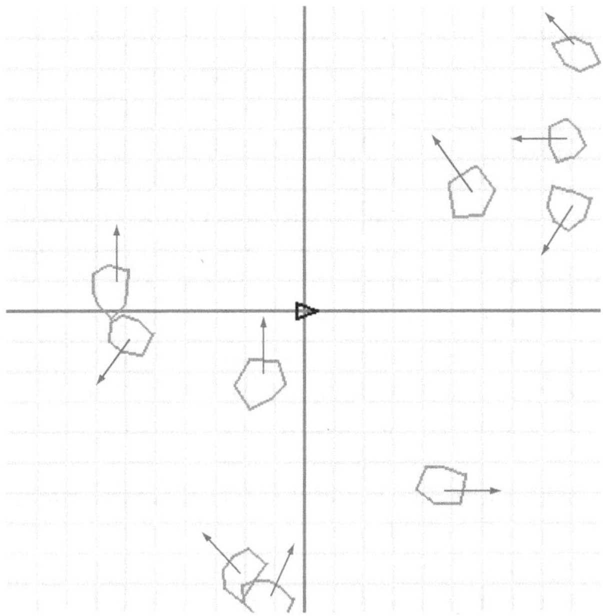

Figure 9.2 With the preceding code included, each asteroid moves with a random, constant velocity.

With asteroids that move, the ship is now in danger—it needs to move to avoid them. But even moving at a constant velocity won't save the ship as it will likely run into an asteroid at some point. The player needs to change the velocity of the ship, meaning both its speed and its direction. Next, we look at how to simulate change in velocity, which is known as *acceleration*.

9.1.4 Exercises

Exercise 9.1 An asteroid has the velocity vector $\mathbf{v} = (v_x, v_y) = (-3, 1)$. Which direction is it moving on the screen?

1. Up and to the right
2. Up and to the left
3. Down and to the left
4. Down and to the right

Solution Because $x'(t) = v_x = -3$ at this moment in time, the asteroid is moving in the negative x direction, or to the left. Because $y'(t) = v_y = 1$, the asteroid is moving in the positive y direction at this moment, which is upward. Therefore, answer *b* is correct.

9.2 Simulating acceleration

Figure 9.3 Schematic of how a rocket propels itself

Let's imagine our spaceship is equipped with a thruster that burns rocket fuel, and the expanding gasses push the spaceship in the direction it's pointed (figure 9.3).

We'll assume that when the rocket is firing its thruster, it accelerates at a constant rate in the direction it's pointed. Because acceleration is defined as the derivative of velocity, constant acceleration values mean that the velocities change at a constant rate in both directions with respect to time. When acceleration is nonzero, the velocities v_x and v_y are not constant; they are the functions $v_x(t)$ and $v_y(t)$ that change over time. Our assumption that acceleration is constant means that there are two numbers, a_x and a_y, so that $v'_x(t) = ax$ and $v'_y(t) = a_y$. As a vector, we denote acceleration by $\mathbf{a} = (a_x, a_y)$.

Our goal is to give the Python spaceship a pair of properties representing a_x and a_y and to have it accelerate and move across the screen according to those values. When the user is not pressing any buttons, the spaceship should have zero acceleration in both directions, and when the user presses the up arrow key, the acceleration values should instantly be updated so that (a_x, a_y) is a non-zero vector pointing in the direction the spaceship is headed. While the user holds down the up arrow key, the spaceship's velocity and position should both change realistically, causing it to move frame-by-frame.

9.2.1 *Accelerating the spaceship*

Regardless of the direction the spaceship is pointing, we want it to appear to accelerate at the same rate. That means that while the thruster is firing, the magnitude of the vector (a_x, a_y) should have a fixed value. By trial and error, I discovered that an acceleration magnitude of 3 makes the ship sufficiently maneuverable. Let's include this constant in our game code:

```
acceleration = 3
```

Thinking of the distance units in our game as meters, this represents a value of 3 meters per second per second (m/s/s). If the spaceship starts at a standstill and the player holds down the up arrow key, the spaceship increases its speed by 3 m/s every second in the direction it's pointing. PyGame works in milliseconds, so the relevant speed change will be 0.003 m/s every millisecond, or 0.003 meters per second per millisecond.

Let's figure out how to calculate the acceleration vector $\mathbf{a} = (a_x, a_y)$ while the up arrow key is pressed. If the ship is pointing at a rotation angle θ, then we need to use trigonometry to find the vertical and horizontal components of the acceleration from the magnitude $|\mathbf{a}| = 3$. By the definition of sine and cosine, the horizontal and vertical components are $|\mathbf{a}| \cdot \cos(\theta)$ and $|\mathbf{a}| \cdot \sin(\theta)$, respectively (figure 9.4). In other words, the acceleration vector is the pair of components $(|\mathbf{a}| \cdot \cos(\theta), |\mathbf{a}| \cdot \sin(\theta))$. Incidentally, you could also use the `from_polar` function we wrote in chapter 2 to get these components from the magnitude and direction of acceleration.

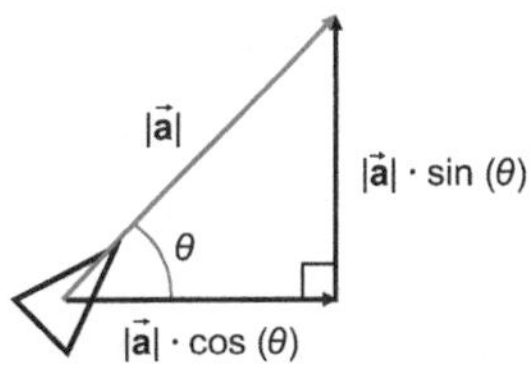

Figure 9.4 Using trigonometry to find the components of acceleration from its magnitude and direction

During each iteration of the game loop, we can update the velocity of the ship before it moves. Over an elapsed time Δt, the update to v_x will be $a_x \cdot \Delta t$ and the update to v_y will be $a_y \cdot \Delta t$. In code, we need to add the appropriate changes in velocity to the ship's vx and vy properties:

```
while not done:
    ...
        if keys[pygame.K_UP]:
            ax = acceleration * cos(ship.rotation_angle)
            ay = acceleration * sin(ship.rotation_angle)
            ship.vx += ax * milliseconds/1000
            ship.vy += ay * milliseconds/1000

        ship.move(milliseconds)
```

Detects whether the up arrow key is pressed

Calculates the values of a_x and a_y based on the fixed magnitude of acceleration and the angle the ship is pointing

Updates the x and y velocities by $a_x \cdot \Delta t$ and $a_y \cdot \Delta t$, respectively

Moves the spaceship, using the updated velocities to update positions

That's it! With this added code, the spaceship should accelerate when you press the up arrow. The code to rotate the spaceship with the left and right arrow keys is similar and included in the source code, but I won't go into it here. With the left, right, and up arrow functionality implemented, you can point the ship in whatever direction to accelerate when you want to avoid asteroids.

This is a slightly more advanced application of Euler's method where we have *second derivatives*: $x''(t) = v_x'(t) = a_x$ and $y''(t) = v_y'(t) = a_y$. At each step, we first update the velocities, then we use the updated velocities in the `move` method to determine the updated positions. We're done with our game programming for this chapter, but in the next sections, we take a closer look at Euler's method and evaluate how well it approximates motion.

9.3 Digging deeper into Euler's method

The core idea of Euler's method is to start with an initial value of a quantity (like position) and an equation describing its derivatives (like velocity and acceleration). The derivatives then tell us how to update the quantity. Let's review how we did this by walking through an example, one step at a time.

Say an object starts at time $t = 0$ at position $(0, 0)$ with an initial velocity $(1, 0)$ and a constant acceleration $(0, 0.2)$. (For notational clarity, I'll leave out units in this section, but you can continue to think in seconds, meters, meters per second, and so on.) This initial velocity points in the positive x direction, and the acceleration points in the positive y direction. This means if we look at the plane, the object starts by moving directly to the right, but it veers upward over time.

Our task is to find the values of the position vector every two seconds from $t = 0$ to $t = 10$ using Euler's method. First, we'll do it by hand and then we'll do the identical computation in Python. Equipped with the resulting positions, we'll draw them in the x,y plane to show the path the spaceship follows.

9.3.1 Carrying out Euler's method by hand

We will continue to think of position, velocity, and acceleration as functions of time: at any given time, the object will have some vector value for each of these quantities. I'll call these vector-valued functions: $\mathbf{s}(t)$, $\mathbf{v}(t)$ and $\mathbf{a}(t)$ where $\mathbf{s}(t) = (x(t), y(t))$, $\mathbf{v}(t) = (x'(t), y'(t))$, and $\mathbf{a}(t) = (x''(t), y''(t))$. Here are the initial values given in a table at time $t = 0$:

t	$\mathbf{s}(t)$	$\mathbf{v}(t)$	$\mathbf{a}(t)$
0	$(0, 0)$	$(1, 0)$	$(0, 0.2)$

In our asteroid game, PyGame dictated how many milliseconds elapsed between each calculation of position. In this example, to make it quick, let's reconstruct the position from time $t = 0$ to $t = 10$ in 2-second increments. The table we need to complete is as follows:

t	$\mathbf{s}(t)$	$\mathbf{v}(t)$	$\mathbf{a}(t)$
0	$(0, 0)$	$(1, 0)$	$(0, 0.2)$
2			$(0, 0.2)$
4			$(0, 0.2)$
6			$(0, 0.2)$
8			$(0, 0.2)$
10			$(0, 0.2)$

I already filled out the acceleration column for us because we've stipulated that the acceleration is constant. What happens in the 2-second period between $t = 0$ and $t = 2$? The velocities change according to the acceleration as calculated in the following pair of equations. In these equations, we again use the Greek letter Δ (delta) to mean the change in a variable on the interval we're considering. For instance, Δt is the change in time, so $\Delta t = 2$ seconds for each of the 5 intervals. The velocity components at time 2 are therefore:

$$v_x(2) = v_x(0) + a_x(0) \cdot \Delta t = 1 + 0 = 1$$

$$v_y(2) = v_y(0) + a_y(0) \cdot \Delta t = 0.2 \cdot 2 = 0.4$$

The new vector value of the velocity at time $t = 2$ is $\mathbf{v}(2) = (v_x(2), v_y(2)) = (1, 0.4)$. The position changes as well, according to the velocity $\mathbf{v}(0)$:

$$x(2) = x(0) + v_x(0) \cdot \Delta t = 0 + 1 \cdot 2 = 2$$

$$y(2) = y(0) + v_y(0) \cdot \Delta t = 0 + 0 \cdot 2 = 0$$

Its updated value is $\mathbf{s} = (x, y) = (2, 0)$. That gives us the second row of the table:

t	$\mathbf{s}(t)$	$\mathbf{v}(t)$	$\mathbf{a}(t)$
0	$(0, 0)$	$(1, 0)$	$(0, 0.2)$
2	$(2, 0)$	$(1, 0)$	$(0, 0.2)$
4			$(0, 0.2)$
6			$(0, 0.2)$
8			$(0, 0.2)$
10			$(0, 0.2)$

Between $t = 2$ and $t = 4$, the acceleration stays the same so the velocity increases by the same amount, $\mathbf{a} \cdot \Delta t = (0, 0.2) \cdot 2 = (0, 0.4)$, to a new value, $\mathbf{v}(4) = (1, 0.8)$. The position increases according to the velocity $\mathbf{v}(2)$:

$$\Delta \mathbf{s} = \mathbf{v}(2) \cdot \Delta t = (1, 0.4) \cdot 2 = (2, 0.8)$$

This increases the position to $s(4) = (4, 0.8)$. We now have three rows of the table completed, and we've calculated two of the five positions we wanted:

t	$\mathbf{s}(t)$	$\mathbf{v}(t)$	$\mathbf{a}(t)$
0	$(0,0)$	$(1,0)$	$(0,0.2)$
2	$(2,0)$	$(1,0)$	$(0,0.2)$
4	$(4,0.8)$	$(1,0.8)$	$(0,0.2)$
6			$(0,0.2)$
8			$(0,0.2)$
10			$(0,0.2)$

We could keep going like this, but it will be more pleasant if we let Python do the work for us—that's our next step. But first, let's pause for a moment. I've taken us through quite a bit of arithmetic in the past few paragraphs. Did any of my assumptions seem suspect to you? I'll give you a hint: it's not quite legal to use the equation $\Delta\mathbf{s} = \mathbf{v} \cdot \Delta t$ as I did here, so the positions in the table are only approximately correct. If you don't see where I snuck in approximations yet, don't worry. It will be clear soon, once we've plotted the position vectors on a graph.

9.3.2 Implementing the algorithm in Python

Describing this procedure in Python isn't too much work. We first need to set the initial values of time, position, velocity, and acceleration:

```
t = 0
s = (0,0)
v = (1,0)
a = (0,0.2)
```

The other values we need to specify are the moments in time we're interested in: 0, 2, 4, 6, 8, and 10 seconds. Rather than list all of these, we can use the fact that $t = 0$ to begin with and specify a constant $\Delta t = 2$ for each time step with 5 time steps in total:

```
dt = 2
steps = 5
```

Finally, we need to update time, position, and velocity once for every time step. As we go, we can store the positions in an array for later use:

```
from vectors import add, scale
positions = [s]
for _ in range(0,5):
    t += 2
    s = add(s, scale(dt,v))
    v = add(v, scale(dt,a))
    positions.append(s)
```

Updates the position by adding the change in position $\Delta\mathbf{s} = \mathbf{v}\cdot\Delta t$ to the current position s. (I used the scale and add functions from chapter 2.)

Updates the velocity by adding the change in velocity $\Delta\mathbf{v} = \mathbf{a}\cdot\Delta t$ to the current velocity v

If we run this code, the positions list is populated with six values of the vector s, corresponding to the times $t = 0, 2, 4, 6, 8, 10$. Now that we have the values in code, we can plot them and picture the object's motion. If we plot them in 2D using the drawing module from chapters 2 and 3, we can see the object initially moving to the right and then veering upward as expected (figure 9.5). Here's the Python code, and the plot it generates:

```
from draw2d import *
draw2d(Points2D(*positions))
```

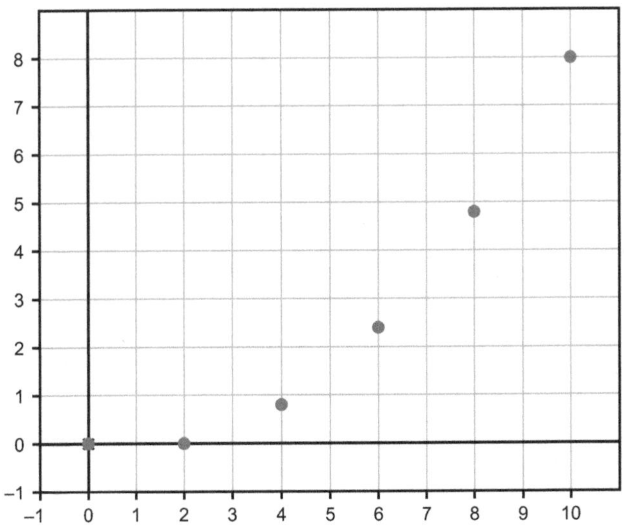

Figure 9.5 Points on the object's trajectory according to our calculation with Euler's method

In our approximation, it's as if the object moved in five straight lines at a different velocity on each of the five time intervals (figure 9.6).

The object is supposed to be accelerating the whole time, so you might expect it to move in a smooth curve instead of in straight lines. Now that we have Euler's method implemented in Python, we can quickly rerun it with different parameters to assess the quality of the approximation.

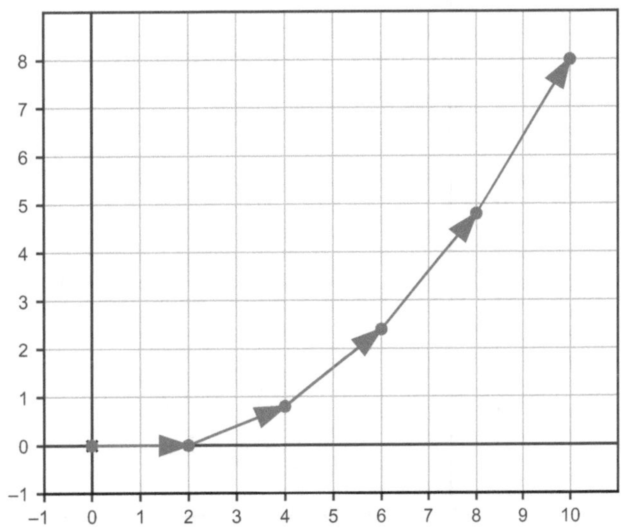

Figure 9.6 The five displacement vectors connecting the points on the trajectory by straight lines.

9.4 Running Euler's method with smaller time steps

We can rerun the calculation again using twice as many time steps by setting `dt = 1` and `steps = 10`. This still simulates 10 seconds of motion, but instead, models it with 10 straight line paths (figure 9.7).

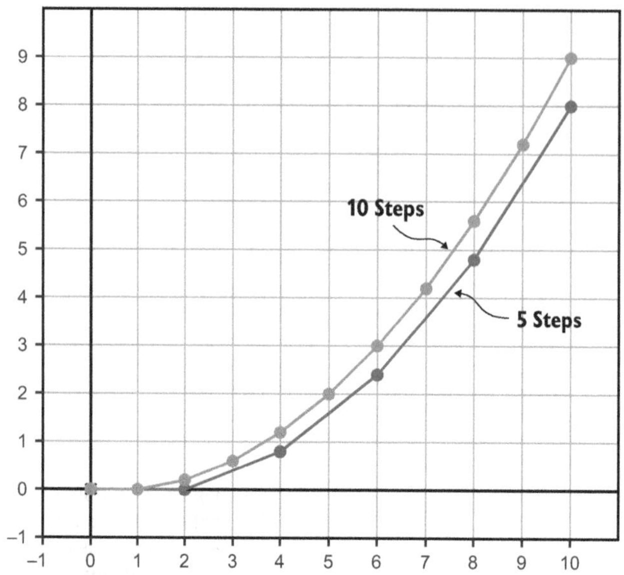

Figure 9.7 Euler's method produces different results with the same initial values and different numbers of steps.

Trying again with 100 steps and `dt = 0.1`, we see yet another trajectory in the same 10 seconds (figure 9.8).

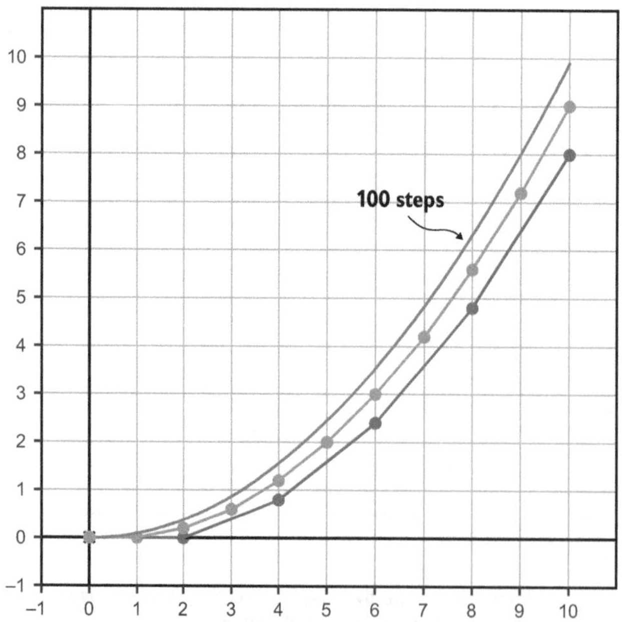

Figure 9.8 With 100 steps instead of 5 or 10, we get yet another trajectory. Dots are omitted for this trajectory because there are so many of them.

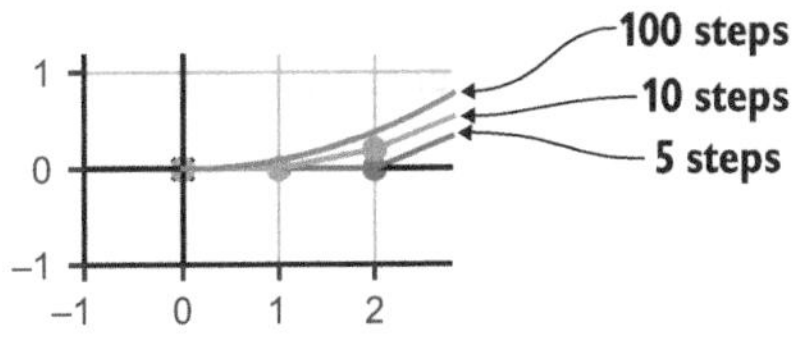

Figure 9.9 Looking closely at the first two segments, the 100-step approximation is the largest because its velocity updates most frequently.

Why do we get different results even though the same equations went into all three calculations? It seems like the more time steps we use, the bigger the *y*-coordinates get. We can see the problem if we look closely at the first two seconds.

In the 5-step approximation, there's no acceleration; the object is still traveling along the *x*-axis. In the 10-step approximation, the object has had one chance to update its velocity, so it has risen above the *x*-axis. Finally, the 100-step approximation has 19 velocity updates between $t = 0$ and $t = 1$, so its velocity increase is the largest (figure 9.9).

This is what I swept under the rug earlier. The equation $\Delta\mathbf{s} = \mathbf{v} \cdot \Delta t$ is only correct when velocity is constant. Euler's method is a good approximation when you use a lot of time steps because on smaller time intervals, velocity doesn't change that much. To confirm this, you can try some large time steps with small values for *dt*. For example, with 100 steps of 0.1 seconds each, the final position is

```
(9.99999999999998, 9.900000000000006)
```

and with 100,000 steps of 0.0001 seconds each, the final position is

```
(9.999999999990033, 9.999899999993497)
```

The exact value of the final position is (10.0, 10.0), and as we add more and more steps to our approximation with Euler's method, our results appear to *converge* to this value. You'll have to trust me for now that (10.0, 10.0) is the exact value. We'll cover how to do exact integrals in the next chapter to prove it. Stay tuned!

9.4.1 Exercises

Exercise 9.2—Mini Project Create a function that carries out Euler's method automatically for a constantly accelerating object. You need to provide the function with an acceleration vector, initial velocity vector, initial position vector, and perhaps other parameters.

Solution I also included the total time and number of steps as parameters to make it easy to test various answers in the solution.

```
def eulers_method(s0,v0,a,total_time,step_count):
    trajectory = [s0]
    s = s0
    v = v0
    dt = total_time/step_count
    for _ in range(0,step_count):
        s = add(s,scale(dt,v))
        v = add(v,scale(dt,a))
        trajectory.append(s)
    return trajectory
```

The duration of each time step dt is the total time elapsed divided by the number of time steps.

For each step, updates the position and velocity and adds the latest position as the next position in the trajectory (list of positions)

Exercise 9.3—Mini Project In the calculation of section 9.4, we under approximated the *y*-coordinate of position because we updated the *y* component of the velocity at the end of each time interval. Update the velocity at the beginning of each time interval and show that you over approximate the *y* position over time.

Solution We can tweak our implementation of the `eulers_method` function from mini-project 9.2 with the only modification being switching the update order of s and v:

```
def eulers_method_overapprox(s0,v0,a,total_time,step_count):
    trajectory = [s0]
    s = s0
    v = v0
    dt = total_time/step_count
    for _ in range(0,step_count):
        v = add(v,scale(dt,a))
        s = add(s,scale(dt,v))
        trajectory.append(s)
    return trajectory
```

With the same inputs, this indeed gives a higher approximation of the *y*-coordinate than the original implementation. If you look closely at the trajectory in the following figure, you can see it is already moving in the *y* direction in the first time step.

```
eulers_method_overapprox((0,0),(1,0),(0,0.2),10,10)
```

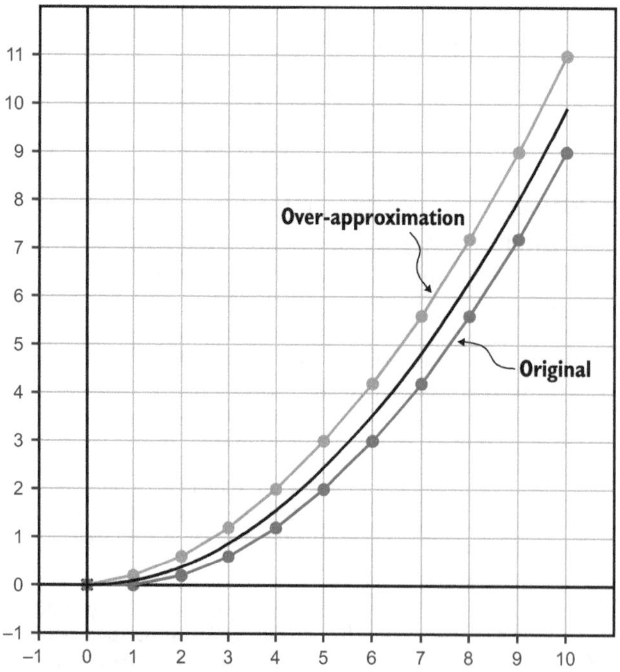

The original Euler's method trajectory and the new one. The exact trajectory is shown in black for comparison.

Exercise 9.4—Mini Project Any projectile like a thrown baseball, a bullet, or an airborne snowboarder experiences the same acceleration vector: 9.81 m/s/s toward the earth. If we think of the *x*-axis of the plane as flat ground with the positive *y*-axis pointing upward, that amounts to an acceleration vector of (0, 9.81). If a baseball is thrown from shoulder height at $x = 0$, we could say its initial position is (0, 1.5). Assume it's thrown at an initial speed of 30 m/s at an angle of 20° up from the positive *x* direction and simulate its trajectory with Euler's method. Approximately how far does the baseball go in the *x* direction before hitting the ground?

Solution The initial velocity is (30 · cos(20°), 30 · sin(20°)). We can use the `eulers_method` function from mini-project 9.2 to simulate the baseball's motion over a few seconds:

```
from math import pi,sin,cos
angle = 20 * pi/180
s0 = (0,1.5)
v0 = (30*cos(angle),30*sin(angle))
a = (0,-9.81)

result = eulers_method(s0,v0,a,3,100)
```

Plotting the resulting trajectory, this figure shows that the baseball makes an arc in the air before returning to the earth at about the 67-meter mark on the *x*-axis. The trajectory continues underground because we didn't tell it to stop.

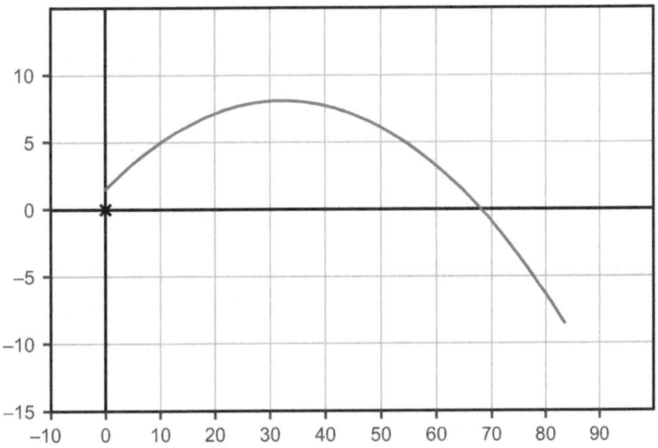

Exercise 9.5—Mini Project Rerun the Euler's method simulation from the previous mini-project with the same initial speed of 30 but using an initial position of (0, 0) and trying various angles for the initial velocity. What angle makes the baseball go the farthest before hitting the ground?

(continued)

Solution To simulate different angles, you can package this code as a function. Using a new starting position of (0, 0), you can see various trajectories in the following figure. It turns out that the baseball makes it the farthest at an angle of 45°. (Notice that I've filtered out the points on the trajectory with negative y components to consider only the motion before the baseball hits the ground.)

```
def baseball_trajectory(degrees):
    radians = degrees * pi/180
    s0 = (0,0)
    v0 = (30*cos(radians),30*sin(radians))
    a = (0,-9.81)
    return [(x,y) for (x,y) in eulers_method(s0,v0,a,10,1000) if y>=0]
```

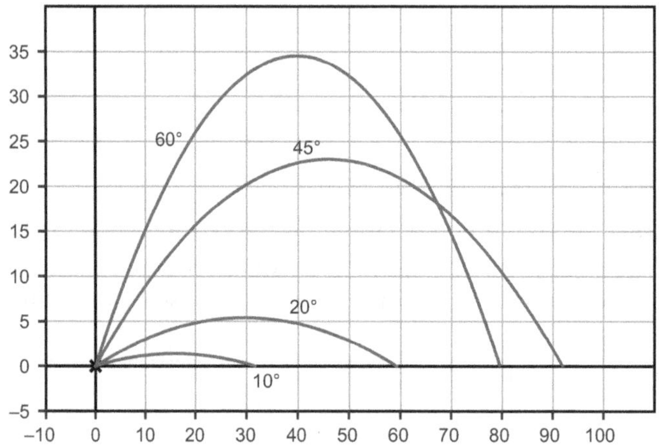

Throwing a baseball at 30 m/s at various angles

Exercise 9.6—Mini Project An object moving in 3D space has an initial velocity of (1, 2, 0) and has a constant acceleration vector of (0, –1, 1). If it starts at the origin, where is it after 10 seconds? Plot its trajectory in 3D using the drawing functions from chapter 3.

Solution It turns out our `eulers_method` implementation can already handle 3D vectors! The figure following the code snippet shows the trajectory in 3D.

```
from draw3d import *
traj3d = eulers_method((0,0,0), (1,2,0), (0,-1,1), 10, 10)
draw3d(
    Points3D(*traj3d)
)
```

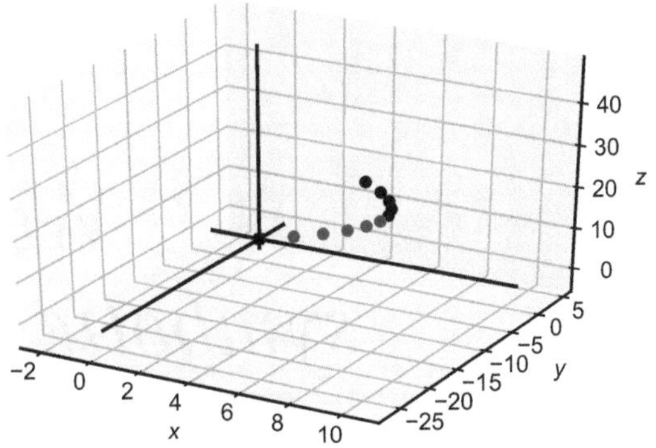

Running with 1,000 steps for improved accuracy, we can find the last position:

```
>>> eulers_method((0,0,0), (1,2,0), (0,-1,1), 10, 1000)[-1]
(9.999999999999831, -29.949999999999644, 49.94999999999933)
```

It's close to (10, –30, 50), which turns out to be the exact position.

Summary

- Velocity is the derivative of position with respect to time. It is a vector consisting of the derivatives of each of the position functions. In 2D, with position functions $x(t)$ and $y(t)$, we can write the position *vector* as a function $\mathbf{s}(t) = (x(t), y(t))$ and the velocity vector as a function $\mathbf{v}(t) = (x'(t), y'(t))$.
- In a video game, you can animate an object moving at a constant velocity by updating its position in each frame. Measuring the time between frames and multiplying it by the object's velocity gives you the change in position for the frame.
- Acceleration is the derivative of velocity with respect to time. It is a vector whose components are the derivatives of the components of velocity, for instance, $\mathbf{a}(t) = (v'_x(t), v'_y(t))$.
- To simulate an accelerating object in a video game, you need to not only update the position with each frame but also update the velocity.
- If you know the rate at which a quantity changes with respect to time, you can compute the value of a quantity itself over time by calculating the quantity's change over many small time intervals. This is called *Euler's method*.

10 Working with symbolic expressions

This chapter covers

- Modeling algebraic expressions as data structures
- Writing code to analyze, transform, or evaluate algebraic expressions
- Finding the derivative of a function by manipulating the expression that defines it
- Writing a Python function to compute derivative formulas
- Using the SymPy library to compute integral formulas

If you followed all of the code examples and did all the exercises in chapter 8 and chapter 9, you already have a solid grasp of the two most important concepts in calculus: the derivative and the integral. First, you learned how to approximate the derivative of a function at a point by taking slopes of smaller and smaller secant lines. You then learned how to approximate an integral by estimating the area under a graph with skinny rectangles. Lastly, you learned how to do calculus with vectors by simply doing the relevant calculus operations in each coordinate.

It might seem like an audacious claim, but I really do hope to have given you the most important concepts you'd learn in a year-long college calculus class in just a few chapters of this book. Here's the catch: because we're working in Python, I'm skipping the most laborious piece of a traditional calculus course, which is doing a lot of formula manipulation by hand. This kind of work enables you to take the formula for a function like $f(x) = x^3$ and figure out an *exact* formula for its derivative, $f'(x)$. In this case, there's a simple answer, $f'(x) = 3x^2$, as shown in figure 10.1.

There are infinitely many formulas you might want to know the derivative of, and you can't memorize derivatives for all of them, so what you end up doing in a calculus class is learning a small set of rules and how to systematically apply them to transform a function into its derivative. By and large, this isn't that useful of a skill for a programmer. If you want to know the exact formula for a derivative, you can use a specialized tool called a *computer algebra system* to compute it for you.

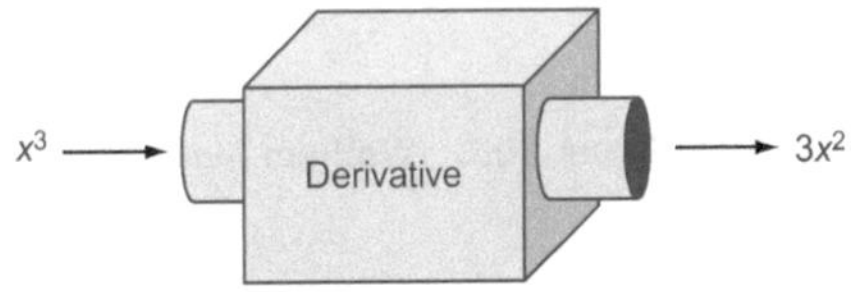

Figure 10.1 The derivative of the function $f(x) = x^3$ has an exact formula, namely $f'(x) = 3x^2$.

10.1 Finding an exact derivative with a computer algebra system

One of the most popular computer algebra systems is called *Mathematica*, and you can use its engine for free online at a website called Wolfram Alpha (wolframalpha.com). In my experience, if you want an exact formula for a derivative for a program you're writing, the best approach is to consult Wolfram Alpha. For instance, when we build a neural network in chapter 16, it will be useful to know the derivative of the function

$$f(x) = \frac{1}{1 + e^{-x}}$$

To find a formula for the derivative of this function, you can simply go to wolframalpha.com and enter the formula in the input box (figure 10.2). Mathematica has its own syntax for mathematical formulas, but Wolfram Alpha is impressively forgiving and understands most simple formulas that you enter (even in Python syntax!).

Figure 10.2 Entering a function in the input box at wolframalpha.com

When you press Enter, the Mathematica engine powering Wolfram Alpha computes a number of facts about this function, including its derivative. If you scroll down, you'll see a formula for the derivative of the function (figure 10.3).

Figure 10.3 Wolfram Alpha reports a formula for the derivative of the function.

For our function $f(x)$, its instantaneous rate of change at any value of x is given by

$$f'(x) = \frac{e^{-x}}{(1+e^{-x})^2}$$

If you understand the concept of a "derivative" and of an "instantaneous rate of change," learning to punch formulas into Wolfram Alpha is a more important skill than any other single skill you'll learn in a calculus class. I don't mean to be cynical; there's plenty to learn about the behavior of specific functions by taking their derivatives by hand. It's just that in your life as a professional software developer, you'll probably never need to figure out the formula for a derivative or integral when you have a free tool like Wolfram Alpha available.

That said, your inner nerd may be asking, "How does Wolfram Alpha do it?" It's one thing to find a crude estimate of a derivative by taking approximate slopes of the graph at various points, but it's another to produce an exact formula. Wolfram Alpha successfully interprets the formula you type in, transforms it with some algebraic manipulations, and outputs a new formula. This kind of approach, where you work with formulas themselves instead of numbers, is called *symbolic programming*.

The pragmatist in me wants to tell you to "just use Wolfram Alpha," while the math enthusiast in me wants to teach you how to take derivatives and integrals by hand, so in this chapter I'm going to split the difference. We do some symbolic programming in Python to manipulate algebraic formulas directly and, ultimately, figure out the formulas for their derivatives. This gets you acquainted with the process of finding derivative formulas, while still letting the computer do most of the work for you.

10.1.1 Doing symbolic algebra in Python

Let me start by showing you how we'll represent and manipulate formulas in Python. Say we have a mathematical function like

$$f(x) = (3x^2 + x)\sin(x)$$

The usual way to represent it in Python is as follows:

```
from math import sin
def f(x):
    return (3*x**2 + x) * sin(x)
```

While this Python code makes it easy to evaluate the formula, it doesn't give us a way to compute facts *about* the formula. For instance, we could ask

- Does the formula depend on the variable *x*?
- Does it contain a trigonometric function?
- Does it involve the operation of division?

We can look at these questions and quickly decide that the answers are yes, yes, and no. There's no simple, reliable way to write a Python program to answer these questions for us. For instance, it's difficult, if not impossible, to write a function `contains_division(f)` that takes the function *f* and returns true if it uses the operation of division in its definition.

Here's where this would come in handy. In order to invoke an algebraic rule, you need to know what operations are being applied and in what order. For instance, the function $f(x)$ is a product of $\sin(x)$ with a sum, and there's a well-known algebraic process for expanding a product of a sum as visualized in figure 10.4.

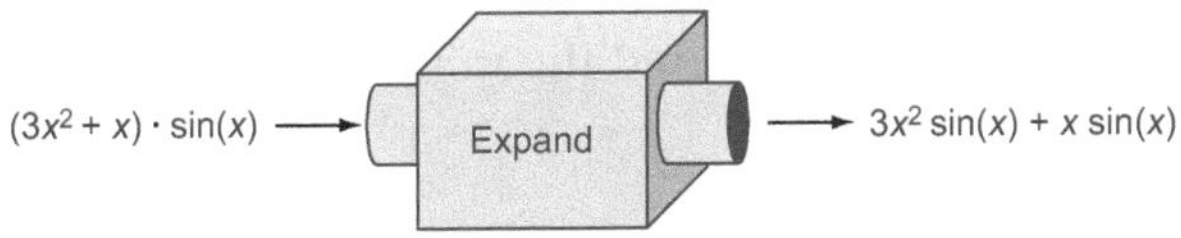

Figure 10.4 Because $(3x^2+x)$ $\sin(x)$ is a product of a sum, it can be expanded.

Our strategy is to model algebraic expressions as data structures rather than translating them directly to Python code, and then they're more amenable to manipulation. Once we can manipulate functions symbolically, we can automate the rules of calculus.

Most functions expressed by simple formulas also have simple formulas for their derivatives. For instance, the derivative of x^3 is $3x^2$, meaning at any value of *x*, the derivative of $f(x) = x^3$ is given by $3x^2$. By the time we're done in this chapter, you'll be able to write a Python function that takes an algebraic expression and gives you an expression for its derivative. Our data structure for an algebraic formula will be able to represent variables, numbers, sums, differences, products, quotients, powers, and special functions like sine and cosine. If you think about it, we can represent a huge variety of different formulas with that handful of building blocks, and our derivative will work on all of them (figure 10.5).

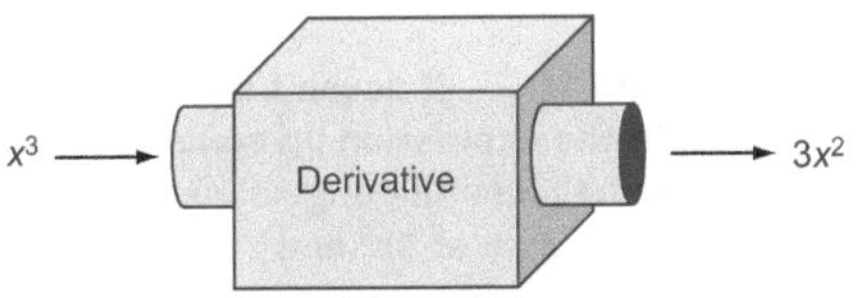

Figure 10.5 A goal is to write a derivative function in Python that takes an expression for a function and returns an expression for its derivative.

We'll get started by modeling expressions as data structures instead of functions in Python code. Then, to warm up, we can do some simple computations with the data structures to do things like plugging in numbers for variables or expanding products of sums. After that, I'll teach you some of the rules for taking derivatives of formulas, and we'll write our own derivative function and perform them automatically on our symbolic data structures.

10.2 Modeling algebraic expressions

Let's focus on the function $f(x) = (3x^2+x)\sin(x)$ for a bit and see how we can break it down into pieces. This is a good example function because it contains a lot of different building blocks: a variable x, as well as numbers, addition, multiplication, a power, and a specially named function, $\sin(x)$. Once we have a strategy for breaking this function down into conceptual pieces, we can translate it into a Python data structure. This data structure is a *symbolic* representation of the function as opposed to a string representation like `"(3*x**2 + x) * sin(x)"`.

A first observation is that f is an arbitrary name for this function. For instance, the right-hand side of this equation expands the same way regardless of what we call it. Because of this, we can focus only on the expression that defines the function, which in this case is $(3x^2 + x)\sin(x)$. This is called an expression in contrast to an equation, which must contain an equals sign (=). An *expression* is a collection of mathematical symbols (numbers, letters, operations, and so on) combined in some valid ways. Our first goal, therefore, is to model these symbols and the valid means of composing this expression in Python.

10.2.1 Breaking an expression into pieces

We can start to model algebraic expressions by breaking them up into smaller expressions. There is only one meaningful way to break up the expression $(3x^2 + x)\sin(x)$. Namely, it's the product of $(3x^2 + x)$ and $\sin(x)$ as shown in figure 10.6.

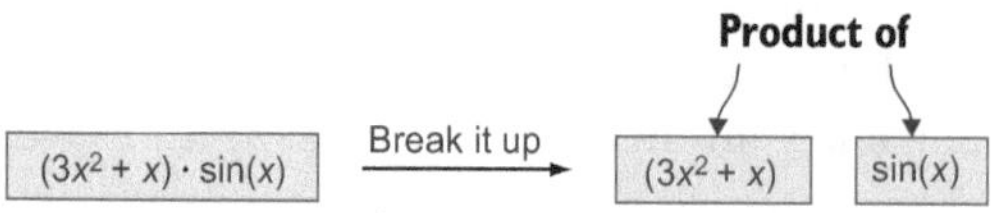

Figure 10.6 A meaningful way to break up an algebraic expression into two smaller expressions

By contrast, we can't split this expression around the plus sign. We could make sense of the expressions on either side of the plus sign if we tried, but the result is not equivalent to the original expression (figure 10.7).

Figure 10.7 It doesn't make sense to split the expression up around the plus sign because the original expression is *not* the sum of $3x^2$ and $x \cdot \sin(x)$.

If we look at the expression $3x^2 + x$, it can be broken up into a sum: $3x^2$ and x. Likewise, the conventional order of operations tells us that $3x^2$ is the product of 3 and x^2, not $3x$ raised to the power of 2.

In this chapter, we'll think of operations like multiplication and addition as ways to take two (or more) algebraic expressions and stick them together side by side to make a new, bigger algebraic expression. Likewise, operators are valid places to break up an existing algebraic expression into smaller ones.

In the terminology of functional programming, functions combining smaller objects into bigger ones like this are often called *combinators*. Here are some of the combinators implied in our expression:

- $3x^2$ is the *product* of the expressions 3 and x^2.
- x^2 is a *power:* one expression x raised to the power of another expression 2.
- The expression $\sin(x)$ is a *function application*. Given the expression sin and the expression x, we can build a new expression $\sin(x)$.

A variable x, a number 2, or a function named sin can't be broken down further. To distinguish these from combinators, we call them *elements*. The lesson here is that while $(3x^2 + x) \sin(x)$ is just a bunch of symbols printed on this page, the symbols are combined in certain ways to convey some mathematical meaning. To bring this concept home, we can visualize how this expression is built from its underlying elements.

10.2.2 Building an expression tree

The elements 3, x, 2, and sin, along with the combinators of adding, multiplying, raising to a power, and applying a function are sufficient to rebuild the whole of the expression $(3x^2 + x) \sin(x)$. Let's go through the steps and draw the structure we'll end up building. One of the first constructions we can put together is x^2, which combines x and 2 with the power combinator (figure 10.8).

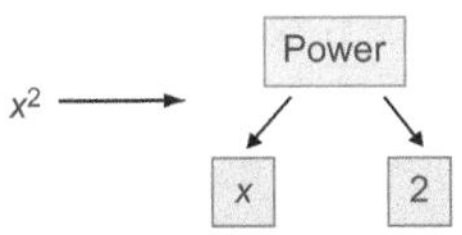

Figure 10.8 Combining *x* and 2 with the power combinator to represent the bigger expression x^2

A good next step is to combine x^2 with the number 3 via the product combinator to get the expression $3x^2$ (figure 10.9).

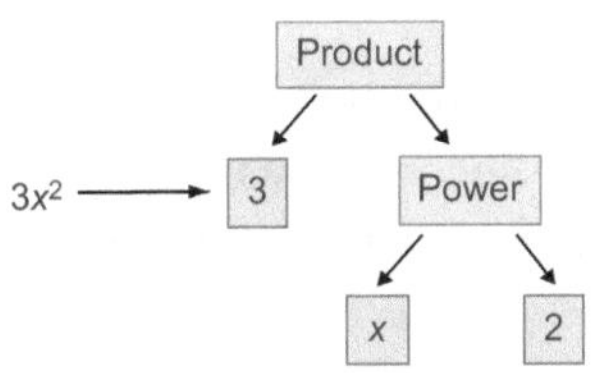

Figure 10.9 Combining the number 3 with a power to model the product $3x^2$

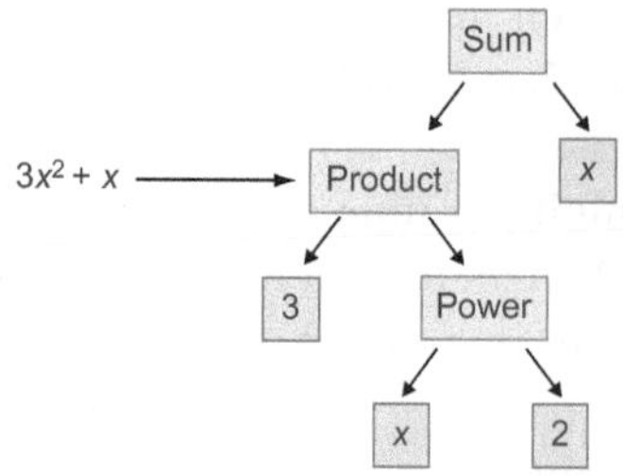

Figure 10.10 Combining the expression $3x^2$ with the element x and the sum combinator to get $3x^2 + x$

This construction is two layers deep: one expression that inputs to the product combinator is itself a combinator. As we add more of the terms of the expression, it gets even deeper. The next step is adding the element x to $3x^2$ using the sum combinator (figure 10.10), which represents the operation of addition.

Finally, we need to use the function application combinator to apply sin to x and then the product combinator to combine $\sin(x)$ with what we've built thus far (figure 10.11).

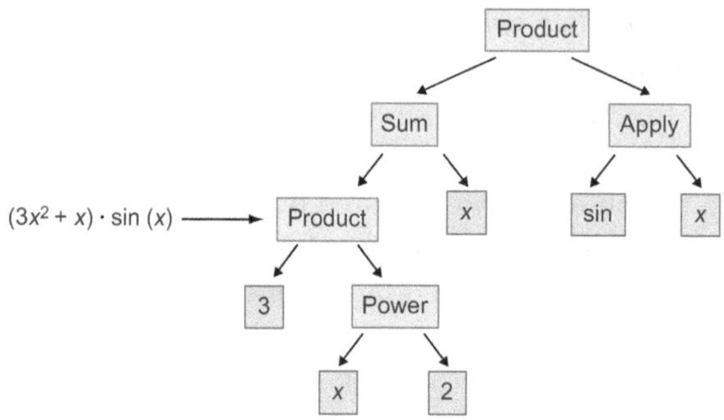

Figure 10.11 A completed picture showing how to build $(3x^2 + x)$ sin(x) from elements and combinators

You may recognize the structure we've built as a *tree*. The root of the tree is the product combinator with two branches coming out of it: `Sum` and `Apply`. Each combinator appearing further down the tree adds additional branches, until you reach the elements that are leaves and have no branches. Any algebraic expression built with numbers, variables, and named functions as elements and operations that are combinators correspond to a distinctive tree that reveals its structure. The next thing we can do is to build the same tree in Python.

10.2.3 Translating the expression tree to Python

When we've built this tree in Python, we'll have achieved our goal of representing the expression as a data structure. I'll use Python classes covered in appendix B to represent each kind of element and each combinator. As we go, we'll revise these classes to give them more and more functionality. You can follow the walk-through Jupyter notebook for chapter 10 if you want to follow the text, or you can skip to a more complete implementation in the Python script file expressions.py.

In our implementation, we model combinators as containers that hold all of their inputs. For instance, a power x to the 2, or x^2, has two pieces of data: the *base* x and the power 2. Here's a Python class that's designed to represent a power expression:

```
class Power():
    def __init__(self,base,exponent):
```

```
        self.base = base
        self.exponent = exponent
```

We could then write `Power("x",2)` to represent the expression x^2. But rather than using raw strings and numbers, I'll create special classes to represent numbers and variables. For example,

```
class Number():
    def __init__(self,number):
        self.number = number

class Variable():
    def __init__(self,symbol):
        self.symbol = symbol
```

This might seem like unnecessary overhead, but it will be useful to be able to distinguish `Variable("x")`, which means the letter x considered as a variable from the string `"x"`, which is merely a string. Using these three classes, we can model the expression x^2 as

```
Power(Variable("x"),Number(2))
```

Each of our combinators can be implemented as an appropriately named class that stores the data of whatever expressions it combines. For instance, a product combinator can be a class that stores two expressions that are meant to be multiplied together:

```
class Product():
    def __init__(self, exp1, exp2):
        self.exp1 = exp1
        self.exp2 = exp2
```

The product $3x^2$ can be expressed using this combinator:

```
Product(Number(3),Power(Variable("x"),Number(2)))
```

After introducing the rest of the classes we need, we can model the original expression as well as an infinite list of other possibilities. (Note that we allow any number of input expressions for the `Sum` combinator, and we could have done this for the Product combinator as well. I restricted the Product combinator to two inputs to keep our code simpler when we start calculating derivatives in section 10.3.)

```
class Sum():
    def __init__(self, *exps):          # Allows a Sum of any number of terms so we
        self.exps = exps                # can add two or more expressions together

class Function():                       # Stores a string that is the
    def __init__(self,name):            # function's name (like "sin")
        self.name = name

class Apply():                          # Stores a function and the
    def __init__(self,function,argument):   # argument it is applied to
        self.function = function
        self.argument = argument
```

```
f_expression = Product(>
               Sum(
                   Product(
                       Number(3),
                       Power(
                           Variable("x"),
                           Number(2))),
                   Variable("x")),
               Apply(
                   Function("sin"),
                   Variable("x")))
```

I use extra whitespace to make the structure of the expression clearer to see.

This is a faithful representation of the original expression $(3x^2 + x)\sin(x)$. By that I mean, we could look at this Python object and see that it describes the algebraic expression and not a different one. For another expression like

```
Apply(Function("cos"),Sum(Power(Variable("x"),Number("3")), Number(-5)))
```

we can read it carefully and see that it represents a different expression: $\cos(x^3 + -5)$. In the exercises that follow, you can practice translating some algebraic expressions to Python and vice versa. You'll see it can be tedious to type out the whole representation of an expression. The good news is that once you get it encoded in Python, the manual work is over. In the next section, we see how to write Python functions to automatically work with our expressions.

10.2.4 Exercises

Exercise 10.1 You may have met the *natural logarithm*, a special mathematical function written $\ln(x)$. Draw the expression $\ln(yz)$ as a tree built from the elements and combinators described in the previous section.

Solution The outermost combinator is an Apply. The function being applied is ln, the natural logarithm, and the argument is *yz*. In turn, *yz* is a power with base *y* and exponent *z*. The result looks like this:

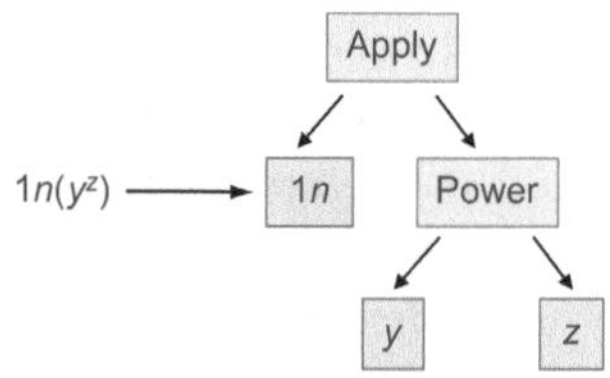

Exercise 10.2 Translate the expression from the previous exercise to Python code, given that the natural logarithm is calculated by the Python function `math.log`. Write it both as a Python function and as a data structure built from elements and combinators.

Solution You can think of $\ln(y^z)$ as a function of two variables y and z. It translates directly to Python, where ln is called `log`:

```
from math import log
def f(y,z):
    return log(y**z)
```

The expression tree is built like this:

```
Apply(Function("ln"), Power(Variable("y"), Variable("z")))
```

Exercise 10.3 What is the expression represented by `Product(Number(3), Sum(Variable("y"),Variable("z")))`?

Solution This expression represents $3 \cdot (y + z)$. Notice that the parentheses are necessary because of the order of operations.

Exercise 10.4 Implement a `Quotient` combinator representing one expression divided by another. How do you represent the following expression?

$$\frac{a+b}{2}$$

Solution A `Quotient` combinator needs to store two expressions: the top expression is called the *numerator* and the bottom is called the *denominator*:

```
class Quotient():
    def __init__(self,numerator,denominator):
        self.numerator = numerator
        self.denominator = denominator
```

The sample expression is the quotient of the sum $a + b$ with the number 2:

```
Quotient(Sum(Variable("a"),Variable("b")),Number(2))
```

Exercise 10.5 Implement a `Difference` combinator representing one expression subtracted from another. How can you represent the expression $b^2 - 4ac$?

Solution The `Difference` combinator needs to store two expressions, and it represents the second subtracted from the first:

```
class Difference():
    def __init__(self,exp1,exp2):
        self.exp1 = exp1
        self.exp2 = exp2
```

(continued)

The expression $b^2 - 4ac$ is the difference of the expressions b^2 and $4ac$ and is represented as follows:

```
Difference(
    Power(Variable('b'),Number(2)),
    Product(Number(4),Product(Variable('a'), Variable('c'))))
```

Exercise 10.6 Implement a `Negative` combinator representing the negation of an expression. For example, the negation of $x^2 + y$ is $-(x^2+y)$. Represent the latter expression in code using your new combinator.

Solution The `Negative` combinator is a class that holds one expression:

```
class Negative():
    def __init__(self,exp):
        self.exp = exp
```

To negate $x^2 + y$, we pass it to the `Negative` constructor:

```
Negative(Sum(Power(Variable("x"),Number(2)),Variable("y")))
```

Exercise 10.7 Add a function called `Sqrt` that represents a square root and use it to encode the following formula:

$$\frac{-b \pm \sqrt{b^2 - 4ac}}{2a}$$

Solution To save some typing, we can name our variables and square root function up front:

```
A = Variable('a')
B = Variable('b')
C = Variable('c')
Sqrt = Function('sqrt')
```

Then it's just a matter of translating the algebraic expression into the appropriate structure of elements and combinators. At the highest level, you can see this is a quotient of a sum (on top) and a product (on the bottom):

```
Quotient(
    Sum(
        Negative(B),
        Apply(
            Sqrt,
            Difference(
                Power(B,Number(2)),
                Product(Number(4), Product(A,C))))),
    Product(Number(2), A))
```

Exercise 10.8—Mini Project Create an abstract base class called `Expression` and make all of the elements and combinators inherit from it. For instance, `class Variable()` would become `class Variable(Expression)`. Then overload the Python arithmetic operations +, –, *, and / so that they produce `Expression` objects. For instance, the code `2*Variable("x")+3` should yield `Sum(Product(Number(2),Variable("x")),Number(3))`.

Solution See the file expressions.py in the source code for this chapter.

10.3 Putting a symbolic expression to work

For the function we've studied so far, $f(x) = (3x^2 + x) \sin(x)$, we wrote a Python function that computes it:

```
def f(x):
    return (3*x**2 + x)*sin(x)
```

As an entity in Python, this function is only good for one thing: returning an output value for a given input value x. The value f in Python does not make it particularly easy to programmatically answer the questions we asked at the beginning of the chapter: whether f depends on its input, whether f contains a trigonometric function, or what the body of f would look like if it were expanded algebraically. In this section, we see that once we translate the expression into a Python data structure built from elements and combinators, we can answer all of these questions and more!

10.3.1 Finding all the variables in an expression

Let's write a function that takes an expression and returns a list of distinct variables that appear in it. For instance, $h(z) = 2z + 3$ is defined using the input variable z, while the definition of $g(x) = 7$ contains no variables. We can write a Python function, `distinct_variables`, that takes an expression (meaning any of our elements or combinators) and returns a Python set containing the variables.

If our expression is an element, like z or 7, the answer is clear. An expression that is just a variable contains one distinct variable, while an expression that is just a number contains no variables at all. We expect our function to behave accordingly:

```
>>> distinct_variables(Variable("z"))
{'z'}
>>> distinct_variables(Number(3))
set()
```

The situation is more complicated when the expression is built from some combinators like $y \cdot z + x^z$. It's easy for a human to read all the variables, y, z, and x, but how do we extract these from the expression in Python? This is actually a Sum combinator representing the sum of $y \cdot z$ and x^z. The first expression in the sum contains y and z, while the second has x and z. The sum then contains all of the variables in these two expressions.

This suggests we should use a recursive solution: the `distinct_variables` for a combinator are the collected `distinct_variables` for each of the expressions it contains. The end of the line has the variables and numbers, which obviously contain either one or zero variables. To implement the `distinct_variables` function, we need to handle the case of every element and combinator that make up a valid expression:

```
def distinct_variables(exp):
    if isinstance(exp, Variable):
        return set(exp.symbol)
    elif isinstance(exp, Number):
        return set()
    elif isinstance(exp, Sum):
        return set().union(*[distinct_variables(exp) for exp in exp.exps])
    elif isinstance(exp, Product):
        return
distinct_variables(exp.exp1).union(distinct_variables(exp.exp2))
    elif isinstance(exp, Power):
        return
distinct_variables(exp.base).union(distinct_variables(exp.exponent))
    elif isinstance(exp, Apply):
        return distinct_variables(exp.argument)
    else:
        raise TypeError("Not a valid expression.")
```

This code looks hairy, but it is just a long if/else statement with one line for each possible element or combinator. Arguably, it would be better coding style to add a `distinct_variables` method to each element and combinator class, but that makes it harder to see the logic in a single code listing. As expected, our `f_expression` contains only the variable x:

```
>>> distinct_variables(f_expression)
{'x'}
```

If you're familiar with the tree data structure, you'll recognize this as a recursive traversal of the expression tree. By the time this function completes, it has called `distinct_variables` on every expression contained in the target expression, which are all of the nodes in the tree. That ensures that we see every variable and that we get the correct answers that we expect. In the exercises at the end of this section, you can use a similar approach to find all of the numbers or all of the functions.

10.3.2 Evaluating an expression

Now, we've got two representations of the same mathematical function $f(x)$. One is the Python function `f`, which is good for evaluating the function at a given input value of x. The new one is this tree data structure that describes the structure of the expression defining $f(x)$. It turns out the latter representation has the best of both worlds; we can use it to evaluate $f(x)$ as well, with only a little more work.

Mechanically, evaluating a function $f(x)$ at, say, $x = 5$ means plugging in the value of 5 for x everywhere and then doing the arithmetic to find the result. If the expression were just $f(x) = x$, plugging in $x = 5$ would tell us $f(5) = 5$. Another simple example would be $g(x) = 7$, where plugging in 5 in place of x has no effect; there are no appearances of x on the right-hand side, so the result of $g(5)$ is just 7.

The code to evaluate an expression in Python is similar to the code we just wrote to find all variables. Instead of looking at the set of variables that appear in each subexpression, we need to evaluate each subexpression, then the combinators tell us how to combine these results to get the value of the whole expression.

The starting data we need is what values to plug-in and which variables these replace. An expression with two different variables like $z(x, y) = 2xy^3$ will need two values to get a result; for instance, $x = 3$ and $y = 2$. In computer science terminology, these are called *variable bindings.* With these, we can evaluate the subexpression y^3 as $(2)^3$, which equals 8. Another subexpression is $2x$, which evaluates to $2 \cdot (3) = 6$. These two are combined with the Product combinator, so the value of the whole expression is the product of 6 and 8, or 48.

As we translate this procedure into Python code, I'm going to show you a slightly different style than in the previous example. Rather than having a separate evaluate function, we can add an `evaluate` method to each class representing an expression. To enforce this, we can create an abstract `Expression` base class with an abstract `evaluate` method and have each kind of expression inherit from it. If you need a review of abstract base classes in Python, take a moment to review the work we did with the `Vector` class in chapter 6 or the overview in appendix B. Here's an `Expression` base class, complete with an `evaluate` method:

```
from abc import ABC, abstractmethod

class Expression(ABC):
    @abstractmethod
    def evaluate(self, **bindings):
        pass
```

Because an expression can contain more than one variable, I set it up so you can pass in the variable bindings as keyword arguments. For instance, the bindings `{"x":3,"y":2}` mean substitute 3 for x and 2 for y. This gives us some nice syntactic sugar when evaluating an expression. If z represents the expression $2xy^3$, then once we're done, we'll be able to execute the following:

```
>>> z.evaluate(x=3,y=2)
48
```

So far, we've only an abstract class. Now we need to have all of our expression classes inherit from `Expression`. For example, a `Number` instance is a valid expression as a number on its own, like 7. Regardless of the variable bindings provided, a number evaluates to itself:

```
class Number(Expression):
    def __init__(self,number):
        self.number = number
    def evaluate(self, **bindings):
        return self.number
```

For instance, evaluating `Number(7).evaluate(x=3,y=6,q=-15)`, or any other evaluation for that matter, returns the underlying number 7.

Handling variables is also simple. If we're looking at the expression `Variable("x")`, we only need to consult the bindings and see what number the variable x is set to. When we're done, we should be able to run `Variable("x").evaluate(x=5)` and get 5 as a result. If we can't find a binding for x, then we can't complete the evaluation, and we need to raise an exception. Here's the updated definition of the `Variable` class:

```
class Variable(Expression):
    def __init__(self,symbol):
        self.symbol = symbol
    def evaluate(self, **bindings):
        try:
            return bindings[self.symbol]
        except:
            raise KeyError("Variable '{}' is not bound.".format(self.symbol))
```

With these elements handled, we need to turn our attention to the combinators. (Note that we won't consider a `Function` object an `Expression` on its own because a function like sine is not a standalone expression. It can only be evaluated when it's given an argument in the context of an `Apply` combinator.) For a combinator like `Product`, the rule to evaluate it is simple: evaluate both expressions contained in the product and then multiply the results together. No substitution needs to be performed in the product, but we'll pass the bindings along to both subexpressions in case either contains a `Variable`:

```
class Product(Expression):
    def __init__(self, exp1, exp2):
        self.exp1 = exp1
        self.exp2 = exp2
    def evaluate(self, **bindings):
        return self.exp1.evaluate(**bindings) *
self.exp2.evaluate(**bindings)
```

With these three classes updated with evaluate methods, we can now evaluate any expression built from variables, numbers, and products. For instance,

```
>>> Product(Variable("x"), Variable("y")).evaluate(x=2,y=5)
10
```

Similarly, we can add an `evaluate` method to the `Sum`, `Power`, `Difference`, or `Quotient` combinators (as well as any other combinators you may have created as exercises). Once we evaluate their subexpressions, the name of the combinator tells us which operation we can use to get the overall result.

The `Apply` combinator works a bit differently, so it deserves some special attention. We need to dynamically look at a function name like sin or sqrt and figure out how to compute its value. There are a few possible ways to do this, but I chose keeping a dictionary of known functions as data on the `Apply` class. As a first pass, we can make our evaluator aware of three named functions:

```
_function_bindings = {
    "sin": math.sin,
    "cos": math.cos,
    "ln": math.log
}

class Apply(Expression):
    def __init__(self,function,argument):
        self.function = function
        self.argument = argument
    def evaluate(self, **bindings):
        return
_function_bindings[self.function.name](self.argument.evaluate(**bindings))
```

You can practice writing the rest of the evaluate methods yourself or find them in the source code for this book. Once you get all of them fully implemented, you'll be able to evaluate our `f_expression` from section 10.1.3:

```
>>>  f_expression.evaluate(x=5)
-76.71394197305108
```

The result here isn't important, only the fact that it's the same as what the ordinary Python function $f(x)$ gives us:

```
>>> f(5)
-76.71394197305108
```

Equipped with the evaluate function, our `Expression` objects can do the same work as their corresponding ordinary Python functions.

10.3.3 Expanding an expression

There are many other things we can do with our expression data structures. In the exercises, you can try your hand at building a few more Python functions that manipulate expressions in different ways. I'll show you one more example for now, which I mentioned at the beginning of this chapter: expanding an expression. What I mean by this is taking any product or power of sums and carrying it out.

The relevant rule of algebra is the *distributive property* of sums and products. This rule says that a product of the form $(a + b) \cdot c$ is equal to $ac + bc$ and, similarly, that $x(y + z) = xy + xz$. For instance, our expression $(3x^2 + x) \sin(x)$ is equal to $3x^2 \sin(x) + x \sin(x)$, which is called the expanded form of the first product. You can use this rule several times to expand more complicated expressions, for instance:

$$
\begin{aligned}
(x+y)^3 &= (x+y)(x+y)(x+y) \\
&= x(x+y)(x+y) + y(x+y)(x+y) \\
&= x^2(x+y) + xy(x+y) + yx(x+y) + y^2(x+y) \\
&= x^3 + x^2y + x^2y + xy^2 + yx^2 + y^2x + y^2x + y^3 \\
&= x^3 + 3x^2y + 3y^2x + y^3
\end{aligned}
$$

As you can see, expanding a short expression like $(x + y)^3$ can be a lot of writing. In addition to expanding this expression, I also simplified the result a bit, rewriting some products that would have looked like xyx or xxy as x^2y, for instance. This is possible because order does not matter in multiplication. Then I further simplified by *combining* like terms, noting that there were three summed copies each of x^2y and y^2x and grouping those together into $3x^2y$ and $3y^2x$. In the following example, we only look at how to do the expanding; you can implement the simplification as an exercise.

We can start by adding an abstract expand method to the Expression base class:

```
class Expression(ABC):
    ...
    @abstractmethod
    def expand(self):
        pass
```

If an expression is a variable or number, it is already expanded. For these cases, the expand method returns the object itself. For instance,

```
class Number(Expression):
    ...
    def expand(self):
        return self
```

Sums are already considered to be expanded expressions, but the individual terms of a sum cannot be expanded. For example, $5 + a(x + y)$ is a sum in which the first term 5 is fully expanded, but the second term $a(x + y)$ is not. To expand a sum, we need to expand each of the terms and sum them:

```
class Sum(Expression):
    ...
    def expand(self):
        return Sum(*[exp.expand() for exp in self.exps])
```

The same procedure works for function application. We can't expand the Apply function itself, but we can expand its arguments. This would expand an expression like $\sin(x(y + z))$ to $\sin(xy + xz)$:

```
class Apply(Expression):
    ...
    def expand(self):
        return Apply(self.function, self.argument.expand())
```

The real work comes when we expand products or powers, where the structure of the expression changes completely. As an example, $a(b + c)$ is a product of a variable with a sum of two variables, while its expanded form is $ab + ac$, the sum of two products of two variables each. To implement the distributive law, we have to handle three cases:

the first term of the product might be a sum, the second term might be a sum, or neither of them might be sums. In the latter case, no expanding is necessary:

```
class Product(Expression):
    ...
    def expand(self):
        expanded1 = self.exp1.expand()
        expanded2 = self.exp2.expand()
        if isinstance(expanded1, Sum):
            return Sum(*[Product(e,expanded2).expand()

                        for e in expanded1.exps])
        elif isinstance(expanded2, Sum):
            return Sum(*[Product(expanded1,e)

                        for e in expanded2.exps])
        else:
            return Product(expanded1,expanded2)
```

Expands both terms of the product

If the first term of the product is a Sum, it takes the product with each of its terms multiplied by the second term of the product, then calls expand on the result in case the second term of the product is also a Sum.

If the second term of the product is a Sum, it multiplies each of its terms by the first term of the product.

Otherwise, neither term is a Sum, and the distributive property doesn't need to be invoked.

With all of these methods implemented, we can test the `expand` function. With an appropriate implementation of `__repr__` (see the exercises), we can see a string representation of the results clearly in Jupyter or in an interactive Python session. It correctly expands $(a + b)\ (x + y)$ to $ax + ay + bx + by$:

```
Y = Variable('y')
Z = Variable('z')
A = Variable('a')
B = Variable('b')

>>> Product(Sum(A,B),Sum(Y,Z))
Product(Sum(Variable("a"),Variable("b")),Sum(Variable("x"),Variable("y")))
>>> Product(Sum(A,B),Sum(Y,Z)).expand()
Sum(Sum(Product(Variable("a"),Variable("y")),Product(Variable("a"),
Variable("z"))),Sum(Product(Variable("b"),Variable("y")),
Product(Variable("b"),Variable("z"))))
```

And our expression, $(3x^2 + x)\ \sin(x)$, expands correctly to $3x^2 \sin(x) + x \sin(x)$:

```
>>> f_expression.expand()
Sum(Product(Product(3,Power(Variable("x"),2)),Apply(Function("sin"),Variable(
"x"))),Product(Variable("x"),Apply(Function("sin"),Variable("x"))))
```

At this point, we've written some Python functions that really do algebra for us, not just arithmetic. There are a lot of exciting applications of this type of programming (called *symbolic programming*, or more specifically, *computer algebra*), and we can't afford to cover all of them in this book. You should try your hand at a few of the following exercises and then we move on to our most important example: finding the formulas for derivatives.

10.3.4 Exercises

Exercise 10.9 Write a function `contains(expression, variable)` that checks whether the given expression contains any occurrence of the specified variable.

Solution You could easily check whether the variable appears in the result of `distinct_variables`, but here's the implementation from scratch:

```
def contains(exp, var):
    if isinstance(exp, Variable):
        return exp.symbol == var.symbol
    elif isinstance(exp, Number):
        return False
    elif isinstance(exp, Sum):
        return any([contains(e,var) for e in exp.exps])
    elif isinstance(exp, Product):
        return contains(exp.exp1,var) or contains(exp.exp2,var)
    elif isinstance(exp, Power):
        return contains(exp.base, var) or contains(exp.exponent, var)
    elif isinstance(exp, Apply):
        return contains(exp.argument, var)
    else:
        raise TypeError("Not a valid expression.")
```

Exercise 10.10 Write a `distinct_functions` function that takes an expression as an argument and returns the distinct, named functions (like sin or ln) that appear in the expression.

Solution The implementation looks a lot like the `distinct_variables` function from section 10.3.1:

```
def distinct_functions(exp):
    if isinstance(exp, Variable):
        return set()
    elif isinstance(exp, Number):
        return set()
    elif isinstance(exp, Sum):
        return set().union(*[distinct_functions(exp) for exp in
exp.exps])
    elif isinstance(exp, Product):
        return
distinct_functions(exp.exp1).union(distinct_functions(exp.exp2))
    elif isinstance(exp, Power):
        return
distinct_functions(exp.base).union(distinct_functions(exp.exponent))
    elif isinstance(exp, Apply):
        return
set([exp.function.name]).union(distinct_functions(exp.argument))
    else:
        raise TypeError("Not a valid expression.")
```

Exercise 10.11 Write a function `contains_sum` that takes an expression and returns `True` if it contains a `Sum`, and `False` otherwise.

Solution

```
def contains_sum(exp):
    if isinstance(exp, Variable):
        return False
    elif isinstance(exp, Number):
        return False
    elif isinstance(exp, Sum):
        return True
    elif isinstance(exp, Product):
        return contains_sum(exp.exp1) or contains_sum(exp.exp2)
    elif isinstance(exp, Power):
        return contains_sum(exp.base) or contains_sum(exp.exponent)
    elif isinstance(exp, Apply):
        return contains_sum(exp.argument)
    else:
        raise TypeError("Not a valid expression.")
```

Exercise 10.12—Mini Project Write a `__repr__` method on the `Expression` classes so that they appear legibly in an interactive session.

Solution See the walk-through notebook for chapter 10 or see appendix B for a discussion of `__repr__` and other special methods on Python classes.

Exercise 10.13—Mini Project If you know how to encode equations using the LaTeX language, write a `_repr_latex_` method on the `Expression` classes that returns LaTeX code representing the given expression. You should see nicely typeset renderings of your expressions in Jupyter after adding the method:

```
In [41]:  1 Product(Power(Variable("x"),Number(2)),Apply(Function("sin"),Variable("y")))
Out[41]:
```

$x^2 \sin(y)$

Adding a `_repr_latex_` method causes Jupyter to render equations nicely in the REPL.

Solution See the walk-through notebook for chapter 10.

Exercise 10.14—Mini Project Write a method to generate the Python code representing an expression. Use the Python `eval` function to turn this into an executable Python function. Compare the result with the evaluate method. For instance, `Power(Variable("x"),Number(2))` represents the expression x^2. This should produce the Python code `x**2`. Then use Python's `eval` function to execute this code and show how it matches the result of the evaluate method.

Solution See the walk-through notebook for implementation. When complete, you can run the following:

```
>>> Power(Variable("x"),Number(2))._python_expr()
'(x) ** (2)'
>>> Power(Variable("x"),Number(2)).python_function(x=3)
9
```

10.4 Finding the derivative of a function

It might not seem obvious, but there is often a clean algebraic formula for the derivative of a function. For instance, if $f(x) = x^3$, then its derivative $f'(x)$, which measures the instantaneous rate of change in f at any point x, is given by $f'(x) = 3x^2$. If you know a formula like this, you can get an exact result such as $f'(2) = 12$ without the numerical issues associated with using small secant lines.

If you took calculus in high school or college, chances are you spent a lot of time learning and practicing how to find formulas for derivatives. It's a straightforward task that doesn't require much creativity, and it can be tedious. That's why we'll briefly spend time covering the rules and then focus on having Python do the rest of the work for us.

10.4.1 Derivatives of powers

Without knowing any calculus, you can find the derivative of a linear function of the form $f(x) = mx + b$. The slope of any secant on this line, no matter how small, is the same as the slope of the line m; therefore, $f'(x)$ doesn't depend on x. Specifically, we can say $f'(x) = m$. This makes sense: a linear function $f(x)$ changes at a constant rate with respect to its input x, so its derivative is a constant function. Also, the constant b has no effect on the slope of the line, so it doesn't appear in the derivative (figure 10.12).

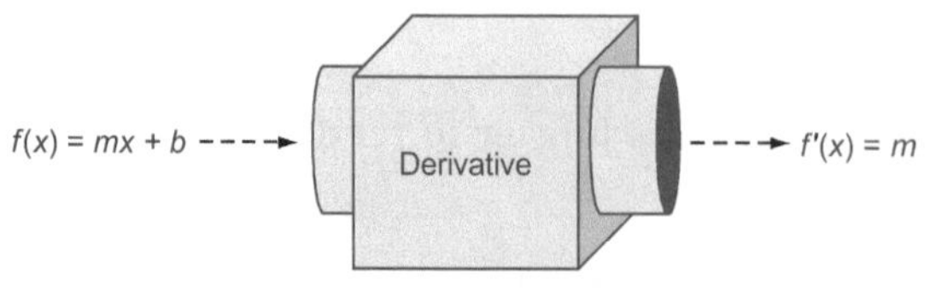

Figure 10.12 The derivative of a linear function is a constant function.

It turns out the derivative of a quadratic function is a linear function. For instance, $q(x) = x^2$ has derivative $q'(x) = 2x$. This also makes sense if you plot the graph of $q(x)$. The slope of $q(x)$ starts negative, increases, and eventually becomes positive after $x = 0$. The function $q'(x) = 2x$ agrees with this qualitative description.

As another example, I showed you that x^3 has derivative $3x^2$. All of these facts are special cases of a general rule: when you take the derivative of a function $f(x)$, which is a power of x, you get back a function that is *one lower* power. Specifically, figure 10.13 shows the derivative of a function of the form ax^n is nax^{n-1}.

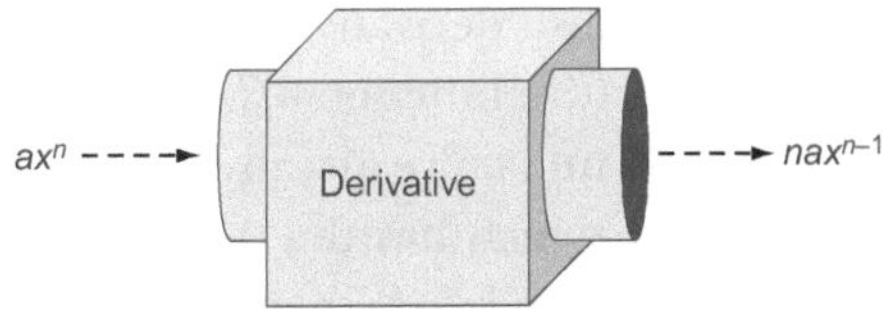

Figure 10.13 A general rule for derivatives of powers: taking the derivative of a function $f(x)$, a power of x, returns a function that is one power lower.

Let's break this down for a specific example. If $g(x) = 5x^4$, then this has the form ax^n with $a = 5$ and $n = 4$. The derivative is nax^{n-1}, which becomes $4 \cdot 5 \cdot x^{4-1} = 20x^3$. Like any other derivative we've covered in this chapter, you can double-check this by plotting it alongside the result from our numerical derivative function from chapter 9. The graphs should coincide exactly.

A linear function like $f(x)$ is a power of x: $f(x) = mx^1$. The power rule applies here as well: mx^1 has a derivative $1 \cdot mx^0$ because $x^0 = 1$. By geometric considerations, adding a constant b does not change the derivative; it moves the graph up and down, but it doesn't change the slope.

10.4.2 Derivatives of transformed functions

Adding a constant to a function never changes its derivative. For instance, the derivative of x^{100} is $100x^{99}$, and the derivative of $x^{100} - \pi$ is also $100x^{99}$. But some modifications of a function *do* change the derivative. For example, if you put a negative sign in front of a function, the graph flips upside down and so does the graph of any secant line. If the slope of the secant line is m before the flip, it is $-m$ after; the change in x is the same as before, but the change in $y = f(x)$ is now in the opposite direction (figure 10.14).

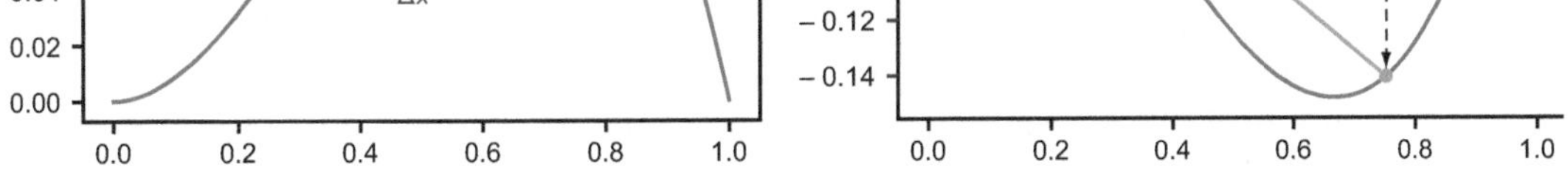

Figure 10.14 For any secant line on f(x), the secant line on the same x interval of −f(x) has the opposite slope.

Because derivatives are determined by the slopes of secant lines, the derivative of a negative function $-f(-x)$ is equal to the negative derivative $-f'(x)$. This agrees with the formula we've already seen: if $f(x) = -5x^2$ then $a = -5$ and $f'(x) = -10x$ (as compared to $5x^2$, which has the derivative $+10x$). Another way to put this is that if you multiply a function by –1, then its derivative is multiplied by –1 as well.

The same turns out to be true for any constant. If you multiply $f(x)$ by 4 to get $4f(x)$, figure 10.15 shows that this new function is four times steeper at every point and, therefore, its derivative is $4f'(x)$.

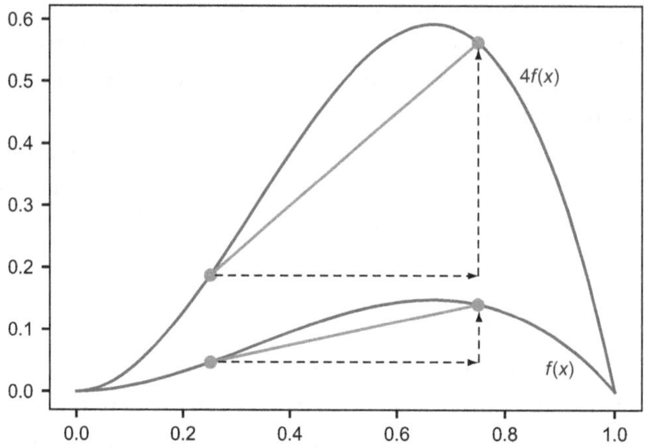

Figure 10.15 Multiplying a function by 4 makes every secant line four times steeper.

This agrees with the power rule for derivatives I showed you. Knowing the derivative of x^2 is $2x$, you also know that the derivative of $10x^2$ is $20x$, the derivative of $-3x^2$ is $-6x$, and so on. We haven't covered it yet, but if I tell you the derivative of $\sin(x)$ is $\cos(x)$, you'll know right away that the derivative of $1.5 \cdot \sin(x)$ is $1.5 \cdot \cos(x)$.

A final transformation that's important is adding two functions together. If you look at the graph of $f(x) + g(x)$ for any pair of functions f and g in figure 10.16, the vertical change for any secant line is the sum of the vertical changes in f and g on that interval.

When we're working with formulas, we can take the derivative of each term in a sum independently. If we know that the derivative of x^2 is $2x$, and the derivative of x^3 is $3x^2$, then the derivative of $x^2 + x^3$ is $2x + 3x^2$. This rule gives a more precise reason why the derivative of $mx + b$ is m; the derivatives of the terms are m and 0, respectively, so the derivative of the whole formula is $m + 0 = m$.

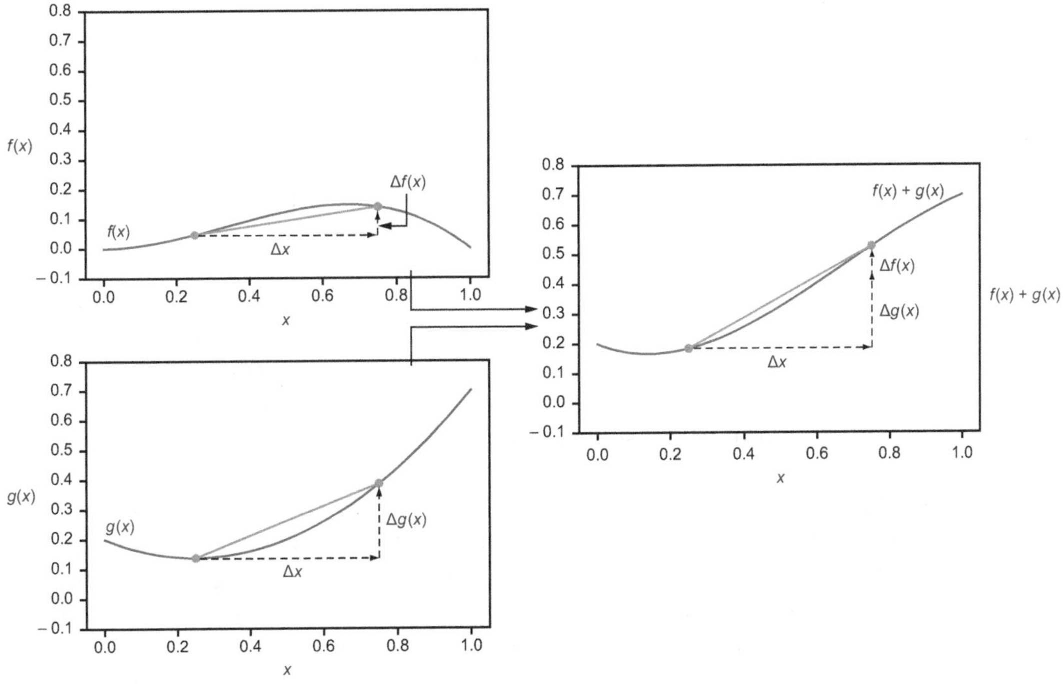

Figure 10.16 The vertical change in $f(x)$ on some x interval is the sum of the vertical change in $f(x)$ and in $g(x)$ on that interval.

10.4.3 Derivatives of some special functions

There are plenty of functions that can't be written in the form *axn* or even as a sum of terms of this form. For example, trigonometric functions, exponential functions, and logarithms all need to be covered separately. In a calculus class, you learn how to figure out the derivatives of these functions from scratch, but that's beyond the scope of this book. My goal is to show you how to take derivatives so that when you meet them in the wild, you'll be able to solve the problem at hand. To that end, I give you a quick list of some other important derivative rules (table 10.1).

Table 10.1 Some basic derivatives

Function name	Formula	Derivative
Sine	$\sin(x)$	$\cos(x)$
Cosine	$\cos(x)$	$-\sin(x)$
Exponential	e^x	e^x
Exponential (any base)	a^x	$\ln(a) \cdot a^x$

Table 10.1 Some basic derivatives *(continued)*

Function name	Formula	Derivative
Natural logarithm	ln(x)	$\frac{1}{x}$
Logarithm (any base)	$\log_a x$	$\frac{1}{\ln(a) \cdot x}$

You can use this table along with the previous rules to figure out more complicated derivatives. For instance, let $f(x) = 6x + 2\sin(x) + 5e^x$. The derivative of the first term is 6, by the power rule from section 10.4.1. The second term contains $\sin(x)$, whose derivative is $\cos(x)$, and the factor of two doubles the result, giving us $2\cos(x)$. Finally, e^x is its own derivative (a very special case!), so the derivative of $5e^x$ is $5e^x$. All together the derivative is $f'(x) = 6 + 2\cos(x) + 5e^x$.

You have to be careful to *only* use the rules we've covered so far: the power law (section 10.4.1), the rules in the table 10.1, and the rules for sums and scalar multiples. If your function is $g(x) = \sin(\sin(x))$, you might be tempted to write $g'(x) = \cos(\cos(x))$, substituting in the derivative for sine in both of its appearances. But this is not correct! Nor can you infer that the derivative of the product $e^x \cos(x)$ is $-e^x \sin(x)$. When functions are combined in other ways than addition and subtraction, we need new rules to take their derivatives.

10.4.4 Derivatives of products and compositions

Let's look at a product like $f(x) = x^2 \sin(x)$. This function can be written as a product of two other functions: $f(x) = g(x) \cdot h(x)$, where $g(x) = x^2$ and $h(x) = \sin(x)$. As I just warned you, $f'(x)$ is *not* equal to $g'(x) \cdot h'(x)$ here. Fortunately, there's another formula that is true, and it's called the *product rule* for derivatives.

THE PRODUCT RULE If $f(x)$ can be written as the product of two other functions g and h as in $f(x) = g(x) \cdot h(x)$, then the derivative of $f(x)$ is given by:

$$f'(x) = g'(x) \cdot h(x) + g(x) \cdot h'(x)$$

Let's practice applying this rule to $f(x) = x^2 \sin(x)$. In this case, $g(x) = x^2$ and $h(x) = \sin(x)$, so $g'(x) = 2x$ and $h'(x) = \cos(x)$ as I showed you previously. Plugging these into the product rule formula $f'(x) = g'(x) \cdot h(x) + g(x) \cdot h'(x)$, we get $f'(x) = 2x\sin(x) + x^2\cos(x)$. That's all there is to it!

You can see that this product rule is compatible with the power rule from section 10.4.1. If you rewrite x^2 as the product of $x \cdot x$, the product rule tells you its derivative is $1 \cdot x + x \cdot 1 = 2x$.

Another important rule tells us how to take derivatives of composed functions like $\ln(\cos(x))$. This function has the form $f(x) = g(h(x))$, where $g(x) = \ln(x)$ and $h(x) = \cos(x)$. We can't just plug in the derivatives where we see the functions, getting

$-1/\sin(x)$; the answer is a bit more complicated. The formula for the derivative of a function of the form $f(x) = g(h(x))$ is called the *chain rule.*

THE CHAIN RULE If $f(x)$ is a composition of two functions, meaning it can be written in the form $f(x) = g(h(x))$ for some functions g and h, then the derivative of f is given by:

$$f'(x) = h'(x) \cdot g'(h(x))$$

In our case, $g'(x) = 1/x$ and $h'(x) = -\sin(x)$ both read from table 10.1. Then plugging into the chain rule formula, we get the result:

$$f'(x) = h'(x) \cdot g'(h(x)) = -\sin(x) \cdot \frac{1}{\cos(x)} = -\frac{\sin(x)}{\cos(x)}$$

You might remember that $\sin(x)/\cos(x) = \tan(x)$, so we could write even more concisely that the derivative of $\ln(\cos(x)) = \tan(x)$. I'll give you a few more opportunities to practice the product and chain rule in the exercises, and you can also turn to any calculus book for abundant examples of calculating derivatives. You don't need to take my word for these derivative rules; you should get a result that looks the same if you find a formula for the derivative or if you use the derivative function from chapter 9. In the next section, I'll show you how to turn the rules for derivatives into code.

10.4.5 Exercises

Exercise 10.15 Show that the derivative of $f(x) = x^5$ is indeed $f'(x) = 5x^4$ by plotting the numerical derivative (using the derivative function from chapter 8) alongside the symbolic derivative $f'(x) = 5x^4$.

Solution

```
def p(x):
    return x**5
plot_function(derivative(p), 0, 1)
plot_function(lambda x: 5*x**4, 0, 1)
```

The two graphs overlap exactly.

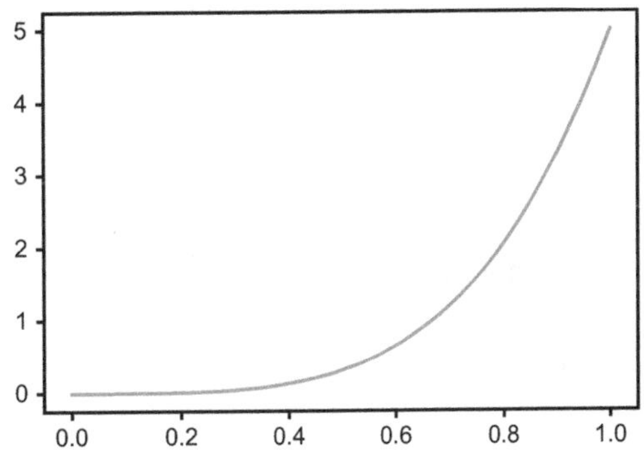

The graph of $5x^4$ and the (numerical) derivative of x^5

Exercise 10.16—Mini Project Let's think again of the functions of one variable as a vector space as we did in chapter 6. Explain why the rules for taking derivatives mean the derivative is a linear transformation of this vector space. (To be specific, you have to restrict your attention to the functions that have derivatives everywhere.)

Solution Thinking of functions f and g as vectors, we can add and multiply them by scalars. Remember that $(f + g)(x) = f(x) + g(x)$ and $(c \cdot f)(x) = c \cdot f(x)$. A *linear transformation* is one that preserves vector sums and scalar multiples.

If we write the derivative as a function D, we can think of it as taking a function as an input and returning its derivative as an output. For instance, $Df = f'$. The derivative of a sum of two functions is the sum of the derivatives

$$D(f + g) = Df + Dg$$

The derivative of a function multiplied by a number c is c times the derivative of the original function:

$$D(c \cdot f) = c \cdot Df$$

These two rules mean that D is a linear transformation. Note, in particular, that the derivative of a linear combination of functions is the same linear combination of their derivatives:

$$D(a \cdot f + b \cdot g) = a \cdot Df + b \cdot Dg$$

Exercise 10.17—Mini Project Find a formula for the derivative of a quotient: $f(x) / g(x)$.

Hint Use the fact that

$$\frac{f(x)}{g(x)} = f(x) \cdot \frac{1}{g(x)} = f(x) \cdot g(x)^{-1}$$

The power law holds for negative exponents; for instance, x^{-1} has the derivative $-x^{-2} = -1/x^2$.

Solution The derivative of $g(x)^{-1}$ is $-g(x)^{-2} \cdot g'(x)$ by the chain rule or

$$-\frac{g'(x)}{g(x)^2}$$

With this information, the derivative of the quotient $f(x)/g(x)$ is equal to the derivative of the product $f(x)/g(x)^{-1}$, which is given by the product rule:

$$f'(x)g(x)^{-1} - \frac{g'(x)}{g(x)^2}f(x) = \frac{f'(x)}{g(x)} - \frac{f(x)g'(x)}{g(x)^2}$$

Multiplying the first term by $g(x)/g(x)$ gives both terms the same denominator, so we can add them:

$$\frac{f'(x)}{g(x)} - \frac{f(x)g'(x)}{g(x)^2} = \frac{f'(x)g(x)}{g(x)^2} - \frac{f(x)g'(x)}{g(x)^2} = \frac{f'(x)g(x) - f(x)g'(x)}{g(x)^2}$$

Exercise 10.18 What is the derivative of $\sin(x) \cdot \cos(x) \cdot \ln(x)$?

Solution There are two products here, and fortunately, we can take the product rule in any order and get the same result. The derivative of $\sin(x) \cdot \cos(x)$ is $\sin(x) \cdot -\sin(x) + \cos(x) \cdot \cos(x) = \cos(x)^2 - \sin(x)^2$. The derivative of $\ln(x)$ is $1/x$, so the product rule tells us that the derivative of the whole product is

$$\ln(x)\left(\cos(x)^2 - \sin(x)^2\right) + \frac{\sin(x)\cos(x)}{x}$$

Exercise 10.19 Assume we know the derivatives of three functions f, g, and h, which are written f', g', and h'. What is the derivative of $f(g(h(x)))$ with respect to x?

Solution We need to apply the chain rule twice here. One term is $f'(g(h(x)))$, but we need to multiply it by the derivative of $g(h(x))$. That derivative is $g'(h(x))$ times the derivative of the inside function $h(x)$. Because the derivative of $g(h(x))$ is $h'(x) \cdot g'(h(x))$, the derivative of $f(g(h(x)))$ is $f'(x) \cdot g'(h(x)) \cdot f'(g(h(x)))$.

10.5 *Taking derivatives automatically*

Even though I taught you only a few rules for taking derivatives, you're now prepared to handle any of an infinite collection of possible functions. As long as a function is built from sums, products, powers, compositions, trigonometric functions, and exponential functions, you are equipped to figure out its derivative using the chain rule, product rule, and so on.

This parallels the approach we used to build algebraic expressions in Python. Even though there are infinitely many possibilities, they are all formed from the same set of building blocks and a handful of predefined ways to assemble them together. To take derivatives automatically, we need to match each case of a representable expression, be it an element or combinator, with the appropriate rule for taking its derivative. The end result is a Python function that takes one expression and returns a new expression representing its derivative.

10.5.1 Implementing a derivative method for expressions

Once again, we can implement the derivative function as a method on each of the `Expression` classes. To enforce that they all have this method, we can add an abstract method to the abstract base class:

```
class Expression(ABC):
    ...
    @abstractmethod
    def derivative(self,var):
        pass
```

The method needs to take a parameter, `var`, indicating which variable we're taking a derivative with respect to. For instance, $f(y) = y^2$ would need a derivative taken with respect to y. As a trickier example, we've worked with expressions like ax^n, where a and n represent constants and only x is the variable. From this perspective, the derivative is nax^{n-1}. However, if we think of this instead as a function of a, as in $f(a) = ax^n$, the derivative is x^{n-1}, a constant to a constant power. We get yet another result if we consider it a function of n: if $f(n) = axn$, then $f'(n) = a \ln(n)\ x^n$. To avoid confusion, we'll consider all expressions as functions of the variable x in the following discussion.

As usual, our easiest examples are the elements: `Number` and `Variable` objects. For `Number`, the derivative is always the expression 0, regardless of the variable passed in:

```
class Number(Expression):
    ...
    def derivative(self,var):
        return Number(0)
```

If we're taking the derivative of $f(x) = x$, the result is $f'(x) = 1$, which is the slope of the line. Taking the derivative of $f(x) = c$ should give us 0 as c represents a constant here, rather than the argument of the function f. For that reason, the derivative of a variable is 1 only if it's the variable we're taking the derivative with respect to; otherwise, the derivative is 0:

```
class Variable(Expression):
    ...
    def derivative(self, var):
        if self.symbol == var.symbol:
            return Number(1)
        else:
            return Number(0)
```

The easiest combinator to take derivatives of is Sum; the derivative of a `Sum` function is just the sum of the derivatives of its terms:

```
class Sum(Expression):

    ...
    def derivative(self, var):
        return Sum(*[exp.derivative(var) for exp in self.exps])
```

With these methods implemented, we can do some basic examples. For instance, the expression `Sum(Variable("x"),Variable("c"),Number(1))` represents $x + c + 1$, and thinking of that as a function of x, we can take its derivative with respect to x:

```
>>> Sum(Variable("x"),Variable("c"),Number(1)).derivative(Variable("x"))
Sum(Number(1),Number(0),Number(0))
```

This correctly reports the derivative of $x + c + 1$ with respect to x as $1 + 0 + 0$, which is equal to 1. This is a clunky way to report the result, but at least we got it right.

I encourage you to do the mini-project for writing a simplify method that gets rid of extraneous terms, like added zeros. We could add some logic to simplify expressions as we compute the derivatives, but it's better to separate our concerns and focus on getting the derivative right for now. Keeping that in mind, let's cover the rest of the combinators.

10.5.2 Implementing the product rule and chain rule

The product rule turns out to be the easiest of the remaining combinators to implement. Given the two expressions that make up a product, the derivative of the product is defined in terms of those expressions and their derivatives. Remember, if the product is $g(x) \cdot h(x)$, the derivative is $g'(x) \cdot h(x) + g(x) \cdot h'(x)$. That translates to the following code, which returns the result as the sum of two products:

```
class Product(Expression):
    ...
    def derivative(self,var):
        return Sum(
            Product(self.exp1.derivative(var), self.exp2),
            Product(self.exp1, self.exp2.derivative(var)))
```

Again, this gives us correct (albeit unsimplified) results. For instance, the derivative of cx with respect to x is

```
>>> Product(Variable("c"),Variable("x")).derivative(Variable("x"))
Sum(Product(Number(0),Variable("x")),Product(Variable("c"),Number(1)))
```

That result represents $0 \cdot x + c \cdot 1$, which is, of course, c.

Now we've got the `Sum` and `Product` combinators handled, so let's look at `Apply`. To handle a function application like $\sin(x^2)$, we need to encode both the derivative of the sine function *and* the use of the chain rule because of the x^2 inside the parentheses.

First, let's encode the derivatives of some of the special functions in terms of a placeholder variable unlikely to be confused with any we use in practice. The derivatives are stored as a dictionary mapping from function names to expressions giving their derivatives:

```
_var = Variable('placeholder variable')

_derivatives = {
    "sin": Apply(Function("cos"), _var),
    "cos": Product(Number(-1), Apply(Function("sin"), _var)),
    "ln": Quotient(Number(1), _var)
}
```

Creates a placeholder variable designed so that it's not confused with any symbol (like x or y) that we might actually use

Records that the derivative of sine is cosine, with cosine expressed as an expression using the placeholder variable

The next step is to add the `derivative` method to the `Apply` class, looking up the correct derivative from the `_derivatives` dictionary and appropriately applying the chain rule. Remember that the derivative of $g(h(x))$ is $h'(x) \cdot g'(h(x))$. If, for example, we're looking at $\sin(x^2)$, then $g(x) = \sin(x)$ and $h(x) = x^2$. We first go to the dictionary to get the derivative of sin, which we get back as cos with a placeholder value. We need to plug in $h(x) = x^2$ for the placeholder to get the $g'(h(x))$ term from the chain rule. This requires a substitute function, which replaces all instances of a variable with an expression (a mini-project from earlier in the chapter). If you didn't do that mini-project, you can see the implementation in the source code. The derivative method for `Apply` looks like this:

```
class Apply(Expression):
    ...
    def derivative(self, var):
        return Product(
                self.argument.derivative(var),
                _derivatives[self.function.name].substitute(_var,
self.argument))
```

Returns h'(x) in h'(x) · g'(h(x)) of the chain rule formula

This is the g'(h(x)) of the chain rule formula, where the _derivatives dictionary looks up g' and h(x) is substituted in.

For $\sin(x^2)$, for example, we have

```
>>> Apply(Function("sin"),Power(Variable("x"),Number(2))).derivative(x)
Product(Product(Number(2),Power(Variable("x"),Number(1))),Apply(Function("cos
"),Power(Variable("x"),Number(2))))
```

Literally, that result translates to $(2x^1) \cdot \cos(x^2)$, which is a correct application of the chain rule.

10.5.3 Implementing the power rule

The last kind of expression we need to handle is the Power combinator. There are actually three derivative rules we need to include in the `derivative` method for the

`Power` class. The first is the rule I called the power rule, which tells us that x^n has derivative nx^{n-1}, when n is a constant. The second is the derivative of the function a^x, where the base, a, is assumed to be constant while the exponent changes. This function has the derivative $\ln(a) \cdot a^x$ with respect to x.

Finally, we need to handle the chain rule here because there could be an expression involved in either the base or the exponent, like $\sin(x)^8$ or $15^{\cos(x)}$. There's yet another case where *both* the base and the exponent are variables like x^x or $\ln(x)^{\sin(x)}$. In all my years taking derivatives, I've never seen a real application where this case comes up, so I'll skip it and raise an exception instead.

Because x^n, $g(x)^n$, a^x, and $a^{g(x)}$ are all represented in Python in the form `Power(expression1, expression2)`, we have to do some checks to find out what rule to use. If the exponent is a number, we use the x^n rule, but if the base is a number, we use the a^x rule. In both cases, I use the chain rule by default. After all, x^n is a special case of $f(x)^n$, where $f(x) = x$. Here's what the code looks like:

```
class Power(Expression):
    ...
    def derivative(self,var):
        if isinstance(self.exponent, Number):
            power_rule = Product(
                    Number(self.exponent.number),
                    Power(self.base, Number(self.exponent.number - 1)))
            return Product(self.base.derivative(var),power_rule)
        elif isinstance(self.base, Number):
            exponential_rule = Product(
                Apply(Function("ln"),
                Number(self.base.number)
            ),
            self)
            return Product(
                self.exponent.derivative(var),
                exponential_rule)
        else:
            raise Exception(
            "can't take derivative of power {}".format(
            self.display()))
```

If the exponent is a number, uses the power rule

The derivative of f(x)^n is f'(x) · nf(x)$^{n-1}$, so here we multiply the factor of f'(x) according to the chain rule.

Checks if the base is a number; if so, we use the exponential rule.

Multiplies in a factor of f'(x) if we're trying to take the derivative of a$^{f(x)}$, again according to the chain rule

In the final case, where neither the base or the exponent is a number, we raise an error. With that final combinator implemented, you have a complete derivative calculator! It can handle (nearly) any expression built out of your elements and combinators. If you test it on our original expression, $(3x^2 + x)\ \sin(x)$, you'll get back the verbose, but correct, result of:

$$\left(0 \cdot x^2 + 3 \cdot 1 \cdot 2 \cdot x^1 + 1\right) \cdot \sin(x) + \left(e \cdot x^2 + x\right) \cdot 1 \cdot \cos(x)$$

This reduces to $(6x + 1)\sin(x) + (3x^2 + x)\cos(x)$ and shows a correct use of the product and the power rules. Coming into this chapter, you knew how to use Python to do arithmetic, then you learned how to have Python do algebra as well. Now, you can really say, you're doing calculus in Python too! In the final section, I'll tell you a bit about taking integrals symbolically in Python, using an off-the-shelf Python library called SymPy.

10.5.4 Exercises

Exercise 10.20 Our code already handles the case where one expression making up a product is constant, meaning a product of the form $c \cdot f(x)$ or $f(x) \cdot c$ for some expression $f(x)$. Either way, the derivative is $c \cdot f'(x)$. You don't need the second term of the product rule, which is $f(x) \cdot 0 = 0$. Update the code taking the derivative of a product to handle this case directly, rather than expanding the product rule and including a zero term.

Solution We could check whether either expression in a product is an instance of the `Number` class. The more general approach is to see whether either term of the product contains the variable we're taking the derivative with respect to. For instance, the derivative of $(3 + \sin(5^a))\, f(x)$ with respect to x doesn't require the product rule because the first term contains no appearance of x. Therefore, its derivative (with respect to x) is 0. We can use the `contains(expression, variable)` function from a previous exercise to do the check for us:

```
class Product(Expression):
    ...
    def derivative(self,var):
        if not contains(self.exp1, var):
            return Product(self.exp1, self.exp2.derivative(var))
        elif not contains(self.exp2, var):
            return Product(self.exp1.derivative(var), self.exp2)
        else:
            return Sum(
                Product(self.exp1.derivative(var), self.exp2),
                Product(self.exp1, self.exp2.derivative(var)))
```

If the first expression has no dependence on the variable, returns the first expression times the derivative of the second

Otherwise, if the second expression has no dependence on the variable, returns the derivative of the first expression times the unmodified second expression

Otherwise, uses the general form of the product rule

Exercise 10.21 Add the square root function to the dictionary of known functions and take its derivative automatically.

Hint The square root of x is equal to $x^{1/2}$.

Solution Using the power law, the derivative of the square root of x with respect to x is $\frac{1}{2} \cdot x^{-1/2}$, which can also be written as:

$$\frac{1}{2} \cdot \frac{1}{x^{1/2}} = \frac{1}{2\sqrt{x}}$$

We can encode that derivative formula as an expression like so:

```
_function_bindings = {
    ...
    "sqrt": math.sqrt
}

_derivatives = {
    ...
    "sqrt": Quotient(Number(1), Product(Number(2),
Apply(Function("sqrt"), _var))
}
```

10.6 Integrating functions symbolically

The other calculus operation we learned about in the last two chapters is integration. While a derivative takes a function and returns a function describing its rate of change, an integral does the opposite—it reconstructs a function from its rate of change.

10.6.1 Integrals as antiderivatives

For instance, when $y = x^2$, the derivative tells us that the instantaneous rate of change in y with respect to x is $2x$. If we started with $2x$, the indefinite integral answers the question: what function of x has an instantaneous rate of change equal to $2x$? For this reason, indefinite integrals are also referred to as *antiderivatives.*

One possible answer for the indefinite integral of $2x$ with respect to x is x^2, but other possibilities are $x^2 - 6$ or $x^2 + \pi$. Because the derivative is 0 for any constant term, the indefinite integral doesn't have a unique result. Remember, even if you know what a car's speedometer reads for the entire trip, it won't tell you where the car started or ended its journey. For that reason, we say that x^2 is *an* antiderivative of $2x$, but not *the* antiderivative.

If we want to talk about *the* antiderivative or *the* indefinite integral, we have to add an unspecified constant, writing something like $x^2 + C$. The C is called the constant of integration, and it has some infamy in calculus classes; it seems like a technicality, but it's important, and most teachers deduct points if students forget this.

Some integrals are obvious if you've practiced enough derivatives. For instance, the integral of $\cos(x)$ with respect to x is written

$$\int \cos(x)dx$$

And the result is $\sin(x) + C$ because for any constant C, the derivative of $\sin(x) + C$ is $\cos(x)$. If you've got the power rule fresh in your head, you might be able to solve the integral:

$$\int 3x^2 dx$$

The expression $3x^2$ is what you get if you apply the power rule to x^3, so the integral is

$$\int 3x^2 dx = x^3 + C$$

There are some harder integrals like

$$\int \tan(x) dx$$

which don't have obvious solutions. You need to invoke more than one derivative rule in reverse to find the answer. A lot of time in calculus courses is dedicated to figuring out tricky integrals like this. What makes the situation worse is that some integrals are *impossible*. Famously, the function

$$f(x) = e^{x^2}$$

is one where it's not possible to find a formula for its indefinite integral (at least without making up a new function to represent it). Rather than torture you with a bunch of rules for integration, let me show you how to use a pre-built library with an `integrate` function so Python can handle integrals for you.

10.6.2 Introducing the SymPy library

The SymPy (*Sym*bolic *Py*thon) library is an open source Python library for symbolic math. It has its own expression data structures, much like the ones we built, along with overloaded operators, making them look like ordinary Python code. Here you can see some SymPy code that looks like what we've been writing:

```
>>> from sympy import *
>>> from sympy.core.core import *
>>> Mul(Symbol('y'),Add(3,Symbol('x')))
y*(x + 3)
```

The `Mul`, `Symbol`, and `Add` constructors replace our `Product`, `Variable`, and `Sum` constructors, but have similar results. SymPy also encourages you to use shorthand; for instance,

```
>>> y = Symbol('y')
>>> x = Symbol('x')
>>> y*(3+x)
y*(x + 3)
```

creates an equivalent expression data structure. You can see that it's a data structure by our ability to substitute and take derivatives:

```
>>> y*(3+x).subs(x,1)
4*y
>>> (x**2).diff(x)
2*x
```

To be sure, SymPy is a much more robust library than the one we've built in this chapter. As you can see, the expressions are automatically simplified.

The reason I'm introducing SymPy is to show you its powerful symbolic integration function. You can find the integral of an expression like $3x^2$ like this:

```
>>> (3*x**2).integrate(x)
x**3
```

That tells us that

$$\int 3x^2 dx = x^3 + C$$

In the next few chapters, we'll continue putting derivatives and integrals to work.

10.6.3 Exercises

Exercise 10.22 What is the integral of $f(x) = 0$? Confirm your answer with SymPy, remembering that SymPy does not automatically include a constant of integration.

Solution Another way of asking this question is what function has a derivative zero? Any constant valued function has a zero slope everywhere, so it has a derivative zero. The integral is

$$\int f(x)dx = \int 0\,dx = C$$

In SymPy, the code `Integer(0)` gives you the number 0 as an expression, so the integral with respect to a variable x is

```
>>> Integer(0).integrate(x)
0
```

Zero, as a function, is one antiderivative of zero. Adding a constant of integration, we get $0 + C$ or just C, matching what we came up with. Any constant function is an antiderivative of the constant, zero function.

Exercise 10.23 What is the integral of $x\cos(x)$?

Hint Look at the derivative of $x\sin(x)$. Confirm your answer with SymPy.

Solution Let's start with the hint—the derivative of $x \sin(x)$ is $\sin(x) + x \cos(x)$ by the product rule. That's almost what we want, but for an extra $\sin(x)$ term. If we had a $-\sin(x)$ term appearing in the derivative, it would cancel this extra $\sin(x)$ out, and the derivative of $\cos(x)$ is $-\sin(x)$. That is, the derivative of $x \sin(x) + \cos(x)$ is $\sin(x) + x \cos(x) - \sin(x) = x \cos(x)$. This was the result we are looking for, so the integral is

$$\int x \cos(x) dx = x \sin(x) + \cos(x) + C$$

Our answer checks out in SymPy:

```
>>> (x*cos(x)).integrate(x)
x*sin(x) + cos(x)
```

This approach of reverse engineering the derivative as one term of a product is called *integration by parts* and is a favorite trick of calculus teachers everywhere.

Exercise 10.24 What is the integral of x^2? Confirm your answer with SymPy.

Solution If $f'(x) = x^2$ then $f(x)$ probably contains x^3 because the power law reduces powers by one. The derivative of x^3 is $3x^2$, so we want a function that gives us a third of that result. What we want is $x^3/3$, which has derivative x^2. In other words,

$$\int x^2 dx = \frac{x^3}{3} + C$$

SymPy confirms this:

```
>>> (x**2).integrate(x)
x**3/3
```

Summary

- Modeling algebraic expressions as data structures rather than as strings of code lets you write programs to answer more questions about the expressions.
- The natural way to model an algebraic expression in code is as a *tree*. The nodes of the tree can be divided into elements (variables and numbers) that are standalone expressions, and combinators (sums, products, and so on) that contain two or more expressions as subtrees.
- By recursively traversing an expression tree, you can answer questions about it, such as what variables it contains. You can also evaluate or simplify the expression, or translate it to another language.

- If you know the expression defining a function, there are a handful of rules you can apply to transform it into the expression for the derivative of the function. Among these are the product rule and the chain rule, which tell you how to take derivatives of products of expressions and compositions of functions, respectively.
- If you program the derivative rule corresponding to each combinator in your Python expression tree, you get a Python function that automatically finds expressions for derivatives.
- SymPy is a robust Python library for working with algebraic expressions in Python code. It has built-in simplification, substitution, and derivative functions. It also has a symbolic integration function that tells you the formula for the indefinite integral of a function.

11 Simulating force fields

This chapter covers

- Modeling forces like gravity using scalar and vector fields
- Calculating force vectors using the gradient
- Taking the gradient of a function in Python
- Adding gravitational force to the asteroid game
- Calculating gradients and working with vector fields in higher dimensions

There has just been a catastrophic event in the universe of our asteroid game: a black hole has appeared in the center of the screen! As a result of this new object in our game, shown in figure 11.1, the spaceship and all of the asteroids will feel a "gravitational pull" toward the middle of the screen. This makes the game even more challenging, and it gives us a new mathematical challenge as well—understanding *force fields*.

Gravity is a familiar example of a force that acts at a distance, meaning that you don't have to be touching an object to feel its gravitational pull. For instance, when you're flying on an airplane, you can still walk around normally because, even at

30,000 feet, the Earth is pulling you downward. Magnetism and static electricity are other familiar forces that act at a distance. In physics, we picture sources of these kinds of forces, like magnets or statically charged balloons, as generating an invisible force field around themselves. Anywhere in the Earth's gravitational force field, called its gravitational field, an object will feel a pull toward the Earth.

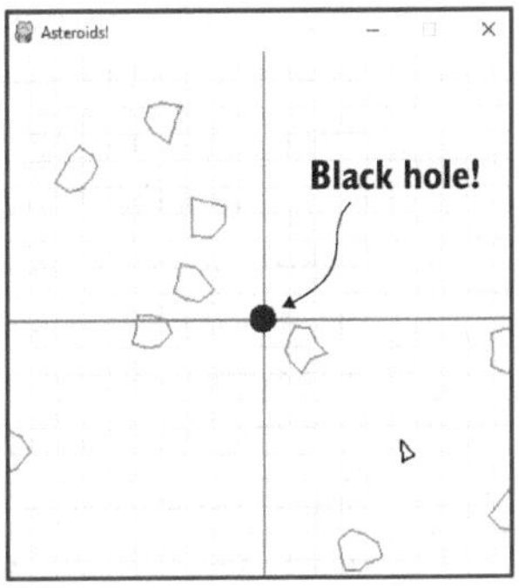

Figure 11.1 Oh no, a black hole!

Our central coding challenge for this chapter is adding a gravitational field to the asteroid game, and once we're done with that, we'll cover the math in more generality. Namely, force fields are modeled with mathematical functions called *vector fields*. Vector fields often arise as outputs from a calculus operation called the *gradient*, which is a key tool in the machine learning examples we cover in part 3.

The math and code in this chapter aren't particularly hard, but there are a lot of new concepts to get familiar with. For that reason, I want to show you the story arc for the chapter before we dig in in earnest.

11.1 Modeling gravity with a vector field

A *vector field* is an assignment of a vector at every point in space. A *gravitational field* is a vector field telling us how strongly gravity pulls and in what direction from any given point. You can visualize a vector field by picking a bunch of points and drawing the vector assigned to each point as an arrow starting from that point. For instance, the gravitational field caused by the black hole in our asteroid game might look like figure 11.2.

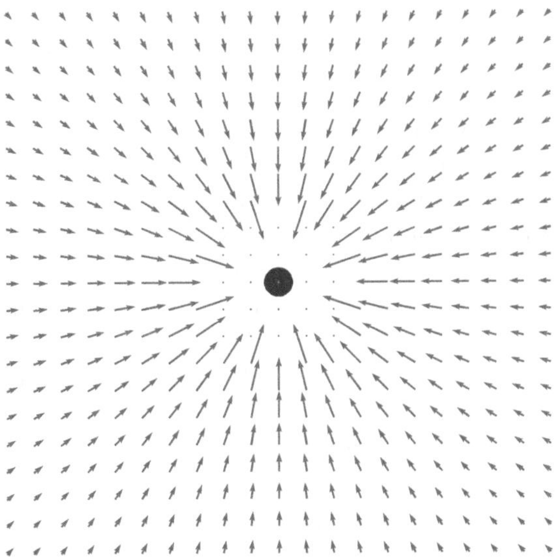

Figure 11.2 Picturing the gravitational field created by the black hole in our asteroid game

Figure 11.2 agrees with our intuition about gravity; all the arrows around the black hole point toward it, so that any object placed in this region feels pulled toward the black hole. Closer to the black hole, it's gravitational pull is stronger, so the arrows are longer.

The first thing we do in this chapter is to model gravitational fields as functions, taking a point in space to tell us the magnitude and direction of the force an object would feel at that point. In Python, a 2D vector field is a function that takes a 2D vector representing a point and returns a 2D vector that is the force at that point.

Once we build that function, we use it to add a gravitational field to our asteroid game. It will tell us what gravitational forces are felt by the spaceship and the asteroids, depending on their locations and, therefore, what their rate and direction of acceleration should be. Once we implement the acceleration, we'll see the objects in the asteroid game accelerate toward the black hole.

11.1.1 Modeling gravity with a potential energy function

After modeling the gravitational field, we'll look at a second, equivalent mental model for force at a distance, called *potential energy*. You can think of potential energy as stored energy, ready to be converted into motion. For instance, a bow and arrow have no potential energy to begin with, but when you stretch the bow it gains potential energy. When the bow is released, this energy is converted into motion (figure 11.3).

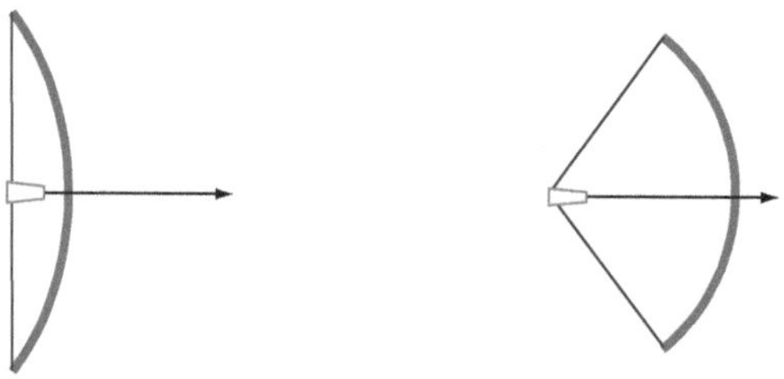

Figure 11.3 On the left, the bow has no potential energy. On the right, it has a lot of potential energy, ready to be spent to put the arrow in motion.

You can picture pulling a spaceship away from the black hole as pulling back an imaginary bow and arrow. The further you pull the spaceship from the black hole, the more potential energy it has, and the faster it ends up going after it is released. We'll model potential energy as another Python function, taking the 2D position vector of an object in the game world and returning a number measuring its potential energy at that point. An assignment of a number (instead of a vector) to every point in space is called a *scalar field*.

With a potential energy function, we'll use several Matplotlib visualizations to see what it looks like. One important example is called a *heatmap*, which uses darker and brighter colors to show how the value of a scalar field changes over a 2D space (figure 11.4).

As figure 11.4 shows, the further away from the black hole you get on this heatmap, the brighter the colors get, meaning the potential energy is greater. The scalar field representing potential energy is a different mathematical model than the vector

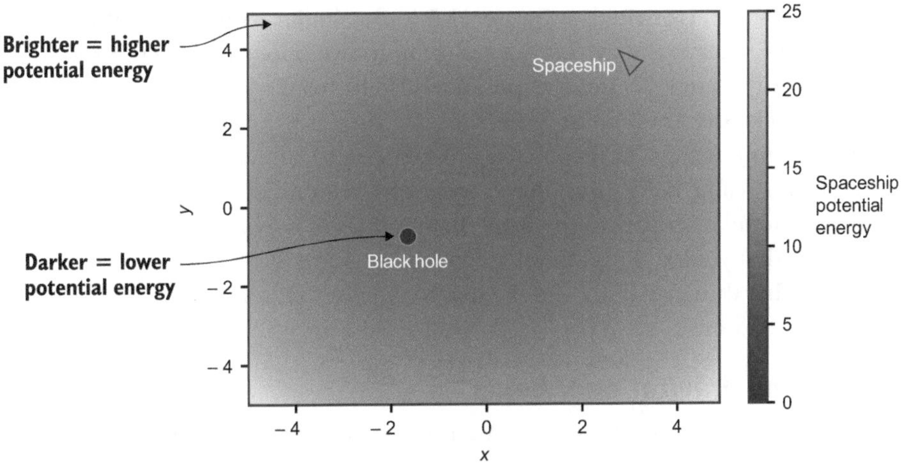

Figure 11.4 A heatmap of potential energy, using brighter colors to represent higher potential energy values

field representing gravitational force, but they represent the same physics. They are also mathematically connected by an operation called the *gradient.*

The gradient of a scalar field is a vector field, which tells us the direction and magnitude of steepest increase in the scalar field. In our example, potential energy increases as you move away from the black hole, so the gradient of potential energy is a vector field pointing outward at every point. Superimposing the gradient vector field on the potential energy heatmap, figure 11.5 shows us that arrows point in the direction that the potential energy increases.

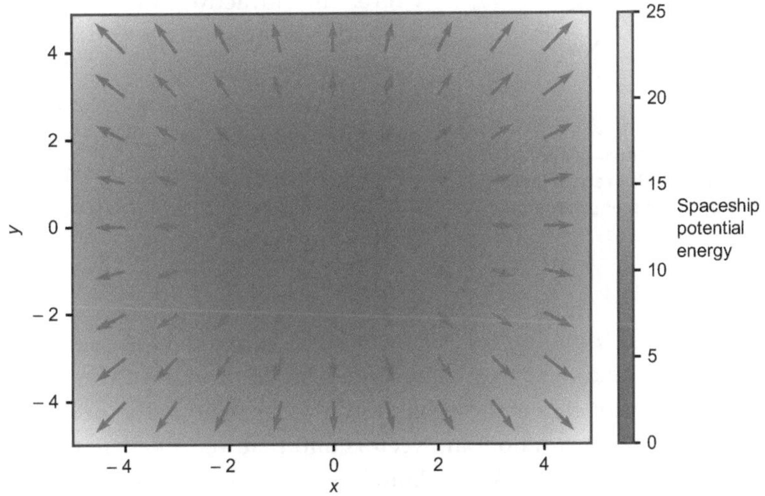

Figure 11.5 Potential energy function plotted as a heatmap with its gradient, a vector field, superimposed. The gradient points in the direction of increasing potential energy.

The gradient vector field in figure 11.5 looks similar to the gravitational field of the black hole, except the arrows are pointing in the opposite directions and the magnitudes are reversed. To get a gravitational field from a potential energy function, we need to take the gradient and then reverse the directions of the force field vectors by adding a minus sign. At the end of this chapter, I'll show you how to calculate the gradient of a scalar field using derivatives, which allow us to switch from the potential energy model of gravity to the force field model of gravity.

Now that you have a sense of where we're going in this chapter, we're ready to dig in. The first thing we'll do is look closer at vector fields and see how to turn them into Python functions.

11.2 Modeling gravitational fields

A vector field is an assignment of a vector to every point in a space—for instance, a gravitational force vector for every location in our asteroid game. We'll look exclusively at 2D vector fields that assign 2D vectors to every point in a 2D space. The first thing we'll do is build concrete representations of vector fields as Python functions taking 2D vectors as inputs and returning 2D vectors as outputs. I've given you a function, `plot_vector_field`, in the source code that takes such a function as an argument and draws a picture of it by drawing the output vectors at a large number of input points in 2D.

Then, we'll write code to add a black hole to our asteroid game. For our purposes, a black hole just means a black circle that exerts an attractive force on all objects around it as shown in figure 11.6.

To make this work, we implement a `BlackHole` class, define its corresponding gravitational field as a function and then update our game loop so that the spaceship and asteroids respond to the forces according to Newton's laws.

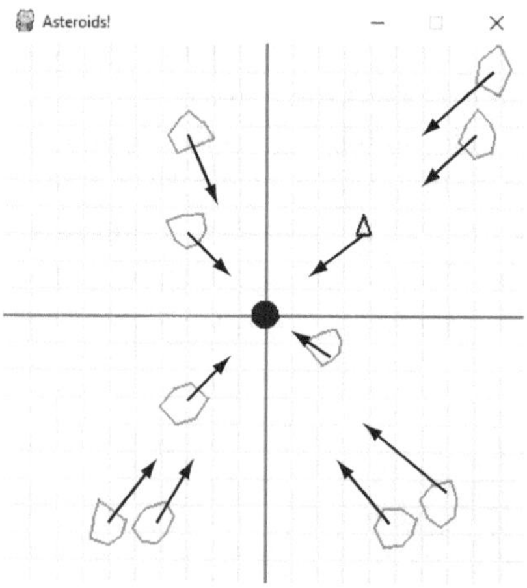

Figure 11.6 The black hole in our asteroid game is a black circle with every object in the game feeling the pull of a force toward it.

11.2.1 Defining a vector field

Let's cover a bit of basic notation for vector fields. A vector field in the 2D plane is a function $\mathbf{F}(x, y)$ that takes a vector represented by its two coordinates, x and y. It returns another 2D vector that is the value of the vector field at the point (x, y). The bold **F** signifies that its return values are vectors and that we can say that **F** is a vector-valued function. When we're talking about vector fields, we usually interpret the inputs as points in the plane and the outputs as arrows. Figure 11.7 shows a schematic for the vector field $\mathbf{F}(x, y) = (-2y, x)$.

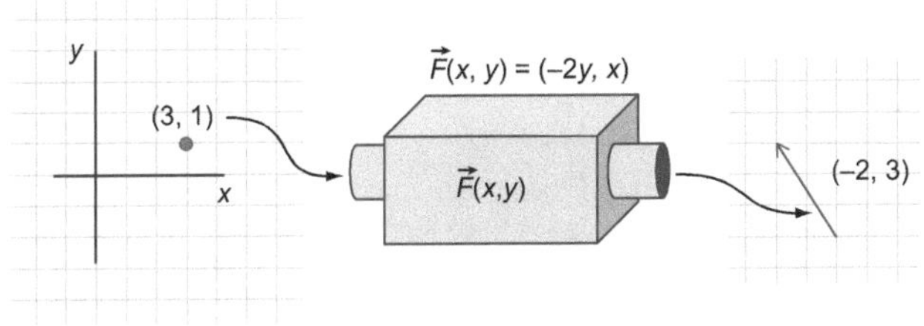

Figure 11.7 The vector field F(x, y) = (–2y, x) takes the point (3, 1) as input and produces the arrow (–2, 3) as output.

It's usual to draw the output vector as an arrow starting from the point in the plane that was the input vector, so that the output vector is "attached" to the input point (figure 11.8).

If you calculate several values of **F**, you can start to picture the vector field by drawing a number of arrows attached to points at once. Figure 11.9 shows three more points, (–2, 2), (–1, –2), and (–1, –2) with arrows attached to them representing the values of **F** at the points. The results are (–4, –2), (4, –1), and (4, 3), respectively.

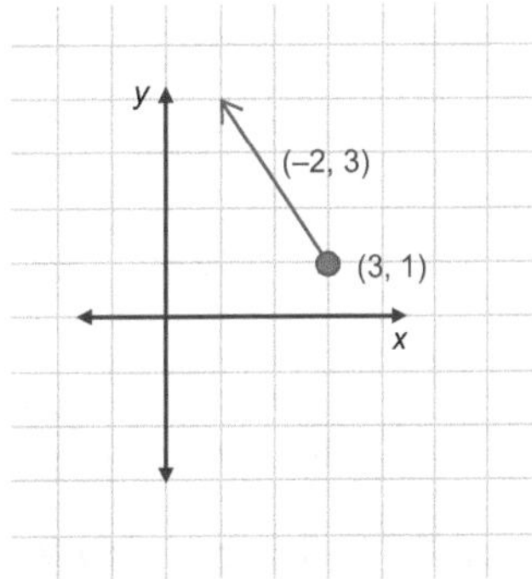

Figure 11.8 Attaching the vector (–2, 3) to the point (3, 1)

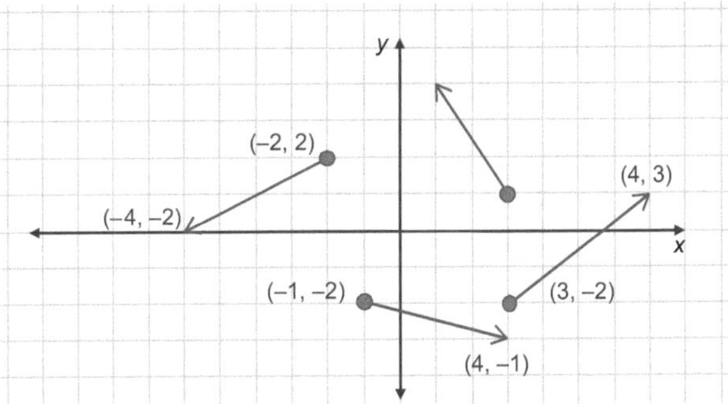

Figure 11.9 Arrows attached to points, representing more values of the vector field F(x, y) = (–2y, x)

If we draw a lot more arrows, they start to overlap and the diagram becomes illegible. To avoid this, we typically scale down the lengths of vectors by a constant factor. I've included a wrapper function called `plot_vector_field` on top of Matplotlib, and you can use it as follows to generate a visualization of a vector field. You can see that the vector field **F**(x, y) circulates in a counterclockwise direction around the origin (figure 11.10):

```
def f(x,y):
    return (-2*y, x)

plot_vector_field(f, -5,5,-5,5)
```

The first argument is the vector field; the next arguments are the x bounds followed by the y bounds for the plot.

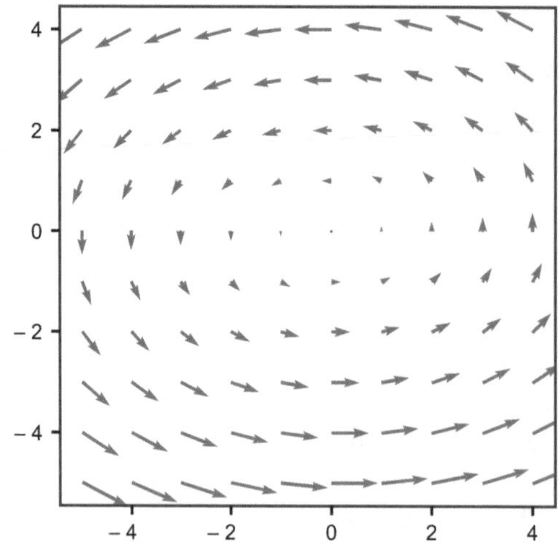

Figure 11.10 A plot of F(x, y) as vectors emanating from (x, y) points that are generated by Matplotlib

One of the big ideas of physics is how some kinds of forces are modeled by vector fields. The example we focus on next is a simplified model of gravity.

11.2.2 Defining a simple force field

As you might expect, gravitational forces get stronger as you get closer to their sources. Even though the sun has stronger gravity than the Earth, you are much closer to the Earth, so you only feel the Earth's gravity. For simplicity, we won't use a realistic gravitational field. Instead, we'll use the vector field $\mathbf{F}(\mathbf{r}) = -\mathbf{r}$, which is $\mathbf{F}(x, y) = (-x, -y)$ in the plane. Here's what it looks like in code, and figure 11.11 shows what it looks like on a plot:

```
def f(x,y):
    return (-x,-y)

plot_vector_field(f,-5,5,-5,5)
```

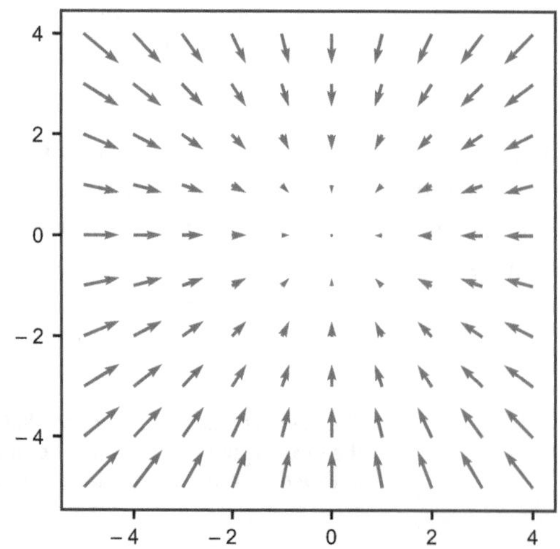

Figure 11.11 A visualization of the vector field F(x, y) = (–x, –y)

This vector field is like a gravitational field in that it points toward the origin everywhere, but it has the advantage that the field gets stronger as you go further away. That guarantees that we can't have a simulated object reach escape velocity and completely disappear from view; every wayward object eventually gets to a point where the force field is big enough to slow it down and pull it back toward the origin. Let's confirm this by implementing this gravitational field for a black hole in our asteroid game.

11.3 Adding gravity to the asteroid game

The black hole in our game is a `PolygonModel` object with 20 vertices at equal distances, so it will be approximately circular. We specify the strength of the gravitational pull of the black hole by one number, which we'll call its gravity. This number is passed to the constructor for the black hole:

```
class BlackHole(PolygonModel):
    def __init__(self,gravity):
        vs = [vectors.to_cartesian((0.5, 2 * pi * i / 20))
                for i in range(0,20)]
        super().__init__(vs)
        self.gravity = gravity #<2>
```

Defines the vertices of the BlackHole as a PolygonModel

Note that the 20 vertices in our `BlackHole` are all 0.5 units from the origin at equally spaced angles, so the black hole appears approximately circular. Adding this line

```
black_hole = BlackHole(0.1)
```

creates a `BlackHole` object with `gravity` valued at `0.1`, which is positioned at the origin by default. To make the black hole appear onscreen (figure 11.12), we need to

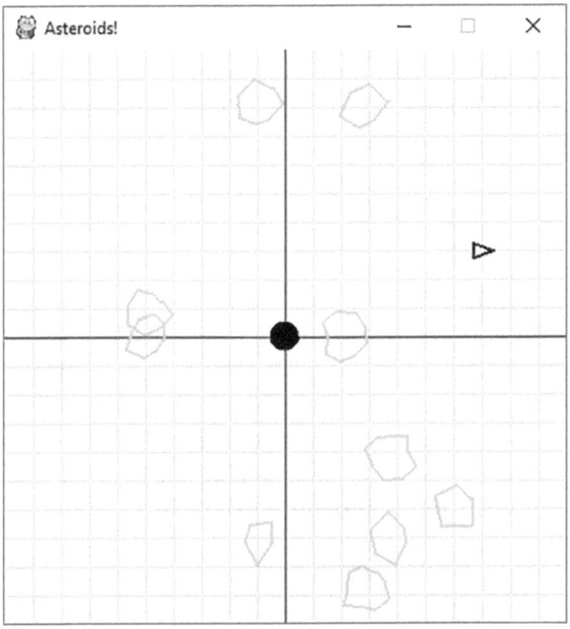

Figure 11.12 Making the black hole show up in the center of our game screen

draw it with each iteration of the game loop. In the following, I add a `fill` keyword argument to the `draw_poly` function to fill in the black hole and make it (appropriately) black:

```
draw_poly(screen, black_hole, fill=True)
```

The gravitational field created by our black hole is inspired by the vector field $\mathbf{F}(x, y) = (-x, -y)$, which points to the origin. If the black hole is centered at (x_{bh}, y_{bh}), the vector field $\mathbf{g}(x, y) = (x_{bh} - x, y_{bh} - y)$ points in the direction from (x, y) to (x_{bh}, y_{bh}). That means that as an arrow attached to the point (x, y), it points toward the center of the black hole. To make the strength of the force field scale with the gravity of the black hole, we can multiply the vectors of the vector field by the gravity value:

```
def gravitational_field(source, x, y):
    relative_position = (x - source.x, y - source.y)
    return vectors.scale(- source.gravity, relative_position)
```

In this function, `source` is a `BlackHole` object, and its `x` and `y` properties indicate its center as a `PolygonModel`, while its `gravity` property is the value passed to it in its constructor. The equivalent force field in mathematical notation would be written like this:

$$\mathbf{g}(x, y) = G_{bh} \cdot (x - x_{bh}, y - y_{bh})$$

Here, G_{bh} represents the made-up `gravity` property of the black hole, and (x_{bh}, y_{bh}), once again, represents its position. The next step is to use this gravitational field to decide how objects should move.

11.3.1 Making game objects feel gravity

If this vector field works like a gravitational field, it tells us the force per unit mass on an object at position (x, y). In other words, the force on an object of mass m will be $\mathbf{F}(x, y) = m \cdot \mathbf{g}(x, y)$. If this is the only force our object feels, we can figure out its acceleration using Newton's second law of motion:

$$\mathbf{a} = \frac{\mathbf{F}_{\text{net}}(x, y)}{m} = \frac{m \cdot \mathbf{g}(x, y)}{m}$$

This last expression for acceleration has the mass m in both the numerator and denominator, so they cancel out. It turns out that the gravitational field vector is equal to the acceleration vector caused by gravity—it has nothing to do with the object's mass. This calculation works for real gravitational fields as well, and it is why objects of different masses all fall at the same rate of about 9.81 meters per second near the Earth's surface. In one iteration of the game loop taking an elapsed time Δt, the change in velocity of a spaceship or asteroid is determined by its (x, y) position as

$$\Delta \mathbf{v} = \mathbf{a} \cdot \Delta t = \mathbf{g}(x, y) \cdot \Delta t$$

We need to add some code to update the velocity of the spaceship as well as each asteroid in each iteration of the game loop. There are a few choices as to how to organize our code, and the one I'll choose is to encapsulate all of the physics into the `move` method of `PolygonModel` objects. You might also remember that instead of having objects fly off the screen, we teleported them to the other side. Another small change I've made here is to add a global `bounce` flag that says whether objects teleport or simply bounce off the sides of the screen. I've done this because if objects teleport, they instantly feel a different gravitational field; if they bounce instead, we get more intuitive physics. Here's the new `move` method:

```
def move(self, milliseconds,
         thrust_vector, gravity_source):
    tx, ty = thrust_vector
    gx, gy = gravitational_field(src, self.x, self.y)
    ax = tx + gx
    ay = ty + gy
    self.vx += ax * milliseconds/1000
    self.vy += ay * milliseconds/1000

    self.x += self.vx * milliseconds / 1000.0
    self.y += self.vy * milliseconds / 1000.0

    if bounce:
        if self.x < -10 or self.x > 10:
            self.vx = - self.vx
        if self.y < -10 or self.y > 10:
            self.vy = - self.vy
    else:
        if self.x < -10:
            self.x += 20
        if self.y < -10:
            self.y += 20
        if self.x > 10:
            self.x -= 20
        if self.y > 10:
            self.y -=20
```

Passes in the thrust vector, which can be (0,0), and the gravity source (black hole) as parameters to the move method

Here the net force is the sum of the thrust vector and gravitational force vector. Assume the mass = 1 and the acceleration is the sum of the thrust and gravitational field.

Updates the velocity as before, using $\Delta v = a*\Delta t$

Updates the position vector as before, using $\Delta s = v*\Delta t$

If the global bounce flag is true, flips the x component of velocity when the object is about to leave the screen on the left or right, or flips the y component of velocity when the object is about to leave the screen through the top or bottom

Otherwise, uses the same teleportation effect as before when the object is about to leave the screen

The remaining work is to call this `move` method for the spaceship as well as each asteroid in the game loop:

```
while not done:
    ...
    for ast in asteroids:
        ast.move(milliseconds, (0,0), black_hole)

    thrust_vector = (0,0)
```

For each asteroid, calls its move method with a thrust vector of 0

The ship's thrust vector is also (0,0) by default.

```
if keys[pygame.K_UP]:
    thrust_vector=vectors.to_cartesian((thrust, ship.rotation_angle))

elif keys[pygame.K_DOWN]:
    thrust_vector=vectors.to_cartesian((-thrust, ship.rotation_angle))

ship.move(milliseconds, thrust_vector, black_hole)
```

Calls the move method for the spaceship to cause it to move

If the up or down arrow is pressed, calculates the thrust_vector using the direction of the spaceship and the fixed thrust scalar value

Running the game now, you'll see that objects start to be attracted by the black hole, and starting with zero velocity, the spaceship falls straight into it! Figure 11.13 shows a time-lapse photo of the ship's acceleration.

With any other starting velocity and no thrust, the ship begins to orbit the black hole, tracing out the shape of an ellipse, or stretched out circle (figure 11.14).

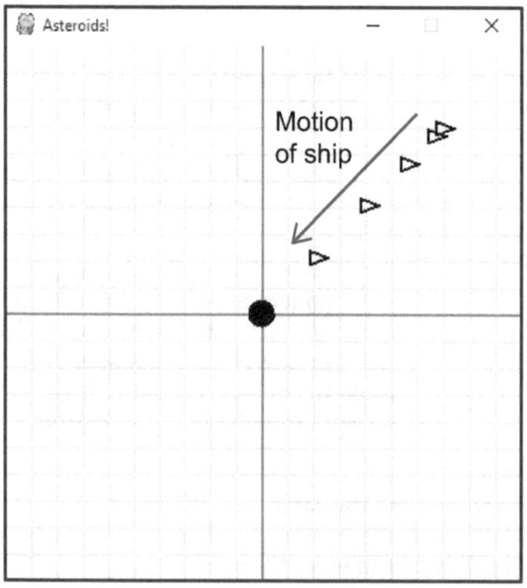

Figure 11.13 With no initial velocity, the spaceship falls into the black hole.

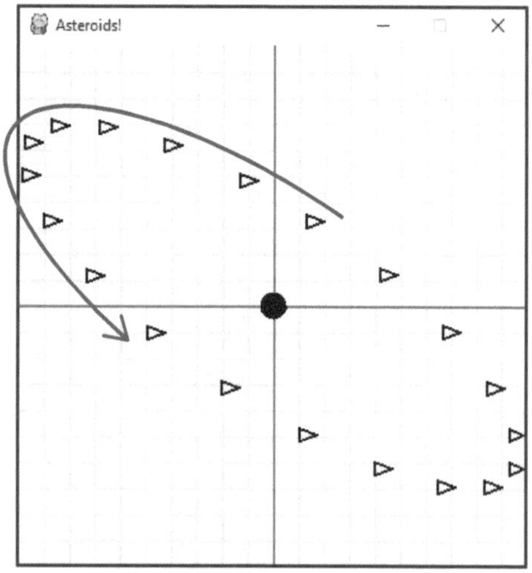

Figure 11.14 With some initial velocity perpendicular to the black hole, the spaceship begins an elliptical orbit.

It turns out that any object feeling no forces other than the gravity of the black hole either fall right in or enter into an elliptical orbit. Figure 11.15 shows a randomly initialized asteroid along with the spaceship. You can see that its trajectory is a different ellipse.

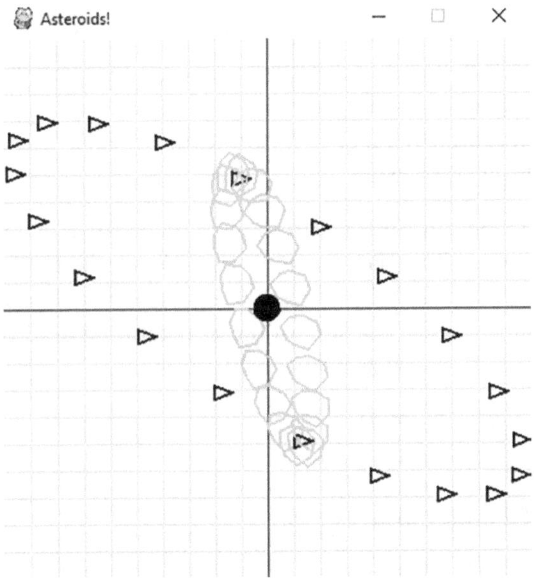

Figure 11.15 An asteroid in another elliptical orbit around our black hole

You can try adding all of the asteroids back in. You'll see that with 11 simultaneously accelerating objects, the game has become much more interesting!

11.3.2 Exercises

Exercise 11.1 Where do all of the vectors in the vector field (–2 – *x*, 4 – *y*) point? Plot the vector field to confirm your answer.

Solution This vector field is the same as the displacement vector (–2, 4) – (*x*, *y*), which is a vector pointing from a point (*x*, *y*) to (–2, 4). Therefore, we expect every vector in this vector field to point toward (–2, 4). Drawing this vector field confirms this.

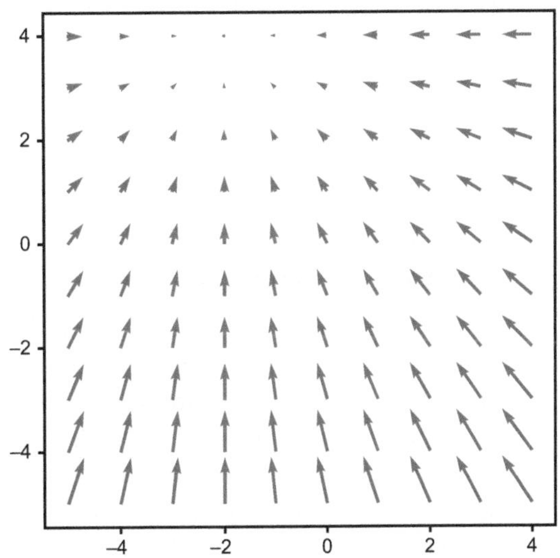

Exercise 11.2—Mini Project Suppose we have *two* black holes, both having gravity 0.1, and positioned at (–3, 4) and (2, 1), respectively. The gravitational fields are $\mathbf{g}_1(x, y) = 0.1 \cdot (-3 - x, 4 - y)$ and $\mathbf{g}_2(x, y) = 0.1 \cdot (2 - x, 1 - y)$. Calculate a formula for the total gravitational field $\mathbf{g}(x, y)$ due to both black holes. Is it equivalent to a single black hole? If so, why?

Solution At every position (x, y), an object with mass m feels two gravitational forces: $m \cdot \mathbf{g}_1(x, y)$ and $m \cdot \mathbf{g}_2(x, y)$. The vector sum of these forces is $m(\mathbf{g}_1(x, y) + \mathbf{g}_2(x, y))$. Per unit of mass, the force felt will be $\mathbf{g}_1(x, y) + \mathbf{g}_2(x, y)$, which confirms that the total gravitational field vector is the sum of the gravitational field vectors due to each of the black holes. This total gravitational field is

$$\begin{aligned}\mathbf{g}(x, y) &= \mathbf{g}_1(x, y) + \mathbf{g}_2(x, y) \\ &= 0.1 \cdot (-3 - x, 4 - y) + 0.1 \cdot (2 - x, 1 - y)\end{aligned}$$

We can divide out a factor of 2 and rewrite as

$$\begin{aligned}\mathbf{g}(x, y) &= 0.1 \cdot 2 \cdot (0.5 - x, 2.5 - y) \\ &= 0.2 \cdot (0.5 - x, 2.5 - y)\end{aligned}$$

This is the same as a single black hole with gravity 0.2, positioned at (0.5, 2.5).

Exercise 11.3—Mini Project In the asteroid game, add two black holes and allow these to feel each other's gravity. Then move while these both also exert gravity on the asteroids and the spaceship.

Solution For a full implementation, see the source code. The key addition calls the `move` method for each black hole in each iteration of the game loop, passing it the list of all other black holes as sources of gravity:

```
for bh in black_holes:
    others = [other for other in black_holes if other != bh]
    bh.move(milliseconds, (0,0), others)
```

11.4 Introducing potential energy

Now that we've seen the behavior of the spaceship and asteroids in our gravitational field, we can build our second model of how they behave using *potential energy*. We already have black holes working in the asteroid game, so the purpose of the rest of the chapter is to broaden your perspective on the underlying math. Vector fields, including gravitational fields, often arise as results of the calculus operation called the gradient, which is a critical tool in the remaining chapters of the book.

The basic idea is as follows: instead of picturing gravity as a force vector at every point, pulling objects toward a source, we can think of objects in a gravitational field like marbles rolling around in a bowl. The marbles may roll back and forth, but they are always "pulled" back toward the bottom of the bowl as they roll away. A potential energy function essentially defines the shape of this bowl. You can see a preview of what the bowl looks like in the center image of figure 11.16.

We'll write potential energy as a function, taking a point (x, y) and returning a single number, representing the gravitational potential energy at the point (x, y). In terms of the bowl analogy, this is something like the height of the bowl at a given point. Once we implement a potential energy function in Python, we can visualize it in three ways: as a heatmap, which you saw at the beginning of this chapter; as a 3D graph; and as a contour map as shown in figure 11.16.

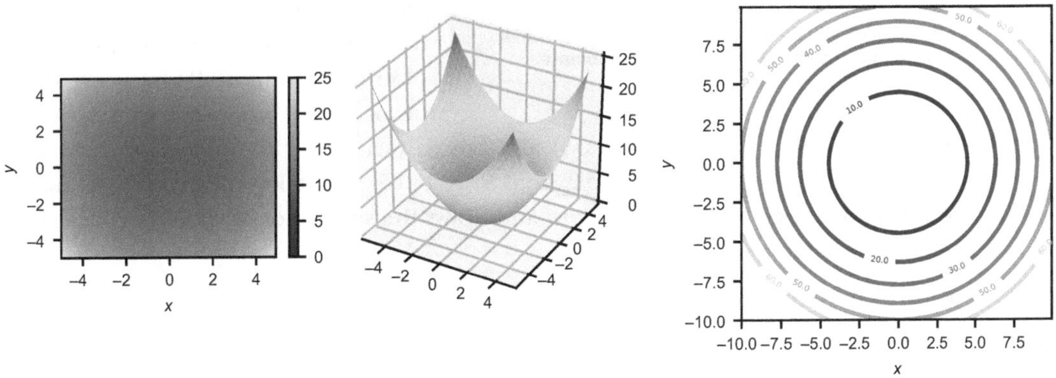

Figure 11.16 Three pictures of a scalar field: a heatmap, a graph, and a contour map

These visualizations will help us picture potential energy functions in the final section of this chapter and in the remaining chapters of the book.

11.4.1 Defining a potential energy scalar field

Like a vector field, we can think of a scalar field as a function that takes (x, y) points as inputs. Instead of vectors, however, the outputs of this function are scalars. For instance, let's work with the function $U(x, y) = ½(x^2 + y^2)$, which defines a scalar field. Figure 11.17 shows that you can plug in a 2D vector, and the output is some scalar determined by the formula for $U(x, y)$.

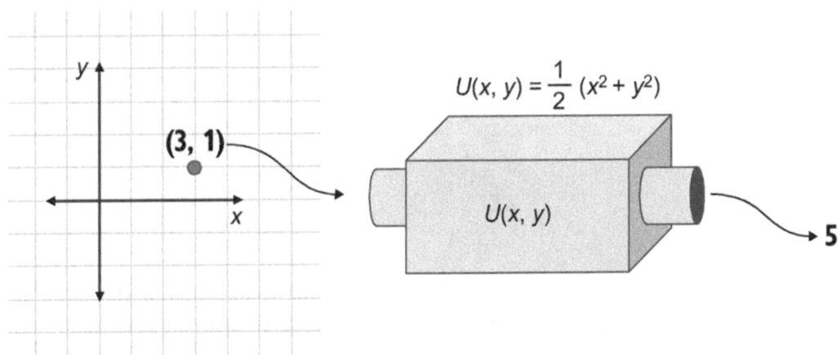

Figure 11.17 As a function, a scalar field takes a point in the plane and produces a corresponding number. In this case, where $(x, y) = (3, 1)$, the value of $U(x, y)$ is $½ \cdot (3^2 + 1^2) = 5$.

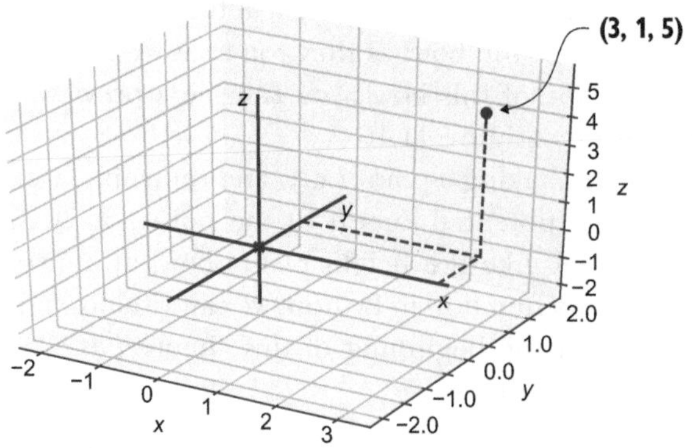

Figure 11.18 To plot one point of $U(x, y) = ½(x^2 + y^2)$, use $(x, y) = (3, 1)$, then use $U(3, 1) = 5$ as the z-coordinate.

The function $U(x, y)$ turns out to be the potential energy function corresponding to the vector field $\mathbf{F}(x, y) = (-x, -y)$. I'll need to do some more work to explain this mathematically, but we can confirm it qualitatively by picturing the scalar field $U(x, y)$.

One way to picture $U(x, y)$ is to draw a 3D plot like figure 11.18, where $U(x, y)$ is a surface of (x, y, z) points and where $z = U(x, y)$. For instance, $U(3, 1) = 5$, so we would plot a first point above the point (3, 1) in the x,y plane at a z-coordinate of 5.

Plotting a point in 3D for every single value of (x, y) gives us a whole surface representing the scalar field $U(x, y)$ and how it varies over the plane. In the source code, you'll find a function, `plot_scalar_field`, that takes a function defining a scalar field, as well as *x* and *y* bounds, and draws the surface of 3D points representing the field:

```
def u(x,y):
    return 0.5 * (x**2 + y**2)

plot_scalar_field(u, -5, 5, -5, 5)
```

While there are several ways to visualize a scalar field, I'll refer to the graph of the function $U(x, y)$ as shown in figure 11.19.

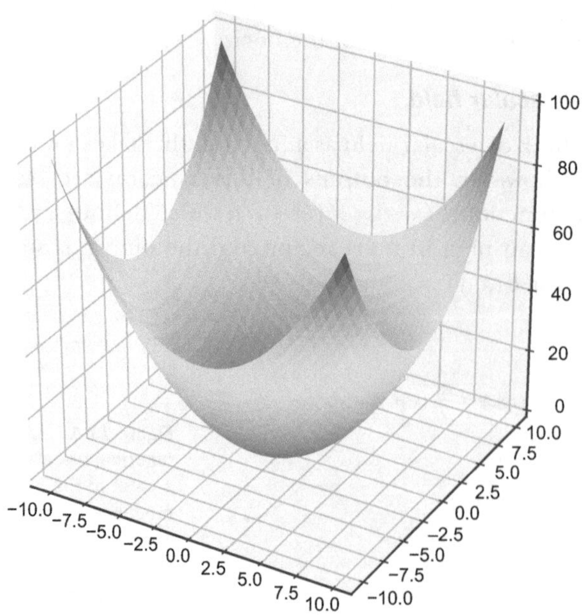

Figure 11.19 A graph of the potential energy scalar field $U(x, y) = ½(x^2 + y^2)$

This is the "bowl" from the previous analogy. It turns out this potential energy function gives the same model of gravity as the vector field $\mathbf{F}(x, y) = (-x, -y)$. We'll see exactly why this is in section 11.5, but for now, we can confirm that the potential energy increases as the distance from the origin (0, 0) increases. In all radial directions, the height of the graph increases, meaning the value of U increases.

11.4.2 Plotting a scalar field as a heatmap

Another way to visualize a scalar function is to draw a heatmap. Instead of using a z-coordinate to visualize the value of $U(x, y)$, we can use a color scheme. This allows us to plot the scalar field without leaving 2D. By including a color legend on the side (as in figure 11.20), we can see the approximate scalar value at (x, y) from the color at that point on the plot.

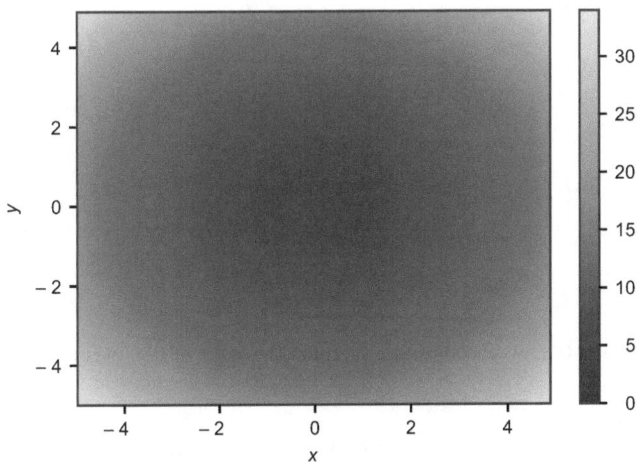

Figure 11.20 A heatmap of the function *U*(*x*, *y*)

In the center of the plot in figure 11.20, near (0, 0), the color is darker, meaning the values of $U(x, y)$ are lower. Toward the edges, the color is lighter, meaning the values of $U(x, y)$ are higher. You can plot the potential energy function using the `scalar_field_heatmap` function you'll find in the source code.

11.4.3 Plotting a scalar field as a contour map

Similar to a heatmap is a *contour map*. You may have seen a contour map before as the format of a topographical map, a map which shows the elevation of terrain over a geographical area. This kind of map consists of paths where the elevation is constant, so if you're walking along the path shown in the map, you're neither walking uphill nor downhill. Figure 11.21 shows the analogous contour map for $U(x, y)$, showing paths in the x,y plane where $U(x, y)$ is equal to 10, 20, 30, 40, 50, and 60.

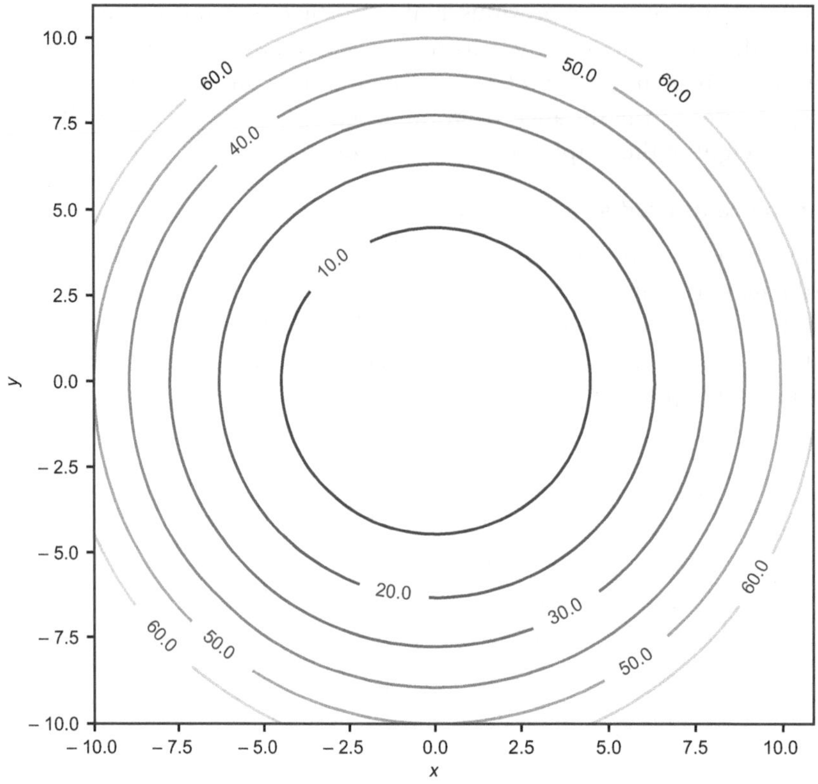

Figure 11.21 A contour map of $U(x, y)$, showing curves where the value of $U(x, y)$ is constant

You can see that the curves are all circular and that they get closer together as they go outward. We can interpret this to mean that $U(x, y)$ gets steeper as we get further from the origin. For instance, $U(x, y)$ increases from 30 to 40 over a shorter distance than it takes to increase from 10 to 20. You can plot the scalar field U as a contour map using the `scalar_field_contour` function from the source code.

11.5 Connecting energy and forces with the gradient

This notion of *steepness* is important—the steepness of a potential energy function tells us how much energy an object has to exert to move in that direction. As you might expect, the exertion required to move in a given direction is a measure of the force *in the opposite direction*. In the remainder of this section, we'll get to a precise and quantitative version of this statement.

As I mentioned in the introduction to this chapter, the gradient is an operation that takes a scalar field like potential energy and produces a vector field like a gravitational field. At every location (x, y) in the plane, the gradient vector field at that location points in the direction of fastest increase in the scalar field. In this section, I will show you how to take the gradient of a scalar field $U(x, y)$, which requires taking a derivative of U with respect to x and, separately, taking a derivative with respect to y. We'll be able to show that the gradient of the potential energy function $U(x, y)$ we've been working with is $-\mathbf{F}(x, y)$, where $\mathbf{F}(x, y)$ is the gravitational field we implemented in our asteroid game. We'll make extensive use of the gradient in the remaining chapters of this book.

11.5.1 Measuring steepness with cross sections

There's one more way of visualizing the function $U(x, y)$ that makes it easy to see how steep it is at various points. Let's focus on a specific point: $(x, y) = (-5, 2)$. On a contour map like that shown in figure 11.22, this point is between the $U = 10$ and $U = 20$ curves, and in fact, $U(-5, 2) = 14.5$. If we move in the $+x$ direction, we hit the $U = 10$ curve, meaning that U decreases in the $+x$ direction. If we instead move in the $+y$ direction, we hit the $U = 20$ curve, meaning that U increases in this direction.

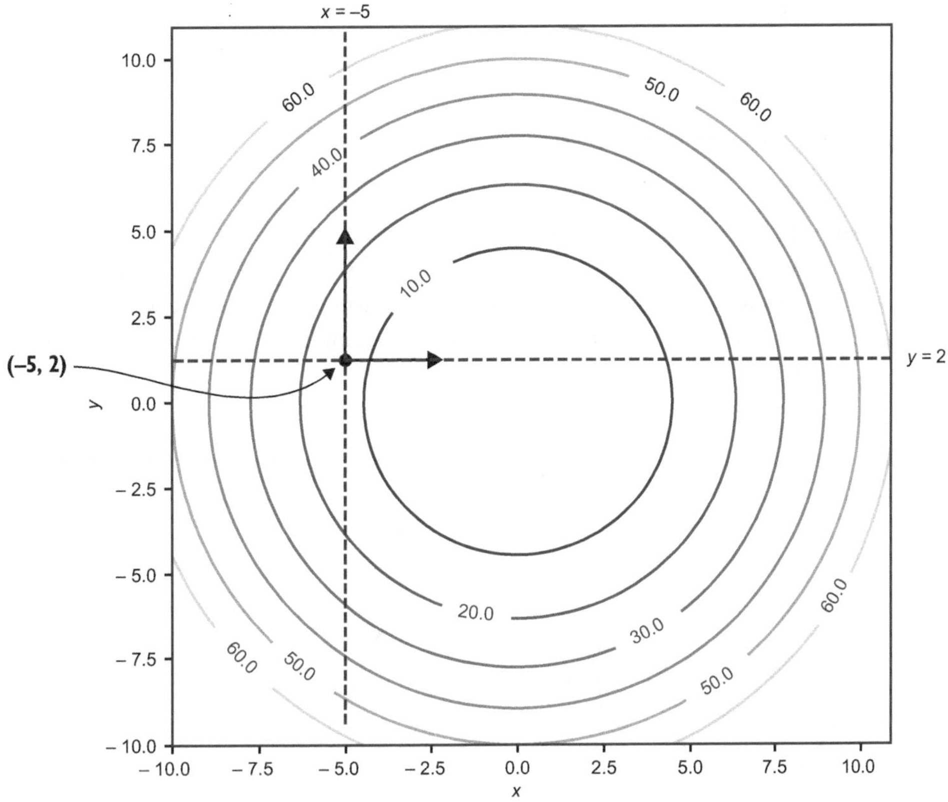

Figure 11.22 Exploring the value of $U(x, y)$ in the $+x$ and $+y$ directions from (–5, 2)

Figure 11.22 shows that the steepness of $U(x, y)$ depends on the direction. We can picture this by plotting the cross sections of $U(x, y)$, where $x = -5$ and $y = 2$. *Cross sections* are slices of the graph of $U(x, y)$ at fixed values of x or y. For example, figure 11.23 shows that the cross section of $U(x, y)$ at $x = -5$ is a slice of $U(x, y)$ in the plane $x = -5$.

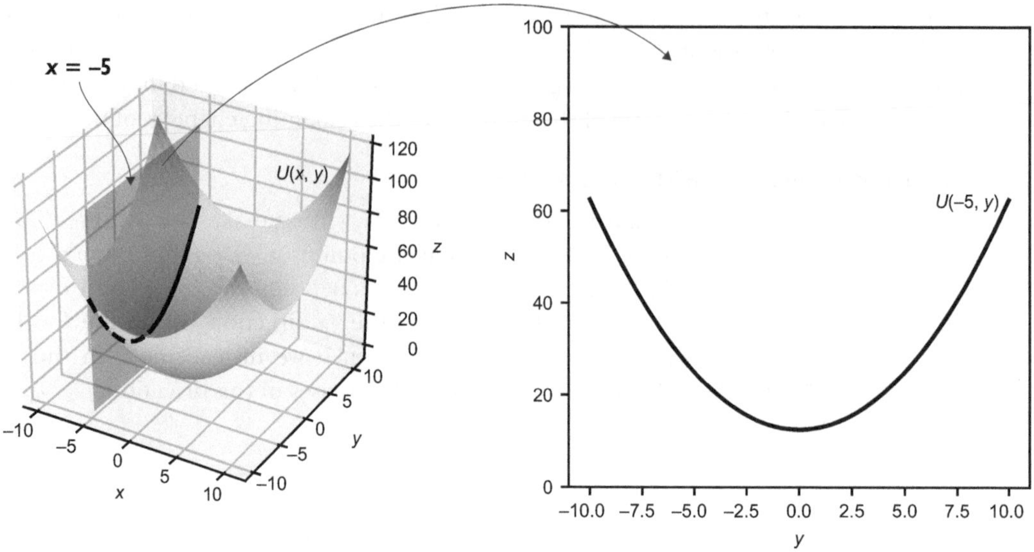

Figure 11.23 The cross section of *U*(*x*, *y*) at *x* = –5

Using the functional programming terminology from chapter 4, we can partially apply U with $x = -5$ to get a function that accepts a single number y and returns the value of U. There's also a cross section in the y direction at (5, 2). This is the cross section of $U(x, y)$, where $y = 2$. Figure 11.24 shows its shape as a graph of $U(x, y)$ after partially applying with $y = 2$.

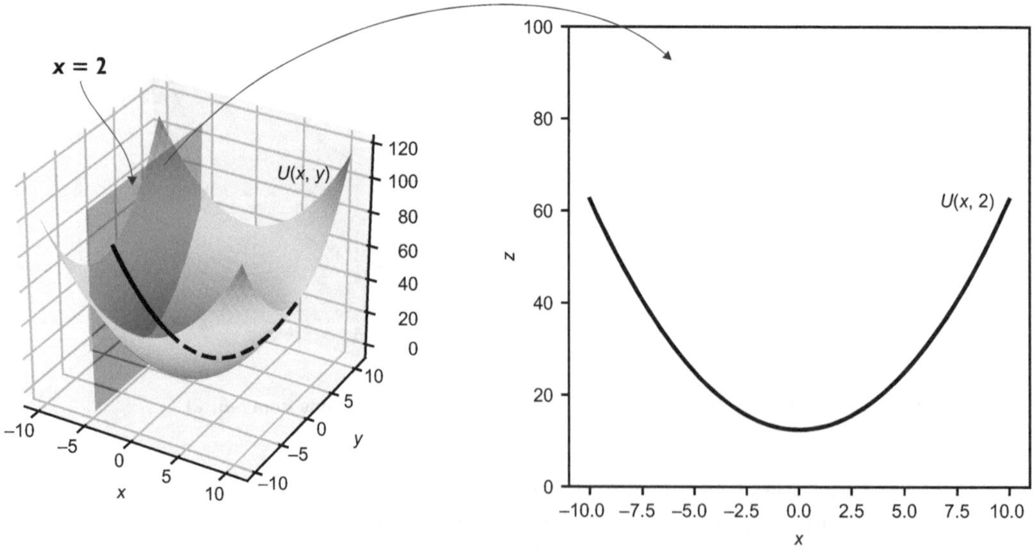

Figure 11.24 The cross section of *U*(*x*, *y*) at *y* = 2

Together, these cross sections tell us how *U* is changing at (–5, 2) in both the *x* and *y* directions. The slope of *U*(*x*, 2) at *x* = –5 is negative, telling us that moving in the +*x* direction from (–5, 2) causes *U* to decrease. Likewise, the slope of *U*(–5, *y*) at *y* = 2 is positive, telling us that moving in the +*y* direction from (–5, 2) causes *U* to increase (figure 11.25).

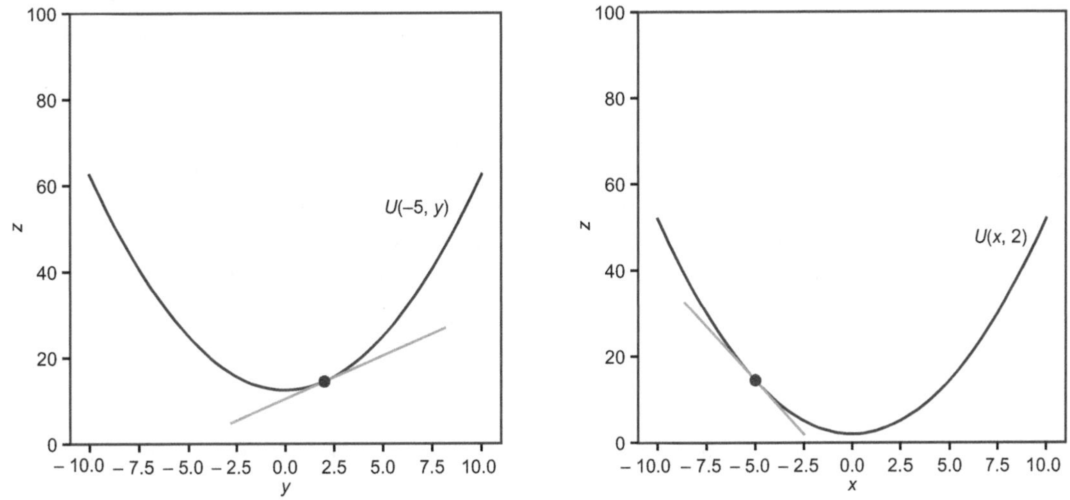

Figure 11.25 Cross sections show us that *U*(*x*, *y*) is increasing in the +*y* direction and decreasing in the +*x* direction.

We haven't found the slope of the scalar field *U*(*x*, *y*) at this point, but we have found what we could call the slope in the *x* direction and the slope in the *y* direction. These values are called *partial derivatives* of *U*.

11.5.2 Calculating partial derivatives

You already know everything you need to know to find the previous slopes. Both *U*(–5, *y*) and *U*(*x*, 2) are functions of one variable, so you could approximate their derivatives by calculating the slope of small secant lines.

For instance, if we want to find the partial derivative of *U*(*x*, *y*) with respect to *x* at the point (–5, 2), we're asking for the slope of *U*(*x*, 2) at *x* = –5. That is, we want to know how fast *U*(*x*, *y*) is changing in the *x* direction at the point (*x*, *y*) = (–5, 2). We could approximate this by plugging in a small value of Δx into the following slope calculation:

$$\frac{U(-5+\Delta x, 2) - U(-5, 2)}{\Delta x}$$

We can also calculate the derivative exactly by writing the formula for $U(x, 2)$. Because $U(x, y) = ½(x^2 + y^2)$, we have $U(x, 2) = ½(x^2 + 2^2) = ½(x^2 + 4) = 2 + (x^2/2)$. Using the power rule for derivatives, the derivative of $U(x, 2)$ with respect to x is $0 + 2x/2 = x$. At $x = -5$, the derivative is –5.

Notice that in both the slope approximation and the symbolic derivative process, the variable y doesn't appear. Instead, we're working with the constant value 2. This is to be expected because when we're thinking about a partial derivative in the x direction, y isn't changing. The general way to calculate partial derivatives symbolically is to take the derivative as if only one symbol (like x) is a variable, while all the other symbols (like y) are constants.

Using this method, the partial derivative of $U(x, y)$ with respect to x is $½(2x + 0) = x$, and the partial derivative with respect to y is $½(0 + 2y) = y$. By the way, the notation $f'(x)$ we previously preferred for the derivative of a function $f(x)$ is insufficient to extend to partial derivatives. When taking partial derivatives, you can take the derivative with respect to different variables, and you need to specify which one you're working with. There's another equivalent notation for the derivative of $f'(x)$:

$$f'(x) \equiv \frac{df}{dx}$$

(I use the ≡ sign to indicate that these notations are equivalent; they represent the same concept.) This is reminiscent of the slope formula $\Delta f/\Delta x$, but in this notation, df and dx instead represent *infinitesimal* changes in the value of f and x. The notation df/dx means the same thing as $f'(x)$, but it makes it clearer that the derivative is taken with respect to x. For a partial derivative of a function like $U(x, y)$, we can take the derivative with respect to either x or y. It's traditional to use different shaped d's to indicate that we're not taking an ordinary derivative (called a *total* derivative). The partial derivatives of U with respect to x and y are (respectively) written as follows:

$$\frac{\partial U}{\partial x} = x \quad \text{and} \quad \frac{\partial U}{\partial y} = y$$

Here's another example with a function $q(x, y) = x\sin(xy) + y$. If we treat y as a constant and take the derivative with respect to x, we need to use the product rule and the chain rule. The result is the partial derivative with respect to x:

$$\frac{\partial q}{\partial x} = \sin(xy) + xy\cos(xy)$$

To take the partial derivative with respect to y, we treat x as a constant, and we need to use the chain rule and the sum rule:

$$\frac{\partial q}{\partial y} = x^2\cos(xy) + 1$$

It's true that each of the partial derivatives only tells part of the story of how a function like $U(x, y)$ changes at any point. Next, we combine them to gain a full understanding, analogous to the total derivative for a function of one variable.

11.5.3 Finding the steepness of a graph with the gradient

Let's zoom in on the point (–5, 2) on the graph of $U(x, y)$ (figure 11.26). Just as any smooth function $f(x)$ looks like a straight line on a sufficiently small range of x values, it turns out the graph of a smooth scalar field looks like a plane in a small enough vicinity of the x,y plane.

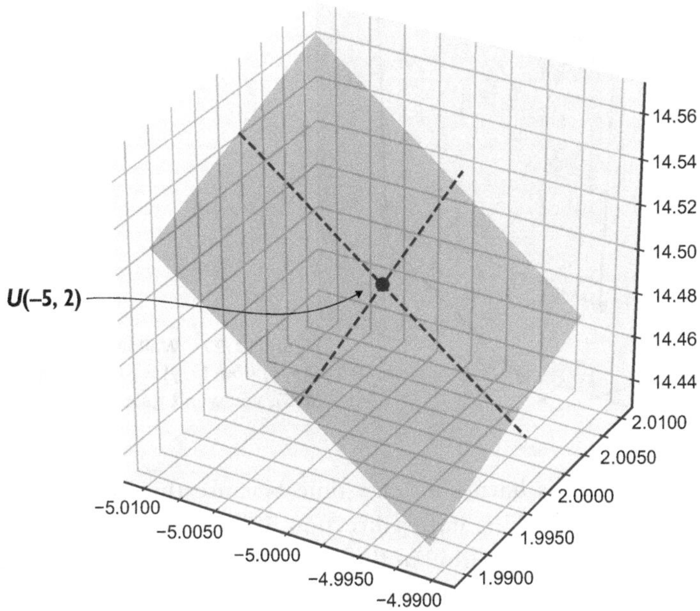

Figure 11.26 Up close, the region of the graph of *U*(*x*, *y*) near (*x*, *y*) = (–5, 2) looks like a plane.

Just as the derivative df/dx tells us about the slope of the line that approximates $f(x)$ at a given point, the partial derivatives $\partial U/\partial x$ and $\partial U/\partial y$ tell us about a plane that approximates $U(x, y)$ at a given point. The dotted lines in figure 11.26 show the x and y cross sections of $U(x, y)$ at this point. In this window, they are approximately straight lines, and their slopes in the x,z and y,z planes are close to the partial derivatives $\partial U/\partial x$ and $\partial U/\partial y$.

I haven't proven it, but suppose there *is* a plane that best approximates $U(x, y)$ near (–5, 2), and because we can't distinguish it, we can pretend the graph in figure 11.26 is that plane for a moment. The partial derivatives tell us how much it is slanted in the x and y directions. On a plane, there are actually two better directions to think about. First of all, there's a direction on the plane you could walk without gaining or losing elevation. In other words, this is the line in the plane that is parallel to the x,y plane. For the plane that approximates $U(x, y)$ at (–5, 2), that turns out to be in the direction (2, 5) as shown by figure 11.27.

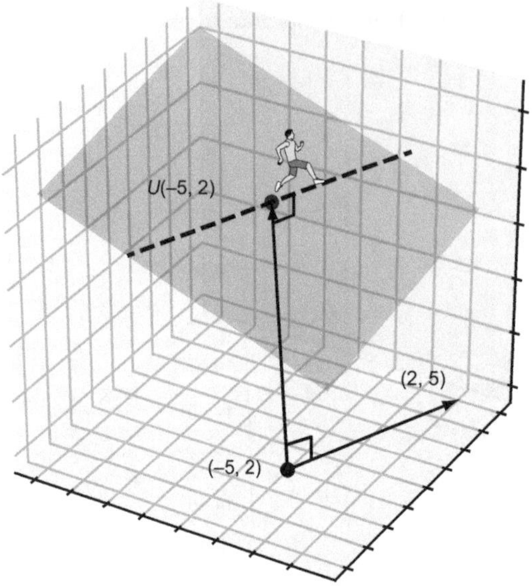

Figure 11.27 Walking along the graph of $U(x, y)$ from $(x, y) = (-5,2)$ in the direction of (2,5), you won't gain or lose elevation.

The walker in figure 11.27 is having an easy time because they're not climbing or descending the plane when walking in this direction. If, however, the walker turns 90° to the left, they would be walking uphill in the steepest direction possible. That is the direction (–5, 2), which is perpendicular to (2, 5).

This direction of steepest ascent happens to be a vector whose components are the partial derivatives of U at a given point. I gave one illustration of this instead of proving it, but this fact is true in general. For a function $U(x, y)$, the vector of its partial derivatives is called its *gradient* and written ∇U. It gives the magnitude and direction of steepest ascent of U at a given point:

$$\nabla U(x, y) = \left(\frac{\partial U}{\partial x}, \frac{\partial U}{\partial y} \right)$$

Because we have formulas for the partial derivatives, we can tell, for instance, that $\nabla U(x, y) = (x, y)$ for our function. The function ∇U, which is the gradient of U, is an assignment of a vector to every point in the plane, so it is indeed a vector field! The plot of ∇U shows us at every point (x, y) which direction is uphill on the graph of $U(x, y)$, as well as how steep it is (figure 11.28).

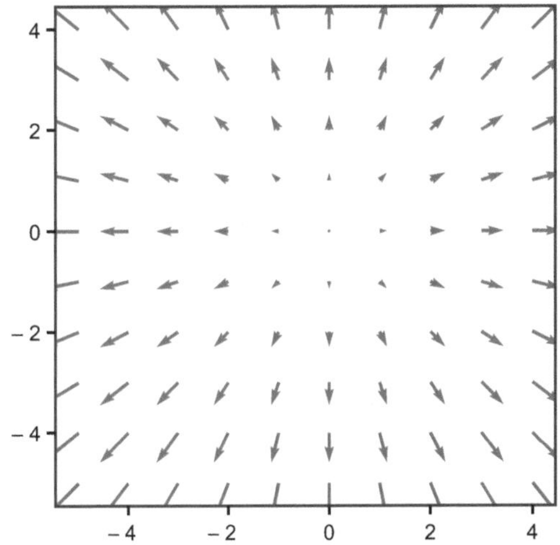

Figure 11.28 The gradient ∇*U* is a vector field telling us the magnitude and direction of steepest ascent on the graph of *U* at any point (*x*, *y*).

The gradient is a way of connecting a scalar field with a vector field. It turns out that this gives the connection between potential energy and force.

11.5.4 Calculating force fields from potential energy with the gradient

The gradient is the best analogy of the ordinary derivative for scalar fields. It has all the information needed to find the direction of steepest ascent of the scalar field, the slope along the x or y directions, or the plane of best approximation. But from the perspective of physics, the direction of steepest ascent is not what we're looking for. After all, there's no object in nature that spontaneously moves uphill.

Neither the spaceship in the asteroid game nor the ball rolling around in the bowl would feel forces that impel them toward regions of higher potential energy. As we've discussed, they would need to apply a force or sacrifice some kinetic energy to gain more potential energy. For that reason, the right description of the force an object feels is the *negative* gradient of potential energy, which points in the direction of steepest *descent* rather than steepest *ascent*. If $U(x, y)$ represents the scalar field of potential energy, then the associated force field $\mathbf{F}(x, y)$ can be calculated by

$$\mathbf{F}(x, y) = -\nabla U(x, y)$$

Let's try a fresh example. What kind of force field would be produced by the following potential energy function?

$$V(x, y) = 1 + y^2 - 2x^2 + x^6$$

We can get a sense for how this function behaves by plotting it.

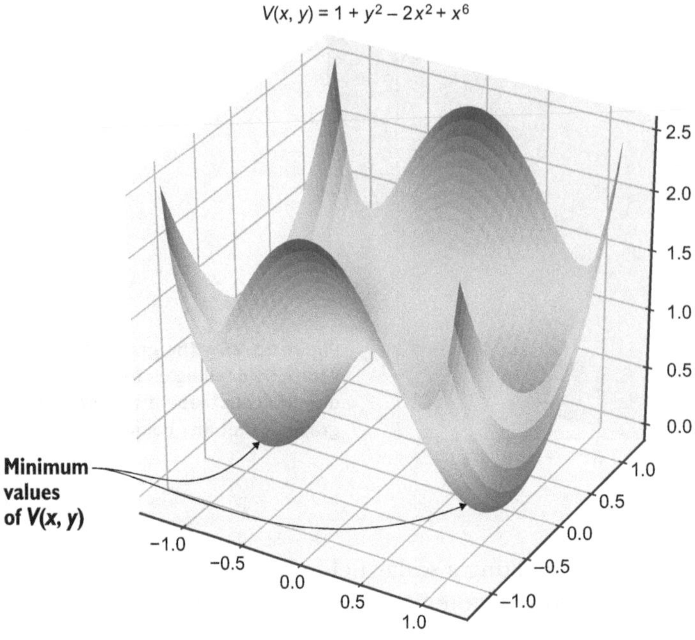

Figure 11.29 The potential energy function *V*(x, y) shown in 3D

Figure 11.29 illustrates that this potential energy function has a double bowl shape with two minimum points and a hump between them. What does the force field associated with this potential energy function look like? To find out, we need to take the negative gradient of V:

$$\mathbf{F}(x, y) = -\nabla V(x, y) = -\left(\frac{\partial V}{\partial x}, \frac{\partial V}{\partial y}\right)$$

We can get the partial derivative of V with respect to x by treating y like a constant, so the terms 1 and y^2 don't contribute. The result is just the derivative of $-2x^2 + x^6$ with respect to x, which is $-4x + 6x^5$.

For the partial derivative of V with respect to y, we treat x like a constant, so the only term that contributes to the result is y^2 having a derivative $2y$. The negative gradient of $V(x, y)$ is therefore

$$\mathbf{F}(x, y) = -\nabla V(x, y) = (4x - 6x^5, -2y)$$

Plotting this vector field, figure 11.30 shows that the force field points toward the points of lowest potential energy. An object feeling this force field would experience these two points as exerting an attractive force.

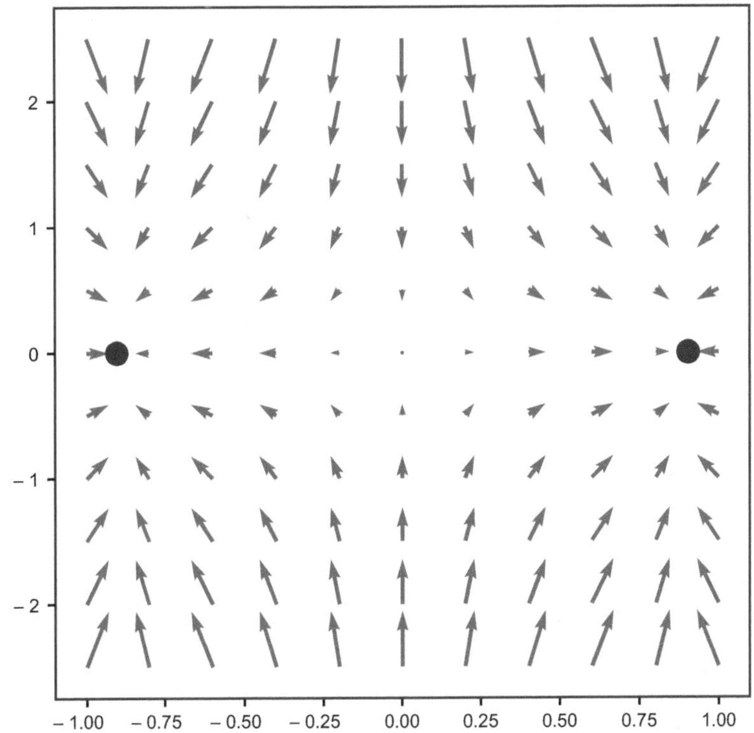

Figure 11.30 A plot of the vector field $-\nabla V(x, y)$, the force field associated with the potential energy function $V(x, y)$. This is an attractive force toward the two points shown.

The negative gradient of potential energy is the direction nature prefers; it is the direction to release stored energy. Objects are naturally pushed toward states where their potential energy is minimized. The gradient is an important tool for finding optimum values of scalar fields, as we'll see in the next chapters. Specifically, in the last part of this book, I'll show you how following the negative gradient in search of an optimal value mimics the process of "learning" in certain machine learning algorithms.

11.5.5 Exercises

Exercise 11.4 Plot the cross section of $h(x, y) = e^y \sin(x)$, where $y = 1$. Then plot the cross section of $h(x, y)$, where $x = \pi/6$.

Solution The cross section of $h(x, y)$ where $y = 1$ is a function of only x: $h(x, 1) = e^1 \sin(x) = e \cdot \sin(x)$ as shown here:

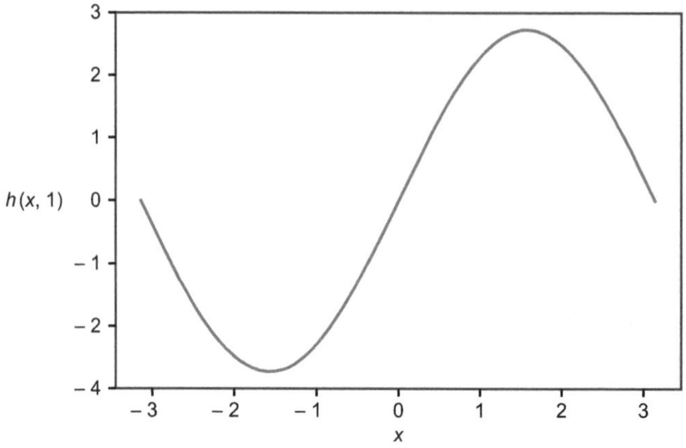

Where $x = \pi/6$, the value of $h(x, y)$ depends only on y. That is, $h(\pi/6, y) = e^y \sin(\pi/6) = e^y/2$. The graph is

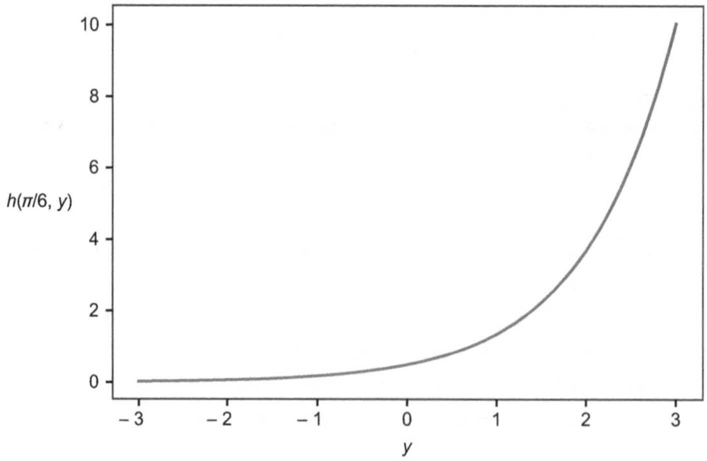

Exercise 11.5 What are the partial derivatives of the function $h(x, y)$ from the first exercise? What is the gradient? What is the value of the gradient at $(x, y) = (\pi/6, 1)$?

Solution The partial derivative of $e^y \sin(x)$ with respect to x is obtained by treating y as a constant. e^y is therefore treated as a constant as well. The result is

$$\frac{\partial h}{\partial x} = e^y \cos(x)$$

Likewise, we get the partial derivative with respect to y by treating x and, therefore, $\sin(x)$ as constants:

$$\frac{\partial h}{\partial y} = e^y \sin(x)$$

The gradient $\nabla h(x, y)$ is the vector field whose components are the partial derivatives:

$$\nabla h(x, y) = \left(\frac{\partial h}{\partial x}, \frac{\partial h}{\partial y}\right) = (e^y \cos(x), e^y \sin(x))$$

At $(x, y) = (\pi/6, 1)$, this vector field evaluates as follows:

$$\nabla h\left(\frac{\pi}{6}, 1\right) = \left(e^1 \cos\left(\frac{\pi}{6}\right), e^1 \sin\left(\frac{\pi}{6}\right)\right) = \frac{e}{2} \cdot (\sqrt{3}, 1)$$

Exercise 11.6 Prove $(-5, 2)$ is perpendicular to $(2, 5)$.

Solution This is a review from chapter 2. These two vectors are perpendicular because their dot product is zero: $(-5, 2) \cdot (2, 5) = -10 + 10 = 0$.

Exercise 11.7—Mini Project Let $z = p(x, y)$ be the equation of the plane that best approximates $U(x, y)$ at $(-5, 2)$. Figure out (from scratch!) an equation for $p(x, y)$ and the line contained in p and passing through $(-5, 2)$, which is parallel to the x, y plane. This line should be parallel to the vector $(2, 5, 0)$ as I claimed in the previous exercise.

Solution Remember that the formula for $U(x, y)$ is $\frac{1}{2}(x^2 + y^2)$. The value of $U(-5, 2)$ is 14.5, so the point $(x, y, z) = (-5, 2, 14.5)$ is on the graph of $U(x, y)$ in 3D.

(continued)

Before we think about the formula for the plane of best approximation for $U(x, y)$, let's review how we got the line of best approximation for a function $f(x)$. The line that best approximates a function $f(x)$ at a point x_0 is the line that passes through the point $(x_0, f(x_0))$ and has a slope $f'(x_0)$. Those two facts ensure that both the value and the derivative of $f(x)$ agree with the line that approximates it.

Following this model, let's look for the plane $p(x, y)$, whose value and *both* partial derivatives match at $(x, y) = (-5, 2)$. That means we must have $p(-5, 2) = 14.5$, while $\partial p/\partial x = -5$ and $\partial p/\partial y = 2$. As a plane, $p(x, y)$ has the form $p(x, y) = ax + by + c$ for some numbers a and b (do you remember why?). The partial derivatives are

$$\frac{\partial p}{\partial x} = a \quad \text{and} \quad \frac{\partial p}{\partial y} = b$$

To make them match, the formula must be $p(x, y) = -5x + 2y + c$, and to satisfy $p(-5, 2) = 14.5$, it must be that $c = -14.5$. Therefore, the formula for the plane of best approximation is $p(x, y) = -5x + 2y - 14.5$.

Now, let's look for the line in the plane $p(x, y)$ passing through $(-5, 2)$, which is parallel to the x,y plane. This is the set of points (x, y) such that $p(x, y) = p(-5, 2)$, meaning that there is no elevation change between $(-5, 2)$ and (x, y).

If $p(x, y) = p(-5, 2)$, then $-5x + 2y - 14.5 = -5 \cdot -5 + 2 \cdot 2 - 14.5$. That simplifies to the equation of a line: $-5x + 2y = 29$. This line is equivalent to the set of vectors $(-5, 2, 14.5) + r \cdot (2, 5, 0)$, where r is a real number, so it is indeed parallel to $(2, 5, 0)$.

Summary

- A vector field is a function that takes a vector both as input and output. Specifically, we picture it as an assignment of an arrow vector to every point in a space.
- Gravitational force can be modeled by a vector field. The value of the vector field at any point in space tells you how strong and in what direction an object is pulled by the force of gravity.
- To simulate the motion of an object in a vector field, you need to use its position to calculate the strength and direction of the force field where it is. In turn, the value of the force field tells you the force on the object, and Newton's second law tells you the resulting acceleration.
- *Potential energy* is stored energy that has the potential to create motion. The potential energy for an object in a force field is determined by where the object is.

- Potential energy can be modeled as a scalar field: an assignment of a number to every point in space, which is the amount of potential energy an object would have at that point.
- There are several ways to picture a scalar field in 2D: as a 3D surface, a heatmap, a contour map, or a pair of cross-section graphs.
- The partial derivatives of a scalar field give the rate of change in the value of the field with respect to the coordinates. For instance, if $U(x, y)$ is a scalar field in 2D, there are partial derivatives with respect to x and y.
- Partial derivatives are the same as the derivatives of a cross section of the scalar field. You can calculate the partial derivative with respect to one variable by treating the other variables as constants.
- The gradient of a scalar field U is a vector whose components are partial derivatives of U with respect to each of the coordinates. The gradient points in the direction of steepest ascent for U or the direction in which U increases most rapidly.
- The negative gradient of a potential energy function corresponding to a force field tells us the vector value of the force field at that point. This means that objects are pushed towards regions of lower potential energy.

12 Optimizing a physical system

This chapter covers

- Building and visualizing a simulation for a projectile
- Finding maximal and minimal values for a function using derivatives
- Tuning simulations with parameters
- Visualizing spaces of input parameters for simulations
- Implementing gradient ascent to maximize functions of several variables

For most of the last few chapters, we've focused on a physical simulation for a video game. This is a fun and simple example to work with, but there are far more important and lucrative applications. For any big feat of engineering like sending a rocket to Mars, building a bridge, or drilling an oil well, it's important to know that it's going to be safe, successful, and on budget before you attempt it. In each of these projects, there are quantities you want to optimize. For instance, you may want to minimize the travel time for your rocket, minimize the amount or cost of concrete in a bridge, or maximize the amount of oil produced by your well.

To learn about optimization, we'll focus on the simple example of a projectile—namely, a cannonball being fired from a cannon. Assuming the cannonball comes out of the barrel at the same speed every time, the launch angle will decide the trajectory (figure 12.1).

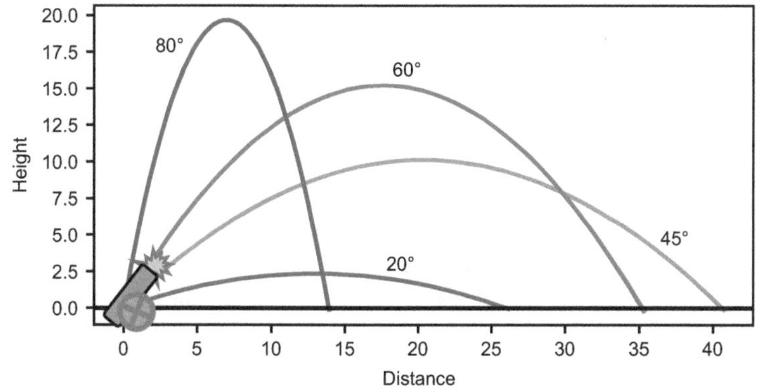

Figure 12.1 Trajectories for a cannonball fired at four different launch angles

As you can see in figure 12.1, four different launch angles produce four different trajectories. Among these, 45° is the launch angle that sends the cannonball the furthest, while 80° is the angle that sends it the highest. These are only a few angles of all of the possible values between 0 and 90°, so we can't be sure they are the best. Our goal is to systematically explore the range of possible launch angles to be sure we've found the one that optimizes the range of the cannon.

To do this, we first build a simulator for the cannonball. This simulator will be a Python function that takes a launch angle as input, runs Euler's method (as we did in chapter 9) to simulate the moment-by-moment motion of the cannonball until it hits the ground, and outputs a list of positions of the cannonball over time. From the result, we'll extract the final horizontal position of the cannonball, which will be the landing position or range. Putting these steps together, we implement a function that takes a launch angle and returns the range of the cannonball at that angle (figure 12.2).

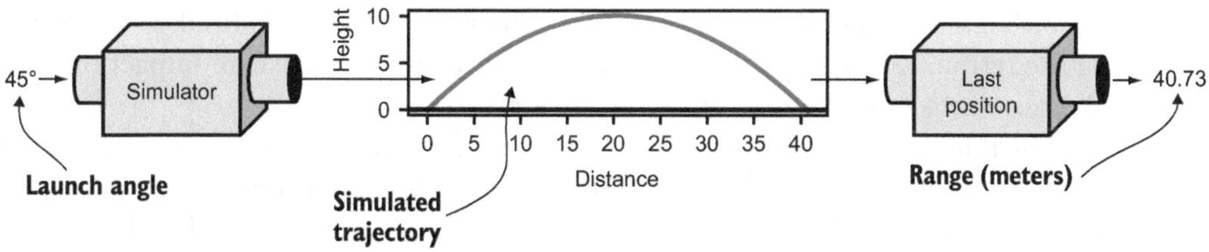

Figure 12.2 Computing the range of a projectile using a simulator

Once we have encapsulated all of this logic in a single Python function called `landing_position`, which computes the range of a cannonball as a function of its launch angle, we can think about the problem of finding the launch angle that maximizes the range. We can do this in two ways: first, we make a graph of the range versus the launch angle and look for the largest value (figure 12.3).

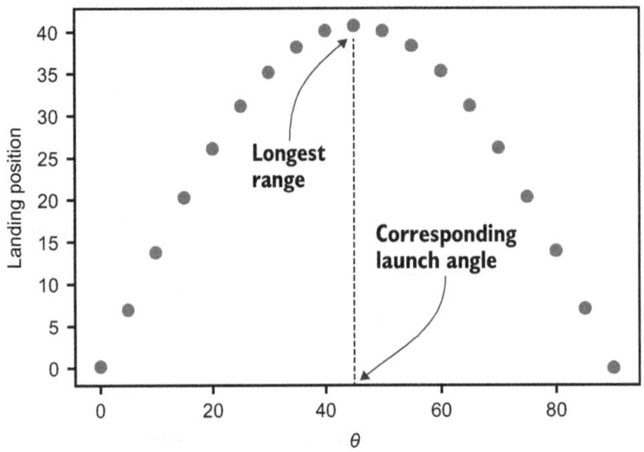

Figure 12.3 Looking at a plot of range vs. launch angle, we can see the approximate value of the launch angle that produces the longest range.

The second way we can find the optimal launch angle is to set our simulator aside and find a formula for the range $r(\theta)$ of the projectile as a function of the launch angle θ. This should produce identical results as the simulation, but because it is a mathematical formula, we can take its derivative using the rules from chapter 10. The derivative of the landing position with respect to the launch angle tells us how much increase in the range we'll get for small increases in the launch angle. At some angle, we can see that we get diminishing returns—increasing the launch angle causes the range to *decrease* and we'll have passed our optimal value. Before this, the derivative of $r(\theta)$ will instantaneously be zero, and the value of θ where the derivative is zero happens to be the maximum value.

Once we've warmed up using both of these optimization techniques on our 2D simulation, we can try a more challenging 3D simulation, where we can control the upward angle of the cannon as well as the lateral direction it is fired. If the elevation of the terrain varies around the cannon, the direction can have an impact on how far the cannonball flies before hitting the ground (figure 12.4).

For this example, let's build a function $r(\theta, \phi)$, taking two input angles θ and ϕ and outputting a landing position. The challenge is to find the pair (θ, ϕ) that maximizes the range of the cannon. This example lets us cover our third and most important optimization technique: *gradient ascent.*

As we learned in the last chapter, the gradient of $r(\theta, \phi)$ at a point (θ, ϕ) is a vector pointing in the direction that causes r to increase most rapidly. We'll write a Python

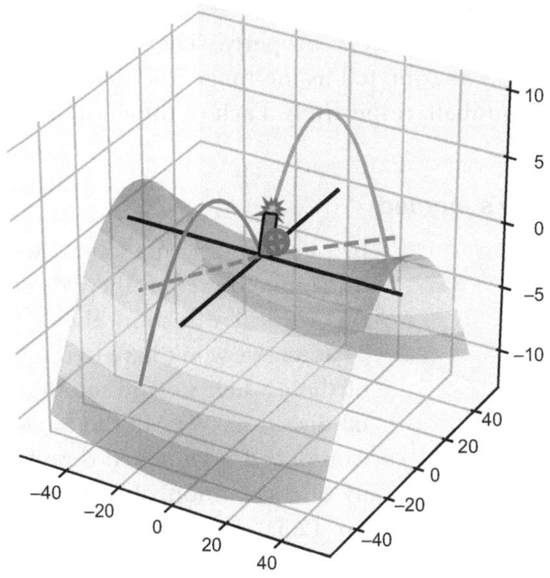

Figure 12.4 With uneven terrain, the direction we fire the cannon can affect the range of the cannonball as well.

function called `gradient_ascent` that takes as input a function to optimize, along with a pair of starting inputs, and uses the gradient to find higher and higher values until it reaches the optimal value.

The mathematical field of optimization is a broad one, and I hope to give you a sense of some basic techniques. All of the functions we'll work with are smooth, so you will be able to make use of the many calculus tools you've learned so far. Also, the way we approach optimization in this chapter sets the stage for optimizing computer "intelligence" in machine learning algorithms, which we turn to in the final chapters of the book.

12.1 Testing a projectile simulation

Our first task is to build a simulator that computes the flight path of the cannonball. The simulator will be a Python function called `trajectory`. It takes the launch angle, as well as a few other parameters that we may want to control, and returns the positions of the cannonball over time until it collides with Earth. To build this simulation, we turn to our old friend from chapter 9—Euler's method.

As a reminder, we can simulate motion with Euler's method by advancing through time in small increments (we'll use 0.01 seconds). At each moment, we'll know the position of the cannonball, as well as its derivatives: velocity and acceleration. The velocity and acceleration let us approximate the change in position to the next moment, and we'll repeat the process until the cannonball hits the ground. As we go, we can save the time, *x* and *y* positions of the cannonball, at each step and output them as the result of the `trajectory` function.

Finally, we'll write functions that take the results we get back from the `trajectory` function and measure one numerical property. The functions `landing_position`, `hang_time`, and `max_height` tell us the range, the time in the air, and the maximum height of the cannonball, respectively. Each of these will be a value we can subsequently optimize for.

12.1.1 Building a simulation with Euler's method

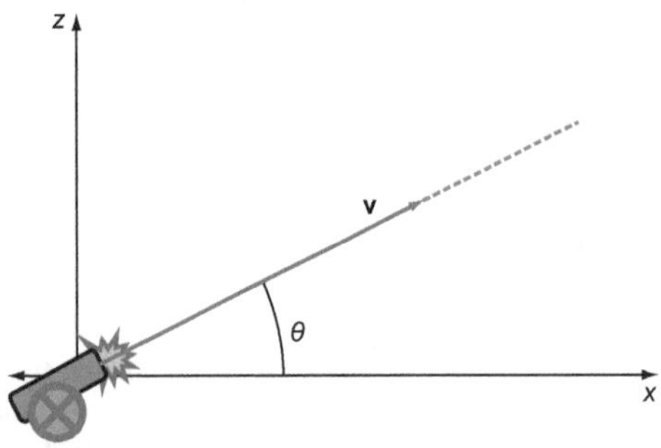

Figure 12.5 The variables in our projectile simulation

In our first 2D simulation, we call the horizontal direction the x direction and the vertical direction the z direction. That way we won't have to rename either of these when we add another horizontal direction. We call the angle that the cannonball is launched θ and the velocity of the cannonball **v** as shown in figure 12.5.

The *speed*, v, of a moving object is defined as the magnitude of its velocity vector, so $v = |\mathbf{v}|$. Given the launch angle θ, the x and z components of the cannonball's velocity are $v_x = |\mathbf{v}| \cdot \cos(\theta)$ and $v_z = |\mathbf{v}| \cdot \sin(\theta)$. I'll assume that the cannonball leaves the barrel of the cannon at time $t = 0$ and with (x, z) coordinates $(0, 0)$, but I'll also include a configurable launch height. Here's the basic simulation using Euler's method:

```
def trajectory(theta,speed=20,height=0,
               dt=0.01,g=-9.81):          # Additional inputs: the time step dt, gravitational field strength g, and angle theta (in degrees)
    vx = speed * cos(pi * theta / 180)    # Calculates the initial x and z components of velocity, converting the input angle from degrees to radians
    vz = speed * sin(pi * theta / 180)
    t,x,z = 0, 0, height
    ts, xs, zs = [t], [x], [z]            # Initializes lists that hold all the values of time and the x and z positions over the course of the simulation
    while z >= 0:                         # Runs the simulation only while the cannonball is above ground
        t += dt
        vz += g * dt                      # Updates time, z velocity, and position. There are no forces acting in the x direction, so the x velocity is unchanged.
        x += vx * dt
        z += vz * dt
        ts.append(t)
        xs.append(x)
        zs.append(z)
    return ts, xs, zs                     # Returns the list of t, x, and z values, giving the motion of the cannonball
```

You'll find a `plot_trajectories` function in the source code for this book that takes the outputs of one or more results of the `trajectory` function and passes them to Matplotlib's plot function, drawing curves that show the path of each cannonball. For instance, figure 12.6 shows plotting a 45° launch next to a 60° launch, which is done using the following code:

```
plot_trajectories(
    trajectory(45),
    trajectory(60))
```

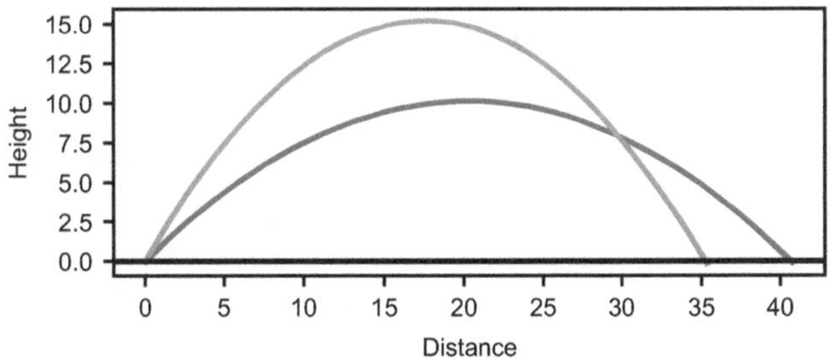

Figure 12.6 An output of the plot_trajectories function showing the results of a 45° and 60° launch angle.

We can already see that the 45° launch angle produces a greater range and that the 60° launch angle produces a greater maximum height. To be able to optimize these properties, we need to measure them from the trajectories.

12.1.2 *Measuring properties of the trajectory*

It's useful to keep the raw output of the trajectory in case we want to plot it, but sometimes we'll want to focus on one number that matters most. For instance, the range of the projectile is the last x-coordinate of the trajectory, which is the last x position before the cannonball hits the ground. Here's a function that takes the result of the trajectory function (parallel lists with time and the x and z positions), and extracts the range or landing position. For the input trajectory, traj, traj[1] lists the x-coordinates, and traj[1][-1] is the last entry in the list:

```
def landing_position(traj):
    return traj[1][-1]
```

This is the main metric of a projectile's trajectory that interests us, but we can also measure some other ones. For instance, we might want to know the hang time (or how long the cannonball stays in the air) or its maximum height. We can easily create other Python functions that measure these properties from simulated trajectories; for example,

```
def hang_time(traj):
    return traj[0][-1]

def max_height(traj):
    return max(traj[2])
```

Total time in the air is equal to the last time value, the time on the clock when the projectile hits the ground.

The maximum height is the maximum among the z positions, the third list in the trajectory output.

To find an optimal value for any of these metrics, we need to explore how the parameters (namely, the launch angle) affect them.

12.1.3 Exploring different launch angles

The `trajectory` function takes a launch angle and produces the full time and position data for the cannonball over its flight. A function like `landing_position` takes this data and produces a single number. Composing these two together (figure 12.7), we get a function for landing position in terms of the launch angle, where all other properties of the simulation are assumed constant.

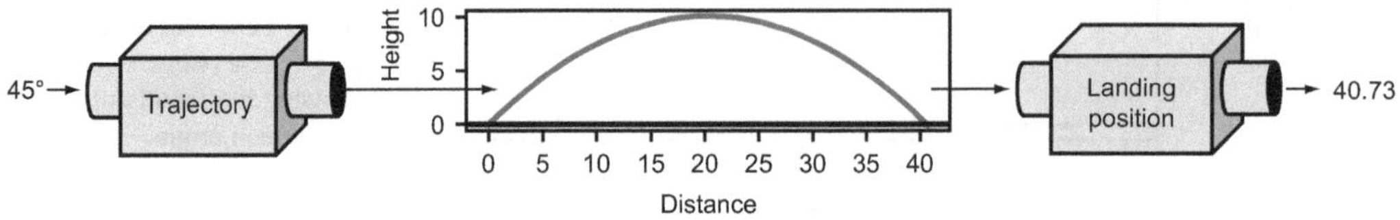

Figure 12.7 Landing position as a function of the launch angle

One way to test the effect of the launch angle on a landing position is to make a plot of the resulting landing position for several different values of the launch angle (figure 12.8). To do this, we need to calculate the result of the composition `landing_position(trajectory(theta))` for several different values of `theta` and pass these to Matplotlib's `scatter` function. Here, for example, I use `range(0,95,5)` as the launch angles. This is every angle from zero to 90 in increments of 5:

```
import matplotlib.pyplot as plt
angles = range(0,90,5)
landing_positions = [landing_position(trajectory(theta))
                     for theta in angles]
plt.scatter(angles,landing_positions)
```

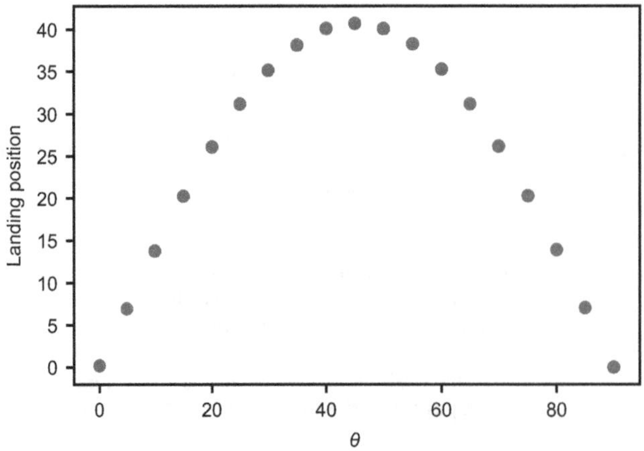

Figure 12.8 A plot of the landing position vs. the launch angle for the cannon for several different values of the launch angle

From this plot, we can guess what the optimal value is. At a launch angle of 45°, the landing position is maximized at a little over 40 meters from the launch position. In this case, 45° turns out to be the *exact* value of the angle that maximizes the landing position. In the next section, we'll use calculus to confirm this maximum value without having to do any simulation.

12.1.4 Exercises

Exercise 12.1 How far does the cannonball go when fired at an angle of 50° from an initial height of zero? How about if it is fired at an angle of 130°?

Solution At 50°, the cannonball goes about 40.1 meters in the positive direction, while at 130°, it goes 40.1 meters in the negative direction:

```
>>> landing_position(trajectory(50))
40.10994684444007
>>> landing_position(trajectory(130))
-40.10994684444007
```

This is because 130° from the positive x-axis is the same as 50° from the negative x-axis.

Exercise 12.2—Mini Project Enhance the `plot_trajectories` function to draw a large dot on the trajectory graph at each passing second so we can see the passing of time on the plot.

Solution Here are the updates to the function. It looks for the index of the nearest time after each whole second and makes a scatter plot of (x, z) values at each of these indices:

```
def plot_trajectories(*trajs,show_seconds=False):
    for traj in trajs:
        xs, zs = traj[1], traj[2]
        plt.plot(xs,zs)
        if show_seconds:
            second_indices = []
            second = 0
            for i,t in enumerate(traj[0]):
                if t>= second:
                    second_indices.append(i)
                    second += 1
            plt.scatter([xs[i] for i in second_indices],
                        [zs[i] for i in second_indices])
  ...
```

(continued)

As a result, you can picture the elapsed time for each of the trajectories you plot; for example,

```
plot_trajectories(
    trajectory(20),
    trajectory(45),
    trajectory(60),
    trajectory(80),
    show_seconds=True)
```

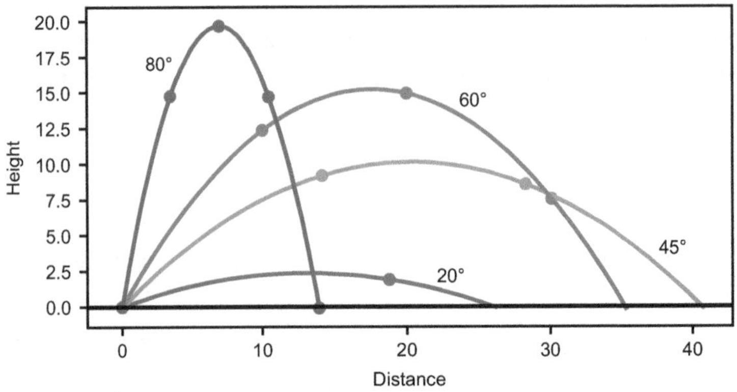

Plots of four trajectories with dots showing their positions at each whole number of seconds.

Exercise 12.3 Make a scatter plot of hang time versus angle for angles between 0 and 180°. Which launch angle produces the maximum hang time?

Solution

```
test_angles = range(0,181,5)
hang_times = [hang_time(trajectory(theta)) for theta in test_angles]
plt.scatter(test_angles, hang_times)
```

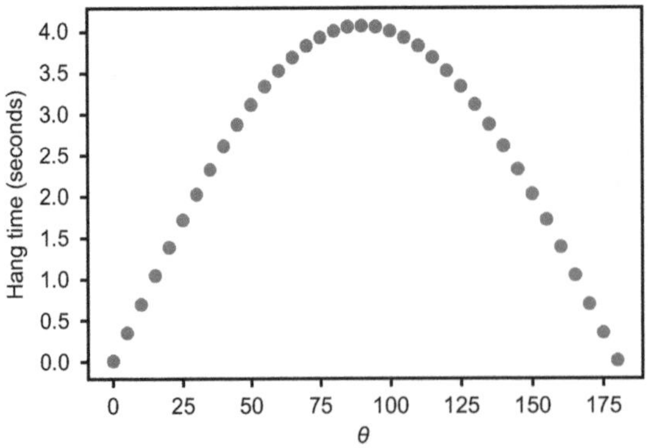

A plot of the hang time of the cannonball as a function of the launch angle

It appears that a launch angle of roughly 90° yields the longest hang time of just about 4 seconds. This makes sense because $\theta = 90°$ yields the initial velocity with the largest vertical component.

Exercise 12.4—Mini Project Write a function `plot_trajectory_metric` that plots the result of any metric we want over a given set of theta (θ) values. For instance,

```
plot_trajectory_metric(landing_position,[10,20,30])
```

makes a scatter plot of landing positions versus launch angle for the launch angles 10°, 20°, and 30°.

As a bonus, pass the keyword arguments from `plot_trajectory_metric` to the internal calls of the `trajectory` function, so you can rerun the test with a different simulation parameter. For instance, this code makes the same plot but simulated with a 10-meter initial launch height:

```
plot_trajectory_metric(landing_position,[10,20,30], height=10)
```

Solution

```
def plot_trajectory_metric(metric,thetas,**settings):
    plt.scatter(thetas,
                [metric(trajectory(theta,**settings))
                 for theta in thetas])
```

We can make the plot from the previous exercise by running the following:

```
plot_trajectory_metric(hang_time, range(0,181,5))
```

Exercise 12.5—Mini Project What is the approximate launch angle that yields the greatest range for the cannonball with a 10-meter initial launch height?

Solution Using the `plot_trajectory_metric` function from the preceding mini-project, we can simply run

```
plot_trajectory_metric(landing_position,range(0,90,5), height=10)
```

(continued)

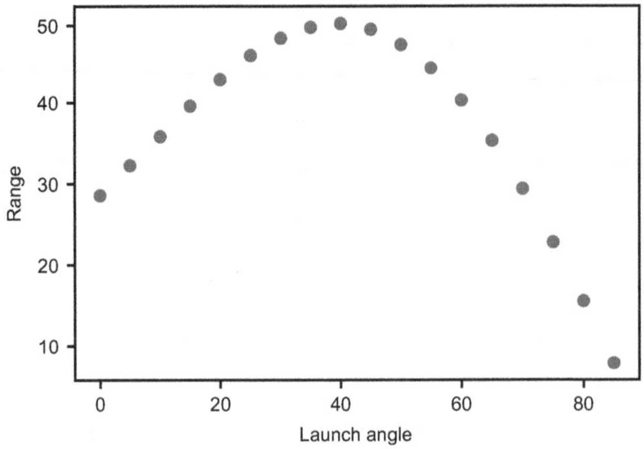

A plot of range of the cannonball vs. launch angle with a 10 meter launch height

The optimal launch angle from a height of 10 meters is about 40°.

12.2 Calculating the optimal range

Using calculus, we can compute the maximum range for the cannon, as well as the launch angle that produces it. This actually takes two separate applications of calculus. First, we need to come up with an exact function that tells us the range r as a function of the launch angle θ. As a warning, this will take quite a bit of algebra. I'll carefully walk you through all the steps, so don't worry if you get lost; you'll be able to jump ahead to the final form of the function $r(\theta)$ and continue reading.

Then I show you a trick using derivatives to find the maximum value of this function $r(\theta)$, and the angle θ that produces it. Namely, a value of θ that makes the derivative $r'(\theta)$ equal zero is also the value of θ that yields the maximum value of $r(\theta)$. It might not be immediately obvious why this works, but it will become clear once we examine the graph of $r(\theta)$ and study its changing slope.

12.2.1 Finding the projectile range as a function of the launch angle

The horizontal distance traveled by the cannonball is actually pretty simple to calculate. The x component of the velocity v_x is constant for its entire flight. For a flight of total time Δt, the projectile travels a total distance of $r = v_x \cdot \Delta t$. The challenge is finding the exact value of that elapsed time Δt.

That time, in turn, depends on the z position of the projectile over time, which is a function $z(t)$. Assuming the cannonball is launched from an initial height of zero, the first time that $z(t) = 0$ is when it's launched at $t = 0$. The second time is the elapsed time we're looking for. Figure 12.9 shows the graph of $z(t)$ from the simulation with θ

= 45°. Note that its shape looks a lot like the trajectory, but the horizontal axis (t) now represents time.

```
trj = trajectory(45)
ts, zs = trj[0], trj[2]
plt.plot(ts,zs)
```

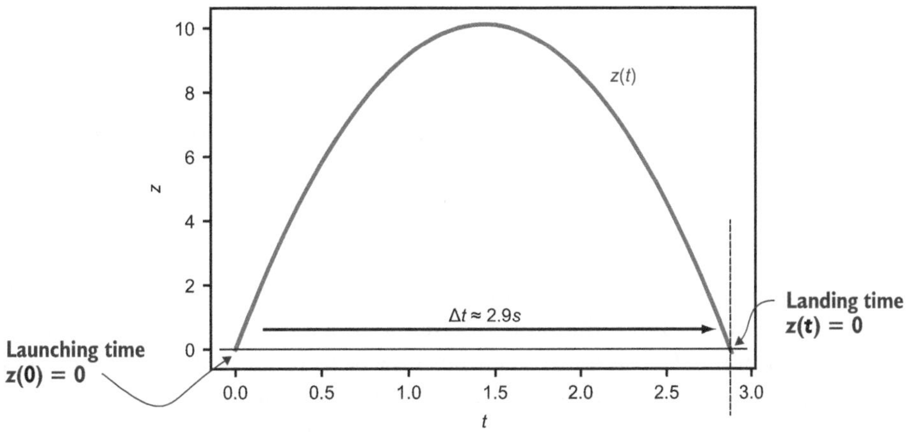

Figure 12.9 A plot of *z*(*t*) for the projectile showing the launching and landing times where *z* = 0. We can see from the graph that the elapsed time is about 2.9 seconds.

We know $z''(t) = g = -9.81$, which is the acceleration due to gravity. We also know the initial z velocity, $z'(0) = |\mathbf{v}| \cdot \sin(\theta)$, and the initial z position, $z(0) = 0$. To recover the position function $z(t)$, we need to integrate the acceleration $z''(t)$ twice. The first integral gives us velocity:

$$z'(t) = z'(0) + \int_0^t g d\tau = |\mathbf{v}| \cdot \sin(\theta) + gt$$

The second integral gives us position:

$$z(t) = z(0) + \int_0^t z'(\tau)\, d\tau = \int_0^t |\mathbf{v}| \cdot \sin(\theta) + g\tau\, d\tau = |\mathbf{v}| \cdot \sin(\theta) \cdot t + \frac{g}{2}t^2$$

We can confirm that this formula matches the simulation by plotting it (figure 12.10). It is nearly indistinguishable from the simulation.

```
def z(t):
    return 20*sin(45*pi/180)*t + (-9.81/2)*t**2

plot_function(z,0,2.9)
```

A direct translation of the result of the integral, z(t), into Python code

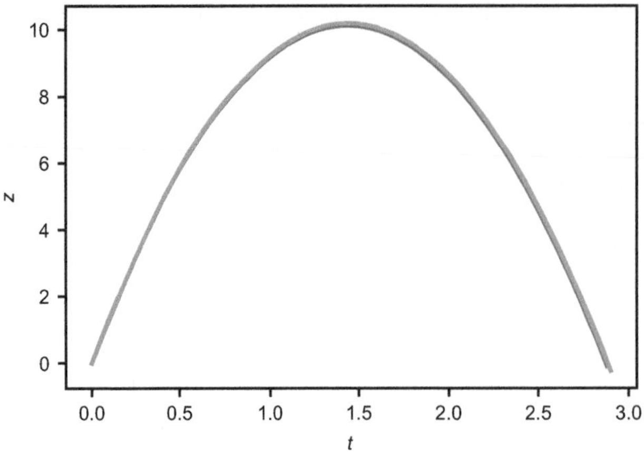

Figure 12.10 Plotting the exact function $z(t)$ on top of the simulated values

For notational simplicity, let's write the initial velocity $|\mathbf{v}| \cdot \sin(\theta)$ as v_z so that $z(t) = v_z t + gt^2/2$. We want to find the value of t that makes $z(t) = 0$, which is the total hang time for the cannonball. You may remember how to find that value from high school algebra, but if not, let me remind you quickly. If you want to know what value of t solves an equation $at^2 + bt + c = 0$, all you have to do is plug the values a, b, and c into the *quadratic formula*:

$$t = \frac{-b \pm \sqrt{b^2 - 4ac}}{2a}$$

An equation like $at^2 + bt + c = 0$ can be satisfied twice; both times when our projectile hits $z = 0$. The symbol ± is shorthand to let you know that using a + or – at this point in the equation gives you two different (but valid) answers.

In the case of solving $z(t) = v_z t + gt^2/2 = 0$, we have $a = g/2$, $b = v_z$ and $c = 0$. Plugging into the formula, we find

$$t = \frac{-v_z \pm \sqrt{v_x^2}}{g} = \frac{-v_z \pm v_z}{g}$$

Treating the ± symbol as a + (plus), the result is $t = (-v_z + v_z)/g = 0$. This says that $z = 0$ when $t = 0$, which is a good sanity check; it confirms that the cannonball starts at $z = 0$. The interesting solution is when we treat ± as a – (minus). In this case, the result is $t = (-v_z - v_z)/g = -2v_z/g$.

Let's confirm the result makes sense. With an initial speed of 20 meters per second and a launch angle of 45° as we used in the simulation, the initial z velocity, v_z, is $-2 \cdot (20 \cdot \sin(45°))/-9.81 \approx 2.88$. This closely matches the result of 2.9 seconds that we read from the graph.

This gives us confidence in calculating the hang time Δt as $\Delta t = -2v_z/g$ or $\Delta t = -2|\mathbf{v}|\sin(\theta)/g$. Because the range is $r = v_x \cdot \Delta t = |\mathbf{v}|\cos(\theta) \cdot \Delta t$, the full expression for the range r as a function of the launch angle θ is

$$r(\theta) = -\frac{2|\mathbf{v}|^2}{g}\sin(\theta)\cos(\theta)$$

We can plot this side by side with the simulated landing positions at various angles as in figure 12.11 and see that it agrees.

```
def r(theta):
    return (-2*20*20/-9.81)*sin(theta*pi/180)*cos(theta*pi/180)

plot_function(r,0,90)
```

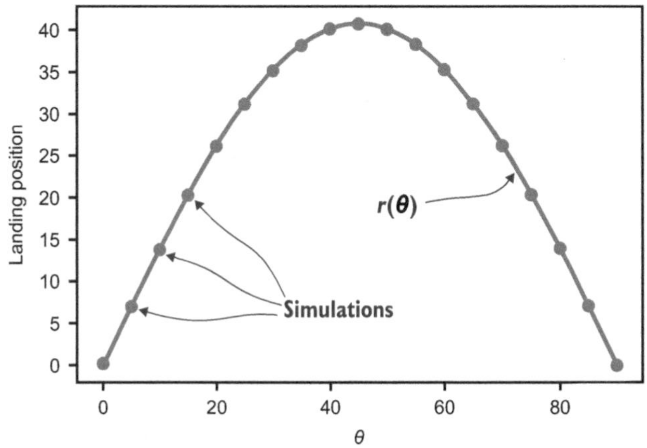

Figure 12.11 Our calculation of projectile range as a function of the launch angle $r(\theta)$, which matches our simulated landing positions

Having a function $r(\theta)$ is a big advantage over repeatedly running the simulator. First of all, it tells us the range of the cannon at *every* launch angle, not just a handful of angles that we simulated. Second, it is much less computationally expensive to evaluate this one function than to run hundreds of iterations of Euler's method. For more complicated simulations, this could make a big difference. Additionally, this function gives us the exact result rather than an approximation. The final benefit, which we'll make use of next, is that the function $r(\theta)$ is smooth, so we can take its derivatives. This gives us an understanding of how the range of the projectile changes with respect to the launch angle.

12.2.2 Solving for the maximum range

Looking at the graph of $r(\theta)$ in figure 12.12, we can set our expectations for what the derivative $r'(\theta)$ will look like. As we increase the launch angle from zero, the range increases as well for a while but at a decreasing rate. Eventually, increasing the launch angle begins to decrease the range.

The key observation to make is that while $r'(\theta)$ is positive, the range is increasing with respect to θ. Then the derivative $r'(\theta)$ crosses below zero, and the range decreases from there. It is precisely at this angle (where the derivative is zero) that the function $r(\theta)$ achieves its maximum value. You can visualize this by seeing that the graph of $r(\theta)$ in figure 12.12 hits its maximum when the slope of the graph is zero.

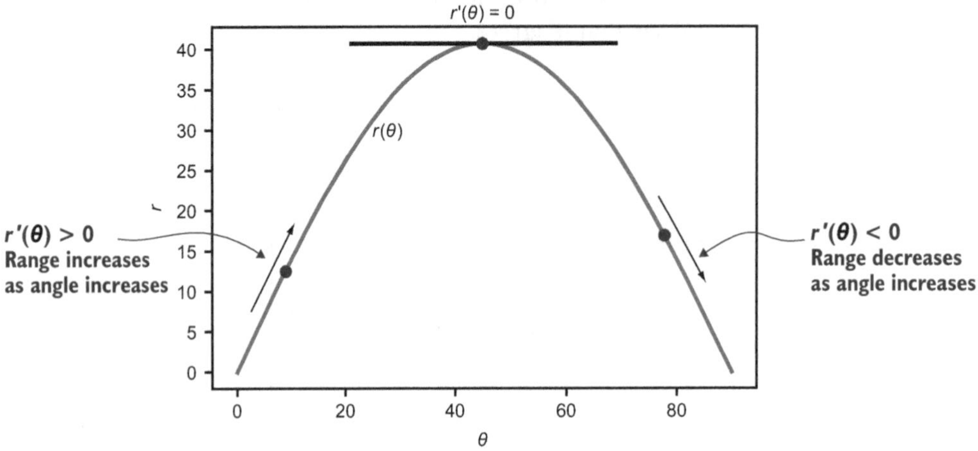

Figure 12.12 The graph of $r(\theta)$ hits its maximum when the derivative is zero and, therefore, the slope of the graph is zero.

We should be able to take the derivative of $r(\theta)$ symbolically, find where it equals zero between 0° and 90°, and this should agree with the rough maximum value of 45°. Remember that the formula for r is

$$r(\theta) = -\frac{2|\mathbf{v}|^2}{g}\sin(\theta)\cos(\theta)$$

Because $-2|\mathbf{v}|^2/g$ is constant with respect to θ, the only hard work is using the product rule on $\sin(\theta)\cos(\theta)$. The result is

$$r'(\theta) = \frac{2|\mathbf{v}|^2}{g}\left(\cos^2(\theta) - \sin^2(\theta)\right)$$

Notice that I factored out the minus sign. If you haven't seen this notation before, $\sin^2(\theta)$ means $(\sin(\theta))^2$. The value of the derivative $r'(\theta)$ is zero when the expression $\sin^2(\theta) - \cos^2(\theta)$ is zero (in other words, we can ignore the constants). There are a few ways to figure out where this expression is zero, but a particularly nice one is to use the trigonometric identity, $\cos(2\theta) = \cos^2(\theta) - \sin^2(\theta)$, which reduces our problem even further. Now we need to figure out where $\cos(2\theta) = 0$.

The cosine function is zero at $\pi/2$ plus any multiple of π, or 90° plus any multiple of 180° (that is, 90°, 270°, 430°, and so on). If 2θ is equal to these values, θ could be half of any of these values: 45°, 135°, 215°, and so on.

Of these, there are two interesting results. First, $\theta = 45°$ is the solution between $\theta = 0$ and $\theta = 90°$, so it is both the solution we expected and the solution we're looking for! The second interesting solution is 135° because this is the same as shooting the cannonball at 45° in the opposite direction (figure 12.13).

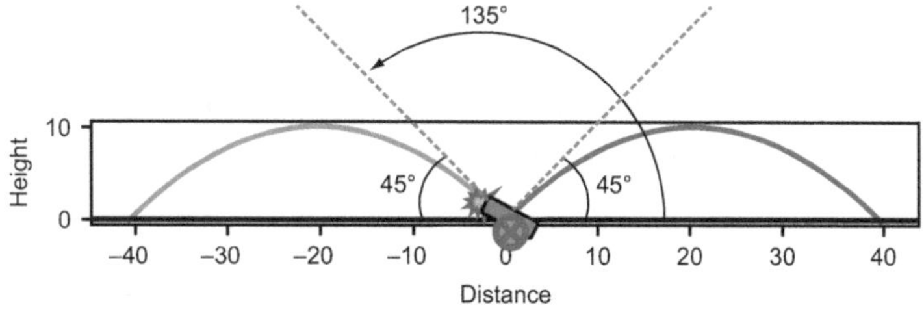

Figure 12.13 In our model, shooting the cannonball at 135° is like shooting at 45° in the opposite direction.

At angles of 45° and 135°, the resulting ranges are

```
>>> r(45)
40.774719673802245
>>> r(135)
-40.77471967380224
```

It turns out that these are the extremes of where the cannonball can end up, with all other parameters equal. A launch angle of 45° produces the maximum landing position, while a launch angle of 135° produces the minimum landing position.

12.2.3 Identifying maxima and minima

To see the difference between the maximum range at 45° and the minimum range at 135°, we can extend the plot of $r(\theta)$. Remember, we found both of these angles because they were at places where the derivative $r'(\theta)$ was zero (figure 12.14).

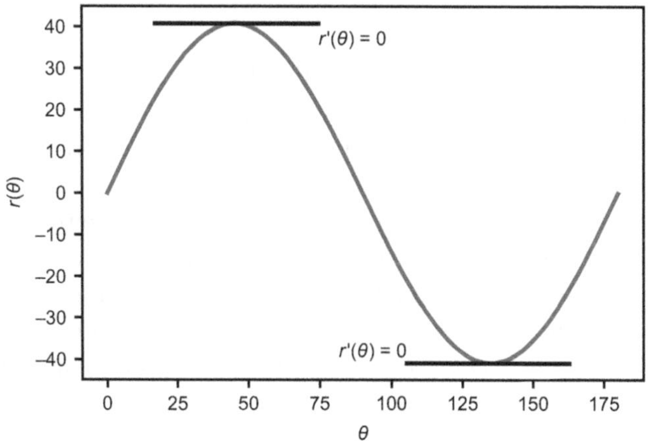

Figure 12.14 The angles $\theta = 45°$ and $\theta = 135°$ are the two values between 0 and 180 where $r'(\theta) = 0$.

While *maxima* (the plural of "maximum") of smooth functions occur where the derivative is zero, the converse is not always true; not every place where the derivative is zero yields a maximum value. As we see in figure 12.14 at $\theta = 135°$, it can also yield a *minimum* value of a function.

You need to be cautious of the global behavior of functions as well, because the derivative can be zero at what's called a *local* maximum or minimum, where the function briefly obtains a maximum or minimum value, but it's real, *global* maximum or minimum values lie elsewhere. Figure 12.15 shows a classic example: $y = x^3 - x$. Zooming in on the region where $-1 < x < 1$, there are two places where the derivative is zero, which look like a maximum and minimum, respectively. When you zoom out, you see that neither of these is the maximum or minimum value for the whole function because it goes off to infinity in both directions.

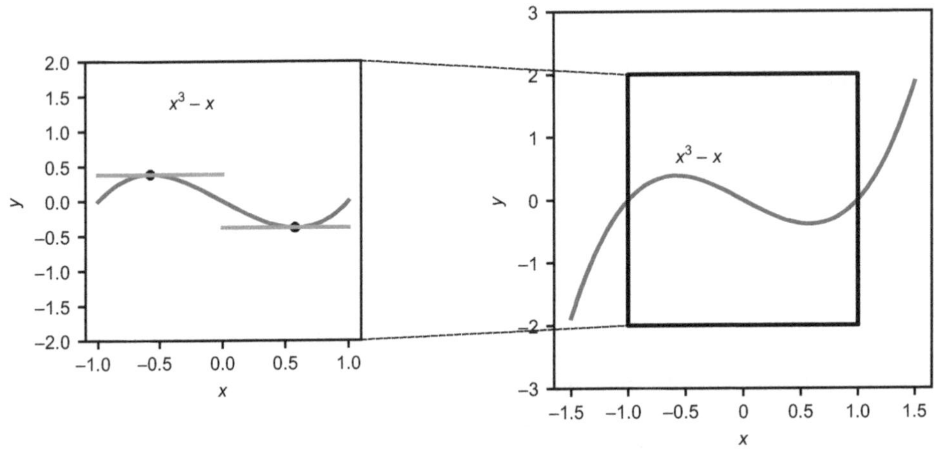

Figure 12.15 Two points that are a *local minimum* and *local maximum*, but neither is the minimum or maximum value for the function

As another confounding possibility, a point where the derivative is zero may not even be a local minimum or maximum. For instance, the function $y = x^3$ has a derivative of zero at $x = 0$ (figure 12.16). This point just happens to be a place where the function x^3 stops increasing momentarily.

I won't go into the technicalities of telling whether a point with a zero derivative is a minimum, maximum, or neither, or how to distinguish local minima and maxima from global ones. The key idea is that you need to fully understand the behavior of a function before you can confidently say you've found an optimal value. With this in mind, let's move on to some more complicated functions to optimize and some new techniques for optimizing them.

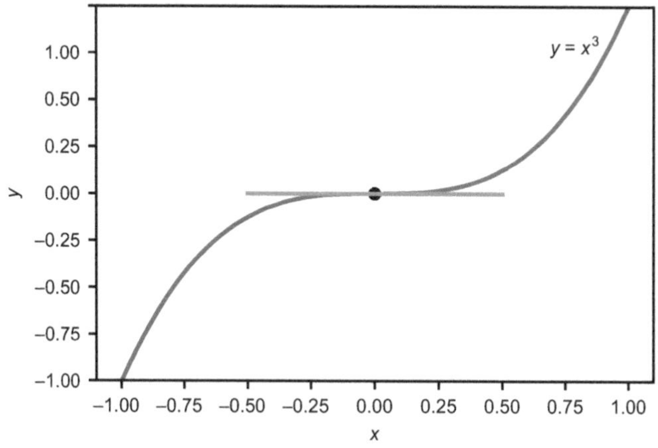

Figure 12.16 For $y = x^3$, the derivative is zero at $x = 0$, but this is not a minimum or maximum value.

12.2.4 Exercises

Exercise 12.6 Use the formula for elapsed time, Δt, in terms of the launch angle θ to find the angle that maximizes the hang time of the cannonball.

Solution The time in the air is $t = 2v_z/g = 2v\sin(\theta)/g$ where the initial speed of the cannonball is $v = |\mathbf{v}|$. This is maximized when $\sin(\theta)$ is maximized. We don't need calculus for this; the maximum value of $\sin(\theta)$ for $0 \leq \theta \leq 180°$ occurs at $\theta = 90°$. In other words, with all other parameters constant, the cannonball stays in the air longest when fired directly upward.

Exercise 12.7 Confirm that the derivative of $\sin(x)$ is zero at $x = 11\pi/2$. Is this a maximum or minimum value of $\sin(x)$?

Solution The derivative of $\sin(x)$ is $\cos(x)$, and

$$\cos\left(\frac{11\pi}{2}\right) = \cos\left(\frac{3\pi}{2} + 4\pi\right) = \cos\left(\frac{3\pi}{2}\right) = 0$$

so the derivative of $\sin(x)$ is indeed zero at $x = 11\pi/2$. Because $\sin(11\pi/2) = \sin(3\pi/2) = -1$ and the sine function ranges between −1 and 1, we can be sure this is a local maximum. Here's a plot of $\sin(x)$ to confirm that:

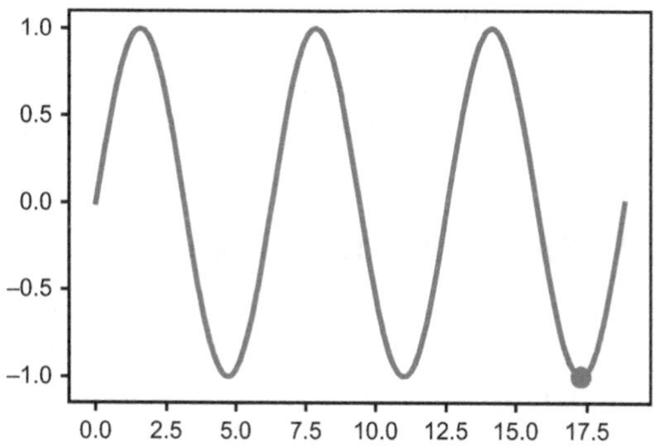

Exercise 12.8 Where does $f(x) = x^3 - x$ have its local maximum and minimum values? What are the values?

Solution You can see from plotting the function that $f(x)$ hits a local minimum value at some $x > 0$ and a local maximum value at some $x < 0$. Let's find these two points.

The derivative is $f'(x) = 3x^2 - 1$, so we want to find where $3x^2 - 1 = 0$. We could use the quadratic formula to solve for *x*, but it's simple enough to eyeball a solution. If $3x^2 - 1 = 0$ then $x^2 = 1/3$, so $x = -1/$ or $x = 1/$. These are the *x* values where $f(x)$ hits its local minimum and maximum values.

The local maximum value is

$$f\left(\frac{-1}{\sqrt{3}}\right) = \frac{-1}{3\sqrt{3}} - \frac{-1}{\sqrt{3}} = \frac{2}{3\sqrt{3}}$$

and the local minimum value is

$$f\left(\frac{1}{\sqrt{3}}\right) = \frac{1}{3\sqrt{3}} - \frac{1}{\sqrt{3}} = \frac{-2}{3\sqrt{3}}$$

Exercise 12.9—Mini Project The graph of a quadratic function $q(x) = ax^2 + bx + c$ with $a \neq 0$ is a *parabola*, an arch shape that either has a single maximum value or a single minimum value. Based on the numbers *a*, *b*, and *c*, what is the *x* value where $q(x)$ is maximized or minimized? How can you tell if this point is a minimum or maximum?

Solution The derivative $q'(x)$ is given by $2ax + b$. This is zero when $x = -b/2a$.

If *a* is positive, the derivative starts negative at some low *x* value, then hits zero at $x = -b/2a$ and is positive from then on. That means *q* is decreasing before $x = -b/2a$ and increasing thereafter; this describes a *minimum* value of $q(x)$.

You can tell the opposite story if *a* is negative. Therefore, $x = -b/2a$ is a minimum value of $q(x)$ if *a* is positive and a maximum value if *a* is negative.

12.3 Enhancing our simulation

As your simulator becomes more complicated, there can be multiple parameters governing its behavior. For our original cannon, the launch angle θ was the only parameter we were playing with. To optimize the range of the cannon, we worked with a function of one variable: $r(\theta)$. In this section, we'll make our cannon fire in 3D, meaning that we need to vary two launch angles as parameters to optimize the range of the cannonball.

12.3.1 Adding another dimension

The first thing is to add a *y* dimension to our simulation. We can now picture the cannon sitting at the origin of the *x*,*y* plane, shooting the cannonball up into the *z* direction at some angle θ. In this version of the simulator, you can control the angle θ as well as a second angle, which we'll call ϕ (the Greek letter phi). This measures how far the cannon is rotated laterally from the $+x$ direction (figure 12.17).

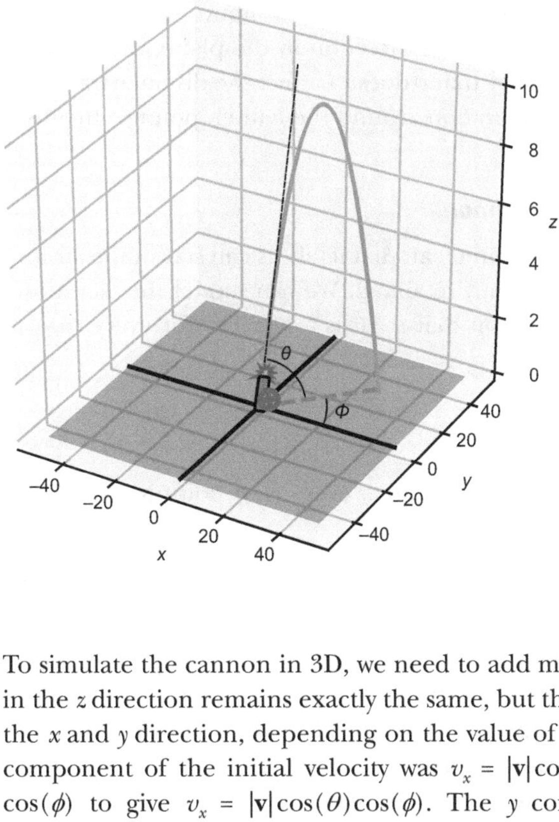

Figure 12.17 Picturing the cannon firing in 3D. Two angles θ and ϕ determine the direction the cannon is fired.

To simulate the cannon in 3D, we need to add motion in the *y* direction. The physics in the *z* direction remains exactly the same, but the horizontal velocity is split between the *x* and *y* direction, depending on the value of the angle ϕ. Whereas the previous *x* component of the initial velocity was $v_x = |\mathbf{v}|\cos(\theta)$, it's now scaled by a factor of $\cos(\phi)$ to give $v_x = |\mathbf{v}|\cos(\theta)\cos(\phi)$. The *y* component of initial velocity is $v_y = |\mathbf{v}|\cos(\theta)\sin(\phi)$. Because gravity doesn't act in the *y* direction, we don't have to update v_y over the course of the simulation. Here's the updated trajectory function:

```
def trajectory3d(theta,phi,speed=20,
                 height=0,dt=0.01,g=-9.81):
    vx = speed * cos(pi*theta/180)*cos(pi*phi/180)
    vy = speed * cos(pi*theta/180)*sin(pi*phi/180)
    vz = speed * sin(pi*theta/180)
    t,x,y,z = 0, 0, 0, height
```

The lateral angle ϕ is the input parameter of the simulation.

Calculates the initial y velocity

```
    ts, xs, ys, zs = [t], [x], [y], [z]
    while z >= 0:
        t += dt
        vz += g * dt
        x += vx * dt
        y += vy * dt
        z += vz * dt
        ts.append(t)
        xs.append(x)
        ys.append(y)
        zs.append(z)
    return ts, xs, ys, zs
```

Stores the values of time and the x, y, positions throughout the simulation

Updates the y position in each iteration

If this simulation is successful, we don't expect it to change the angle θ that yields the maximum range. Whether you fire a projectile at 45° above the horizontal in the $+x$ direction, the $-x$ direction, or any other direction in the plane, the projectile should go the same distance. That is to say that ϕ doesn't affect the distance traveled. Next, we add the terrain with a variable elevation around the launch point so the distance traveled changes.

12.3.2 Modeling terrain around the cannon

Hills and valleys around the cannon mean that its shots can stay in the air for different durations depending on where they're aimed. We can model the elevation above or below the plane $z = 0$ by a function that returns a number for every (x,y) point. For instance,

```
def flat_ground(x,y):
    return 0
```

represents flat ground, where the elevation at every (x,y) point is zero. Another function we'll use is a ridge between two valleys:

```
def ridge(x,y):
    return (x**2 - 5*y**2) / 2500
```

On this ridge, the ground slopes upward from the origin in the positive and negative x directions, and it slopes downward in the positive and negative y directions. (You can plot the cross sections of this function at $x = 0$ and $y = 0$ to confirm this.)

Whether we want to simulate the projectile on flat ground or on the ridge, we have to adapt the `trajectory3d` function to terminate when the projectile hits the ground, not just when its altitude is zero. To do this, we can pass the elevation function defining the terrain as a keyword argument, defaulting to flat ground, and revise the test for whether the projectile is above ground. Here are the changed lines in the function:

```
def trajectory3d(theta,phi,speed=20,height=0,dt=0.01,g=-9.81,
                elevation=flat_ground):
    ...
    while z >= elevation(x,y):
        ...
```

In the source code, I also provide a function called `plot_trajectories_3d`, which plots the result of `trajectory3D` as well as the specified terrain. To confirm our simulation works, we see the trajectory end below $z = 0$ when the cannonball is fired downhill and above $z = 0$ when it is fired uphill (figure 12.18):

```
plot_trajectories_3d(
    trajectory3d(20,0,elevation=ridge),
    trajectory3d(20,270,elevation=ridge),
    bounds=[0,40,-40,0],
    elevation=ridge)
```

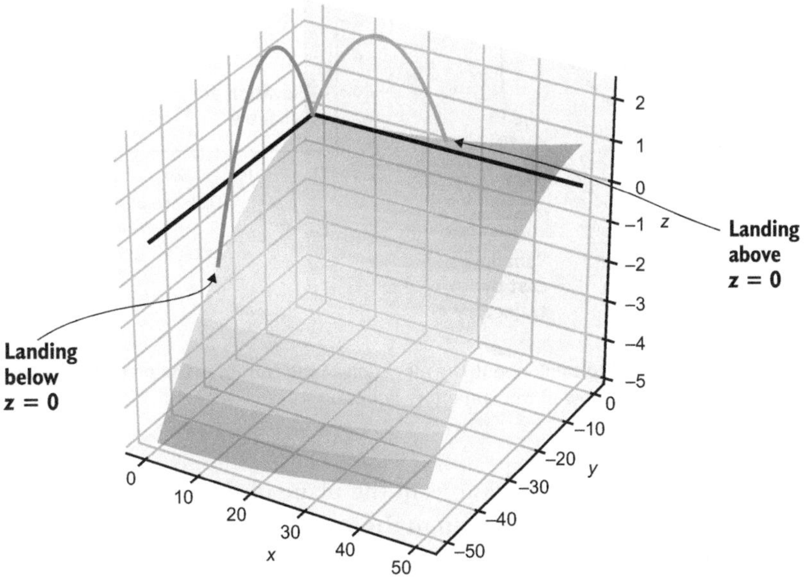

Figure 12.18 A projectile fired downhill lands below $z = 0$ and a projectile fired uphill lands above $z = 0$.

If you had to guess, it seems reasonable that the maximum range for the cannon is attained in the downhill direction rather than the uphill direction. On its way down, the cannonball has further to fall, taking more time and allowing it to travel further. It's not clear if the vertical angle θ will yield the optimal range because our calculation of 45° made the assumption that the ground was flat. To answer this question, we need to write the range r of the projectile as a function of θ and ϕ.

12.3.3 Solving for the range of the projectile in 3D

Even though the cannonball is fired in 3D space in our latest simulation, its trajectory lies in a vertical plane. As such, given an angle ϕ, we only need to work with the slice of the terrain in the direction the cannonball is fired. For instance, if the cannonball is fired at an angle $\phi = 240°$, we only need to think about terrain values when (x, y) lies along a line at 240° from the origin. This is like thinking about the elevation of the terrain only at the points in the shadow cast by the trajectory (figure 12.19).

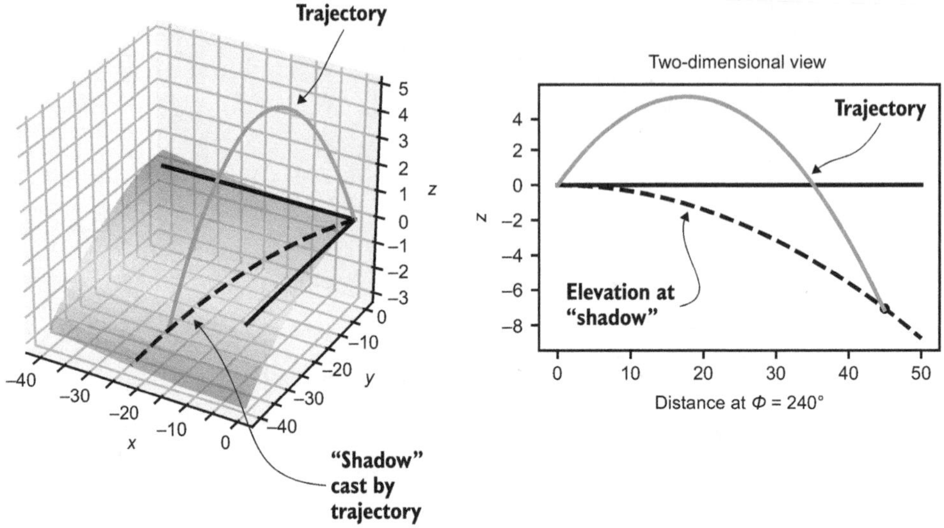

Figure 12.19 We only need to think about the elevation of the terrain in the plane where the projectile is fired. This is where the shadow of the trajectory is cast.

Our goal is to do all of our calculations in the plane of the shadow's trajectory, working with the distance d from the origin in the x,y plane as our coordinate, rather than x and y themselves. At some distance, the trajectory of the cannonball and the elevation of the terrain will have the same z value, which is where the cannonball stops. This distance is the range that we want to find an expression for.

Let's keep calling the height of the projectile z. As a function of time, the height is exactly the same as in our 2D example

$$z(t) = v_z \cdot t + \frac{1}{2}gt^2$$

where v_z is the z component of the initial velocity. The x and y positions are also given as simple functions of time $x(t) = v_x t$ and $y(t) = v_y t$ because no forces act in the x or y direction.

On the ridge, the elevation is given as a function of the x,y position by $(x^2 - 5y^2)/2500$. We can write this elevation as $h(x, y) = Bx^2 - Cy^2$ where $B = 1/2500 = 0.0004$ and $C = 5/2500 = 0.002$. It's useful to know the elevation of the terrain directly under the projectile at a given time t, which we can call $h(t)$. We can calculate the value of h under the projectile at any point in time t because the projectile's x and y positions are given by $v_x t$ and $v_y t$, and the elevation at the same (x, y) point will be $h(v_x t, v_y t) = Bv_x^2 t^2 - Cv_y^2 t^2$.

The altitude of the projectile above the ground at a time t is the difference between $z(t)$ and $h(t)$. The time of impact is the time when this difference is zero, that is $z(t) - h(t) = 0$. We can expand that condition in terms of the definitions of $z(t)$ and $h(t)$:

$$\left(v_z \cdot t + \frac{1}{2}gt^2\right) - (Bv_x^2 t^2 - Cv_y^2 t^2) = 0$$

Once again, we can reshape this into the form $at^2 + bt + c = 0$:

$$\left(\frac{g}{2} - Bv_x^2 + Cv_y^2\right) t^2 - v_z t = 0$$

Specifically, $a = g/2 - Bv_x^2 + Cv_y^2$, $b = v_z$ and $c = 0$. To find the time t that satisfies this equation, we can use the quadratic formula:

$$t = \frac{-b \pm \sqrt{b^2 - 4ac}}{2a}$$

Because $c = 0$, the form is even simpler:

$$t = \frac{-b \pm b}{2a}$$

When we use the + operator, we find $t = 0$, confirming that the cannonball is at ground level at the moment it is launched. The interesting solution is obtained using the – operator, which is the time the projectile lands. This time is $t = (-b - b)/2a = -b/a$. Plugging in the expressions for a and b, we get an expression for landing time in terms of quantities we know how to calculate:

$$t = \frac{-v_z}{\frac{g}{2} - Bv_x^2 + Cv_y^2}$$

The distance in the (x,y) plane that the projectile lands is $\sqrt{x(t)^2 + y(t)^2}$ for this time t. That expands to $\sqrt{(v_x t)^2 + (v_y t)^2} = t\sqrt{v_x^2 + v_y^2}$. You can think of $\sqrt{v_x^2 + v_y^2}$ as the component of the initial velocity parallel to the x,y plane, so I'll call this number $v_x y$. The distance at landing is

$$d = \frac{-v_z \cdot v_{xy}}{\frac{g}{2} - Bv_x^2 + Cv_y^2}$$

All of these numbers in the expression are either constants that I specified or are computed in terms of the initial speed $v = |\mathbf{v}|$ and the launch angles θ and ϕ. It's straightforward (albeit a bit tedious) to translate this to Python, where it becomes clear exactly how we can view the distance as a function of θ and ϕ:

```
B = 0.0004
C = 0.005
v = 20
g = -9.81

def velocity_components(v,theta,phi):
    vx = v * cos(theta*pi/180) * cos(phi*pi/180)
    vy = v * cos(theta*pi/180) * sin(phi*pi/180)
    vz = v * sin(theta*pi/180)
    return vx,vy,vz

def landing_distance(theta,phi):
    vx, vy, vz = velocity_components(v, theta, phi)
    v_xy = sqrt(vx**2 + vy**2)
    a = (g/2) - B * vx**2 + C * vy**2
    b = vz
    landing_time = -b/a
    landing_distance = v_xy * landing_time
    return landing_distance
```

Constants for the shape of the ridge, launch speed, and acceleration due to gravity

A helper function that finds the x, y, and z components of the initial velocity

The horizontal component of initial velocity (parallel to the x,y plane)

The constants a and b

Solves the quadratic equation for landing time, which is –b/a

The horizontal distance traveled

The horizontal distance traveled is the horizontal velocity times the elapsed time. Plotting this point alongside the simulated trajectory, we can verify that our calculated value for the landing position matches the simulation with Euler's method (figure 12.20).

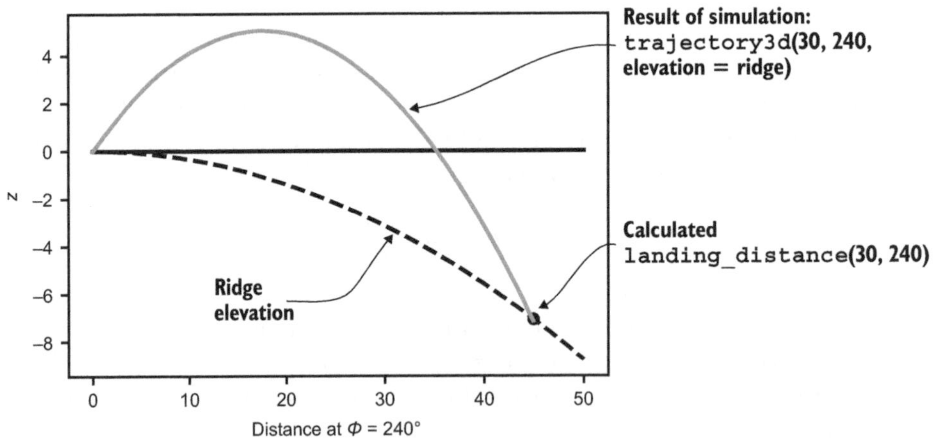

Figure 12.20 Comparing the calculated landing point with the result of the simulation for $\theta = 30°$ and $\phi = 240°$

Now that we have a function $r(\theta, \phi)$ for the range of the cannon in terms of the launch angles θ and ϕ, we can turn our attention to finding the angles that optimize the range.

12.3.4 Exercises

Exercise 12.10 If $|\mathbf{v}| = v$ is the initial speed of the cannonball, verify that the initial velocity vector has a magnitude equal to v. In other words, show that the vector ($v \cos \theta \cos \phi$, $v \cos \theta \sin \phi$, $v \sin \theta$) has length v.

Hint By the definitions of sine and cosine and the Pythagorean theorem, $\sin^2 x + \cos^2 x = 0$ for any value of x.

Solution The magnitude of ($v \cos \theta \cos \phi$, $v \cos \theta \sin \phi$, $v \sin \theta$) is given by

$$\begin{aligned}
&\sqrt{v^2 \cos^2\theta \cos^2\phi + v^2 \cos^2\theta \sin^2\phi + v^2 \sin^2\theta} \\
&= \sqrt{v^2(\cos^2\theta \cos^2\phi + \cos^2\theta \sin^2\phi + \sin^2\theta)} \\
&= \sqrt{v^2(\cos^2\theta(\cos^2\phi + \sin^2\phi) + \sin^2\theta)} \\
&= \sqrt{v^2(\cos^2\theta \cdot 1 + \sin^2\theta} \\
&= \sqrt{v^2 \cdot 1} \\
&= v
\end{aligned}$$

Exercise 12.11 Explicitly write out the formula for the range of the cannonball on the ridge with elevation $Bx^2 - Cy^2$ as a function of θ and ϕ. The constants that appear are B and C, as well as the initial launch speed v and the acceleration due to gravity g.

Solution Starting with the formula

$$d = \frac{-v_z \cdot v_{xy}}{\frac{g}{2} - Bv_x^2 + Cv_y^2}$$

we can plug in $v_z = v \sin \theta$, $v_x y = v \cos \theta$, $v_y = v \cos \theta \sin \phi$, and $v_x = v \cos \theta \cos \phi$ to get

$$d(\theta, \phi) = \frac{-v^2 \sin\theta \cos\theta}{\frac{g}{2} - Bv^2 \cos^2\theta \cos^2\phi + Cv^2 \cos^2\theta \sin^2\phi}$$

With a little simplification in the denominator, this becomes

$$d(\theta, \phi) = \frac{-v^2 \sin\theta \cos\theta}{\frac{g}{2} + v^2 \cos^2\theta \cdot (C \sin^2\phi - B \cos^2\phi)}$$

Exercise 12.12—Mini Project When an object like a cannonball moves quickly through the air, it experiences frictional force from the air, called *drag*, which pushes it in the opposite direction it's moving. The drag force depends on a lot of factors, including the size and shape of the cannonball and the density of the air, but for simplicity, let's assume it works as follows. If v is the cannonball's velocity vector at any point, the drag force, $\mathbf{F}_d$, is

$$\mathbf{F}_d = -\alpha\mathbf{v}$$

where α (the Greek letter alpha) is a number giving the magnitude of drag felt by a particular object in the air. The fact that the drag force is proportional to the velocity means that as an object speeds up, it feels more and more drag. Figure out how to add a drag parameter to the cannonball simulation and show that drag causes the cannonball to slow down.

Solution We want to add to our simulation is an acceleration based on drag. The force will be $-\alpha v$, so the acceleration it causes is $-\alpha v/m$. Because we're not varying the mass of the cannonball, we can use a single drag constant, which is α/m. The components of the acceleration due to drag is $v_x\alpha/m$, $v_y\alpha/m$ and $v_z\alpha/m$. Here's the updated section of the code:

```
def trajectory3d(theta,phi,speed=20,height=0,dt=0.01,g=-9.81,
                 elevation=flat_ground, drag=0):
    ...
    while z >= elevation(x,y):
        t += dt
        vx -= (drag * vx) * dt
        vy -= (drag * vy) * dt
        vz += (g - (drag * vz)) * dt
        ...
    return ts, xs, ys, zs
```

Reduces both vx and vy in proportion to the drag force

Changes the z velocity (vz) by the effects of gravity and drag

You can see that a small drag constant of 0.1 slows down the cannonball noticeably, causing it to fall short of the trajectory without drag.

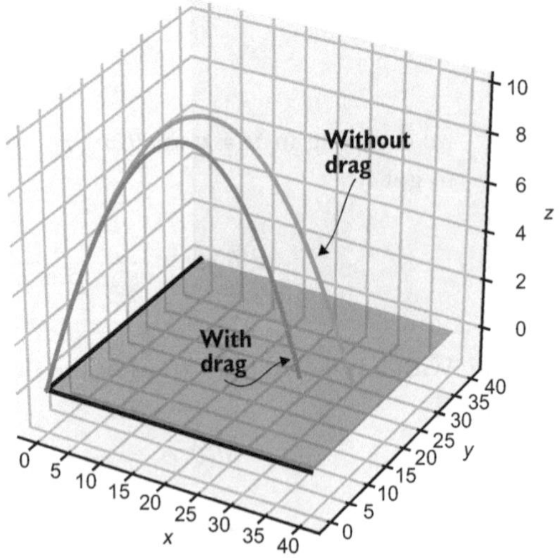

Trajectories of the cannonball with `drag = 0` and `drag = 0.1`

12.4 Optimizing range using gradient ascent

Let's continue to assume that we're firing the cannon on the ridge terrain with some launch angles θ and ϕ, and all other launch parameters set to their defaults. In this case, the function $r(\theta, \phi)$ tells us what the range of the cannon is at these launch angles. To get a qualitative sense of how the angles affect the range, we can plot the function *r*.

12.4.1 Plotting range versus launch parameters

I showed you a few different ways to plot a function of two variables in the last chapter. My preference for plotting $r(\theta, \phi)$ is to use a heatmap. On a 2D canvas, we can vary θ in one direction and vary ϕ in the other and then use color to indicate the corresponding range of the projectile (figure 12.21).

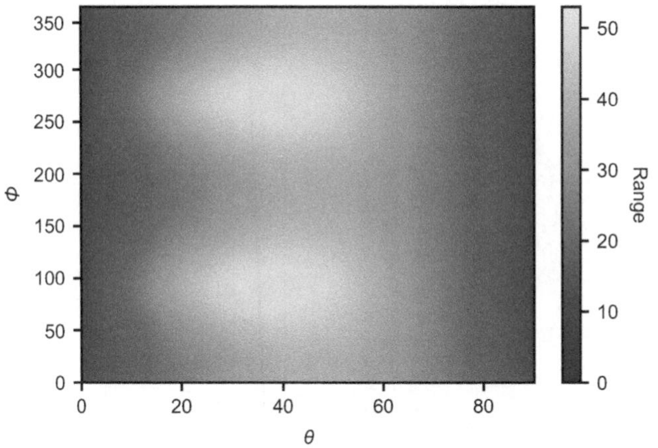

Figure 12.21 A heatmap of the range of the cannon as a function of the launch angles θ and ϕ

This 2D space is an abstract one, having coordinates θ and ϕ. That is to say that this rectangle isn't a drawing of a 2D slice of the 3D world we've modelled. Rather, it's just a convenient way to show how the range r varies as the two parameters change.

On the graph in figure 12.22, brighter values indicate higher ranges, and there appear to be two brightest points. These are possible maximum values of the range of the cannon.

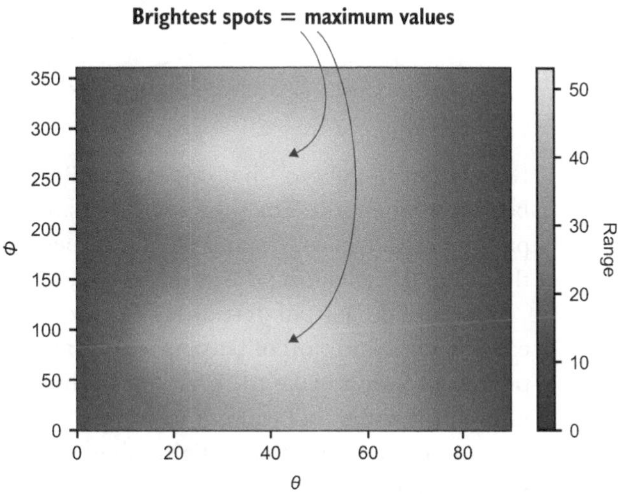

Figure 12.22 The brightest spots occur when the range of the projectile is maximized.

These spots occur at around $\theta = 40$, $\phi = 90$, and $\phi = 270$. The ϕ values make sense because they are the downhill directions in the ridge. Our next goal is to find the exact values of θ and ϕ to maximize the range.

12.4.2 The gradient of the range function

Just as we used the derivative of a function of one variable to find its maximum, we'll use the gradient $\nabla r(\theta, \phi)$ of the function $r(\theta, \phi)$ to find its maximum values. For a smooth function of one variable, $f(x)$, we saw that $f'(x) = 0$ when f attained its maximum value. This is when the graph of $f(x)$ was momentarily flat, meaning the slope of $f(x)$ was zero, or more precisely, that the slope of the line of best approximation at the given point was zero. Similarly, if we make a 3D plot of $r(\theta, \phi)$, we can see that it is flat at its maximum points (figure 12.23).

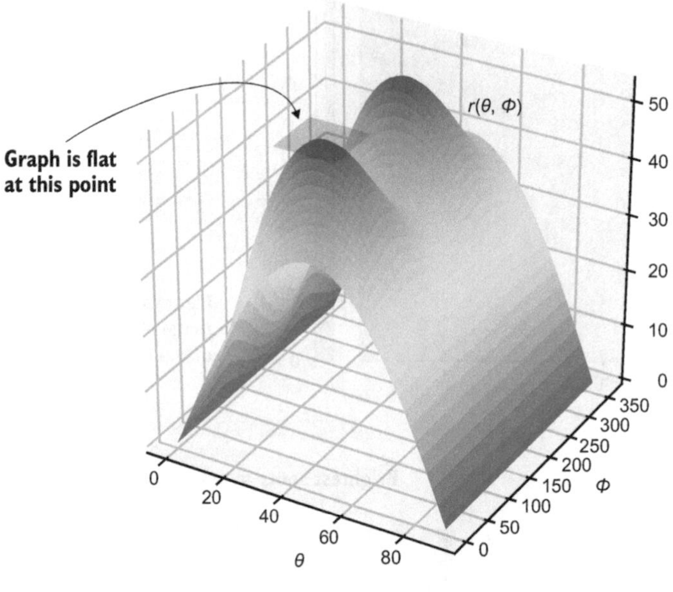

Figure 12.23 The graph of $r(\theta, \phi)$ is flat at its maximum points.

Let's be precise about what this means. Because $r(\theta, \phi)$ is smooth, there is a plane of best approximation. The slopes of this plane in the θ and ϕ directions are given by the partial derivatives $\partial r/\partial\theta$ and $\partial r/\partial\phi$, respectively. Only when both of these are zero is the plane flat, meaning the graph of $r(\theta, \phi)$ is flat.

Because the partial derivatives of r are defined to be the components of the gradient of r, this condition of flatness is equivalent to saying that $\nabla r(\theta, \phi) = 0$. To find such points, we have to take the gradient of the full formula for $r(\theta, \phi)$ and then solve for values of θ and ϕ that cause it to be zero. Taking these derivatives and solving them is a lot of work and not that enlightening, so I'll leave it as an exercise for you. Next, I'll show you a way to follow an *approximate* gradient up the slope of the graph toward the maximum point, which won't require any algebra.

Before I move on, I want to reiterate a point from the previous section. Just because you've found a point on the graph where the gradient is zero doesn't mean that it's a maximum value. For instance, on the graph of $r(\theta, \phi)$, there's a point in between the two maxima where the graph is flat and the gradient is zero (figure 12.24).

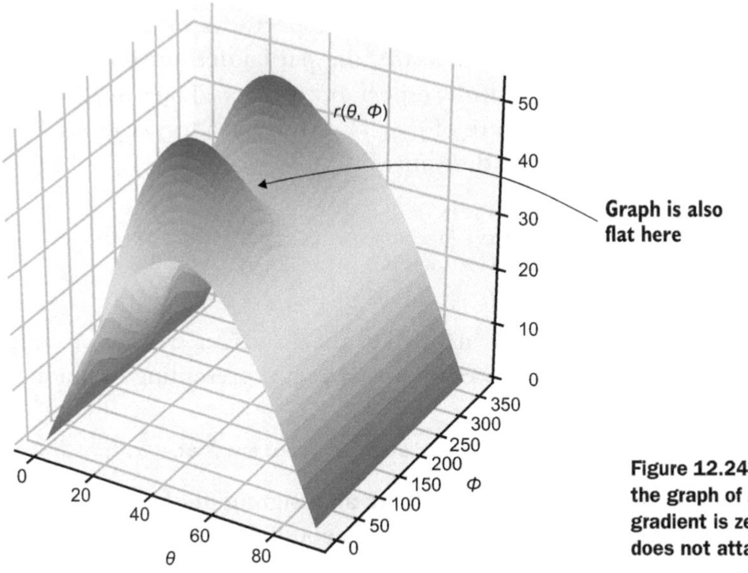

Figure 12.24 A point (θ, ϕ) where the graph of $r(\theta, \phi)$ is flat. The gradient is zero, but the function does not attain a maximum value.

This point isn't meaningless, it happens to tell you the best angle θ when you're shooting the projectile at $\phi = 180°$, which is the worst possible direction because it is the steepest uphill direction. A point like this is called a *saddle point*, where the function simultaneously hits a maximum with respect to one variable and a minimum with respect to another. The name comes from the fact that the graph kind of looks like a saddle.

Again, I won't go into the details of how to identify maxima, minima, saddle points, or other kinds of places where the gradient is zero, but be warned: with more dimensions, there are weirder ways that a graph can be flat.

12.4.3 Finding the uphill direction with the gradient

Rather than take the partial derivatives of the complicated function $r(\theta, \phi)$ symbolically, we can find approximate values for the partial derivatives. The direction of the gradient that they give us tells us, for any given point, which direction the function increases the most quickly. If we jump to a new point in this direction, we should move uphill and towards a maximum value. This procedure is called *gradient ascent*, and we'll implement it in Python.

The first step is to be able to approximate the gradient at any point. To do that, we use the approach I introduced in chapter 9: taking the slopes of small secant lines. Here are the functions as a reminder:

```
def secant_slope(f,xmin,xmax):
    return (f(xmax) - f(xmin)) / (xmax - xmin)
```

Finds the slope of a secant line, f(x), between x values of xmin and xmax

```
def approx_derivative(f,x,dx=1e-6):
    return secant_slope(f,x-dx,x+dx)
```

The approximate derivative is a secant line between x – 10 – 6 and x + 10 – 6.

To find the approximate partial derivative of a function $f(x, y)$ at a point (x_0, y_0), we want to fix $x = x_0$ and take the derivative with respect to y, or fix $y = y_0$ and take the derivative with respect to x. In other words, the partial derivative $\partial f/\partial x$ at (x_0, y_0) is the ordinary derivative of $f(x, y_0)$ with respect to x at $x = x_0$. Likewise, the partial derivative $\partial f/\partial y$ is the ordinary derivative of $f(x_0, y)$ with respect to y at $y = y_0$. The gradient is a vector (tuple) of these partial derivatives:

```
def approx_gradient(f,x0,y0,dx=1e-6):
    partial_x = approx_derivative(lambda x: f(x,y0), x0, dx=dx)
    partial_y = approx_derivative(lambda y: f(x0,y), y0, dx=dx)
    return (partial_x,partial_y)
```

In Python, the function $r(\theta, \phi)$ is encoded as the `landing_distance` function, and we can store a special function, `approx_gradient`, representing its gradient:

```
def landing_distance_gradient(theta,phi):
    return approx_gradient(landing_distance_gradient, theta, phi)
```

This, like all gradients, defines a vector field: an assignment of a vector to every point in space. In this case, it tells us the vector of steepest increase in r at any point (θ, ϕ). Figure 12.25 shows the plot of the `landing_distance_gradient` on top of the heatmap for $r(\theta, \phi)$.

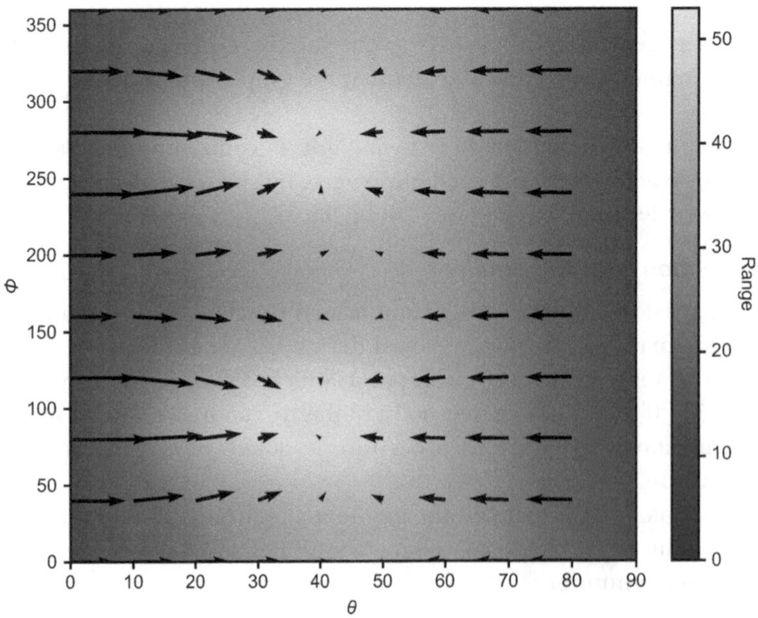

Figure 12.25 A plot of the gradient vector field $\nabla r(\theta, \phi)$ on top of the heatmap of the function $r(\theta, \phi)$. The arrows point in the direction of increase in r, toward brighter spots on the heatmap.

It's even clearer that the gradient arrows converge on maximum points for the function if you zoom in (figure 12.26).

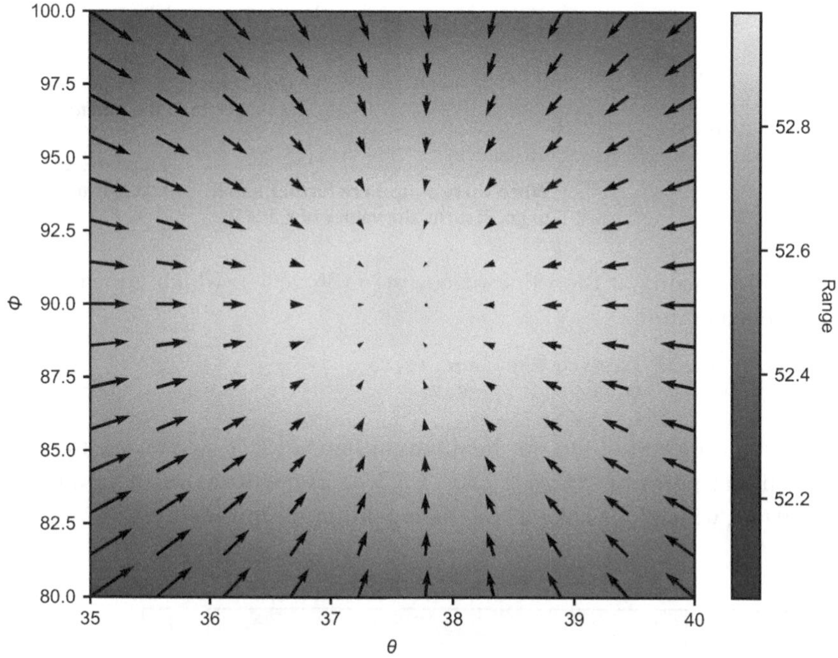

Figure 12.26 The same plot as in figure 12.25 near (θ, ϕ) = (37.5°, 90°), which is the approximate location of one of the maxima

The next step is to implement the *gradient ascent* algorithm, where we start at an arbitrarily chosen point (θ, ϕ) and follow the gradient field until we arrive at a maximum.

12.4.4 Implementing gradient ascent

The gradient ascent algorithm takes as inputs the function we're trying to maximize, as well as a starting point where we'll begin our exploration. Our simple implementation calculates the gradient at the starting point and adds it to the starting point, giving us a new point some distance away from the original in the direction of the gradient. Repeating this process, we can move to points closer and closer to a maximum value.

Eventually, as we approach a maximum, the gradient will get close to zero as the graph reaches a plateau. When the gradient is near zero, we have nowhere further uphill to go, and the algorithm should terminate. To make this happen, we can pass in a *tolerance*, which is the smallest value of the gradient that we should follow. If the gradient is smaller, we can be assured the graph is flat, and we've arrived at a maximum for the function. Here's the implementation:

```
def gradient_ascent(f,xstart,ystart,tolerance=1e-6):
    x = xstart
    y = ystart
    grad = approx_gradient(f,x,y)
    while length(grad) > tolerance:
        x += grad[0]
        y += grad[1]
        grad = approx_gradient(f,x,y)
    return x,y
```

- **Sets the initial values of (x, y) to the input values**
- **Tells us how to move uphill from the current (x, y) value**
- **Only steps to a new point if the gradient is longer than the minimum length**
- **Updates (x, y) to (x, y) + ∇f(x, y)**
- **Updates the gradient at this new point**
- **When there's nowhere further uphill to go, returns the values of x and y**

Let's test this, starting at the value of $(\theta, \phi) = (36°, 83°)$, which appears to be fairly close to the maximum:

```
>>> gradient_ascent(landing_distance,36,83)
(37.58114751557887, 89.99992616039857)
```

This is a promising result! On our heatmap in figure 12.27, we can see the movement from the initial point of $(\theta, \phi) = (36°, 83°)$ to a new location of roughly $(\theta, \phi) = (37.58, 90.00)$, which looks like it has the maximum brightness.

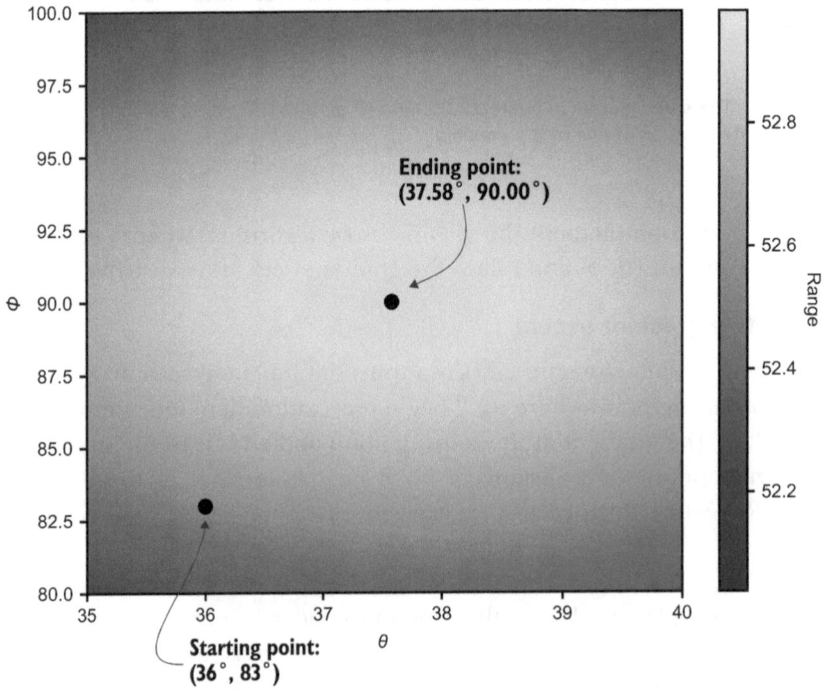

Figure 12.27 The starting and ending points for the gradient ascent

To get a better sense of how the algorithm works, we can track the trajectory of the gradient ascent through the θ, ϕ plane. This is similar to how we tracked the time and position values as we iterated through Euler's method:

```
def gradient_ascent_points(f,xstart,ystart,tolerance=1e-6):
    x = xstart
    y = ystart
    xs, ys = [x], [y]
    grad = approx_gradient(f,x,y)
    while length(grad) > tolerance:
        x += grad[0]
        y += grad[1]
        grad = approx_gradient(f,x,y)
        xs.append(x)
        ys.append(y)
    return xs, ys
```

With this implemented, we can run

```
gradient_ascent_points(landing_distance,36,83)
```

and we get back two lists, consisting of the θ values and the ϕ values at each step of the ascent. These lists both have 855 numbers, meaning that this gradient ascent took 855 steps to complete. When we plot the θ and ϕ points on the heatmap (figure 12.28), we can see the path that our algorithm took to ascend the graph.

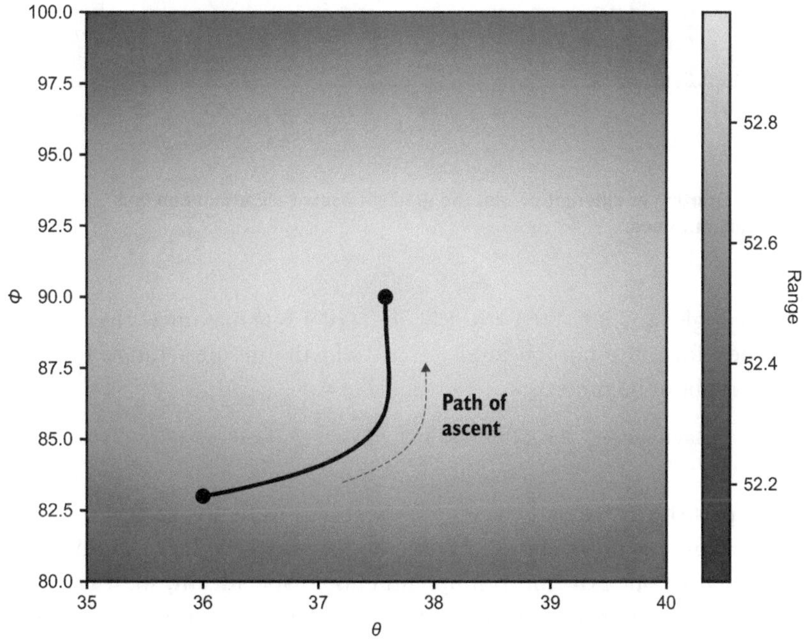

Figure 12.28 The path that the gradient ascent algorithm takes to reach the maximum value of the range function

Note that because there are two maximum values, the path *and* the destination depend on our choice of the initial point. If we start close to $\phi = 90°$, we're likely to hit that maximum, but if we're closer to $\phi = 270°$, our algorithm finds that one instead (figure 12.29).

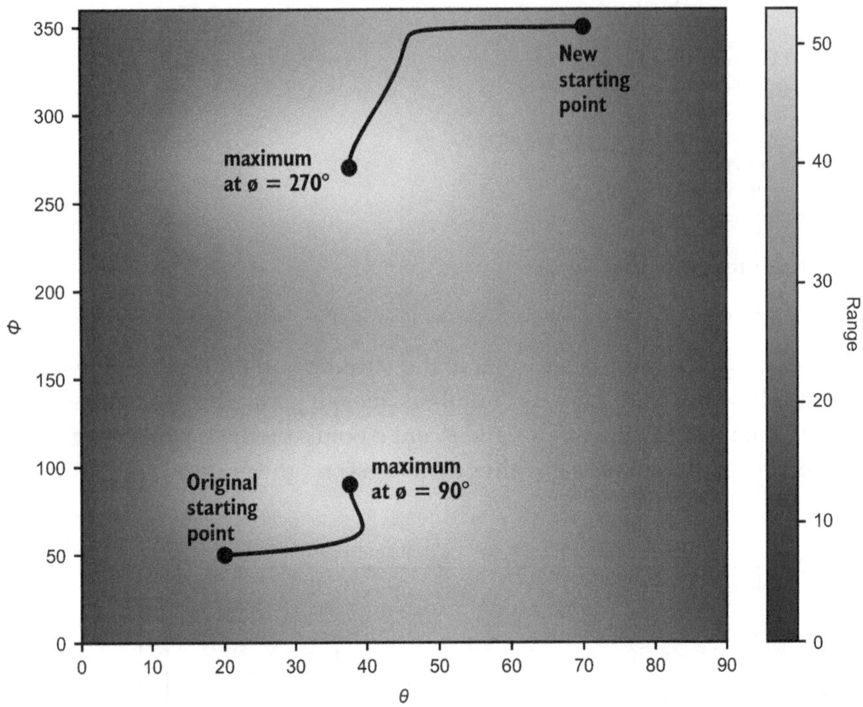

Figure 12.29 Starting at different points, the gradient ascent algorithm can find different maximum values.

The launch angles (37.58°, 90°) and (37.58°, 270°) *both* maximize the function $r(\theta, \phi)$ and are, therefore, the launch angles that yield the greatest range for the cannon. That range is about 53 meters

```
>>> landing_distance(37.58114751557887, 89.99992616039857)
52.98310689354378
```

and we can plot the associated trajectories as shown in figure 12.30.

As we explore some machine learning applications, we'll continue to rely on the gradient to figure out how to optimize functions. Specifically, we'll use the counterpart to gradient ascent, called *gradient descent*. This finds the minimum values for

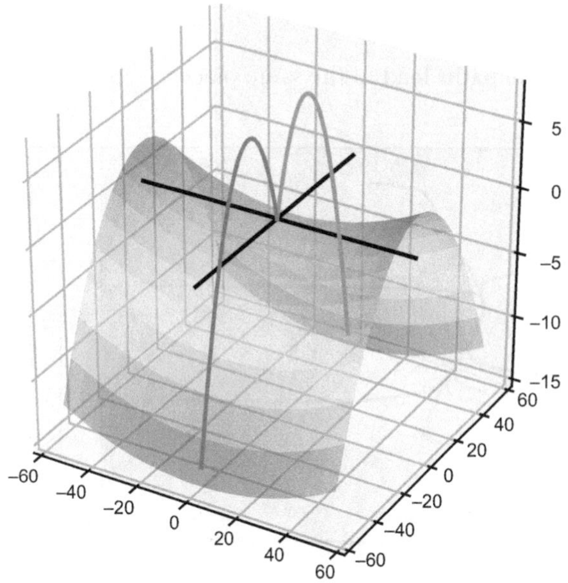

Figure 12.30 The trajectories for the cannon having maximum range

functions by exploring the parameter space in the direction *opposite* the gradient, thereby moving downhill instead of uphill. Because gradient ascent and descent can be performed automatically, we'll see they give a way for machines to learn optimal solutions to problems on their own.

12.4.5 Exercises

Exercise 12.13 On the heatmap, simultaneously plot the paths of gradient ascent from 20 randomly chosen points. All of the paths should end up at one of the two maxima.

Solution With a heatmap already plotted, we can run the following to execute and plot 20 random gradient ascents:

```
from random import uniform
for x in range(0,20):
    gap = gradient_ascent_points(landing_distance,
                                 uniform(0,90),
                                 uniform(0,360))
    plt.plot(*gap,c='k')
```

(continued)

The result shows that all of the paths lead to the same places.

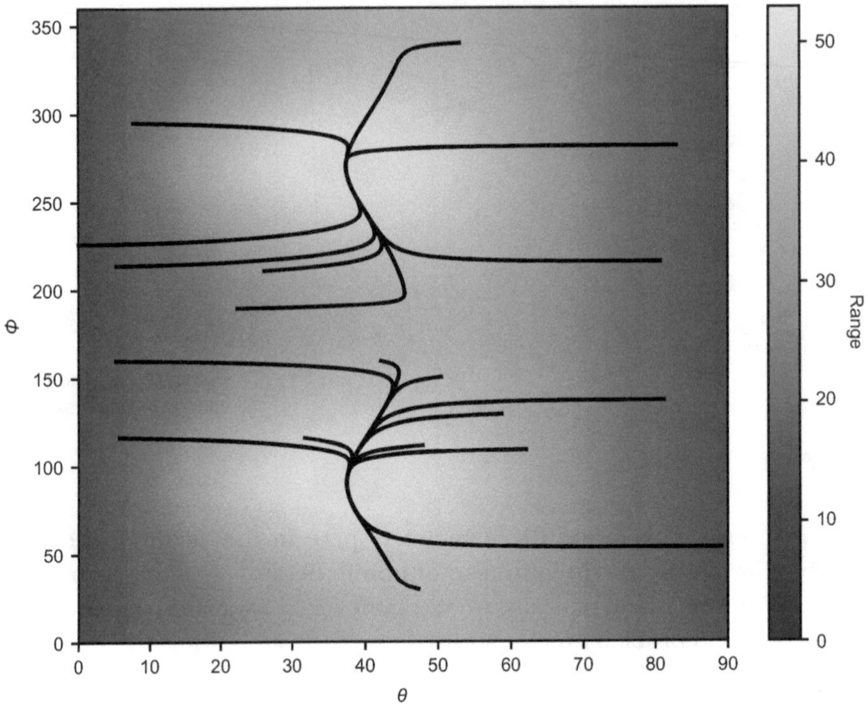

The paths of gradient ascents from 20 random initial points

Exercise 12.14—Mini Project Find the partial derivatives $\partial r/\partial\theta$ and $\partial r/\partial\phi$ symbolically and write a formula for the gradient $\nabla r(\theta, \phi)$.

Exercise 12.15 Find the point on $r(\theta, \phi)$ where the gradient is zero, but the function is not maximized.

Solution We can trick the gradient ascent by starting it with $\phi = 180°$. By the symmetry of the setup, we can see that $\partial r/\partial\phi = 0$ wherever $\phi = 180°$, so the gradient ascent never has a reason to leave the line where $\phi = 0$:

```
>>> gradient_ascent(landing_distance,0,180)
(46.122613357930206, 180.0)
```

This is the optimal launch angle if you fix $\phi = 0$ or $\phi = 180°$, which is the worst angle because you're firing uphill.

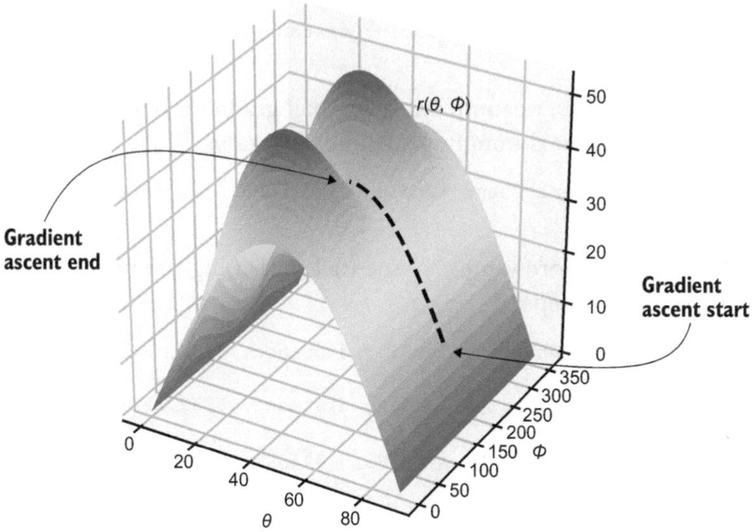

Tricking gradient ascent by initializing it on a cross section where $\partial r/\partial \phi = 0$

Exercise 12.16 How many steps does it take for gradient ascent to reach the origin from (36, 83)? Instead of jumping one gradient, jump 1.5 gradients. Show that you get there in fewer steps. What happens if you jump even further in each step?

Solution Let's introduce a parameter `rate` to the gradient ascent calculation, which indicates how fast the ascent tries to go. The higher the rate, the more we trust the current calculated gradient and jump in that direction:

```
def gradient_ascent_points(f,xstart,ystart,rate=1,tolerance=1e-6):
    ...
    while length(grad) > tolerance:
        x += rate * grad[0]
        y += rate * grad[1]
        ...
    return xs, ys
```

Here's a function that counts the number of steps that a gradient ascent process takes to converge:

```
def count_ascent_steps(f,x,y,rate=1):
    gap = gradient_ascent_points(f,x,y,rate=rate)
    print(gap[0][-1],gap[1][-1])
    return len(gap[0])
```

(continued)

It takes 855 steps to perform our original ascent, with the `rate` parameter equal to 1:

```
>>> count_ascent_steps(landing_distance,36,83)
855
```

With `rate=1.5`, we jump one and a half gradients in each step. Not surprisingly, we get to the maximum faster, in only 568 steps:

```
>>> count_ascent_steps(landing_distance,36,83,rate=1.5)
568
```

Trying some more values, we see that increasing the rate gets us to the solution in even fewer steps:

```
>>> count_ascent_steps(landing_distance,36,83,rate=3)
282
>>> count_ascent_steps(landing_distance,36,83,rate=10)
81
>>> count_ascent_steps(landing_distance,36,83,rate=20)
38
```

Don't get too greedy though! When we use a rate of 20, we get the answer in fewer steps, but some steps appear to overshoot the answer and the next step doubles back. If you set the rate too high, the algorithm can get further and further from the solution; in which case, it is said to diverge rather than converge.

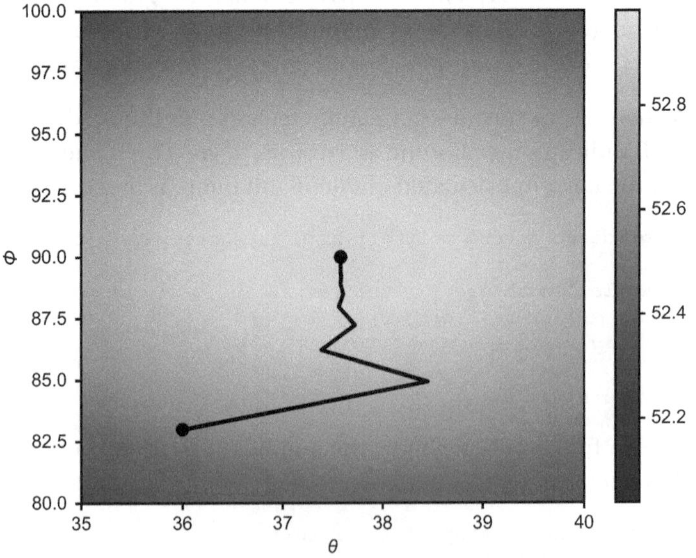

A gradient ascent with a rate of 20. The algorithm initially overshoots the maximum θ value and has to double back.

If you up the rate to 40, your gradient ascent won't converge. Each jump overshoots further than the last, and the exploration of the parameter space runs off into infinity.

Exercise 12.17 What happens when you try to run `gradient_ascent` directly using simulated results for r as a function of θ and ϕ instead of calculated results?

Solution The result is not pretty. This is because the simulated results depend on numerical estimations (like deciding when the projectile hits the ground), so these fluctuate rapidly for small changes in the launch angles. Here's a plot of the cross section $r(\theta, 270°)$ that our derivative approximator would consider when calculating the partial derivative $\partial r/\partial\theta$:

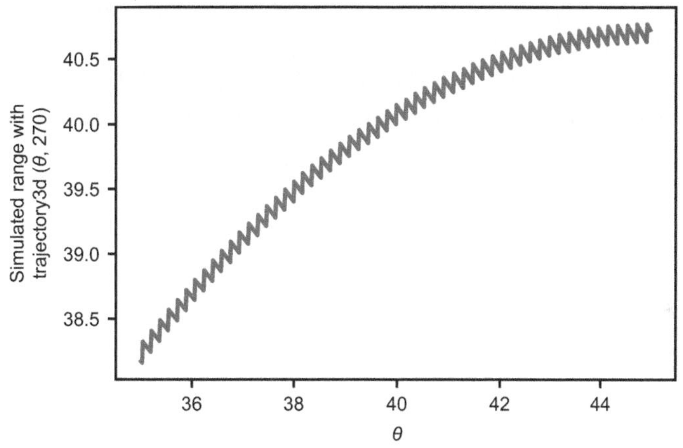

A cross section of simulated trajectories shows that our simulator doesn't produce a smooth function $r(\theta, \phi)$.

The value of the derivative fluctuates wildly, so the gradient ascent moves in random directions.

Summary

- We can simulate a moving object's trajectory by using Euler's method and recording all of the times and positions along the way. We can compute facts about the trajectory, like final position or elapsed time.
- Varying a parameter of our simulation, like the launch angle of the cannon, can lead to different results—for instance, a different range for the cannonball. If

we want to find the angle that maximizes range, it helps to write range as a function of the angle $r(\theta)$.

- Maximum values of a smooth function $f(x)$ occur where the derivative $f'(x)$ is zero. You need to be careful, though, because when $f'(x) = 0$, the function f might be at a maximum value, or it could also be a minimum value or a point where function f has temporarily stopped changing.
- To optimize a function of two variables, like the range r as a function of the vertical launch angle θ and the lateral launch angle ϕ, you need to explore the 2D space of all possible inputs (θ, ϕ) and figure out which pair produces the optimal value.
- Maximum and minimum values of a smooth function of two variables $f(x, y)$ occur when *both* partial derivatives are zero; that is, $\partial f/\partial x = 0$ and $\partial f/\partial y = 0$, so $\nabla f(x, y) = 0$ as well (by definition). If the partial derivatives are zero, it might also be a *saddle point*, which minimizes the function with respect to one variable while maximizing it with respect to the other.
- The gradient ascent algorithm finds an approximate maximum value for a function $f(x,y)$ by starting at an arbitrarily chosen point in 2D and moving in the direction of the gradient $\nabla f(x, y)$. Because the gradient points in the direction of most rapid increase in the function f, this algorithm finds (x, y) points with increasing f values. The algorithm terminates when the gradient is near zero.

13 Analyzing sound waves with a Fourier series

This chapter covers

- Defining and playing sound waves with Python and PyGame
- Turning sinusoidal functions into playable musical notes
- Combining two sounds by adding their sound waves as functions
- Decomposing a sound wave function into its Fourier series to see its musical notes

For a lot of part 2, we've focused on using calculus to simulate moving objects. In this chapter, I'll show you a completely different application: working with audio data. Digital audio data is a computer representation of *sound waves,* which are repeating changes of pressure in the air that our ears perceive as sound. We'll think of sound waves as functions that we can add and scale as vectors, and then we can use integrals to understand what kinds of sounds they represent. As a result, our exploration of sound waves combines a lot of what you've learned about both linear algebra and calculus in earlier chapters.

I won't go too deep into the physics of sound waves, but it's useful to understand how they work at a basic level. What we perceive as sound isn't air pressure itself, but rather rapid changes in air pressure that cause our eardrums to vibrate. For instance, if you play a violin, you drag the bow across one of the strings and cause the string to vibrate. The vibrating string causes the air around it to rapidly change pressure, and the changes in pressure propagate through the air as sound waves until they reach your ear. At that point, your eardrum vibrates at the same rate, and you perceive a sound (figure 13.1).

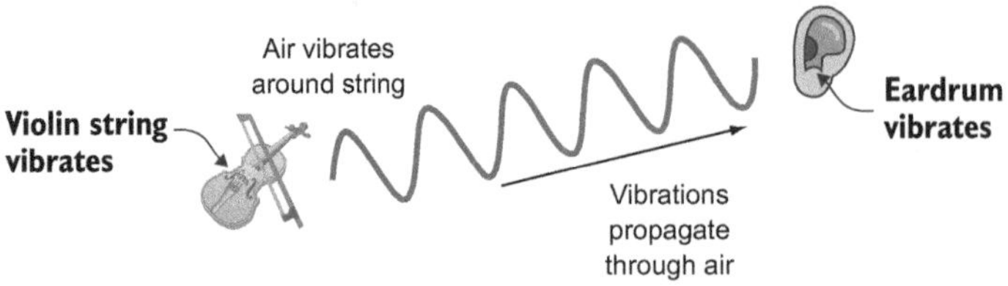

Figure 13.1 Schematic diagram of the sound of a violin reaching an eardrum

You can think of a digital audio file as a function describing a vibration over time. Audio software interprets the function and instructs your speakers to vibrate accordingly, producing sound waves of a similar shape in the air around the speakers. For our purposes, it doesn't matter exactly what the function represents, but you can interpret it loosely as describing air pressure over time (figure 13.2).

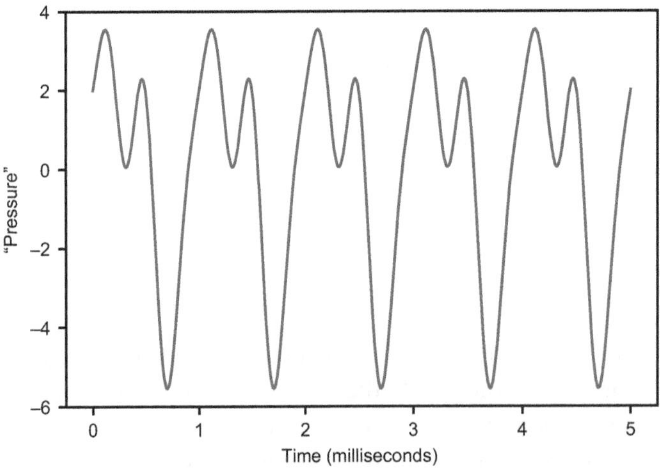

Figure 13.2 Thinking of sound waves as functions, loosely interpreted as representing pressure over time

Interesting sounds like musical notes have sound waves with repeating patterns, like the one shown in figure 13.2. The rate at which the function repeats itself is called the *frequency* and tells how high or low the musical note sounds. The quality, or *timbre*, of the sound is controlled by the shape of the repeating pattern, for instance, whether it sounds more like a violin, a trumpet, or a human voice.

13.1 Combining sound waves and decomposing them

Throughout this chapter, we do mathematical operations on functions and use Python to play them as actual sounds. The two main things we'll do are combining existing sound waves to make new ones and then decomposing complex sound waves into simpler ones. For instance, we can combine several musical notes into a chord and then we can decompose a chord to see its musical notes.

Before we do that, however, we need to cover the basic building blocks: sound waves and musical notes. I start by showing you how to use Python to turn a sequence of numbers, representing a sound wave, into a real sound coming out of your speakers. To make a sound corresponding to a function, we extract some *y* values from the graph of the function and pass these to the audio library as an array. This is a process called *sampling* (figure 13.3).

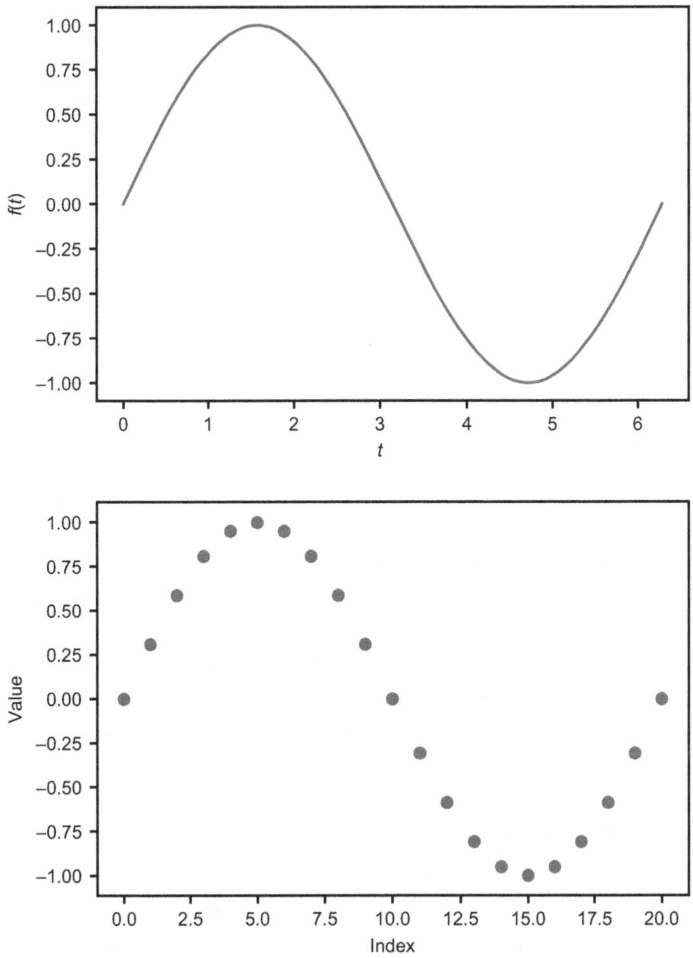

Figure 13.3 Starting with the graph of a function *f*(*t*) (top) and sampling some of the *y* values (bottom) to send to an audio library

The main sound wave functions we'll use are *periodic functions*, whose graphs are built from the same repeating shape. Specifically, we'll use *sinusoidal functions*, a family of periodic functions including sine and cosine that produce natural-sounding musical notes. After sampling them to turn them into sequences of numbers, we'll build Python functions to play musical notes.

Once we can produce individual notes, we'll write Python code to help us combine different notes to create chords and other complex sounds. We'll do this by adding the functions defining each of the sound waves together. We'll see that combining a few musical notes can make a chord, and combining dozens of musical notes together can produce some quite interesting and qualitatively dissimilar sounds.

Our last goal will be to decompose a function representing any sound wave into a sum of (pure) musical notes and their corresponding volumes, which make up the sound wave (figure 13.4). Such a decomposition into a sum is called a *Fourier series* (pronounced *FOR-ee-yay*). Once we've found the sound waves making up a Fourier series, we can play them together and get the original sound.

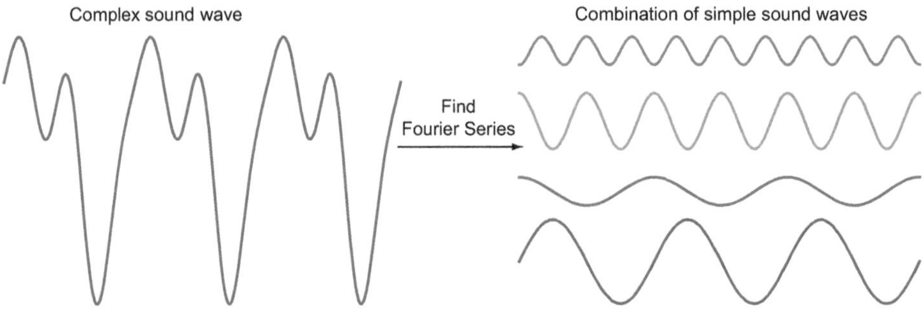

Figure 13.4 Decomposing a sound wave function into a combination of simpler ones using a Fourier series

Mathematically, finding a Fourier series means writing a function as a sum or, more specifically, a linear combination of sine and cosine functions. This procedure and its variants are among the most important algorithms of all time. Methods similar to the ones we'll cover are used in common applications like MP3 compression, as well as in more grandiose ones like the recent Nobel prize-winning detection of gravitational waves.

It's one thing to look at these sound waves as graphs, but it's another to actually hear them coming out of your speakers. Let's make some noise!

13.2 Playing sound waves in Python

To play sounds in Python, we turn to the PyGame library that we used in a few of the preceding chapters. A particular function in this library takes an array of numbers as input and plays a sound as a result. As a first step, we use a random sequence of numbers in Python and write code to interpret and play these sounds with PyGame. This will just be *noise* (yes, that's a technical term!) rather than beautiful music at this point, but we need to start somewhere.

After producing some noise, we'll make a slightly more appealing sound by running the same process on a sequence of numbers that have repeating patterns, rather than just being completely random. This sets us up for the next section, where we'll get a sequence of repeating numbers by sampling a periodic function.

13.2.1 Producing our first sound

Before we pass PyGame an array of numbers representing a sound, we need to tell it how the numbers should be interpreted. There are several technical details about audio data here, and I'll explain them so you know how PyGame thinks about that, but these details won't be critical to the rest of the chapter.

In this application, we use conventions typically used in CD audio. Specifically, we'll represent one second of audio with an array of 44,100 values, each of which is a 16-bit integer (between –32,768 and 32,767). These numbers roughly represent the intensity of the sound at every step of time, with 44,100 steps in a second. This is not unlike how we represented an image in chapter 6. Instead of an array of values giving the brightness of pixels, we have an array of values giving the intensity of a sound wave at different moments in time. Eventually, we'll get these numbers as the *y*-coordinates of points on a sound wave graph, but for now, we're going to pick them randomly to make *some* noise.

We also use a single *channel*, meaning we only play one sound wave as opposed to *stereo* audio, which produces two sound waves simultaneously, one in the left speaker and one in the right. The other thing we configure is the bit depth of the sound. While frequency is analogous to the resolution of an image, *bit depth* is like the number of allowable pixel colors, more bit depth means a more refined range of sound intensities. We used three numbers between 0 and 256 for the color of a pixel, but here, we use a single 16-bit number to represent the sound intensity at a moment in time. With these parameters selected, the first step in our code is to import PyGame and initialize the sound library:

```
>>> import pygame, pygame.sndarray
>>> pygame.mixer.init(frequency=44100,
                      size=-16,
                      channels=1)
```

–16 indicates a bit depth of 16 and an input of 16-bit signed integers ranging from –32,768 to 32,767

To start with the simplest possible example, we can generate one second of audio by creating a NumPy array of 44,100 random integers between –32,768 and 32,767. We can do this in one line with NumPy's `randint` function:

```
>>> import numpy as np
>>> arr = np.random.randint(-32768, 32767, size=44100)
>>> arr
array([-16280, 30700, -12229, ..., 2134, 11403, 13338])
```

To interpret this array as a sound wave, we can plot its first few values on a scatter graph. I've included a `plot_sequence` function in the source code for this book to help you quickly plot an array of integer values like this. If you run `plot_sequence (arr,max=100)`, you get a picture of the first 100 values of this array. As compared to numbers sampled from a smooth function, these numbers are all over the place (figure 13.5).

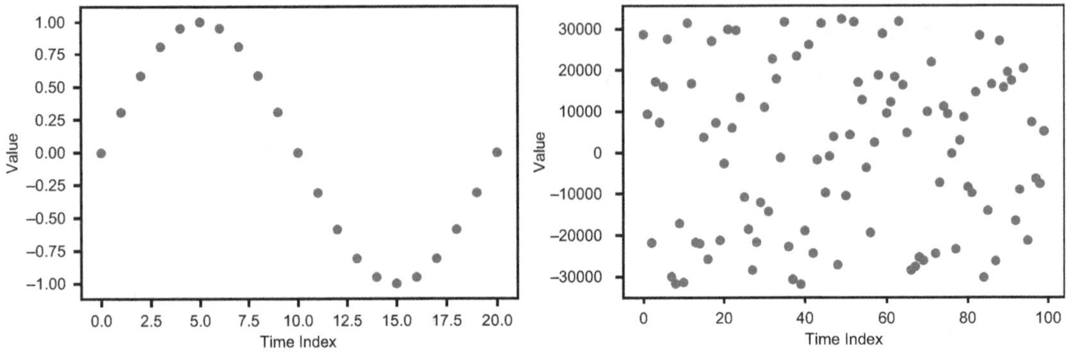

Figure 13.5 Sampled values from a sound wave (left) vs. our random values (right)

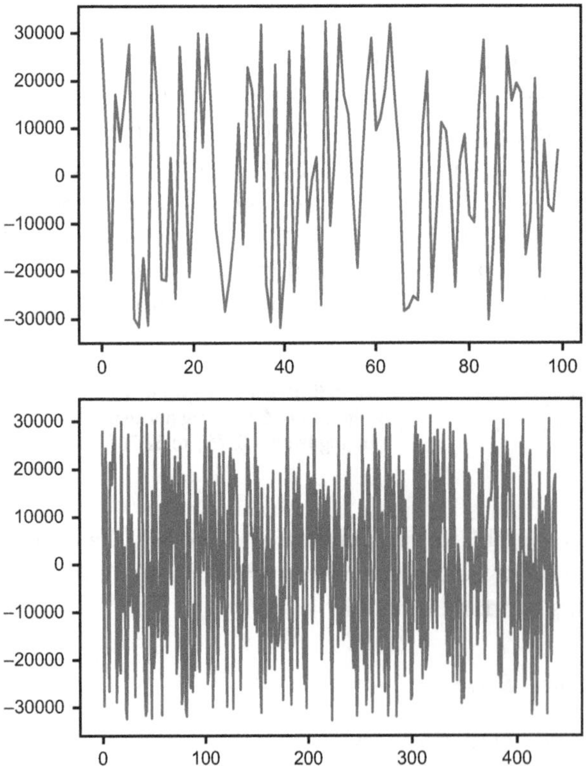

Figure 13.6 The first 100 values (top) and the first 441 values (bottom) connected to define a function

If you connect the dots, you can picture them as defining a function over this time period. Figure 13.6 shows two graphs of the array of numbers with the dots connected, showing the first 100 and 441 numbers, respectively. This data is completely random, so there's nothing particularly interesting to see, but this will be the first sound wave we play.

Because 44,100 values define the whole second of sound, the 441 values on the bottom define the sound during the first one-hundredth of a second. Next, we can play the sound using a library call.

CAUTION Before you run the next few lines of Python code, make sure your speaker volume isn't too high. The first sound we've made won't be that pleasant, not to mention, you don't want to hurt your ears!

To play the sound, you can run:

```
sound = pygame.sndarray.make_sound(arr)
sound.play()
```

The result should sound like one second of static, as if you turned on the radio without tuning it to a station. A sound wave like this, consisting of random values over time, is called *white noise*.

About the only thing you can adjust about white noise is the volume. The human ear responds to changes in pressure, and the bigger the sound wave, the bigger the changes in pressure, and the louder the perceived sound. If this white noise was unpleasantly loud for you, you can create a quieter version by generating sound data consisting of smaller numbers. For instance, this white noise is generated by numbers ranging from –10,000 to 10,000:

```
arr = np.random.randint(-10000, 10000, size=44100)
sound = pygame.sndarray.make_sound(arr)
sound.play()
```

This sound should be nearly identical to the first white noise you played, except that it is quieter. The loudness of a sound wave depends on how big the function values are, and the measure of this is called the *amplitude* of the wave. In this case, because the values vary 10,000 units from the average value of 0, the amplitude is said to be 10,000.

Although some people find white noise soothing, it's not very interesting. Let's produce a more interesting sound, namely a musical note.

13.2.2 Playing a musical note

When we hear a musical note, our ears are detecting a pattern in the vibrations in contrast to the randomness of white noise. We can put together a series of 44,100 numbers with an obvious pattern, and you'll hear them produce a musical note. Specifically, let's start by repeating the number 10,000 fifty times and then repeating the number –10,000 fifty times. I picked 10,000 because we just saw it's a big enough amplitude to make the sound wave audible. Figure 13.7 shows the plot for the first 100 numbers returned from the following code snippet:

```
form = np.repeat([10000,-10000],50)    <-- Repeats each value in the list
plot_sequence(form)                        the specified number of times
```

If we repeat this sequence of 100 numbers 441 times, we have 44,100 total values that define one second of audio. To achieve this, we can use another handy NumPy function, called `tile`, which repeats a given array a specified number of times:

```
arr = np.tile(form,441)
```

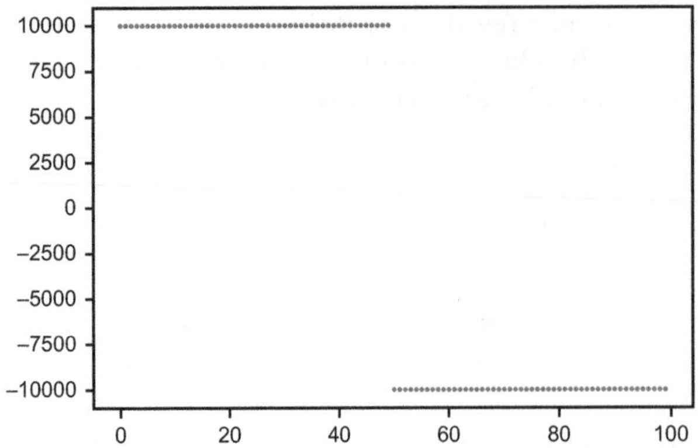

Figure 13.7 A plot of the sequence, consisting of the number 10,000 repeated 50 times followed by the number –10,000 repeated 50 times.

Figure 13.8 shows the plot of the first 1,000 values of the array with the "dots" connected. You can see that it jumps back and forth between 10,000 and –10,000 every 50 numbers. That means the pattern repeats every 100 numbers.

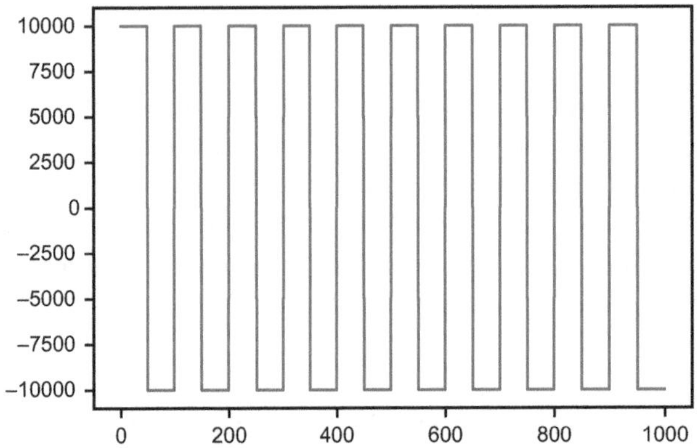

Figure 13.8 A plot of the first 1,000 of 44,100 numbers shows the repeating pattern.

This waveform is called a *square wave* because its graph has sharp, 90° corners. (Note that the vertical lines are only there because MatPlotLib connects all of the dots; there are no values of the sequence between 10,000 and –10,000, just a dot at 10,000 connected to a dot at –10,000.)

The 44,100 numbers represent one second, so the 1,000 numbers graphed in figure 13.8 represent 1/44.1 seconds (or 0.023 seconds) of audio. Playing this sound data using the following lines produces a clear musical note. This is approximately the note A (or A_4 in scientific pitch notation). You can listen to it with the same `play()` method as used in section 13.2.1:

```
sound = pygame.sndarray.make_sound(arr)
sound.play()
```

The rate of repetition (in this case, 441 repetitions per second) is called the *frequency* of the sound wave, and it determines the *pitch* of the note, or how high or low the note sounds. Frequencies of repetition are measured in units of *hertz*, abbreviated Hz, where 441 Hz means the same thing as 441 per second. The most common definition for the pitch A is 440 Hz, but 441 is close enough, and it conveniently divides the CD sampling rate of 44,100 values per second.

Interesting sound waves come from *periodic* functions, which repeat themselves on fixed intervals like the square wave in figure 13.8. The repeated sequence for the square wave consists of 100 numbers, and we repeat it 441 times to get 44,100 numbers giving one second of audio. That's a repetition rate of 441 Hz or once every 0.0023 seconds. What our ear detects as a musical note is this rate of repetition. In the next section, we'll play sounds corresponding to the most important periodic functions, sine and cosine, at different frequencies.

13.2.3 Exercises

Exercise 13.1 Our musical note A was a pattern that repeated 441 times in one second. Create a similar pattern that repeats 350 times in one second, which produces the musical note F.

Solution Fortunately, the frequency of 44,100 Hz is divisible by 350: 44,100 / 350 = 126. With 63 values of 10,000 and 63 values of –10,000, we can repeat that sequence 350 times to create one second of audio. The resulting note sounds lower than the A and is indeed an F:

```
form = np.repeat([10000,-10000],63)
arr = np.tile(form,350)
sound = pygame.sndarray.make_sound(arr)
sound.play()
```

13.3 Turning a sinusoidal wave into a sound

The sound we played with the square wave was a recognizable musical note, but it wasn't very natural sounding. That's because in nature, things don't usually vibrate in square waves. More often, vibrations are *sinusoidal*, meaning if we measure and graph these, we get results that look like the graphs of the sine or cosine functions. These functions turn out to be mathematically natural as well, so we can use them as the building blocks for the music we're going to make. After sampling the notes and passing them to PyGame, you'll be able to hear the difference between a square wave and a sinusoidal wave.

13.3.1 Making audio from sinusoidal functions

The sinusoidal functions sine and cosine, which we've used several times already in this book, are intrinsically periodic functions. That's because their inputs are interpreted as angles; if you rotate 360° or 2π radians, you're back where you started, and the sine and cosine functions return the same values. Therefore, $\sin(t)$ and $\cos(t)$ repeat themselves every 2π units as shown in figure 13.9.

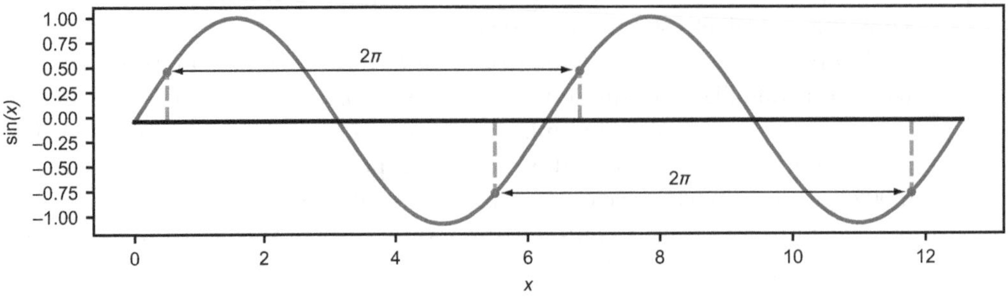

Figure 13.9 Every 2π units, the function sin(t) repeats the same value.

This interval of repetition is called the *period* of the periodic function, so for both sine and cosine, the period is 2π. When you graph them (figure 13.10), you can see that they look the same between 0 and 2π as they do between 2π and 4π, or between 4π and 6π, and so on.

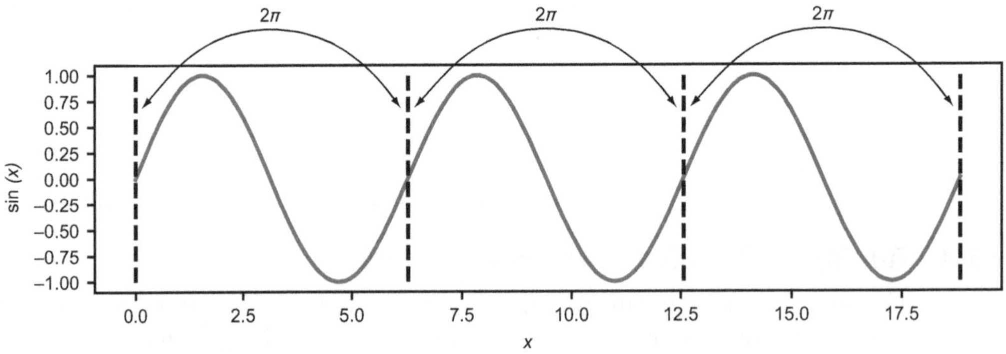

Figure 13.10 Because the sine function is periodic with period 2π, its graph has the same shape over every 2π interval.

The only difference for the cosine function is that the graph is shifted by $\pi/2$ units to the left, but it still repeats itself every 2π units (figure 13.11).

For the purposes of audio, one repetition every 2π seconds is a frequency of $1/2\pi$ or about 0.159 Hz, which turns out to be too small to be audible by the human ear. The amplitude of 1.0 also turns out to be too small to hear in 16-bit audio. To solve this, let's write a Python function, `make_sinusoid(frequency,amplitude)`, which

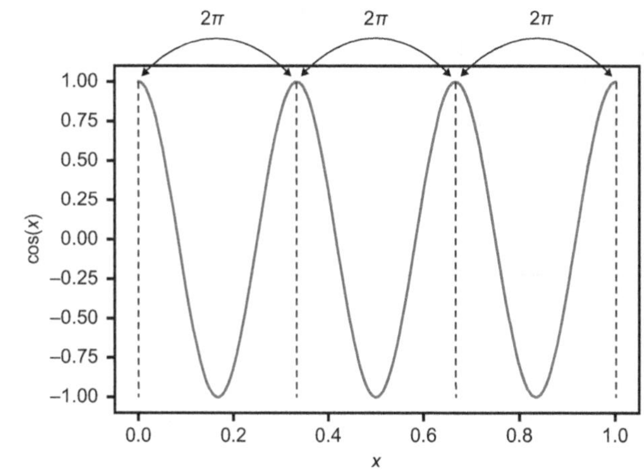

Figure 13.11 The graph of the cosine function has the same shape as the graph of the sine function, but it is shifted to the left. It also repeats itself every 2π units.

produces a sine function that is stretched or compressed vertically and horizontally to have a more desirable frequency and amplitude. A frequency of 441 Hz and an amplitude of 10,000 should represent an audible sound wave.

Once we've produced that function, we want to extract 44,100 evenly spaced values of the function to pass to PyGame. The process of extracting function values like this is called *sampling*, so we can write a function called `sample(f,start,end,count)` that gets the specified count number of values of $f(t)$ in the range of t values between `start` and `end`. Once we have our desired sinusoid function, we can run `sample(sinusoid,0,1,44100)` to get an array of 44,100 samples to pass to PyGame, and we'll hear what a sinusoidal wave sounds like.

13.3.2 Changing the frequency of a sinusoid

As a first example, let's create a sinusoid with a frequency of 2, meaning a function shaped like a sine graph but repeating itself twice between zero and one. The period of the sine function is 2π, so by default, it takes 4π units to repeat itself twice (figure 13.12).

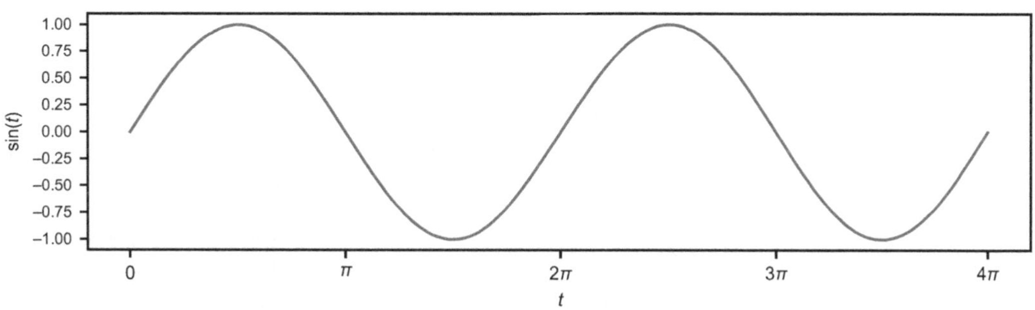

Figure 13.12 The sine function repeats itself twice between $t = 0$ and $t = 4\pi$.

To get two periods of the sine function graph, we need the sine function to receive values from 0 to 4π as inputs, but we want the input variable t to vary from 0 to 1. To achieve that, we can use the function $\sin(4\pi t)$. From $t = 0$ to $t = 1$, all of the values between 0 and 4π are passed to the sine function. The plot of $\sin(4\pi t)$ in figure 13.13 has the same graph as figure 13.12 but with two full periods of the sine function squished into the first 1.0 units.

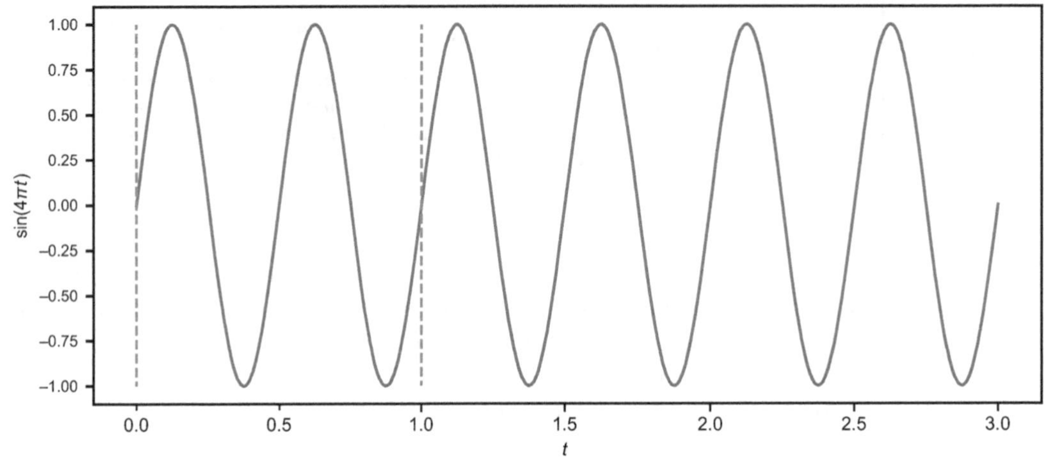

Figure 13.13 The graph of sin(4πt) is sinusoidal, repeating itself twice in every unit of t for a frequency of 2.

The period of the function $\sin(4\pi t)$ is ½ instead of 2π, so the "squishing factor" is 4π. That is, the original period was 2π, and the reduced period is 4π times shorter. In general, for any constant k, a function of the form $f(t) = \sin(kt)$ has a period shrunk by a factor of k to $2\pi/k$. The frequency is increased by a factor of k from the usual value of $1/(2\pi)$ to $k/2\pi$.

If we want a sinusoidal function with a frequency of 441, the appropriate value of k would be $441 \cdot 2 \cdot \pi$. That gives us a frequency of

$$\frac{441 \cdot 2 \cdot \pi}{2\pi} = 441$$

Increasing the amplitude of the sinusoid is simpler by comparison. All you need to do is multiply the sine function by a constant factor, and the amplitude increases by the same factor. With that, we have what we need to define our `make_sinusoid` function:

```
def make_sinusoid(frequency,amplitude):
    def f(t):                                          # Defines f(t)—the sinusoidal function that is returned
        return amplitude * sin(2*pi*frequency*t)       # Multiplies the input t by 2 · π times the frequency, then multiplies the output of the sine function by the amplitude
    return f
```

We can test this, for example, by making a sinusoidal function with a frequency of 5 and an amplitude of 4, and plotting it (figure 13.14) from $t = 0$ to $t = 1$:

```
>>> plot_function(make_sinusoid(5,4),0,1)
```

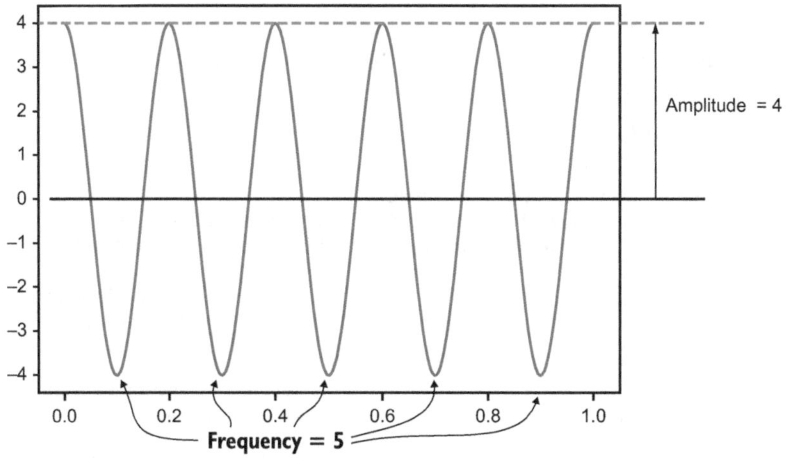

Figure 13.14 The graph of `make_sinusoid(5,4)` has a height (amplitude) of 4 and repeats itself 5 times from *t* = 0 to *t* = 5, so it has a frequency of 5.

Next, we work with the sound wave function that is the result of `make_sinusoid (441,8000)` having a frequency of 441 Hz and an mplitude of 8,000.

13.3.3 Sampling and playing the sound wave

To play the sound wave mentioned in the last section, we need to sample it to get the array of numbers that are playable by PyGame. Let's set

```
sinusoid = make_sinusoid(441,8000)
```

so the `sinusoid` function from $t = 0$ to $t = 1$ represents 1 second of a sound wave we try to play. We pick 44,100 values of t, evenly spaced between 0 and 1, and the resulting function values are the corresponding values of sinusoid(t).

We can use the NumPy function `np.arange`, which provides evenly spaced numbers on a given interval. For instance, `np.arange(0,1,0.1)` gives 10 evenly spaced values, starting from 0 and below 1 at 0.1 unit intervals:

```
>>> np.arange(0,1,0.1)
array([0. , 0.1, 0.2, 0.3, 0.4, 0.5, 0.6, 0.7, 0.8, 0.9])
```

For our application, we want to use 44,100 time values between 0 and 1, which are evenly spaced by 1/44100 units:

```
>>> np.arange(0,1,1/44100)
array([0.00000000e+00, 2.26757370e-05, 4.53514739e-05, ...,
       9.99931973e-01, 9.99954649e-01, 9.99977324e-01])
```

We want to apply the sinusoid function to every entry of this array to produce another NumPy array as a result. The NumPy function `np.vectorize(f)` takes a Python function `f` and produces a new one that applies the same operation to *every* entry of an array. So for us, `np.vectorize(sinusoid)(arr)` applies the sinusoid function to every entry of an array.

This is almost a complete procedure for sampling a function. The last detail we need to include is converting the outputs to 16-bit integer values using the `astype` method on NumPy arrays. Putting these steps together, we can build the following general sampling function:

```
def sample(f,start,end,count):
    mapf = np.vectorize(f)
    ts = np.arange(start,end,(end-start)/count)
    values = mapf(ts)
    return values.astype(np.int16)
```

Inputs are the function f to sample the start and end of the range and the number of values we want.

Creates a version of f that can be applied to a NumPy array

Creates the evenly spaced input values for the function over the desired range

Applies the function to every value in the NumPy array

Converts the resulting array to 16-bit integer values and returns it

Equipped with the following function, you can hear the sound of a 441 Hz sinusoidal wave:

```
sinusoid = make_sinusoid(441,8000)
arr = sample(sinusoid, 0, 1, 44100)
sound = pygame.sndarray.make_sound(arr)
sound.play()
```

If you play this alongside the 441 Hz square wave, you'll notice that it plays the same note; in other words, it has the same pitch. However, the quality of the sound is much different; the sinusoidal wave plays a much smoother sound. It sounds almost like it could be coming out of a flute rather than out of an old-school video game. This quality of sound is called *timbre* (pronounced *TAM-ber).*

For the rest of the chapter, we focus on sound waves that are built as combinations of sinusoids. It turns out that with the right combination, you can approximate any shape of wave and, therefore, any timbre you want.

13.3.4 Exercises

Exercise 13.2 Plot the tangent function $\tan(t) = \sin(t)/\cos(t)$. What is its period?

Solution The tangent function gets infinitely big in every period, so it helps to plot it with a restricted range of *y* values:

```
from math import tan
plot_function(tan,0,5*pi)
plt.ylim(-10,10)
```

Limits the graph window to a y range of –10 < y < 10

A graph of $\tan(x)$, which is periodic, looks like this:

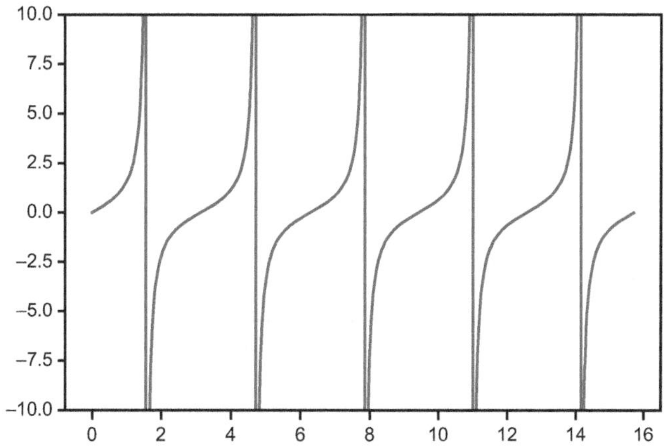

Because $\tan(t)$ depends only on the values of $\cos(t)$ and $\sin(t)$, it should repeat itself at least every 2π units. In fact, it repeats itself *twice* every 2π units; we can see on the graph that its period is π.

Exercise 13.3 What is the frequency of $\sin(3\pi t)$? What is the period?

Solution The frequency of $\sin(t)$ is $1/(2\pi)$ and multiplying the argument by 3π increases this frequency by a factor of 3π. The resulting frequency is $(3\pi)/(2\pi) = 3/2$. The period is the reciprocal of this value, which is $2/3$.

Exercise 13.4 Find the value of k such that $\cos(kt)$ has a frequency of 5. Plot the resulting function $\cos(kt)$ from zero to one and show that it repeats itself 5 times.

(continued)

Solution The default frequency of $\cos(t)$ is $1/2\pi$, so $\cos(kt)$ has a frequency of $k/2\pi$. If we want this value to equal 5, we need to have $k = 10\pi$. The resulting function is $\cos(10\pi t)$:

```
>>> plot_function(lambda t: cos(10*pi*t),0,1)
```

Here is its graph, where it repeats itself five times between the values $t = 0$ to $t = 1$.

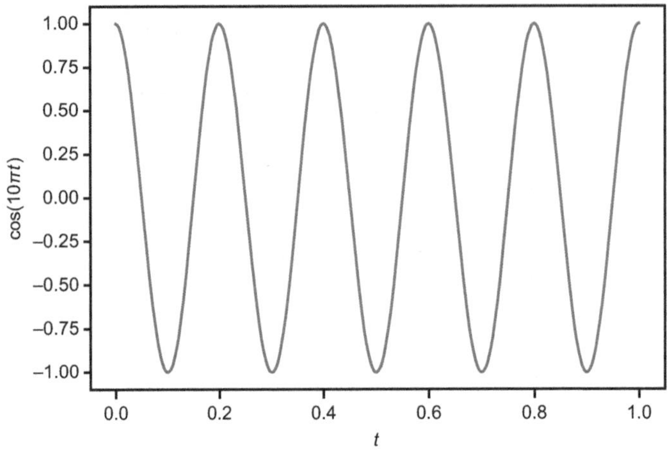

13.4 Combining sound waves to make new ones

In chapter 6, you learned that functions can be treated like vectors; you can add functions or multiply them by scalars to produce new functions. When you create linear combinations of functions defining sound waves, you can create new, interesting sounds.

The simplest way to combine two sound waves in Python is to sample both and then add the corresponding values of the two arrays to create a new one. We start by writing some Python code to add sampled sound waves of different frequencies, and the result they produce will sound like a musical chord, just as if you strummed several strings of a guitar at once.

Once we do that, we can do a more advanced and more surprising example—we'll add together several dozen sinusoidal sound waves of different frequencies in a prescribed linear combination, and the result will look and sound like the square wave from before.

13.4.1 Adding sampled sound waves to build a chord

NumPy arrays can be added using the ordinary + operator in Python, making the job of adding sampled sound waves easy. Here's a small example showing that NumPy does addition by adding the corresponding values of each array to build a new array:

```
>>> np.array([1,2,3]) + np.array([4,5,6])
array([5, 7, 9])
```

It turns out that doing this operation with two sampled sound waves produces the same sound as if you played both at once. Here are two samples: our sinusoid at 441 Hz and a second sinusoid at 551 Hz, approximately 5/4 of the frequency of the first:

```
sample1 = sample(make_sinusoid(441,8000),0,1,44100)
sample2 = sample(make_sinusoid(551,8000),0,1,44100)
```

If you ask PyGame to start one and immediately start playing the next, it plays the two sounds almost simultaneously. If you run the following code, you should hear a chord consisting of two different musical notes. If you run either of the last two lines on its own, you hear one of the two individual notes:

```
sound1 = pygame.sndarray.make_sound(sample1)
sound2 = pygame.sndarray.make_sound(sample2)
sound1.play()
sound2.play()
```

Now, using NumPy, we can add the two sample arrays to produce a new one and play it with PyGame. When `sample1` and `sample2` are added, a new array of length 44,100 is created, containing the sums of entries from `sample1` and `sample2`. If you play the result, it sounds exactly like playing the previous sounds:

```
chord = pygame.sndarray.make_sound(sample1 + sample2)
chord.play()
```

13.4.2 Picturing the sum of two sound waves

Let's see what this looks like in terms of the graphs of the sound waves. Here are the first 400 points of `sample1` (441 Hz) and `sample2` (551 Hz). In figure 13.15, you can see that sample 1 makes it through four periods, while sample 2 makes it through five periods.

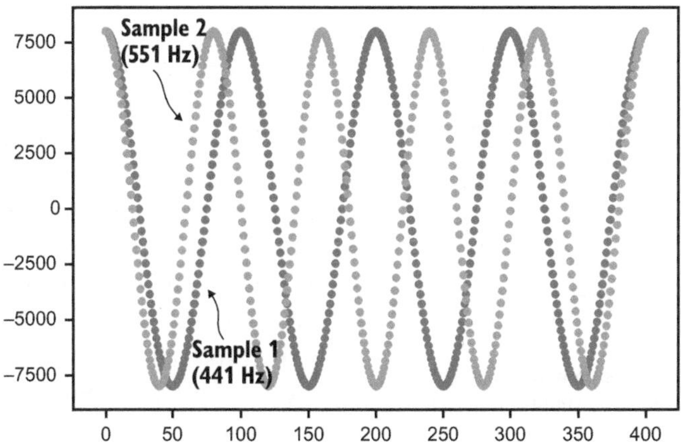

Figure 13.15 Plotting the first 400 points of `sample1` and `sample2`

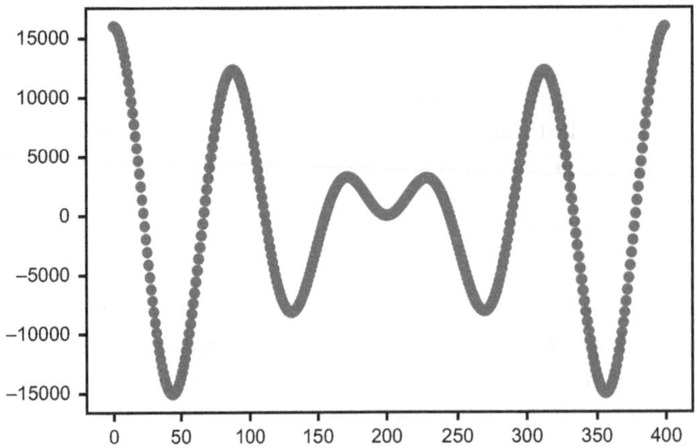

Figure 13.16 Plotting the sum of the two waves, `sample1 + sample2`

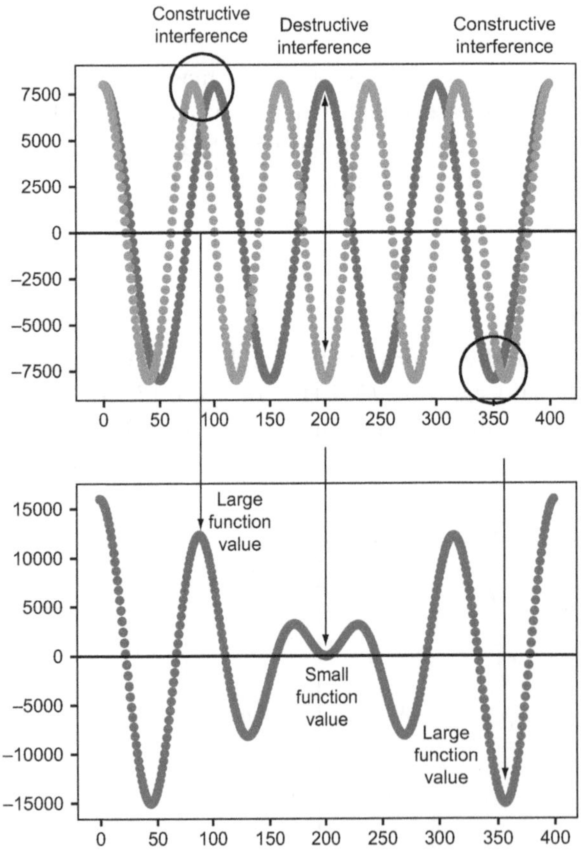

Figure 13.17 The absolute value of the sum wave is large where there is constructive interference and small where there is destructive interference.

It might come as a surprise that the sum of `sample1` and `sample2` doesn't produce a sinusoid even though it's built out of two sinusoids. Instead, the sequence `sample1 + sample2` traces a wave whose amplitude seems to fluctuate. Figure 13.16 shows what the sum looks like.

Let's look closely at the summation to see how we got this shape. Near the 85th point of the sample, the waves are both large and positive, so the 85th point of the sum is also large and positive. Around the 350th point, both waves have large, negative values and so does their sum. When two waves align, their sum is even bigger (and louder), which is called *constructive interference*.

There's an interesting effect in figure 13.17, where the values are opposite (at the 200th point). For example, `sample1` is large and positive while `sample2` is large and negative. This causes their sum to be close to zero even though neither wave on its own is close to zero. When two waves cancel each other out like this, it is called *destructive interference*.

Because the waves have different frequencies, they go in and out of sync with each other, alternating between constructive and destructive interference. As a consequence, the sum of the waves is not a sinusoid; rather, it appears to change amplitude over time. Figure 13.17 displays the two graphs lined up, showing the relationship between the two samples and their sum.

As you can see, the relative frequencies of summed sinusoids have an influence on the shape of the resulting graph. Next, I show you an even more extreme example of this as we build a linear combination with several dozen sinusoidal functions.

13.4.3 Building a linear combination of sinusoids

Let's start with a big collection of sinusoids of different frequencies. We can make a list (as long as we want) of sine functions, starting with:

$$\sin(2\pi t),\ \sin(4\pi t),\ \sin(6\pi t),\ \sin(8\pi t),\ \ldots$$

These functions have the frequencies 1, 2, 3, 4, and so on. Likewise, the list of cosine functions, starting with

$$\cos(2\pi t),\ \cos(4\pi t),\ \cos(6\pi t),\ \cos(8\pi t),\ \ldots$$

has the respective frequencies 1, 2, 3, 4, and so on. The idea is that with so many different frequencies at our disposal, we can create a wide variety of different shapes by taking linear combinations of these functions. For reasons we'll see later, I'll also include a constant function $f(x) = 1$ in the linear combination. If we pick some highest frequency N, the most general linear combination of the sines, cosines, and a constant is given by figure 13.18.

Constant function

Cosine functions

$$\begin{aligned} &a_0 \\ &+ a_1\cos(2\pi t) + a_2\cos(4\pi t) + a_3\cos(6\pi t) + a_4\cos(8\pi t) + \cdots + a_N\cos(2\pi N t) \\ &+ b_1\sin(2\pi t) + b_2\sin(4\pi t) + b_3\sin(6\pi t) + b_4\sin(8\pi t) + \cdots + b_N\sin(2\pi N t) \end{aligned}$$

Sine functions

Figure 13.18 The sine and cosine functions in our linear combination

This linear combination is a Fourier series, and it is, itself, a function of the variable t. It is specified by $2N + 1$ numbers: the constant term a_0, the coefficients a_1 through a_N on the cosine functions, and the coefficients b_1 through b_N on the sine functions. We can evaluate the function by plugging a given t value into every sine and cosine, and adding the linear combination of results. Let's do this in Python, so we can easily test out a few different Fourier series.

The `fourier_series` function takes a single constant a_0, and lists a and b containing the coefficients $a_1, \ldots, a_N$ and $b_1, \ldots, b_N$, respectively. This function works even if the arrays are different lengths; it's as if the unspecified coefficients are zero. Note that the sine and cosine frequencies start from one, while Python's enumerate starts with zero, so $(n + 1)$ is the frequency corresponding to the coefficient at index n in either of the arrays:

```
def const(n):
    return 1

def fourier_series(a0,a,b):
    def result(t):
        cos_terms = [an*cos(2*pi*(n+1)*t)
            for (n,an) in enumerate(a)]
        sin_terms = [bn*sin(2*pi*(n+1)*t)
            for (n,bn) in enumerate(b)]
        return a0*const(t) + \
            sum(cos_terms) + sum(sin_terms)
    return result
```

Creates a constant function that returns 1 for any input

Evaluates all cosine terms with their respective constants and adds the results

Evaluates the sine terms with their respective constants and adds the results

Adds both of the results with the constant coefficient a_0 times the value of the constant function (1)

Here's an example for calling this function with $b_4 = 1$ and $b_5 = 1$, and all other constants are 0. This is a very short Fourier series, $\sin(8\pi t) + \sin(10\pi t)$, whose plot is shown in figure 13.19. Because the ratio of the frequencies is 4 : 5, the shape of the result should look like the last graph we plotted (figure 13.17):

```
>>> f = fourier_series(0,[0,0,0,0,0],[0,0,0,1,1])
>>> plot_function(f,0,1)
```

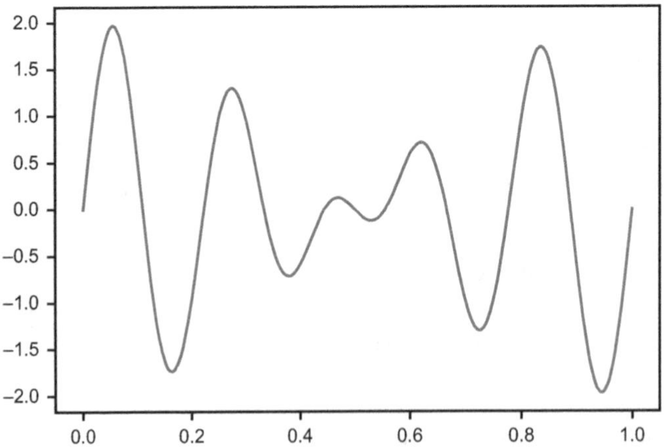

Figure 13.19 The graph of the Fourier series $\sin(8\pi t) + \sin(10\pi t)$

This is a good test to see if our function is working, but it doesn't show the full power of the Fourier series yet. Next, we try a Fourier series with more terms.

13.4.4 Building a familiar function with sinusoids

Let's create a Fourier series that still has no constant and no cosine terms, but a lot more sine terms. Specifically, we use the following sequence of values for b_1, b_2, b_3, and so on:

$$b_1 = \frac{4}{\pi}$$
$$b_2 = 0$$
$$b_3 = \frac{4}{3\pi}$$
$$b_4 = 0$$
$$b_5 = \frac{4}{5\pi}$$
$$b_6 = 0$$
$$b_7 = \frac{4}{7\pi}$$
$$\ldots$$

Or $b_n = 0$ for every even n, while $b_n = 4/(n\pi)$ when n is odd. This gives us a base to make a Fourier series with as many terms as we want. For instance, the first non-zero term is

$$\frac{4}{\pi}\sin(2\pi t)$$

and with the next term that gets added, the series becomes

$$\frac{4}{\pi}\sin(2\pi t) + \frac{4}{3\pi}\sin(6\pi t).$$

Here is the code, and figure 13.20 shows the graphs of these two functions plotted simultaneously.

```
>>> f1 = fourier_series(0,[],[4/pi])
>>> f3 = fourier_series(0,[],[4/pi,0,4/(3*pi)])
>>> plot_function(f1,0,1)
>>> plot_function(f3,0,1)
```

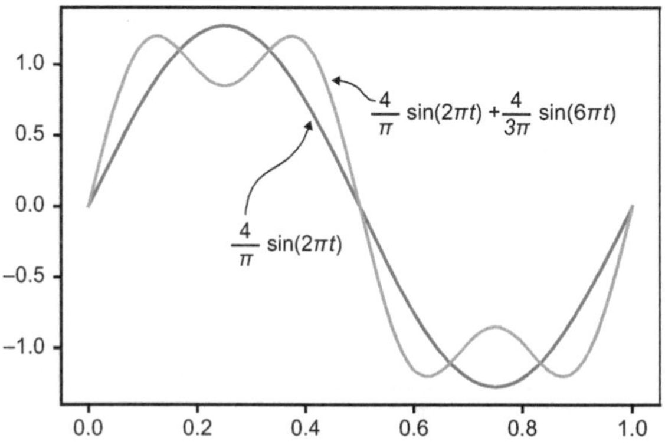

Figure 13.20 A plot of the first term and then the first two terms of the Fourier series

Using a list comprehension, we can make a much longer list of the coefficients, b_n, and construct the Fourier series programmatically. We can leave the list of cosine coefficients empty, and it will be as if all of the a_n values are set to 0:

```
b = [4/(n * pi)
     if n%2 != 0 else 0 for n in range(1,10)]
f = fourier_series(0,[],b)
```

Lists the values of $b_n = 4/n\pi$ for odd values of n and $b_n = 0$, otherwise

This list covers $1 \leq n < 10$, so the non-zero coefficients are b_1, b_3, b_5, b_7, and b_9. With these terms, the graph of the series looks like figure 13.21.

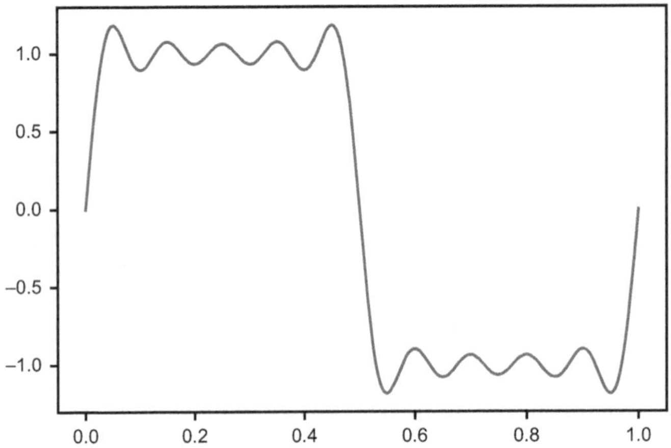

Figure 13.21 A sum of the first 5 non-zero terms of the Fourier series

This is an interesting pattern of constructive and destructive interference! Around $t = 0$ and $t = 1$, all of the sine functions are simultaneously increasing, while around $t = 0.5$, they are all simultaneously decreasing. This constructive interference is the dominant effect, while alternating constructive and destructive interference keeps the

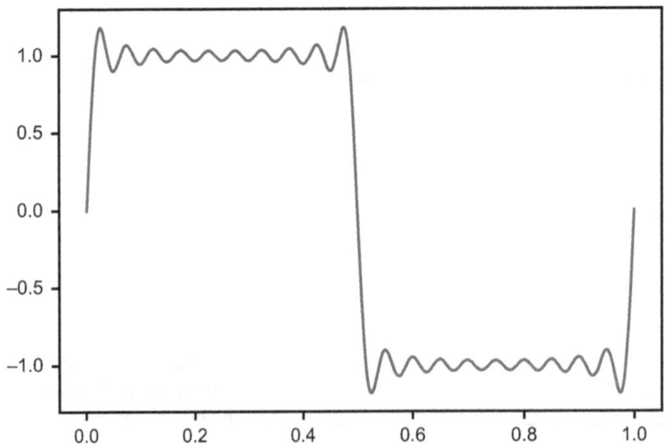

Figure 13.22 The first 10 non-zero terms of the Fourier series

graph relatively flat in the other regions. With n ranging up to 19, as shown in figure 13.22, there are 10 non-zero terms and this effect is even more striking.

```
>>> b = [4/(n * pi) if n%2 != 0 else 0 for n in range(1,20)]
>>> f = fourier_series(0,[],b)
```

If we let n range all the way up to 99, we get a sum of 50 sine functions, and the function becomes nearly flat outside of a few big jumps (figure 13.23).

```
>>> b = [4/(n * pi) if n%2 != 0 else 0 for n in range(1,100)]
>>> f = fourier_series(0,[],b)
```

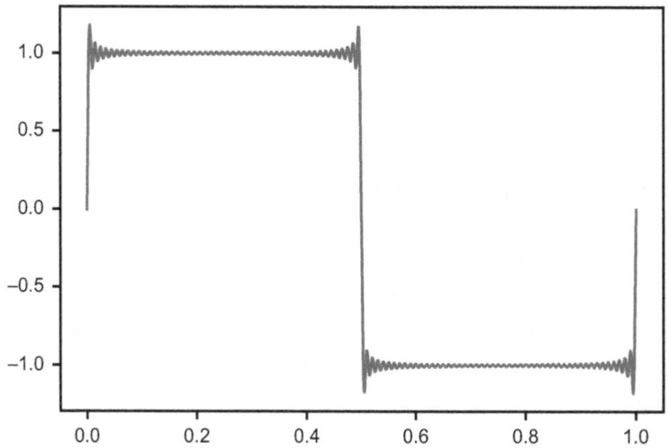

Figure 13.23 With 99 terms, the graph of the Fourier series is nearly flat, apart from big steps at 0, 0.5, and 1.0.

If you zoom out, you can see that this Fourier series comes close to the square wave we plotted at the beginning of the chapter (figure 13.24).

What we've done here is to build an approximation of the square wave function as a linear combination of sinusoids. It's counterintuitive that we can do this! After all, all

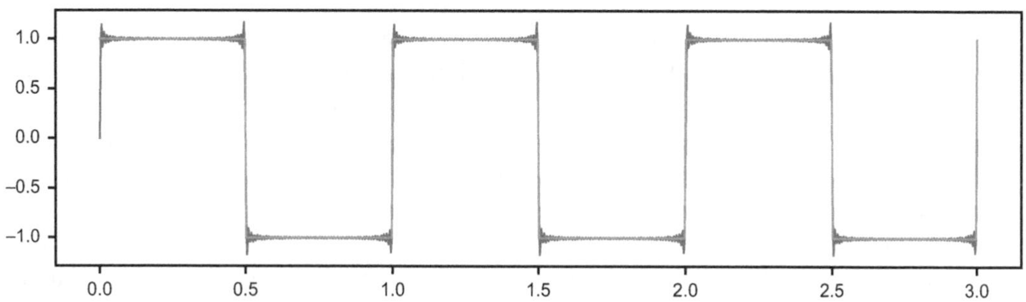

Figure 13.24 The first 50 non-zero terms of the Fourier series are close to a square wave, like the first function we met in this chapter.

of the sinusoids in the Fourier series are round and smooth, and the square wave is flat and jagged. We'll conclude this chapter by showing how to reverse engineer this approximation, starting with any periodic function and recovering the coefficients for a Fourier series that approximates it.

13.4.5 Exercises

Exercise 13.5—Mini Project Create a manipulated version of the square wave Fourier series so that its frequency is 441 Hz, then sample it and confirm that it doesn't just look like the square wave. It should sound like the square wave as well.

13.5 Decomposing a sound wave into its Fourier series

Our last goal is to take an arbitrary periodic function, like the square wave, and figure out how to write it (or at least an approximation of it) as a linear combination of sinusoidal functions. This means breaking any sound wave into a combination of pure notes. As a basic example, we'll look at a sound wave defining a chord and identify which notes make up the chord. More profoundly, we can break any sound into musical notes: a person talking, a dog barking, or a car revving its engine. Behind this result are some elegant mathematical ideas, and you now have all the background you need to understand them.

The process of breaking a function into its Fourier series is analogous to writing a vector as a linear combination of basis vectors as we did in part 1. Here's how the analogy works. We'll work in the vector space of functions and think of a function like the square wave, as a function of interest. Then, we'll think of our basis as the set of functions $\sin(2\pi t)$, $\sin(4\pi t)$, $\sin(6\pi t)$, and so on. In section 13.3, we approximated the square wave as a linear combination beginning with

$$\frac{4}{\pi}\sin(2\pi t) + \frac{4}{3\pi}\sin(6\pi t) + \cdots$$

You can picture two of the basis vectors, $\sin(2\pi t)$ and $\sin(6\pi t)$, as defining two perpendicular directions in the infinite-dimensional space of functions with many other directions defined by the other basis vectors. The square wave has a component of length $4/\pi$ in the $\sin(2\pi t)$ direction and a component of length $4/3\pi$ in the $\sin(6\pi t)$ direction. These are the first two in what would be an infinite list of coordinates for the square wave in this basis (figure 13.25).

We can write a `fourier_coefficients(f,N)` function that takes a function f, which is periodic with period one, and a number N of desired coefficients. The function treats the constant function, as well as the functions $\cos(2n\pi t)$ and $\sin(2n\pi t)$ from $1 \leq n < N$, as directions in the vector space of functions and find the components of f in those directions. It returns the Fourier coefficient a_0, representing the constant

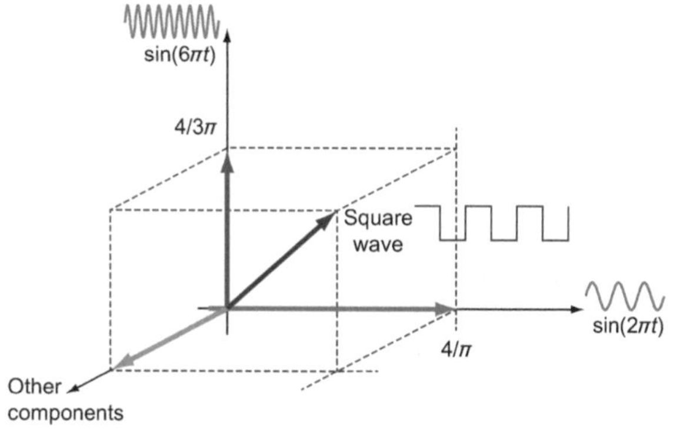

Figure 13.25 You can think of the square wave as a vector in the space of functions with a component length of $4/\pi$ in the sin($2\pi t$) direction and component length of $4/3\pi$ in the sin($6\pi t$) direction. The square wave has infinitely many more components beyond these two.

function, a list of Fourier coefficients a_1, a_2, ..., a_N, and a list of Fourier coefficients b_1, b_2, ..., b_N as a result.

13.5.1 Finding vector components with an inner product

In chapter 6, we covered how to do vector sums and scalar multiples with functions in analogy with the operations with 2D and 3D vectors. Another tool we need is an analogy for the dot product. The dot product is one example of an *inner product*, a way of multiplying two vectors to get a scalar that measures how aligned are the two vectors.

Let's think back to the 3D world for a moment and show how to use the dot product to find components of a 3D vector, then we'll do the same thing to find components of a function in the basis of sinusoidal functions. Suppose our goal is to find the components of the vector $\mathbf{v} = (3, 4, 5)$ in the directions of the standard basis vectors, $\mathbf{e}_1 = (1, 0, 0)$, $\mathbf{e}_2 = (0, 1, 0)$, and $\mathbf{e}_3 = (0, 0, 1)$. This question is so obvious that we never put much thought into it. The components are 3, 4, and 5, respectively; that's what the coordinates (3, 4, 5) mean!

Here, I'll show you another way to find the components of $\mathbf{v} = (3, 4, 5)$ using the dot product. It's going to be overkill because we already have the answer, but it will be useful for the case of function vectors. Notice that each of the dot products of $\mathbf{v}$ with a standard basis vector gives us back one of the components:

$$\mathbf{v} \cdot \mathbf{e}_1 = (3,4,5) \cdot (1,0,0) = 3 + 0 + 0 = 3$$
$$\mathbf{v} \cdot \mathbf{e}_2 = (3,4,5) \cdot (0,1,0) = 0 + 4 + 0 = 4$$
$$\mathbf{v} \cdot \mathbf{e}_3 = (3,4,5) \cdot (0,0,1) = 0 + 0 + 5 = 5$$

These dot products immediately tell us how to build v as a linear combination of the standard basis: $\mathbf{v} = 3\mathbf{e}_1 + 4\mathbf{e}_2 + 5\mathbf{e}_3$. Be careful. This only works because the dot product agrees with our definitions of lengths and angles. Any pair of perpendicular standard basis vectors has zero dot product:

$$\mathbf{e}_1 \cdot \mathbf{e}_2 = \mathbf{e}_2 \cdot \mathbf{e}_3 = \mathbf{e}_3 \cdot \mathbf{e}_1 = 0$$

And the dot products of standard basis vectors with themselves yield their (squared) lengths of one:

$$\mathbf{e}_1 \cdot \mathbf{e}_1 = \mathbf{e}_2 \cdot \mathbf{e}_2 = \mathbf{e}_3 \cdot \mathbf{e}_3 = |\mathbf{e}_1|^2 = |\mathbf{e}_2|^2 = |\mathbf{e}_3|^2 = 1$$

Another way to look at these relationships is that, according to the dot product, none of the standard basis vectors have components in the direction of the other standard basis vectors. Furthermore, each standard basis vector has component 1 in its own direction. If we want to invent an inner product to calculate components of functions, we need our basis to have the same desirable properties. In other words, we need to know that our basis functions, like $\sin(2\pi t)$, $\cos(2\pi t)$, and so on, are all perpendicular and have length 1. We'll create an inner product for functions and test these facts.

13.5.2 Defining an inner product for periodic functions

Suppose $f(t)$ and $g(t)$ are two functions defined on the interval from $t = 0$ to $t = 1$, and that these repeat themselves every one unit of t. We can write the inner product of f and g as $\langle f, g\rangle$ and define it by a definite integral:

$$\langle f, g\rangle = 2 \cdot \int_0^1 f(t)g(t)\,dt.$$

Let's implement this in Python code, approximating the integral as a Riemann sum (as we did in chapter 8), so you can get a sense for how this inner product works like the familiar dot product. This Riemann sum defaults to 1,000 time steps as shown here:

```
def inner_product(f,g,N=1000):
    dt = 1/N
    return 2*sum([f(t)*g(t)*dt
                  for t in np.arange(0,1,dt)])
```

The dt size defaults to 1/1000 = 0.001.

For each time step, the contribution to the integral is f(t) * g(t) * dt. The integral's result is multiplied by 2, according to the formula.

Like the dot product, this integral approximation is a sum of products of values from the input vectors. Instead of being a sum of products of coordinates, it is a sum of products of function values. You can think of a function's values as a set of infinitely many coordinates, and this inner product as being a kind of "infinite dot product" over these coordinates.

Let's take this inner product for a spin. For convenience, let's define some Python functions to create the nth sine and cosine functions in our basis, and then we can test them with the `inner_product` function. These functions are like simplified versions of the `make_sinusoid` function from section 13.3.2:

```
def s(n):
    def f(t):
        return sin(2*pi*n*t)
    return f
```

s(n) takes a whole number n and returns the function sin(2nπt).

```
def c(n):
    def f(t):
        return cos(2*pi*n*t)
    return f
```

c(n) takes a whole number n and returns the function cos(2nπt).

A dot product of two 3D vectors like (1, 0, 0) and (0, 1, 0) returns zero, confirming they are perpendicular. Our inner product shows that all of our pairs' basis functions are (approximately) perpendicular. For instance,

```
>>> inner_product(s(1),c(1))
4.2197487366314734e-17
>>> inner_product(s(1),s(2))
-1.4176155163484784e-18
>>> inner_product(c(3),s(10))
-1.7092447249233977e-16
```

These numbers are extremely close to zero, confirming that $\sin(2\pi t)$ and $\cos(2\pi t)$ are perpendicular, and $\sin(2\pi t)$ and $\sin(4\pi t)$ are perpendicular, as well as $\cos(6\pi t)$ and $\cos(20\pi t)$. Using exact integration formulas, which we won't cover here, it's possible to *prove* that for any whole numbers n and m:

$$\langle \sin(2n\pi t), \cos(2m\pi t) \rangle = 0$$

And for any pair of distinct whole numbers n and m, both

$$\langle \sin(2n\pi t), \sin(2m\pi t) \rangle = 0$$

and

$$\langle \cos(2n\pi t), \cos(2m\pi t) \rangle = 0$$

This is a way of saying that with respect to this inner product, all of our sinusoidal basis functions are perpendicular; none has a component in the direction of another. The other thing we need to check is that the inner product implies our basis vectors have components of 1 in their own directions. Indeed, within numerical error, this looks to be true:

```
>>> inner_product(s(1),s(1))
1.0000000000000002
>>> inner_product(c(1),c(1))
0.9999999999999999
>>> inner_product(c(3),c(3))
1.0
```

Even though we won't go through it here, using integral formulas makes it possible to prove directly that for any whole number n

$$\langle \sin(2n\pi t), \sin(2n\pi t) \rangle = 1$$

and

$$\langle \cos(2n\pi t), \cos(2n\pi t) \rangle = 1$$

The last bit of tidying up we need to do is to include our constant function in this discussion. I promised before that I'd explain why we need to include a constant term in the Fourier series, and now I can give an initial explanation. The constant function is required to build a complete basis of functions; not including this would be like omitting $\mathbf{e}_2$ from the basis for 3D space and going forward with $\mathbf{e}_1$ and $\mathbf{e}_3$. If you did that, there'd be functions you simply couldn't build out of basis vectors.

Any constant function is perpendicular to every sine and cosine function in our basis, but we need to pick the value of the constant function so that it has component 1 in its own direction. That is, if we implement a Python function `const(t)`, we should find that `inner_product(const,const)` returns 1. The right constant value for `const` to return turns out to be $1/\sqrt{2}$ (and you can check in the following exercise that this value makes sense!):

```
from math import sqrt

def const(n):
    return 1 /sqrt(2)
```

With this defined, we can confirm the constant function has the right properties:

```
>>> inner_product(const,s(1))
-2.2580204307905138e-17
>>> inner_product(const,c(1))
-3.404394821604484e-17
>>> inner_product(const,const)
1.0000000000000007
```

We now have the tools we need to find the Fourier coefficients of a periodic function. These coefficients are nothing more than components of the function in the basis we've defined.

13.5.3 Writing a function to find Fourier coefficients

In the 3D example, we saw that the dot product of a vector **v** with a basis vector $\mathbf{e}_i$ gave us the component of **v** in the direction of $\mathbf{e}_i$. We'll use the same process for a periodic function *f*.

The coefficients a_n for $n \geq 1$ tell us the components of *f* in the direction of the basis function $\cos(2n\pi t)$. They are computed as the inner products of *f* with these basis functions:

$$a_n = \langle f, \cos(2n\pi t) \rangle, \quad n \geq 1$$

Likewise, every Fourier coefficient *bn* tells us the component of *f* in the direction of a basis function $\sin(2n\pi t)$ and can also be computed with an inner product:

$$b_n = \langle f, \sin(2n\pi t) \rangle$$

Finally, the number a_0 is the inner product of *f* with the constant function, whose value is $1/\sqrt{2}$. All of these Fourier coefficients can be computed with Python functions we've already written, so we're ready to assemble the `fourier_coefficients` function we set out to write. Remember, the first argument to the function is the function we want to analyze, and the second argument is the maximum number of sine and cosine terms we want:

```
def fourier_coefficients(f,N):
    a0 = inner_product(f,const)
    an = [inner_product(f,c(n))
          for n in range(1,N+1)]
    bn = [inner_product(f,s(n))
          for n in range(1,N+1)]
    return a0, an, bn
```

The constant term a_0 is the inner product of f with the constant basis function.

The coefficients a_n are given by inner products of f with cos($2n\pi t$) for $1 < n < N + 1$.

The coefficients b_n are given by inner products of f with sin($2n\pi t$) for $1 \leq n < N + 1$.

As a sanity check, a Fourier series should give back its own coefficients. For instance

```
>>> f = fourier_series(0,[2,3,4],[5,6,7])
>>> fourier_coefficients(f,3)
(-3.812922200197022e-15,
 [1.9999999999999887, 2.999999999999999, 4.0],
 [5.000000000000002, 6.000000000000001, 7.0000000000000036])
```

> **NOTE** If you want the inputs and outputs to match non-zero constant terms, you need to revise the const function to be $f(t) = 1/\sqrt{2}$ instead of $f(t) = 1$. See exercise 13.8.

Now that we can automatically compute Fourier coefficients, we can conclude our exploration by building some Fourier approximations of interestingly shaped periodic functions.

13.5.4 Finding the Fourier coefficients for the square wave

We saw in the last section that the Fourier coefficients for the square wave were all zero except for the b_n coefficients for odd n values. That is, the Fourier series is built as a linear combination of the function $\sin(2n\pi t)$ for odd values of n. For odd n, the coefficient was $b_n = 4/n\pi$. I didn't explain why those were the coefficients, but now we can check our work.

To make a square wave that repeats itself every unit of t, we can use the value `t%1` in Python, which computes the fractional part of t. Because, for example, `2.3 % 1` is `0.3` and `0.3 % 1` is `0.3`, a function written in terms of `t % 1` is automatically periodic with the period 1. The square wave has a value of +1 when `t % 1 < 0.5` and –1 otherwise

```
def square(t):
    return 1 if (t%1) < 0.5 else -1
```

Let's look at the first 10 Fourier coefficients for this square wave. Run

```
a0, a, b = fourier_coefficients(square,10)
```

and you'll see that a_0 and the entries of a are all small, as with every other entry of b. The values of b_1, b_3, b_5, and so on are represented by `b[0]`, `b[2]`, `b[4]`, ..., because Python arrays are zero-indexed. These are all close to the expected values:

```
>>> b[0], 4/pi
(1.273235355942202, 1.2732395447351628)
>>> b[2], 4/(3*pi)
(0.4244006151333577, 0.4244131815783876)
>>> b[4], 4/(5*pi)
(0.2546269646514865, 0.25464790894703254)
```

We already saw that a Fourier series with these coefficients is a solid approximation of the square wave graph. Let's conclude this section by looking at two example functions we haven't seen before and plotting the Fourier series alongside the original functions to show that the approximation works.

13.5.5 Fourier coefficients for other waveforms

Next, we consider more functions beyond the square wave graph that can be modeled using a Fourier transform. Figure 13.26 shows a new, interestingly shaped waveform called a *sawtooth wave*.

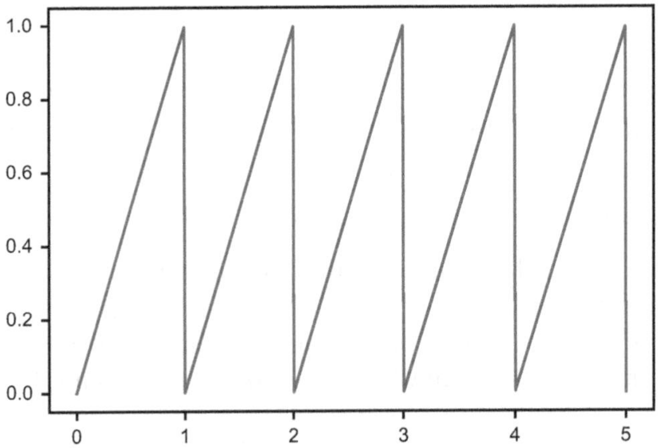

Figure 13.26 A sawtooth wave plotted over five periods

On the intervals from $t = 0$ to $t = 1$, the sawtooth wave is identical to the function $f(t) = t$ and then it repeats itself every one unit. To define the sawtooth wave as a Python function, we can simply write

```
def sawtooth(t):
    return t%1
```

To see its Fourier series approximation with up to 10 sine and cosine terms, we can plug the Fourier coefficients directly into our Fourier series function. Plotting it alongside the sawtooth, as shown in figure 13.27, we can see it has a good fit.

```
>>> approx = fourier_series(*fourier_coefficients(sawtooth,10))
>>> plot_function(sawtooth,0,5)
>>> plot_function(approx,0,5)
```

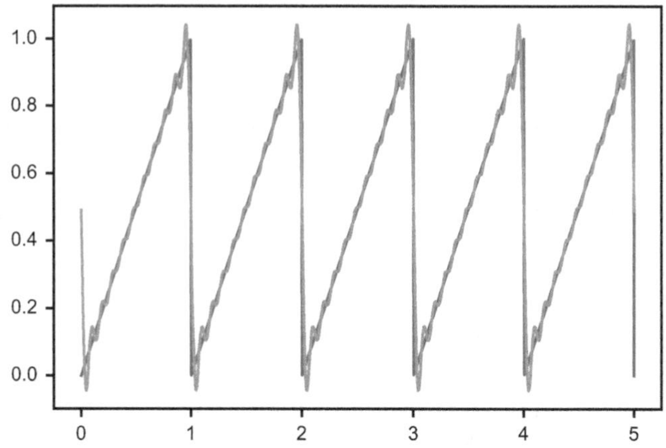

Figure 13.27 The original sawtooth wave from figure 13.26 with its Fourier series approximation

Once again, it's striking how close we can come to a function with sharp corners using only a linear combination of smooth sine and cosine waves. This function happens to have a non-zero constant coefficient a_0. That's required because this function only has values above zero, while sine and cosine functions contribute negative values.

As a final example, take a look at the following function defined as `speedbumps(t)` in the source code for this book. Figure 13.28 shows the graph.

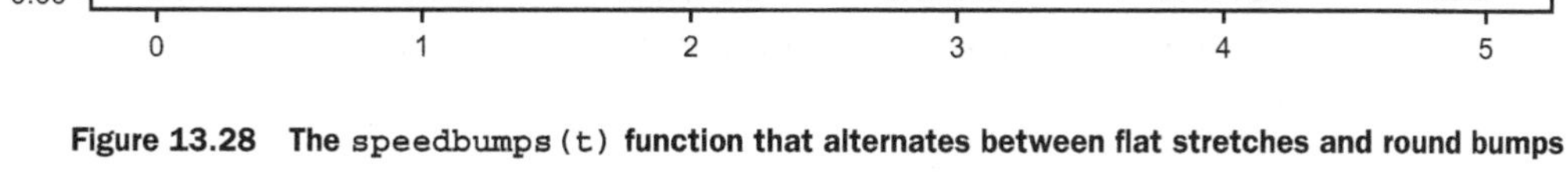

Figure 13.28 The `speedbumps(t)` function that alternates between flat stretches and round bumps

The implementation of this function isn't important, but this one is an interesting example because it has non-zero coefficients for the cosine functions and all zeros for the sines. Even with 10 terms, we get a good approximation. Figure 13.29 shows the graph of the Fourier series with a_0 and ten cosine terms (the coefficients b_n are all zero).

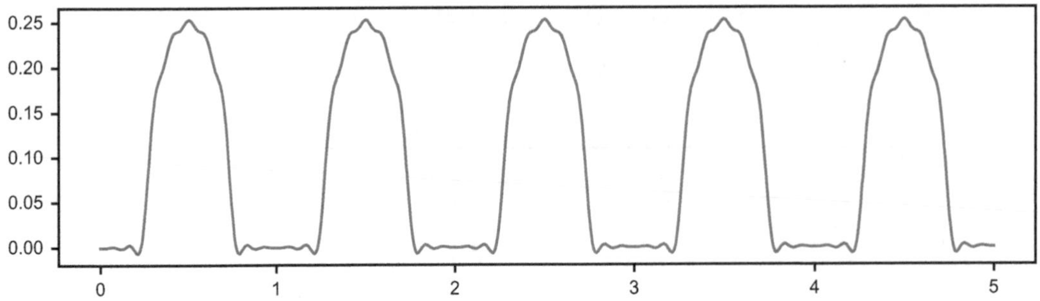

Figure 13.29 The constant term and first 10 cosine terms for the Fourier series of the `speedbumps(t)` function

You can see some wobbles when we graph these approximations, but when these waveforms are translated to sound, the Fourier series can be good enough. Because we are able to transform waveforms of all shapes to lists of their Fourier coefficients, we can store and transmit audio files efficiently.

13.5.6 Exercises

Exercise 13.6 The vectors $\mathbf{u}_1 = (2, 0, 0)$, $\mathbf{u}_2 = (0, 1, 1)$, and $\mathbf{u}_3 = (1, 0, -1)$ form a basis for $\mathbb{R}^3$. For a vector $\mathbf{v} = (3, 4, 5)$, compute three dot products $a_1 = \mathbf{v} \cdot \mathbf{u}_1$, $a_2 = \mathbf{v} \cdot \mathbf{u}_2$, and $a_3 = \mathbf{v} \cdot \mathbf{u}_3$. Show that $\mathbf{v}$ is *not* equal to $a_1\mathbf{u}_1 + a_2\mathbf{u}_2 + a_3\mathbf{u}_3$. Why aren't they equal?

Solution The dot products are

$$a_1 = \mathbf{v} \cdot \mathbf{u}_1 = (3,4,5) \cdot (2,0,0) = 6$$
$$a_2 = \mathbf{v} \cdot \mathbf{u}_2 = (3,4,5) \cdot (0,1,1) = 9$$
$$a_3 = \mathbf{v} \cdot \mathbf{u}_3 = (3,4,5) \cdot (1,0,-1) = -2$$

That makes the linear combination $6 \cdot (2, 0, 0) + 9 \cdot (0, 1, 1) - 2 \cdot (1, 0, -1) = (16, 9, 2)$, which is not equal to $(3, 4, 5)$. This approach does not give the correct result because these basis vectors do not have length 1 and are not perpendicular to each other.

Exercise 13.7—Mini Project Suppose $f(t)$ is constant, meaning $f(t) = k$. Use the integral formula for the inner product to find a value k making $\langle f, f \rangle = 1$. (Yes, I've already told you that $k = 1/\sqrt{2}$ but see if you can get to that value yourself!)

Solution If $f(t) = k$, then $\langle f, f \rangle$ is given by the integral:

$$2 \cdot \int_0^1 f(t) \cdot f(t)\, dt = 2 \cdot \int_0^1 k \cdot k\, dt = 2k^2$$

(The area under the constant function k^2 from 0 to 1 is k^2.) If we want $2k^2$ to equal 1, then $k^2 =$ and $k = \sqrt{\frac{1}{2}} = 1/\sqrt{2}$.

Exercise 13.8 Update the `fourier_series` function to use $f(t) = 1/\sqrt{2}$ for the constant function instead of $f(t) = 1$.

Solution

Multiplies the coefficient a_0 by the constant function $f(t) = 1/\sqrt{2}$ in the linear combination, contributing $a_0/\sqrt{2}$ to the Fourier series result regardless of the value of t

```
def fourier_series(a0,a,b):
    def result(t):
        cos_terms = [an*cos(2*pi*(n+1)*t) for (n,an) in enumerate(a)]
        sin_terms = [bn*sin(2*pi*(n+1)*t) for (n,bn) in enumerate(b)]
        return a0/sqrt(2) + sum(cos_terms) + sum(sin_terms)
    return result
```

Exercise 13.9—Mini Project Play a sawtooth wave at 441 Hz and compare it with the square and sinusoidal waves you played at that frequency.

Solution We can create a modified sawtooth wave function with amplitude 8,000 and frequency 441 and then sample it to pass to PyGame:

```
def modified_sawtooth(t):
    return 8000 * sawtooth(441*t)
arr = sample(modified_sawtooth,0,1,44100)
sound = pygame.sndarray.make_sound(arr)
sound.play()
```

People often compare the sound of a sawtooth wave to that of a string instrument, like a violin.

Summary

- Sound waves are pressure changes over time that propagate through the air to our ears where we perceive these as sounds. We can represent a sound wave as a function that loosely represents the change in air pressure over time.
- PyGame and most other digital audio systems used *sampled* audio. Rather than a function defining a sound wave, these systems use arrays of values of the function

taken at uniform intervals. For instance, CD audio commonly uses 44,100 values for each second of audio.

- Sound waves with random shapes sound like noise, while waves with shapes that repeat on fixed intervals produce well-defined musical notes. A function that repeats its values on a certain interval is called a *periodic function.*
- The sine and cosine functions are periodic functions, and their graphs repeat curved shapes called *sinusoids.*
- Sine and cosine repeat their values every 2π units. That value is called their *period.* The *frequency* of a periodic function is the reciprocal of the period, which is $1/(2\pi)$ for sine and cosine.
- A function of the form $\sin(2n\pi t)$ or $\cos(2n\pi t)$ has frequency n. High frequency sound wave functions produce high-pitched musical notes.
- The maximum height of a periodic function is called its *amplitude.* Multiplying a sine or cosine function by a number increases the amplitude of the function and the volume of the corresponding sound wave.
- To create the effect of two sounds playing at once, you can add the functions that define their corresponding sound waves to create a new function and a new sound wave. Generally, you can take any linear combination of existing sound waves to create a new sound wave.
- A linear combination of a constant function along with functions of the form $\sin(2n\pi t)$ and $\cos(2n\pi t)$ for various values of n is called a *Fourier series.* Despite being built out of smooth sine and cosine functions, Fourier series can be good approximations for any periodic functions, even those with sharp corners like square waves.
- You can think of the constant function along with the sines and cosines at different frequencies as a basis for the vector space of periodic functions. The linear combination of these basis vectors that best approximate a given function are called *Fourier coefficients.*
- We can use the dot product of a 2D or 3D vector with a standard basis vector to find its component in the direction of that basis vector.
- Analogously, we can take a special inner product of a periodic function with a sine or cosine function to find a component associated with that function. The inner product for periodic functions is a definite integral taken over a specified range, in our case, from zero to one.

Part 3

Machine learning applications

In part 3, we apply what you've learned about mathematical functions, vectors, and calculus to implement some machine learning algorithms. We hear a lot of hype around machine learning, so it's worth being precise about what it actually is. Machine learning is part of the field of *artificial intelligence*, or AI, which studies how to write computer programs to accomplish tasks intelligently. If you've ever played a video game against a computer adversary, you've interacted with an artificial intelligence. Such an adversary is programmed (usually) with a set of rules which help it destroy you, outmaneuver you, or otherwise defeat you.

For an algorithm to be classified as *machine learning*, it must not only operate autonomously and intelligently, but it must learn from experience. That means that the more data it receives, the better it performs at the task at hand. The next three chapters focus on a specific kind of machine learning called *supervised learning*. When we write supervised learning algorithms, we give them *training* data sets with pairs of inputs and corresponding outputs, and the algorithms should then be able to look at new inputs and come up with correct outputs on their own. In this sense, the result of training a machine learning algorithm is a new mathematical function that can effectively map some kind of input data to some kind of decision as an output.

In chapter 14, we cover a simple supervised learning algorithm called *linear regression* and use it to predict the price of a used car based on its mileage. The training data set consists of the known mileage and price for many used cars and, without any prior knowledge about how cars are valued, our algorithm learns to put a price on a car based on its mileage. The linear regression algorithm works by taking pairs (x, p) of a mileage x and a price p and finding the linear function that best approximates them. This amounts to finding the equation of the line that comes closest to all of the known (x, p) points in 2D. Much of our work is figuring out what the word "closest" should mean!

In chapters 15 and 16, we cover a different type of supervised learning problem called *classification.* For any numeric input data point, we want to answer a yes/no or multiple-choice question about it. In chapter 15, we create an algorithm that looks at mileage and price data for two different models of cars, and it will try to correctly identify the model of car based on new data it sees. Again, this amounts to finding a function that comes "closest" to the values in a training data set, and we have to decide what closeness means for a function answering a yes or no question.

In chapter 16, the classification problem is harder. The input data set are images of hand-drawn digits (from 0 to 9), and the desired output will be the digit that is drawn. As we saw in chapter 6, an image is made up of a lot of data; we can think of images as living in vector spaces with a lot of different dimensions. To deal with this complexity, we use a special kind of mathematical function called a *multilayer perceptron.* This is a specific kind of *artificial neural network*, one of the most talked-about machine learning algorithms today.

While you may not be a machine learning expert by the end of these three short chapters, I hope you'll have a solid foundation for further exploration in the subject. Specifically, these chapters should demystify the subject of machine learning for you. We won't magically imbue our computers with human-like sentience. Instead we'll process real-world data in Python and then creatively apply the math we've seen so far.

14 Fitting functions to data

This chapter covers

- Measuring how closely a function models a data set
- Exploring spaces of functions determined by constants
- Using gradient descent to optimize the quality of "fit"
- Modeling data sets with different kinds of functions

The calculus techniques you learned in part 2 require well-behaved functions to be applicable. For a derivative to exist, a function needs to be sufficiently smooth, and to calculate an exact derivative or integral, you need a function to have a simple formula. For most real-world data, we aren't so lucky. Due to randomness or measurement error, we rarely come across perfectly smooth functions in the wild. In this chapter, we cover how to take messy data and model it with a simple mathematical function—a task called *regression*.

I'll walk you through an example on a real data set, consisting of 740 used cars listed for sale on the website CarGraph.com. These cars are all Toyota Priuses, and they all have mileage and sale price reported. Plotting this data on a scatter plot, figure 14.1 shows that we can see there's a downward trend in price as mileage increases. This reflects that cars lose value as they are driven. Our goal is to come up with a simple function that describes how the price of a used Prius changes as its mileage increases.

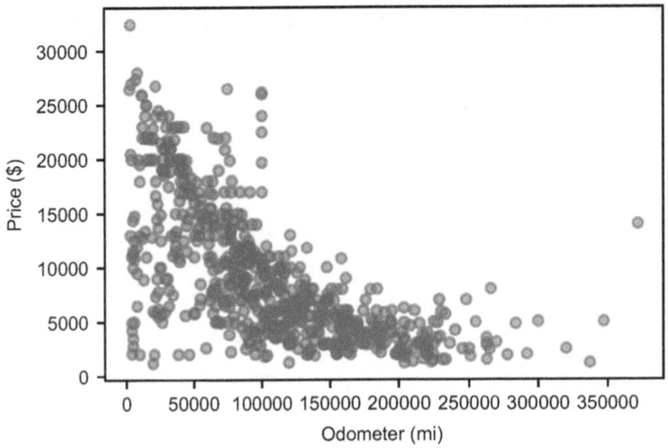

Figure 14.1 A plot of price vs. mileage for used Toyota Priuses listed for sale on CarGraph.com

We can't draw a smooth function that passes through all of these points, and even if we could, it would be nonsensical. Many of these points are *outliers* and probably erroneous (for instance, the handful of nearly new cars that are selling for less than $5,000 in figure 14.1). And there are certainly other factors that affect the resale price of a used car. We shouldn't expect mileage alone to put an exact value on the price.

What we can do is find a function that approximates the trend of this data. Our function $p(x)$ takes mileage x as an input and returns a typical price for a Prius with that given mileage. To do this, we need to make a *hypothesis* about what kind of function this will be. We can start with the simplest possible example: a linear function.

We looked at linear functions in many forms in chapter 7, but in this chapter, we'll write these in the format $p(x) = ax + b$, where x is the mileage of a car, p is its price, and a and b are numbers that determine the shape of the function. With a choice of a and b, this function is an imaginary machine that takes the mileage of a Toyota Prius and predicts its price as shown in figure 14.2.

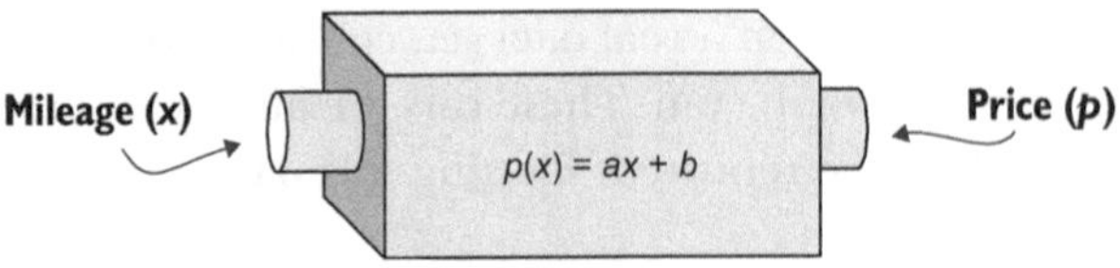

Figure 14.2 A schematic of a linear function predicting price *p* from mileage *x*

Remember, a is the slope of the line and b is its value at zero. With values like $a = -0.05$ and $b = 20{,}000$, the graph of the function becomes a line starting at a price of \$20,000 and decreasing by \$0.05 every time the mileage increases by one mile (figure 14.3).

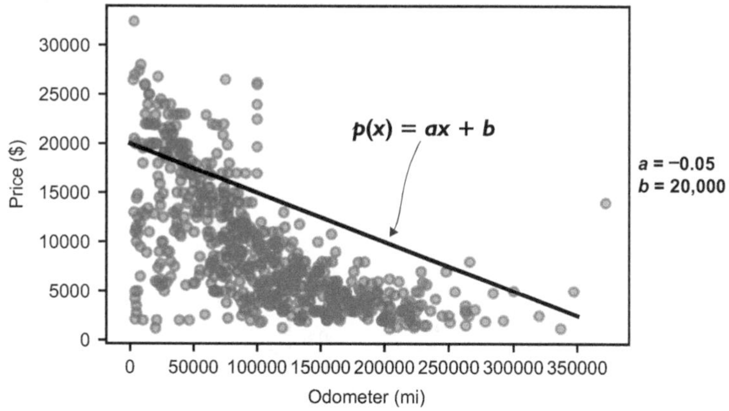

Figure 14.3 Predicting the price of a Prius based on the mileage, using a function of the form $p(x) = ax + b$ with $a = -0.05$ and $b = 20{,}000$

This choice of prediction function implies a new Prius is worth \$20,000, and that it depreciates or loses value at a rate of \$0.05 per mile. These values may or may not be correct; in fact, we have reason to believe they aren't perfect because the graph of this line doesn't come close to most of the data. The task of finding the values of a and b so that $p(x)$ follows the trend of the data as well as possible is called *linear regression.* Once we find the best values, we can say $p(x)$ is the *line of best fit.*

If $p(x)$ is going to come close to the real data, it seems reasonable that slope a should be negative so that the predicted price decreases as mileage increases. We won't have to assume that, however, because we can implement an algorithm that figures this out directly from the raw data. This is why regression is a simple example of a machine learning algorithm; based on data alone, it infers a trend and can then make predictions about new data points.

The only real constraint we impose is that our algorithm looks for linear functions. A *linear function* assumes that the rate of depreciation is constant—that the loss in dollar value of the car in its first 1,000 miles is the same as the loss in value during 100,000 to 101,000 miles. Conventional wisdom says this is not the case, and in fact, that cars lose a good portion of their value the moment they are driven off the lot. Our goal will not be to find the perfect model, but to find a simple model that still performs well.

The first thing we need to do is be able to measure how well a given linear function, meaning a given choice of a and b, predicts the price of a Prius from its mileage. To do this, we write a function in Python, called a *cost function,* which takes a function $p(x)$ as input and returns a number telling us how far it is from the raw data. For any pair of numbers a and b, we can then measure how well the function $p(x) = ax + b$ fits the data set using the cost function. There's one linear function for every pair (a, b), so we can think of our task as exploring the 2D space of such pairs and evaluating the linear function they imply.

Figure 14.4 shows that picking positive values of a and b yield a line, sloping upwards. If that were our price function, it would imply that a car gains value as it is driven, which is not likely.

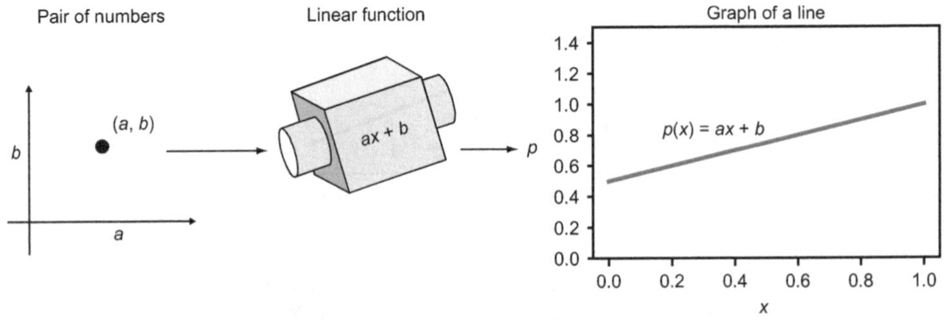

Figure 14.4 A pair of numbers (*a*, *b*) define a linear function that we can plot on a graph as a line. For positive values of *a*, the graph slopes upward.

Our cost function compares a line like this to the actual data and returns a big number, indicating that the line is far away from the data. The closer the line gets to the data, the lower the cost and the better the fit.

What we want are values of a and b that don't just make the cost function small, but that make it the *exact* smallest function possible. The second major function we'll write is called `linear_regression`, and it automatically finds these best values of a and b. This, in turn, tell us the line of best fit. To implement this, we build a function telling us the cost for any values of a and b and minimize it using the technique of gradient descent from chapter 12. Let's get started by implementing a cost function in Python to measure how well a function fits a data set.

14.1 Measuring the quality of fit for a function

We'll write our cost function so that it can work on any data set, not just our collection of used cars. That allows us to test it out on simpler (made-up) data sets, so we can see how it works. With that in mind, the cost function is a Python function taking two inputs. One of these is the Python function $f(x)$ that we want to test, and the second is the data set to test against, a collection of (x, y) pairs. For the used car example, our $f(x)$ might be a linear function giving a cost in dollars for any mileage, and the (x, y) pairs are the actual values of mileage and price from the data set.

The output of the cost function is a single number, measuring how far the values of $f(x)$ are from the correct y values. If $y = f(x)$, for every x, the function is a perfect fit for the data, so the cost function returns zero. More realistically, the function won't agree exactly with all the data points, and it will return some positive number. We'll actually write two cost functions to compare them and give you a sense of how cost functions work:

- `sum_error`—Adds the distance from $f(x)$ to y for every (x, y) value in the data set
- `sum_square_error`—Adds up the squares of these distances

The second of these is the cost function most commonly used in practice, and you'll see why shortly.

14.1.1 Measuring distance from a function

In the source code for this book, you'll find a made-up data set called `test_data`. It's a Python list of (x, y) values, where the x values range from –1 to 1. I've intentionally chosen y values so that the points lie close to the line $f(x) = 2x$. Figure 14.5 shows a scatter plot of the `test_data` data set next to that line.

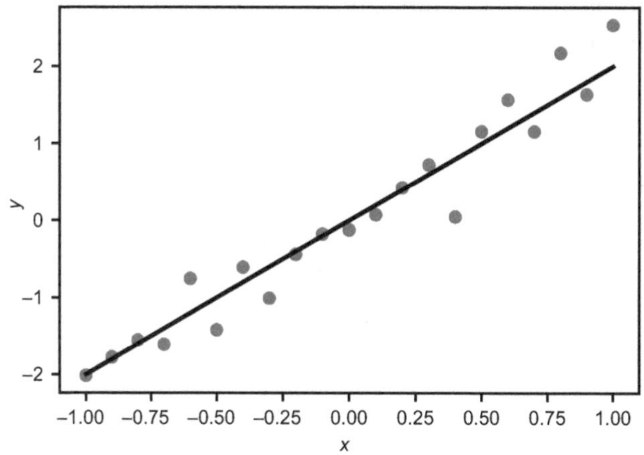

Figure 14.5 A set of randomly generated data that intentionally stays close to the line *f*(*x*) = 2*x*

The fact that $f(x) = 2x$ stays close to the data set means that for any x value in the data set, $2x$ is a pretty good guess for the corresponding y value. For instance, the point

$$(x, y) = (0.2, 0.427)$$

is an actual value from the data set. Given only the value $x = 0.2$, our $f(x) = 2x$ would have predicted $y = 0.4$. The absolute value of the difference $|f(0.2) - 0.4|$ tells us the size of this error, which is about 0.027.

An error value, which is the difference between the actual y value and the one predicted by the function $f(x)$, can be pictured as the vertical distance from the actual (x, y) point to the graph of f. Figure 14.6 shows the error distances drawn as vertical lines.

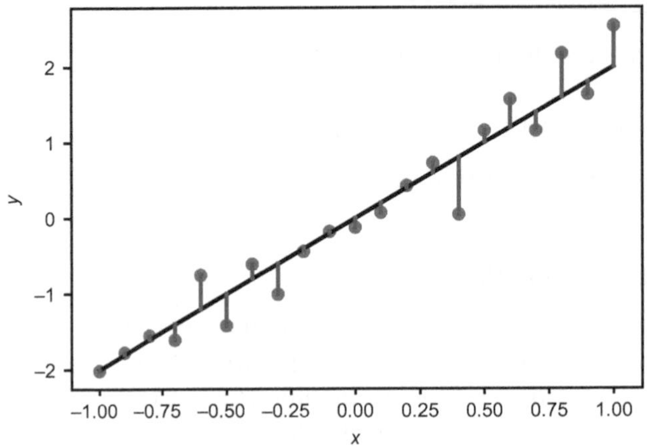

Figure 14.6 The error values are the differences between the function *f*(*x*) and the actual *y* values.

Some of these errors are smaller than others, but how can we quantify the quality of the fit? Let's compare this to a picture of a function, $g(x) = 1 - x$, which is obviously a bad fit (figure 14.7).

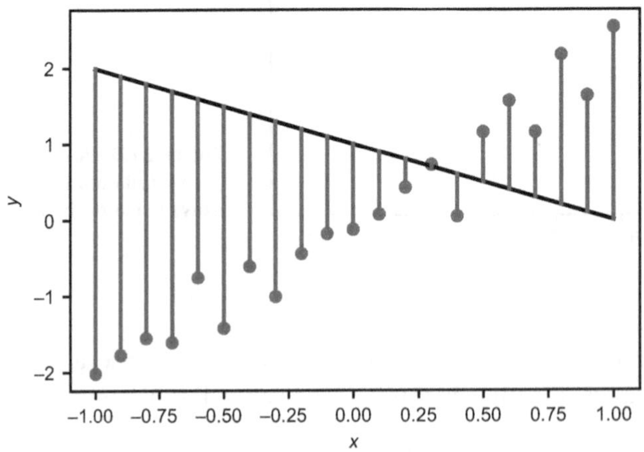

Figure 14.7 Picturing a function with larger error values

Our function, $g(x) = 1 - x$, happens to come close to one of the points, but the total of the errors is much larger. For that reason, we can write our first cost function by adding all of the errors. A larger sum of errors means a worse fit, while a lower value means a better fit. To implement this function, we simply iterate over the (x, y) pairs, take the absolute value of the difference between $f(x)$ and y, and sum the results:

```
def sum_error(f,data):
    errors = [abs(f(x) - y) for (x,y) in data]
    return sum(errors)
```

To test this function, we can translate our $f(x)$ and $g(x)$ into code:

```
def f(x):
    return 2*x

def g(x):
    return 1-x
```

As expected, the summed error for $f(x) = 2x$ is lower than for $g(x) = 1 - x$:

```
>>> sum_error(f,test_data)
5.021727176394801
>>> sum_error(g,test_data)
38.47711311130152
```

The exact values of these outputs don't matter; what matters is the comparison between them. Because the sum of the error for $f(x)$ is lower than for $g(x)$, we can conclude that $f(x)$ is a better fit for the given data.

14.1.2 Summing the squares of the errors

The `sum_error` function might be the most obvious way to measure the distance from the line to the data, but in practice, we'll use a cost function that sums up the *squares* of all the errors. There are a few good reasons for this. The simplest is because a squared distance function is smooth, so we can use derivatives to minimize it, while an absolute value function is not smooth, so we can't take its derivative everywhere. Keep in mind the pictures of the graphs of functions $|x|$ and x^2 (figure 14.8), both of which return bigger values when x is further from 0, but only the latter is smooth at $x = 0$ and has a derivative there.

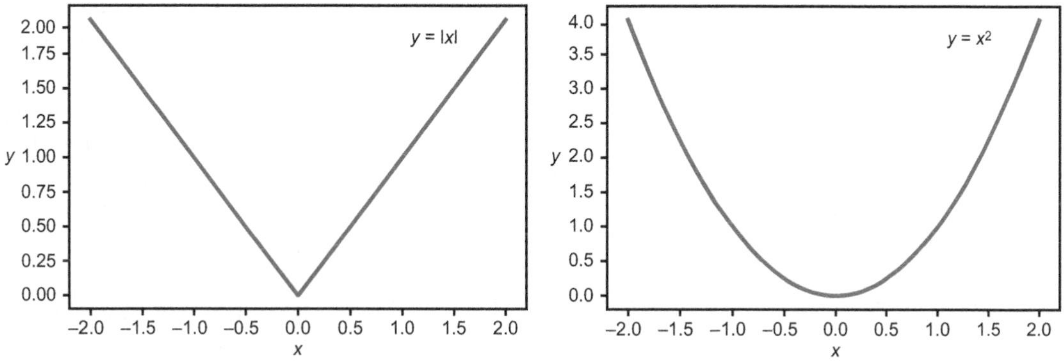

Figure 14.8 The graph of $y = |x|$ is not smooth at $x = 0$, but the graph of $y = x^2$ is.

Given a test function $f(x)$, we can look at every (x, y) pair and add the value of $(f(x) - y)^2$ to the cost. The `sum_squared_error` function does this, and its implementation doesn't look much different than `sum_error`. We only need to square the error instead of taking its absolute value:

```
def sum_squared_error(f,data):
    squared_errors = [(f(x) - y)**2 for (x,y) in data]
    return sum(squared_errors)
```

We can also visualize this cost function. Instead of looking at the vertical distances between points and the graph of the function, we can think of those distances as edges of squares. The area of each square is the squared error for that data point, and the total area of all squares is the result of `sum_squared_error`. The total area of squares in figure 14.9 shows the sum of the squared error between `test_data` and $f(x) = 2x$. (Note that the squares don't look like squares because the x- and y-axes have different scales!)

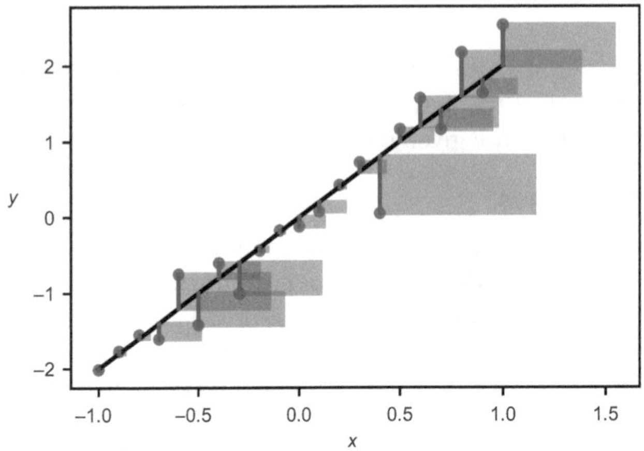

Figure 14.9 Picturing the sum of the squared error between a function and a data set

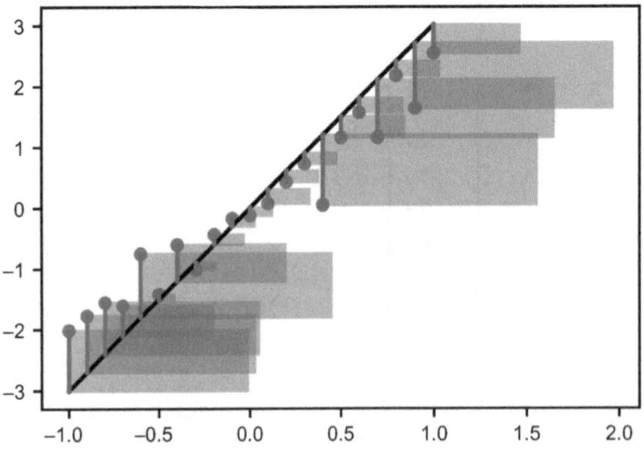

Figure 14.10 Picturing the `sum_squared_error` for $h(x) = 3x$ relative to the test data

A y value, which is twice as far from the graph in figure 14.9, contributes to the sum squared error by a factor of four. One of the reasons to prefer this cost function is that it penalizes poor fits more aggressively. For $h(x) = 3x$, you can see that the squares are quite a bit bigger (figure 14.10).

It's not worth drawing the squared errors for $g(x) = 1 - x$ because the squares are so big they fill almost the whole chart area and overlap each other significantly. You can see, though, that the difference in value of the `sum_squared_error` is even more drastic for $f(x)$ and $g(x)$ than for the difference in `sum_error`:

```
>>> sum_squared_error(f,test_data)
2.105175107540148
>>> sum_squared_error(g,test_data)
97.1078879283203
```

The graph of $y = x^2$ in figure 14.8 is clearly smooth, and it turns out that if you move the line by varying the parameters a and b that define it, the cost function changes "smoothly" as well. For this reason, we'll continue using the `sum_squared_error` cost function.

14.1.3 Calculating cost for car price functions

I'll start by making an educated guess about how Priuses depreciate as they gain mileage. There are several different models of Toyota Priuses, but I would guess the average retail price is about \$25,000. To make the calculation simple, our first, naïve model assumes they stay on the road for 125,000 miles, after which they are worth exactly \$0. That means that the cars depreciate at an average rate of \$0.20 per mile. That implies the price p of a Prius in terms of its mileage x is given by subtracting $0.2x$ dollars of depreciation from the starting price of \$25,000, and this means $p(x)$ is a linear function because it has the familiar form, $p(x) = ax + b$, with $a = -0.2$ and $b = 25{,}000$:

$$p(x) = -0.2x + 25{,}000$$

Let's see how this function looks on the graph by plotting it next to the CarGraph data (figure 14.11). You can find the data and the Python code to plot it in the source code for this chapter.

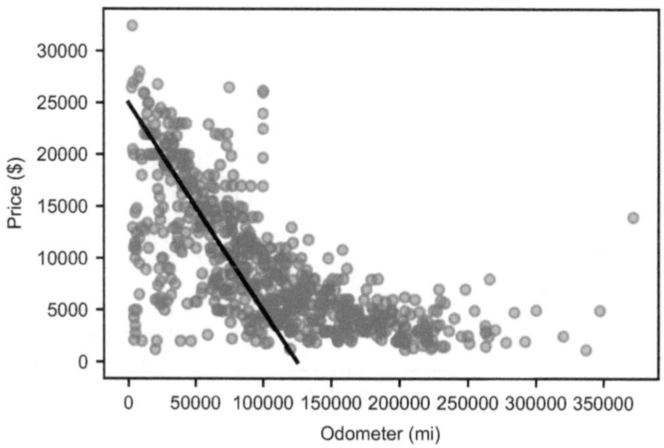

Figure 14.11 The scatter plot of price and mileage for used Priuses with my hypothetical depreciation function

Clearly, many of the cars in the data set have made it past my guess of 125,000 miles. That can mean that my guess of the depreciation rate was too high. Let's try a depreciation rate of $0.10 per mile, implying a pricing function:

$$p(x) = -0.10x + 25{,}000$$

This isn't perfect either. We can see on the graph in figure 14.12 that this function overestimates the price of the majority of the cars.

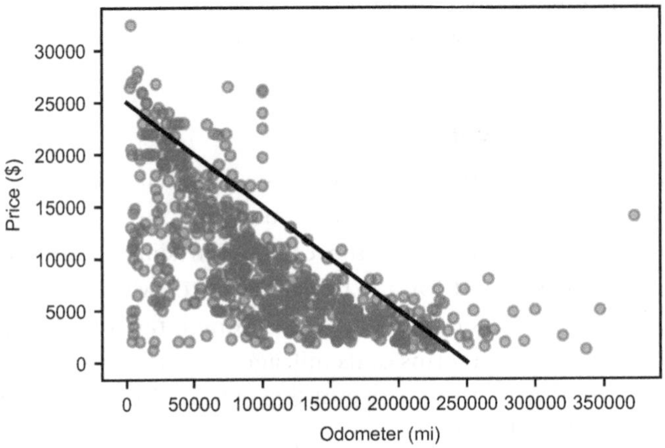

Figure 14.12 Plotting a different function that assumes a depreciation of $0.10 per mile

We can also experiment with the starting value, which we've assumed is $25,000. Anecdotally, a car loses much of its value the moment it drives off the lot, so a price of $25,000 might be an overestimation for a used car with very few miles on it. If the car loses 10% of its value when it drives off the lot, a price of $22,500 at zero mileage might give us better results (figure 14.13).

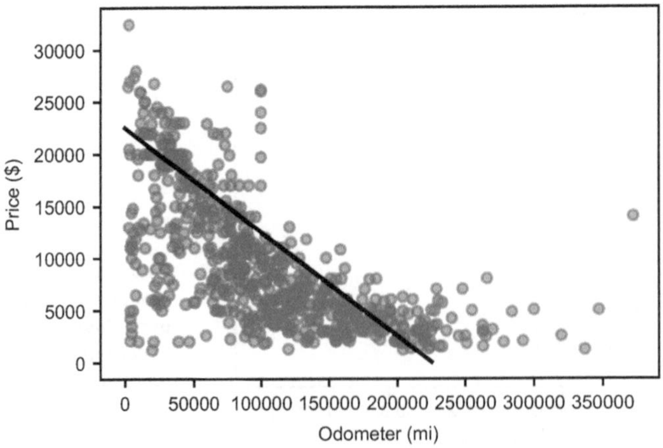

Figure 14.13 Testing a starting value of $22,500 for used Toyota Priuses

We can spend a lot of time speculating about what the best linear function is to fit the data, but to see if our speculations are improving, we need to use a cost function. Using the `sum_squared_error` function, we can measure which of our educated guesses is the closest to the data. Here are three pricing functions translated to Python code:

```
def p1(x):
    return 25000 - 0.2 * x

def p2(x):
    return 25000 - 0.1 * x

def p3(x):
    return 22500 - 0.1 * x
```

The `sum_squared_error` function takes a function as well as a list of pairs of numbers representing the data. In this case, we want pairs of mileages and prices:

```
prius_mileage_price = [(p.mileage, p.price) for p in priuses]
```

Using the `sum_squared_error` function for each of the three pricing functions, we can compare their quality of fit to the data:

```
>>> sum_squared_error(p1, prius_mileage_price)
88782506640.24002
>>> sum_squared_error(p2, prius_mileage_price)
34723507681.56001
>>> sum_squared_error(p3, prius_mileage_price)
22997230681.560013
```

These are some big values, roughly 88.7 billion, 34.7 billion, and 22.9 billion, respectively. Once again, the values don't matter, only their relative sizes. Because the last one is the lowest, we can conclude that `p3` is the best of the three pricing functions. Given how unscientific I was when making up these functions, it seems likely I could keep guessing and find a linear function making the cost even lower. Rather than guessing and checking, we'll look at how to explore the space of possible linear functions systematically.

14.1.4 Exercises

Exercise 14.1 Create a set of data points lying on a line and demonstrate that the `sum_error` and `sum_squared_error` cost functions both return exactly zero for the appropriate linear function.

Solution Here is a linear function and some points that lie on its graph:

```
def line(x):
    return 3*x-2
points = [(x,line(x)) for x in range(0,10)]
```

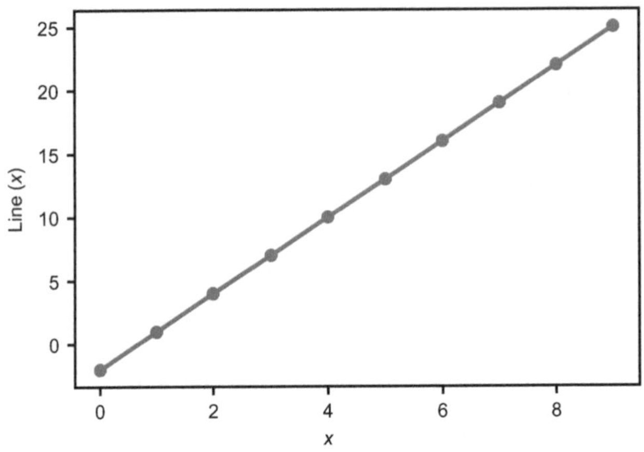

Both `sum_error(line,points)` and `sum_squared_error(line,points)` return zero because there is no distance from any of the points to the line.

Exercise 14.2 Calculate the value of the cost for the two linear functions, $x + 0.5$ and $2x - 1$. Which one produces a lower sum squared error relative to `test_data`, and what does that say about the quality of the fits?

Solution

```
>>> sum_squared_error(lambda x:2*x-1,test_data)
23.1942461283472
>>> sum_squared_error(lambda x:x+0.5,test_data)
16.607900877665685
```

The function $x + 0.5$ produces a lower value for `sum_squared_error`, so it is a better fit to the `test_data`.

Exercise 14.3 Find a linear function `p4` that fits the data even better than `p1`, `p2`, or `p3`. Demonstrate that it is a better fit by showing the cost function is lower than for `p1`, `p2`, or `p3`.

Solution The best fit we found so far is `p3`, represented by $p(x) = 22{,}500 - 0.1 \cdot x$. To get an even better fit, you can try tweaking the constants in this formula until the cost is reduced. One observation that you might make is that `p3` was a better fit because we reduced the b value from 25,000 to 22,500. If we reduce it slightly further, the fit gets even better. If we define a new function `p4` with a b value of 20,000

```
def p4(x):
    return 20000 - 0.1 * x
```

it turns out the `sum_squared_error` is even lower:

```
>>> sum_squared_error(p4, prius_mileage_price)
18958453681.560005
```

This is lower than the values for any of the three previous functions, demonstrating that it is a better fit to the data.

14.2 Exploring spaces of functions

We ended the last section by guessing some pricing functions of the form $p(x) = ax + b$, where x represents the mileage on a used Toyota Prius and p is a prediction of its price. By choosing different values of a and b and graphing the resulting function, $p(x)$, we could tell which choices were better than others. The cost function gave us a way of measuring how close the functions came to the data, rather than eyeballing it. Our goal in this section is to systematize the process of trying different values of a and b to make the cost function as small as possible.

If you did the last exercise from section 14.1 and searched manually for a better fit, you might have noticed that part of the challenge is tuning *both* a and b simultaneously. If you remember from chapter 6, the collection of all functions like $p(x) = ax + b$ form a 2D vector space. When you guess and check, you're blindly picking points in various directions in this 2D space and hoping the cost function decreases.

In this section, we'll try to understand the 2D space of possible linear functions by graphing the `sum_squared_error` cost function with respect to the parameters a and b, which define a linear function. Specifically, we can plot the cost as a function of the two parameters a and b, which define a choice of $p(x)$ (figure 14.14).

The actual function we'll plot takes two numbers, a and b, and returns one number, which is the cost of the function $p(x) = ax + b$. We call this function `coefficient_cost(a,b)` because the numbers a and b are *coefficients*. To plot such a function, we use a heatmap like we did in chapter 12.

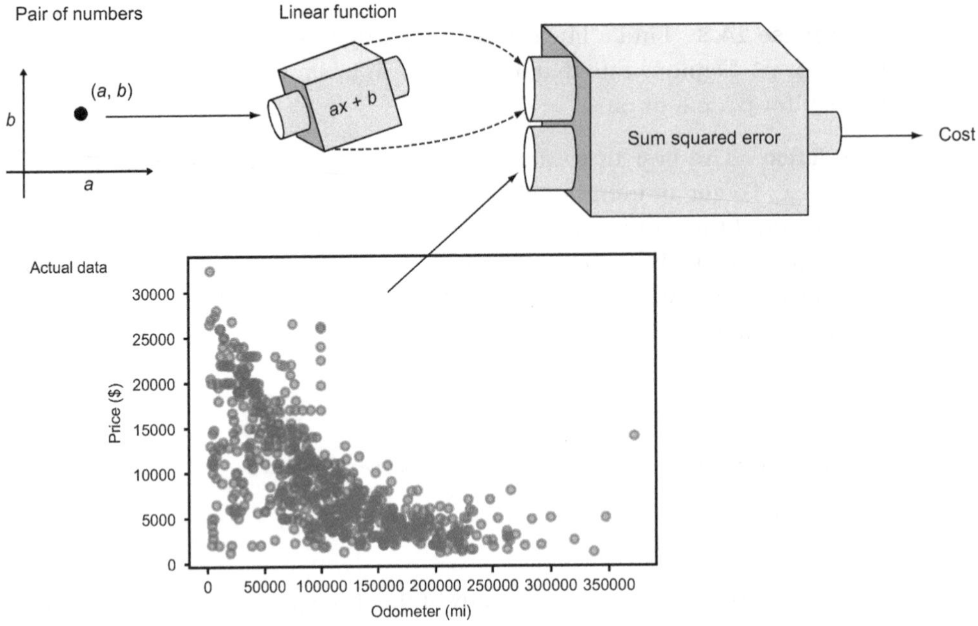

Figure 14.14 A pair of numbers, (*a*, *b*), define a linear function. Comparing it to the fixed actual data produces the cost as a single number.

As a warm up, we can try to fit a function $f(x) = ax$ to the `test_data` data set we used before. This is a simpler problem because `test_data` doesn't have as many data points, and because we've only got one parameter to tune: $f(x) = ax$ is a linear function with the value of b fixed at zero. The graph of a function of this form is a line through the origin and the coefficient a controls its slope. That means that there's only one dimension to explore, and we can plot the value of the sum squared error versus the value of a, which is an ordinary function graph.

14.2.1 Picturing cost for lines through the origin

Let's use the same `test_data` data set as before and compute the `sum_squared_error` from functions of the form $f(x) = ax$. We can then write a function `test_data_coefficient_cost`, taking the parameter a (the slope) and returning the cost for $f(x) = ax$. To do this, we first create the function f from the value of the input a and then we can pass it and the test data into the `sum_squared_error` cost function:

```
def test_data_coefficient_cost(a):
    def f(x):
        return a * x
    return sum_squared_error(f,test_data)
```

Each value of this function corresponds to a choice of the slope, a, and, therefore, tells us the cost of a line we could plot alongside `test_data`. Figure 14.15 shows a

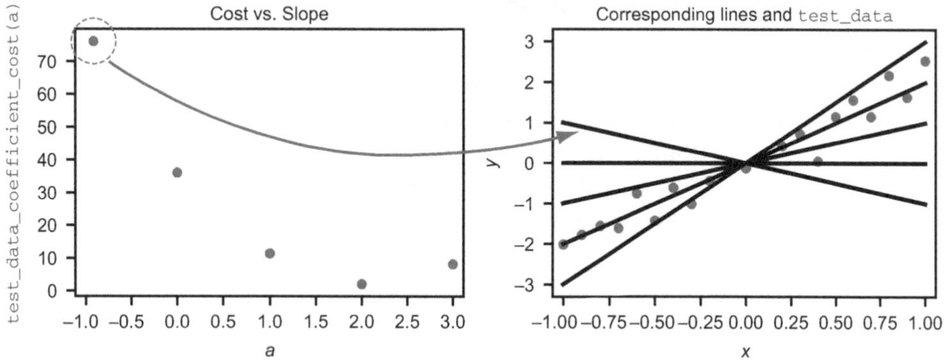

Figure 14.15 Costs for various values of the slope *a* and the lines represented by each

scatter plot of a few values of a and their corresponding lines. I've drawn attention to the slope of $a = -1$, which produces the highest cost and the line with the worst fit.

The `test_data_coefficient_cost` function turns out to be a smooth function that we can plot over a range of a values. The graph in figure 14.16 shows us that the cost gets lower and lower until it hits a minimum around $a = 2$ and then it starts increasing.

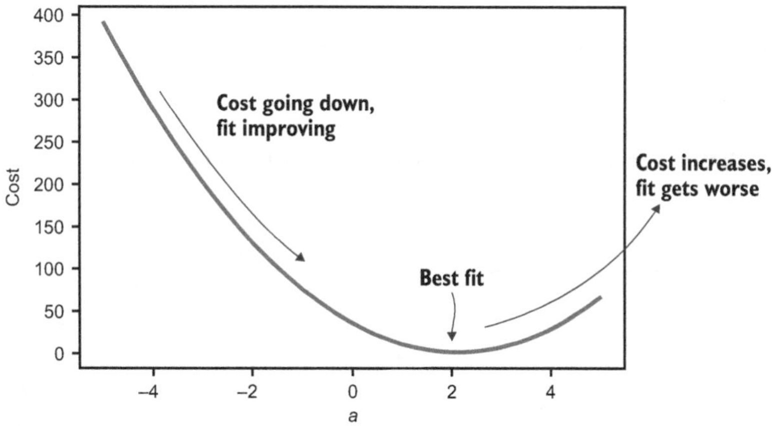

Figure 14.16 A graph of cost vs. the slope *a*, showing the quality of fit for different slope values

The graph in figure 14.16 tells us the line through the origin producing the lowest cost and, therefore, the *best fit*, which has roughly a slope of 2 (we'll find the exact value shortly). To find the best linear function fitting the used car data, let's look at the cost over a space with one more dimension.

14.2.2 The space of all linear functions

We're looking for a function $p(x) = ax + b$ that comes closest to predicting the price of a Prius based on its mileage as measured by the `sum_squared_error` function. To evaluate different choices of the coefficients a and b, we need to first write a function `coefficient_cost(a,b)` that gives the sum squared error for $p(x) = ax + b$ relative to the car data. This looks like the `test_data_coefficient_cost` function, except there are two parameters and we use a different data set:

```
def coefficient_cost(a,b):
    def p(x):
        return a * x + b
    return sum_squared_error(p,prius_mileage_price)
```

Now, there's a 2D space of pairs of coefficients (a,b), any of which gives us a different candidate function $p(x)$ to compare with the price data. Figure 14.17 shows two points in the a,b plane and the corresponding lines on the graph.

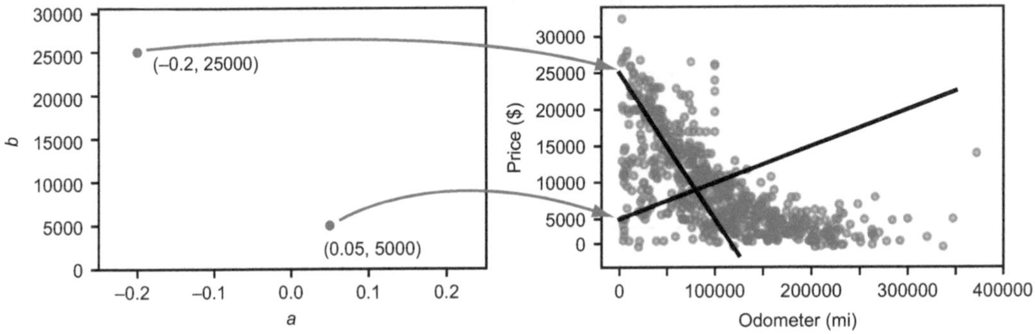

Figure 14.17 Different pairs of numbers (*a*, *b*) correspond to different price functions

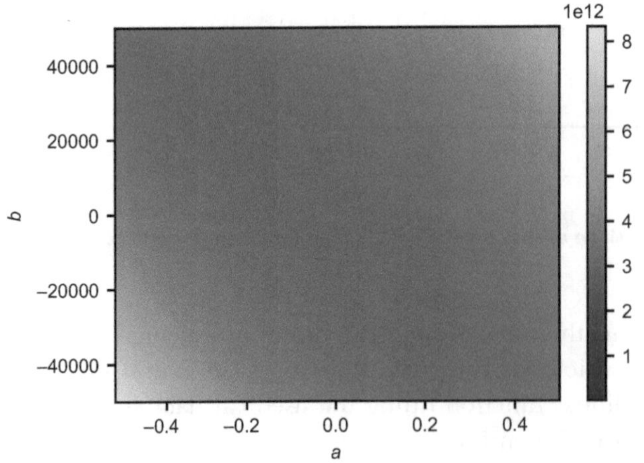

For every pair (a, b) and corresponding function $p(x) = ax + b$, we can compute the `sum_squared_error` function; that's what the `coefficient_cost` function does for us in one step. This gives us a cost value for every point in the a,b plane that we can plot as a heat map (figure 14.18).

Figure 14.18 Cost for the linear function as a heatmap over values of *a* and *b*

On this heatmap, you can see that when (a, b) are at their extremes, the cost function is high. The heatmap is darkest in the middle, but it's not visually clear whether there's a minimum value for the cost or exactly where it occurs. Fortunately, we have a method for finding where in the (a, b) plane the cost function is minimized—gradient descent.

14.2.3 Exercises

Exercise 14.4 Find the exact formula for a line through the origin that passes through one point (3, 4). Do this by finding the function $f(x) = ax$, which minimizes the sum squared error relative to this one-point data set.

Solution There is one coefficient we need to find, which is a. The sum of squared error is the squared difference between $f(3) = a \cdot 3$ and 4. This is $(3a - 4)^2$, which expands to $9a^2 - 24a + 16$. We can think of this as a cost function in terms of a, that is, $c(a) = 9a^2 - 24a + 16$.

The best value of a is the one that minimizes this cost. That value of a causes the derivative of the cost function to be zero. Using the derivative rules from chapter 10, we find $c'(a) = 18a - 24$. This is solved when $a = 4/3$, meaning our line of best fit is

$$f(x) = \frac{4}{3}x$$

This clearly contains the origin and the point (4, 3).

Exercise 14.5 Suppose we use a linear function to model the price of a sports car with respect to its mileage with coefficients $(a, b) = (-0.4, 80000)$. In English, what does that say about how the car depreciates over time?

Solution The value of $ax + b$ when $x = 0$ is just $b = 80{,}000$. That means that at a mileage of 0, we can expect the car to sell for \$80,000. The value a of -0.4 means that the function value $ax + b$ decreases at a rate of 0.4 units for every one unit increase in x. That means that the car's value decreases, on average, by 40 cents for every one mile it is driven.

14.3 Finding the line of best fit using gradient descent

In chapter 12, we used the gradient descent algorithm to minimize a smooth function of the form $f(x, y)$. In simpler terms, that meant finding the values of x and y that made the value of $f(x, y)$ as small as possible. Because we have a `gradient_descent` function implemented, we can simply pass it a Python version of a function we want to minimize, and it automatically finds the inputs that minimize it.

Now, we want to find the values of a and b that make the cost of $p(x) = ax + b$ as small as possible, in other words, minimizing the Python function `coefficient_cost(a,b)`. When we plug `coefficient_cost` into our `gradient_descent` function, we get back the pair (a, b) such that $p(x) = ax + b$ is the line of best fit. We can use the values of a and b that we find to plot the line $ax + b$ and visually confirm that it's a good fit to the data.

14.3.1 Rescaling the data

There's one last tricky detail we need to deal with before applying gradient descent. The numbers we've been working with have drastically different sizes: the depreciation rates are between 0 and –1, the prices are in the tens of thousands, and the cost function returned results in the hundreds of billions. If we don't specify otherwise, our derivative approximation is taken using a `dx` value of 10^{-6}. Because these numbers differ so greatly in magnitude, we can run into numerical errors if we try to run the gradient descent as is.

NOTE I won't go into the details of the numerical issues that come up; my goal is to show you how to apply the math concepts, not to write robust numerical code. Instead, I'll just show you how to get around this issue by reshaping the data we're using.

Based on our intuition about the data set, we can figure out some conservative bounds on the values of a and b, producing the line of best fit. The value a represents the depreciation, so the best value probably has a magnitude greater than 0.5 or 50 cents per mile. The b value represents the price of a Prius with zero miles on it and should be safely below $50,000.

If we define new variables c and d by $a = 0.5 \cdot c$ and $b = 50{,}000 \cdot d$, then when c and d have a magnitude less than one, a and b should have a magnitude less than 0.5 and 50,000, respectively. For values of a and b smaller than these values, the cost function goes no higher than 10^{13}. If we divide the cost function result by 10^{13} and express it in terms of c and d, we have a new version of the cost function whose inputs and outputs all have absolute value between zero and one:

```
def scaled_cost_function(c,d):
    return coefficient_cost(0.5*c,50000*d)/1e13
```

If we find the values of c and d that minimize this scaled cost function, we can find the a and b values minimizing the original function, using the facts that $a = 0.5 \cdot c$ and $b = 50{,}000 \cdot d$.

This is a somewhat back-of-the-envelope approach, and there are more scientific ways to scale data to make it more numerically mangeable, one of which we'll see in chapter 15. If you want to learn more, the usual process is called *feature scaling* in machine learning literature. For now, we've got what we need—a function we can plug into the gradient descent algorithm.

14.3.2 Finding and plotting the line of best fit

The function we're going to optimize is `scaled_cost_function`, and we can expect the minimum to occur at a point (c, d), where $|c| < 1$ and $|d| < 1$. Because the optimal c

and d are reasonably close to the origin, we can start the gradient descent at (0,0). The following code finds the minimum, although it may take a while to run, depending on what kind of machine you're using:

```
c,d = gradient_descent(scaled_cost_function,0,0)
```

When it runs, it finds the following values for c and d:

```
>>> (c,d)
(-0.12111901781176426, 0.31495422888049895)
```

To recover a and b, we need to multiply by the respective factors:

```
>>> a = 0.5*c
>>> b = 50000*d
>>> (a,b)
(-0.06055950890588213, 15747.711444024948)
```

And, at last, we have our coefficients that we've been looking for! Rounding, we can say that the price function

$$p(x) = -0.0606 \cdot x + 15{,}700$$

is the linear function that (approximately) minimizes the sum squared error over the whole data set of cars. It implies that the price of a Toyota Prius with zero miles on it is, on average, \$15,700, and that Priuses depreciate at an average rate of just over 6 cents per mile. Figure 14.19 shows what the line looks like on a graph.

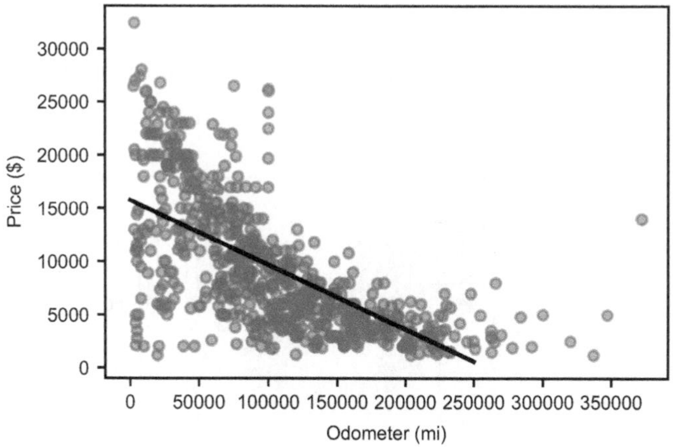

Figure 14.19 The line of best fit for the car price data

This looks at least as good as the other linear functions $p_1(x)$, $p_2(x)$, and $p_3(x)$ that we tried, if not better. We can be sure that it's a better fit to our data as measured by the cost function:

```
>>> coefficient_cost(a,b)
14536218169.403479
```

Having automatically found a line of best fit that minimizes the cost function, we can say that our algorithm "learned" how to value Priuses based on their mileage, and we achieved the main goal of this chapter.

There are a number of ways to compute linear regression to get this line of best fit, including a number of optimized Python libraries. Regardless of the methodology, they should get you to the same linear function that minimizes the sum of the squared error. I picked this specific methodology using gradient descent, both because it is a great application of a number of concepts we covered in part 1 and part 2, and also because it is highly generalizable. In the last section, I'll show you one more application of gradient descent for regression, and we'll make use of gradient descent and regression in the next two chapters as well.

14.3.3 Exercises

Exercise 14.6 Use gradient descent to find the linear function that best fits the test data. Your resulting function should be close to $2x + 0$, but not exactly, because the data was randomly generated around that line.

Solution First, we need to write a function that computes the cost of $f(x) = ax + b$ relative to the test data in terms of the coefficients a and b:

```
def test_data_linear_cost(a,b):
    def f(x):
        return a*x+b
    return sum_squared_error(f,test_data)
```

The values of a and b that minimize this function give us the linear function of best fit. We expect a and b to be close to 2 and 0, respectively, so we can plot a heat map around those points to understand the function we're minimizing:

```
scalar_field_heatmap(test_data_linear_cost,-0,4,-2,2)
```

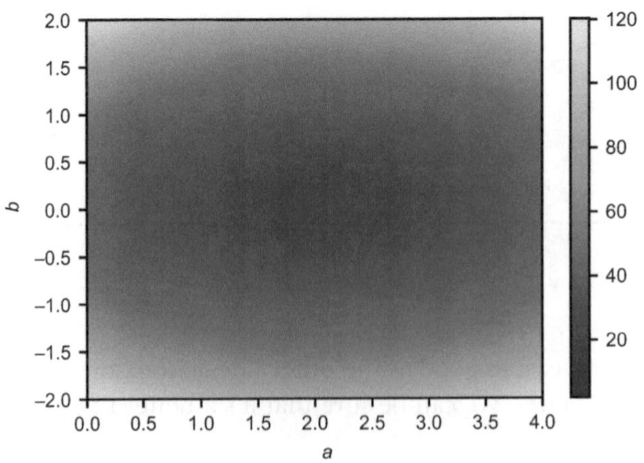

The cost of *ax* + *b* relative to the test data as a function of *a* and *b*

It looks like there's a minimum to this cost function in the vicinity of $(a, b) = (2,0)$, as expected. Using gradient descent to minimize this function, we can find the exact values:

```
>>> gradient_descent(test_data_linear_cost,1,1)
(2.103718204728344, 0.0021207385859157535)
```

This means the line of best fit to the test data is approximately $2.10372 \cdot x + 0.00212$.

14.4 Fitting a nonlinear function

In the work we've done so far, there was no step that *required* the price function $p(x)$ to be linear. A linear function was a good choice because it was simple, but we can apply the same method with any function of one variable defined by two constants. As an example, let's find the exponential function of best fit having the form $p(x) = qe^{rx}$ and minimizing the sum of the squared error relative to the car data. In this equation, e is the special constant 2.71828 . . ., and we'll find the values of q and r that give us the best fit.

14.4.1 Understanding the behavior of exponential functions

In case it's been a while since you've worked with exponential functions, let's do a quick review of how they work. You can recognize a function $f(x)$ as exponential when the argument x is in an exponent. For instance, $f(x) = 2^x$ is an exponential function, while $f(x) = x^2$ is not. In fact, $f(x) = 2^x$ is one of the most familiar exponential functions. The value of 2^x at each whole number of x is 2 multiplied by itself x times. Table 14.1 gives us some values of 2^x.

Table 14.1 Values of the familiar exponential function 2^x

x	0	1	2	3	4	5	6	7	8	9
2^x	1	2	4	8	16	32	64	128	256	512

The number raised to the x power is called the *base*, so in the case of 2^x, the base is 2. If the base is greater than one, the function increases as x increases, but if it is less than one, it decreases as x increases. For instance, in $(½)^x$, each whole number value is half the previous as shown in table 14.2.

Table 14.2 Values of the decreasing exponential function (½)x

x	0	1	2	3	4	5	6	7	8	9
$(½)^x$	1	0.5	0.25	0.125	~0.06	~0.03	~0.015	~0.008	~0.004	~0.002

This is called an *exponential decay*, and it's more like what we want for our car depreciation model. An exponential decay means that the value of the function decreases by the same ratio over every x interval of a fixed size. Once we have our model, it can tell us that a Prius loses half its value every 50,000 miles, implying that it's worth is ¼ of its original price at 100,000 miles, and so on.

Intuitively, this could be a better way to model depreciation. Toyotas, which are reliable cars that last a long time, retain some value as long as they are driveable. Our linear model, by comparison, suggests that their value becomes negative after a long time (figure 14.20).

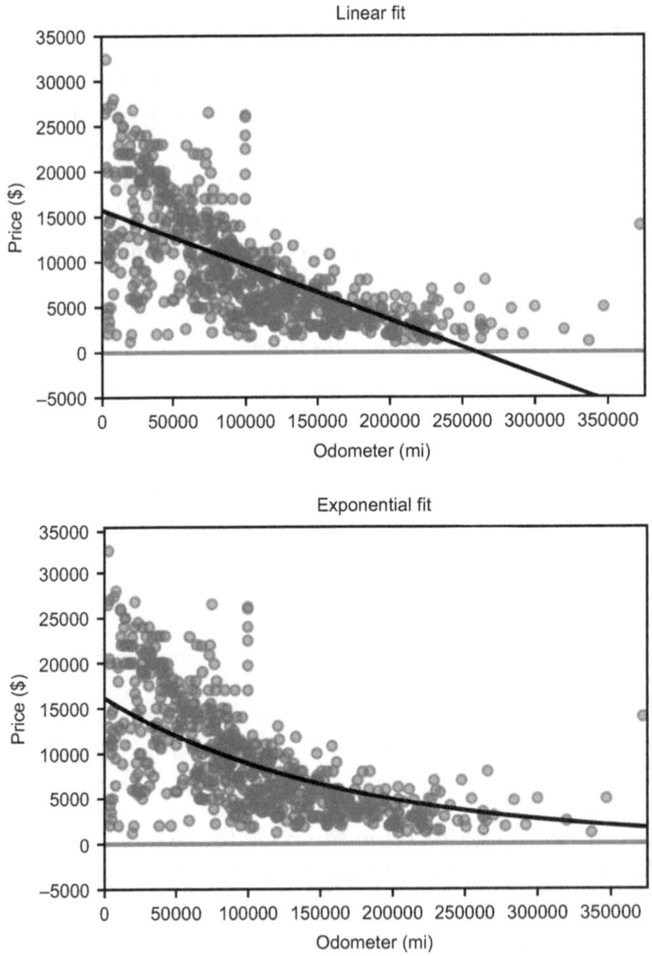

Figure 14.20 A linear model predicts negative values for Priuses, as compared with an exponential model that shows a positive value at any mileage.

The form of exponential function we can use is $p(x) = qe^{rx}$, where $e = 2.71828\ldots$ is the fixed base, and r and q are the coefficients we can adjust. (It might seem arbitrary or even inconvenient to use the base e, but e^x is the standard exponential function, so it's worth getting used to.) In the case of exponential decay, the value of r is negative.

Because $e^{r \cdot 0} = e^0 = 1$, we have $p(0) = qe^{r \cdot 0} = q$, so q still models the price of the Prius with zero miles. The constant r decides the rate of depreciation.

14.4.2 Finding the exponential function of best fit

With the formula $p(x) = qe^{rx}$ in mind, we can use our methodology from the previous sections to find the exponential function of best fit. The first step is to write a function that takes the coefficients q and r and returns the cost of the corresponding function:

```
def exp_coefficient_cost(q,r):
    def f(x):
        return q*exp(r*x)
    return sum_squared_error(f,prius_mileage_price)
```

Python's exp function computes the exponential function e^x.

The next thing we need to do is choose a reasonable range for the coefficients q and r, which set the starting price and the depreciation rate, respectively. For q, we expect it to be close to the value of b from our linear model because both q and b represent the price of the car with zero miles on it. I'll use a range from \$0 to \$30,000 to be safe.

The value of r, which controls the depreciation rate, is a bit trickier to understand and set limits on. The equation $p(x) = qe^{rx}$ with a negative r value implies that every time x increases by $-1/r$ units, the price decreases by a *factor* of e, meaning it is multiplied by $1/e$ or about 0.36. (I've added an exercise at the end of the section to help you convince yourself of this!)

To be conservative, let's say a car is reduced in price by a factor of $1/e$, or to 36% of its original value, after 10,000 miles at the earliest. That would give us $r = 10^{-4}$. A smaller r value would mean a slower rate of depreciation. These benchmark magnitudes show us how to rescale the function, and if we divide by 10^{11}, the cost values stay small as well. Here's the implementation of the scaled cost function, and figure 14.21 shows a heat map of its output:

```
def scaled_exp_coefficient_cost(s,t):
    return exp_coefficient_cost(30000*s,1e-4*t) / 1e11

scalar_field_heatmap(scaled_exp_coefficient_cost,0,1,-1,0)
```

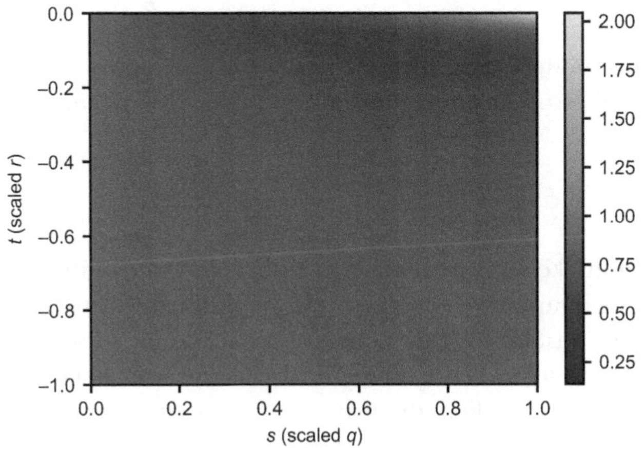

Figure 14.21 Cost as a function of the rescaled values of *q* and *r*, called s and *t*, respectively

The dark region at the top of the heatmap in figure 14.21 shows that the lowest cost occurs at a small value of t and a value of s somewhere in the middle of the range 0 to 1. We're ready to plug the scaled cost function into the gradient descent algorithm. The outputs of the gradient descent function are the s and t values minimizing the cost function, and we can undo the scaling to get q and r:

```
>>> s,t = gradient_descent(scaled_exp_coefficient_cost,0,0)
>>> (s,t)
(0.6235404892859356, -0.07686877731125034)
>>> q,r = 30000*s,1e-4*t
>>> (q,r)
(18706.214678578068, -7.686877731125035e-06)
```

This implies that the exponential function that best predicts the price of a Prius in terms of its mileage is approximately

$$p(x) = 18{,}700 \cdot e^{0.00000768 \cdot x}$$

Figure 14.22 shows the graph with the actual price data.

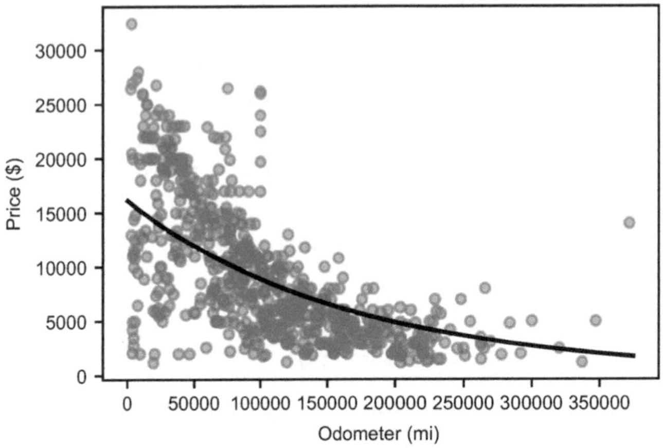

Figure 14.22 The exponential function of best fit for a Prius and its mileage

You could argue that this model is even better than our linear model because it produces a lower sum squared error, meaning it fits the data (slightly) better according to the cost function:

```
>>> exp_coefficient_cost(q,r)
14071654468.28084
```

Using a nonlinear function like an exponential function is just one of the many variations on this regression technique. We could use other nonlinear functions, functions defined by more than two constants or data fit in more than 2 dimensions. In the next two chapters, we'll continue to use cost functions to measure the quality of fit for regression models and then use gradient descent to make the fit as good as possible.

14.4.3 Exercises

Exercise 14.7 Confirm by choosing a sample value of r that e^{-rx} decreases by a factor of e every time x increases by $1/r$ units.

Solution Let's take $r = 3$, so our test function is e^{-3x}. We want to confirm that this function decreases by a *factor* of e every time x increases by ⅓ units. Defining the function in Python as follows

```
def test(x):
    return exp(-3*x)
```

we can see that it starts at a value of 1 at $x = 0$ and decreases by a factor of e for every ⅓ we add to x:

```
>>> test(0)
1.0
>>> from math import e
>>> test(1/3), test(0)/e
(0.36787944117144233, 0.36787944117144233)
>>> test(2/3), test(1/3)/e
(0.1353352832366127, 0.1353352832366127)
>>> test(1), test(2/3)/e
(0.049787068367863944, 0.04978706836786395)
```

In each of these cases, adding ⅓ to the input of `test` yields the same result as dividing the previous result by e.

Exercise 14.8 According to the exponential function of best fit, what percentage of the value of a Prius is lost every 10,000 miles?

Solution The price function is $p(x) = 18{,}700 \cdot e^{-0.00000768 \cdot x}$, where the value q = \$18,700, which represents the initial price and not how fast the price is decreasing. We can focus on the term $erx = e^{-0.00000768 \cdot x}$ and see how much it changes over 10,000 miles. For $x = 0$, the value of this expression is 1, and for $x = 10{,}000$, the value is

```
>>> exp(r * 10000)
0.9422186306357088
```

This means that after 10,000 miles, the Prius is worth only 94.2% of its original price, a decrease of 5.8%. Given how the exponential function behaves, this will be the case over *any* 10,000-mile increase in the mileage.

Exercise 14.9 Asserting that the retail price (the price at zero miles) is \$25,000, what is the exponential function that best fits the data? In other words, fixing $q = 25{,}000$, what is the value of r yielding the best fit for qe^{rx}?

Solution We can write a separate function that gives the cost of the exponential function in terms of the single unknown coefficient r:

```
def exponential_cost2(r):
    def f(x):
        return 25000 * exp(r*x)
    return sum_squared_error(f,prius_mileage_price)
```

The following plot confirms that there's a value of r between -10^{-4} and 0, which minimizes the cost function:

```
plot_function(exponential_cost2,-1e-4,0)
```

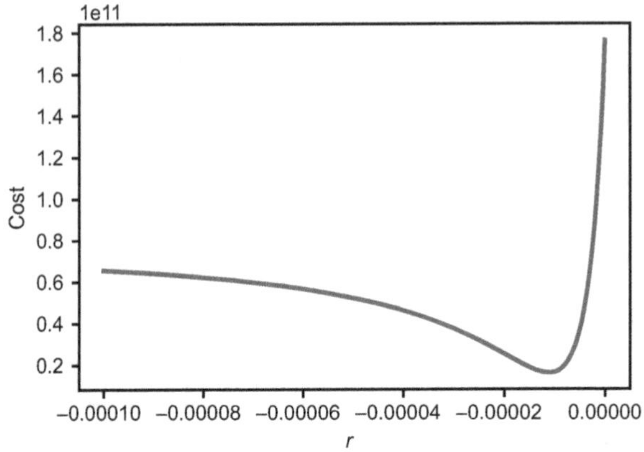

It looks like an approximate value of $r = -10^{-5}$ minimizes the cost function. To automatically minimize this function, we need to write a one-dimensional version of the gradient descent or use another minimization algorithm. You can try that approach if you like, but because there's only one parameter, we can simply guess and check to see that $r = -1.12 \cdot 10^{-5}$ is approximately the r value yielding the minimum cost. This implies the best fit function is $p(x) = 25{,}000 \cdot e^{-0.0000112 \cdot x}$. Here's the graph of the new exponential fit, plotted with the raw price data:

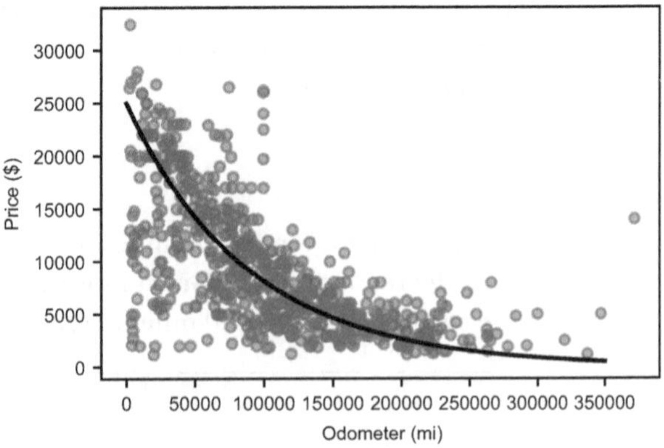

Summary

- *Regression* is the process of finding a model to describe the relationship between various data sets. In this chapter, we use linear regression to approximate the price of a car from its mileage as a linear function.
- For a set of many (x, y) data points, there is likely no line that passes through all of the points.
- For a function $f(x, y)$ modeling the data, you can measure how close it comes to the data by taking the distance between $f(x)$ and y for the specified points (x, y).
- A function measuring how well a model fits a data set is called a *cost function.* A commonly used cost function is the sum of squared distances from (x, y) points to the corresponding model value $f(x)$. The function that best fits the data has the lowest cost function.
- Considering linear functions of the form $f(x)$, every pair of coefficients (a, b) defines a unique linear function. There is a 2D space of such pairs and, therefore, a 2D space of lines to explore.
- Writing a function that takes a pair of coefficients (a, b) and computes the cost of $ax + b$ gives a function taking a 2D point and returning a number. Minimizing this function gives the coefficients defining the line of best fit.
- Whereas a linear function $p(x)$ increases or decreases by a constant amount for constant changes in x, an exponential function decreases or increases by a constant ratio for constant changes in x.
- To fit an exponential equation to data, you can follow the same procedure as for a linear equation; you need to find the pair (q, r) yielding an exponential function qe^{rx} minimizing the cost function.

15 Classifying data with logistic regression

This chapter covers

- Understanding classification problems and measuring classifiers
- Finding decision boundaries to classify two kinds of data
- Approximating classified data sets with logistic functions
- Writing a cost function for logistic regression
- Carrying out gradient descent to find a logistic function of best fit

One of the most important classes of problems in machine learning is *classification*, which we'll focus on in the last two chapters of this book. A classification problem is one where we've got one or more pieces of raw data, and we want to say what *kind* of object each one represents. For instance, we might want an algorithm to look at the data of all email messages entering our inbox and classify each one as an interesting message or as unwanted spam. As an even more impactful example, we could write a classification algorithm to analyze a data set of medical scans and decide whether they contain benign or malevolent tumors.

We can build machine learning algorithms for classification where the more real data our algorithm sees, the more it learns, and the better it performs at the classification task. For instance, every time an email user flags an email as spam or a radiologist identifies a malignant tumor, this data can be passed back to the algorithm to improve its calibration.

In this chapter, we look at the same simple data set as in the last chapter: mileages and prices of used cars. Instead of using data for a single model of car like in the last chapter, we'll look at two car models: Toyota Priuses and BMW 5 series sedans. Based only on the numeric data of the car's mileage and price, and a reference data set of known examples, we want our algorithm to give us a yes or no answer as to whether the car is a BMW. As opposed to a regression model that takes in a number and produces another number, the classification model will take in a vector and produce a number between zero and one, representing the confidence that the vector represents a BMW instead of a Prius (figure 15.1).

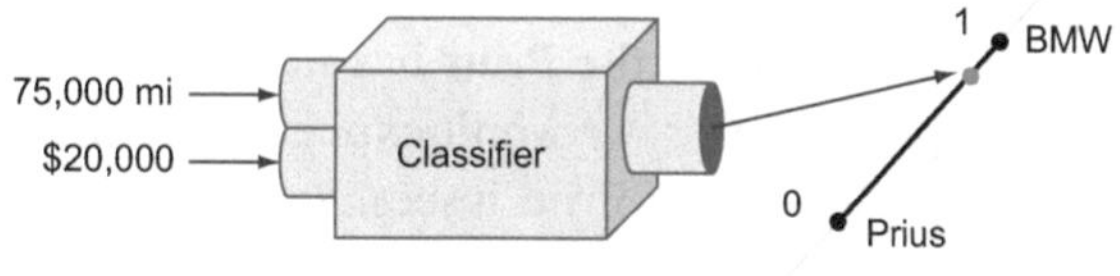

Figure 15.1 Our classifier takes a vector of two numbers, the mileage and price of a used car, and returns a number representing its confidence that the car is a BMW.

Even though classification has different inputs and outputs than regression, it turns out we can build our classifier using a type of regression. The algorithm we'll implement in this chapter is called *logistic regression*. To train this algorithm, we start with a known data set of used car mileages and prices, labeled with a 1 if they are BMWs and a 0 if they are Priuses. Table 15.1 shows sample points in this data set that we use to train our algorithm.

Table 15.1 Sample data points used to train the algorithm

Mileage (mi)	Price ($)	Is BMW?
110,890.0	13,995.00	1
94,133.0	13,982.00	1
70,000.0	9,900.00	0
46,778.0	14,599.00	1
84,507.0	14,998.00	0
...	...	...

We want a function that takes the values in the first two columns and produces a result that is between zero and one, and hopefully, close to the correct choice of car. I'll introduce you to a special kind of function called a *logistic function*, which takes a pair of input numbers and produces a single output number that is always between zero and one. Our classification function is the logistic function that "best fits" the sample data we provide.

Our classification function won't always get the answer right, but then again neither would a human. BMW 5 series sedans are luxury cars, so we would expect to get a lower price for a Prius than a BMW with the same mileage. Defying our expectations, the last two rows of the data in table 5.1 show a Prius and BMW at roughly the same price, where the Prius has nearly twice the mileage of the BMW. Due to fluke examples like this, we won't expect the logistic function to produce exactly one or zero for each BMW or Prius it sees. Rather it can return 0.51, which is the function's way of telling us it's not sure, but the data is slightly more likely to represent a BMW.

In the last chapter, we saw that the linear function we chose was determined by the two parameters a and b in the formula $f(x) = ax + b$. The logistic functions we'll use in this chapter are parametrized by three parameters, so the task of logistic regression boils down to finding three numbers that get the logistic function as close as possible to the sample data provided. We'll create a special cost function for the logistic function and find the three parameters that minimize the cost function using gradient descent. There's a lot of steps here, but fortunately, they all parallel what we did in the last chapter, so it will be a useful review if you're learning about regression for the first time.

Coding the logistic regression algorithm to classify the cars is the meat of the chapter, but before doing that, we spend a bit more time getting you familiar with the process of classification. And before we train a computer to do the classification, let's measure how well we can do the task. Then, once we build our logistic regression model, we can evaluate how well it does by comparison.

15.1 Testing a classification function on real data

Let's see how well we can identify BMWs in our data set using a simple criterion. Namely, if a used car has a price above \$25,000, it's probably too expensive to be a Prius (after all, you can get a brand new Prius for near that amount). If the price is above \$25,000, we'll say that it is a BMW; otherwise, we'll say that it's a Prius. This classification is easy to build as a Python function:

```
def bmw_finder(mileage,price):
    if price > 25000:
        return 1
    else:
        return 0
```

The performance of this classifier might not be that great because it's conceivable that BMWs with a lot of miles might sell for less than \$25,000. But we don't have to speculate: we can measure how well this classifier does on actual data.

In this section, we measure the performance of our algorithm by writing a function called `test_classifier`, which takes a classification function like `bmw_finder` as well as the data set to test. The data set is an array of tuples of mileages, prices, and a `1` or `0`, indicating whether the car is a BMW or a Prius. Once we run the `test_classifier` function with real data, it returns a percent value, telling us how many of the cars it identifies correctly. At the end of the chapter when we've implemented logistic regression, we can instead pass in our logistic classification function to `test_classifier` and see its relative performance.

15.1.1 Loading the car data

It is easier to write the `test_classifier` function if we first load the car data. Rather than fuss with loading the data from CarGraph.com or from a flat file, I've made it easy for you by providing a Python file called cardata.py in the source code for the book. It contains two arrays of data: one for Priuses and one for BMWs. You can import the two arrays as follows:

```
from car_data import bmws, priuses
```

If you inspect either the BMW or Prius raw data in the car_data.py file, you'll see that this file contains more data than we need. For now, we're focusing on the mileage and price of each car, and we know what car it is, based on the list it belongs to. For instance, the BMW list begins like this:

```
[('bmw', '5', 2013.0, 93404.0, 13999.0, 22.09145859494213),
 ('bmw', '5', 2013.0, 110890.0, 13995.0, 22.216458611342592),
 ('bmw', '5', 2013.0, 94133.0, 13982.0, 22.09145862741898),
 ...
```

Each tuple represents one car for sale, and the mileage and price are given by the fourth and fifth entries of the tuple, respectively. Within car_data.py, these are converted to `Car` objects, so we can write `car.price` instead of `car[4]`, for example. We can make a list, called `all_car_data`, of the shape we want by pulling the desired entries from the BMW tuples and Prius tuples:

```
all_car_data = []
for bmw in bmws:
    all_car_data.append((bmw.mileage,bmw.price,1))
for prius in priuses:
    all_car_data.append((prius.mileage,prius.price,0))
```

Once this is run, `all_car_data` is a Python list starting with the BMWs and ending with the Priuses, labeled with 1's and 0's, respectively:

```
>>> all_car_data
[(93404.0, 13999.0, 1),
 (110890.0, 13995.0, 1),
 (94133.0, 13982.0, 1),
 (46778.0, 14599.0, 1),
 ....
(45000.0, 16900.0, 0),
(38000.0, 13500.0, 0),
(71000.0, 12500.0, 0)]
```

15.1.2 Testing the classification function

With the data in a suitable format, we can now write the `test_classifier` function. The job of the `bmw_finder` is to look at the mileage and price of a car and tell us whether these represent a BMW. If the answer is yes, it returns a 1; otherwise, it returns a 0. It's likely that `bmw_finder` will get some answers wrong. If it predicts that a car is a BMW (returning 1), but the car is actually a Prius, we'll call that a *false positive.* If it predicts the car is a Prius (returning 0), but the car is actually a BMW, we'll call that a *false negative.* If it correctly identifies a BMW or a Prius, we'll call that a *true positive* or *true negative*, respectively.

To test a classification function against the all_car_data data set, we need to run the classification function on each mileage and price in that list, and see whether the result of 1 or 0 matches the given value. Here's what that looks like in code:

```
def test_classifier(classifier, data):
    trues = 0
    falses = 0
    for mileage, price, is_bmw in data:
        if classifier(mileage, price) == is_bmw:     <-- Adds 1 to the trues counter if the classification is correct
            trues += 1
        else:
            falses += 1                              <-- Otherwise, adds 1 to the falses counter
    return trues / (trues + falses)
```

If we run this function with the `bmw_finder` classification function and the all_car_data data set, we see that it has 59% accuracy:

```
>>> test_classifier(bmw_finder, all_car_data)
0.59
```

That's not too bad; we got most of the answers right. But we'll see we can do much better than this! In the next section, we plot the data set to understand what's qualitatively wrong with the `bmw_finder` function. This helps us to see how we can improve the classification with our logistic classification function.

15.1.3 Exercises

Exercise 15.1 Update the `test_classifier` function to print the number of true positives, true negatives, false positives, and false negatives. Printing these for the `bmw_finder` classifier, what can you tell about the performance of the classifier?

Solution Rather than just keeping track of correct and incorrect predictions, we can track true and false positives and negatives separately:

```
def test_classifier(classifier, data, verbose=False):    <-- Specifies whether to print the data (we might not want to print it every time).
    true_positives = 0        <-- We now have 4 counters to keep track of.
    true_negatives = 0
    false_positives = 0
    false_negatives = 0

    for mileage, price, is_bmw in data:
        predicted = classifier(mileage,price)
```

```
        if predicted and is_bmw:
            true_positives += 1
        elif predicted:
            false_positives += 1
        elif is_bmw:
            false_negatives += 1
        else:
            true_negatives += 1

    if verbose:
        print("true positives %f" % true_positives)
        print("true negatives %f" % true_negatives)
        print("false positives %f" % false_positives)
        print("false negatives %f" % false_negatives)

    total = true_positives + true_negatives

    return total / len(data)
```

Depending on whether the car is a Prius or BMW and whether it's classified correctly, increments 1 of 4 counters

Prints the results of each counter

Returns the number of correct classifications (true positives or negatives) divided by the length of the data set

For the bmw_finder function, this prints the following text:

```
true positives 18.000000
true negatives 100.000000
false positives 0.000000
false negatives 82.000000
```

Because the classifier returns no false positives, this tells us it always correctly identifies when the car is *not* a BMW. But we can't be too proud of our function yet, because it says most of the cars are not BMWs, including many that are! In the next exercise, you can relax the constraint to get a higher overall success rate.

Exercise 15.2 Find a way to update the bmw_finder function to improve its performance and use the test_classifier function to confirm that your improved function has better than 59% accuracy.

Solution If you solved the last exercise, you saw that bmw_finder was too aggressive in saying that cars were not BMWs. We can lower the price threshold to $20,000 and see if it makes a difference:

```
def bmw_finder2(mileage,price):
    if price > 20000:
        return 1
    else:
        return 0
```

Indeed, by lowering this threshold, bmw_finder improved the success rate to 73.5%:

```
>>> test_classifier(bmw_finder2, all_car_data)
0.735
```

15.2 Picturing a decision boundary

Before we implement the logistic regression function, let's look at one more way to measure our success at classification. Because two numbers, mileage and price, define our used car data points, we can think of these as 2D vectors and plot them as points on a 2D plane. This plot gives us a better sense of where our classification function "draws the line" between BMWs and Priuses, and we can see how to improve it. It turns out that using our `bmw_finder` function is equivalent to drawing a literal line in the 2D plane, calling any point above the line a BMW and any point below it a Prius.

In this section, we use Matplotlib to draw our plot and see where `bmw_finder` places the dividing line between BMWs and Priuses. This line is called the *decision boundary*, because what side of the line a point lies on, helps us decide what class it belongs to. After looking at the car data on a plot, we can figure out where to draw a better dividing line. This lets us define an improved version of the `bmw_finder` function, and we can measure exactly how much better it performs.

15.2.1 Picturing the space of cars

All of the cars in our data set have mileage and price values, but some of them represent BMWs and some represent Priuses, depending on whether they are labeled with a 1 or with a 0. To make our plot readable, we want to make a BMW and a Prius visually distinct on the scatter plot.

The `plot_data` helper function in the source code takes the whole list of car data and automatically plots the BMWs with X's and the Priuses with circles. Figure 15.2 shows the plot.

```
>>> plot_data(all_car_data)
```

In general, we can see that the BMWs are more expensive than the Priuses; most BMWs are higher on the price axis. This justifies our strategy of classifying the more expensive cars as BMWs. Specifically, we drew the line at a price of $25,000 (figure

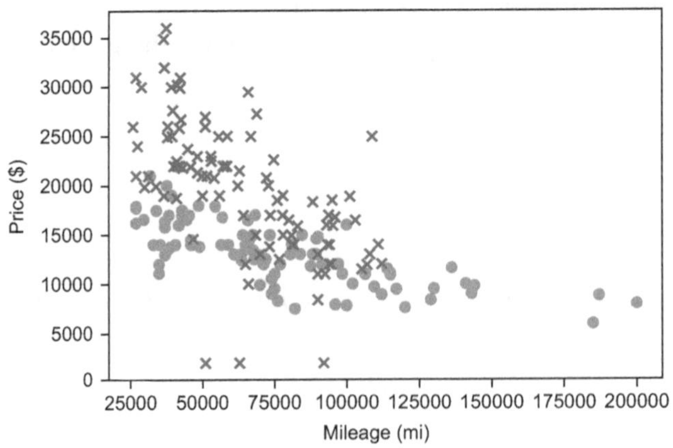

Figure 15.2 A plot of price vs. mileage for all cars in the data set with each BMW represented by an X and each Prius represented with a circle

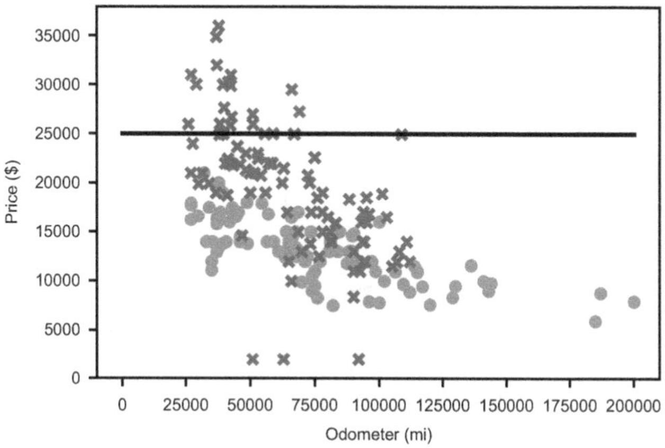

Figure 15.3 Shows the decision line with car data plotted

15.3). On the plot, this line separates the top of the plot with more expensive cars from the bottom with less expensive cars.

This is our decision boundary. Every X above the line was correctly identified as a BMW, while every circle below the line was correctly identified as a Prius. All other points were classified incorrectly. It's clear that if we move this decision boundary, we can improve our accuracy. Let's give it a try.

15.2.2 Drawing a better decision boundary

Based on the plot in figure 15.3, we could lower the line and correctly identify a few more BMWs, while not incorrectly identifying any Priuses. Figure 15.4 shows what the decision boundary looks like if we lower the cut-off price to $21,000.

The $21,000 cut-off might be a good boundary for low-mileage cars, but the higher the mileage, the lower the threshold. For instance, it looks like most BMWs with 75,000 miles or more are below $21,000. To model this, we can make our cut-off price

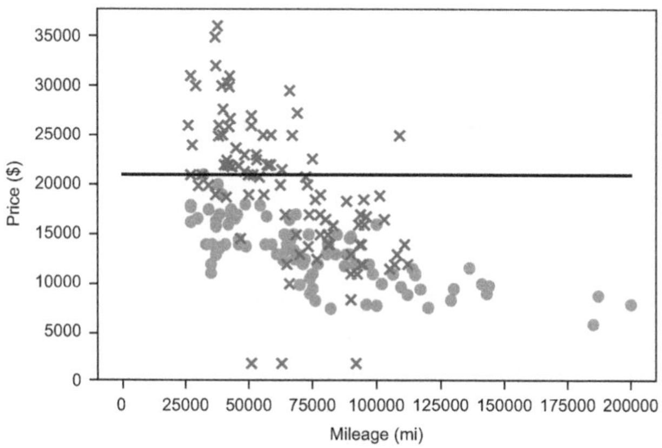

Figure 15.4 Lowering the decision boundary line appears to increase our accuracy.

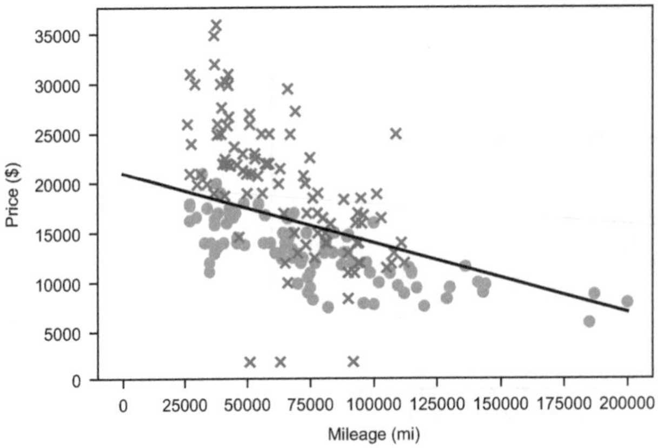

Figure 15.5 Using a downward-sloping decision boundary

mileage dependent. Geometrically that means drawing a line that slopes downward (figure 15.5).

This line is given by the function $p(x) = 21{,}000 - 0.07 \cdot x$, where p is price and x is mileage. There is nothing special about this equation; I just played around with the numbers until I plotted a line that looked reasonable. But it looks like it correctly identifies even more BMWs than before, with only a handful of false positives (Priuses incorrectly classified as BMWs). Rather than just eyeballing these decision boundaries, we can turn them into classifier functions and measure their performance.

15.2.3 Implementing the classification function

To turn this decision boundary into a classification function, we need to write a Python function that takes a car mileage and price, and returns one or zero depending on whether the point falls above or below the line. That means taking the given mileage, plugging it into the decision boundary function, $p(x)$, to see what the threshold price is and comparing the result to the given price. This is what it looks like:

```
def decision_boundary_classify(mileage,price):
    if price > 21000 - 0.07 * mileage:
        return 1
    else:
        return 0
```

Testing this out, we can see it is much better than our first classifier; 80.5% of the cars are correctly classified by this line. Not bad!

```
>>> test_classifier(decision_boundary_classify, all_car_data)
0.805
```

You might ask why we can't just do a gradient descent on the parameters defining the decision boundary line. If 20,000 and 0.07 don't give the most accurate decision boundary, maybe some pair of numbers near them do. This isn't a crazy idea. When we implement logistic regression, you'll see that under the hood, it moves the decision boundary around using gradient descent until it finds the best one.

There are two important reasons we'll implement the more sophisticated logistic regression algorithm rather than doing a gradient descent on the parameters a and b of the decision boundary function, $ax + b$. The first is that if the decision boundary is close to vertical at any step in the gradient descent, the numbers a and b could get very large and cause numerical issues. The other is that there isn't an obvious cost function. In the next section, we see how logistic regression takes care of both of these issues so we can search for the best decision boundary using gradient descent.

15.2.4 Exercises

Exercise 15.3—Mini Project What is the decision boundary of the form p = *constant* that gives the best classification accuracy on the test data set?

Solution The following function builds a classifier function for any specified, constant cut-off price. In other words, the resulting classifier returns true if the test car has price above the cutoff and false otherwise:

```
def constant_price_classifier(cutoff_price):
    def c(x,p):
        if p > cutoff_price:
            return 1
        else:
            return 0
    return c
```

The accuracy of this function can be measured by passing the resulting classifier to the `test_classify` function. Here's a helper function to automate this check for any price we want to test as a cut-off value:

```
def cutoff_accuracy(cutoff_price):
    c = constant_price_classifier(cutoff_price)
    return test_classifier(c,all_car_data)
```

The best cut-off price is between two of the prices in our list. It's sufficient to check each price and see if it is the best cut-off price. We can do that quickly in Python using the `max` function. The keyword argument `key` lets us choose what function we want to maximize by. In this case, we want to find the price in the list that is the best cut-off, so we can maximize by the `cutoff_accuracy` function:

```
>>> max(all_prices,key=cutoff_accuracy)
17998.0
```

This tells us that according to our data set, $17,998 is the best price to use as a cut-off when deciding whether a car is a BMW 5 series or a Prius. It turns out to be quite accurate for our data set, with 79.5% accuracy:

```
>>> test_classifier(constant_price_classifier(17998.0), all_car_data)
0.795
```

15.3 Framing classification as a regression problem

The way that we can reframe our classification task as a regression problem is to create a function that takes in the mileage and price of a car, and returns a number measuring how likely it is to be a BMW instead of a Prius. In this section, we implement a function called `logistic_classifier` that, from the outside, looks a lot like the classifiers we've built so far; it takes a mileage and a price, and outputs a number telling us whether the car is a BMW or a Prius. The only difference is that rather than outputting one or zero, it outputs a value between zero and one, telling us how likely it is that the car is a BMW.

You can think of this number as the probability that the mileage and price describe a BMW, or more abstractly, you can think of it as giving the "BMWness" of the data point (figure 15.6). (Yes, that's a made-up word, which I pronounce "bee-em-doubleyou-ness." It means how much it looks like a BMW. Maybe we could call the antonym "Priusity.")

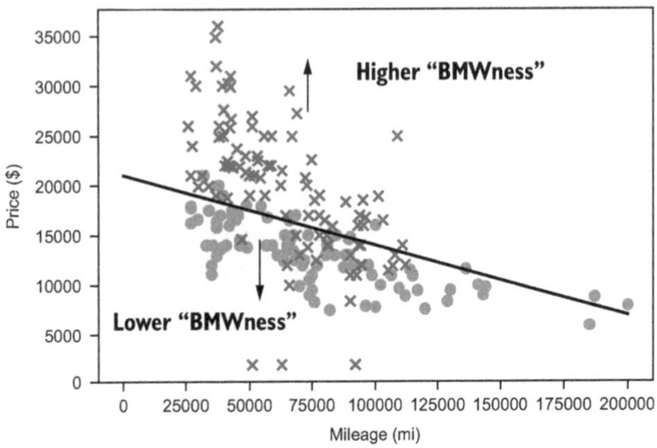

Figure 15.6 The concept of "BMWness" describes how much like a BMW a point in the plane is.

To build the logistic classifier, we start with a guess of a good decision boundary line. Points above the line have high "BMWness," meaning these are likely to be BMWs and the logistic function should return values close to one. Data points below the line have a low "BMWness," meaning these are more likely to be Priuses and our function should return values close to zero. On the decision boundary, the "BMWness" value will be 0.5, meaning a data point there is equally as likely to be a BMW as it is to be a Prius.

15.3.1 Scaling the raw car data

There's a chore we need to take care of at some point in the regression process, so we might as well take care of it now. As we discussed in the last chapter, the large values of mileage and price can cause numerical errors, so it's better to rescale them to a small, consistent size. We should be safe if we scale all of the mileages and the prices linearly to values between zero and one.

We need to be able to scale and unscale each of mileage and price, so we need four functions in total. To make this a little bit less painful, I've written a helper function that takes a list of numbers and returns functions to scale and unscale these linearly, between zero and one, using the maximum and minimum values in the list. Applying this helper function to the whole list of mileages and of prices gives us the four functions we need:

```
def make_scale(data):
    min_val = min(data)
    max_val = max(data)
    def scale(x):
        return (x-min_val) / (max_val - min_val)
    def unscale(y):
        return y * (max_val - min_val) + min_val
    return scale, unscale

price_scale, price_unscale =\
    make_scale([x[1] for x in all_car_data])
mileage_scale, mileage_unscale =\
    make_scale([x[0] for x in all_car_data])
```

The maximum and minimum provide the current range of the data set.

Puts the data point at the same fraction of the way between 0 and 1 as it was from min_val to max_val

Puts the scaled data point at the same fraction of the way from min_val to max_val as it was from 0 to 1

Returns the scale and unscale functions (closures, if you're familiar with that term) to use when we want to scale or unscale members of this data set.

Returns two sets of functions, one for price and one for mileage

We can now apply these scaling functions to every car data point in our list to get a scaled version of the data set:

```
scaled_car_data = [(mileage_scale(mileage), price_scale(price), is_bmw)
                    for mileage,price,is_bmw in all_car_data]
```

The good news is that the plot looks the same (figure 15.7), except that the values on the axes are different.

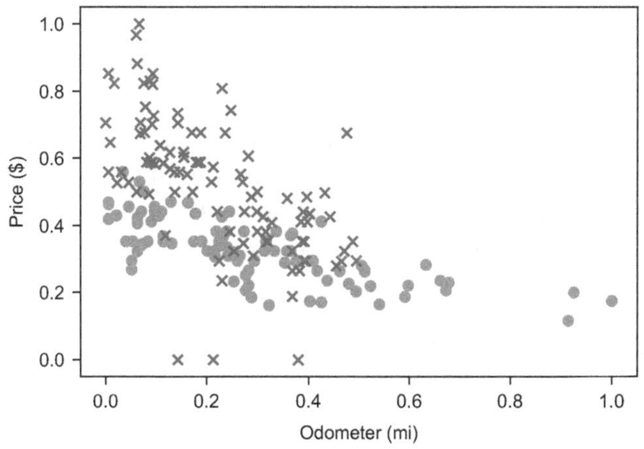

Figure 15.7 The mileage and price data scaled so that all values are between zero and one. The plot looks the same as before, but our risk of numerical error has decreased.

Because the geometry of the scaled data set is the same, it should give us confidence that a good decision boundary for this scaled data set translates to a good decision boundary for the original data set.

15.3.2 Measuring the "BMWness" of a car

Let's start with a decision boundary that looks similar to the one from the last section. The function $p(x) = 0.56 - 0.35 \cdot x$ gives price at the decision boundary as a function of mileage. This is pretty close to the one I found by eyeballing in the last section, but it applies to the scaled data set instead (figure 15.8).

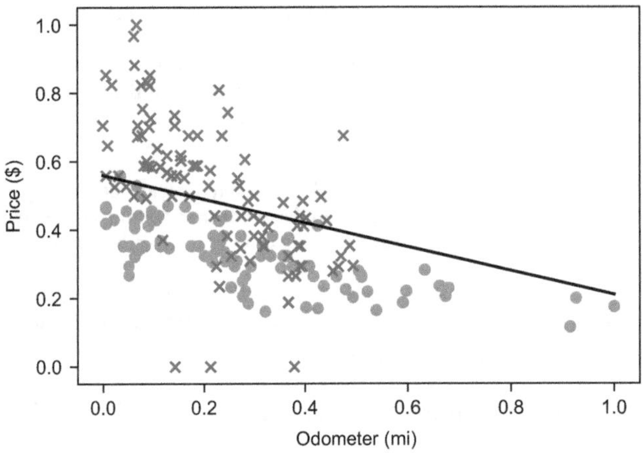

Figure 15.8 The decision boundary $p(x) = 0.56 - 0.35 \cdot x$ on the scaled data set

We can still test classifiers on the scaled data set with our `test_classifier` function; we just need to take care to pass in the scaled data instead of the original. It turns out this decision boundary gives us a 78.5% accurate classification of the data.

It also turns out that this decision boundary function can be rearranged to give a measure of the "BMWness" of a data point. To make our algebra easier, let's write the decision boundary as

$$p = ax + b$$

where p is price, x is still mileage, and a and b are the slope and intercept of the line (in this case, $a = -0.35$ and $b = 0.56$), respectively. Instead of thinking of this as a function, we can think of it as an equation satisfied by points (x, p) on the decision boundary. If we subtract $ax + b$ from both sides of the equation, we get another correct equation:

$$p - ax - b = 0$$

Every point (x, p) on the decision boundary satisfies this equation as well. In other words, the quantity $p - ax - b$ is zero for every point on the decision boundary.

Here's the point of this algebra: the quantity $p - ax - b$ is a measure of the "BMWness" of the point (x, p). If (x, p) is above the decision boundary, it means p is too big, relative to x, so $p - ax - b > 0$. If, instead, (x, p) is below the decision boundary, it means p is too small relative to x, then $p - ax - b < 0$. Otherwise, the expression $p - ax - b$ is exactly zero, and the point is right at the threshold of being interpreted as a Prius or a BMW. This might be a little bit abstract on the first read, so table 15.2 lists the three cases.

Table 15.2 Summary of the possible cases

(x, p) *above* decision boundary	$p - ax - b > 0$	Likely to be a BMW
(x, p) *on* decision boundary	$p - ax - b = 0$	Could be either car model
(x, p) *below* decision boundary	$p - ax - b < 0$	Likely to be a Prius

If you're not convinced that $p - ax - b$ is a measure of "BMWness" compatible with the decision boundary, an easier way to see this is to look at the heat map of $f(x, p) = p - ax - b$, together with the data (figure 15.9). When $a = -0.35$ and $b = 0.56$, the function is $f(x, p) = p - 0.35 \cdot x - 0.56$.

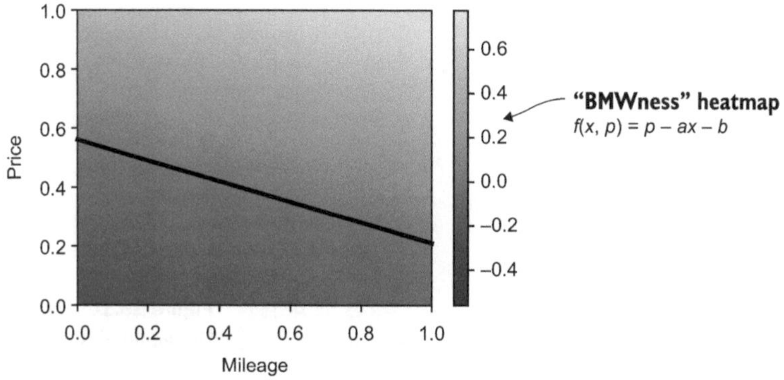

Figure 15.9 A plot of the heatmap and decision boundary showing that the bright values (positive "BMWness") are above the decision boundary and dark values (negative "BMWness") occur below the decision boundary

The function, $f(x, p)$, *almost* meets our requirements. It takes a mileage and a price, and it outputs a number that is higher if the numbers are likely to represent a BMW, and lower if the values are likely to represent a Prius. The only thing missing is that the output numbers aren't constrained to be between zero and one, and the cutoff is at a value of zero rather than at a value of 0.5 as desired. Fortunately, there's a handy kind of mathematical helper function we can use to adjust the output.

15.3.3 Introducing the sigmoid function

The function $f(x, p) = p - ax - b$ is linear, but this is not a chapter on linear regression! The topic at hand is *logistic regression*, and to do logistic regression, you need to use a logistic function. The most basic logistic function is the one that follows, which is often called a *sigmoid* function:

$$\sigma(x) = \frac{1}{1 + e^{-x}}$$

We can implement this function in Python with the exp function, which stands in for e^x, where $e = 2.71828...$ and is the constant we've used for exponential bases before:

```
from math import exp
def sigmoid(x):
    return 1 / (1+exp(-x))
```

Figure 15.10 shows its graph.

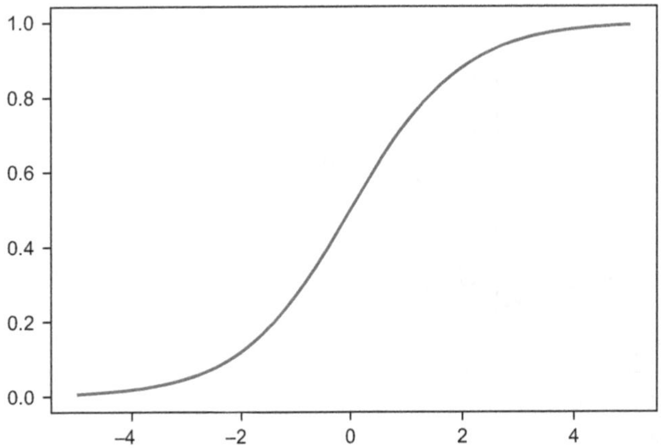

Figure 15.10 The graph of the sigmoid function $\sigma(x)$

In the function, we use the Greek letter σ (sigma) because σ is the Greek version of the letter *S*, and the graph of $\sigma(x)$ looks a bit like the letter *S*. Sometimes the words *logistic function* and *sigmoid function* are used interchangeably to mean a function like the one in figure 15.10, which smoothly ramps up from one value to another. In this chapter (and the next), when I refer to *the* sigmoid function, I'll be talking about this specific function: $\sigma(x)$.

You don't need to worry too much about how this function is defined, but you do need to understand the shape of the graph and what it means. This function sends any input number to a value between zero and one, with big negative numbers yielding results closer to zero, and big positive numbers yielding results closer to one. The result of $\sigma(0)$ is 0.5. We can think of σ as translating the range from $-\infty$ to ∞ to the more manageable range from zero to one.

15.3.4 Composing the sigmoid function with other functions

Returning to our function $f(x, p) = p - ax - b$, we saw that it takes a mileage value and a price value and returns a number measuring how much the values look like a BMW rather than a Prius. This number could be large or positive or negative, and a value of zero indicates that it is on the boundary between being a BMW and being a Prius.

What we want our function to return is a value between zero and one (with values close to zero and one), representing cars likely to be Priuses or BMWs, respectively, and a value of 0.5, representing a car that is equally likely to be either a Prius or a BMW. All we have to do to adjust the outputs of $f(x, p)$ to be in the expected range is to pass through the sigmoid function $\sigma(x)$ as shown in figure 15.11. That is, the function we want is $\sigma(f(x, p))$, where x and p are the mileage and price.

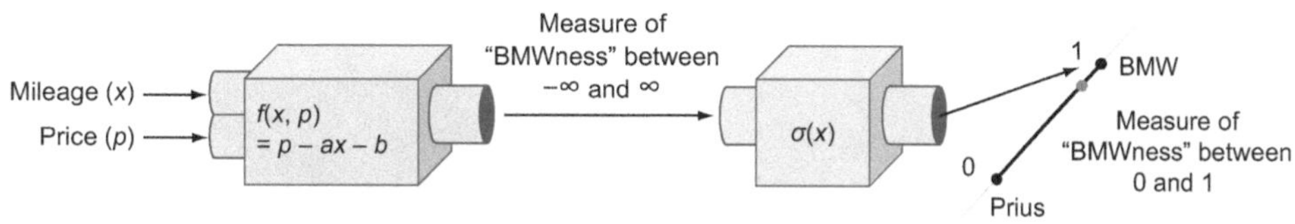

Figure 15.11 Schematic diagram of composing the "BMWness" function $f(x, p)$ with the sigmoid function $\sigma(x)$

Let's call the resulting function $L(x, p)$, so in other words, $L(x, p) = \sigma(f(x, p))$. Implementing the function $L(x, p)$ in Python and plotting its heatmap (figure 15.12), we can see that it increases in the same direction as $f(x, p)$, but its values are different.

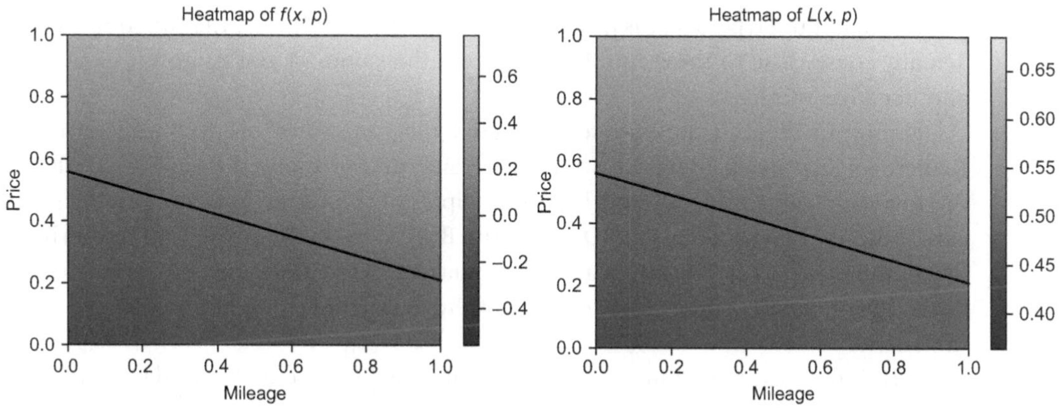

Figure 15.12 The heatmaps look basically the same, but the values of the function are slightly different.

Based on this picture, you might wonder why we went through the trouble of passing the "BMWness" function through the sigmoid. From this perspective, the functions look mostly the same. However, if we plot their graphs as 2D surfaces in 3D (figure 15.13), you can see that the curvy shape of the sigmoid has an effect.

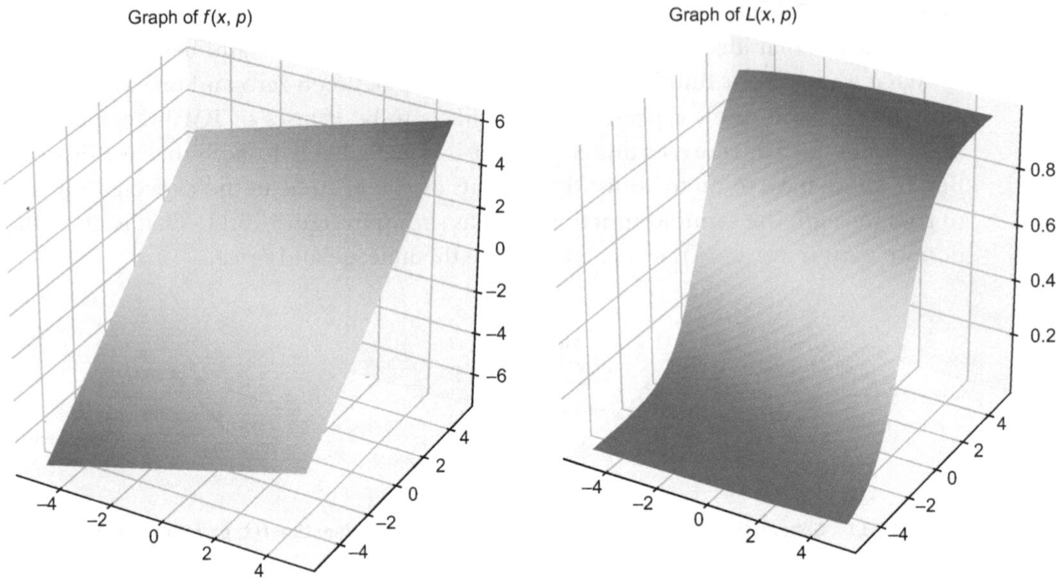

Figure 15.13 While $f(x, p)$ slopes upward linearly, $L(x, p)$ curves up from a minimum value of 0 to a maximum value of 1.

In fairness, I had to zoom out a bit in (x, p) space to make the curvature clear. The point is that if the type of car is indicated by a 0 or 1, the values of the function $L(x, p)$ actually come close to these numbers, whereas the values of $f(x, p)$ go off to positive and negative infinity!

Figure 15.14 illustrates two exaggerated diagrams to show you what I mean. Remember that in our data set, scaled_car_data, we represented Priuses as triples of the form (mileage, price, 0) and BMWs as triples of the form (mileage, price, 1). We can interpret these as points in 3D where the BMWs live in the plane $z = 1$ and Priuses live in the plane $z = 0$. Plotting scaled_car_data as a 3D scatter plot, you can see that a linear function can't come close to many of the data points in the same way as a logistic function.

With functions shaped like $L(x, p)$, we can actually hope to *fit* the data, and we'll see how to do that in the next section.

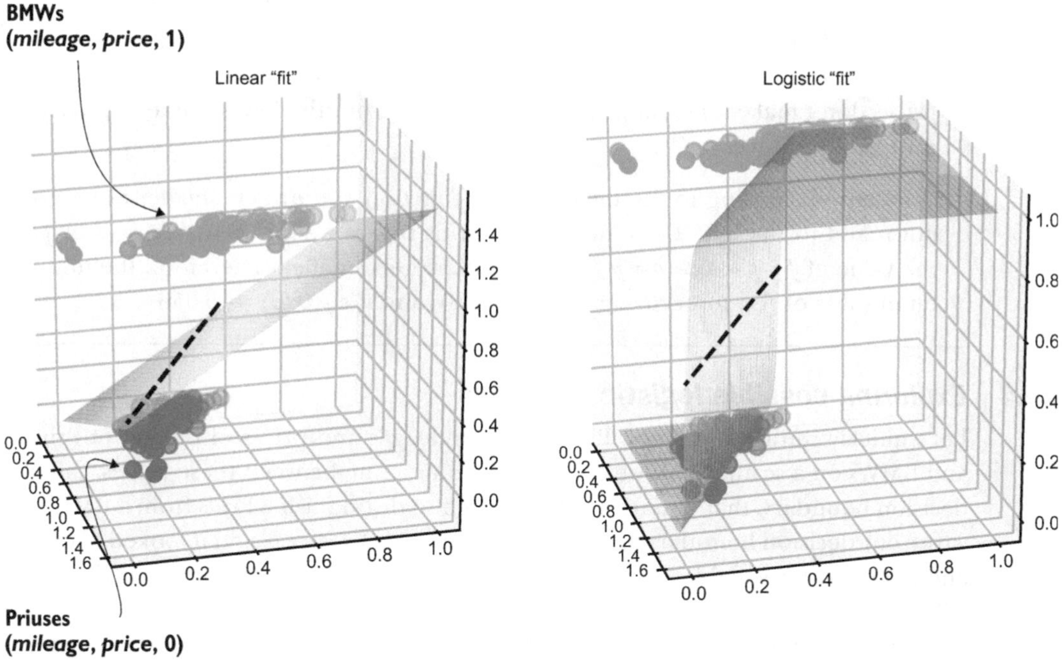

Figure 15.14 The graph of a linear function in 3D can't come as close to the data points as the graph of a logistic function.

15.3.5 Exercises

Exercise 15.4 Find a function $h(x)$ such that large positive values of x cause $h(x)$ to be close to 0, large negative values of x cause $h(x)$ to be close to 1, and $h(3) = 0.5$.

Solution The function $y(x) = 3 - x$ has $y(3) = 0$ and it goes off to positive infinity when x is large and negative and off to negative infinity when x is large and positive. That means passing the result of $y(x)$ into our sigmoid function gives us a function with the desired properties. Specifically, $h(x) = \sigma(y(x)) = \sigma(3 - x)$ works, and its graph is shown here to convince you:

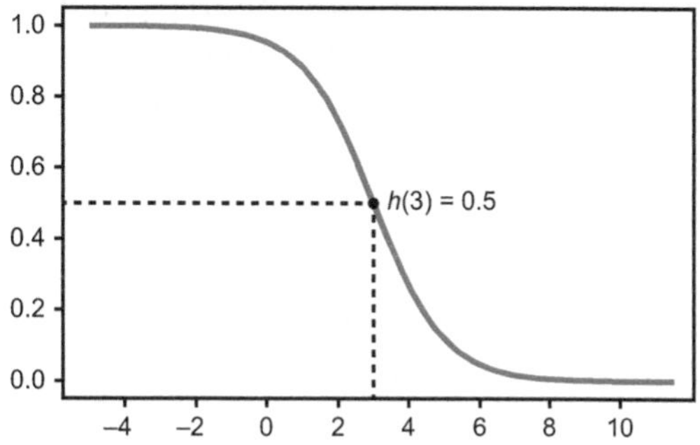

Exercise 15.5—Mini Project There is actually a lower bound on the result of $f(x, p)$ because x and p are not allowed to be negative (negative mileages and prices don't make sense, after all). Can you figure out the lowest value of f that a car could produce?

Solution According to the heatmap, the function $f(x, p)$ gets smaller as we go down and to the left. The equation confirms this as well; if we decrease x or p, the value of $f = p - ax - b = p + 0.35 \cdot x - 0.56$ gets smaller. Therefore, the minimum value of $f(x, p)$ occurs at $(x, p) = (0, 0)$, and it's $f(0, 0) = -0.056$.

15.4 Exploring possible logistic functions

Let's quickly retrace our steps. Plotting the mileages and prices of our set of Priuses and BMWs on a scatter plot, we could try to draw a line between these values, called a decision boundary, that defines a rule by which to distinguish a Prius from a BMW. We wrote our decision boundary as a line in the form $p(x) = ax + b$, and it looked like -0.35 and 0.56 were reasonable choices for a and b, giving us a classification that was about 80% correct.

Rearranging this function, we found that $f(x, p) = p - ax - b$ was a function taking a mileage and price (x, p) and returning a number that was greater than zero on the BMW side of the decision boundary and smaller than zero on the Prius side. On the decision boundary, $f(x, p)$ returned zero, meaning a car would be equally likely to be a BMW or a Prius. Because we represent BMWs with a 1 and Priuses with a 0, we wanted a version of $f(x, p)$ that returned values between zero and one, where 0.5 would represent a car equally likely to be a BMW or a Prius. Passing the result of f into a sigmoid function σ, we got a new function $L(x, p) = \sigma(f(x, p))$, satisfying that requirement.

But we don't want the $L(x, p)$ I made by eyeballing the best decision boundary—we want the $L(x, p)$ that *best fits the data.* On our way to doing that, we'll see that there are three parameters we can control to write a general logistic function that takes 2D vectors and returns numbers between zero and one, and also has a decision boundary $L(x, p) = 0.5$, which is a straight line. We'll write a Python function, `make_logistic(a,b,c)`, that takes in three parameters a, b, and c, and returns the logistic function they define. As we explored a 2D space of (a, b) pairs to choose a linear function in chapter 14, we'll explore a 3D space of values (a, b, c) to define our logistic function (figure 15.15).

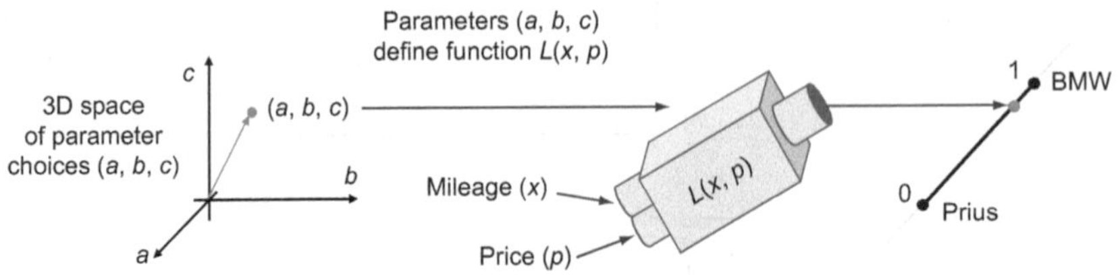

Figure 15.15 Exploring a 3D space of parameter values (a, b, c) to define a function L(x, p)

Then we'll create a cost function, much like the one we created for linear regression. The cost function, which we'll call `logistic_cost(a,b,c)`, takes the parameters *a*, *b*, and *c*, which define a logistic function and produce one number, measuring how far the logistic function is from our car data set. The `logistic_cost` function needs to be implemented in such a way that the lower its value, the better the predictions from the associated logistic function.

15.4.1 Parameterizing logistic functions

The first task is to find the general form of a logistic function $L(x, p)$, whose values range from zero to one and whose decision boundary $L(x, p) = 0.5$ is a straight line. We got close to this in the last section, starting with the decision boundary $p(x) = ax + b$ and reverse engineering a logistic function from that. The only problem is that a linear function of the form $ax + b$ can't represent any line in the plane. For instance, figure 15.16 shows a data set where a vertical decision boundary, $x = 0.6$, makes sense. Such a line can't be represented in the form $p = ax + b$, however.

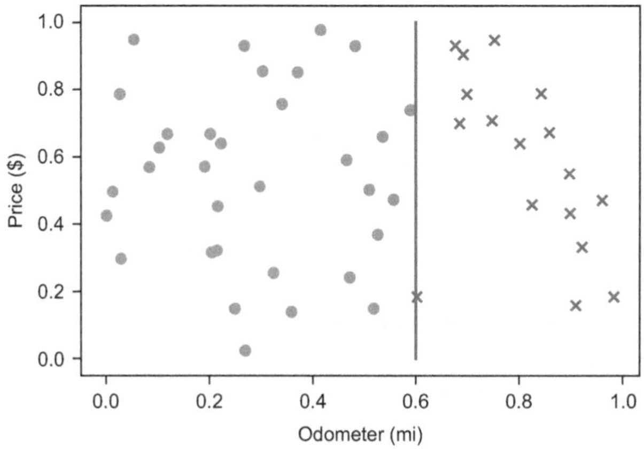

Figure 15.16 A vertical decision boundary might make sense, but it can't be represented in the form $p = ax + b$.

The general form of a line that does work is the one we met in chapter 7: $ax + by = c$. Because we're calling our variables x and p, we'll write $ax + bp = c$. Given an equation like this, the function $z(x, p) = ax + bp - c$ is zero on the line with positive values on one side and negative values on the other. For us, the side of the line where $z(x, p)$ is positive is the BMW side, and the side where $z(x, p)$ is negative is the Prius side.

Passing $z(x, p)$ through the sigmoid function, we get a general logistic function $L(x, p) = \sigma(z(x, p))$, where $L(x, p) = 0.5$ on the line where $z(x, p) = 0$. In other words, the function $L(x, p) = \sigma(ax + bp - c)$ is the general form we're looking for. This is easy to translate to Python, giving us a function of *a*, *b*, and *c* that returns a corresponding logistic function $L(x, p) = \sigma(ax + bp - c)$:

```
def make_logistic(a,b,c):
    def l(x,p):
        return sigmoid(a*x + b*p - c)
    return l
```

The next step is to come up with a measure of how close this function comes to our scaled_car_data dataset.

15.4.2 Measuring the quality of fit for a logistic function

For any BMW, the `scaled_car_data` list contains an entry of the form $(x, p, 1)$, and for every Prius, it contains an entry of the form $(x, p, 0)$, where x and p denote (scaled) mileage and price values, respectively. If we apply a logistic function, $L(x, p)$, to the x and p values, we'll get a result between zero and one.

A simple way to measure the error or cost of the function L is to find how far off it is from the correct value, which is either zero or one. If you add up all of these errors, you'll get a total value telling you how far the function $L(x, p)$ comes from the data set. Here's what that looks like in Python:

```
def simple_logistic_cost(a,b,c):
    l = make_logistic(a,b,c)
    errors = [abs(is_bmw-l(x,p))
              for x,p,is_bmw in scaled_car_data]
    return sum(errors)
```

This cost reports the error reasonably well, but it isn't good enough to get our gradient descent to converge to a best value of a, b, and c. I won't go into a full explanation of why this is, but I'll try to quickly give you the general idea.

Suppose we have two logistic functions, $L_1(x, p)$ and $L_2(x, p)$, and we want to compare the performance of both. Let's say they both look at the same data point $(x, p, 0)$, meaning a data point representing a Prius. Then let's say $L_1(x, p)$ returns 0.99, which is greater than 0.5, so it predicts incorrectly that the car is a BMW. The error for this point is $|0–0.99| = 0.99$. If another logistic function, $L_2(x, p)$, predicts a value of 0.999, the model predicts with more certainty that the car is a BMW, and is even more wrong. That said, the error would be only $|0–0.999| = 0.999$, which is not much different.

It's more appropriate to think of L_1 as reporting a 99% chance the data point represents a BMW and a 1% chance that it represents a Prius, with L_2 reporting a 99.9% chance it is a BMW and a 0.1% chance it is a Prius. Instead of thinking of this as a 0.09% worse Prius prediction, we should really think of it as being *ten times* worse! We can, therefore, think of L_2 as being ten times more wrong than L_1.

We want a cost function such that if $L(x, p)$ is *really sure* of the wrong answer, then the cost of L is high. To get that, we can look at the difference between $L(x, p)$ and the wrong answer, and pass it through a function that makes tiny values big. For instance, $L_1(x, p)$ returned 0.99 for a Prius, meaning it was 0.01 units from the wrong answer, while $L_2(x, p)$ returned 0.999 for a Prius, meaning it was 0.001 units from the wrong answer. A good function to return big values from tiny ones is $-\log(x)$, where log is the special natural logarithm function. It's not critical that you know what the –log function does, only that it returns big numbers for small inputs. Figure 15.17 shows the plot of $-\log(x)$.

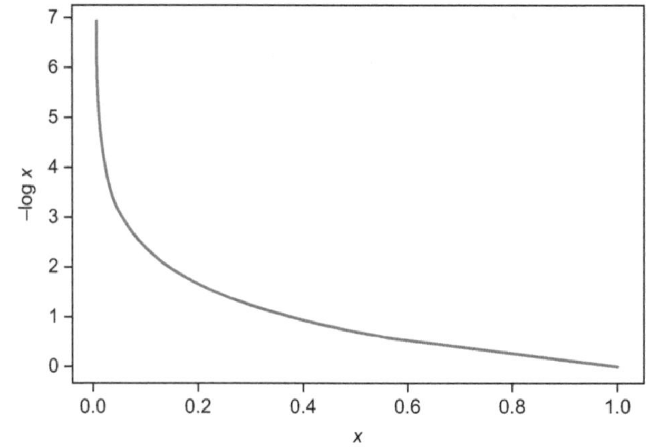

Figure 15.17 The function −log(x) returns big values for small inputs, and −log(**1**) = **0**.

To familiarize yourself with −log(x), you can test it with some small inputs. For $L_1(x, p)$, which was 0.01 units from the wrong answer, we get a smaller cost than $L_2(x, p)$, which was 0.001 units from the wrong answer:

```
from math import log
>>> -log(0.01)
4.605170185988091
>>> -log(0.001)
6.907755278982137
```

By comparison, if $L(x, p)$ returns zero for a Prius, it would be giving the correct answer. That's one unit away from the wrong answer, and −log(1) = 0, so there is zero cost for the right answer.

Now we're ready to implement the `logistic_cost` function that we set out to create. To find the cost for a given point, we calculate how close the given logistic function comes to the wrong answer and then take the negative logarithm of the result. The total cost is the sum of the cost at every data point in the `scaled_car_data` data set:

```
def point_cost(l,x,p,is_bmw):      <-- Determines the cost of a single data point
    wrong = 1 - is_bmw
    return -log(abs(wrong - l(x,p)))

def logistic_cost(a,b,c):
    l = make_logistic(a,b,c)
    errors = [point_cost(l,x,p,is_bmw)      <-- The overall cost of the logistic function is the same as before, except that we use the new point_cost function for each data point instead of just the absolute value of the error.
              for x,p,is_bmw in scaled_car_data]
    return sum(errors)
```

It turns out, we get good results if we try to minimize the `logistic_cost` function using gradient descent. But before we do that, let's do a sanity check and confirm that `logistic_cost` returns lower values for a logistic function with an (obviously) better decision boundary.

15.4.3 Testing different logistic functions

Let's try out two logistic functions with different decision boundaries, and confirm if one has an obviously better decision boundary than if it has a lower cost. As our two examples, let's use $p = 0.56 - 0.35 \cdot x$, my best-guess decision boundary, which is the same as $0.35 \cdot x + 1 \cdot p = 0.56$, and also an arbitrarily selected one, say $x + p = 1$. Clearly, the former is a better dividing line between the Priuses and the BMWs.

In the source code, you'll find a `plot_line` function to draw a line based on the values a, b, and c in the equation $ax + by = c$ (and as an exercise at the end of the section, you can try implementing this function yourself). The respective values of (a, b, c) are (0.35, 1, 0.56) and (1, 1, 1). We can plot them alongside the scatter plot of car data (shown in figure 15.18) with these three lines:

```
plot_data(scaled_car_data)
plot_line(0.35,1,0.56)
plot_line(1,1,1)
```

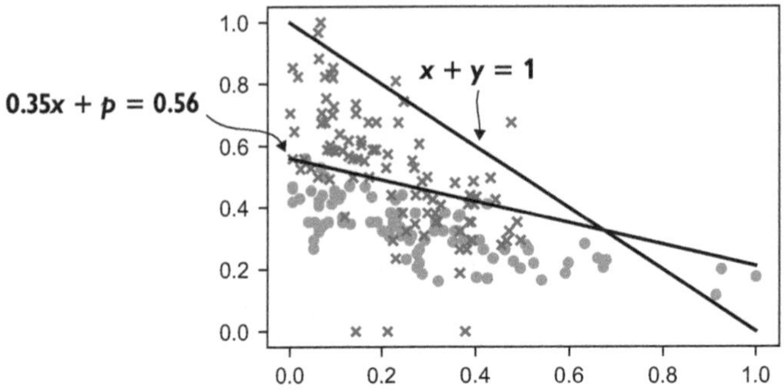

Figure 15.18 The graphs of two decision boundary lines. One is clearly better than the other at separating Priuses from BMWs.

The corresponding logistic functions are $\sigma(0.35 \cdot x + p - 0.56)$ and $\sigma(x + p - 1)$, and we expect the first one has a lower cost with respect to the data. We can confirm this with the `logistic_cost` function:

```
>>> logistic_cost(0.35,1,0.56)
130.92490748700456
>>> logistic_cost(1,1,1)
135.56446830870456
```

As expected, the line $x + p = 1$ is a worse decision boundary, so the logistic function $\sigma(x + p - 1)$ has a higher cost. The first function $\sigma(0.35 \cdot x + p - 0.56)$ has a lower cost and a better fit. But is it the best fit? When we run gradient descent on the `logistic_cost` function in the next section, we'll find out.

15.4.4 Exercises

Exercise 15.6 Implement the function `plot_line(a,b,c)` referenced in section 15.4.3 that plots the line $ax + by = c$, where $0 \leq x \leq 1$ and $0 \leq y \leq 1$.

Solution Note that I used different names other than *a*, *b*, and *c* for the function arguments because `c` is a keyword argument that sets the color of the plotted line for Matplotlib's `plot` function, which I commonly make use of:

```
def plot_line(acoeff,bcoeff,ccoeff,**kwargs):
    a,b,c = acoeff, bcoeff, ccoeff
    if b == 0:
        plt.plot([c/a,c/a],[0,1])
    else:
        def y(x):
            return (c-a*x)/b
        plt.plot([0,1],[y(0),y(1)],**kwargs)
```

Exercise 15.7 Use the formula for the sigmoid function σ to write an expanded formula for $\sigma(ax + by - c)$.

Solution Given that

$$\sigma(x) = \frac{1}{1 + e^{-x}}$$

we can write

$$\sigma(ax + by + c) = \frac{1}{1 + e^{c-ax-by}}$$

Exercise 15.8—Mini Project What does the graph of $k(x, y) = \sigma(x^2 + y^2 - 1)$ look like? What does the decision boundary look like, meaning the set of points where $k(x, y) = 0.5$?

Solution We know that $\sigma(x^2 + y^2 - 1) = 0.5$, wherever $x^2 + y^2 - 1 = 0$ or where $x^2 + y^2 = 1$. You can recognize the solutions to this equation as the points of distance one from the origin or a circle of radius 1. Inside the circle, the distance from the origin is smaller, so $x^2 + y^2 < 1$ and $\sigma(x^2 + y^2) < 0.5$, while outside the circle $x^2 + y^2 > 1$, so $\sigma(x^2 + y^2 - 1) > 0.5$. The graph of this function approaches 1

(continued)

as we move further away from the origin in any direction, while it decreases inside the circle to a minimum value of about 0.27 at the origin. Here's the graph:

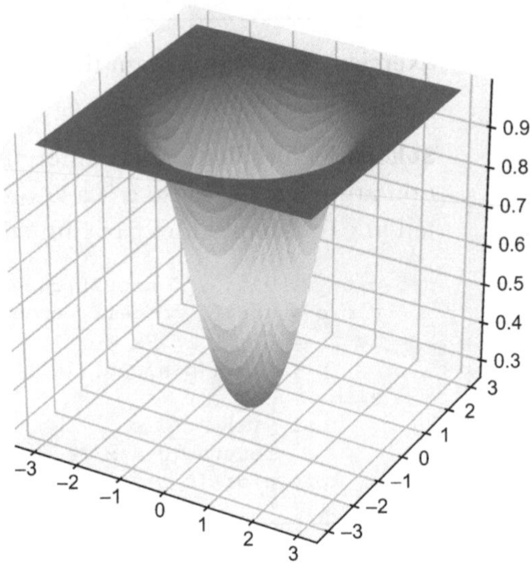

A graph of $\sigma(x^2 + y^2 - 1)$. Its value is less than 0.5 inside the circle of a radius of 1, and it increases to a value of 1 in every direction outside that circle.

Exercise 15.9—Mini Project Two equations, $2x + y = 1$ and $4x + 2y = 2$, define the same line and, therefore, the same decision boundary. Are the logistic functions $\sigma(2x + y - 1)$ and $\sigma(4x + 2y - 2)$ the same?

Solution No, they aren't the same function. The quantity $4x + 2y - 2$ increases more rapidly with respect to increases in x and y, so the graph of the latter function is steeper:

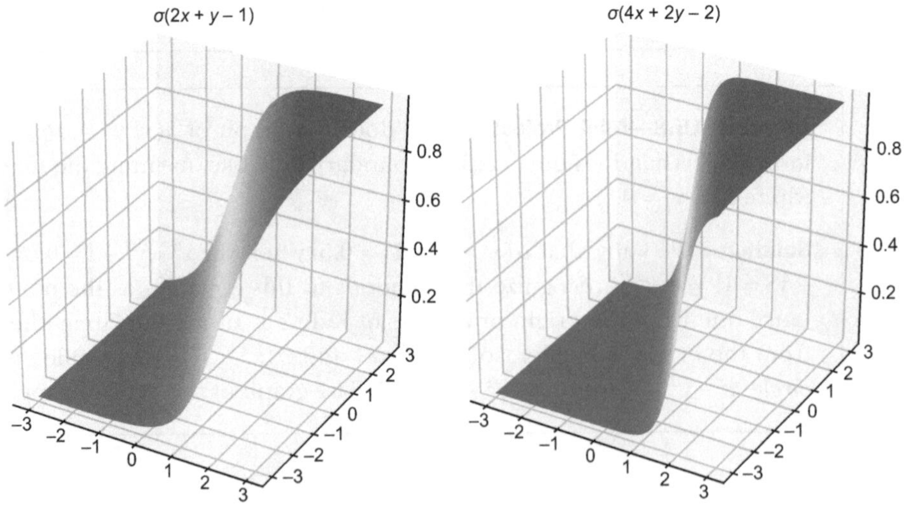

The graph of the second logistic function is steeper than the graph of the first.

Exercise 15.10—Mini Project Given a line $ax + by = c$, it's not as easy to define what is above that line and what is below. Can you describe which side of the line the function $z(x, y) = ax + by - c$ returns positive values?

Solution The line $ax + by = c$ is the set of points where $z(x, y) = ax + by - c = 0$. As we saw for equations of this form in chapter 7, the graph of $z(x, y) = ax + by - c$ is a plane, so it increases in one direction from the line and decreases in the other direction. The gradient of $z(x, y)$ is $\nabla z(x, y) = (a, b)$, so $z(x, y)$ increases most rapidly in the direction of the vector (a, b) and decreases most rapidly in the opposite direction $(-a, -b)$. Both of these directions are perpendicular to the direction of the line.

15.5 Finding the best logistic function

We now have a straightforward minimization problem to solve; we'd like to find the values a, b, and c that make the `logistic_cost` function as small as possible. Then the corresponding function, $L(x, p) = \sigma(ax + bp - c)$ will be the best fit to the data. We can use that resulting function to build a classifier by plugging in the mileage x and price p for an unknown car and labeling it as a BMW if $L(x, p) > 0.5$ and as a Prius, otherwise. We'll call this classifier `best_logistic_classifier(x,p)`, and we can pass it to `test_classifier` to see how well it does.

The only major work we have to do here is upgrading our `gradient_descent` function. So far, we've only done gradient descent with functions that take 2D vectors and return numbers. The `logistic_cost` function takes a 3D vector (a, b, c) and outputs a number, so we need a new version of gradient descent. Fortunately, we covered 3D analogies for every 2D vector operation we've used, so it won't be too hard.

15.5.1 Gradient descent in three dimensions

Let's look at our existing gradient calculation that we used to work with functions of two variables in chapters 12 and 14. The partial derivatives of a function $f(x, y)$ at a point (x_0, y_0) are the derivatives with respect to x and y individually, while assuming the other variable is a constant. For instance, plugging in y_0 into the second slot of $f(x, y)$, we get $f(x, y_0)$, which we can treat as a function of x alone and take its ordinary derivative. Putting the two partial derivatives together as components of a 2D vector gives us the gradient:

```
def approx_gradient(f,x0,y0,dx=1e-6):
    partial_x = approx_derivative(lambda x:f(x,y0),x0,dx=dx)
    partial_y = approx_derivative(lambda y:f(x0,y),y0,dx=dx)
    return (partial_x,partial_y)
```

The difference for a function of three variables is that there's one other partial derivative we can take. If we look at $f(x, y, z)$ at some point (x_0, y_0, z_0), we can look at $f(x, y_0, z_0)$, $f(x_0, y, z_0)$, and $f(x_0, y_0, z)$ as functions of x, y, and z, respectively, and take their ordinary derivatives to get three partial derivatives. Putting these three partial derivatives together in a vector, we get the 3D version of the gradient:

```
def approx_gradient3(f,x0,y0,z0,dx=1e-6):
    partial_x = approx_derivative(lambda x:f(x,y0,z0),x0,dx=dx)
    partial_y = approx_derivative(lambda y:f(x0,y,z0),y0,dx=dx)
    partial_z = approx_derivative(lambda z:f(x0,y0,z),z0,dx=dx)
    return (partial_x,partial_y,partial_z)
```

To do the gradient descent in 3D, the procedure is just as you'd expect; we start at some point in 3D, calculate the gradient, and step a small amount in that direction to arrive at a new point, where hopefully, the value of $f(x, y, z)$ is smaller. As one additional enhancement, I've added a `max_steps` parameter so we can set a maximum number of steps to take during the gradient descent. With that parameter set to a reasonable limit, we won't have to worry about our program stalling even if the algorithm doesn't converge to a point within the tolerance. Here's what the result looks like in Python:

```
def gradient_descent3(f,xstart,ystart,zstart,
                      tolerance=1e-6,max_steps=1000):
    x = xstart
    y = ystart
    z = zstart
    grad = approx_gradient3(f,x,y,z)
    steps = 0
    while length(grad) > tolerance and steps < max_steps:
        x -= 0.01 * grad[0]
        y -= 0.01 * grad[1]
        z -= 0.01 * grad[2]
        grad = approx_gradient3(f,x,y,z)
        steps += 1
    return x,y,z
```

All that remains is to plug in the `logistic_cost` function, and the `gradient_descent3` function finds inputs that minimize it.

15.5.2 Using gradient descent to find the best fit

To be cautious, we can start by using a small number of `max_steps`, like 100:

```
>>> gradient_descent3(logistic_cost,1,1,1,max_steps=100)
(0.21114493546399946, 5.04543972557848, 2.1260122558655405)
```

If we allow it to take 200 steps instead of 100, we see that it has further to go after all:

```
>>> gradient_descent3(logistic_cost,1,1,1,max_steps=200)
(0.884571531298388, 6.657543188981642, 2.955057286988365)
```

Remember, these results are the parameters required to define the logistic function, but they are also the parameters (a, b, c) defining the decision boundary in the form $ax + bp = c$. If we run gradient descent for 100 steps, 200 steps, 300 steps, and so on, and plot the corresponding lines with `plot_line`, we can see the decision boundary converging as in figure 15.19.

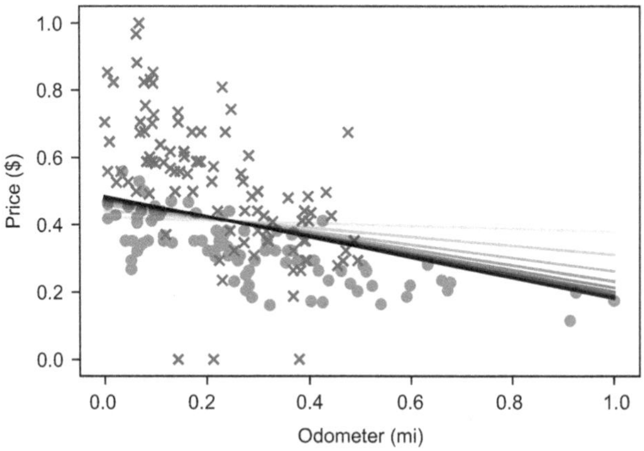

Figure 15.19 With more and more steps, the values of (*a*, *b*, *c*) returned by gradient descent seem to be settling on a clear decision boundary.

Somewhere between 7,000 and 8,000 steps, the algorithm actually converges, meaning it finds a point where the length of the gradient is less than 10^{-6}. Approximately speaking, that's the minimum point we're looking for:

```
>>> gradient_descent3(logistic_cost,1,1,1,max_steps=8000)
(3.7167003153580045, 11.422062409195114, 5.596878367305919)
```

We can see what this decision boundary looks like relative to the one we've been using (figure 15.20 shows the result):

```
plot_data(scaled_car_data)
plot_line(0.35,1,0.56)
plot_line(3.7167003153580045, 11.422062409195114, 5.596878367305919)
```

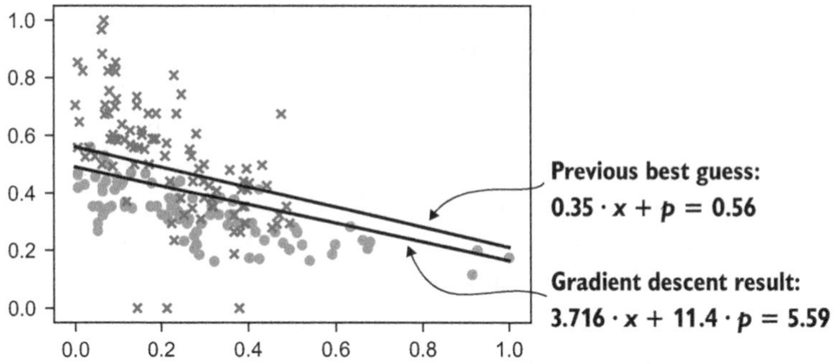

Figure 15.20 Comparing our previous best-guess decision boundary to the one implied by the result of gradient descent

This decision boundary isn't too far off from our guess. The result of the logistic regression appears to have moved the decision boundary slightly downward from our guess, trading off a few false positives (Priuses that are now incorrectly above the line in figure 15.20) for a few more true positives (BMWs that are now correctly above the line).

15.5.3 Testing and understanding the best logistic classifier

We can easily plug these values for (*a*, *b*, *c*) into a logistic function and then use it to make a car classification function:

```
def best_logistic_classifier(x,p):
    l = make_logistic(3.7167003153580045, 11.422062409195114, 
5.596878367305919)
    if l(x,p) > 0.5:
        return 1
    else:
        return 0
```

Plugging this function into the `test_classifier` function, we can see its accuracy rate on the test data set is about what we got from our best attempts, 80% on the dot:

```
>>> test_classifier(best_logistic_classifier,scaled_car_data)
0.8
```

The decision boundaries are fairly close, so it makes sense that the performance is not too far off of our guess from section 15.2. That said, if what we had previously was close, why did the decision boundary converge so decisively where it did?

It turns out logistic regression does more than simply find the optimal decision boundary. In fact, we saw a decision boundary early in the section that outperformed this best fit logistic classifier by 0.5%, so the logistic classifier doesn't even maximize accuracy on the test data set. Rather, logistic regression looks holistically at the data set and finds the model that is most likely to be accurate given all of the examples. Rather than moving the decision boundary slightly to grab one or two more percentage points of accuracy on the test set, the algorithm orients the decision boundary based on a holistic view of the data set. If our data set is representative, we can trust our logistic classifier to do well on data it hasn't seen yet, not just the data in our training set.

The other information that our logistic classifier has is an amount of certainty about every point it classifies. A classifier based only on a decision boundary is 100% certain that a point above that boundary is a BMW and that a point below that is a Prius. Our logistic classifier has a more nuanced view; we can interpret the values it returns between zero and one as a probability a car is a BMW rather than a Prius. For real-world applications, it can be valuable to know not only the best guess from your machine learning model, but also how trustworthy it considers itself to be. If we were classifying benign tumors from malignant ones based on medical scans, we might act much differently if the algorithm told us it was 99% sure, as opposed to 51% sure, if a tumor was malignant.

The way certainty comes through in the shape of the classifier is the magnitude of the coefficients (*a*, *b*, *c*). For instance, you can see that the ratio between (*a*, *b*, *c*) in our guess of (0.35, 1, 0.56) is similar to the ratio in the optimal values of (3.717, 11.42, 5.597). The optimal values are approximately ten times bigger than our best guess. The biggest difference that causes this change is the steepness of the logistic function. The optimal logistic function is much more certain of the decision boundary than the

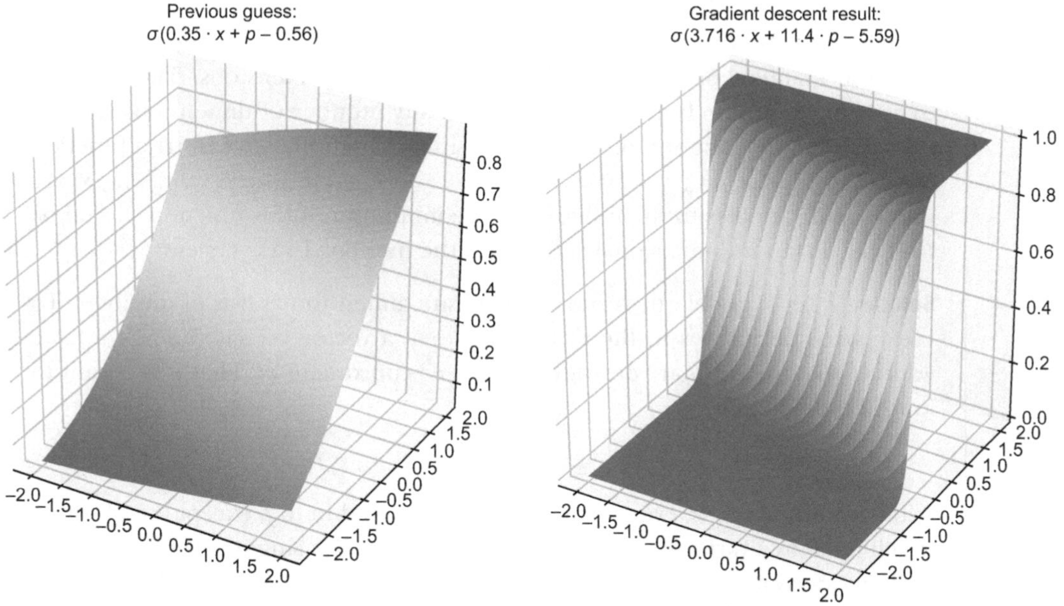

Figure 15.21 The optimized logistic function is much steeper, meaning its certainty that a car is a BMW rather than a Prius increases rapidly as you cross the decision boundary.

first. It tells us that as soon as you cross the decision boundary, certainty of the result increases significantly as figure 15.21 shows.

In the final chapter, we'll continue to use sigmoid functions to produce certainties of results between zero and one as we implement classification using neural networks.

15.5.4 Exercises

Exercise 15.11 Modify the `gradient_descent3` function to print the total number of steps taken before it returns its result. How many steps does the gradient descent take to converge for `logistic_cost`?

Solution All you need to do is add the line `print(steps)` right before `gradient_descent3` to return its result:

```
def gradient_descent3(f,xstart,ystart,zstart,tolerance=1e-
6,max_steps=1000):
    ...
    print(steps)
    return x,y,z
```

Running the following gradient descent

```
gradient_descent3(logistic_cost,1,1,1,max_steps=8000)
```

the number printed is `7244`, meaning the algorithm converges in 7,244 steps.

Exercise 15.12—Mini Project Write an `approx_gradient` function that calculates the gradient of a function in any number of dimensions. Then write a `gradient_descent` function that works in any number of dimensions. To test your `gradient_descent` on an n-dimensional function, you can try a function like $f(x_1, x_2, \ldots, x_n) = (x_1 - 1)^2 + (x_2 - 1)^2 + \ldots + (x_n - 1)^2$, where $x_1, x_2, \ldots, x_n$ are the n input variables to the function f. The minimum of this function should be $(1, 1, \ldots, 1)$, an n-dimensional vector with the number 1 in every entry.

Solution Let's model our vectors of arbitrary dimension as lists of numbers. To take partial derivatives in the ith coordinate at a vector $\mathbf{v} = (v_1, v_2, \ldots, v_n)$, we want to take the ordinary derivative of the ith coordinate *xi*. That is, we want to look at the function:

$$f(v_1, v_2, \ldots, v_{i-1}, x_i, v_{i+1}, \ldots, v_n)$$

that is, in other words, every coordinate of **v** plugged in to f, except the ith entry, which is left as a variable x_i. This gives us a function of a single variable x_i, and its ordinary derivative is the ith partial derivative. The code for partial derivatives looks like this:

```
def partial_derivative(f,i,v,**kwargs):
    def cross_section(x):
        arg = [(vj if j != i else x) for j,vj in enumerate(v)]
        return f(*arg)
    return approx_derivative(cross_section, v[i], **kwargs)
```

Note that our coordinates are zero-indexed, and the dimension of input to f is inferred from the length of **v**.

The rest of the work is easy by comparison. To build the gradient, we just take the n partial derivatives and put them in order in a list:

```
def approx_gradient(f,v,dx=1e-6):
    return [partial_derivative(f,i,v) for i in range(0,len(v))]
```

To do the gradient descent, we replace all of the manipulations of named coordinate variables, like x, y, and z, with list operations on the list vector of coordinates called **v**:

```
def gradient_descent(f,vstart,tolerance=1e-6,max_steps=1000):
    v = vstart
    grad = approx_gradient(f,v)
    steps = 0
    while length(grad) > tolerance and steps < max_steps:
        v = [(vi - 0.01 * dvi) for vi,dvi in zip(v,grad)]
        grad = approx_gradient(f,v)
        steps += 1
    return v
```

To implement the suggested test function, we can write a generalized version of it that takes any number of inputs and returns the sum of their squared difference from one:

```
def sum_squares(*v):
    return sum([(x-1)**2 for x in v])
```

This function can't be lower than zero because it's a sum of squares, and a square cannot be less than zero. The value zero is obtained if every entry of the input vector **v** is one, so that's the minimum. Our gradient descent confirms this (with only a small numerical error), so everything looks good! Note that because the starting vector **v** is 5D, all vectors in the computation are automatically 5D.

```
>>> v = [2,2,2,2,2]
>>> gradient_descent(sum_squares,v)
[1.0000002235452137,
 1.0000002235452137,
 1.0000002235452137,
 1.0000002235452137,
 1.0000002235452137]
```

Exercise 15.13—Mini Project Attempt to run the gradient descent with the `simple_logistic_cost` cost function. What happens?

Solution It does not appear to converge. The values of a, b, and c continue increasing without bound even though the decision boundary stabilizes. This means as the gradient descent explores more and more logistic functions, these are staying oriented in the same direction but becoming infinitely steep. It is incentivized to become closer and closer to most of the points, while neglecting the ones it has already mislabeled. As I mentioned, this can be solved by penalizing the incorrect classifications for which the logistic function is the most confident, and our `logistic_cost` function does that well.

Summary

- Classification is a type of machine learning task where an algorithm is asked to look at unlabeled data points and identify each one as a member of a class. In our examples for this chapter, we looked at mileage and price data for used cars and wrote an algorithm to classify them either as 5 series BMWs or Toyota Priuses.
- A simple way to classify vector data in 2D is to establish a decision boundary; that means drawing a literal boundary in the 2D space where your data lives, where points on one side of the boundary are classified in one class and points

on the other side are classified in another. A simple decision boundary is a straight line.

- If our decision boundary line takes the form $ax + by = c$, then the quantity $ax + by - c$ is positive on one side of the line and negative on the other. We can interpret this value as a measure of how much the data point looks like a BMW. A positive value means that the data point looks like a BMW, while a negative value means that it looks more like a Prius.
- The sigmoid function, defined as follows, takes numbers between $-\infty$ and ∞ and crunches them into the finite interval from zero to one:

$$\sigma(x) = \frac{1}{1 + e^{-x}}$$

- Composing the sigmoid with the function $ax + by - c$, we get a new function $\sigma(ax + by - c)$ that also measures how much the data point looks like a BMW, but it only returns values between zero and one. This type of function is a logistic function in 2D.
- The value between zero and one that a logistic classifier outputs can be interpreted as how confident it is that a data point belongs to one class versus another. For instance, return values of 0.51 or 0.99 would both indicate that the model thinks we're looking at a BMW, but the latter would be a much more confident prediction.
- With an appropriate cost function that penalizes confident, incorrect classifications, we can use gradient descent to find the logistic function of best fit. This is the best logistic classifier according to the data set.

16 Training neural networks

This chapter covers

- Classifying images of handwritten digits as vector data
- Designing a type of neural network called a multilayer perceptron
- Evaluating a neural network as a vector transformation
- Fitting a neural network to data with a cost function and gradient descent
- Calculating partial derivatives for neural networks in backpropagation

In the final chapter of this book, we combine almost everything you've learned so far to introduce one of the most famous machine learning tools used today: artificial neural networks. *Artificial neural networks*, or neural networks for short, are mathematical functions whose structure is loosely based on the structure of the human brain. These are called artificial to distinguish from the "organic" neural networks that exist in the brain. This might sound like a lofty and complex goal, but it's all based on a simple metaphor for how the brain works.

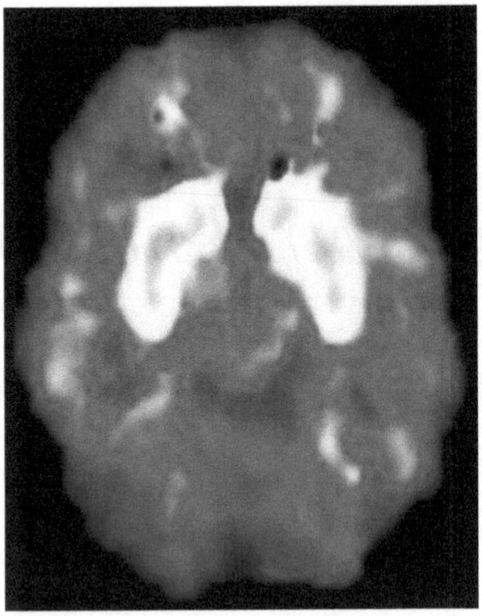

Figure 16.1 Different kinds of brain activity cause different neurons to electrically activate, showing bright areas in a brain scan.

Before explaining the metaphor, I'll preface this discussion by reminding you that I'm not a neurologist. The rough idea is that the brain is a big clump of interconnected cells called *neurons* and, when you think certain thoughts, what's actually happening is electrical activity at specific neurons. You can see this electrical activity in the right kind of brain scan where various parts of the brain light up (figure 16.1).

As opposed to the billions of neurons in the human brain, the neural networks we build in Python have only a few dozen neurons, and the degree to which a specific neuron is turned on is represented by a single number called its *activation*. When a neuron activates in the brain or in our artificial neural network, it can cause adjacent, connected neurons to turn on as well. This allows one idea to lead to another, which we can loosely see as creative thinking.

Mathematically, the activation of a neuron in our neural network is a function of the numerical activation values of neurons it is connected to. If a neuron connects to four others with the activation values a_1, a_2, a_3, and a_4, then its activation will be some mathematical function applied to those four values, say $f(a_1, a_2, a_3, a_4)$.

Figure 16.2 shows a schematic diagram with all of the neurons drawn as circles. I've shaded the neurons differently to indicate that they have different levels of activation, kind of like the brighter or darker areas of a brain scan.

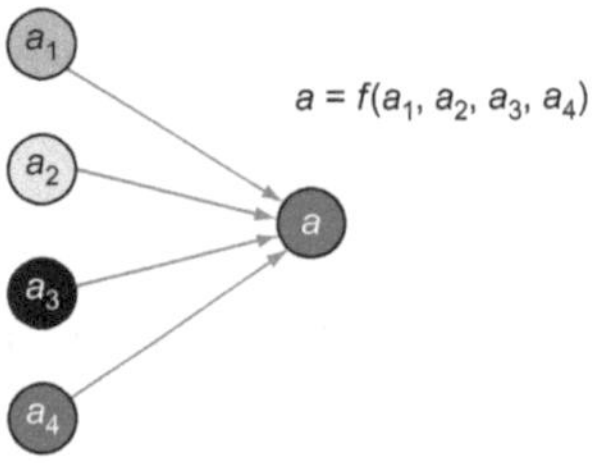

Figure 16.2 Picturing neuron activation as a mathematical function, where a_1, a_2, a_3, and a_4 are the activation values applied to the function *f*.

If each of a_1, a_2, a_3, and a_4 depend on the activation of other neurons, the value of a could depend on even more numbers. With more neurons and more connections, you can build an arbitrarily complicated mathematical function, and the goal is to model arbitrarily complicated ideas.

The explanation I've just given you is a somewhat philosophical introduction to neural networks, and it's definitely not enough for you to start coding. In this chapter, I show you, in detail, how to run with these ideas and build your own neural network. As in the last chapter, the problem we'll solve with neural networks is *classification*. There are many steps in building a neural network and training it to perform well on classification, so before we dive in, I'll lay out the plan.

16.1 Classifying data with neural networks

In this section, I focus on a classic application of neural networks: classifying images. Specifically, we'll use low resolution images of handwritten digits (numbers from 0 to 9), and we want our neural network to identify which digit is shown in a given image. Figure 16.3 shows some example images for these digits.

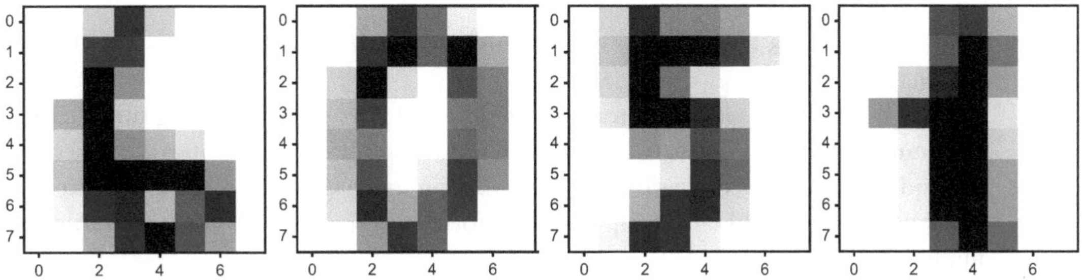

Figure 16.3 Low resolution images of some handwritten digits

If you identified the digits in figure 16.3 as 6, 0, 5, and 1, then congratulations! Your organic neural network (that is, your brain) is well trained. Our goal here is to build an artificial neural network that looks at such an image and classifies it as one of ten possible digits, perhaps as well as a human could.

In chapter 15, the classification problem amounted to looking at a 2D vector and classifying it in one of two classes. In this problem, we look at 8x8 pixel grayscale images, where each of the 64 pixels is described by one number that tells us its brightness. Just as we treated images as vectors in chapter 6, we'll treat the 64-pixel brightness values as a 64-dimensional vector. We want to put each 64-dimensional vector in one of ten classes, indicating which digit it represents. Thus, our classification function will have more inputs and more outputs than the one in chapter 15.

Concretely, the neural network classification function we'll build in Python will look like a function with 64 inputs and 10 outputs. In other words, it's a (non-linear!) vector transformation from $\mathbb{R}^{64}$ to $\mathbb{R}^{10}$. The input numbers are the pixel darkness values, scaled from 0 to 1, and the ten output values represent how likely the image is to be any of the ten digits. The index of the largest output number is the answer. In the following case (shown by figure 16.4), an image of a 5 is passed in, and the neural

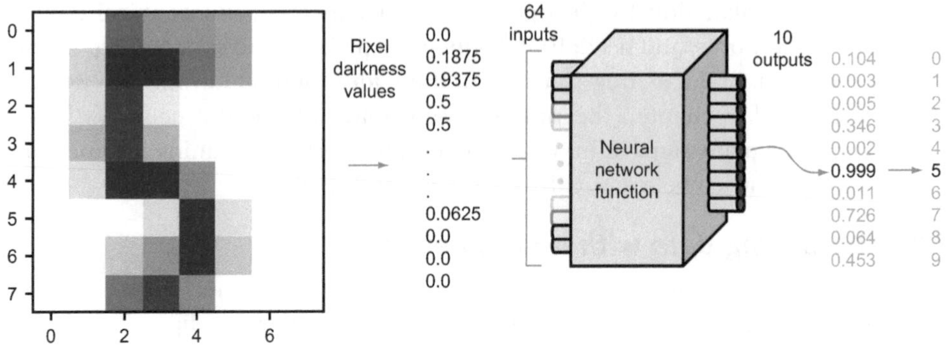

Figure 16.4 How our Python neural network function classifies images of digits.

network returns its largest value in the fifth slot, so it correctly identifies the digit in the image.

The neural network function in the middle of figure 16.4 is nothing more than a mathematical function. Its structure will be more complex than the ones we've seen so far, and in fact, the formula defining it is too long to write on paper. Evaluating a neural network is more like carrying out an algorithm. I'll show you how to do this and implement it in Python.

Just as we tested many different logistic functions in the previous chapter, we could try many different neural networks and see which one has the best predictive accuracy. Once again, the systematic way to do this is a gradient descent. While a linear function is determined by the two constants a and b in the formula $f(x) = ax + b$, a neural network of a given shape can have thousands of constants determining how it behaves. That's a lot of partial derivatives to take! Fortunately, due to the form of the functions connecting neurons in our neural network, there's a shortcut algorithm for taking the gradient, which is called *backpropagation*.

It's possible to derive the backpropagation algorithm from scratch and implement it using only the math we've covered so far, but unfortunately, that's too big of a project to fit in this book. Instead, I'll show you how to use a famous Python library called scikit-learn ("sci" pronounced as in "science") to do the gradient descent for us, so it automatically trains the neural network to predict as well as possible for our data set. Finally, I'll leave you with a teaser of the math behind backpropagation. I hope this will be just the starting point for your prolific career in machine learning.

16.2 Classifying images of handwritten digits

Before we start implementing our neural network, we need to prepare the data. The digit images I use are among the extensive, free test data that comes with the scikit-learn data. Once we download those, we need to convert them into 64-dimensional vectors with values scaled between zero and one. The data set also comes with the correct answers for each digit image, represented as Python integers from zero to nine.

Then we build two Python functions to practice the classification. The first is a fake digit identification function called `random_classifier`, which takes 64 numbers representing an image and (randomly) outputs 10 numbers representing the certainty that the image represents each digit from 0 to 9. The second is a function called `test_digit_classify`, which takes a classifier and automatically plugs in every image in the data set, returning a count of how many correct answers come out. Because our `random_classifier` produces random results, it should only guess the right answer 10% of the time. This sets the stage for improvement when we replace it with a real neural network.

16.2.1 Building the 64-dimensional image vectors

If you're working with the Anacondas Python distribution as described in appendix A, you should already have the scikit-learn library available as `sklearn`. If not, you can install it with pip. To open `sklearn` and import the digits data set, you need the following code:

```
from sklearn import datasets
digits = datasets.load_digits()
```

Each entry of digits is a 2D NumPy array (a matrix), giving the pixel values of one image. For instance, `digits.images[0]` gives the pixel values of the first image in the data set, which is an 8-by-8 matrix of values:

```
>>> digits.images[0]
array([[ 0.,  0.,  5., 13.,  9.,  1.,  0.,  0.],
       [ 0.,  0., 13., 15., 10., 15.,  5.,  0.],
       [ 0.,  3., 15.,  2.,  0., 11.,  8.,  0.],
       [ 0.,  4., 12.,  0.,  0.,  8.,  8.,  0.],
       [ 0.,  5.,  8.,  0.,  0.,  9.,  8.,  0.],
       [ 0.,  4., 11.,  0.,  1., 12.,  7.,  0.],
       [ 0.,  2., 14.,  5., 10., 12.,  0.,  0.],
       [ 0.,  0.,  6., 13., 10.,  0.,  0.,  0.]])
```

You can see that the range of grayscale values is limited. The matrix consists only of whole numbers from 0 to 15.

Matplotlib has a useful built-in function called `imshow`, which shows the entries of a matrix as an image. With the correct grayscale specification, the zeroes in the matrix appear as white and the bigger non-zero values appear as darker shades of gray. For instance, figure 16.5 shows the first image in the data set, which looks like a zero, resulting from `imshow`:

```
import matplotlib.pyplot as plt
plt.imshow(digits.images[0], cmap=plt.cm.gray_r)
```

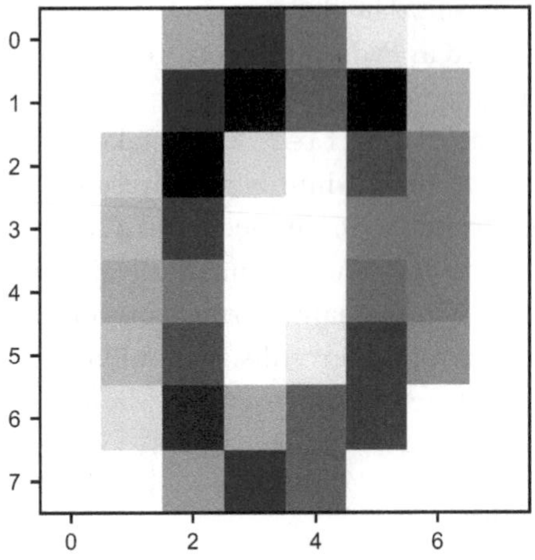

Figure 16.5 The first image in sklearn's digit data set, which looks like a zero

To emphasize once more how we're going to think of this image as a 64-dimensional vector, figure 16.6 shows a version of the image with each of the 64-pixel brightness values overlaid on the corresponding pixels.

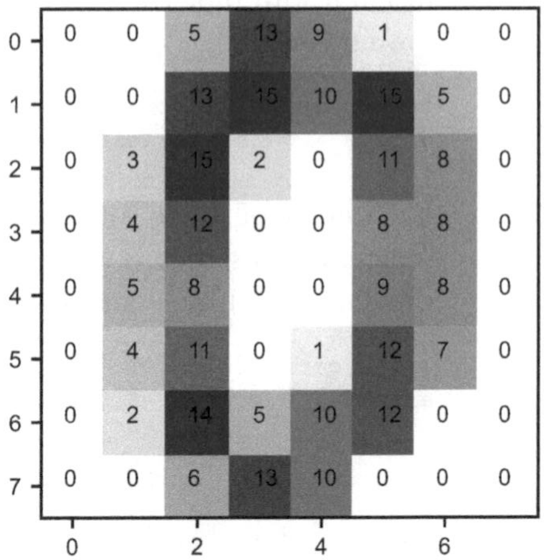

Figure 16.6 An image from the digit data set with brightness values overlaid on each pixel.

To turn this 8-by-8 matrix of numbers into a single 64-entry vector, we can use a built-in NumPy function called `np.matrix.flatten`. This function builds a vector starting with the first row of the matrix, followed by the second row, and so on, giving us a vector representation of an image similar to the one we used in chapter 6. Flattening the first image matrix indeed gives us a vector with 64 entries:

```
>>> import numpy as np
>>> np.matrix.flatten(digits.images[0])
array([ 0.,  0.,  5., 13.,  9.,  1.,  0.,  0.,  0.,  0., 13., 15., 10.,
       15.,  5.,  0.,  0.,  3., 15.,  2.,  0., 11.,  8.,  0.,  0.,  4.,
       12.,  0.,  0.,  8.,  8.,  0.,  0.,  5.,  8.,  0.,  0.,  9.,  8.,
        0.,  0.,  4., 11.,  0.,  1., 12.,  7.,  0.,  0.,  2., 14.,  5.,
       10., 12.,  0.,  0.,  0.,  0.,  6., 13., 10.,  0.,  0.,  0.])
```

To keep our analysis numerically tidy, we'll once again scale our data so that the values are between 0 and 1. Because all the pixel values for every entry in this data set are between 0 and 15, we can scalar multiply these vectors by 1 / 15 to get scaled versions.

NumPy overloads the `*` and `/` operators to automatically work as scalar multiplication (and division) of vectors, so we can simply type

```
np.matrix.flatten(digits.images[0]) / 15
```

and we'll get a scaled result. Now we can build a sample digit classifier to plug these values into.

16.2.2 Building a random digit classifier

The input to the digit classifier is a 64-dimensional vector, like the ones we just constructed, and the output is a 10-dimensional vector with each entry value between 0 and 1. For our first example, the output vector entries can be randomly generated, but we interpret them as the classifier's certainty that the image represents each of the ten digits.

Because we're okay with random outputs for now, this is easy to implement; NumPy has a function, `np.random.rand`, that produces an array of random numbers between 0 and 1 of a specified size. For instance, `np.random.rand(10)` gives us a NumPy array of 10 random numbers between 0 and 1. Our `random_classifier` function takes an input vector, ignores it, and returns a random vector:

```
def random_classifier(input_vector):
    return np.random.rand(10)
```

To classify the first image in the data set, we can run the following:

```
>>> v = np.matrix.flatten(digits.images[0]) / 15.
>>> result = random_classifier(v)
>>> result
array([0.78426486, 0.42120868, 0.47890909, 0.53200335, 0.91508751,
       0.1227552 , 0.73501115, 0.71711834, 0.38744159, 0.73556909])
```

The largest entry of this output is about `0.915`, occurring at index 4. Returning this vector, our classifier tells us that there's some chance that the image represents any of the digits and that it is most likely a 4. To get the index of a maximum value programmatically, we can use the following Python code:

```
>>> list(result).index(max(result))
4
```

Here, `max(result)` finds the largest entry of the array, and `list(result)` treats the array as an ordinary Python list. Then we can use the built-in `list` index function to find the index of the maximum value. The return value of 4 is incorrect; we saw previously that the picture is a 0, and we can check the official result as well.

The correct digit for each image is stored at the corresponding index in the `digits.target` array. For the image `digits.images[0]`, the correct value is `digits.target[0]`, which is zero as we expected:

```
>>> digits.target[0]
0
```

Our random classifier predicted the image to be a 4 when in fact it was a 0. Because it is guessing at random, it should be wrong 90% of the time, and we can confirm this by testing it on a lot of test examples.

16.2.3 Measuring performance of the digit classifier

Now we'll write the function `test_digit_classify`, which takes a classifier function and measures its performance on a large set of digit images. Any classifier function will have the same shape; it takes a 64-dimensional input vector and returns a 10-dimensional output vector. The `test_digit_classify` function goes through all of the test images and known correct answers and sees if the classifier produces the right answer:

```
def test_digit_classify(classifier,test_count=1000):
    correct = 0
    for img, target in zip(digits.images[:test_count],
digits.target[:test_count]):
        v = np.matrix.flatten(img) / 15.
        output = classifier(v)
        answer = list(output).index(max(output))
        if answer == target:
            correct += 1
    return (correct/test_count)
```

Starts the counter of correct classifications at 0

Loops over pairs of images in the test set with corresponding targets, giving the correct answer for the digit

Flattens the image matrix into a 64D vector and scales it appropriately

Passes the image vector through the classifier to get a 10D result

Finds the index of the largest entry in this result, which is the classifier's best guess

If this matches our answer, increments the counter

Returns the number of correct classifications as a fraction of the total number of test data points

We expect our random classifier to get about 10% of the answers right. Because it acts randomly, it might do better on some trials than others, but because we're testing on so many images, the result should be somewhere close to 10% every time. Let's give it a try:

```
>>> test_digit_classify(random_classifier)
0.107
```

In this test, our random classifier did slightly better than expected at 10.7%. This isn't too interesting on its own, but now we've got our data organized and a baseline example to beat so we can start building our neural network.

16.2.4 Exercises

Exercise 16.1 Suppose a digit classifier function outputs the following NumPy array. What digit has it concluded is in the image?

```
array([5.00512567e-06, 3.94168539e-05, 5.57124430e-09, 9.31981207e-09,
       9.98060276e-01, 9.10328786e-07, 1.56262695e-03, 1.82976466e-04,
       1.48519455e-04, 2.54354113e-07])
```

Solution The largest number in this array is `9.98060276e-01`, or approximately 0.998, which appears fifth, or in index 4. Therefore, this output says the image is classified as a 4.

Exercise 16.2—Mini Project Find the average of all the images of 9's in the data set in the same way we took averages of the images in chapter 6. Plot the resulting image. What does it look like?

Solution This code takes an integer *i* and averages the images in the data set that represent the digit *i*. Because the digit images are represented as NumPy arrays, which support addition and scalar multiplication, we can average them using the ordinary Python `sum` function and division operator:

```
def average_img(i):
    imgs = [img for img,target in zip(digits.images[1000:],
digits.target[1000:]) if target==i]
    return sum(imgs) / len(imgs)
```

With this code, `average_img(9)` computes an 8-by-8 matrix representing the average of all the images of 9's, and it looks like this:

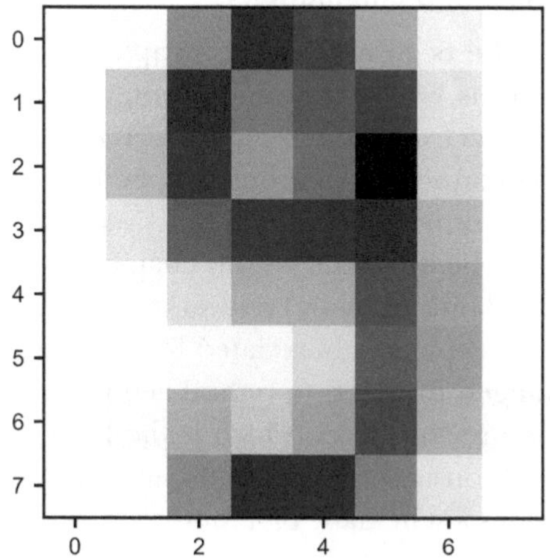

Exercise 16.3—Mini Project Build a better classifier than the random one by finding the average image of each kind of digit in the test data set and comparing a target image with all of the averages. Specifically, return a vector of the dot products of the target image with each average digit image.

Solution

```
avg_digits = [np.matrix.flatten(average_img(i)) for i in range(10)]
def compare_to_avg(v):
    return [np.dot(v,avg_digits[i]) for i in range(10)]
```

Testing this classifier, we get 85% of the digits correct in the test data set. Not bad!

```
>>> test_digit_classify(compare_to_avg)
0.853
```

16.3 Designing a neural network

In this section, I show you how to think of a neural network as a mathematical function and how you can expect it to behave, depending on its structure. That sets us up for the next section, where we implement our first neural network as a Python function in order to classify digit images.

For our image classification problem, our neural network has 64 input values and 10 output values, and requires hundreds of operations to evaluate. For that reason, in this section, I stick with a simpler neural network with three inputs and two outputs. This makes it possible to picture the whole network and walk through every step of its evaluation. Once we cover this, it will be easy to write the evaluation steps that work on a neural network of any size in general Python code.

16.3.1 Organizing neurons and connections

As I described in the beginning of this chapter, the model for a neural network is a collection of neurons, where a given neuron activates, depending on how much its connected neurons activate. Mathematically, turning on a neuron is a function of the activations of the connected neurons. Depending on how many neurons are used, which neurons are connected, and the functions that connect them, the behavior of a neural network can be different. In this chapter, we'll restrict our attention to one of the simplest useful kinds of neural networks—a *multilayer perceptron.*

A multilayer perceptron, abbreviated MLP, consists of several columns of neurons called *layers*, arranged from left to right. Each neuron's activation is a function of the activations in the previous layer, which is the layer immediately to the left. The leftmost layer depends on no other neurons, and its activation is based on training data. Figure 16.7 provides a schematic of a four-layer MLP.

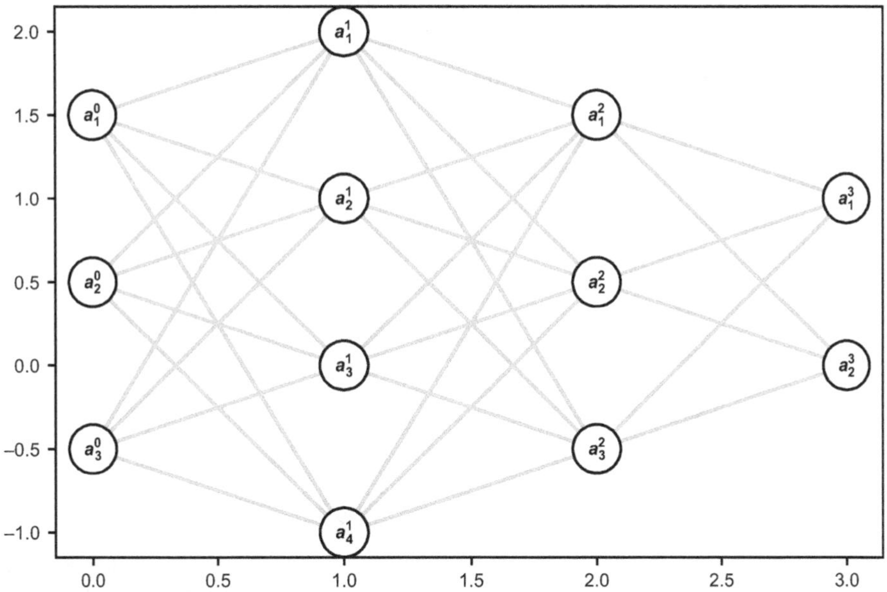

Figure 16.7 A schematic of a multilayer perceptron (MLP), consisting of several layers of neurons

In figure 16.7, each circle is a neuron, and lines between the circles show connected neurons. Turning on a neuron depends only on the activations of neurons from the previous layer, and it influences the activations of every neuron in the next layer. I arbitrarily chose the number of neurons in each layer, and in this particular schematic, the layers consist of three, four, three, and two neurons, respectively.

Because there are 12 total neurons, there are 12 total activation values. Often there can be many more neurons (we'll use 90 for digit classification), so we can't give a letter variable name to every neuron. Instead, we represent all activations with the letter a and index them with superscripts and subscripts. The superscript indicates the layer, and the subscript indicates which neuron we're talking about within the layer. For instance, a_2^2 is a number representing the activation of the second neuron in the second layer.

16.3.2 Data flow through a neural network

To evaluate a neural network as a mathematical function, there are three basic steps, which I describe in terms of the activation values. I'll walk through them conceptually, and then I'll show you the formulas. Remember, a neural network is just a function that takes an input vector and produces an output vector. The steps in between are just a recipe for getting to the output from the given input. Here's the first step in the pipeline.

STEP 1: SET THE INPUT LAYER ACTIVATIONS TO THE ENTRIES OF THE INPUT VECTOR

The *input* layer is another word for the first or leftmost layer. The network in figure 16.7 has three neurons in the input layer, so this neural network can take 3D vectors as inputs. If our input vector is (0.3, 0.9, 0.5), then we can perform this first step by setting $a_1^0 = 0.3$, $a_2^0 = 0.9$, and $a_3^0 = 0.5$. That fills in 3 of the 12 total neurons in the network (figure 16.8).

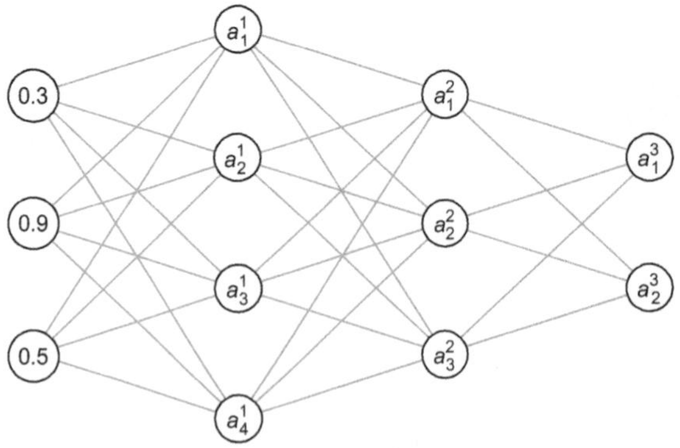

Figure 16.8 Setting the input layer activations to the entries of the input vector (left)

Each activation value in layer one is a function of the activations in layer zero. Now we have enough information to calculate them, so that's step 2.

STEP 2: CALCULATE EACH ACTIVATION IN THE NEXT LAYER AS A FUNCTION OF ALL OF THE ACTIVATIONS IN THE INPUT LAYER

This step is the meat of the calculation, and I'll return to it once I've gone through all of the steps conceptually. The important thing to know for now is that each activation in the next layer is usually given by a *distinct function* of the previous layer activations. Say we want to calculate a_0^1. This activation is some function of a_1^0, a_2^0 and a_3^0, which we can simply write as $a_1^1 = f(a_1^0, a_2^0, a_3^0)$ for now. Suppose, for instance, we calculate $f(0.3, 0.9, 0.5)$ and the answer is 0.6. Then the value of a_1^1 becomes 0.6 in our calculation (figure 16.9).

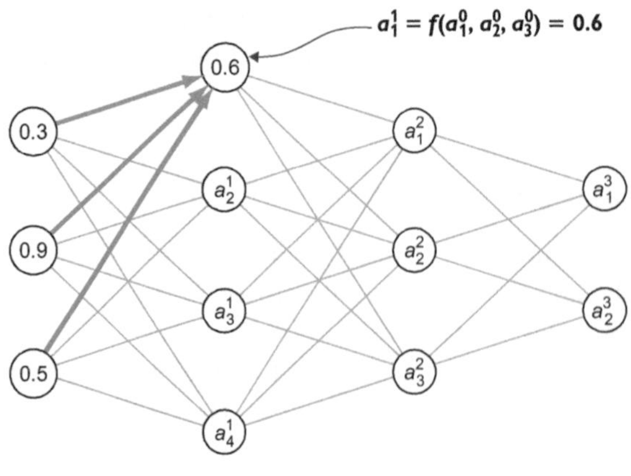

Figure 16.9 Calculating an activation in layer one as some function of the activations in layer zero

When we calculate the next activation in layer one, a_2^1, it is also a function of the input activations a_1^0, a_2^0, and a_3^0, but in general, it is a different function, say $a_2^1 = g(a_1^0, a_2^0, a_3^0)$. The result still depends on the same inputs, but as a different function, it's likely we'll get a different result. Let's say, $g(0.3, 0.9, 0.5) = 0.1$, then that's our value for a_2^1 (figure 16.10).

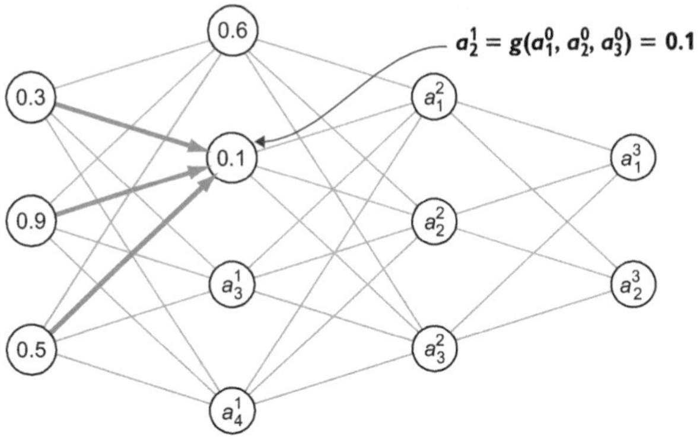

Figure 16.10 Calculating another activation in layer one with another function of the input layer activations

I used *f* and *g* because those are simple placeholder function names. There are two more distinct functions for a_3^1 and a_4^1 in terms of the input layer. I won't keep naming these functions, because we'll quickly run out of letters, but the important point is that each activation has a special function of the previous layer activations. Once we calculate all of the activations in layer one, we've 7 of the 12 total activations filled in. The numbers here are still made up, but the result might look something like that shown in figure 16.11.

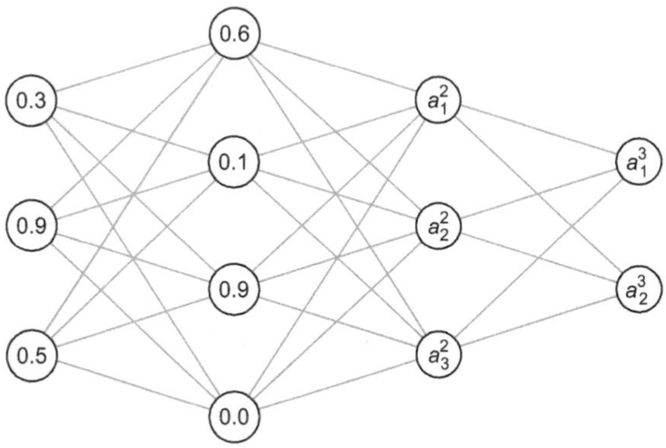

Figure 16.11 Two layers of activations for our multilayer perceptron (MLP) calculated.

From here on out, we repeat the process until we've calculated the activation of every neuron in the network, which is step 3.

STEP 3: REPEAT THIS PROCESS, CALCULATING THE ACTIVATIONS OF EACH SUBSEQUENT LAYER BASED ON THE ACTIVATIONS IN THE PRECEDING LAYER

We start by calculating a_1^2 as a function of the layer one activations, a_1^1, a_2^1, a_3^1, and a_4^1. Then we move on to a_2^2 and a_3^2, which are given by their own functions. Finally, we calculate a_1^3 and a_2^3 as their own functions of the layer two activations. At this point, we have an activation for every neuron in the network (figure 16.12).

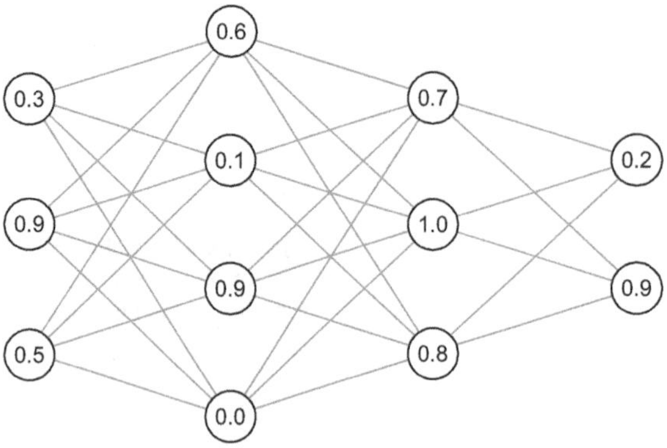

Figure 16.12 An example of an MLP with all activations calculated

At this point, our calculation is done. We have activations calculated for the middle layers, called *hidden layers*, and the final layer, called the *output layer*. All we need to do now is to read off the activations of the output layer to get our result and that's step 4.

STEP 4: RETURN A VECTOR WHOSE ENTRIES ARE THE ACTIVATIONS OF THE OUTPUT LAYER

In this case, the vector is (0.2, 0.9), so evaluating our neural network as a function of the input vector (0.3, 0.9, 0.5) produces the output vector (0.2, 0.9).

That's all there is to it! The only thing I didn't cover is how to calculate individual activations, and these are what make the neural network distinct. Every neuron, except for those in the input layer, has its own function, and the parameters defining those functions are the numbers we'll tweak to make the neural network do what we want.

16.3.3 Calculating activations

The good news is that we'll use a familiar form of function to calculate the activations in one layer as a function of those in the previous layer: logistic functions. The tricky part is that our neural network has 9 neurons outside the input layer, so there are 9 distinct functions to keep track of. What's more, there are several constants to determine the behavior of each logistic function. Most of the work will be keeping track of all of these constants.

To focus on a specific example, we noted that in our sample MLP, we have the activation depend on the three input layer activations: a_1^0, a_2^0, and a_3^0. The function giving a_1^1 is a linear function of these inputs (including a constant) passed into a sigmoid function. There are four free parameters here, which I name *A*, *B*, *C*, and *D* for the moment (figure 16.13).

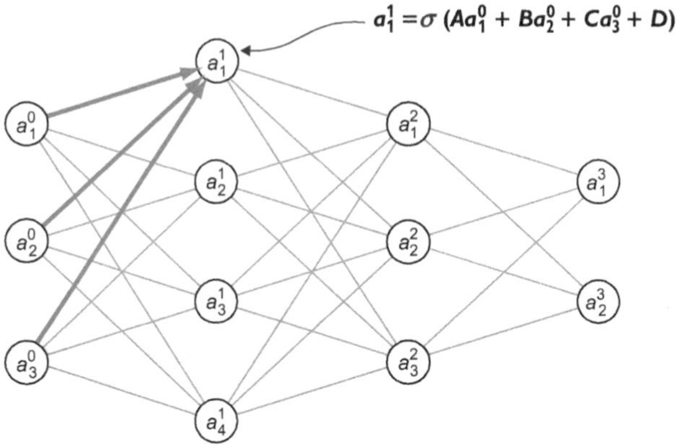

Figure 16.13 The general form of the function to calculate a_1^1 as a function of the input layer activations

We need to tune the variables *A*, *B*, *C*, and *D* to make a_1^1 respond appropriately to inputs. In chapter 15, we thought of logistic functions as taking in several numbers and making a yes-or-no decision about them, reporting the answer as a certainty of "yes" from zero to one. In that sense, you can think of the neurons in the middle of the network as breaking the overall classification problem into smaller yes-or-no classifications.

For every connection in the network, there is a constant telling us how strongly the input neuron activation affects the output neuron activation. In this case, the constant *A* tells us how strongly a_1^0 affects a_1^1, while *B* and *C* tell us how strongly a_2^0 and a_3^0 affect a_1^1, respectively. These constants are called *weights* for the neural network, and there is one weight for every line segment in the neural network general diagram used throughout this chapter.

The constant *D* doesn't affect the connection, but instead, independently increases or decreases the value of a_1^1, which is not dependent on an input activation. This is appropriately named the *bias* for the neuron because it measures the inclination to make a decision without any input. The word *bias* sometimes comes with a negative connotation, but it's an important part of any decision-making process; it helps avoid outlier decisions unless there is strong evidence.

As messy as it might look, we need to index these weights and biases rather than giving them names like *A*, *B*, *C*, and *D*. We'll write the weights in the form w_{ij}^l, where l is the layer on the right of the connection, i is the index of the previous neuron in layer $l-1$, and j is the index of the target neuron in layer l. For instance, the weight *A*, which impacts the first neuron of layer one based on the value of the first neuron of layer zero is denoted by w^1_{11}. The weight connecting the second neuron of layer three to the first neuron of the previous layer is w^3_{21} (figure 16.14).

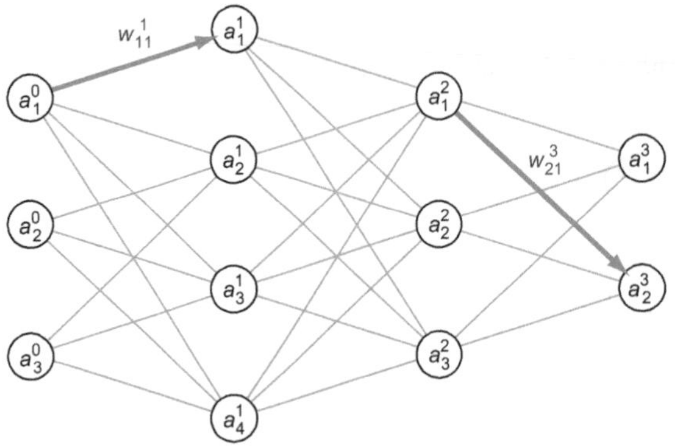

Figure 16.14 Showing the connections corresponding to weights w^1_{11} and w^3_{21}

The biases correspond to neurons, not pairs of neurons, so there is one bias for each neuron: b^l_j for the bias of the jth neuron in the lth layer. In terms of these naming conventions, we could write the formula for a^1_1 as

$$a^1_1 = \sigma(w^1_{11}a^0_1 + w^2_{12}a^0_2 + w^3_{13}a^0_3 + b^1_1)$$

or the formula for a^2_3 as

$$a^2_3 = \sigma(w^2_{31}a^1_1 + w^2_{32}a^1_2 + w^2_{33}a^1_3 + w^2_{34}a^1_4 + b^2_3)$$

As you can see, computing activations to evaluate an MLP is not difficult, but the number of variables can make it a tedious and error-prone process. Fortunately, we can simplify the process and make it easier to implement using the notation of matrices we covered in chapter 5.

16.3.4 Calculating activations in matrix notation

As nasty as it could be, let's do a concrete example and write the formula for the activations of a whole layer of the network, and then we'll see how to simplify it in matrix notation and write a reusable formula. Let's take layer two. The formulas for the three activations are as follows:

$$a^2_1 = \sigma(w^2_{11}a^1_1 + w^2_{12}a^1_2 + w^2_{13}a^1_3 + w^2_{14}a^1_4 + b^2_1)$$
$$a^2_2 = \sigma(w^2_{21}a^1_1 + w^2_{22}a^1_2 + w^2_{23}a^1_3 + w^2_{24}a^1_4 + b^2_2)$$
$$a^2_3 = \sigma(w^2_{31}a^1_1 + w^2_{32}a^1_2 + w^2_{33}a^1_3 + w^2_{34}a^1_4 + b^2_3)$$

It turns out to be useful to name the quantities inside the sigmoid function. Let's denote the three quantities z_1^2, z_2^2, and z_3^2, so that by definition

$$a_1^2 = \sigma(z_1^2)$$
$$a_2^2 = \sigma(z_2^2)$$

and

$$a_3^2 = \sigma(z_3^2)$$

The formulas for these *z* values are nicer because they are all linear combinations of the previous layer activations, plus a constant. That means we can write them in matrix vector notation. Starting with

$$\begin{aligned}
z_1^2 &= w_{11}^2 a_1^1 + w_{12}^2 a_2^1 + w_{13}^2 a_3^1 + w_{14}^2 a_4^1 + b_1^2 \\
z_2^2 &= w_{21}^2 a_1^1 + w_{22}^2 a_2^1 + w_{23}^2 a_3^1 + w_{24}^2 a_4^1 + b_2^2 \\
z_3^2 &= w_{31}^2 a_1^1 + w_{32}^2 a_2^1 + w_{33}^2 a_3^1 + w_{34}^2 a_4^1 + b_3^2
\end{aligned}$$

we can write all three equations as a vector

$$\begin{pmatrix} z_1^2 \\ z_2^2 \\ z_3^2 \end{pmatrix} = \begin{pmatrix} w_{11}^2 a_1^1 + w_{12}^2 a_2^1 + w_{13}^2 a_3^1 + w_{14}^2 a_4^1 + b_1^2 \\ w_{21}^2 a_1^1 + w_{22}^2 a_2^1 + w_{23}^2 a_3^1 + w_{24}^2 a_4^1 + b_2^2 \\ w_{31}^2 a_1^1 + w_{32}^2 a_2^1 + w_{33}^2 a_3^1 + w_{34}^2 a_4^1 + b_3^2 \end{pmatrix}$$

and then pull out the biases as a vector sum:

$$\begin{pmatrix} z_1^2 \\ z_2^2 \\ z_3^2 \end{pmatrix} = \begin{pmatrix} w_{11}^2 a_1^1 + w_{12}^2 a_2^1 + w_{13}^2 a_3^1 + w_{14}^2 a_4^1 \\ w_{21}^2 a_1^1 + w_{22}^2 a_2^1 + w_{23}^2 a_3^1 + w_{24}^2 a_4^1 \\ w_{31}^2 a_1^1 + w_{32}^2 a_2^1 + w_{33}^2 a_3^1 + w_{34}^2 a_4^1 \end{pmatrix} + \begin{pmatrix} b_1^2 \\ b_2^2 \\ b_3^2 \end{pmatrix}$$

This is just a 3D vector addition. Even though the big vector in the middle looks like a larger matrix, it is just a column of three sums. This big vector, however, can be expanded into a matrix multiplication as follows:

$$\begin{pmatrix} z_1^2 \\ z_2^2 \\ z_3^2 \end{pmatrix} = \begin{pmatrix} w_{11}^2 & w_{12}^2 & w_{13}^2 & w_{14}^2 \\ w_{21}^2 & w_{22}^2 & w_{23}^2 & w_{24}^2 \\ w_{31}^2 & w_{32}^2 & w_{33}^2 & w_{34}^2 \end{pmatrix} \begin{pmatrix} a_1^1 \\ a_2^1 \\ a_3^1 \\ a_4^1 \end{pmatrix} + \begin{pmatrix} b_1^2 \\ b_2^2 \\ b_3^2 \end{pmatrix}$$

The activations in layer two are then obtained by applying σ to every entry of the resulting vector. This is nothing more than a notational simplification, but it is useful psychologically to pull out the numbers w_{ij}^l and b_j^l into their own matrices. These are the numbers that define the neural network itself, as opposed to the activations a_j^l that are the incremental steps in the evaluation.

To see what I mean, you can compare evaluating a neural network to evaluating the function $f(x) = ax + b$. The input variable is x, and by contrast, a and b are the constants that define the function; the space of possible linear functions is defined by the choice of a and b. The quantity ax, even if we relabeled it something like q, is merely an incremental step in the calculation of $f(x)$. The analogy is that, once you've decided the number of neurons per layer in your MLP, the matrices of weights and vectors of biases for each layer are really the data defining the neural network. With that in mind, we can implement the MLP in Python.

16.3.5 Exercises

Exercise 16.4 What neuron and layer is represented by the activation a_2^3? What value does this activation have in the following image? (Neurons and layers are indexed as throughout the previous sections.)

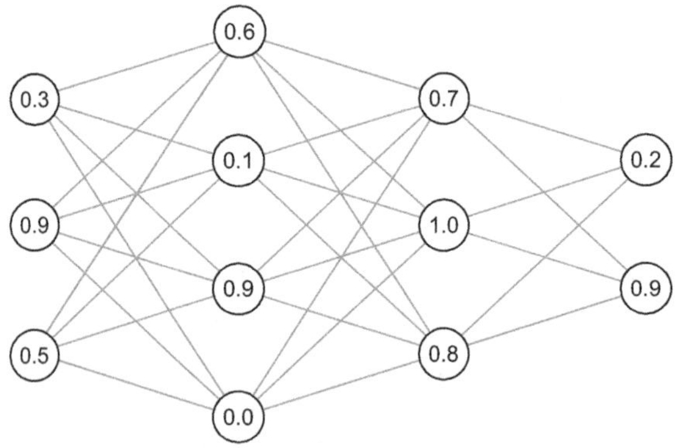

Solution The superscript indicates the layer, and the subscript indicates the neuron within the layer. The activation a_2^3, therefore, corresponds to the second neuron in layer 3. In the image, it has an activation value of 0.9.

Exercise 16.5 If layer 5 of a neural network has 10 neurons and layer 6 has 12 neurons, how many total connections are there between neurons in layers 5 and 6?

Solution Each of the 10 neurons in layer 5 is connected to each of the 12 neurons in layer 6. That's 120 total connections.

Exercise 16.6 Suppose we have an MLP with 12 layers. What are the indices *l*, *i*, and *j* of the weight w^l_{ij}, connecting the third neuron of layer 4 to the seventh neuron of layer 5?

Solution Remember that *l* is the destination layer of the connection, so $l = 5$ in this case. The indices *i* and *j* refer to the neurons in layers *l* and $l - 1$, respectively, so $i = 7$ and $j = 3$. The weight is labeled w^5_{73}.

Exercise 16.7 Where is the weight w^3_{31} in the network used throughout the section?

Solution There is no such weight. This would connect to a third neuron in layer three, the output layer, but there are only two neurons in this layer.

Exercise 16.8 In the neural network from this section, what's a formula for a^3_1 in terms of the activations of layer 2 and the weights and biases?

Solution The previous layer activations are a^2_1, a^2_2, and a^3_2, and the weights connecting them to a^3_1 are w^3_{11}, w^3_{12}, and w^3_{13}. The bias for activation a^3_1 is denoted b^3_1, so the formula is as follows:

$$a^3_1 = \sigma(w^3_{11}a^2_1 + w^3_{12}a^2_2 + w^3_{13}a^2_3 + b^3_1)$$

Exercise 16.9—Mini Project Write a Python function `sketch_mlp(*layer_sizes)` that takes layer sizes of a neural network and outputs a diagram like the ones used throughout this section. Show all of the neurons with labels and draw their connections with straight lines. Calling `sketch_mlp(3,4,3,2)` should produce the example from the diagram we have used to represent the neural net throughout.

Solution See the source code for this book for an implementation.

16.4 Building a neural network in Python

In this section, I show you how to take the procedure for evaluating an MLP that I explained in the last section and implement it in Python. Specifically, we'll implement a Python class called `MLP` that stores weights and biases (randomly generated at first), and provides an `evaluate` method that takes a 64-dimensional input vector and returns the output 10-dimensional vector. This code is a somewhat rote translation of the MLP design I described in the last section into Python, but once we're done with the implementation, we can test it at the task of classifying handwritten digits.

As long as the weights and biases are randomly selected, it probably won't do better than the random classifier we built to start with. But once we have the structure of a neural network to predict for us, we can tune the weights and biases to make it more predictive. We'll turn to that problem in the next section.

16.4.1 Implementing an MLP class in Python

If we want our class to represent an MLP, we need to specify how many layers we want and how many neurons we want per layer. To initialize our MLP with the structure we want, our constructor can take a list of numbers, representing the number of neurons in each layer.

The data we need to evaluate the MLP are the weights and biases for every layer after the input layer. As we just covered, we can store the weights as a matrix (a NumPy array) and the biases as a vector (also a NumPy array). To start, we can use random values for all of the weights and biases, and then when we train the network, we can gradually replace these values with more meaningful ones.

Let's quickly review the dimensions of the weight matrices and bias vectors that we want. If we pick a layer with m neurons, and the previous layer has n neurons, then our weights describe the linear part of the transformation from an n-dimensional vector of activations to an m-dimensional vector of activations. That's described by an m-by-n matrix, in other words, one with m rows and n columns. To see this, we can return to the example from section 16.3, where the weights connecting a layer with four neurons to a layer with three neurons made up a 4-by-3 matrix as shown in figure 16.15.

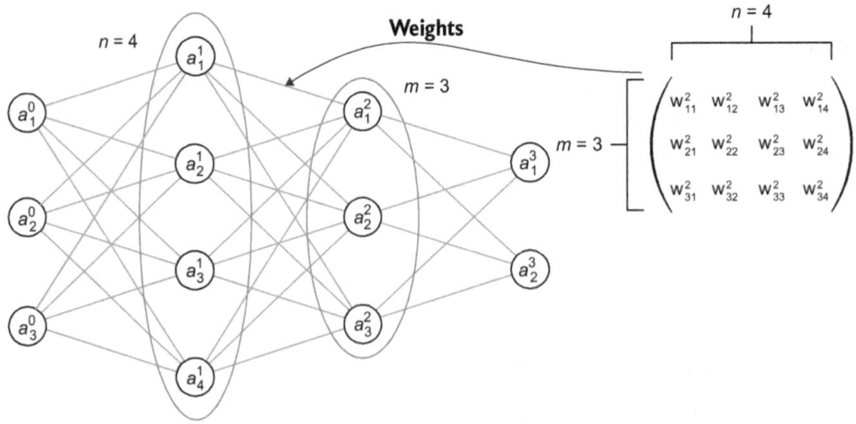

Figure 16.15 The weight matrix connecting a four neuron layer to a three neuron layer is a 3-by-4 matrix.

The biases for a layer of *m* neurons simply make up a vector with *m* entries, one for each neuron. Now that we've reminded ourselves how to find the size of the weight matrix and bias vector for each layer, we're ready to have our class constructor create them. Notice that we iterate over `layer_sizes[1:]`, which gives us the sizes of layers in the MLP, skipping the input layer which comes first:

```
class MLP():
    def __init__(self,layer_sizes):
        self.layer_sizes = layer_sizes
        self.weights = [
            np.random.rand(n,m)
            for m,n in zip(layer_sizes[:-1],
                           layer_sizes[1:])
        ]
        self.biases = [np.random.rand(n)
                       for n in layer_sizes[1:]]
```

Initializes the MLP with a list of layer sizes, giving the number of neurons for each layer

The weight matrices are n-by-m matrices with random entries ...

... where m and n are the number of neurons of adjacent layers in the neural network.

The bias for each layer (skipping the first) is a vector with one entry per neuron in the layer.

With this implemented, we can double-check that a two-layer MLP has exactly one weight matrix and one bias vector, and the dimensions match. Let's say the first layer has two neurons and the second layer has three neurons. Then we can run this code:

```
>>> nn = MLP([2,3])
>>> nn.weights
[array([[0.45390063, 0.02891635],
        [0.15418494, 0.70165829],
        [0.88135556, 0.50607624]])]
>>> nn.biases
[array([0.08668222, 0.35470513, 0.98076987])]
```

This confirms that we've a single 3-by-2 weight matrix and a single 3D bias vector, both populated with random entries.

The number of neurons in the input layer and output layer should match the dimensions of vectors we want to pass in and receive as output. Our problem of image classification calls for a 64D input vector and a 10D output vector. For this chapter, I stick with a 64-neuron input layer, a 10-neuron output layer, and a single 16-neuron layer in between. There is some combination of art and science to picking the right number of layers and layer sizes to get a neural network to perform well on a given task, and that's the kind of thing machine learning experts get paid big bucks for. For the purpose of this chapter, I say this structure is sufficient to get us a good, predictive model.

Our neural network can be initialized as `MLP([64,16,10])`, and it is much bigger than any of the ones we've drawn so far. Figure 16.16 shows what it looks like.

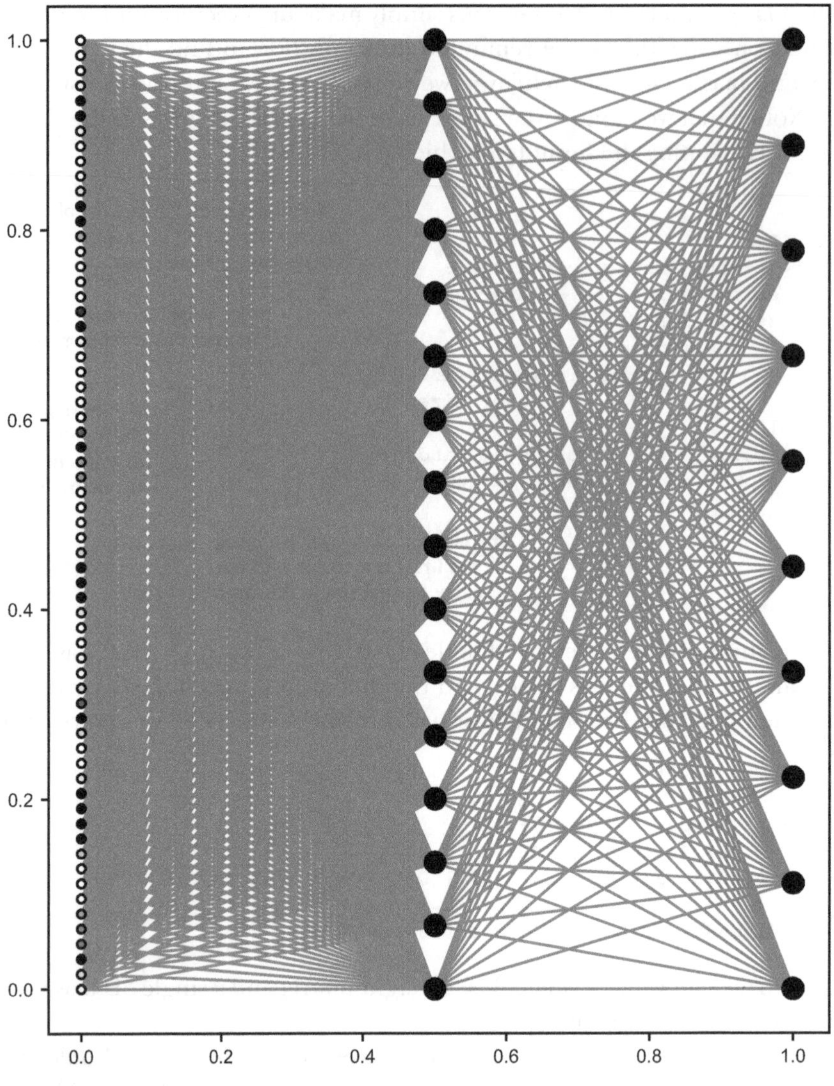

Figure 16.16 An MLP with three layers of 64, 16, and 10 neurons, respectively

Fortunately, once we implement our evaluation method, it's no harder for us to evaluate a big neural network than a small one. That's because Python does all of the work for us!

16.4.2 Evaluating the MLP

An evaluation method for our `MLP` class should take a 64D vector as input and return a 10D vector as output. The procedure to get from the input to the output is based on calculating the activations layer-by-layer from the input layer all the way to the output layer. As you'll see when we discuss backpropagation, it's useful to keep track of all of the activations as we go, even for the hidden layers in the middle of the network. For that reason, I'll build the `evaluate` function in two steps: first, I'll build a method to calculate all of the activations, and then I'll build another one to pull the last layer activation values and produce the results.

I call the first method `feedforward`, which is a common name for the procedure of calculating activations layer-by-layer. The input layer activations are given, and to get to the next layer, we need to multiply the vector of these activations by the weight matrix, add the next layer biases, and pass the coordinates of the result through the sigmoid function. We repeat this process until we get to the output layer. Here's what it looks like:

```
class MLP():
    ...
    def feedforward(self,v):
        activations = []
        a = v
        activations.append(a)
        for w,b in zip(self.weights, self.biases):
            z = w @ a + b
            a = [sigmoid(x) for x in z]
            activations.append(a)
        return activations
```

- `activations = []`: **Initializes with an empty list of activations**
- `activations.append(a)`: **The first layer activations are exactly the entries of the input vector; we append those to the list of activations.**
- `for w,b in zip(...)`: **Iterates over the layers with one weight matrix and bias vector per layer**
- `z = w @ a + b`: **The vector z is the matrix product of the weights with the previous layer activations plus the bias vector.**
- `a = [sigmoid(x) for x in z]`: **Takes the sigmoid function of every entry of z to get the activation**
- `activations.append(a)` (in the loop): **Adds the new computed activation vector to the list of activations**

The last layer activations are the results we want, so an evaluate method for the neural network simply runs the `feedforward` method for the input vector and then extracts the last activation vector like this:

```
class MLP():
    ...
    def evaluate(self,v):
        return np.array(self.feedforward(v)[-1])
```

That's it! You can see that the matrix multiplication saved us a lot of loops over neurons we'd otherwise be writing to calculate the activations.

16.4.3 Testing the classification performance of an MLP

With an appropriately sized MLP, it can now accept a vector for a digit image and output a result:

```
>>> nn = MLP([64,16,10])
>>> v = np.matrix.flatten(digits.images[0]) / 15.
>>> nn.evaluate(v)
array([0.99990572, 0.9987683 , 0.99994929, 0.99978464, 0.99989691,
       0.99983505, 0.99991699, 0.99931011, 0.99988506, 0.99939445])
```

That's passing in a 64-dimensional vector representing an image and returning a 10-dimensional vector as an output, so our neural network is behaving as a correctly shaped vector transformation. Because the weights and biases are random, these numbers should not be a good prediction of what digit the image is likely to be. (Incidentally, the numbers are all close to 1 because all of our weights, biases, and input numbers are positive, and the sigmoid sends big positive numbers to values close to 1.) Even so, there is a *biggest* entry in this output vector, which happens to be the number at index 2. This (incorrectly) predicts that image 0 in the data set represents the number 2.

The randomness suggests that our MLP only guesses 10% of the answers correctly. We can confirm this with a `test_digit_classify` function. For the random MLP I initialized, it gave exactly 10%:

```
>>> test_digit_classify(nn.evaluate)
0.1
```

This may not seem like much progress, but we can pat ourselves on the back for getting the classifier working, even if it's not good at its task. Evaluating a neural network is much more involved than evaluating a simple function like $f(x) = ax + b$, but we'll see the payoff soon as we *train* the neural network to classify images more accurately.

16.4.4 Exercises

> **Exercise 16.10—Mini Project** Rewrite the `feedforward` method using explicit loops over the layers and weights rather than using NumPy matrix multiplication. Confirm that your result matches exactly with the previous implementation.

16.5 Training a neural network using gradient descent

Training a neural network might sound like an abstract concept, but it just means finding the best weights and biases that makes the neural network do the task at hand as well as possible. We can't cover the whole algorithm here, but I show you how it works conceptually and how to use a third-party library to do it automatically. By the end of this section, we'll have adjusted the weights and biases of our neural network to predict which digit is represented by an image to a high degree of accuracy. We can then run it through `test_digit_classify` again and measure how well it does.

16.5.1 Framing training as a minimization problem

In the previous chapters for a linear function $ax + b$ or a logistic function $\sigma(ax + by + c)$, we created a cost function that measured the failure of the linear or logistic function, depending on the constants in the formula, to match the data exactly. The constants for the linear function were the slope and y-intercept a and b, so the cost function had the form $C(a, b)$. The logistic function had the constants a, b, and c (to be determined), so its cost function had the form $C(a, b, c)$. Internally, both of these cost functions depended on *all* of the training examples. To find the best parameters, we'll use gradient descent to minimize the cost function.

The big difference for an MLP is that its behavior can depend on hundreds or thousands of constants: all of its weights w_{ij}^l and biases b_j^l for every layer l and valid neuron indices i and j. Our neural network with 64, 16, and 10 neurons and its three layers have $64 \cdot 16 = 1{,}024$ weights between the first two layers and $16 \cdot 10 = 160$ weights between the second two. It has 16 biases in the hidden layer and 10 biases in the output layer. All in all, that's 1,210 constants we need to tune. You can picture our cost function as a function of these 1,210 values, which we need to minimize. If we write it out, it would look something like this:

$$C(w_{11}^1, w_{12}^1, \ldots, b_1^1, b_2^1, \ldots)$$

In the equation, where I've written the ellipses, there are over a thousand more weights and 24 more biases I didn't write out. It's worth thinking briefly about how to create the cost function, and as a mini-project, you can try implementing it yourself.

Our neural network outputs vectors, but we consider the answer to the classification problem to be the digit represented by the image. To resolve this, we can think of the correct answer as the 10-dimensional vector that a perfect classifier would have as output. For instance, if an image clearly represents the digit 5, we would like to see 100% certainty that the image is a 5 and 0% certainty that the image is any other digit. That means a 1 in the fifth index and 0's elsewhere (figure 16.17).

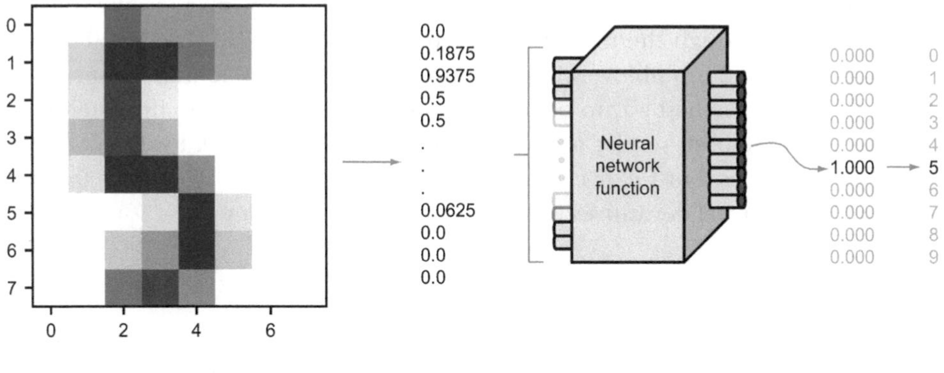

Figure 16.17 Ideal output from a neural network: 1.0 in the correct index and 0.0 elsewhere

Just as our previous attempts at regression never fit the data exactly, neither will our neural network. To measure the error from our 10-dimensional output vector to the ideal output vector, we can use the square of their distance in 10 dimensions.

Suppose the ideal output is written $y = (y_1, y_1, y_2, ..., y_{10})$. Note that I'm following the math convention for indexing from 1 rather than the Python convention of indexing from 0. That's actually the same convention I used for neurons within a layer, so the output layer (layer two) activations are indexed $(a_1^2, a_2^2, a_3^2, ..., a_{10}^1)$. The squared distance between these vectors is the sum:

$$(y_1 - a_1^2)^2 + (y_2 - a_2^2)^2 + (y_3 - a_3^2)^2 + \cdots + (y_{10} - a_{10}^2)^2$$

As another potentially confusing point, the superscript 2 above the *a* values indicates the output layer is layer two in our network, while the 2 outside the parentheses means squaring the quantity. To get a total cost relative to the data set, you can evaluate the neural network for all of the sample images and take the average squared distance. At the end of the section, you can try the mini-project for implementing this yourself in Python.

16.5.2 Calculating gradients with backpropagation

With the cost function $C(w_{11}^1, w_{12}^1, ..., b_1^1, b_2^1, ...)$ coded in Python, we could write a 1,210-dimensional version of gradient descent. This would mean taking 1,210 partial derivatives in each step to get a gradient. That gradient would be the 1,210-dimensional vector of the partial derivatives at the point, having this form

$$\nabla C(w_{11}^1, w_{12}^1, \ldots, b_1^1, b_2^1, \ldots) = \left(\frac{\partial C}{\partial w_{11}^1}, \frac{\partial C}{\partial w_{12}^1}, \ldots, \frac{\partial C}{\partial b_1^1}, \frac{\partial C}{\partial b_2^1}, \ldots \right)$$

Estimating so many partial derivatives would be computationally expensive because each would require evaluating *C* twice to test the effect of tweaking one of its input variables. In turn, evaluating *C* requires looking at every image in the training set and passing it through the network. It might be possible to do this, but the computation time would be prohibitively long for most real-world problems like ours.

Instead, the best way to calculate the partial derivatives is to find their exact formulas using methods similar to those we covered in chapter 10. I won't completely cover how to do this, but I'll give you a teaser in the last section. The key is that while there are 1,210 partial derivatives to take, they all have the form:

$$\frac{\partial C}{w_{ij}^l} \quad \text{or} \quad \frac{\partial C}{b_j^l}$$

for some choice of indices *l*, *i*, and *j*. The algorithm of *backpropagation* calculates all of these partial derivatives recursively, working backward from the output layer weights and biases, all the way to layer one.

If you're interested in learning more about backpropagation, stay tuned for the last section of the chapter. For now, I'll turn to the scikit-learn library to calculate costs, carry out backpropagation, and complete the gradient descent automatically.

16.5.3 Automatic training with scikit-learn

We don't need any new concepts to train an MLP with scikit-learn; we just need to tell it to set up the problem the same way we have and then find the answer. I won't explain everything the scikit-learn library can do, but I will step you through the code to train the MLP for digit classification.

The first step is to put all of our training data (in this case, the digit images as 64-dimensional vectors) into a single NumPy array. Using the first 1,000 images in the data set gives us a 1,000-by-64 matrix. We'll also put the first 1,000 answers in an output list:

```
x = np.array([np.matrix.flatten(img) for img in digits.images[:1000]]) / 15.0
y = digits.target[:1000]
```

Next, we use the `MLP` class that comes with scikit-learn to initialize an MLP. The sizes of the input and output layers are determined by the data, so we only need to specify the size of our single hidden layer in the middle. Additionally, we include parameters telling the MLP how we want it to be trained. Here's the code:

```
from sklearn.neural_network import MLPClassifier

mlp = MLPClassifier(hidden_layer_sizes=(16,),
                    activation='logistic',
                    max_iter=100,
                    verbose=10,
                    random_state=1,
                    learning_rate_init=.1)
```

Specifies that we want a hidden layer with 16 neurons (`hidden_layer_sizes=(16,)`)

Specifies that we want to use logistic (ordinary sigmoid) activation functions in the network (`activation='logistic'`)

Sets the maximum number of gradient descent steps to take in case there are convergence issues (`max_iter=100`)

Selects that the training process provides verbose logs (`verbose=10`)

Initializes the MLP with random weights and biases (`random_state=1`)

Decides the learning rate or what multiple of the gradient to move in each iteration of the gradient descent (`learning_rate_init=.1`)

Once this is done, we can train the neural network to the input data *x* and corresponding output data *y* in one line:

```
mlp.fit(x,y)
```

When you run this line of code, you'll see a bunch of text print in the terminal window as the neural network trains. This logging shows how many gradient descent steps it takes and the value of the cost function, which scikit-learn calls "loss" instead of "cost."

```
Iteration 1, loss = 2.21958598
Iteration 2, loss = 1.56912978
Iteration 3, loss = 0.98970277
...
Iteration 58, loss = 0.00336792
Iteration 59, loss = 0.00330330
Iteration 60, loss = 0.00321734
Training loss did not improve more than tol=0.000100 for two consecutive
epochs. Stopping.
```

At this point, after 60 iterations of gradient descent, a minimum has been found and the MLP is trained. You can test it on image vectors using the `_predict` method. This method takes an array of inputs, meaning an array of 64-dimensional vectors, and returns the output vectors for all of them. For instance, `mlp._predict(x)` gives the 10-dimensional output vectors for all 1,000 image vectors stored in *x*. The result for the zeroth training example is the zeroth entry of the result:

```
>>> mlp._predict(x)[0]
array([9.99766643e-01, 8.43331208e-11, 3.47867059e-06, 1.49956270e-07,
       1.88677660e-06, 3.44652605e-05, 6.23829017e-06, 1.09043503e-04,
       1.11195821e-07, 7.79837557e-05])
```

It takes some squinting at these numbers in scientific notation, but the first one is 0.9998, while the others are all less than 0.001. This correctly predicts that the zeroth training example is a picture of the digit 0. So far so good!

With a small wrapper, we can write a function that uses this MLP to do *one* prediction, taking a 64-dimensional image vector and outputting a 10-dimensional result. Because scikit-learn's MLP works on collections of input vectors and produces arrays of results, we just need to put our input vector in a list before passing it to `mlp._predict`:

```
def sklearn_trained_classify(v):
    return mlp._predict([v])[0]
```

At this point, the vector has the correct shape to have its performance tested by our `test_digit_classify` function. Let's see what percentage of the test digit images it correctly identifies:

```
>>> test_digit_classify(sklearn_trained_classify)
1.0
```

That's an astonishing 100% accuracy! You might be skeptical of this result; after all, we're testing on the same data set that the neural network used to train. In theory, when storing 1,210 numbers, the neural net could have just memorized every example in the training set. If you test the images the neural network hasn't seen before, you'll see this isn't the case; it still does an impressive job classifying the images correctly as digits. I found that it had 96.2% accuracy on the next 500 images in the data set, and you can test this yourself in an exercise.

16.5.4 Exercises

Exercise 16.11 Modify the `test_digit_classify` function to work on a custom range of examples in the test set. How does it do on the next 500 examples after the 1,000 training examples?

Solution Here I've added a `start` keyword argument to indicate which test example to start with. The `test_count` keyword argument still indicates the number of examples to test:

```
def test_digit_classify(classifier,start=0,test_count=1000):
    correct = 0
    end = start + test_count                                    <-- Calculates the end index for test data we want to consider
    for img, target in zip(digits.images[start:end],
digits.target[start:end]):                                      <-- Loops only over the test data between the start and end indices
        v = np.matrix.flatten(img) / 15
        output = classifier(v)
        answer = list(output).index(max(output))
        if answer == target:
            correct += 1
    return (correct/test_count)
```

My trained MLP identifies 96.2% of these fresh digit images correctly:

```
>>>
test_digit_classify(sklearn_trained_classify,start=1000,test_count=500)
0.962
```

Exercise 16.12 Using the squared distance cost function, what is the cost of your randomly generated MLP for the first 1,000 training examples? What is the cost of the scikit-learn MLP?

Solution First, we can write a function to give us the ideal output vector for a given digit. For instance, for the digit 5, we'd like an output vector y, which is all zeros except for a one in the fifth index.

```
def y_vec(digit):
    return np.array([1 if i == digit else 0 for i in range(0,10)])
```

The cost of one test example is the sum of squared distance from what the classifier outputs to the ideal result. That's the sum of squared differences in the coordinates added up:

```
def cost_one(classifier,x,i):
    return sum([(classifier(x)[j] - y_vec(i)[j])**2 for j in range(10)])
```

The total cost for a classifier is the average cost over all of the 1,000 training examples:

```
def total_cost(classifier):
    return sum([cost_one(classifier,x[j],y[j]) for j in range(1000)])/1000.
```

(continued)

As expected, a randomly initialized MLP with only 10% predictive accuracy has a much higher cost than a 100% accurate MLP produced by scikit-learn:

```
>>> total_cost(nn.evaluate)
8.995371023185067
>>> total_cost(sklearn_trained_classify)
5.670512721637246e-05
```

Exercise 16.13—Mini Project Extract the `MLPClassifier` weights and biases using its properties `coefs_` and `intercepts_`, respectively. Plug these weights and biases into the `MLP` class we built from scratch earlier in this chapter and show that your resulting MLP performs well on digit classification.

Solution If you try this, you'll notice one problem; where we expect the weight matrices to be 16-by-64 and 10-by-16, the `coefs_` property of `MLPClassifier` gives a 64-by-16 matrix and a 16-by-10 matrix. It looks like scikit-learn uses a convention that stores columns of the weight matrices versus our convention that stores rows. There's a quick way to fix this.

NumPy arrays have a `T` property returning the *transpose* of a matrix (a matrix obtained by pivoting the matrix so that the rows become the columns of the result). With this trick in mind, we can plug the weights and biases into our neural network and test it:

```
>>> nn = MLP([64,16,10])
>>> nn.weights = [w.T for w in mlp.coefs_]
>>> nn.biases = mlp.intercepts_
>>> test_digit_classify(nn.evaluate,
                        start=1000,
                        test_count=500) 0.962
```

Sets our weight matrices to the ones from the scikit-learn MLP, after transposing them to agree with our convention

Sets our network's biases to the ones from the scikit-learn MLP

Tests the performance of our neural network at the classification task with new weights and biases

This is 96.2% accurate on the 500 images after the training data set, just like the MLP produced by scikit-learn directly.

16.6 Calculating gradients with backpropagation

This section is completely optional. Frankly, because you know how to train an MLP using scikit-learn, you're ready to solve real-world problems. You can test neural networks of different shapes and sizes on classification problems and experiment with their design to improve classification performance. Because this is the last section in the book, I wanted to give you some final, challenging (but doable!) math to chew on—calculating partial derivatives of the cost function by hand.

The process of calculating partial derivatives of an MLP is called *backpropagation* because it's efficient to start with the weights and biases of the last layer and work backwards. Backpropagation can be broken into four steps: calculating the derivatives with respect to the last layer weights, last layer biases, hidden layer weights, and hidden layer biases. I'll show you how to get the partial derivatives with respect to the weights in the last layer, and you can try running with this approach to do the rest.

16.6.1 Finding the cost in terms of the last layer weights

Let's call the index of the last layer of the MLP L. That means that the last weight matrix consists of the weights w_{ij}^L, where $l = L$, in other words, the weights w_{ij}^L. The biases in this layer are b_j^L and the activations are labeled a_j^L.

The formula to get the jth neuron's activation in the last layer a_j^L is a sum of the contribution from every neuron in layer $L - l$, indexed by i. In a made-up notation, it becomes

$$a_j^L = \sigma(b_j^L + \text{sum of } [w_{ij}^L a_i^{L-1}] \text{ for every value of } i)$$

The sum is taken over all values of i from one to the number of neurons in layer $L - l$. Let's write the number of neurons in layer l as n_i, with i ranging from l to n_{L-1} in our sum. In proper mathematical summation notation, this sum is written:

$$\sum_{i=1}^{n_L} w_{ij}^L a_i^{L-1}$$

The English translation of this formula is "fixing values of L and j by adding up the values of the expression $w_{ij}^L a_i^{L-1}$ for every i from one to n_L." This is nothing more than the formula for matrix multiplication written as a sum. In this form, the activation is as follows:

$$a_j^L = \sigma\left(b_j^L + \sum_{i=1}^{n_{L-1}} w_{ij}^L a_i^{L-1}\right)$$

Given an actual training example, we can have some ideal output vector **y** with a 1 in the correct slot and 0's elsewhere. The cost is the squared distance between the activation vector a_j^L and the ideal output values y_j. That is,

$$C = \sum_{j=1}^{n_L} (a_j^L - y_j)$$

The impact of a weight w_{ij}^L on C is indirect. First, it is multiplied by an activation from the previous layer, added to the bias, passed through a sigmoid, and then passed through the quadratic cost function. Fortunately, we covered how to take derivatives of compositions of functions in chapter 10. This example is a bit more complicated, but you should be able to recognize it as the same chain rule we saw before.

16.6.2 Calculating the partial derivatives for the last layer weights using the chain rule

Let's break it down into three steps to get from w_{ij}^L to C. First, we can calculate the value to be passed into the sigmoid, which we called z_j^L earlier in the chapter:

$$z_j^L = b_j^L + \sum_{i=1}^{n_{L-1}} w_{ij}^L a_i^{L-1}$$

Then we can pass z_j^L into the sigmoid function to get the activation a_j^L:

$$a_j^L = \sigma(z_j^L)$$

And finally, we can compute the cost:

$$C = \sum_{j=1}^{n_L} (a_j^L - y_j)^2$$

To find the partial derivative of C with respect to w_{ij}^L, we multiply the derivatives of these three "composed" expressions together. The derivative of z_j^L with respect to *one* particular w_{ij}^L is the specific activation a_j^{L-1} that it's multiplied by. This is similar to the derivative of $y(x) = ax$ with respect to x, which is the constant a. The partial derivative is

$$\frac{\partial z_j^L}{\partial w_{ij}^L} = a_i^{L-1}$$

The next step is applying the sigmoid function, so the derivative of a_j^L with respect to z_j^L is the derivative of σ. It turns out, and you can confirm this as an exercise, that the derivative of $\sigma(x)$ is $\sigma(x)(1 - \sigma(x))$. This nice formula follows in part from the fact that e^x is its own derivative. That gives us

$$\frac{da_j^L}{dz_j^L} = \sigma'(z_j^L) = \sigma(z_j^L)(1 - \sigma(z_j^L))$$

This is an ordinary derivative, not a partial derivative, because a_j^L is a function of only one input: z_j^L. Finally, we need the derivative of C with respect to a_j^L. Only one term of the sum depends on w_{ij}^L, so we just need the derivative of $(a_j^L - y_j)^2$ with respect to a_j^L. In this context, y_j is a constant, so the derivative is $2a_j^L$. This comes from the power rule, telling us that if $f(x) = x^2$, then $f'(x) = 2x$. For our last derivative, we need

$$\frac{\partial C}{\partial a_j^L} = 2(a_j^L - y_j)$$

The multivariable version of the chain rule says the following:

$$\frac{\partial C}{\partial w_{ij}^L} = \frac{\partial C}{\partial a_j^L} \frac{da_j^L}{dz_j^L} \frac{\partial z_j^L}{w_{ij}^L}$$

This looks a little bit different from the version we saw in chapter 10, which covered only composition of two functions of one variable. The principle is the same here though: with C written in terms of a_j^L, a_j^L written in terms of z_j^L, and z_j^L written in terms of w_{ij}^L, we have C written in terms of w_{ij}^L. What the chain rule says is that to get the derivative of the whole chain, we multiply together the derivatives of each step. Plugging in the derivatives, the result is

$$\frac{\partial C}{\partial w_{ij}^L} = 2(a_j^L - y_j) \cdot \sigma(z_j^L)(1 - \sigma(z_j^L)) \cdot a_i^{L-1}$$

This formula is one of the four we need to find the whole gradient of C. Specifically, this gives us the partial derivative for any weight in the last layer. There are 16 × 10 of these, so we've covered 160 of the 1,210 total partial derivatives we need to have the complete gradient.

The reason I'll stop here is because derivatives of other weights require more complicated applications of the chain rule. An activation influences every subsequent activation in the neural network, so every weight influences every subsequent activation. This isn't beyond your capabilities, but I feel I'd owe you a better explanation of the multivariable chain rule before digging in. If you're interested in going deeper, there are excellent resources online that walk through all the steps of backpropagation in gory detail. Otherwise, you can stay tuned for the (fingers crossed) sequel to this book. Thanks for reading!

16.6.3 Exercises

Exercise 16.14—Mini Project Use SymPy or your own code from chapter 10 to automatically find the derivative of the sigmoid function

$$\sigma(x) = \frac{1}{1 + e^{-x}}$$

Show that the answer you get is equal to $\sigma(x)(1 - \sigma(x))$.

(continued)

Solution In SymPy, we can quickly get a formula for the derivative:

```
>>> from sympy import *
>>> X = symbols('x')
>>> diff(1 / (1+exp(-X)),X)
exp(-x)/(1 + exp(-x))**2
```

In math notation, that's

$$\frac{e^{-x}}{(1+e^{-x})^2} = \frac{e^{-x}}{1+e^{-x}} \cdot \frac{1}{1+e^{-x}} == \frac{e^{-x}}{1+e^{-x}} \cdot \sigma(x)$$

The computation to show this expression equals $\sigma(x)(1 - \sigma(x))$ and requires a bit of rote algebra, but it's worth it to convince yourself that this formula is valid. Multiplying the top and bottom by e^x and noting that $e^x \cdot e^{-x} = 1$, we get

$$\begin{aligned}
\frac{e^{-x}}{(1+e^{-x})^2} &= \frac{1}{e^x+1} \cdot \sigma(x) \\
&= \frac{e^{-x}}{e^{-x}} \cdot \frac{1}{e^x+1} \cdot \sigma(x) \\
&= \frac{e^{-x}}{1+e^{-x}} \cdot \sigma(x) \\
&= \left(\frac{1+e^{-x}}{1+e^{-x}} - \frac{1}{1+e^{-x}}\right) \cdot \sigma(x) \\
&= \left(1 - \frac{1}{1+e^{-x}}\right) \cdot \sigma(x) \\
&= (1-\sigma(x)) \cdot \sigma(x)
\end{aligned}$$

Summary

- An artificial neural network is a mathematical function whose computation mirrors the flow of signals in the human brain. As a function, it takes a vector as input and returns another vector as output.
- A neural network can be used to classify vector data: for instance, images converted to vectors of grayscale pixel values. The output of the neural network is a vector of numbers that express confidence that the input vector should be classified in any of the possible classes.
- A multilayer perceptron (MLP) is a particular kind of artificial neural network consisting of several ordered layers of neurons, where the neurons in each layer

are connected to and influenced by the neurons in the previous layer. During evaluation of the neural network, each neuron gets a number that is its activation. You can think of an activation as an intermediate yes-or-no answer on the way to solving the classification problem.

- To evaluate a neural network, the activations of the first layer of neurons are set to the entries of the input vector. Each subsequent layer of activations is calculated as a function of the previous layer. The final layer of activations is treated as a vector and returned as the result vector of the calculation.
- The activation of a neuron is based on a linear combination of the activations of all neurons in the previous layer. The coefficients in the linear combination are called *weights*. Each neuron also has a *bias*, a number which is added to the linear combination. This value is passed through a sigmoid function to get the activation function.
- Training a neural network means tuning the values of all of the weights and biases so that it performs its task optimally. To do this, you can measure the error of the neural network's predictions relative to actual answers from a training data set with a cost function. With a fixed training data set, the cost function depends only on the weights and biases.
- Gradient descent allows us to find the values of weights and biases that minimize the cost function and yield the best neural network.
- Neural networks can be trained efficiently because there are simple, exact formulas for the partial derivatives of the cost function with respect to the weights and biases. These are found using an algorithm called *backpropagation*, which in turn makes use of the chain rule from calculus.
- Python's scikit-learn library has a built in `MLPClassifer` class that can automatically be trained on classified vector data.

appendix A
Getting set up with Python

This appendix covers the basic steps to get Python and related tools installed so you can run the code examples in this book. The main thing to install is Anaconda, which is a popular Python distribution for mathematical programming and data science. Specifically, Anaconda comes with an interpreter that runs Python code, as well as a number of the most popular math and data science libraries and a coding interface called Jupyter. The steps are mostly the same on any Linux, Mac, or Windows machine. I'm going to show you the steps on my Mac.

A.1 Checking for an existing Python installation

It's possible you already have Python installed on your computer, even if you didn't realize it. To check the existing installation, open a terminal window (or CMD or PowerShell on Windows) and type `python`. On a Mac straight from the factory, you should see a Python 2.7 terminal appear. You can press Ctrl-D to exit the terminal.

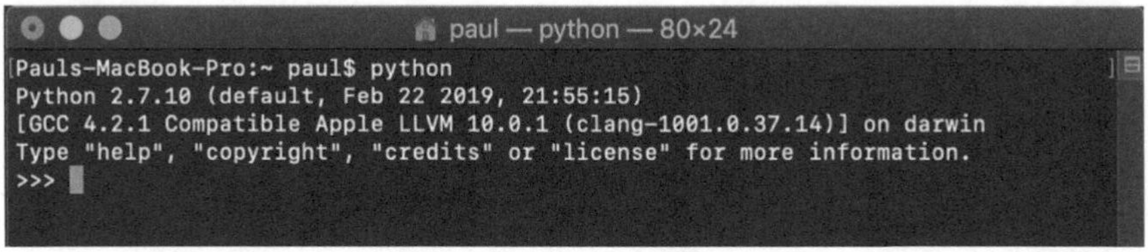

For the examples in this book, I use Python 3, which is becoming the new standard, and specifically, I use the Anaconda distribution. As a warning, if you have an existing Python installation, the following steps might be tricky. If any of the instructions below don't work for you, my best advice is to search Google or StackOverflow with any error messages you see.

If you are a Python expert and don't want to install or use Anaconda, you should be able to find and install the relevant libraries like NumPy, Matplotlib, and Jupyter using the pip package manager. For a beginner, I strongly recommend installing Anaconda as follows.

A.2 Downloading and installing Anaconda

Go to https://www.anaconda.com/distribution/ to download the Anaconda Python distribution. Click Download and choose the Python version beginning with 3 (figure A.1). At the time of writing, this was Python 3.7.

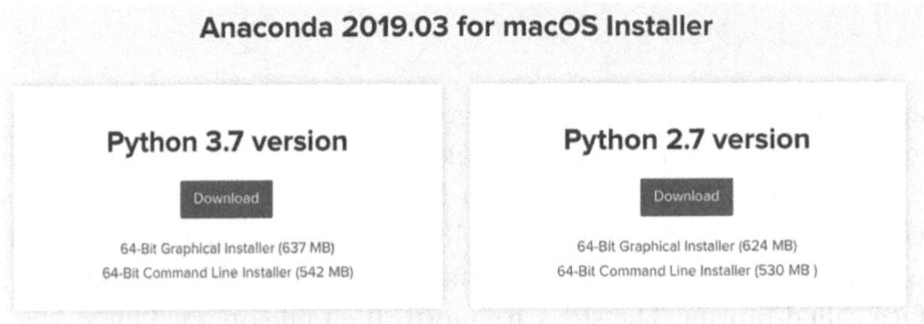

Figure A.1 At the time of writing, here's what I see after clicking Download on the Anaconda website. To install Python, choose the Python 3.x download link.

Open the installer. It walks you through the installation process. The installer dialog box looks different depending on your operating system, but figure A.2 shows what it looks like on my Mac.

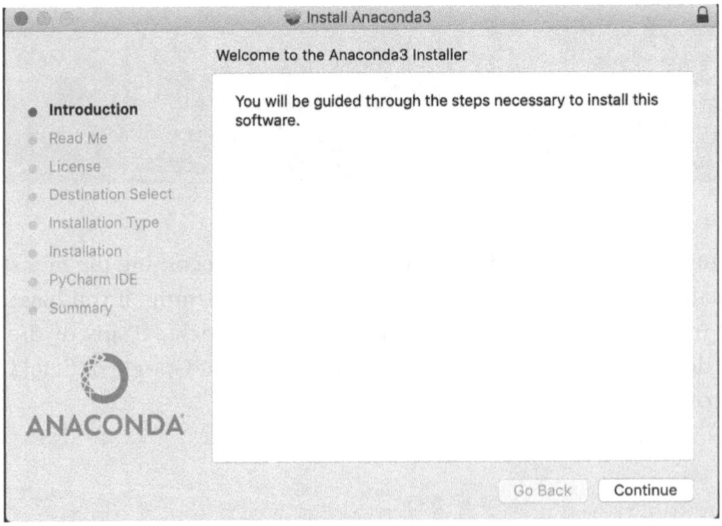

Figure A.2 The Anaconda installer as it appears on my Mac

I used the default installation location and didn't add any optional features like the PyCharm IDE. Once the installation is complete, you should be able to open a fresh terminal. Type `python` to enter a Python 3 session with Anaconda (figure A.3).

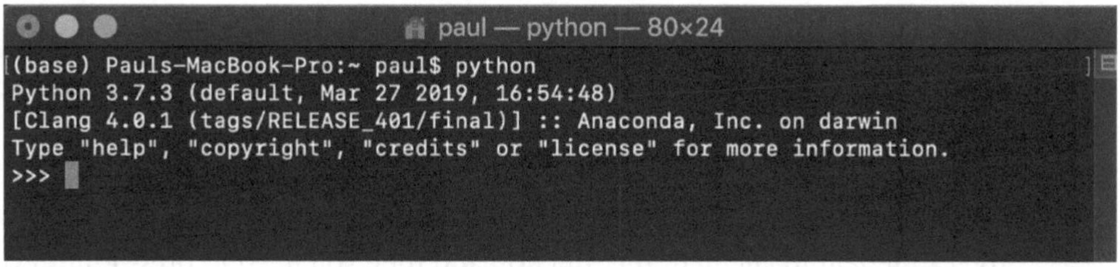

Figure A.3 How the Python interactive session should look once you install Anaconda. Notice the Python 3.7.3 and the Anaconda, Inc. labels that appear.

If you don't see a Python version starting with the number 3 and the word *Anaconda*, it probably means you're stuck on the previous Python installation on your system. You need to edit your PATH environment variables so that your terminal knows which Python version you want when you type `python` in a terminal window. Hopefully, you won't run into this problem, but if so, you can search online for a fix. You can also type `python3` instead of `python` to explicitly use your Python 3 installation.

A.3 Using Python in interactive mode

Three angle brackets (`>>>`) in the terminal window prompt you to enter a line of Python code. When you type `2+2` and press Enter, you should see the Python interpreter's evaluation of this statement, which is `4` (figure A.4).

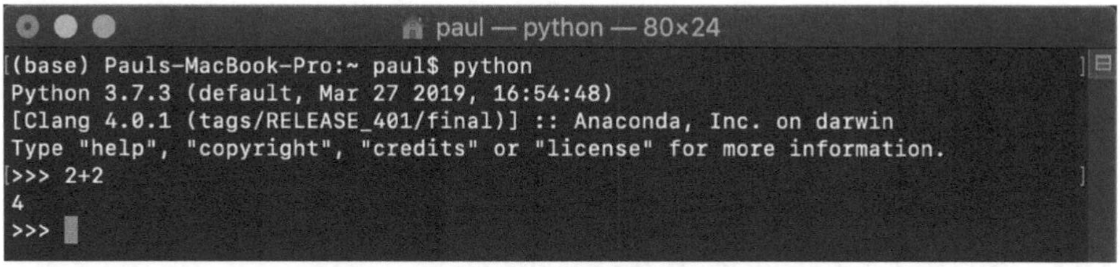

Figure A.4 Entering a line of Python in the interactive session

Interactive mode is also referred to as a REPL, standing for read-evaluate-print loop. The Python session reads in a line of typed code, evaluates it, and prints the result, and this process is repeated as many times as you want in a loop. Pressing Ctrl-D signifies that you're done entering code and sends you back to your terminal session.

Python interactive can usually identify if you enter a multi-line statement. For instance, `def f(x):` is the first line you enter when defining a new Python function called `f`. The Python interactive session shows you `...` to indicate that it expects more input (figure A.5).

```
>>> def f(x):
...
```

Figure A.5 The Python interpreter knows you're not done with your multi-line statement.

You can indent to augment the function and then you need to press Enter twice to let Python know that you've finished the multi-line code input and to implement the function (figure A.6).

```
>>> def f(x):
...     return x * x
...
>>>
```

Figure A.6 Once you've completed your multi-line code, you need to press Enter twice to submit it.

The function `f` is now defined in your interactive session. On the next line, you can give Python input to evaluate (figure A.7).

```
>>> f(5)
25
```

Figure A.7 Evaluating the function defined previously

Beware that any code you write in an interactive session disappears when you exit the session. For that reason, if you're writing a lot of code, it's best to put it in a script file or in a Jupyter notebook. I'll cover both of these methods next.

A.3.1 *Creating and running a Python script file*

You can create Python files with basically any text editor you like. As usual, it's better to use a text editor designed for programming rather than a rich-text editor like Microsoft Word, which could insert invisible or unwanted characters for formatting. My preference is Visual Studio Code, and other popular choices are Atom, which is cross-platform, and Notepad++ for Windows. At your own risk, you can use a terminal-based text editor like Emacs or Vim. All of these tools are free and readily downloadable.

To create a Python script, simply create a new text file in your editor with a .py filename extension. Figure A.8 shows that I've created a file called first.py, which is in my ~/Documents directory. You can also see in figure A.8 that my text editor, Visual Studio Code, comes with syntax highlighting for Python. Keywords, functions, and literal values are colored to make it easy to read the code. Many editors (including Visual Studio Code) have optional extensions you can install to give you more helpful tools like checking for simple errors as you type, for instance.

Figure A.8 shows a few lines of Python code typed into the first.py file. Because this is a math book, we can use an example that's more "mathy" than Hello World. When we run the code, it prints the squares of all digits from 0 to 9.

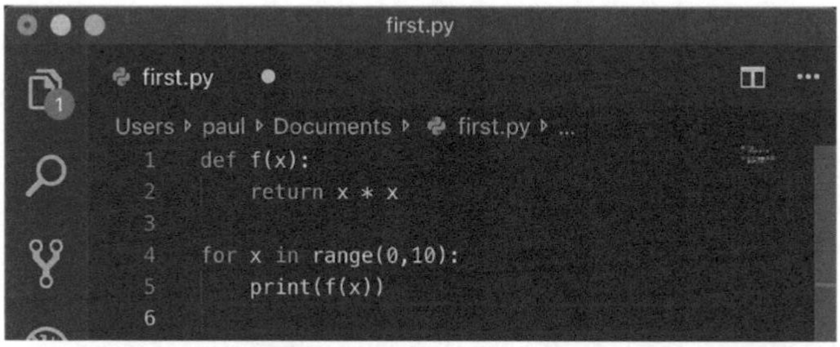

Figure A.8 Some example Python code in a file. This code prints the squares of all digits from 0 to 9.

Back in the terminal, go to the directory where your Python file lives. On my Mac, I go to the ~/Documents directory by typing `cd ~/Documents`. You can type `ls first.py` to confirm that you're in the same directory as your Python script (figure A.9).

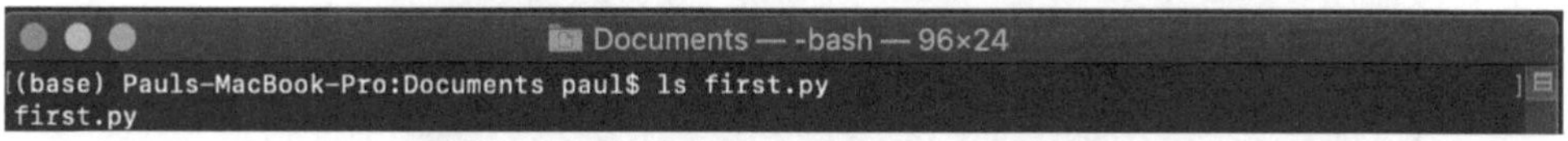

Figure A.9 The `ls` command shows you that the file first.py is in the directory.

To execute the script file, type `python first.py` into the terminal window. This invokes the Python interpreter and tells it to run the first.py file. The interpreter does what we hoped and prints some numbers (figure A.10).

Figure A.10 The result of running a simple Python script from the command line

When you're solving more complicated problems, you might want to break your code up into separate files. Next, I'll show you how to put the function `f(x)` in a different Python file that can be used by first.py. Let's call this new file function.py and save it in the same directory as first.py, then cut and paste the code for `f(x)` into it (figure A.11).

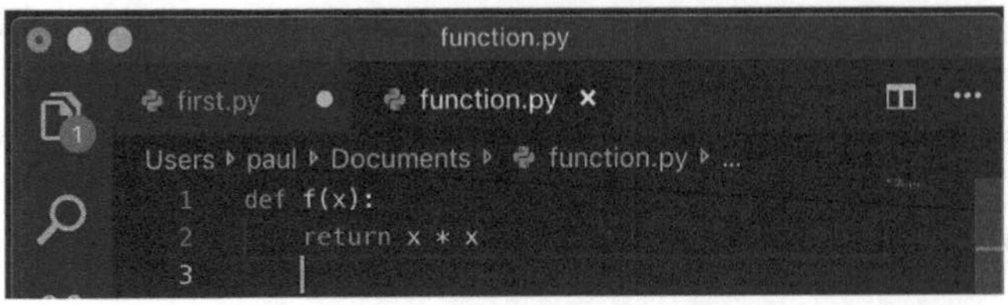

Figure A.11 Putting the code to define the function `f(x)` in its own Python file

To let Python know that you're going to combine multiple files in this directory, you need to add an empty text file called __init__.py in the directory. (That's two underscores before and after the word *init*.)

> **TIP** A quick way to create this empty file on a Mac or Linux machine is to type `touch __init__.py`.

To use the function `f(x)` from function.py in your script first.py, we need to let the Python interpreter know to retrieve it. To do that, we write `from function import f` as the first line in first.py (figure A.12).

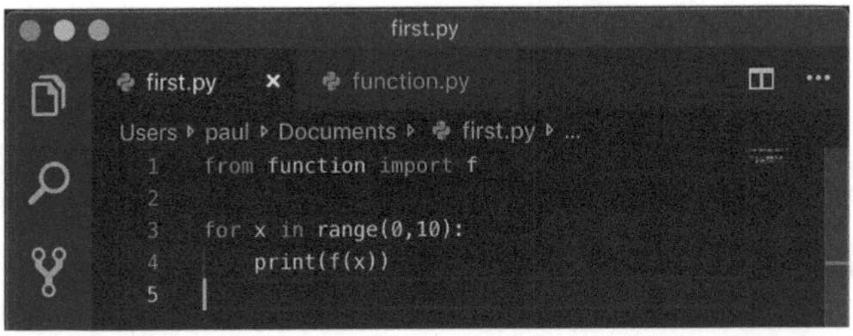

Figure A.12 Rewriting the file first.py to include the function `f(x)`

When you run the command `python first.py` again, you should get the same result as the previous time you ran it. This time, Python is getting the function *f* from function.py in the process.

An alternative to doing all of your work in text files and running them from the command line is to use Jupyter notebooks, which I'll cover next. For this book, I did most of my examples in Jupyter notebooks, but I wrote any reusable code in separate Python files and imported those files.

A.3.2 Using Jupyter notebooks

A Jupyter Notebook is a graphical interface for coding in Python (and other languages as well). As with a Python interactive session, you type lines of code into a Jupyter notebook, and it prints the result. The difference is that Jupyter is a prettier interface than your terminal, and you can save your sessions to resume or re-run later.

Jupyter Notebook should automatically install with Anaconda. If you're using a different Python distribution, you can install Jupyter with pip as well. See https://jupyter.org/install for documentation if you want to do a custom installation.

To open the Jupyter Notebook interface, type `jupyter notebook` or `python -m notebook` in a directory you want to work in. You should see a lot of text stream into the terminal, and your default web browser should open showing you the Jupyter Notebook interface.

Figure A.13 shows what I see in my terminal after typing `python -m notebook`. Again, your mileage may vary, depending on the Anaconda version you have.

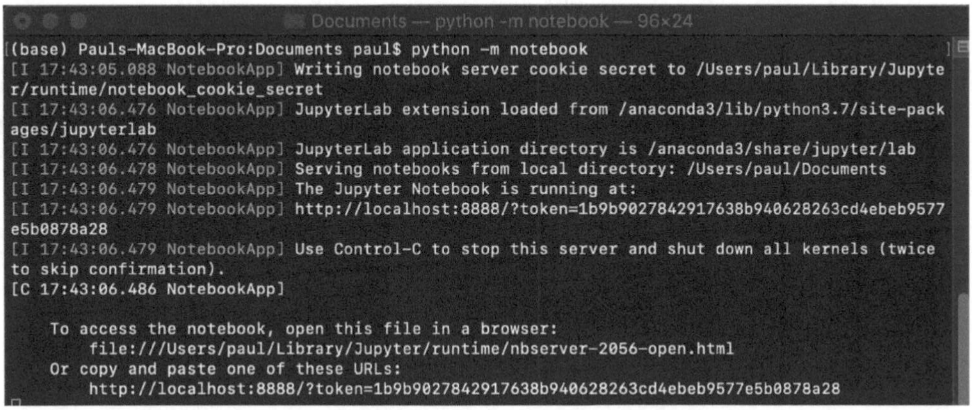

Figure A.13 What the terminal looks like when you open a Jupyter notebook

Your default web browser should open, showing the Jupyter interface. Figure A.14 shows what I see when Jupyter opens in the Google Chrome browser.

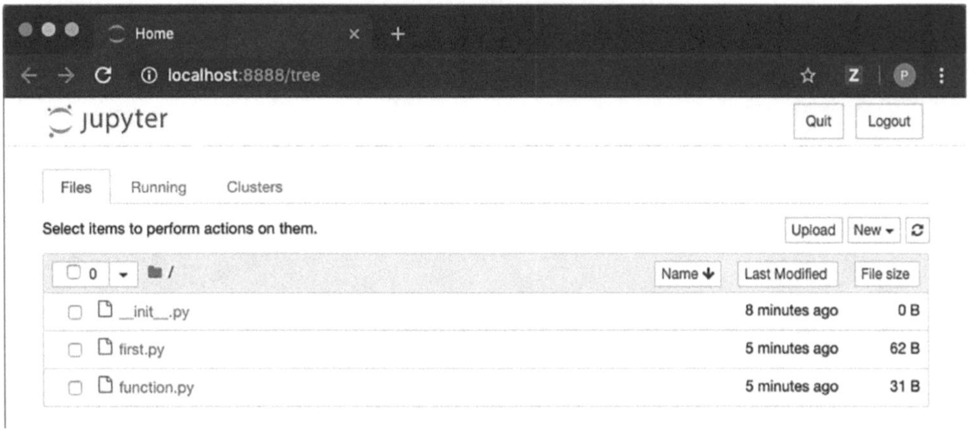

Figure A.14 When you start Jupyter, a browser tab automatically opens, looking something like this.

What's going on here is that the terminal is running Python behind the scenes and also serving a local website at the address localhost:8888. From here on, you only have to think about what's going on in the browser. The browser automatically sends the code you write to the Python process in the terminal via web requests. This Python background process is called a *kernel* in Jupyter terminology.

At the first screen that opens in the browser, you can see all of the files in the directory you're working in. For example, I opened the notebook in my ~/Documents folder, so I can see the Python files we wrote in the last section. If you click one of the files, you'll see you can view and edit it directly in the web browser. Figure A.15 shows what I see when I click first.py.

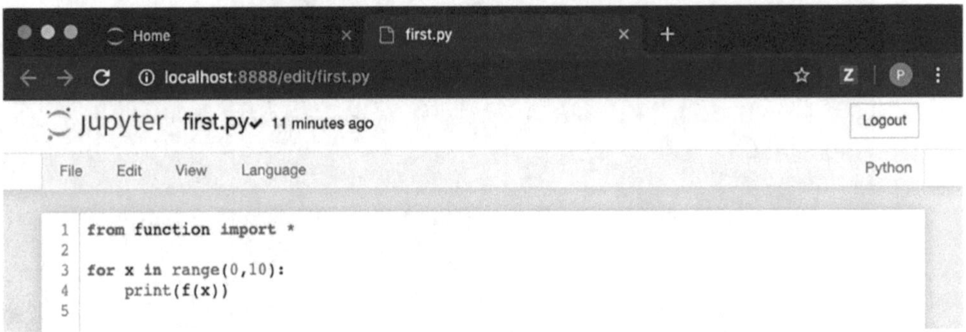

Figure A.15 Jupyter has a basic text editor for Python files. Here, I've opened the first.py file.

This isn't a notebook yet. A notebook is a different kind of file than an ordinary Python file. To create a notebook, return to the main view by clicking the Jupyter logo in the top left corner, then go to the New dropdown menu on the right, and click Python 3 (figure A.16).

Figure A.16 Selecting the menu option to create a new Python 3 notebook

Once you've clicked Python 3, you'll be taken to your new notebook. It should look like the one shown in figure A.17, with one blank input line ready to accept some Python code.

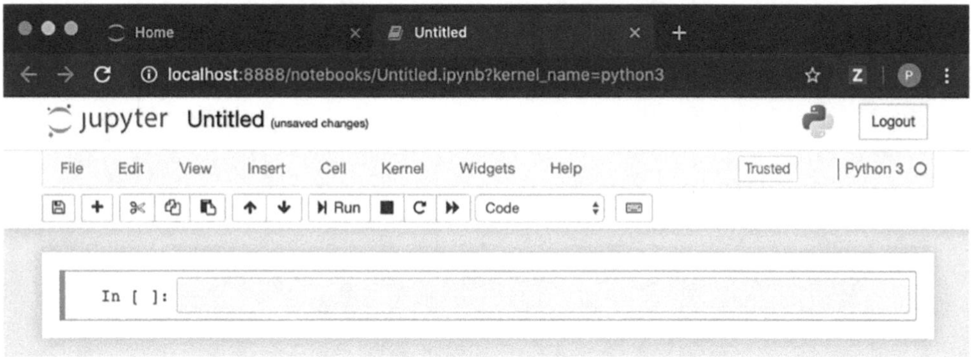

Figure A.17 A new, empty Jupyter notebook ready for coding

You can type a Python expression into the text box and then press Shift-Enter to evaluate it. In figure A.18, I typed `2+2` and then pressed Shift-Enter to see the output, `4`.

Figure A.18 Evaluating 2 + 2 in a Jupyter notebook

As you can see, it works just like an interactive session except it looks nicer. Each input is shown in a box, and the corresponding output is shown below it.

If you just press Enter rather than Shift-Enter, you can add a new line inside your input box. In the interface, variables and functions defined in boxes above can be used by boxes below. Figure A.19 shows what our original example could look like in a Jupyter notebook.

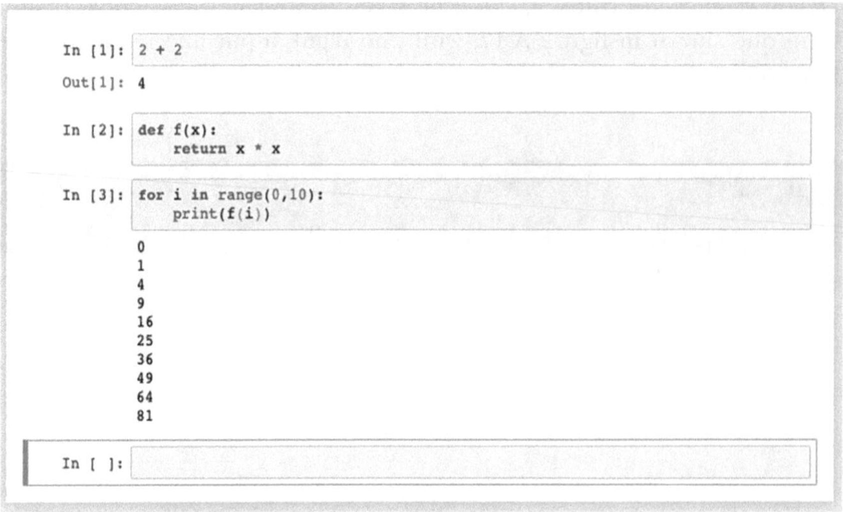

Figure A.19 Writing and evaluating several snippets of Python code in a Jupyter notebook. Notice the input boxes and the resulting output.

Strictly speaking, each box doesn't depend on the boxes above it, but on the boxes you *previously evaluated*. For example, if I redefine the function `f(x)` in the next input box, and then rerun the previous box, I overwrite the previous output (figure A.20).

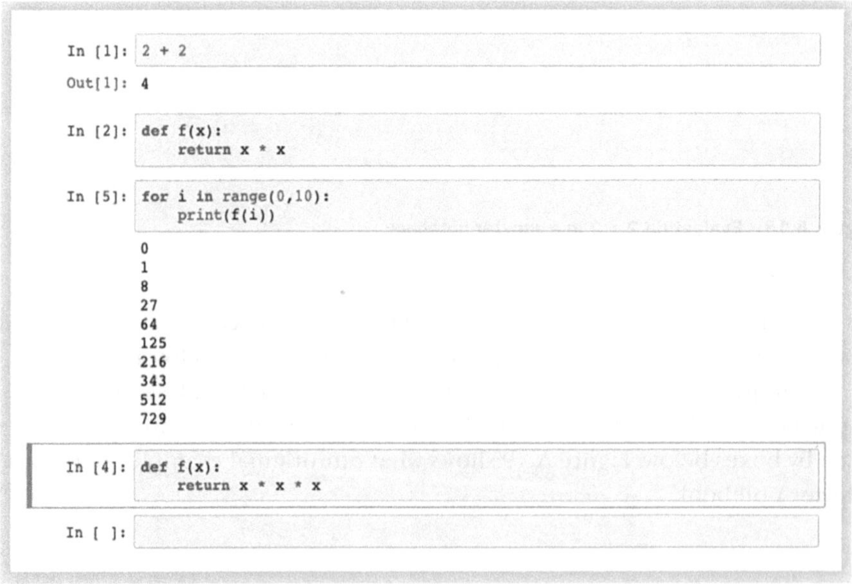

Figure A.20 If you redefine a symbol like `f` below the previous output and then rerun a box above, Python uses the new definition. Compare this figure with figure A.19 to see the new cell.

This can be confusing, but at least Jupyter renumbers the input boxes as you run them. For reproducibility, I suggest you define variables and functions *above* their first usage. You can confirm your code runs correctly from top to bottom by clicking the menu item Kernel > Restart & Run All (figure A.21). This wipes out all of your existing computations, but if you stay organized, you should get the same results back.

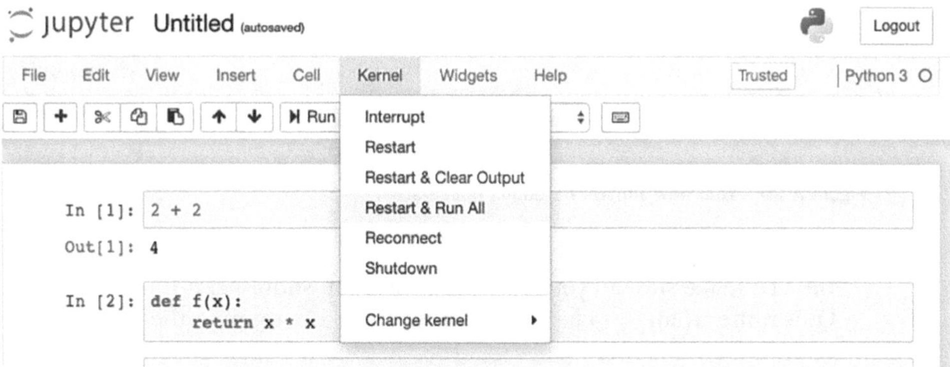

Figure A.21 Use the menu item Restart & Run All to clear the outputs and run all of your inputs from top to bottom.

Your notebook is automatically saved as you go. When you're done coding, you can name the notebook by clicking Untitled at the top of the screen and entering a new name (figure A.22).

Figure A.22 Giving your notebook a name

Then you can click the Jupyter logo once again to return to the main menu, and you should see your new notebook saved as a file with the .ipynb extension (figure A.23). If you want to return to your notebook, you can click its name to open it.

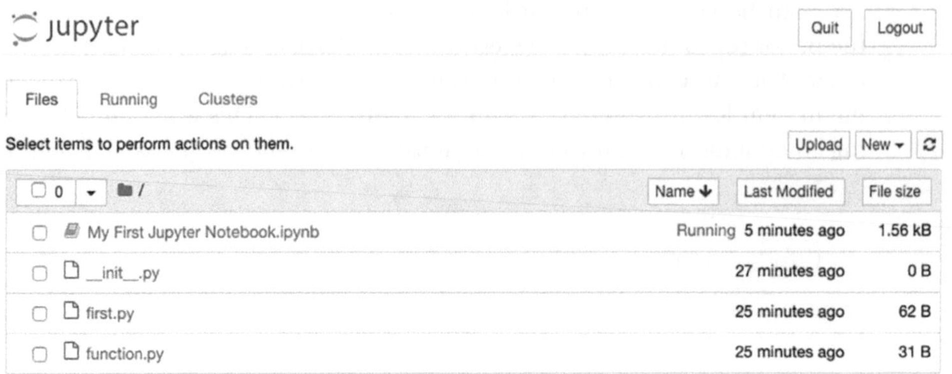

Figure A.23 Your new Jupyter notebook appears.

> **TIP** To make sure all your files are saved, you should exit Jupyter by clicking Quit rather than just closing the browser tab or stopping the interactive process.

For more info on Jupyter notebooks, you can consult the extensive documentation at https://jupyter.org/. At this point, however, you know enough to download and play with the source code for this book, which is organized in a Jupyter notebook for almost all of the chapters.

appendix B
Python tips and tricks

After following the set-up instructions in appendix A, you should be able to run Python code on your computer. If you're new to Python, the next step is to learn some of the language features. If you haven't seen *any* Python before, don't sweat! It's one of the simplest and easiest-to-learn programming languages out there. Plus, there are many excellent online resources and books to help you learn the basics of Python programming, and the website python.org is a great starting point.

This appendix assumes you've tinkered around with Python a bit and are comfortable with the basics: numbers, strings, True and False, if/else statements, and so on. To make this book as accessible as possible, I've avoided using advanced Python language features. This appendix covers some Python features used in this book that are either beyond the "basics" or warrant special attention because of their importance in this book. Don't worry if this is a lot to digest; when these features appear in the book, I often include a quick review of how they work. All of the code in this appendix is covered in a "walkthrough" notebook in the source code.

B.1 Python numbers and math

Like most languages, Python has built-in support for basic math. I'll assume you're already familiar with the basic Python arithmetic operators: `+`, `-`, `*`, and `/`. Note that when you divide integers in Python 3, you can get a fractional value, for instance,

```
>>> 7/2
3.5
```

By contrast, in Python 2, this would have returned `2`, which is the result of whole number division with the remainder of 1 discarded. But sometimes we want to get a remainder, in which case, we can use the `%` operator, called the *modulus* operator. Running `13 % 5` returns 3, which tells us that 13 divided by 5 has remainder 3 (as in $13 = 2 \times 5 + 3$). Note also that the modulus operator works for floating-point numbers. In particular, you can use it to get the fractional part of a number as the remainder when dividing by 1. Running `3.75 % 1` returns `0.75`.

Another useful operator beyond the basic four is the `**` operator, which raises numbers to a given power. For instance, `2 ** 3` represents two to the third power, or two cubed, which is 8. Similarly, `4 ** 2` is four squared, which is 16.

A final thing to keep in mind when doing math in Python is that floating-point arithmetic is not exact. I won't go into why this is, but will show you what it looks like so you aren't caught by surprise. For instance, 1000.1 – 1000.0 is obviously 0.1, but Python doesn't compute this value exactly:

```
>>> 1000.1 - 1000.0
0.10000000000002274
```

Of course, this result is within one-trillionth of the correct answer so it won't cause us problems, but it can lead to results that look wrong at first glance. For instance, we expect (1000.1 – 1000.0) – 0.1 to be zero, but Python gives us a big, hairy-looking result instead:

```
>>> (1000.1 - 1000.0) - 0.1
2.273181642920008e-14
```

This long number is written in scientific notation and is roughly 2.27 times 10^{-14} power. The number 10^{-14} is the same as 1/100,000,000,000,000 (1 over 1 followed by 14 zeros or 1 over 100 trillion), so this number is very close to zero after all.

B.1.1 The math module

Python has a math module with more helpful mathematical values and functions. Like any Python module, you need to import the objects you want to use from it. For instance,

```
from math import pi
```

imports the variable `pi` from the `math` module, which represents the number π. You may remember π from geometry class; it's the ratio of the circumference of a circle to its diameter. With the value `pi` imported, we can use it in subsequent code like any other variable:

```
>>> pi
3.141592653589793
>>> tau = 2 * pi
>>> tau
6.283185307179586
```

Another way to access values from modules in Python is to import the module itself and then access the values as needed. Here I import the math module and then use it to access the number π and another special number e that we'll run into a few times.

```
>>> import math
>>> math.pi
```

```
3.141592653589793
>>> math.e
2.718281828459045
```

The math module also contains a number of important functions we'll use in the book. Among these are the square root function `sqrt`, the trigonometric functions `cos` and `sin`, the exponential function `exp`, and the natural logarithm function `log`. We'll go over each of these functions as needed, but the important thing to know for now is that you call them like ordinary Python functions, providing their input values in parentheses:

```
>>> math.sqrt(25)
5.0
>>> math.sin(pi/2)
1.0
>>> math.cos(pi/3)
0.5000000000000001
>>> math.exp(2)
7.38905609893065
>>> math.log(math.exp(2))
2.0
```

As a quick reminder about exponential functions, `math.exp(x)` is the same as `math.e ** x`, for any value of `x`, and the `math.log` function undoes the effect of `math.exp`. The trigonometric functions are introduced in chapter 2.

B.1.2 Random numbers

Sometimes we want to choose some arbitrary numbers to test our computations, and we can use Python's random number generators to do this. These are stored in the `random` module, so we need to import that first:

```
import random
```

The first important function in this module is `randint`, which returns a randomly selected floating-point value from a given range. If you run `random.randint(0,10)`, you get a randomly selected integer from 0 to 10, and both 0 and 10 are possible outputs:

```
>>> random.randint(0,10)
7
>>> random.randint(0,10)
1
```

The other function we use for generating random numbers is `random.uniform`, which generates a random floating-point number on a specified interval. The following code returns a randomly selected number between 7.5 and 9.5:

```
>>> random.uniform(7.5, 9.5)
8.200084576283352
```

The word *uniform* indicates that no subrange is more likely than any other. By contrast, if you picked people at random and returned their ages, you would get a *non-uniform* distribution of random numbers, meaning you'd find more far people between ages 10–20 than between ages 100–110.

B.2 Collections of data in Python

Throughout this book, we do math involving *collections* of data. These can be ordered pairs of numbers representing points in the plane, lists of numbers representing measured data from the real world, or sets of symbols in an algebraic expression. Python has a number of ways to model collections, and in this section, I'll introduce them and compare them.

B.2.1 Lists

The most basic collection in Python is the list. To create a list, simply enclose some values between square brackets and separate them by commas. Here's a list of three strings, saved as a variable called `months`:

```
months = ["January", "February", "March"]
```

We can retrieve an entry from a list by its index (plural, *indices*) or its numerical position in the list. In Python, lists are *zero-indexed*, meaning the entries are numbered, counting from zero instead of from one. In the `months` list, the three indices are 0, 1, and 2. Therefore, we can get

```
>>> months[0]
'January'
>>> months[1]
'February'
>>> months[2]
'March'
```

Attempting to access an entry of a list outside the range of valid indices returns an error. For example, there's no `months[3]` or `months[17]`. A trick I use in a few places in the book is to use the modulus operator on indices to guarantee a valid entry. For any Python integer `n`, the expression `months[n % 3]` is guaranteed to be valid because `n % 3` always returns `0`, `1`, or `2`.

Another way to access list entries is to *unpack* them. If we are sure there are three entries in the list of months, we can write

```
j, f, m = months
```

which sets the variables `j`, `f`, and `m` equal to the three values from `months` in order. After running this, we have

```
>>> j
'January'
>>> f
'February'
>>> m
'March'
```

Another basic thing we can do with lists is *concatenate* them or combine them in order to make a bigger list. In Python, this is done with the + operator. Concatenating `[1, 2, 3]` and `[4, 5, 6]` gives us a new list consisting of the entries of the first followed by the entries of the second:

```
>>> [1,2,3] + [4,5,6]
[1, 2, 3, 4, 5, 6]
```

More list indexing and slicing

Python also lets you extract a *slice* of a list, which is the list of all values between two indices. For instance,

```
>>> months[1:3]
['February', 'March']
```

gives the slice starting at index 1 and going up to (but not including) index 3. For an even clearer example, we can look at a list whose entries are equal to their corresponding indices:

```
>>> nums = [0,1,2,3,4,5,6,7,8,9,10]
>>> nums[2:5]
[2, 3, 4]
```

The length of a list can be computed with the `len` function:

```
>>> len(months)
3
>>> len(nums)
11
```

Because the entries of a list are indexed starting from zero, the last entry in a list has index one less than the length of the list. To get the last entry of a list (like `nums`), we can write

```
>>> nums[len(nums)-1]
10
```

To get the last entry of a list, you can also use

```
>>> nums[-1]
10
```

Likewise, `nums[-2]` returns the second-to-last entry of the `nums` list, which is `9`. There are many ways to combine positive and negative indices and slices. For instance, `nums[1:]` returns all entries of the list except the first (at index zero), and `nums[3:-1]` returns the entries of `nums` from index 3 up until the second to last entry:

```
>>> nums[1:]
[1, 2, 3, 4, 5, 6, 7, 8, 9, 10]
>>> nums[3:-1]
[3, 4, 5, 6, 7, 8, 9]
```

Make sure not to confuse the slice syntax, which involves two indices, with retrieving an entry from a list of lists, which also involves two indices. With a list like

```
list_of_lists = [[1,2,3],[4,5,6],[7,8,9]]
```

the number 8 is in the third list (index 2) and it's the second entry (index 1) in that list, so if we run `list_of_lists[2][1]` we get `8`.

ITERATING OVER A LIST

Often when we're computing something about a list, we want to use every value in it. That means *iterating* over the list, visiting all of its values. The easiest way to do this in Python is with a *for loop*. The following `for` loop prints a statement for each value in the `months` list:

```
>>> for x in months:
>>>     print('Month: ' + x)
Month: January
Month: February
Month: March
```

It's also possible to build a new list by starting with an empty list and successively adding entries to it using the `append` method. This next code creates an empty list called `squares` and then loops over the `nums` list, adding the square of each number in `nums` to the `squares` list by calling `squares.append`:

```
squares = []
for n in nums:
    squares.append(n * n)
```

By the end of the `for` loop, `squares` contain the square of every number in `nums`:

```
>>> squares
[0, 1, 4, 9, 16, 25, 36, 49, 64, 81, 100]
```

LIST COMPREHENSIONS

Python has a special syntax for iteratively building lists: the *list comprehension*. A list comprehension is essentially a special kind of `for` loop that lives between square brackets, indicating that list entries are being added at each step of the iteration. List comprehensions read like plain English, making it easy to understand what they are doing. For instance, the following list comprehension builds a list consisting of squares of the form `x * x` for every value `x` in the `nums` list:

```
>>> [x * x for x in nums]
[0, 1, 4, 9, 16, 25, 36, 49, 64, 81, 100]
```

It's possible to iterate over multiple source lists in a list comprehension. For instance, the following code loops over all possible values from both a `years` list and a `months` list, making every combination of a year and a month into a string:

```
>>> years = [2018,2019,2020]
>>> [m + " " + str(y) for y in years for m in months]
```

```
['January 2018',
 'February 2018',
 'March 2018',
 'January 2019',
 'February 2019',
 'March 2019',
 'January 2020',
 'February 2020',
 'March 2020']
```

Similarly, we can build a list of lists by putting one comprehension inside of another. Adding one more pair of square brackets, we change the comprehension to return a list for each value in the `months` list:

```
>>> [[m + " " + str(y) for y in years] for m in months]
[['January 2018', 'January 2019', 'January 2020'],
 ['February 2018', 'February 2019', 'February 2020'],
 ['March 2018', 'March 2019', 'March 2020']]
```

B.2.2 Other iterables

In Python, and especially in Python 3.*x*, there are a few other types of collections. In particular, some of these are referred to as *iterables* because we can iterate over them as if they were lists. Probably the most frequently used in this book are *ranges*, which are used to construct sequences of numbers in order. For instance, `range(5,10)` represents the sequence of integers starting from 5 and going up to (but not including) 10. If you evaluate `range(5,10)` on its own in Python, the result is unexciting:

```
>>> range(5,10)
range(5, 10)
```

Even though a range doesn't display the numbers that make it up, we can iterate over it just like a list:

```
>>> for i in range(5,10):
>>>     print(i)
5
6
7
8
9
```

The fact that ranges are not lists allows us to use very big ranges and not iterate over them all at once. For instance, `range(0,1000000000)` defines a range of a billion numbers that we could iterate over, but it doesn't actually store a billion numbers. It only stores the instructions to produce the numbers during an iteration. If you want to turn an iterable like a `range` into a list, all you have to do is convert it with the `list` function:

```
>>> list(range(0,10))
[0, 1, 2, 3, 4, 5, 6, 7, 8, 9]
```

It's useful to get lists of consecutive integers, so we use the `range` function a lot. One more note about the `range` function is that some of its arguments are optional. If you call it with only one input, it automatically starts from zero and goes up to the input number, and if you provide a third argument it counts by that number. For instance, `range(10)` counts from 0 to 9 and `range(0,10,3)` counts from 0 to 9 in increments of 3:

```
>>> list(range(10))
[0, 1, 2, 3, 4, 5, 6, 7, 8, 9]
>>> list(range(0,10,3))
[0, 3, 6, 9]
```

Another example of a function that returns its own special type of iterable is the `zip` function. The `zip` function takes two iterables of the same length and returns an iterable consisting of pairs of corresponding entries from the first and second iterable:

```
>>> z = zip([1,2,3],["a","b","c"])
>>> z
<zip at 0x15fa8104bc8>
>>> list(z)
[(1, 'a'), (2, 'b'), (3, 'c')]
```

Note that not all iterables support indexing; `z[2]` is invalid, so you'll need to convert it to a list first (like `list(z)[2]` instead) to get the third entry of a `zip z`. (Ranges do support indexing, and `range(5,10)[3]` returns 8.) Be careful—once you've iterated through a `zip` it is no longer there! It's a good idea to convert a `zip` to a list immediately if you plan to reuse it.

B.2.3 Generators

Python's *generators* give you a way to create iterables that don't store all of their values at once, but rather store instructions to produce values. This allows us to define large or even infinite sequences of values without storing them in memory. Generators can be created in a few ways, the most basic of which looks like a function with the keyword `yield` instead of `return`. The difference is that a generator can yield many values, where as a function returns at most once and then is done.

Here's a generator representing the infinite sequence of integers beginning 0, 1, 2, 3, and so on. The `while` loop goes forever, and in each loop the variable x is yielded and then incremented by 1.

```
def count():
    x = 0
    while True:
        yield x
        x += 1
```

Even though this represents an infinite sequence, you can run `count()` without blowing up your computer. It just returns a generator object, not a complete list of values:

```
>>> count()
<generator object count at 0x0000015FA80EC750>
```

A `for` loop beginning with `for x in count()` is valid, but runs forever. Here's an example of using this infinite generator in a `for` loop with a `break` to escape instead of iterating forever:

```
for x in count():
    if x > 1000:
        break
    else:
        print(x)
```

Here's a more practical version of the count generator that only yields finitely many values. It works like the `range` function, starting from the first input value and going up to the second:

```
def count(a,b):
    x = a
    while x < b:
        yield x
        x += 1
```

The result of `count(10,20)` is a generator that is like `range(10,20)`; we can't see its values directly, but we can iterate over it, for instance, in a list comprehension:

```
>>> count(10,20)
<generator object count at 0x0000015FA80EC9A8>
>>> [x for x in count(10,20)]
[10, 11, 12, 13, 14, 15, 16, 17, 18, 19]
```

We can create generator comprehensions much like list comprehensions by wrapping the comprehension code in parentheses instead of brackets. For example,

```
(x*x for x in range(0,10))
```

is a generator yielding the squares of numbers from 0 to 9. It's behavior is the same as the generator:

```
def squares():
    for x in range(0,10):
        yield x*x
```

When a generator is finite, you can safely convert it to a list with the `list` function:

```
>>> list(squares())
[0, 1, 4, 9, 16, 25, 36, 49, 64, 81]
```

B.2.4 Tuples

Tuples are iterables that are a lot like lists except that they are *immutable*; you can't change them once they are created. That means there is no `append` method on a tuple. In particular, once you've created a tuple, it always has the same fixed length.

That makes them useful for storing data that comes in pairs or triples. Tuples are created like lists with the only difference being that we use parentheses (or no brackets at all) instead of square brackets:

```
>>> (1,2)
(1, 2)
>>> ("a","b","c")
('a', 'b', 'c')
>>> 1,2,3,4,5
(1, 2, 3, 4, 5)
```

If you look at the `zip` in section B.2.2 again, you'll see its entries are actually tuples. Tuples are in a sense the default collection in Python. If you write `a = 1,2,3,4,5` (without the parentheses), then `a` is automatically interpreted as a tuple of those numbers. Likewise, if you conclude a function with `return a,b`, the output will actually be the tuple `(a,b)`.

Tuples are often short, so we often don't need to iterate over them. There's no such thing as a tuple comprehension, but you can iterate over a tuple in another comprehension and convert the result back to a tuple with the built-in `tuple` function. Here's something that looks like a tuple comprehension, but it's actually a generator comprehension whose result is passed to the `tuple` function:

```
>>> a = 1,2,3,4,5
>>> tuple(x + 10 for x in a)
(11, 12, 13, 14, 15)
```

B.2.5 Sets

Python *sets* are collections where every entry must be distinct, and they do not keep track of order. We won't use sets much in this book, except that turning a list into a set is a quick way to guarantee it has no duplicate values. The `set` function turns an iterable into a set as follows:

```
>>> dups = [1,2,3,3,3,3,4,5,6,6,6,6,7,8,9,9,9]
>>> set(dups)
{1, 2, 3, 4, 5, 6, 7, 8, 9}
>>> list(set(dups))
[1, 2, 3, 4, 5, 6, 7, 8, 9]
```

Python sets are written as a list of entries between curly braces, which incidentally is the same way mathematical sets are written. You can define a set from scratch by listing some entries, separated by commas and enclosing them in curly braces. Because sets don't respect order, sets are equal if they have exactly the same entries:

```
>>> set([1,1,2,2,3]) == {3,2,1}
True
```

B.2.6 NumPy arrays

The final collection we make extensive use of in this book is not a built-in Python collection; it comes from the NumPy package, which is the de-facto standard Python library for numerics (efficient number crunching). This collection is the NumPy *array*, and it's mostly important because of how ubiquitous NumPy is. Many other Python libraries have functions that expect NumPy arrays as inputs.

To use NumPy arrays, make sure you have access to the NumPy library. First, you need to make sure NumPy is installed. If you're using Anaconda as described in appendix A, you should already have it. Otherwise, you'll want to install NumPy with the pip package manager via `pip install numpy` in your terminal. Once NumPy is installed, you need to import it to your Python program. The traditional way to import NumPy is with the name `np`:

```
import numpy as np
```

To create a NumPy array, simply pass an iterable to the `np.array` function:

```
>>> np.array([1,2,3,4,5,6])
array([1, 2, 3, 4, 5, 6])
```

One NumPy function we use is `np.arange`, which is like a floating-point version of the built-in Python `range` function. With two arguments, `np.arange` works the same way as `range`, producing a NumPy array instead of a `range` object:

```
>>> np.arange(0,10)
array([0, 1, 2, 3, 4, 5, 6, 7, 8, 9])
```

With a third argument, you can specify a value to count by, which can be a float. The following code gives us a NumPy array with values from 0 up to 10 in increments of 0.1, which is 100 numbers in total:

```
>>> np.arange(0,10,0.1)
array([0. , 0.1, 0.2, 0.3, 0.4, 0.5, 0.6, 0.7, 0.8, 0.9, 1. , 1.1, 1.2,
       1.3, 1.4, 1.5, 1.6, 1.7, 1.8, 1.9, 2. , 2.1, 2.2, 2.3, 2.4, 2.5,
       2.6, 2.7, 2.8, 2.9, 3. , 3.1, 3.2, 3.3, 3.4, 3.5, 3.6, 3.7, 3.8,
       3.9, 4. , 4.1, 4.2, 4.3, 4.4, 4.5, 4.6, 4.7, 4.8, 4.9, 5. , 5.1,
       5.2, 5.3, 5.4, 5.5, 5.6, 5.7, 5.8, 5.9, 6. , 6.1, 6.2, 6.3, 6.4,
       6.5, 6.6, 6.7, 6.8, 6.9, 7. , 7.1, 7.2, 7.3, 7.4, 7.5, 7.6, 7.7,
       7.8, 7.9, 8. , 8.1, 8.2, 8.3, 8.4, 8.5, 8.6, 8.7, 8.8, 8.9, 9. ,
       9.1, 9.2, 9.3, 9.4, 9.5, 9.6, 9.7, 9.8, 9.9])
>>> len(np.arange(0,10,0.1))
100
```

B.2.7 Dictionaries

Dictionaries are collections that work quite differently than lists, tuples, or generators. Instead of accessing entries of a dictionary by a numeric index, you can label them with another piece of data called a *key*. At least in this book, keys are most frequently strings. The following code defines a dictionary called `dog` with two keys and corresponding values; the key `"name"` is associated with the string `"Melba"` and the key `"age"` is associated with the number `2`:

```
dog = {"name" : "Melba", "age" : 2}
```

To make dictionaries more readable, we often use some extra whitespace and write each key-value pair on its own line. The following is the same `dog` dictionary with extra whitespace:

```
dog = {
    "name" : "Melba",
    "age" : 2
}
```

To access the values of a dictionary, you use a similar syntax as when getting the entries of a list, but instead of passing an index, you pass a key:

```
>>> dog["name"]
'Melba'
>>> dog["age"]
2
```

If you want to get all the values from a dictionary, you can get an iterable of key-value pair tuples using the `items` method on a dictionary. Dictionaries don't order their values, so don't expect the result of items to have any particular order:

```
>>> list(dog.items())
[('name', 'Melba'), ('age', 2)]
```

B.2.8 Useful collection functions

Python comes with a number of useful built-in functions that work with iterables, particularly for iterables of numbers. We already saw the length function `len`, which we will use most frequently, as well as the `zip` function, but there are a few others worth a quick mention. The `sum` function adds up an iterable of numbers, and the `max` and `min` functions return the largest and smallest values, respectively:

```
>>> sum([1,2,3])
6
>>> max([1,2,3])
3
>>> min([1,2,3])
1
```

The `sorted` function returns a list that is a sorted copy of an iterable. It's important to note that `sorted` returns a new list; the order of the original list is unaffected:

```
>>> q = [3,4,1,2,5]
>>> sorted(q)
[1, 2, 3, 4, 5]
>>> q
[3, 4, 1, 2, 5]
```

Similarly, the `reversed` function returns a reversed version of a given iterable, while leaving the original iterable's order unchanged. The result is an iterable but not a list, so you need to convert it to see the result:

```
>>> q
[3, 4, 1, 2, 5]
```

```
>>> reversed(q)
<list_reverseiterator at 0x15fb652eb70>
>>> list(reversed(q))
[5, 2, 1, 4, 3]
```

By contrast, if you do want to sort or reverse a list in place, you can use the sort and reverse methods like `q.sort()` or `q.reverse()`.

B.3 Working with functions

Python functions are like mini-programs that take some input values (or possibly none), do some computations, and possibly produce an output value. We already used some Python functions, like `math.sqrt` and `zip`, and saw the outputs they produce for different input values.

We can define our own Python functions using the `def` keyword. The following code defines a function called `square` that takes an input value called `x`, stores the value `x * x` in a variable called `y`, and returns the value of `y`. Like a `for` loop or `if` statement, we need to use indentation to show which lines belong to the function definition:

```
def square(x):
    y = x * x
    return y
```

The net result of this function is returning the square of the input value:

```
>>> square(5)
25
```

This section covers a few of the more advanced ways we use functions in the book.

B.3.1 Giving functions more inputs

We can define our function to take as many inputs, or *arguments*, as we want. The following function takes three arguments and adds them together:

```
def add3(x,y,z):
    return x + y + z
```

Sometimes it's useful to have a single function accept a variable number of arguments. For instance, we might want to write a single `add` function where `add(2,2)` returns `4`, `add(1,2,3)` returns `6`, and so on. We can do this by adding a star to a single input value, commonly called `args`. The star indicates that we are taking all of the input values and storing them in a tuple called `args`. Then we are free to write logic within our function that iterates over all of the arguments. This `add` function iterates over all arguments it is passed and adds them up, returning the total:

```
def add(*args):
    total = 0
    for x in args:
        total += x
    return total
```

Then add(1,2,3,4,5) returns 1 + 2 + 3 + 4 + 5 = 15 as desired and add() returns 0. Our add function works differently than the sum function from before; sum takes an iterable, while add takes the underlying values directly as arguments. Here's a comparison:

```
>>> sum([1,2,3,4,5])
15
>>> add(1,2,3,4,5)
15
```

The * operator has a second application: you can use it to take a list and convert it into arguments of a function. For instance,

```
>>> p = [1,2,3,4,5]
>>> add(*p)
15
```

This is equivalent to evaluating add(1,2,3,4,5).

B.3.2 Keyword arguments

Using a starred argument to a function is one way to optional parameters. Another way is to pass in named arguments, called *keyword arguments*. Here's an example function with two optional keyword arguments called name and age, which returns a string containing a birthday greeting:

```
def birthday(name="friend", age=None):
    s = "Happy birthday, %s" % name
    if age:
        s += ", you're %d years old" % age
    return s + "!"
```

(This function uses the string formatting operator %, which substitutes occurrences of %s with a given string and occurrences of %d with a given number.) Because they are keyword arguments, both name and age are optional. The name defaults to "friend", so if we call birthday with no arguments, we get a generic greeting:

```
>>> birthday()
'Happy birthday, friend!'
```

We can optionally specify a different name instead. The first argument is understood to be the name, but we can also make it explicit by setting the name argument directly:

```
>>> birthday('Melba')
'Happy birthday, Melba!'
>>> birthday(name='Melba')
'Happy birthday, Melba!'
```

The age argument is also optional and defaults to None. We can specify a name and an age, or just an age. Because age is the second keyword argument, we need to identify it if we don't provide a name. When all arguments are identified, we can pass them in any order. Here are a few examples:

```
>>> birthday('Melba', 2)
"Happy birthday, Melba, you're 2 years old!"
>>> birthday(age=2)
"Happy birthday, friend, you're 2 years old!"
>>> birthday('Melba', age=2)
"Happy birthday, Melba, you're 2 years old!"
>>> birthday(age=2,name='Melba')
"Happy birthday, Melba, you're 2 years old!"
```

If you have a lot of arguments, you can package them as a dictionary and pass them to the function using the ** operator. This is like the * operator, except instead of passing a list of arguments, you pass a dictionary of keyword arguments:

```
>>> dog = {"name" : "Melba", "age" : 2}
>>> dog
{'name': 'Melba', 'age': 2}
>>> birthday(**dog)
"Happy birthday, Melba, you're 2 years old!"
```

When defining your function, you can similarly use the `**` operator to process all of the keywords supplied to the function as a single dictionary. We can rewrite the birthday function as follows, but then we need to specify the names of all of the arguments when calling it:

```
def birthday(**kwargs):
    s = "Happy birthday, %s" % kwargs['name']
    if kwargs['age']:
        s += ", you're %d years old" % kwargs['age']
    return s + "!"
```

Specifically, the `name` and `age` variables are replaced with `kwargs['name']` and `kwargs['age']`, and we can run it either of the following ways:

```
>>> birthday(**dog)
"Happy birthday, Melba, you're 2 years old!"
>>> birthday(age=2,name='Melba')
"Happy birthday, Melba, you're 2 years old!"
```

B.3.3 Functions as data

In Python, functions are *first-class* values, meaning you can assign them to variables, pass them to functions, and return functions as the output values of other functions. In other words, Python functions look like any other piece of data in Python. In the *functional programming* paradigm, which we introduce in chapter 4, it's common to have functions that operate on other functions. The following function takes two inputs, a function `f` and a value `x`, and returns the value `f(x)`:

```
def evaluate(f,x):
    return f(x)
```

Using the `square` function from section B.3, `evaluate(square,10)` should return `square(10)` or `100`:

```
>>> evaluate(square,10)
100
```

A more useful function that takes a function as input is Python's `map` function. The `map` function takes a function and an iterable, and returns a new iterable obtained by applying the function to every entry of the original. For instance, the following `map` applies `square` to every number in `range(10)`. Converting it to a list, we can see the first 10 square numbers:

```
>>> map(square,range(10))
<map at 0x15fb752e240>
>>> list(map(square,range(10)))
[0, 1, 4, 9, 16, 25, 36, 49, 64, 81]
```

The `evaluate` and `map` functions are examples of functions that take other functions as inputs. It's also possible for a function to return another function as output. The following function, for instance, returns a function that raises a number to some power. Notice, in particular, that a full function definition can live inside another function:

```
def make_power_function(power):

    def power_function(x):
        return x ** power

    return power_function
```

With this defined, `make_power_function(2)` returns a function that behaves just like the previous `square` function. Similarly, `make_power_function(3)` returns a function that cubes its input:

```
>>> square = make_power_function(2)
>>> square(2)
4
>>> cube = make_power_function(3)
>>> cube(2)
8
```

When the `make_power_function` finishes evaluating, the `power_function` returned still remembers the `power` variable it was passed, even though variables within a function are usually gone when the function finishes running. Such a function that remembers outside variables used in its definition is called a *closure.*

B.3.4 Lambdas: Anonymous functions

There's another simpler syntax we can use when creating functions on the fly. The `lambda` keyword allows us to create a function without a name, referred to as an *anonymous function* or a *lambda.* This name comes from the Greek letter λ, written lambda and pronounced LAM-duh, which is the symbol computer scientists use for function definitions in the theory of functional programming. To define a function as a lambda, you specify the input variable or variables, separated by commas, then a colon, then the return expression for the function. This lambda defines a function that takes a single input x and adds 2 to it:

```
>>> lambda x: x + 2
<function __main__.<lambda>(x)>
```

You can use a lambda anywhere you use a function, so you can apply it directly to a value as follows:

```
>>> (lambda x: x + 2)(7)
9
```

Here's another lambda function taking two input variables and returning the value of the first plus twice the value of the second. In this case, the first input is 2 and the second is 3, so the output is $2 + 2 \cdot 3 = 8$:

```
>>> (lambda x,y: x + 2 * y)(2,3)
8
```

You can also bind a lambda to a name like any function value in Python, although that somewhat defeats the purpose of using the anonymous function syntax:

```
>>> plus2 = lambda x: x + 2
>>> plus2(5)
7
```

Lambdas should be used sparingly because if a function does anything interesting, it probably deserves a name. One place you might use a lambda is if you're writing a function that returns another function. For instance, the `make_power_function` can be equivalently implemented with a lambda as follows:

```
def make_power_function(p):
    return lambda x: x ** p
```

We can see that this function behaves the same way as the original implementation:

```
>>> make_power_function(2)(3)
9
```

The name of the outside function makes this clear, and not much is gained by naming the return function. It's also possible to use lambdas as inputs to functions. For instance, if you want to add 2 to every number from 0 to 9, you could concisely write

```
map(lambda x: x + 2, range(0,9))
```

To see the data, we again need to convert this result to a list. In most places, however, it's just as concise and more readable to use comprehensions. The equivalent list comprehension is

```
[x+2 for x in range(0,9)]
```

B.3.5 Applying functions to NumPy arrays

NumPy has its own versions of some of the built-in Python math functions, which are useful because they can be applied to every entry of a NumPy array at once. For instance, `np.sqrt` is a square root function that can either take the square root of a number or a whole NumPy array. For instance, `np.sqrt(np.arange(0,10))` produces a NumPy array with the square roots of integers from 0 to 9:

```
>>> np.sqrt(np.arange(0,10))
array([0.        , 1.        , 1.41421356, 1.73205081, 2.        ,
       2.23606798, 2.44948974, 2.64575131, 2.82842712, 3.        ])
```

This is not just a shortcut. NumPy actually has an implementation that is faster than iterating over the array in Python. If you want to apply a custom function to every entry of a NumPy array, you can use the `np.vectorize` function. Here's an example that takes a number and returns another number:

```
def my_function(x):
    if x % 2 == 0:
        return x/2
    else:
        return 0
```

The following code vectorizes the function and applies it to every entry in the NumPy array: `np.arange(0,10)`.

```
>>> my_numpy_function = np.vectorize(my_function)
>>> my_numpy_function(np.arange(0,10))
array([0., 0., 1., 0., 2., 0., 3., 0., 4., 0.])
```

B.4 Plotting data with Matplotlib

Matplotlib is the most popular plotting library in Python. Throughout the book, we use it to create plots of data sets, graphs of functions, and drawings of other geometric figures. To avoid library-specific discussions, I've hidden most of Matplotlib's usage in wrapper functions, so you can mainly use those to do all of the exercises and mini-projects. In case you want to dig deeper into the implementations, here's a quick overview of making plots in Matplotlib. You should have Matplotlib already installed with Anaconda, or you can manually install it with `pip install matplotlib`.

B.4.1 Making a scatter plot

Scatter plots are useful for visualizing a collection of ordered pairs of numbers of the form (x, y) as points in the plane (covered in more detail in chapter 2). To create a scatter plot (or any plot) with Matplotlib, the first step is to install the library and import it into your Python script. It's traditional to import Matplotlib's plotting module, pyplot, with the name `plt`:

```
import matplotlib.pyplot as plt
```

Say we wanted to make a scatter plot of the points (1, 1), (2, 4), (3, 9), (4, 16), and (5, 25), which are pairs of some numbers and their squares. Thinking of these as points of the form (x, y), the x values are 1, 2, 3, 4, and 5, and the y values are 1, 4, 9, 16, and 25. To make a scatter plot, we use the `plt.scatter` function, passing first a list of the x values and then a list of the y values:

```
x_values = [1,2,3,4,5]
y_values = [1,4,9,16,25]
plt.scatter(x_values,y_values)
```

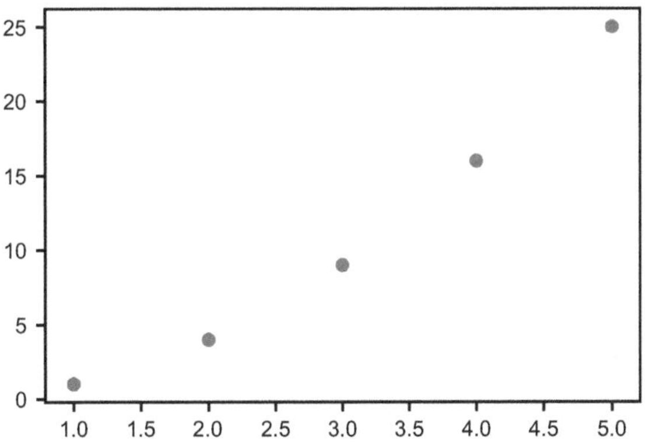

Figure B.1 A scatter plot created with the Matplotlib `plt.scatter` function

The horizontal position tells us an *x* value of a given point and the vertical position tells us a *y* value. Notice that Matplotlib automatically scales the graph area to fit all of the points, so in this case, the *y* scale is bigger than the *x* scale.

There are also a few keyword arguments you can use to customize the appearance of your scatter plot. For instance, the `marker` keyword argument sets the shape of the dots on the plot and the `c` keyword argument sets the color of the dots. The following line plots the same data with a red "x" for each point instead of the default blue circle:

```
plt.scatter(x_values,y_values,marker='x',c='red')
```

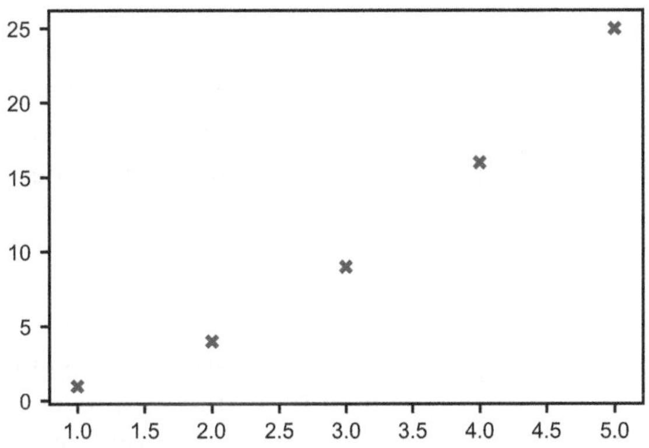

Figure B.2 Customizing the appearance of a Matplotlib scatter plot

The documentation at https://matplotlib.org/ covers all of the possible keyword arguments and customizations that are possible with Matplotlib plots.

B.4.2 Making a line chart

If we use the Matplotlib `plt.plot` function instead of the `plt.scatter` function, our points are connected with lines rather than being marked as dots. This is sometimes called a *line chart.* For example,

```
plt.plot(x_values,y_values)
```

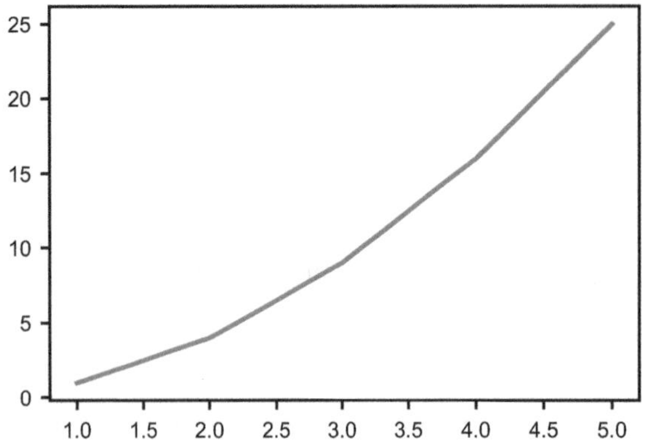

Figure B.3 Creating a line chart with the Matplotlib `plt.plot` function

One useful application of this is specifying only two points to draw a line segment. For instance, we could write a function that takes two input (*x*, *y*) points as tuples and draws the line segment connecting them by extracting their *x* and *y* values and then using `plt.plot`:

```
def plot_segment(p1,p2):
    x1,y1 = p1
    x2,y2 = p2
    plt.plot([x1,x2],[y1,y2],marker='o')
```

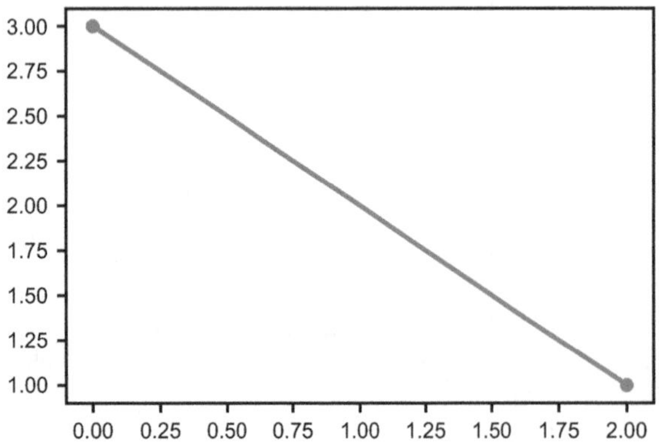

Figure B.4 A function that draws a line segment between two points

This example also shows that you can set a `marker` keyword argument for `plt.plot` to mark the individual points in addition to drawing the line:

```
point1 = (0,3)
point2 = (2,1)
plot_segment(point1,point2)
```

This `draw_segment` function is an example of a wrapper function; we can now use `draw_segment` any time we want to make a line segment between two (*x*, *y*) points, rather than using Matplotlib functions.

Another important use of a line chart is plotting the *graph* of a function. That is, plotting all of the pairs $(x, f(x))$ for some fixed function f over a range of x values. Theoretically, a smooth, continuous graph consists of infinitely many points. We can't use infinitely many points, but the more we use, the more accurate the graph will look. Here's a plot of $f(x) = \sin(x)$ from $x = 0$ to $x = 10$ with 1,000 points:

```
x_values = np.arange(0,10,0.01)
y_values = np.sin(x_values)
plt.plot(x_values,y_values)
```

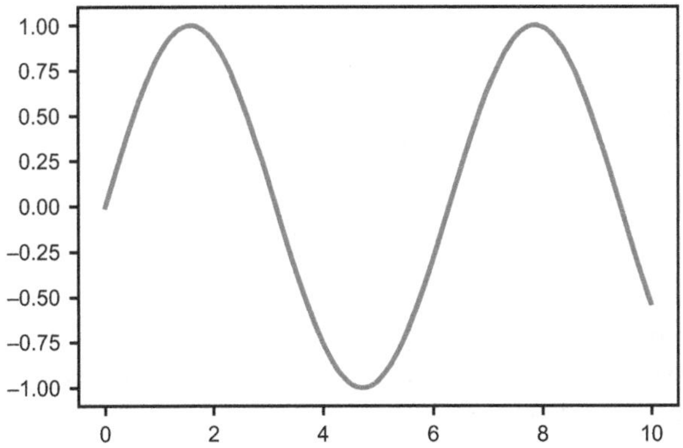

Figure B.5 Using a lot of points, we can approximate a smooth function graph.

B.4.3 More plot customizations

As I mentioned, the best way to learn more about customizing your Matplotlib plots is to search the documentation at matplotlib.org when you want to accomplish something specific. There are a few more important ways to control the appearance of your Matplotlib plots, which I'll mention because they appear frequently in the examples in this book.

The first is setting the scale and size of your plot. You may have noticed that the result of `plot_segment(point1,point2)` was not drawn proportionally. If we want to see our line segment plotted to scale, we could explicitly set the x and y bounds for the graph to be the same. For instance, this code sets both the x and y bounds to the range from 0 to 5:

```
plt.ylim(0,5)
plt.xlim(0,5)
plot_segment(point1,point2)
```

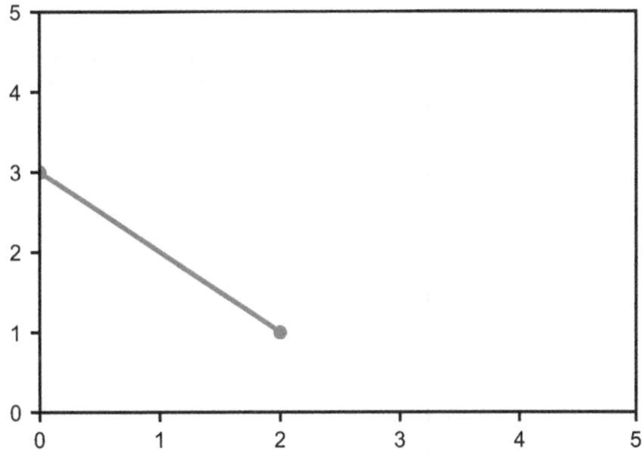

Figure B.6 Updating the *x* and *y* scales on a plot with *x* and *y* bounds

This is still not quite to scale. One unit on the *x*-axis is not the same as one unit on the *y*-axis. To make them the same size visually, we need to make our graph a square. This can be done with the `set_size_inches` method. That method actually belongs to the current "figure" object that Matplotlib is working with, which we can retrieve with the `gcf` (get current figure) method on `plt`. The following code draws the line segment in correct proportion in a 5 in.-by-5 in. plot area. Depending on your display, it could appear another size, but the proportions should be correct:

```
plt.ylim(0,5)
plt.xlim(0,5)
plt.gcf().set_size_inches(5,5)
plot_segment(point1,point2)
```

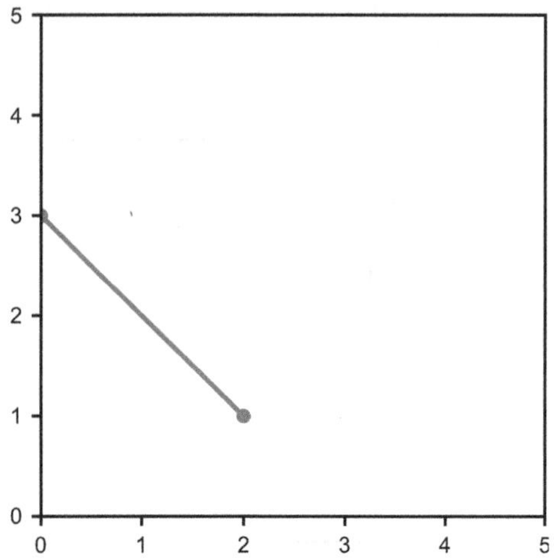

Figure B.7 Drawing with correct proportions by setting the figure size in inches

The other important customization you can add to your graph is setting labels for the axes and for the whole graph. You can add a title to the current graph with the `plt.title` function, and you can add labels to the *x*- and *y*-axes with the `plt.xlabel` and `plt.ylabel` functions, respectively. Here's an example of adding labels to the graph of the sine function:

```
x_values = np.arange(0,10,0.01)
y_values = np.sin(x_values)
plt.plot(x_values,y_values)
plt.title('Graph of sin(x) vs. x',fontsize=16)
plt.xlabel('this is the x value',fontsize=16)
plt.ylabel('the value of sin(x)',fontsize=16)
```

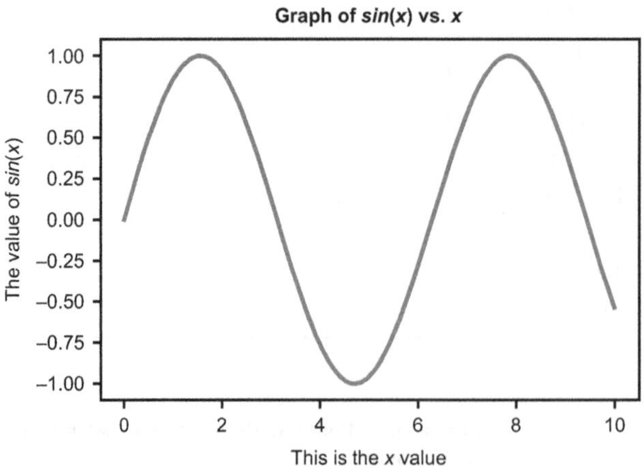

Figure B.8 A Matplotlib plot with a title and axis labels

B.5 Object-oriented programming in Python

Object-oriented programming (OOP) is, roughly speaking, the programming paradigm that emphasizes organizing your program data with *classes.* Classes can store values called *properties* as well as functions called *methods,* which relate data and functionality in a program. You don't need to know much about OOP to appreciate this book, but some mathematical ideas do have an object-oriented flavor. In chapters 6 and 10 especially, we use classes and some object-oriented design principles to help understand math. This section gives a quick introduction to classes and OOP in Python.

B.5.1 Defining classes

Let's work with a concrete example. Suppose you are writing a Python program that deals with geometric shapes like a drawing app, for example. One kind of shape you might want to describe would be a rectangle. To do this, we define a `Rectangle` class in Python that describes properties of rectangles and then we create *instances* of this class that represent specific rectangles.

In Python, a class is defined with the `class` keyword, and the name of the class is typically capitalized, like `Rectangle`. The indented lines below the name of the class describe the properties (values) and methods (functions) associated with the class. The most basic method of a class is its *constructor,* which is a function that allows us to make instances of the class. In Python, constructors are given the special name `__init__`. For a rectangle, we might want to describe it by two numbers, representing its height and width. In that case, the `__init__` function takes three values: the first represents the new instance of the class we're building, and the next two represent the height and width values. The constructor does the work of setting height and width properties of the new instance to the input values:

```
class Rectangle():
    def __init__(self,w,h):
        self.width = w
        self.height = h
```

Having created a constructor, we can use the name of the class like a function that takes two numbers and returns a `Rectangle` object. For instance `Rectangle(3,4)` creates an instance with the `width` property set to 3 and the `height` property set to 4. Even though the constructor is defined with a `self` argument, you don't need to include it when you call the constructor. With that `Rectangle` object created, we can access its height and width properties:

```
>>> r = Rectangle(3,4)
>>> type(r)
__main__.Rectangle
>>> r.width
3
>>> r.height
4
```

B.5.2 Defining methods

A *method* is a function associated with a class that lets you compute something about instances or gives the instances some kind of functionality. For a rectangle, it would make sense to have an `area()` method that computes the area, which is the height times the width. Like the constructor, any method must take a `self` parameter, which represents the current instance. Once again, you don't need to pass a `self` parameter to the method; `self` is automatically taken to be the object the method is being called on:

```
class Rectangle():
    def __init__(self,w,h):
        self.width = w
        self.height = h

    def area(self):
        return self.width * self.height
```

To find the area of a rectangle, we can call the `area` method as follows:

```
>>> Rectangle(3,4).area()
12
```

Note that a `self` parameter is not passed to the function; the instance `Rectangle(3,4)` is automatically taken to be the `self` value. As another example, we could have a `scale` method that takes a number and returns a new `Rectangle` object whose dimensions of height and width are scaled by that factor from the original. (I'll start using "..." as a placeholder on the printed page to stand-in for code in the `Rectangle` class that we've already written.)

```
class Rectangle():
    ...
    def scale(self, factor):
        return Rectangle(factor * self.width, factor * self.height)
```

Calling `Rectangle(2,1)` constructs a rectangle with a width of 2 and a height of 1. If we scale it by a factor of 3, we get a new rectangle with a width of 6 and a height of 3:

```
>>> r = Rectangle(2,1)
>>> s = r.scale(3)
>>> s.width
6
>>> s.height
3
```

B.5.3 Special methods

Some methods are either automatically available or have special effects once implemented. For instance, the `__dict__` method is available by default on every instance of a new class and returns a dictionary of all the properties of the instance. With no additional modifications to the `Rectangle` class, we can write:

```
>>> Rectangle(2,1).__dict__
{'width': 2, 'height': 1}
```

Another special method name is `__eq__`. This method when implemented describes the behavior of the `==` operator on instances of a class, and therefore, decides when two instances are equal. Without a custom equality method implemented, different instances of a class are always unequal, even if they contain the same data:

```
>>> Rectangle(3,4) == Rectangle(3,4)
False
```

For rectangles, we might like to say they are the same if they are geometrically indistinguishable, having the same width and the same height. We can implement the `__eq__` method accordingly. The method takes two arguments, the `self` argument, as usual, and a second argument representing another instance we compare `self` to:

```
class Rectangle():
    ...
    def __eq__(self,other):
        return self.width == other.width and self.height == other.height
```

With this done, `Rectangle` instances are equal if their heights and widths agree:

```
>>> Rectangle(3,4) == Rectangle(3,4)
True
```

Another useful special method is `__repr__`, which produces a default string representation of an object. The following `__repr__` method makes it easier to see the width and height of a rectangle at first glance:

```
class Rectangle():
    ...
    def __repr__(self):
        return 'Rectangle (%r by %r)' % (self.width, self.height)
```

We can see it work:

```
>>> Rectangle(3,4)
Rectangle (3 by 4)
```

B.5.4 Operator overloading

There are more special methods that we can implement to describe how Python operators should behave with instances of a class. Repurposing operators that have an existing meaning to work on objects of a new class is called *operator overloading*. For instance, the `__mul__` and `__rmul__` methods describe how a class behaves with respect to the multiplication operator *, acting on the right and left, respectively. For a `Rectangle` instance `r`, we might want to write `r * 3` or `3 * r` to represent scaling the rectangle by a factor of 3. The following implementations of `__mul__` and `__rmul__` call the `scale` method we already implemented, producing a new rectangle scaled by the given factor:

```
class Rectangle():
    ...
    def __mul__(self,factor):
        return self.scale(factor)
```

```
    def __rmul__(self,factor):
        return self.scale(factor)
```

We can see that either `10 * Rectangle(1,2)` or `Rectangle(1,2) * 10` returns a new `Rectangle` instance with width 10 and height 20:

```
>>> 10 * Rectangle(1,2)
Rectangle (10 by 20)
>>> Rectangle(1,2) * 10
Rectangle (10 by 20)
```

B.5.5 Class methods

Methods are functions that can only be run with an existing instance of a class. Another option is to create a *class method*, which is a function attached to the class itself rather than individual instances. For a `Rectangle` class, a class method would contain some functionality that has to do with rectangles in general, rather than with a specific rectangle.

One typical use of a class method is to create an alternative constructor. For instance, we could create a class method on the `Rectangle` class taking a single number as an argument, returning a rectangle with height and width both equal to that number. In other words, this class method constructs a rectangle that is a square of a given side length. The first argument of a class method represents the class itself and is often abbreviated to `cls`:

```
class Rectangle():
    ...
    @classmethod
    def square(cls,side):
        return Rectangle(side,side)
```

With this class method implemented, we could write `Rectangle.square(5)` to get the same result as `Rectangle(5,5)`.

B.5.6 Inheritance and abstract classes

The last topic in OOP that we use is *inheritance.* If we say class `A` inherits from class `B`, it is like saying instances of class `A` are special cases of class `B`; they work like instances of class `B` but with some additional or modified functionality. In this case, we also say that `A` is a *subclass* of `B`, and that `B` is a *superclass* of `A`. As a simple example, we could create a subclass `Square` from `Rectangle` that represents squares, while keeping most of the same underlying logic from `Rectangle`. Writing `class Square(Rectangle)` means that `Square` is a subclass of `Rectangle`, and the call to `super().__init__` runs the superclass (`Rectangle`) constructor from the `Square` constructor:

```
class Square(Rectangle):

    def __init__(self,s):
        return super().__init__(s,s)

    def __repr__(self):
        return "Square (%r)" % self.width
```

This is all we need to do to define the `Square` class, and we can use any `Rectangle` method out of the box:

```
>> Square(5).area()
25
```

In practice, you might want to re-implement or *override* some more methods, like `scale`, which by default returns the scaled square as a rectangle.

One common pattern in OOP is to have two classes inherit from the same *abstract base class*, a class defining some common methods or code but that you could never have an instance of. As an example, suppose we had a similar class `Circle`, representing a circle of a given radius. Most of the `Circle` class implementation is analogous to the `Rectangle` class:

```
from math import pi

class Circle():
    def __init__(self, r):
        self.radius = r

    def area(self):
        return pi * self.radius * self.radius

    def scale(self, factor):
        return Circle(factor * self.radius)

    def __eq__(self,other):
        return self.radius == other.radius

    def __repr__(self):
        return 'Circle (radius %r)' % self.radius

    def __mul__(self,factor):
        return self.scale(factor)

    def __rmul__(self,factor):
        return self.scale(factor)
```

(Remember from geometry class that the area of a circle of radius r is πr^2.) If we're dealing with a number of different shapes in our program, we could have both `Circle` and `Rectangle` inherit from a common `Shape` class. The concept of a shape is not specific enough that we could create an instance of it, so only some of the methods can be implemented. The others are marked as *abstract methods*, meaning they can't be implemented for a `Shape` on its own but could be implemented for any concrete subclass.

We can use the following code to create an abstract class in Python. *ABC* stands for "abstract base class," and `ABC` is a special base class that any abstract class must inherit from in Python:

```
from abc import ABC, abstractmethod

class Shape(ABC):
    @abstractmethod
    def area(self):
        pass
```

```
    @abstractmethod
    def scale(self, factor):
        pass

    def __eq__(self,other):
        return self.__dict__ == other.__dict__

    def __mul__(self,factor):
        return self.scale(factor)

    def __rmul__(self,factor):
        return self.scale(factor)
```

Equality and the multiplication overloads are fully implemented, with `__eq__` checking that all the properties of two shapes agree. Area and scale are left to be implemented, and their implementations depend on the particular shape we're working with.

If we were to reimplement the `Rectangle` class based on the `Shape` abstract base class, we could start by having it inherit from `Shape`, while giving it its own constructor:

```
class Rectangle(Shape):
    def __init__(self,w,h):
        self.width = w
        self.height = h
```

If we try to instantiate a `Rectangle` with just this code, we run into an error because the `area` and `scale` methods are not implemented:

```
>>> Rectangle(1,3)
TypeError: Can't instantiate abstract class Rectangle with abstract methods
area, scale
```

We can include the earlier implementations:

```
class Rectangle(Shape):
    def __init__(self,w,h):
        self.width = w
        self.height = h

    def area(self):
        return self.width * self.height

    def scale(self, factor):
        return Rectangle(factor * self.width, factor * self.height)
```

Once we have these methods whose behavior is rectangle-specific, we have access to all the functionality from the `Shape` base class. For instance, equality and the multiplication operator overload behave as expected:

```
>>> 3 * Rectangle(1,2) == Rectangle(3,6)
True
```

We can now quickly implement a `Circle` class, a `Triangle` class, or a class for any other 2D shape, all of which would be unified by their area and shape methods and operator overloads.

appendix C
Loading and rendering 3D Models with OpenGL and PyGame

Beyond chapter 3, when we start writing programs that transform and animate graphics, I use OpenGL and PyGame instead of Matplotlib. This appendix provides an overview of how to set up a game loop in PyGame and render 3D models in successive frames. The culmination is an implementation of a `draw_model` function that renders a single image of a 3D model like the teapot we used in chapter 4.

The goal of `draw_model` is to encapsulate the library-specific work, so you don't have to spend a lot of time wrestling with OpenGL. But if you want to understand how the function works, feel free to follow along in this appendix and play with the code yourself. Let's start with our octahedron from chapter 3 and recreate it with PyOpenGL, an OpenGL binding for Python and PyGame.

C.1 Recreating the octahedron from chapter 3

To begin working with the PyOpenGL and PyGame libraries, you need to install them. I recommend using pip as follows:

```
> pip install PyGame
> pip install PyOpenGL
```

The first thing I'll show you is how to use these libraries to recreate work we've already done, rendering a simple 3D object.

In a new Python file called octahedron.py (which you can find in the source code for appendix C), we start with a bunch of imports. The first few come from the two new libraries, PyGame and PyOpenGL, and the rest should be familiar from chapter 3. In particular, we'll continue to use all of the 3D vector arithmetic functions we already built, organized in the file vectors.py in the source code for this book. Here are the import statements:

```
import pygame
from pygame.locals import *
from OpenGL.GL import *
from OpenGL.GLU import *
import matplotlib.cm
from vectors import *
from math import *
```

While OpenGL has automatic shading capabilities, let's continue to use our shading mechanism from chapter 3. We can use a blue color map from Matplotlib to compute colors for the shaded sides of the octahedron:

```
def normal(face):
    return(cross(subtract(face[1], face[0]), subtract(face[2], face[0])))

blues = matplotlib.cm.get_cmap('Blues')

def shade(face,color_map=blues,light=(1,2,3)):
    return color_map(1 - dot(unit(normal(face)), unit(light)))
```

Next, we have to specify the geometry of the octahedron and the light source. Again, this is the same as in chapter 3:

```
light = (1,2,3)
faces = [
    [(1,0,0), (0,1,0), (0,0,1)],
    [(1,0,0), (0,0,-1), (0,1,0)],
    [(1,0,0), (0,0,1), (0,-1,0)],
    [(1,0,0), (0,-1,0), (0,0,-1)],
    [(-1,0,0), (0,0,1), (0,1,0)],
    [(-1,0,0), (0,1,0), (0,0,-1)],
    [(-1,0,0), (0,-1,0), (0,0,1)],
    [(-1,0,0), (0,0,-1), (0,-1,0)],
]
```

Now it's time for some unfamiliar territory. We're going to show the octahedron as a PyGame game window, which requires a few lines of boilerplate. Here, we start the game, set the window size in pixels, and tell PyGame to use OpenGL as the graphics engine:

```
pygame.init()
display = (400,400)
window = pygame.display.set_mode(display,
                                 DOUBLEBUF|OPENGL)
```

Asks PyGame to show our graphics in a 400 × 400 pixel window

Lets PyGame know that we're using OpenGL for our graphics and indicates that we want to use a built-in optimization called double-buffering, which isn't important to understand for our purposes

In our simplified example in section 3.5, we drew the octahedron from the perspective of someone looking from a point far up the *z*-axis. We computed which triangles should be visible to such an observer and projected them to 2D by removing the *z*-axis. OpenGL has built-in functions to configure our perspective even more precisely:

```
gluPerspective(45, 1, 0.1, 50.0)
glTranslatef(0.0,0.0, -5)
glEnable(GL_CULL_FACE)
glEnable(GL_DEPTH_TEST)
glCullFace(GL_BACK)
```

For the purpose of learning math, you don't really need to know what these functions do, but I'll give you a short overview in case you are curious. The call to `gluPerspective` describes our perspective looking at the scene, where we have a 45° viewing angle and an aspect ratio of 1. This means the vertical units and the horizontal units display as the same size. As a performance optimization, the numbers 0.1 and 50.0 put limits on the *z*-coordinates that are rendered: no objects further than 50.0 units from the observer or closer than 0.1 units will show up. Our use of `glTranslatef` indicates that we'll observe the scene from 5 units up the *z*-axis, meaning we move the scene down by vector (0, 0, –5). Calling `glEnable(GL_CULL_FACE)` turns on an OpenGL option that automatically hides polygons oriented away from the viewer, saving us some work we already did in chapter 3, and `glEnable(GL_DEPTH_TEST)` ensures that we render polygons closer to us on top of those further from us. Finally, `glCullFace(GL_BACK)` enables an OpenGL option that automatically hides polygons that are facing us but that are behind other polygons. For the sphere, this wasn't a problem, but for more complex shapes it can be.

Finally, we can implement the main code that draws our octahedron. Because our eventual goal is to animate objects, we'll actually write code that draws the object over and over repeatedly. These successive drawings, like frames of a movie, show the same octahedron over time. And, like any video of any stationary object, the result is indistinguishable from a static picture.

To render a single frame, we loop through the vectors, decide how to shade them, draw them with OpenGL, and update the frame with PyGame. Inside of an infinite `while` loop, this process can be automatically repeated as fast as possible as long as the program runs:

```
clock = pygame.time.Clock()                          ◁ Initializes a clock to measure the advancement of time for PyGame
while True:
    for event in pygame.event.get():                 ◁ In each iteration, checks the events PyGame receives and quits if the user closes the window
        if event.type == pygame.QUIT:
            pygame.quit()
            quit()

    clock.tick()                                     ◁ Indicates to the clock that time should elapse
    glClear(GL_COLOR_BUFFER_BIT|GL_DEPTH_BUFFER_BIT)
    glBegin(GL_TRIANGLES)                            ◁ Instructs OpenGL that we are about to draw triangles
    for face in faces:
        color = shade(face,blues,light)
```

```
        for vertex in face:
            glColor3fv((color[0],
                        color[1],
                        color[2]))
            glVertex3fv(vertex)
    glEnd()
    pygame.display.flip()
```

For each vertex of each face (triangle), sets the color based on the shading

Specifies the next vertex of the current triangle

Indicates to PyGame that the newest frame of the animation is ready and makes it visible

Running this code, we see a 400 × 400 pixel PyGame window appear, containing an image that looks like the one from chapter 3 (figure C.1).

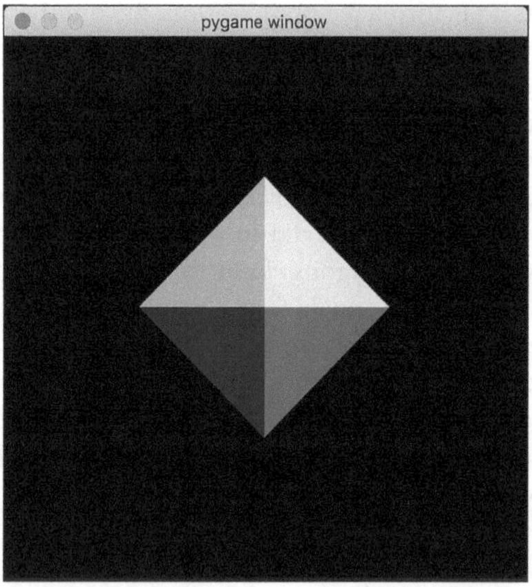

Figure C.1 The octahedron rendered in a PyGame window

If you want to prove that something more interesting is happening, you can include the following line at the end of the `while True` loop:

```
print(clock.get_fps())
```

This prints instantaneous quotes of the rate (in frames per second, or fps) at which PyGame is rendering and re-rendering the octahedron. For a simple animation like this, PyGame should reach or exceed its default maximum frame rate of 60 fps.

But what's the point of rendering so many frames if nothing changes? Once we include a vector transformation with each frame, we see the octahedron move in various ways. For now, we can cheat by moving the "camera" with each frame instead of actually moving the octahedron.

C.2 Changing our perspective

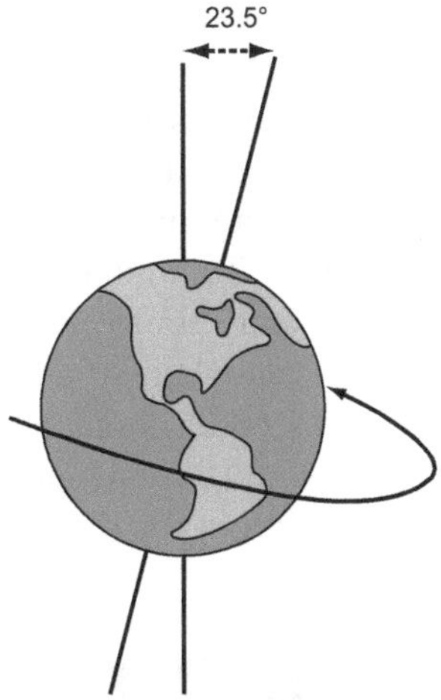

Figure C.2 A familiar example of an object rotating about an axis. The Earth's axis of rotation is tilted at 23.5° relative to its orbital plane.

The `glTranslatef` function in the previous section tells OpenGL the position from which we want to see the 3D scene we're rendering. Similarly, there is a `glRotatef` function that lets us change the angle at which we observe the scene. Calling `glRotatef (theta, x, y, z)` rotates the whole scene by the angle `theta` about an axis specified by the vector (*x*, *y*, *z*).

Let me clarify what I mean by "rotating by an angle about an axis." You can think of the familiar example of the Earth rotating in space. The Earth rotates by 360° every day or 15° every hour. The *axis* is the invisible line that the Earth rotates around; it passes through the North and South poles—the only two points that aren't rotating. For the Earth, the axis of rotation is not directly upright, rather it is tilted by 23.5° (figure C.2).

The vector (0, 0, 1) points along the *z*-axis, so calling `glRotatef(30,0,0,1)` rotates the scene by 30° about the *z*-axis. Likewise, `glRotatef(30,0,1,1)` rotates the scene by 30° but, instead, about the axis (0, 1, 1), which is 45° tilted between the *y*- and *z*-axes. If we call `glRotatef (30,0,0,1)` or `glRotatef(30,0,1,1)` after `glTranslatef(...)` in the octahedron code, we see the octahedron rotated (figure C.3).

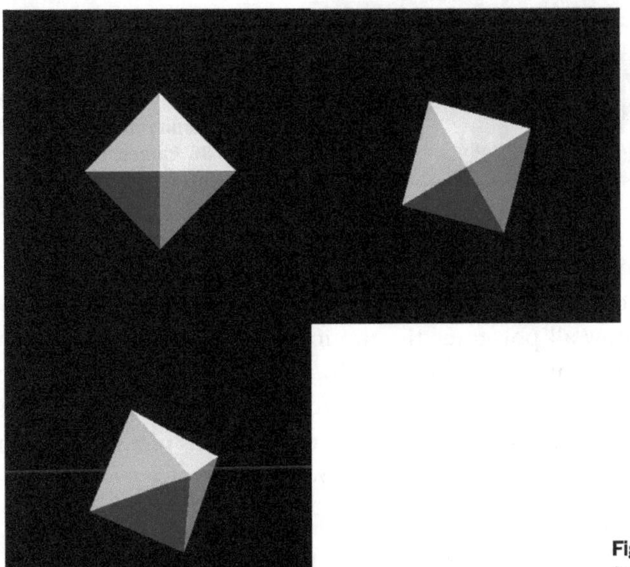

Figure C.3 The octahedron as seen from three different rotated perspectives using the `glRotatef` function from OpenGL

Notice that the shading of the four visible sides of the octahedron in figure C.3 has not changed. This is because none of the vectors change; the vertices of the octahedron and the light source are all the same! We have only changed the position of the "camera" relative to the octahedron. When we actually change the position of the octahedron, we'll see the shading change too.

To animate the rotation of the cube, we can call `glRotate` with a small angle in every frame. For instance, if PyGame draws the octahedron at about 60 fps, and we call `glRotatef(1,x,y,z)` in every frame, the octahedron rotates about 60° every second about the axis (x, y, z). Adding `glRotatef(1,1,1,1)` within the infinite `while` loop before `glBegin` causes the octahedron to rotate by 1° per frame about an axis in the direction (1, 1, 1) as shown in figure C.4.

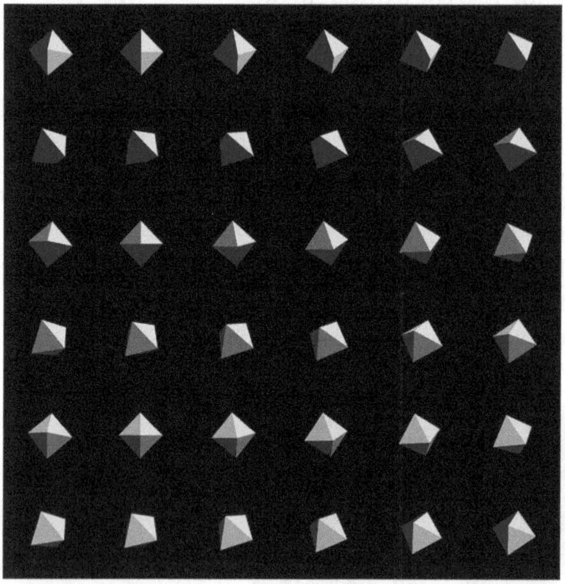

Figure C.4 Every tenth frame of our octahedron rotating at 1° per frame. After 36 frames, the octahedron completes a full rotation.

This rotation rate is only accurate if PyGame draws exactly 60 fps. In the long run, this may not be true; if a complex scene requires more than a sixtieth of a second to compute all vectors and draw all polygons, the motion actually slows down. To keep the motion of the scene constant regardless of the frame rate, we can use PyGame's clock.

Let's say we want our scene to rotate by a full rotation (360°) every 5 seconds. PyGame's clock thinks in milliseconds, which are thousandths of a second. For a thousandth of a second, the angle rotated is divided by 1,000:

```
degrees_per_second = 360./5
degrees_per_millisecond = degrees_per_second / 1000
```

The PyGame clock we created has a `tick()` method that both advances the clock and returns the number of milliseconds since `tick()` was last called. This gives us a reliable number of milliseconds since the last frame was rendered, and lets us compute the angle that the scene should be rotated in that time:

```
milliseconds = clock.tick()
glRotatef(milliseconds * degrees_per_millisecond, 1,1,1)
```

Calling `glRotatef` like this every frame guarantees that the scene rotates exactly 360° every 5 seconds. In the file rotate_octahedron.py in the appendix C source code, you can see exactly how this code is inserted.

With the ability to move our perspective over time, we already have better rendering capabilities than we developed in chapter 3. Now, we can turn our attention to drawing a more interesting shape than an octahedron or a sphere.

C.3 Loading and rendering the Utah teapot

As we manually identified the vectors outlining a 2D dinosaur in chapter 2, we could manually identify the vertices of any 3D object, organize them into triples representing triangles, and build the surface as a list of triangles. Artists who design 3D models have specialized interfaces for positioning vertices in space and then saving them to files. In this section, we use a famous pre-built 3D model: the *Utah teapot*. The rendering of this teapot is the Hello World program for graphics programmers: a simple, recognizable example for testing.

The teapot model is saved in the file teapot.off in the source code, where the .off filename extension stands for Object File Format. This is a plaintext format, specifying the polygons that make up the surface of a 3D object and the 3D vectors that are vertices of the polygon. The teapot.off file looks something like what is shown in this listing.

Listing C.1 A schematic of the teapot.off file

```
OFF                                  <- Indicates that this file follows the Object File Format
480  448  926                        <- Contains the number of vertices, faces, and edges of the 3D model in that order
0  0  0.488037                       <- Specifies 3D vectors for each of the vertices, as x-, y-, and z-coordinate values
0.00390625  0.0421881  0.476326
0.00390625  -0.0421881  0.476326
0.0107422  0  0.575333
...
4 324 306 304 317                    <- Specifies the 448 faces of the model
4 306 283 281 304
4 283 248 246 281
...
```

For the last lines of this file, specifying the faces, the first number of each line tells us what kind of polygon the face is. The number 3 indicates a triangle, 4 a quadrilateral, 5 a pentagon, and so on. Most of the teapot's faces turn out to be quadrilaterals. The next numbers on each line tell us the indices of the vertices from the previous lines that form the corners of the given polygon.

In the file teapot.py in the appendix C source code, you'll find the functions `load_vertices()` and `load_polygons()` that load the vertices and faces (polygons) from the teapot.off file. The first function returns a list of 440 vectors, which are all the vertices for the model. The second returns a list of 448 lists, each one containing vectors that are the vertices of one of the 448 polygons making up the model. Finally, I included a third function, `load_triangles()`, that breaks up the polygons with four or more vertices so that our entire model is built out of triangles.

I've left it as a mini-project for you to dig deeper into my code or to try to load the teapot.off file as well. For now, I'll continue with the triangles loaded by teapot.py, so we can get to drawing and playing with our teapot more quickly. The other step I skip is organizing the PyGame and OpenGL initialization into a function so that we don't have to repeat it every time we draw a model. In draw_model.py, you'll find the following function:

```
def draw_model(faces, color_map=blues, light=(1,2,3)):
...
```

It takes the faces of a 3D model (assumed to be correctly oriented triangles), a color map for shading, and a vector for the light source, and draws the model accordingly. There are also a few more keyword arguments that we introduced in chapters 4 and 5. Like our code to draw the octahedron, it draws whatever model is passed in, over and over in a loop. This listing shows how I put these together in draw_teapot.py.

Listing C.2 Loading the teapot triangles and passing those to `draw_model`

```
from teapot import load_triangles
from draw_model import draw_model

draw_model(load_triangles())
```

The result is an overhead view of a teapot. You can see the circular lid, the handle on the left, and the spout on the right (figure C.5).

Figure C.5 Rendering the teapot

Now that we can render a shape that's more interesting than a simple geometric figure, it's time to play! If you read chapter 4, you learned about the mathematical transformations that you can do on all of the vertices of the teapot to move and distort it in 3D space. Here, I've also left you some exercises if you want to do some guided exploration of the rendering code.

C.4 Exercises

Exercise C.1 Modify the `draw_model` function to display the input figure from any rotated perspective. Specifically, give the `draw_model` function a keyword argument `glRotatefArgs` that provides a tuple of four numbers corresponding to the four arguments of `glRotatef`. With this extra information, add an appropriate call to `glRotatef` within the body of `draw_model` to execute the rotation.

Solution In the source code for this book, see draw_model.py for the solution and draw_teapot_glrotatef.py for an example usage.

Exercise C.2 If we call `glRotatef(1,1,1,1)` in every frame, how many seconds does it take for the scene to complete a full revolution?

Solution The answer depends on the frame rate. This call to `glRotatef` rotates the perspective by 1° each frame. At 60 fps, it would rotate 60° per second and complete a full rotation of 360° in 6 seconds.

Exercise C.3—Mini Project Implement the `load_triangles()` function shown previously, which loads the teapot from the teapot.off file and produces a list of triangles in Python. Each triangle should be specified by three 3D vectors. Then, pass your result to `draw_model()` and confirm that you see the same result.

Solution In the source code, you can find `load _triangles()` implemented in the file teapot.py.

Hint You can turn the quadrilaterals into pairs of triangles by connecting their opposite vertices.

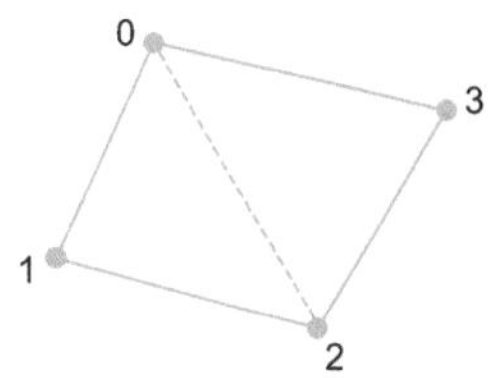

Indexing four vertices of a quadrilateral, two triangles are formed by vertices 0, 1, 2 and 0, 2, 3, respectively.

Exercise C.4—Mini Project Animate the teapot by changing the arguments to `gluPerspective` and `glTranslatef`. This will help you visualize the effects of each of the parameters.

Solution In the file animated_octahedron.py in the source code, an example is given for rotating the octahedron by 360 / 5 = 72° per second by updating the `angle` parameter of `glRotatef` every frame. You can try similar modifications yourself with either the teapot or the octahedron.

index

Symbols

Numerics

A

B

C

D

E

F

G

H

I

N

O

P

T

U

V

W

Z

RELATED MANNING TITLES

Grokking Algorithms
by Aditya Y. Bhargava

ISBN 9781617292231
256 pages, $44.99
May 2016

Real-World Cryptography
by David Wong

ISBN 9781617296710
388 pages (estimated), $59.99
Spring 2021 (estimated)

Algorithms and Data Structures in Action
by Marcello La Rocca

ISBN 9781617295485
746 pages (estimated), $59.99
February 2021 (estimated)

For ordering information go to www.manning.com